T0174354

Quantum Mechanics I

Quantum Mechanics I: The Fundamentals provides a graduate-level account of the behaviour of matter and energy at the molecular, atomic, nuclear, and sub-nuclear levels. It covers basic concepts, mathematical formalism, and applications to physically important systems.

This fully updated new edition addresses many topics not typically found in books at this level, including:

- Bound state solutions of quantum pendulum
- Morse oscillator
- Solutions of classical counterpart of quantum mechanical systems
- A criterion for bound state
- Scattering from a locally periodic potential and reflection-less potential
- Modified Heisenberg relation
- Wave packet revival and its dynamics
- An asymptotic method for slowly varying potentials
- Klein paradox, Einstein-Podolsky-Rosen (EPR) paradox, and Bell's theorem
- Delayed-choice experiments
- Fractional quantum mechanics
- Numerical methods for quantum systems

A collection of problems at the end of each chapter develops students' understanding of both basic concepts and the application of theory to various physically important systems. This book, along with the authors' follow-up *Quantum Mechanics II: Advanced Topics*, provides students with a broad, up-to-date introduction to quantum mechanics.

Quantum Mechanics I
The Fundamentals
Second Edition

S. Rajasekar and R. Velusamy

CRC Press
Taylor & Francis Group
Boca Raton London New York

CRC Press is an imprint of the
Taylor & Francis Group, an **informa** business

A CHAPMAN & HALL BOOK

Second edition published 2023
by CRC Press
4 Park Square, Milton Park, Abingdon, Oxon, OX14 4RN

and by CRC Press
6000 Broken Sound Parkway NW, Suite 300, Boca Raton, FL 33487-2742

First edition published by CRC Press 2015

CRC Press is an imprint of Informa UK Limited

British Library Cataloguing-in-Publication Data
A catalogue record for this book is available from the British Library

---

Library of Congress Cataloging-in-Publication Data

Names: Rajasekar, S. (Shanmuganathan), 1963- author. | Velusamy, R., 1952- author.
Title: Quantum mechanics / S. Rajasekar, R. Velusamy.
Description: Second edition. | Boca Raton : CRC Press, 2022. | Includes bibliographical references and index. | Contents: v. 1. The fundamentals -- v. 2. Advanced topics. | Summary: "Quantum Mechanics I: The Fundamentals provides a graduate-level account of the behavior of matter and energy at the molecular, atomic, nuclear, and sub-nuclear levels. It covers basic concepts, mathematical formalism, and applications to physically important systems. This fully updated new edition addresses many topics not typically found in books at this level, including: Bound state solutions of quantum pendulum Morse oscillator Solutions of classical counterpart of quantum mechanical systems A criterion for bound state Scattering from a locally periodic potential and reflection-less potential Modified Heisenberg relation Wave packet revival and its dynamics An asymptotic method for slowly varying potentials Klein paradox, Einstein-Podolsky-Rosen (EPR) paradox, and Bell's theorem Delayed-choice experiments Fractional quantum mechanics Numerical methods for quantum systems A collection of problems at the end of each chapter develops students' understanding of both basic concepts and the application of theory to various physically important systems. This book, along with the authors' follow-up Quantum Mechanics II: Advanced Topics, provides students with a broad, up-to-date introduction to quantum mechanics. Print Versions of this book also include access to the ebook version"-- Provided by publisher.
Identifiers: LCCN 2022021033 | ISBN 9780367769987 (v. 1 ; hardback) | ISBN 9780367776367 (v. 1 ; paperback) | ISBN 9781003172178 (v. 1 ; ebook) | ISBN 9780367770006 (v. 2 ; hardback) | ISBN 9780367776428 (v. 2 ; paperback) | ISBN 9781003172192 (v. 2 ; ebook)
Subjects: LCSH: Quantum theory.
Classification: LCC QC174.12 .R348 2022 | DDC 530.12--dc23/eng20220518
LC record available at https://lccn.loc.gov/2022021033

ISBN: 978-0-367-76998-7 (hbk)
ISBN: 978-0-367-77636-7 (pbk)
ISBN: 978-1-003-17217-8 (ebk)

DOI: 10.1201/9781003172178

Typeset in CMR10 font
by KnowledgeWorks Global Ltd.

Publisher's note: This book has been prepared from camera-ready copy provided by the authors.

To our wives.

Contents

Preface   xv

About the Authors   xix

1   Why Was Quantum Mechanics Developed?   **1**
   1.1   Introduction 1
   1.2   Black Body Radiation 2
   1.3   Photoelectric Effect 5
   1.4   Hydrogen Spectrum 8
   1.5   Franck–Hertz Experiment 10
   1.6   Stern–Gerlach Experiment 11
   1.7   Correspondence Principle 13
   1.8   Compton Effect 14
   1.9   Specific Heat Capacity 18
   1.10   de Broglie Waves 18
   1.11   Particle Diffraction 19
   1.12   Wave-Particle Duality 20
   1.13   Concluding Remarks 23
   1.14   Bibliography 24
   1.15   Exercises 25

2   Schrödinger Equation and Wave Function   **27**
   2.1   Introduction 27
   2.2   Construction of Schrödinger Equation 27
   2.3   Solution of Time-Dependent Equation 29
   2.4   Physical Interpretation of $\psi^*\psi$ 30
   2.5   Conditions on Allowed Wave Functions 32
   2.6   Box Normalization 33
   2.7   A Special Feature of Occurrence of i in the Schrödinger Equation 34
   2.8   Conservation of Probability 35
   2.9   Expectation Value 36
   2.10   Ehrenfest's Theorem 40
   2.11   Basic Postulates 43
   2.12   Time Evolution of Stationary States 43
   2.13   Conditions for Allowed Transitions 44
   2.14   Orthogonality of Two States 45
   2.15   Phase of the Wave Function 47
   2.16   Classical Limit of Quantum Mechanics 48
   2.17   Concluding Remarks 49
   2.18   Bibliography 50
   2.19   Exercises 51

3 Operators, Eigenvalues and Eigenfunctions   **55**
   3.1   Introduction 55
   3.2   Linear Operators 55
   3.3   Commuting and Noncommuting Operators 58
   3.4   Self-Adjoint and Hermitian Operators 62
   3.5   Discrete and Continuous Eigenvalues 65
   3.6   Meaning of Eigenvalues and Eigenfunctions 67
   3.7   Parity Operator 68
   3.8   Some Other Useful Operators 71
   3.9   Concluding Remarks 73
   3.10   Bibliography 73
   3.11   Exercises 74

4 Exactly Solvable Systems I: Bound States   **79**
   4.1   Introduction 79
   4.2   Classical Probability Distribution 80
   4.3   Free Particle 80
   4.4   Harmonic Oscillator 82
   4.5   Particle in the Potential $V(x) = x^{2k}$, $k = 1, 2, \ldots$ 92
   4.6   Particle in a Box 93
   4.7   Morse Oscillator 99
   4.8   Pöschl–Teller Potentials 101
   4.9   Quantum Pendulum 103
   4.10   Criteria for the Existence of a Bound State 105
   4.11   Time-Dependent Harmonic Oscillator 107
   4.12   Damped and Forced Linear Harmonic Oscillator 108
   4.13   Two-Dimensional Systems 112
   4.14   Rigid Rotator 117
   4.15   Concluding Remarks 121
   4.16   Bibliography 122
   4.17   Exercises 124

5 Exactly Solvable Systems II: Scattering States   **129**
   5.1   Introduction 129
   5.2   Potential Barrier: Tunnel Effect 129
   5.3   Finite Square-Well Potential 137
   5.4   Potential Step 142
   5.5   Locally Periodic Potential 147
   5.6   Reflectionless Potentials 151
   5.7   Dynamical Tunnelling 153
   5.8   Concluding Remarks 154
   5.9   Bibliography 154
   5.10   Exercises 156

6 Matrix Mechanics   **159**
   6.1   Introduction 159
   6.2   Linear Vector Space and Tensor Products 160
   6.3   Matrix Representation of Operators and Wave Function ... 162
   6.4   Unitary Transformation 164
   6.5   Schrödinger Equation and Other Quantities in Matrix Form ... 165
   6.6   Application to Certain Systems 166

6.7 Dirac's Bra and Ket Notations 171
6.8 Dimensions of Kets and Bras 174
6.9 Hilbert Space 176
6.10 Symmetry Operators in Hilbert Space 177
6.11 Projection and Displacement Operators 179
6.12 Quaternionic Quantum Mechanics 181
6.13 Concluding Remarks 184
6.14 Bibliography 184
6.15 Exercises 185

7 Various Pictures and Density Matrix   **189**
7.1 Introduction 189
7.2 Schrödinger Picture 189
7.3 Heisenberg Picture 193
7.4 Interaction Picture 196
7.5 Comparison of Three Representations 197
7.6 Density Matrix for a Single System 197
7.7 Density Matrix for an Ensemble 199
7.8 Time Evolution of Density Operator 201
7.9 Concluding Remarks 202
7.10 Bibliography 202
7.11 Exercises 202

8 Heisenberg Uncertainty Principle   **205**
8.1 Introduction 205
8.2 The Classical Uncertainty Relation 206
8.3 Heisenberg Uncertainty Relation 206
8.4 Condition for Minimum Uncertainty Product 211
8.5 Implications of Uncertainty Relation 212
8.6 Illustration of Uncertainty Relation 213
8.7 Some Extensions of Uncertainty Relation 216
8.8 The Modified Heisenberg Relation 217
8.9 Concluding Remarks 219
8.10 Bibliography 220
8.11 Exercises 221

9 Momentum Representation   **223**
9.1 Introduction 223
9.2 Momentum Eigenfunctions 223
9.3 Schrödinger Equation 225
9.4 Expressions for $\langle \mathbf{X} \rangle$ and $\langle \mathbf{p} \rangle$ 226
9.5 Transformation Between Momentum and Coordinate Representations . . 228
9.6 Operators in Momentum Representation 228
9.7 Momentum Function of Some Systems 230
9.8 Concluding Remarks 233
9.9 Bibliography 233
9.10 Exercises 234

10 Wave Packet   **235**
  10.1 Introduction 235
  10.2 Phase and Group Velocities 235
  10.3 Wave Packets and Uncertainty Principle 238
  10.4 Gaussian Wave Packet 240
  10.5 Wave Packet Revival 245
  10.6 Almost Periodic Wave Packets 248
  10.7 Concluding Remarks 249
  10.8 Bibliography 250
  10.9 Exercises 251

11 Theory of Angular Momentum   **253**
  11.1 Introduction 253
  11.2 Scalar Wave Function Under Rotations 254
  11.3 Orbital Angular Momentum 257
  11.4 Spin Angular Momentum 266
  11.5 Spin-Orbit Coupling 273
  11.6 Addition of Angular Momenta 275
  11.7 Rotational Transformation of a Vector Wave Function and Spin 281
  11.8 Rotational Properties of Vector Operators 283
  11.9 Tensor Operators and the Wigner–Eckart Theorem 284
  11.10 Concluding Remarks 287
  11.11 Bibliography 288
  11.12 Exercises 289

12 Hydrogen Atom   **293**
  12.1 Introduction 293
  12.2 Hydrogen Atom in Three-Dimension 293
  12.3 Hydrogen Atom in D-Dimension 305
  12.4 Field Produced by a Hydrogen Atom 306
  12.5 System in Parabolic Coordinates 309
  12.6 Concluding Remarks 311
  12.7 Bibliography 312
  12.8 Exercises 312

13 Approximation Methods I: Time-Independent Perturbation Theory   **315**
  13.1 Introduction 315
  13.2 Theory for Nondegenerate Case 316
  13.3 Applications to Nondegenerate Levels 322
  13.4 Theory for Degenerate Levels 324
  13.5 First-Order Stark Effect in Hydrogen 327
  13.6 Alternate Perturbation Theories 331
  13.7 Concluding Remarks 333
  13.8 Bibliography 334
  13.9 Exercises 334

14 Approximation Methods II: Time-Dependent Perturbation Theory   **337**
  14.1 Introduction 337
  14.2 Transition Probability 337
  14.3 Constant Perturbation 339
  14.4 Harmonic Perturbation 342

14.5   Adiabatic Perturbation   346
14.6   Sudden Approximation   348
14.7   The Semiclassical Theory of Radiation   349
14.8   Concluding Remarks   352
14.9   Bibliography   352
14.10 Exercises   352

15 Approximation Methods III: WKB and Asymptotic Methods   **355**
15.1   Introduction   355
15.2   Principle of WKB Method   355
15.3   Applications of WKB Method   358
15.4   WKB Quantization with Perturbation   365
15.5   An Asymptotic Method   366
15.6   Concluding Remarks   368
15.7   Bibliography   368
15.8   Exercises   368

16 Approximation Methods IV: Variational Approach   **371**
16.1   Introduction   371
16.2   Calculation of Ground State Energy   371
16.3   Trial Eigenfunctions for Excited States   375
16.4   Application to Hydrogen Molecule   378
16.5   Hydrogen Molecule Ion   381
16.6   Concluding Remarks   383
16.7   Exercises   383

17 Scattering Theory   **385**
17.1   Introduction   385
17.2   Classical Scattering Cross-Section   386
17.3   Centre of Mass and Laboratory Coordinates Systems   387
17.4   Scattering Amplitude   390
17.5   Green's Function Approach   392
17.6   Born Approximation   395
17.7   Partial Wave Analysis   398
17.8   Scattering from a Square-Well System   404
17.9   Phase-Shift of One-Dimensional Case   406
17.10 Inelastic Scattering   407
17.11 Concluding Remarks   407
17.12 Bibliography   408
17.13 Exercises   408

18 Identical Particles   **411**
18.1   Introduction   411
18.2   Permutation Symmetry   411
18.3   Symmetric and Antisymmetric Wave Functions   414
18.4   The Exclusion Principle   416
18.5   Spin Eigenfunctions of Two Electrons   417
18.6   Exchange Interaction   418
18.7   Excited States of the Helium Atom   420
18.8   Collisions Between Identical Particles   422
18.9   Uncertainty Principle for a System of Identical Particles   424

18.10 Concluding Remarks 425
18.11 Bibliography 425
18.12 Exercises 425

19 Relativistic Quantum Theory   427
19.1   Introduction 427
19.2   Klein–Gordon Equation 428
19.3   Dirac Equation for a Free Particle 431
19.4   Minimum Uncertainty Wave Packet 441
19.5   Spin of a Dirac Particle 443
19.6   Particle in a Potential 444
19.7   Klein Paradox 445
19.8   Relativistic Particle in a Box 449
19.9   Relativistic Hydrogen Atom 450
19.10 The Electron in a Field 452
19.11 Spin-Orbit Energy 453
19.12 Relativistic Quaternionic Quantum Mechanics 455
19.13 Concluding Remarks 456
19.14 Bibliography 457
19.15 Exercises 458

20 Mysteries in Quantum Mechanics   461
20.1   Introduction 461
20.2   The Collapse of the Wave Function 462
20.3   Einstein–Podolsky–Rosen (EPR) Paradox 463
20.4   Hidden Variables 467
20.5   The Paradox of Schrödinger's Cat 467
20.6   Bell's Theorem 469
20.7   Violation of Bell's Theorem 470
20.8   Resolving EPR Paradox 474
20.9   Concluding Remarks 475
20.10 Bibliography 475
20.11 Exercises 477

21 Delayed-Choice Experiments   479
21.1   Introduction 479
21.2   Single Slit and Double Slit Experiments 479
21.3   Quantum Mechanical Explanation 482
21.4   Experiment With Mach–Zehnder Interferometer 483
21.5   Delayed-Choice Experiment 484
21.6   Delayed-Choice Quantum Eraser 486
21.7   Concluding Remarks 488
21.8   Bibliography 489
21.9   Exercises 491

22 Fractional Quantum Mechanics   493
22.1   Introduction 493
22.2   Integer and Fractional Diffusion Equations 494
22.3   Wave Function and Kernel 495
22.4   Space Fractional Schrödinger Equation 497
22.5   Solutions of Certain Space Fractional Schrödinger Equations 500

22.6 Fractional Schrödinger Equation in Relativistic Quantum Mechanics . . . 503
22.7 Time Fractional Schrödinger Equation 505
22.8 Solutions of Certain Time Fractional Schrödinger Equations 510
22.9 Space-Time Fractional Schrödinger Equation 512
22.10 Concluding Remarks 516
22.11 Bibliography 516
22.12 Exercises 518

23 Numerical Methods for Quantum Mechanics **521**
23.1 Introduction 521
23.2 Matrix Method for Computing Stationary State Solutions 522
23.3 Finite-Difference Time-Domain Method 529
23.4 Time-Dependent Schrödinger Equation 534
23.5 Quantum Scattering 542
23.6 Electronic Distribution of Hydrogen Atom 547
23.7 Schrödinger Equation with an External Field 549
23.8 Concluding Remarks 551
23.9 Bibliography 553
23.10 Exercises 554

A Calculation of Numerical Values of h and k_{B} **557**

B A Derivation of the Factor $h\nu/(\mathrm{e}^{h\nu/k_{\mathrm{B}}T} - 1)$ **559**

C Bose's Derivation of Planck's Law **561**

D Distinction Between Self-Adjoint and Hermitian Operators **563**

E Proof of Schwarz's Inequality **565**

F Calculation of Eigenvalues of a Symmetric Tridiagonal Matrix–QL Method **567**

G Random Number Generators for Desired Distributions **571**

Solutions to Selected Exercises **575**

Index **581**

Preface

Quantum mechanics is the study of the behaviour of matter and energy at the molecular, atomic, nuclear levels and even at sub-nuclear level. This book is intended to provide a broad introduction to fundamental and advanced topics of quantum mechanics. Volume I is devoted to basic concepts, mathematical formalism and application to physically important systems. Volume II covers most of the advanced topics of current research interest in quantum mechanics. Both the volumes are primarily developed as texts at the graduate level and also as reference books. In addition to worked-out examples, numerous collection of exercises are included at the end of each chapter. Solutions are available to confirmed instructors upon request to the publisher. Some of the exercises serve as a mode of understanding and highlighting the significances of basic concepts while others form application of theory to various physically important systems/problems. Developments made in recent years on various mathematical treatments, theoretical methods, their applications and experimental observations are pointed out wherever necessary and possible and moreover they are quoted with references so that readers can refer them for more details.

Volume I consists of 23 chapters and 7 appendices. Chapter 1 summarizes the needs for the quantum theory and its early development (old quantum theory). Chapters 2 and 3 provide the basic mathematical framework of quantum mechanics. Schrödinger wave mechanics and operator formalism are introduced in these chapters. Chapters 4 and 5 are concerned with the analytical solutions of bound states and scattering states, respectively, of certain physically important microscopic systems. The basics of matrix mechanics, Dirac's notation of state vectors and Hilbert space are elucidated in chapter 6. The next chapter gives the Schrödinger, Heisenberg and interaction pictures of time evolution of quantum mechanical systems. Description of time evolution of ensembles by means of density matrix is also described. Chapter 8 is concerned with Heisenberg's uncertainty principle. A brief account of wave function in momentum space and wave packet dynamics are presented in chapters 9 and 10, respectively. Theory of angular momentum is covered in chapter 11. Chapter 12 is devoted exclusively to the theory of hydrogen atom.

Chapters 13 through 16 are mainly concerned with approximation methods such as time-independent and time-dependent perturbation theories, WKB method and variational method. The elementary theory of elastic scattering is presented in chapter 17. Identical particles are treated in chapter 18. The next chapter presents quantum theory of relativistic particles with specific emphasize on Klein–Gordon equation, Dirac equation and its solution for a free particle, a particle in a box (Klein paradox) and hydrogen atom.

Quantum mechanics has novel concepts like wave-particle duality, the uncertainty principle and wave function collapse due to measurement done. Chapter 20 examines the strange consequences of role of measurement through the paradoxes of EPR and a thought experiment of Schrödinger. A brief sketch of Bell's inequality and the quantum mechanical examples violating it are given. With reference to wave-particle duality modified double slit experiments, referred as delayed-choice experiments, have been proposed and performed to identify whether a microscopic particle decides to behave as a particle or wave at the slits itself irrespective of the later path. These experiments and their outcomes are presented in chapter 21. The next chapter introduces and covers the various features of fractional quantum mechanics where in the Schrödinger equation the space derivative or the time derivative or both can be fractional.

Considering the rapid growth of numerical techniques in solving physical problems and significances of simulation studies in describing complex phenomena, the final chapter is devoted for a detailed description of numerical computation of bound state eigenvalues and eigenfunctions, transmission and reflection probabilities of scattering potentials, transition probabilities of quantum systems in the presence of external fields and electronic distribution of atoms. Some supplementary and background materials are presented in the appendices.

The pedagogic features of volume I of the book, which are not usually found in textbooks at this level, are the presentation of bound state solutions of Morse oscillator, quantum pendulum, Pöschl–Teller potential, damped and forced linear harmonic oscillator, solutions of classical counter part of quantum mechanical systems considered, criterion for bound state, scattering from a locally periodic potential and reflectionless potential, modified Heisenberg relation, quaternionic quantum mechanics, wave packet revival and its dynamics, hydrogen atom in D-dimension, alternate perturbation theories, an asymptotic method for slowly varying potentials, Klein paradox, EPR paradox, Bell's theorem, delayed-choice experiments, fractional quantum mechanics and numerical methods for quantum systems.

The volume II consists of 16 chapters. Chapter 1 describes the basic ideas of both classical and quantum field theories. Quantization of nonrelativistic equation, electromagnetic field, Klein–Gordon equation and Dirac field is given. The formulation of quantum mechanics in terms of path integrals is presented in chapter 2. Application of it to free particle and undamped and damped linear harmonic oscillators are considered. In chapter 3 some illustrations and interpretation of supersymmetric potentials and partners are presented. A simple general procedure to construct all the supersymmetric partners of a given quantum mechanical systems with bound states is described. The method is then applied to a few interesting system. A method to generate complex potentials with real eigenvalues is presented. The next chapter is concerned with coherent and squeezed states. Construction of these states and their characteristic properties are enumerated. Chapter 5 is devoted to Berry's phase, Aharonov–Bohm and Sagnac effects. Their origin, properties, effects and experimental demonstration are presented. The features of Wigner distribution function are elucidated in chapter 6.

In composite quantum systems, entanglement is a kind of correlation between subsystems with no classical counterpart. The definition, detection, classification, quantification and the application of quantum entanglement are covered in chapter 7. The next chapter discusses the concept of quantum decoherence, its models, experimental studies on it and its significance in the interpretation of quantum mechanics. There is a growing interest on quantum computing. Basic aspects of quantum computing are presented in chapter 9. Deutsch–Jozsa algorithm of finding whether a function is constant or not, Grover's search algorithm and Shor's efficient quantum algorithm for integer factorization and evaluation of discrete logarithms are described. Chapter 10 deals with quantum cryptography. Basic principles of classical cryptography and quantum cryptography and features of a few quantum cryptographic systems are discussed.

Chapter 11 presents no-cloning theorem concerned with the impossibility of copying a quantum state without destroying the original state and the features of quantum cloning machines. Learning the state of a quantum system by appropriate measurements is known as quantum tomography. An overview of the relevant theory and techniques of quantum tomography is provided in chapter 12. Simulating or emulating less controllable quantum systems through some controllable quantum systems is the aim of quantum simulation. Chapter 13 is devoted to the theoretical and experimental aspects of quantum simulation. The next chapter is concerned with quantum error correction (QEC), which is foundational in quantum computing and quantum information processing due to the presence of noise and interaction of the systems with the environment and the measurement devices. Basic ideas of QEC, setting and using of QEC codes and practical issues in the implementation of QEC codes are covered.

A brief introduction to other advanced topics such as quantum gravity, quantum cosmology, quantum Zeno effect, quantum teleportation, quantum games, quantum diffusion and quantum chaos is presented in chapter 15. The last chapter gives features of some of the recent technological applications of quantum mechanics. Particularly, promising applications of quantum mechanics in ghost imaging, detection of weak amplitude objects and small displacements, entangled two-photon microscopy, lithography, metrology, teleportation of optical images and quantum sensors, batteries and internet are briefly discussed.

In the second edition of both the volumes, comprehensive revision of the chapters in the first edition and extensive enlargement through the addition of two chapters in volume I and six chapters in volume II have been made. Other chapters in the first edition are updated by including the advancements witnessed in the past several years. Key revision to the volume I includes new sections/subsections on Morse oscillator, damped and forced linear harmonic oscillators, moments of linear harmonic oscillator, two-dimensional exactly solvable systems, quaternionic quantum mechanics, condition for a wave function to possess the minimum uncertainty product, wave packet revival of a particle in a box potential, density matrix of a spin-1/2 system and two new chapters on delayed-choice experiments and fractional quantum mechanics.

In volume II six new chapters on quantum entanglement, quantum decoherence, no-cloning theorem and quantum cloning machines, quantum tomography, quantum simulation and QEC are presented. Further, some notable newly added topics in other chapters are application of path integral formulation to linearly damped systems, generation of complex supersymmetric potentials with real eigenvalues, coherent states of position-dependent mass systems, spin coherent states, Zak phase, cumulants with geometric phases, Aharonov–Bohm effect in electrodynamics, testing quantum computers using Grover's algorithm, multi-photon and mult-stage protocol for cryptography, quantum cheque scheme, quantum cosmology, quantum batteries, quantum sensors and quantum internet. Typographical errors occurred in the first edition of both the volumes are corrected in the second edition.

During the preparation of this book, we have received great supports from many colleagues, students and friends. In particular, we are grateful to Prof. N. Arunachalam, Prof. K.P.N. Murthy, Prof. M. Daniel, Dr. S. Sivakumar, Mr. S. Kanmani, Dr. V. Chinnathambi, Dr. P. Philominathan, Dr. K. Murali, Dr. S.V.M. Sathyanarayana, Dr. K. Thamilmaran, Dr.T. Arivudainambi and Dr.V.S. Nagarathinam for their suggestions and encouragement. It is a great pleasure to thank Dr. V.M. Gandhimathi, Dr. V. Ravichandran, Dr. S. Jeyakumari, Dr. G. Sakthivel, Dr. M. Santhiah, Dr. R. Arun, Dr. C. Jeevarathinam, Dr. R. Jothimurugan, Dr. K. Abirami and Dr. S. Rajamani for typesetting some of the chapters. We thank the senior publishing editor Luna Han, commissioning editor for physics Carolina Antunes, editorial assistants Betsy Byers and Danny Kielty, and Michael Davidson, production editor at Taylor & Francis for various suggestions and careful editing of the manuscript, and their team members for smooth handling of the publication process. Finally, we thank our family members for their unflinching support, cooperation and encouragement during the course of preparation of this work.

Tiruchirapalli   *S. Rajasekar*
July, 2022   *R. Velusamy*

About the Authors

Shanmuganathan Rajasekar was born in Thoothukudi, Tamilnadu, India in 1962. He received his B.Sc. and M.Sc. in Physics, both from St. Joseph's College, Tiruchirapalli. He was awarded Ph.D. degree from Bharathidasan University in 1992 under the supervision of Prof. M. Lakshmanan. In 1993, he joined as a Lecturer at the Department of Physics, Manonmaniam Sundaranar University, Tirunelveli. In 2003, the book on *Nonlinear Dynamics: Integrability, Chaos and Patterns* written by Prof. M. Lakshmanan, and the author was published by Springer. In 2005, he joined as a Professor at the School of Physics, Bharathidasan University. In 2016 Springer has published the book on Nonlinear Resonances written by Prof. Miguel A.F. Sanjuan and the author. In 2021 Professors U.E. Vincent, P.V.E. McClintock, I.A. Khovanov, and the author compiled and edited two issues of Philosophical Transactions of the Royal Society A on the theme Vibrational and Stochastic Resonances in Driven Nonlinear Systems. He has also edited a book on Recent Trends in Chaotic, Nonlinear and Complex Dynamics with Professors Jan Awrejecewicz and Minvydas Ragulskis published by World Scientific in 2022. His recent research focuses on nonlinear dynamics with a special emphasize on nonlinear resonances. He has authored or coauthored more than 120 research papers in nonlinear dynamics.

Ramiah Velusamy was born in Srivilliputhur, Tamilnadu, India in the year 1952. He received his B.Sc. degree in Physics from the Ayya Nadar Janaki Ammal College, Sivakasi in 1972 and M.Sc. in Physics from the P.S.G. Arts and Science College, Coimbatore in 1974. He worked as a demonstrator in the Department of Physics in P.S.G. Arts and Science College during 1974–1977. He received an M.S. Degree in Electrical Engineering at Indian Institute of Technology, Chennai, in the year 1981. In the same year, he joined in Ayya Nadar Janaki Ammal College as an Assistant Professor in Physics. He was awarded M.Phil. degree in Physics in the year 1988. He retired in the year 2010. His research topics are quantum confined systems and wave packet dynamics.

1

Why Was Quantum Mechanics Developed?

1.1 Introduction

Quantum mechanics is the study of matter and radiation at an atomic level. It is a mathematical tool for predicting the behaviours of microscopic particles[1]. Classical physics produced satisfactory results for systems that are much larger than atoms and moving with much slower than the speed of light. But it failed for the systems with small size (atomic size) or with speed near that of light. In the early 20th century, scientists observed that results of certain experiments were not satisfactorily explained by classical physics. Particularly, it failed to account for the observations in the following phenomena.

1. **Black body radiation:**

   In the typical experiments with a black body, the energy emitted per unit time and per unit wavelength at a temperature increases with increase in wavelength (λ), reaches a maximum and then decreases. Classical theory gave incorrect variation of emitted energy.

2. **Photoelectric effect:**

   Electrons emission from a light illuminated metal surface happens only above a threshold frequency of the incident light. This and other associated experimental results were not accounted by the classical theory.

3. **Spectrum of hydrogen:**

   Hydrogen atoms are found to emit radiations of certain wavelengths alone. But classical theory predicted emission of radiations of all wavelengths.

4. **X-ray scattering:**

   The experimentally observed results with X-ray scattering have not been satisfactorily interpreted by the electromagnetic wave theory.

5. **Variation of heat capacity at low temperatures:**

   Classically, the atomic heat of solids is $3R$, a universal constant where $R = 8.3 \times 10^3 \, \text{J} \cdot \text{kilomole}^{-1}\text{K}^{-1}$. In experiments it decreases and approaches to zero as temperature T approaches $0 \, \text{K}$.

In classical physics energy is a continuous variable. To obtain a radiation formula that fit the experimental data, Max Karl Ludwig Planck, in 1901, proposed a revolutionary idea of discrete energy levels of harmonic oscillators. He assumed that the absorption and emission of electromagnetic radiation happen in discrete quanta. His formula agreed remarkably with the experimental result. The importance of Planck's idea was first appreciated by Albert Einstein. He applied it to the photoelectric effect and specific heat. The results of

---

[1]The microscopic scale is one where the lengths are of the order of several angstroms at the most. Macroscopic scale is the one where the length is not less than, say, 10^{-4} cm.

DOI: 10.1201/9781003172178-1

Einstein were subjected to experimental test by Robert Andrews Millikan, Louis Cesar Victor Maurice de Broglie (the elder brother of Louis Victor Pierre Raymond de Broglie) and others. Their observations established the validity of predictions of Einstein. Next, Niels Henrik David Bohr applied the idea of Planck to the atomic structure and spectral lines. He proposed that the electron in hydrogen atom moves in a circular orbit with angular momentum assuming integral multiples of $h/2\pi$, where $h = 6.626 \times 10^{-34}$ J s is Planck constant. In this way, Bohr connected h to the mechanics of the atom. The experiment of James Franck and Gustav Hertz gave the striking evidence for Bohr's postulate, namely, atomic energy states are quantized and the amount of energy transfer to atomic electrons is indeed discrete. Then Arthur Compton applied it to the scattering of X-rays. His results were verified experimentally. In 1924 Louis de Broglie proposed that all matters exhibit wave-like properties. The wave-particle duality is supported by electron diffraction with crystals and double slit experiment. In this chapter, we briefly discuss the above historical background of quantum theory.

1.2   Black Body Radiation

James Clerk Maxwell established that *radiation* is a wave made of electric and magnetic fields and is characterized by its frequency ν and wavelength λ. At a given temperature, different materials emit radiation of different wavelengths. A solid need not be so hot to emit electromagnetic energy. In fact, all objects radiate energy continuously irrespective of their temperatures. However, the predominate frequency depends on the temperature. Most of the radiation at room temperature is in the invisible infrared part of the spectrum.

1.2.1   Black Body Spectrum

In 1859–1860 Gustav Robert Kirchhoff (grandfather of quantum theory) introduced the concept of black body radiation. A *black body* is an object which absorbs and emits all the radiation incident on it. It is difficult to realize a perfect black body. However, a closed oven with a small hole is a good approximation to black body. Here, the radiation is admitted through the hole. The entered radiation is totally absorbed due to repeated reflections inside it. The radiation emitted can be studied using a spectrometer, and the spectral distribution of the radiation can be obtained.

Suppose the walls of the black body are at a constant temperature T. Then a thermodynamic equilibrium, a balance between emission and absorption of radiation, is reached. The spectrum obtained can be characterized by $E(\nu, T)$, where $E(\nu, T)\mathrm{d}\nu$ is the energy contribution by radiation per unit volume of the object in the interval ν to $\nu + \mathrm{d}\nu$. From the thermodynamic arguments $E(\nu, T)$ is shown to depend only on T. Since a black body absorbs all frequencies, we expect that when heated it should radiate all frequencies. But, experimentally, $E(\nu, T)$ was found to be as shown in Fig. 1.1. *Why does the black body spectrum have the shape shown in Fig. 1.1?* Attempts were made to describe this by combining the concepts of thermodynamics and electromagnetic theory.

1.2.2   The Stefan, Wien and Rayleigh–Jeans' Laws

Kirchhoff proved that E must be $E(\nu, T)$ and does not depend on the particular object. In 1879 Josef Stefan calculated the total energy density $u(T)$ of thermal radiation by adding all frequencies and showed that it was $\propto T^4$ and is known as *Stefan's fourth power-law of*

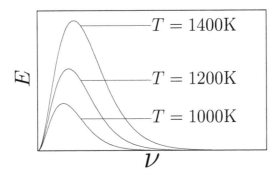

FIGURE 1.1
The spectral distribution of energy density in a black body.

black body:

$$u(T) = \int_0^\infty E(\nu, T)\mathrm{d}\nu = \sigma T^4 \,. \tag{1.1}$$

In 1884, Ludwig Boltzmann presented a proof of this result with the idea of pressure exerted by the radiation. This law did not account correctly for the energy distribution in the spectrum. In 1893, Wilheim Wien, through thermodynamic arguments, obtained the *empirical formula*

$$E(\nu, T) = a\,\nu^3\,e^{-b\nu/T} \,, \tag{1.2}$$

where a and b are constants. It explained the spectrum for high value of ν (visible and ultraviolet regions) but not for low values of ν (infrared region). In 1900, Lord Rayleigh argued that radiation consisted of modes of vibration. Applying the equipartition energy theorem, he calculated the energy of each mode of vibration and the energy distribution. His result after an amendment by James Jeans in the year 1905 read as

$$E(\nu, T) = \frac{8\pi\nu^2 k_B T}{c^3} \,, \tag{1.3}$$

where k_B is called *Boltzmann constant* $(= RN_0^{-1} = 1.38 \times 10^{-23}\,\mathrm{JK^{-1}}$, where $R = 8.3 \times 10^3\,\mathrm{J\,kilomole^{-1}K^{-1}}$ is the universal gas constant and $N_0 = 6 \times 10^{26}\,\mathrm{atoms\,kilomole^{-1}})$ and c is the velocity of light in vacuum. For the calculation of the value of k_B see appendix A. Equation (1.3) is the *Rayleigh–Jeans' law*. It did a good job only for low frequencies and failed at high frequencies. Further, it led to the total energy in the spectrum as

$$\int_0^\infty E\mathrm{d}\nu = \frac{8\pi k_B T}{c^3} \int_0^\infty \nu^2 \mathrm{d}\nu = \infty \,. \tag{1.4}$$

The spectrum diverges at high frequencies. Paul Ehrenfest called this phenomenon *ultraviolet catastrophe*. This divergence was not observed experimentally. The disagreement of theoretical predictions with the experimental observation hinted that *certain basic assumptions of classical physics were somehow wrong.*

1.2.3 Planck's Radiation Formula

Max Planck [1] thought that the assumptions on which Wien's and Rayleigh–Jeans' laws were derived might be incorrect. He assumed that the assumption of continuous emission of radiation in classical mechanics led to incorrect predictions. Therefore, he made the following assumptions:

1. Changes of energy in radiators happen discontinuously and discretely.

2. The walls of the black body are made of harmonic oscillators. They have discrete energies $E_n = nh\nu$, where h is Planck constant[2], $n = 0, 1, \ldots$. (This is the origin of the concept of quantization.) That is, the oscillators have only discrete values of energy and are integral multiples of $h\nu$. (In the words of Born in this postulate, there is an absurdity. Here the energy is that of a localized particle, while frequency ν is usually associated with a wave.)

3. The distribution of oscillators in discrete energy states is governed by the Boltzmann distribution law[3].

If $\langle E(\nu, T)\rangle$ denotes the average energy of an oscillator, then

$$E(\nu, T) = \frac{8\pi\nu^2\langle E(\nu, T)\rangle}{c^3}. \tag{1.5}$$

Planck started with

$$\frac{\mathrm{d}S}{\mathrm{d}\langle E\rangle} = \frac{1}{T} = \text{function of } \langle E(\nu, T)\rangle, \tag{1.6}$$

where S is the entropy[4] of the oscillator. Now, for high ν (or small $\langle E\rangle$) from Eqs. (1.2) and (1.5) we obtain

$$\frac{1}{T} = \frac{1}{b\nu} \ln\left(\frac{a\nu c^3}{8\pi}\right) - \frac{1}{b\nu}\ln\langle E\rangle. \tag{1.7}$$

Then

$$\frac{\mathrm{d}S}{\mathrm{d}\langle E\rangle} = \frac{1}{b\nu}\ln\left(\frac{a\nu c^3}{8\pi}\right) - \frac{1}{b\nu}\ln\langle E\rangle, \quad \frac{\mathrm{d}^2 S}{\mathrm{d}\langle E\rangle^2} = -\frac{1}{b\nu\langle E\rangle}. \tag{1.8}$$

From the Rayleigh–Jeans' law, for small ν (or large $\langle E\rangle$) we get

$$\frac{\mathrm{d}S}{\mathrm{d}\langle E\rangle} = \frac{k_B}{\langle E\rangle}, \quad \frac{\mathrm{d}^2 S}{\mathrm{d}\langle E\rangle^2} = -\frac{k_B}{\langle E\rangle^2}. \tag{1.9}$$

Planck combined the second equations in Eqs. (1.8) and (1.9) as

$$\frac{\mathrm{d}^2 S}{\mathrm{d}\langle E\rangle^2} = -\frac{1}{b\nu\langle E\rangle + (\langle E\rangle^2/k_B)} = -\frac{k_B}{\langle E\rangle\left(\langle E\rangle + k_B b\nu\right)}. \tag{1.10}$$

Integration of Eq. (1.10) gives

$$\frac{\mathrm{d}S}{\mathrm{d}\langle E\rangle} = \frac{1}{b\nu}\ln\left(1 + \frac{k_B b\nu}{\langle E\rangle}\right) + C, \tag{1.11}$$

where C is a constant. Setting $\mathrm{d}S/\mathrm{d}\langle E\rangle = k_B T$ for large $\langle E\rangle$ makes $C = 0$. Comparison of Eqs. (1.6) and (1.11) gives

$$\frac{1}{T} = \frac{k_B}{h\nu}\ln\left(1 + \frac{h\nu}{\langle E\rangle}\right), \tag{1.12}$$

---

[2]This is first occurrence of Planck constant in quantum mechanics. It appears in all quantum phenomena. It occurs in angular momentum, spin, momentum, commutator relations, Heisenberg uncertainty relation, etc. We cannot speak of quantum mechanics without h. h is the ratio of the frequency of radiation and the size of the quanta ($h = 4.2 \times 10^{-15}$ eV or $h = 6.626 \times 10^{-34}$ J s). For a determination of h from black body radiation refer refs. [2–5]. Also see appendix A.

[3]If there are N molecules in the system that can be in different energy states, then the fraction of the molecules in the jth state (N_j/N) is given by the Boltzmann distribution law $N_j/N = \mathrm{e}^{-E_j/k_B T}/\sum_i \mathrm{e}^{-E_i/k_B T}$.

[4]Quantitatively, entropy, S, is defined through the differential quantity $\mathrm{d}S = \delta Q/T$, where δQ is the heat absorbed in an isothermal reversible process when the system undergoes transition from one state to another and T is the absolute temperature.

where $k_B b = h$, Planck constant. From the above equation, we have

$$\langle E \rangle = \frac{h\nu}{e^{h\nu/k_B T} - 1} . \tag{1.13}$$

(For an alternative simple derivation of the above equation see appendix B.) Then Eq. (1.5) becomes

$$E(\nu, T) = \frac{8\pi\nu^2}{c^3} \frac{h\nu}{e^{h\nu/k_B T} - 1} . \tag{1.14}$$

This is *Planck's radiation formula.*

Planck's formula agreed remarkably with the experiments. At low frequencies, $h\nu \ll k_B T$, (1.14) becomes the Rayleigh–Jeans' law (1.3). Planck's discovery led to the formulation of quantum mechanics. Planck received Nobel Prize for physics for the year 1918 for his work. Satyendra Nath Bose derived Planck's law by considering photons as indistinguishable particles obeying *Bose–Einstein statistics*. His derivation is presented in appendix C.

We note that the size of the energy quanta is proportional to the frequency. Radiation of low frequencies is easy to generate because it needs only small quanta of energy. But for high frequencies radiation requires large amounts of energy. This accounts the explanation for the spectrum of black body radiation. It is easy to verify that the ultraviolet catastrophe does not occur in Planck's formula. In the limit $\nu \to \infty$, the exponential term in the denominator of Eq. (1.14) diverges much faster than ν^3 in the numerator and $E \to 0$.

Planck's idea of discrete energy of a harmonic oscillator implies that the frequencies for the absorbed and emitted radiation should be discrete. According to classical physics, a system can absorb or emit any amount of energy. In contrast, in the work of Planck the energy emitted by an oscillator is discrete or quantized. Planck coined the term *quanta* for the packets of energy. Bohr in 1913 proposed that like Planck's oscillator, atoms can exist only in certain energy states. Essentially, *in classical physics energy is a continuous variable while in the new theory it is quantized.* Quantization of energy and certain other quantities such as angular momentum and spin in the new theory led to the name *quantum physics* or *quantum mechanics*. Max Born coined the term *quantum mechanics*.

1.3   Photoelectric Effect

The electrons emitted in the photoelectric effect are called *photoelectrons*, and the current formed by them is called *photoelectric current*. The substance exhibiting photoelectric effect is said to be *photoemissive*. The effect was discovered first by Willoughby Smith in 1873. In 1839, Alexandre Edmond Becquerel observed the photoelectric effect with an electrode in a conductive solution exposed to light. In 1887 Heinrich Rudolf Hertz noticed the photoelectric effect in his experiments with a spark-gap generator. Hertz observed an increase in the sensitivity of the device when it was illuminated with ultraviolet or visible light. In 1897 Joseph John Thomson (Nobel Prize in 1906) showed that the observed increased sensitivity was because of the emission of electrons from the cathode by the incident light. Nikola Tesla described the photoelectric effect in 1901. He regarded such radiations as vibrations of aether of short wavelengths which ionized the atmosphere.

In 1902 Phillip Edward Anton Von Lenard (Nobel Prize in 1905), an assistant of Hertz, studied photoelectric effect. His setup consisted of a photocell connected to a circuit with a power supply, voltmeter and micro-ammeter. He illuminated the photoemissive surface with light of different frequencies and intensities. He found that the kinetic energy of the emitted electron was independent of the intensity. Classically, if a plate was illuminated with a high-intensity light, then the emitted photoelectron should have higher energy.

1.3.1   A Typical Experimental Study

We give briefly a typical experimental study and the common outcomes of the photoelectric effect. Consider a vacuum tube with two electrodes connected to a variable voltage source. Suppose a light incident on the surface of a cathode in the vacuum tube. The electrons released from the metal surface arrive at the other electrode and give rise the flow of current I in the circuit. There would be no flow of current until the potential difference V between the anode and the cathode is raised to a critical value, say, V_0. When V is increased beyond V_0 the current I would increase and reach a limiting value I_s called the *saturation photoelectric current*. At $I = I_s$ all the electrons that escaped from the cathode would reach the anode. The following are some of the important striking observations with the photoelectric effect.

1. Photoelectric effect occurs when the frequency of the incident light is greater than a certain frequency ν_0 called *threshold frequency*. ν_0 differs from metal to metal.

2. The kinetic energy of the photoelectrons emitted increases with increase in frequency, however, remains constant on further increasing the intensity of light.

3. Flow of photoelectric current takes place only above a certain critical applied voltage V_0.

4. The intensity of the incident light has no influence on the maximum kinetic energy of the photoelectrons. But the number of particles emitted decreases with decrease in the intensity of incident light. The velocity of the emitted particles remain the same.

5. There is no delay between absorption of the incident light energy and the emission of photoelectrons. Classically, the electrons would require a certain time to absorb sufficient energy to escape.

The above results were completely unexpected. These observations were not explained by classical physics. Einstein solved these mysteries in 1905 [6].

1.3.2   Einstein's Photoelectric Equation

In 1905, Albert Einstein wrote a paper on photoelectric effect. (In the same year he wrote his paper on the special theory of relativity.) He proposed that light is absorbed by matter in discrete quanta of electromagnetic radiation that were later called *photons*[5]. The electromagnetic radiation itself is quantized. All photons of monochromatic light have the same energy $E_{\mathrm{ph}} = h\nu = hc/\lambda$, where λ is the wavelength of the light. Here E_{ph} *is the energy of the photon (a particle property) and* ν *is its frequency (a wave property)*. A conduction electron gets energy $h\nu$ by absorbing a photon. The electron uses this energy to overcome the binding energy in the metal and to raise its kinetic energy.

Let us denote ϕ_0 as the minimum energy required to impart to an electron to escape it from a metal. ϕ_0 is called *work function* because it is the work done by a photoelectron to escape from the metal. It is h times the critical frequency. The conservation of energy here

---

[5]The term photon was coined by the chemist Gilbert Newton Lewis in 1926 [7].

is

$$h\nu = \frac{1}{2}mv_{\max}^2 + \phi_0 \, , \qquad (1.15)$$

where m and $v_{\max}$ are the mass and maximum velocity of a photoelectron, respectively. When a photon of energy $h\nu$ strikes an atom, $h\nu - \phi_0$ amount of energy is available for imparting kinetic energy (K) to an electron. Then

$$K_{\max} = \frac{1}{2}mv_{\max}^2 = |e|V_0 = h\nu - \phi_0 \, . \qquad (1.16)$$

The above equation is called Einstein's *photoelectric equation*.

The notable features of Einstein's equation are:

1. $K_{\max} = 0$ when $\nu = \nu_0 = \phi_0/h$. If $\nu < \nu_0$ (or $h\nu < \phi_0$) the incident photon will not have sufficient energy to remove the electron from the metal surface. Hence, there exists a threshold frequency (corresponding to $K_{\max} = 0$) of the incident photon above which only the photoelectric effect occurs.

2. $K_{\max}$, $v_{\max}$ and V_0 depend only on the frequency of the incident light and ϕ_0. They are independent of the intensity of the incident light. By considering the instantaneous absorption of light, we account for observed instantaneous emission of photoelectrons.

The above features explained the experimental facts not explained by classical theory. For his work on photoelectric effect Einstein won the Nobel Prize in 1922.

Experiments have been conducted by Robert Andrews Millikan (Nobel Prize in 1923), Arthur Llewelyn Hughes, Owen Willans Richardson (Nobel Prize in 1928), Arthur Holly Compton and Maurice de Broglie to verify Einstein's photoelectric equation. Their experiments at both low and high frequencies ranges confirmed the validity of the equation. Though the predictions of the model were verified by Millikan and others it did not attract interest immediately. This is because the particle model of photon does not explain wave effects. Only when Compton in 1923 verified that X-rays were scattered by electrons as if they were particles the Einstein's model got great interest. The photoelectric effect led to the invention of photoelectric cells, photomultipliers, photovoltaic cells and the image sensors in digital cameras.

In the photoelectric effect, energy is transferred to electrons from photons. *Is the inverse process, that is, conversion of a part of kinetic energy of a electron into a photon possible?* An energetic electron passing through matters will loss a part of its kinetic energy in the form of photons. The process by which an electron emits photons is called *bremsstrahlung*[6]. The inverse photoelectric effect was observed by Wilhelm Konrad Roentgen in 1895. He found that a highly penetrating radiation (later called *X-rays*) is produced when fast electrons impinge on matter. He received Nobel Prize in 1902 for the discovery of X-rays.

Solved Problem 1:

When a light of wavelength 210 nm falls on a metal surface the measured kinetic energy (K.E.) of ejected electrons is 1.914 eV. When a light of 250 nm falls on the same metal surface the maximum kinetic energy is 0.968 eV. Find the value of hc and the work function of the metal.

The photoelectric equation is K.E. $= h\nu - \phi_0 = (hc/\lambda) - \phi_0$. For $\lambda = 210$ nm and K.E. $= 1.914$ eV the above equation is

$$1.914\,\text{eV} = \frac{hc}{210\,\text{nm}} - \phi_0 \, . \qquad (1.17)$$

---

[6]*Bremsstrahlung* is a German word meaning breaking radiation or deceleration radiation.

For $\lambda = 250\,\text{nm}$ and K.E. $= 0.968\,\text{eV}$ the photoelectric equation is

$$0.968\,\text{eV} = \frac{hc}{250\,\text{nm}} - \phi_0 \,. \tag{1.18}$$

From the Eqs. (1.17) and (1.18) we get

$$0.946\,\text{eV} = hc \left(\frac{1}{210\,\text{nm}} - \frac{1}{250\,\text{nm}} \right) \,. \tag{1.19}$$

Then $hc = 1242\,\text{eVnm}$. Substituting the value of hc in Eq. (1.18) we get

$$\phi_0 = \left(\frac{1242}{250} - 0.968 \right) \text{eV} = 4\,\text{eV} \,. \tag{1.20}$$

1.4   Hydrogen Spectrum

Classical mechanics failed to explain the stability of a nuclear atom and the existence of spectral lines. Consider a hydrogen-like atom with electron of mass m_e and charge $-e$ at a distance r from the nucleus of mass M and the charge Ze. The force between proton and electron is electrostatic and is (Coulomb interaction) $-(1/4\pi\epsilon_0)Ze^2/r^2$, where $\epsilon_0 = 8.85 \times 10^{-12}\,\text{C}^2\text{N}^{-1}\text{m}^{-2}$ is the permittivity of free space. Due to the Coulomb force the electron will move towards the nucleus and collide with it in a fraction of second. We know that this is not happening.

To stabilize the atom assume that the electron is orbiting the nucleus like the planets orbiting the sun. In a planetary system the motion of the planets and the centrifugal force prevents them colliding with the sun under the influence of gravity. With this analogy, Niels Bohr assumed that the electrons move along circular orbits around the nucleus. The Coulomb force gives a centrifugal acceleration of the electron given by v^2/r, where v is the velocity of the electron. The centrifugal force is $F = mv^2/r$, where $m = m_e M/(m_e + M)$ is the *reduced mass* and the Coulomb force is $(1/4\pi\epsilon_0)Ze^2/r^2$. The reduced mass is used because the motion of the electron is considered as relative to the centre-of-mass of the atom. In the orbit, these forces are balanced so that

$$F = \frac{1}{4\pi\epsilon_0} \frac{Ze^2}{r^2} = \frac{mv^2}{r} \tag{1.21}$$

or

$$mv^2 = \frac{1}{4\pi\epsilon_0} \frac{Ze^2}{r} \,. \tag{1.22}$$

Now, the energy of the atom is

$$E = \text{K.E.} + \text{P.E.} = \frac{1}{2}mv^2 - \frac{1}{4\pi\epsilon_0} \frac{Ze^2}{r} = -\frac{Ze^2}{8\pi\epsilon_0 r} \,. \tag{1.23}$$

Classically, r can take any positive value and therefore E can take any value in the interval $[-\infty, 0]$. An accelerated charge will emit electromagnetic radiation. Since the electron is revolving around the nucleus it will emit radiation. The emission will reduce the energy of the electron. Hence, the radius of the electron orbit will decrease. The frequencies of the emitted radiation will be continuous. Therefore, we must get a continuous spectrum. Further, the electron will spiral inside the nucleus and become unstable. But the experimentally observed energy spectrum is discrete and an atom is stable.

To explain the discrete spectrum of hydrogen, in 1914 Bohr [8] extended the discreteness of energy to atom and postulated that emission of radiation occurs when the atom jumps from a higher energy state E_i to a lower energy state E_f and the frequency of the emitted radiation is

$$\nu = (E_i - E_f)/h. \tag{1.24}$$

Further, he assumed that an electron moves only in certain allowed orbits called *stationary states* with the orbital angular momentum[7] mvr equal to an integer times $\hbar = h/2\pi$:

$$mvr = nh/(2\pi) = n\hbar, \quad n = 1, 2, \ldots. \tag{1.25}$$

Notice the presence of Planck constant in the above equation. (Here onwards we use $\hbar$ in place of $h/2\pi$). When the condition (1.25) holds, the electron does not radiate energy and the orbit is stable. That is, according to Bohr, atoms in the above-allowed orbits remain in them and could not radiate. There is a minimum angular momentum determining the closest orbit to the nucleus. Consequently, the electron cannot get into the nucleus. From Eq. (1.25) we have $mv^2 = n^2\hbar^2/(mr^2)$. Comparison of this with Eq. (1.22) gives

$$r_n = \frac{n^2 r_0}{Z}, \quad r_0 = \frac{4\pi\epsilon_0\hbar^2}{me^2}. \tag{1.26}$$

r_0 (or a_0) is called *Bohr radius* ($r_0 = 5.29167 \times 10^{-11}$m). The total energy is

$$E_n = \frac{1}{2}mv^2 - \frac{1}{4\pi\epsilon_0}\frac{Ze^2}{r_n} = -\frac{Z^2e^4m}{32\pi^2\epsilon_0^2 n^2\hbar^2}. \tag{1.27}$$

Since n is discrete, the energy levels of the hydrogen atom are thus discrete. *Why is the energy < 0? What does the negative sign of E_n imply?*

In view of Eq. (1.27), we rewrite Eq. (1.24) as

$$\nu = (E_i - E_f)/h = \frac{Z^2e^4m}{8\epsilon_0^2 h^3}\left(\frac{1}{n_f^2} - \frac{1}{n_i^2}\right). \tag{1.28}$$

Or

$$\nu = R_H\left(\frac{1}{n_f^2} - \frac{1}{n_i^2}\right), \quad R_H = \frac{Z^2e^4m}{8\epsilon_0^2 h^3}. \tag{1.29}$$

This expression agreed with the *Rydberg equation*

$$\nu = R_H\left(\frac{1}{n^2} - \frac{1}{m^2}\right), \tag{1.30}$$

where n and m are integers such that $n < m$ and R_H is the Rydberg constant ($= 1.097373 \times 10^7$m^{-1}) deduced from experimental spectrum. This formula was introduced by Johannes Rydberg in 1890. Equation (1.29) gives the frequencies of all transitions in the hydrogen atom. In this way Bohr accounted for the spectrum of hydrogen. Bohr shared the 1922 Nobel Prize (with Einstein) for his work on the investigation of the structure of atoms and of the radiation coming from them. The formula (1.30) with $n = 2$ was proposed few years earlier by Johann Jakob Balmer.

---

[7]In three-dimension the orbital angular momentum $\mathbf{L}$ is $\mathbf{r} \times \mathbf{p}$, where $\mathbf{p} = m\mathbf{v}$.

In 1925 de Broglie gave an explanation for non-decay of an electron from a Bohr orbit. He hypothesized that electrons have wavelengths. If the wavelength of an electron orbiting a nucleus is an integral multiple of $\hbar$ (length of the orbit), a standing electron wave results. de Broglie's equation is $n\lambda = 2\pi r$, $n = 1, 2, \ldots$. Then the quantization of angular momentum means

$$n\lambda = 2\pi r = \frac{2\pi n\hbar}{mv} = \frac{nh}{mv}. \tag{1.31}$$

Thus, $\lambda = h/(mv) = h/p$.

Although the theory of Bohr was developed for hydrogen atom it was successful for hydrogen-like ions such as He^+, Li^{++} which consist of a charged nucleus and a single orbiting electron. However, there were several unanswered questions. For example, some of the spectral lines have fine structures. Difficulties arose when the theory was extended to complex atoms. The model failed to account the effect of a magnetic field on the radiation emitted by certain atoms. It couldn't account for the intensities of the spectral lines.

1.5   Franck–Hertz Experiment

In 1914 James Franck and Gustav Hertz [9] studied experimentally the impact of electrons on atoms. They observed that electrons moving through mercury vapour with an energy equal to or greater than a critical value near 4.9 eV excited the 2536 Å line of the mercury spectrum. Their experiment supported the discrete energy states of atoms proposed by Bohr, and for this the 1925 Nobel Prize was awarded to Franck and Hertz.

The experimental apparatus [9–11] consisted of a glass tube with mercury vapour and an electrode. A grid was placed between the anode and cathode. Electrons emitted from cathode were accelerated toward a positively charged grid by a voltage V. Acceleration gave kinetic energy to the electrons. This energy can be stated in electron volts (eV). For example, an electron accelerated by 5 V will have 5 eV of kinetic energy. A small potential difference V_0 between the grid and the anode disallowed electrons with energies lower than a certain minimum amount reaching the anode. An ammeter connected to the anode measured the flow of current. Franck and Hertz measured the current by changing the accelerating voltage.

In the experiment, as the voltage increased from zero the current increased. At about 4.9 V the current decreased sharply indicating the occurrence of a new phenomenon wherein much energy of electrons were taken away. Consequently, the electrons were unable to reach the anode. Essentially, inelastic collisions took place between the atomic electrons of the mercury atoms and the accelerated electrons. The increase of current up to 4.9 V implied that energy less than 4.9 eV was not absorbed by the mercury. Electrons with energy less than 4.9 eV bounced elastically when they collided with mercury atoms and unable to excite electromagnetic radiation.

The result of Franck–Hertz experiment is explained based on Bohr's idea of quantized energy levels. The energy of first excited state of mercury is 4.9 eV. If the kinetic energy of accelerated electron is lower than 4.9 eV then the available energy becomes insufficient to excite an electron of mercury to next higher energy and thus no absorption of energy. When the kinetic energy of an electron is, say, at least 4.9 eV then it would able to transfer 4.9 eV to an electron in mercury. The mercury can absorb this specific quanta of energy. Now, an electron of mercury jumps to the first excited state and at the same time the electron that has lost 4.9 eV is unable to reach the collector. Consequently, the current is reduced.

In the experiment, when the accelerating voltage was raised just above 4.9 V, the mercury atoms absorbed only 4.9 eV. Hence, the electrons had much energy to reach the anode. The current started increasing steadily. At 9.8 V a second decrease in the current was noticed. At this voltage an electron with energy 9.8 eV lost 4.9 eV because of collision with one mercury atom. This electron moved with the energy of 4.9 eV. When this electron collided with another mercury atom it lost the remaining 4.9 eV. The electrons lost all their 4.9 eV by two inelastic collisions and excited two mercury atoms. Then the electrons were with zero kinetic energy. As a result the current dropped. The above process repeated at intervals of 4.9 V and each time the electrons underwent one more inelastic collision.

Drops in the current occurred at multiples of 4.9 V. This is an indication that the atoms absorbed energy only in specific amounts–quanta! In this process the atomic electrons made an abrupt transition to a discrete higher energy state. If the hitting electrons had energy less than the energy 4.9 eV then no energy was absorbed by the atomic electrons and hence no transition. When they had this specific quanta of energy they lost it to the electrons of mercury. The absorbed energy got stored in the electron excited to higher states. This energy was released as a photon when the electrons came back to the lowest energy state (ground state).

As described above the Franck–Hertz experiment confirmed quantized model of Bohr by demonstrating that atoms could only absorb specific amount of energy, namely those quantities which transfer the atom from one state to another stationary state.

Solved Problem 2:

Atomic hydrogen is bombarded with electrons in a Franck–Hertz type experiment. Excitation potentials of 10.21 V and 12.10 V are measured, and three spectral lines are found in the emission spectrum. Compute the wavelengths of these spectral lines.

Using the relation $E = h\nu = hc/\lambda$, we write

$$\frac{hc}{\lambda_1} = 10.21\,\text{eV}, \quad \frac{hc}{\lambda_2} = 12.10\,\text{eV}, \tag{1.32a}$$

$$\frac{hc}{\lambda_3} = (12.10 - 10.21)\,\text{eV} = 1.89\,\text{eV}. \tag{1.32b}$$

Since $1\text{eV} = 1.6 \times 10^{-19}\text{J}$, we get

$$hc = \frac{6.626 \times 10^{-34}\,\text{eV} \times 3 \times 10^8\,\text{m}}{1.6 \times 10^{-19}} = 1242.375\,\text{eVnm}. \tag{1.33}$$

Then

$$\lambda_1 = \frac{hc}{10.21\,\text{eV}} = 1216.8217\,\text{Å}, \tag{1.34a}$$

$$\lambda_2 = 1026.7562\,\text{Å}, \quad \lambda_3 = 6573.4127\,\text{Å}. \tag{1.34b}$$

1.6   Stern–Gerlach Experiment

Equation (1.25) was realized as a special case of the Bohr–Sommerfeld quantum rule [12]

$$\oint p\,dq = n\hbar, \tag{1.35}$$

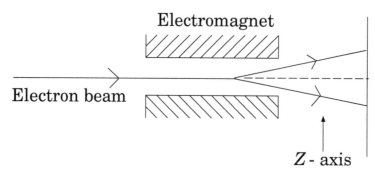

FIGURE 1.2
Splitting of electron beam in Stern–Gerlach experiment.

where q is a generalized coordinate and p is the corresponding canonically conjugate momentum. This condition is satisfied for periodic motion, and the integral is taken over one period. Orbits with a specific direction for the angular momentum vector were alone considered. There is a further degeneracy with the various orientations for an angular momentum vector **L** with respect to any fixed axis. The angular momentum component parallel to the axis must be $m\hbar$ with the quantum number m taking integer values only. This is called *space quantization*. When l is the quantum number characterizing the magnitude of the angular momentum then $m = -l, -l+1, \ldots, l-1, l$.

The first evidence of space quantization was shown experimentally by Otto Stern and Walther Gerlach in 1922 [13]. They demonstrated the magnetic nature of the electron. Suppose a beam of electrons is allowed to pass through a region where the magnetic field points in the z-direction. The strength of field increases with z. Such a field would make a pull upward on the south pole but downward on the north pole of the equivalent magnet of the electron. The total force on the electron will depend on its orientation. If the south pole of the electron is much higher in the field then the force will be upward. When its north pole is higher then the total force will be downward. Hence, an electron moving through this region would experience a downward or upward deflecting force and is directly proportional to S_z (the projection of its spin angular momentum[8] on the z-axis).

If we send an electron beam through a region with a magnetic field of the above system, then the upward or downward forces on the electrons will make the beam to fan-out vertically. The vertical position of an electron, on a display screen, can be used as a direct measurement of S_z of that electron. Because of the random orientation of the incoming electrons we expect the electrons be at the screen to be with a continuous range of vertical positions. But in the experiment electrons were found to reach at only two positions as shown in Fig. 1.2. This implied that S_z could have one of the two possible values. *The spin is thus quantized.*

The point is that the beam is bent by the field into two paths: one above and one below the z-axis. The force in the z-direction on a particle is $F_z = \mu_z \partial B_z / \partial z$, where μ_z is the z-component of the magnetic moment. The separation of beams means that μ_z takes on two equal but opposite values. The atoms from the source are with their magnetic moments oriented at random, but when entered the field they take one of the two values in the z-direction. Because of the relationship between μ_z and L_z (z-component of the orbital angular momentum), L_z can take only one value. For a given value of the orbital quantum

---

[8]The spin is usually said to be non-orbital, intrinsic or inherent angular momentum. For more details see chapter 11.

number l, there are $2l+1$ values of the z-component. Here the two values of the z-component means $l = 1/2$, which is in disagreement with l an integer. George Eugene Uhlenbeck and Samuel Abraham Goudsmit suggested that the electron has an intrinsic angular momentum (spin) with two values of the z-component $\pm 1/2$.

If the electron is like a spinning ball, then its collisions with other electrons would give spin angular momentum with different magnitudes. Hence, more than two parts could occur. But this is not observed. Thus, the spin of an electron appears as one of its intrinsic characteristics like mass or charge.

1.7 Correspondence Principle

Bohr investigated the passage of α-particles in matter and studied the processes of ionization of the atoms. The classical ideas failed. Bohr thought that the new theory (quantum theory) should match with the classical theory for low frequencies or large quantum numbers. This led to the famous *correspondence principle*. This principle is of great interest in statistical mechanics and electrodynamics. It played a significant role in the early formulation of quantum theory [14,15].

Planck stated that

$$\lim_{h \to 0} [\text{quantum physics}] = [\text{classical physics}]. \tag{1.36}$$

He illustrated this principle through the observation that (1.14) reduced to the classical Rayleigh–Jeans' law (1.3) in the limit $h \to 0$. He stated: *The classical theory can simply be characterized by the fact that the quantum of action becomes infinitesimally small.* Another form of this principle was given by Paul Ehrenfest and is known as Ehrenfest's theorem which will be discussed in the next chapter.

In the year 1913, Bohr observed that the frequency spectrum emitted by a hydrogen atom approaches the classical spectrum in the limit of large principal quantum number n. The emission frequency in this limit is the orbiting of the electron classical frequency which approach zero as $n \to \infty$. This is called *Bohr frequency correspondence principle*. This is obtained when the quantum spectrum coalesces with that of the classical.

Let us illustrate the correspondence principle [14,15] for linear harmonic oscillator and hydrogen atom. In Planck's idea energy of harmonic oscillator is discrete, while in classical mechanics it is continuous. To understand this difference we calculate $\Delta E/E$ for any two adjacent energy levels of linear harmonic oscillator. Since $E_n = (n + 1/2)\hbar\omega$, $n = 0, 1, 2, \ldots$ (see sec. 4.4) the quantity $\Delta E/E$ corresponding to n and $n + 1$ levels is

$$\frac{\Delta E}{E} = \frac{\left(n + \dfrac{3}{2}\right)\hbar\omega - \left(n + \dfrac{1}{2}\right)\hbar\omega}{\left(n + \dfrac{3}{2}\right)\hbar\omega} = \frac{1}{\left(n + \dfrac{3}{2}\right)}. \tag{1.37}$$

In the limit of $n \to \infty$, we get $\lim_{n\to\infty}(\Delta E)/E \to 0$. This is in agreement with the correspondence principle.

In hydrogen atom the force on the electron due to the nucleus is $F = -(1/4\pi\epsilon_0)e^2\hat{r}/r^2$. This gives the centripetal force to the electron to follow a circular orbit. The radial acceleration is $a_r = v^2/r$. Applying Newton's second law, we get

$$\frac{mv^2}{r} = \frac{e^2}{4\pi\epsilon_0 r^2} \quad \text{or} \quad v = \left(\frac{e^2}{4\pi\epsilon_0 m r}\right)^{1/2}. \tag{1.38}$$

Then the classical frequency ν_c is

$$\nu_c = \frac{\omega}{2\pi} = \frac{v}{2\pi r} = \left(\frac{e^2}{16\pi^3\epsilon_0 m r^3}\right)^{1/2}. \tag{1.39}$$

We substitute r_n from Eq. (1.26) in (1.39) and obtain

$$\nu_c = \frac{2|E_1|}{hn^3}, \quad E_1 = -\frac{me^4}{32\pi^2\epsilon_0^2\hbar^2}, \tag{1.40}$$

where E_1 is the lowest energy of hydrogen atom (refer Eq. (12.36)). For the transition from $(n+1)$ to nth orbit we get from Eqs. (1.24) and (1.27)

$$\nu_{\text{Bohr}} = \frac{|E_1|}{h}\left(\frac{1}{n^2} - \frac{1}{(n+1)^2}\right) = \frac{|E_1|}{h}\frac{(2n+1)}{n^2(n+1)^2}. \tag{1.41}$$

For large n

$$\nu_{\text{Bohr}} = \frac{2|E_1|}{hn^3} = \nu_c. \tag{1.42}$$

Bohr predicted ν_{Bohr} coincides with ν_c for large n. This led to the correspondence principle: *For large quantum numbers the state of a system is governed by the laws of classical physics and we ignore the quantum effects.* The results of quantum mechanics in the limit of large quantum numbers and $h \to 0$ must be same as that of classical mechanics.

The limit $n \to \infty$ and $h \to 0$ are not the same. Let us point out an example [15]. The energies of a particle of mass m bounded to a cubical box with impenetrable walls of length a are given by

$$E_n = E_1\left(n_x^2 + n_y^2 + n_z^2\right), \quad E_1 = h^2/(8ma^2) \tag{1.43}$$

and $n_{x,y,z} = 1, 2, 3, \ldots$. The frequencies of emission ν_n are obtained from

$$h\nu_n = E_1\left(n_x'^2 + n_y'^2 + n_z'^2\right) - E_1\left(n_x^2 + n_y^2 + n_z^2\right). \tag{1.44}$$

Since $h\nu_n/E_1$ is the difference of integers, it is an integer. Then

$$\nu_n = \left(\frac{E_1}{h}\right)q = \nu_1 q, \quad q - \text{an integer}. \tag{1.45}$$

In any limiting process with n these frequencies do not include the classical continuum.

For high quantum number domain the adjacent frequencies are given by

$$\nu_{n+1} = \nu_n + \frac{h}{I}, \quad I - \text{constant}. \tag{1.46}$$

In the limit $h \to 0$, $\nu_{n+1} \to \nu_n$. $h \to 0$ leads to frequency correspondence but the limit $n \to \infty$ does not.

1.8   Compton Effect

The Franck–Hertz experiment gave an evidence of discrete energy levels of atoms. The particle nature of radiation is established by the Compton effect. In 1923, both theoretically and experimentally, Arthur Holly Compton (Nobel Prize in 1927) showed that during propagation electromagnetic radiation manifests as a stream of particles (photons) [16].

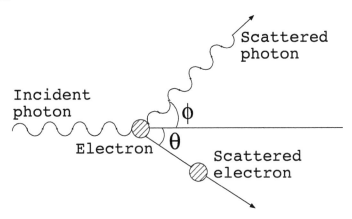

FIGURE 1.3
Scattering of a photon by an electron.

Let us consider a collision between an electron and a photon. Assume that an X-ray photon of frequency ν and energy $E(= h\nu)$ is made to strike a stationary electron. The photon is scattered, and the electron gets an impulse and moves as shown in Fig. 1.3.

Classical theory of X-ray scattering predicts the following:

1. The wavelength of a scattered photon should have a continuous values.

2. The wavelength of incident and scattered photons are the same.

3. Scattering occurs in all directions.

But experimental observations have shown discrepancy from the above results. Compton noticed that when X-rays were scattered by a graphite block the scattered radiation had a longer wavelength in addition to the original wavelength. Further, the longer wavelength component was found to increase with increase in the scattering angle. Considering the scattering as a result of an elastic collision of two particles (photon and electron) Compton accounted the experimental results.

In the scattering process shown in Fig. 1.3, the kinetic energy of the scattered electron has to be from the incident photon. Therefore, the energy E_{sp} of the scattered photon is lower than E_{ip}, the incident photon energy. Thus, the frequency ν' of the scattered photon is lower than the frequency ν of the incident photon. The kinetic energy gained by the scattered electron is thus $h\nu - h\nu'$.

In the collision process momentum must be conserved in the direction of the incident photon as well as in the direction perpendicular to it. The momenta of the photon and the electron before collision are $p = h\nu/c$ and 0, respectively. After collision they become $p' = h\nu'/c$ and P, respectively. From Fig. 1.4 we write the equations of conservation of momenta.

Parallel to the direction of incident photon:

$$\frac{h\nu}{c} + 0 = \frac{h\nu'}{c}\cos\phi + P\cos\theta. \tag{1.47}$$

Perpendicular to the direction of incident photon:

$$0 = \frac{h\nu'}{c}\sin\phi - P\sin\theta. \tag{1.48}$$

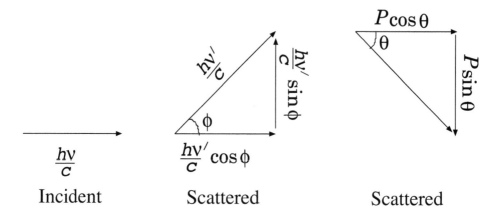

FIGURE 1.4
Vector components of momenta of the incident and scattered photons.

We multiply Eqs. (1.47) and (1.48) by c and obtain

$$Pc \cos\theta = h\nu - h\nu' \cos\phi, \quad Pc \sin\theta = h\nu' \sin\phi. \tag{1.49}$$

θ can be eliminated by squaring the above two subequations and adding them. We obtain

$$P^2 c^2 = (h\nu)^2 - 2h\nu h\nu' \cos\phi + (h\nu')^2. \tag{1.50}$$

The energy conservation equation is

$$E_{\mathrm{ip}} + E_{\mathrm{re}} = E_{\mathrm{sp}} + E_{\mathrm{se}}, \tag{1.51}$$

where $E_{\mathrm{re}} = m_e c^2$ is the energy of the electron at rest and $E_{\mathrm{se}} = \sqrt{P^2 c^2 + m_e^2 c^4}$ is the energy of the scattered electron. Equation (1.51) takes the form

$$h\nu + m_e c^2 = h\nu' + \left(P^2 c^2 + m_e^2 c^4\right)^{1/2}. \tag{1.52}$$

Eliminating square-root in Eq. (1.52) and then substituting Eq. (1.50) for $P^2 c^2$ results in

$$\left[h(\nu - \nu') + m_e c^2\right]^2 = m_e^2 c^4 + (h\nu)^2 - 2h\nu h\nu' \cos\phi + (h\nu')^2. \tag{1.53}$$

Simplifying the above equation we obtain

$$\frac{m_e c}{h}\left(\frac{\nu}{c} - \frac{\nu'}{c}\right) = \frac{\nu}{c}\frac{\nu'}{c}(1 - \cos\phi). \tag{1.54}$$

In terms of $\lambda = c/\nu$ and $\lambda' = c/\nu'$, the above relation becomes

$$\triangle\lambda = \lambda' - \lambda = \frac{h}{m_e c}(1 - \cos\phi). \tag{1.55}$$

The shift in the wavelength $\triangle\lambda$ of the scattered photon is from its collision with the electron at rest and is called *Compton effect*. This shift in the wavelength is not detected by the classical physics. $\triangle\lambda$ depends on h, c, m_e and ϕ. The shift in the wavelength when the scattering is at an angle $\pi/2$ is called the *Compton wavelength* and is given by

$$\lambda_c = \frac{h}{m_e c} = 2.426 \times 10^{-12}\mathrm{m}. \tag{1.56}$$

From Eq. (1.55) we obtain

$$\lambda' = \lambda + \frac{h}{m_e c}(1 - \cos\phi).$$ (1.57)

Substituting $\lambda = c/\nu$, $\lambda' = c/\nu'$ and simplifying the above equation we get

$$\nu' = \frac{\nu}{1 + 2\alpha\sin^2(\phi/2)}, \quad \alpha = \frac{h\nu}{m_e c^2}.$$ (1.58)

Then from Eqs. (1.47) and (1.48) we obtain

$$\tan\theta = \frac{\nu'\sin\phi}{\nu - \nu'\cos\phi} = \frac{1}{(1+\alpha)\tan(\phi/2)}$$ (1.59)

which gives the possible directions of scattering.

From the above analysis the following conclusions were drawn:

1. Equation (1.55) shows that λ' of the scattered radiation (photon) is greater than that of the incident photon.

2. $\triangle\lambda$ is independent of the wavelength of the incident photon and the nature of the target but depends on the angle of scattering.

3. The relation between θ and ϕ, Eq. (1.59), shows that as ϕ increases from 0 to π, θ decreases from $\pi/2$ to 0. That is, the photon is scattered in all directions while the target electron is scattered only in the forward direction.

4. From Eq. (1.56) we write $m_e c^2 = hc/\lambda_c = h\nu_c$. Therefore, λ_c is the wavelength of the light produced if the energy of the electron is converted into a light quanta.

5. Here p (or p') is a particle property while λ is a wave property. There is a paradox in Compton's experiment: *Light manifests both particle-like and wave-like properties.*

Soon after Compton's theoretical work, several scientists including Compton himself carried out experiments with X-rays. Their results confirmed Compton's predictions.

Solved Problem 3:

Express the change in wavelength in terms of Compton wavelength of electron when light is scattered through an angle $\pi/4$ by an electron.

The change in wavelength of scattered light due to a particle of mass m is

$$\triangle\lambda = \lambda' - \lambda = \frac{h}{mc}(1 - \cos\phi) = \lambda_c\frac{m_e}{m}(1 - \cos\phi),$$

where $\lambda_c = h/(m_e c)$. For an electron $m = m_e$ and hence

$$\triangle\lambda = \lambda_c(1 - \cos(\pi/4)) = \lambda_c\left(1 - \frac{1}{\sqrt{2}}\right) = 0.29289\lambda_c.$$

1.9   Specific Heat Capacity

Due to thermal agitations, atoms in solids oscillate about their mean positions. When each atom is considered as a three-dimensional harmonic oscillator then its average thermal energy should be $3k_{\mathrm{B}}T$, where $k_{\mathrm{B}}T$ is that of one-dimensional harmonic oscillator. Then the thermal energy of a solid becomes $3k_{\mathrm{B}}T$ per atom or $3RT = 3Nk_{\mathrm{B}}T$ per gram-atom, where N is the Avogadro number. The rate of increase of thermal energy with T, per gram-atom (called *atomic heat* or specific heat) becomes $3R$ $(= 3 \times 8.3 \times 10^3 \, \mathrm{J \cdot kilomole^{-1} K^{-1}})$, a universal constant. This was confirmed experimentally in solids at normal temperatures. But for sufficiently lower temperature, the specific heat decreases and approach to zero as $T \to 0\mathrm{K}$. In order to explain this, Einstein [17] assumed that the energy of an atom in a solid takes only a discrete set of values: the energy of each component of oscillation is $nh\nu$, $n = 0, 1, \ldots$. The mean energy per atom is obtained as

$$\langle E(\nu) \rangle = \frac{h\nu}{e^{h\nu/k_{\mathrm{B}}T} - 1}. \tag{1.60}$$

Then the atomic heat is given by

$$C = \frac{\mathrm{d}}{\mathrm{d}T} \left(\frac{3Nh\nu}{e^{h\nu/k_{\mathrm{B}}T} - 1} \right) = 3R \left(\frac{h\nu}{k_{\mathrm{B}}T} \right)^2 \frac{e^{h\nu/k_{\mathrm{B}}T}}{\left(e^{h\nu/k_{\mathrm{B}}T} - 1 \right)^2}. \tag{1.61}$$

C given by the above equation decreases with decrease in T. The variation of specific heat of solids is approximately accounted by Eq. (1.61).

1.10   de Broglie Waves

According to special theory of relativity mass and energy are equivalent. Therefore, electromagnetic waves and electrons are considered as different forms of energy. Electromagnetic waves possess particle properties. When the postulates of Planck and Einstein shown that electromagnetic wave posses particle properties, at that time, wave properties of particles were not observed. To have a symmetrical picture Louis de Broglie, a young French prince, in his doctorate thesis (1924), attached waves to all moving particles [18]. For his idea of *matter waves* he was awarded the Nobel Prize in the year 1929.

The matter waves proposed by de Broglie are called *de Broglie waves*. To know the wavelength of a de Broglie wave, let us attach a periodic travelling wave with a moving particle. Denote the wave solution as ϕ and let it be $a_0 \sin(2\pi\nu_0 t_0)$, where a_0 is the amplitude and ν_0 is the frequency observed by an observer at rest with respect to the particle. Suppose the particle's speed is v along the x-axis. Then taking according to the time-dilation

$$t_0 = \frac{t - (\beta x/c)}{\sqrt{1 - \beta^2}}, \quad \beta = v/c \tag{1.62}$$

we have

$$\phi = a_0 \sin \left[\frac{2\pi\nu_0 \left(t - (\beta x/c) \right)}{\sqrt{1 - \beta^2}} \right]. \tag{1.63}$$

Comparison of this with the standard solution

$$\phi = A \sin \left[2\pi\nu \left(t - \frac{x}{u} \right) \right], \tag{1.64}$$

where u is the speed of the wave along the x-direction, gives

$$u = c/\beta = c^2/v, \quad \nu = \nu_0/\sqrt{1-\beta^2}. \tag{1.65}$$

From $E = h\nu_0 = mc^2$ we have $\nu_0 = mc^2/h$. Substituting this expression in Eq. (1.65) we obtain

$$\nu = \frac{mc^2}{h\sqrt{1-\beta^2}}. \tag{1.66}$$

Now, the *(de Broglie)wavelength* λ is $u/\nu = h/p$. λ depends on m and speed of the particle. It is shorter if the particle moves faster. The angular frequency ω and the wave number k of the de Broglie wave are

$$\omega = 2\pi\nu = \frac{2\pi mc^2}{h\sqrt{1-\beta^2}}, \quad k = \frac{2\pi}{\lambda} = \frac{2\pi mv}{h\sqrt{1-\beta^2}}. \tag{1.67}$$

Evidently, the de Broglie wavelength is extremely small for ordinary objects like bullets and balls because of the extreme smallness of h ($= 6.626 \times 10^{-34}$ J s). For example, a cricket ball of 100 g travelling with a speed 120 km/h have a de Broglie wavelength of 1.9878×10^{-34} m which is too small to be detectable. But for an electron accelerated through 150 V, the de Broglie wavelength is $\lambda = \sqrt{150/V} \times 10^{-8}$ cm = 1 Å which is significant in atomic scale.

We note that matter waves are quite different from electromagnetic waves. The latter in vacuum always travels with velocity 3×10^8 m/s while the former do not. de Broglie formula has found several applications. The wavelength of electrons at high speed or high energy is much smaller than those of visible light. Because of this electron microscope have relatively much higher resolutions than those of optical microscopes.

Solved Problem 4:

Find the de Broglie wavelength (λ_d) associated with an electron of energy 20 MeV.

We find $\lambda_d = h/(mv)$, that is,

$$\begin{aligned}
\lambda_d &= \frac{h}{\sqrt{2mE}} \\
&= \frac{6.626 \times 10^{-34}}{(2 \times 9.11 \times 10^{-31} \times 20 \times 10^6 \times 1.6 \times 10^{-19})^{1/2}} \text{ m} \\
&= 0.002744 \text{ Å}.
\end{aligned}$$

1.11 Particle Diffraction

Compton effect shows that light behaves like particles when it collides with electrons. *Can a particle exhibit a wave phenomenon, supporting matter waves concept?* In 1927 Clinton Joseph Davisson, Lester Halbert Germer [19] and George Paget Thomson[9] [20] independently observed the wave nature of particles in their experiments with diffraction of a particle with a single crystal. Then Arthur Jeffrey Dempster in 1927 and Immanuel Estermann and Otto Stern in 1930 obtained diffraction effects with hydrogen and helium by

---

[9]Joseph John Thomson got Nobel Prize for his discovery of electron and showing that it has *particle* character. His son George Paget Thomson received Nobel Prize for showing electron behaves as a *wave*.

reflecting them from crystal gratings. The results of their experiments could be explained only by attributing wave-like character to microscopic particles.

According to classical physics the scattered electrons move in all directions with a moderate dependence of their intensity on angle of scattering and less on the energy of the incident electrons. These results were verified by Davisson and Germer using the target as a polycrystalline nickel solid. They heated the nickel to a high temperature and then cooled it. They got a single crystal with the atoms arranged in a regular lattice. They measured the intensity of scattering in various directions. The intensity of scattering increased from zero in the plane of the target to a highest value in co-latitude $20°$. When the co-latitude and the bombarding potential were fixed and measurement was done in azimuth angle, they observed a variation in the intensity of scattering. At 40 V a slight hump occurred near $60°$ in the co-latitude curve for azimuth 111. This hump then developed rapidly with increasing voltage into a strong spur. The spurs were due to scattered electrons. At 54 V and co-latitude $50°$ the graph of intensity of scattered electron with azimuth angle was shown a series of maxima and minima. When the incident electron beam was replaced by a monochromatic X-rays, phenomena noticed with the electron were observed. If the electrons were particles then the above result cannot be explained. Thus, the experiment of Davisson and Germer shown that a beam of electrons possesses wave-like characteristics. According to de Broglie hypothesis *the electron waves were diffracted by the single crystal.*

The de Broglie wavelength λ of the diffracted electrons obtained from $n\lambda = 2d\sin\theta$ agreed with the electron wavelength $\lambda = h/(mv)$, thereby verifying the wave nature of moving bodies. Beams of neutrons, hydrogen atom and helium atom have produced diffraction patterns.

The wave properties of electrons are utilized in many practical ways. For example, in an electron microscope, electrons replace waves due to their shorter wavelengths compared to visible light and thus able to resolve much finer detail. Further, electron and neutron wave effects are useful to investigate defects in metals. A beam of neutron waves directed towards a target nuclei can be tuned in frequency to resonate with the natural interval frequencies of the nuclei.

1.12   Wave-Particle Duality

In classical mechanics, we have two types of phenomena: *waves* and *particles*. As we know waves spread in space and carry energy with them but do not transport mass. In contrast, particles are localized and have both energy and mass. In quantum mechanics these differences are unclear. In certain cases objects thought of as particles behave like waves. In some cases waves behave like particles. For example, the photoelectric effect can only be explained if the light has a particle nature. On the other hand, electrons can give rise to wave-like diffraction patterns. As suggested by de Broglie, *all objects have both particle and wave aspects.* The different aspects manifest according to the type of process it undergoes. This is known as the principle of *wave-particle duality.* In this section, we demonstrate it by considering the famous Young's double-slit experiment.

In 1803, Thomas Young (1773–1829) demonstrated that when a light beam passes through two narrow slits, the waves coming out from these slits interfere with one another

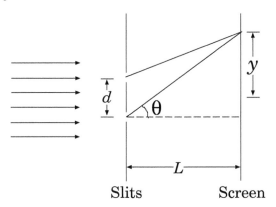

Slits   Screen

FIGURE 1.5
Schematic arrangement of a double slit experiment.

and produce interference patterns on a photographic plate[10]. This experiment exemplifies the wave and particle nature of light.

Here, we take up three experiments similar in concept with a double slit [22, 23]. Suppose a light is sent to a screen through a double slit (Fig. 1.5). A photographic film on the screen is used to extract information created by the incident light. We consider the following three cases:

1. Slit-1 (upper slit) is opened while slit-2 (lower slit) is closed.

2. Slit-2 is opened and slit-1 is closed.

3. Both the slits are opened.

Image patterns are recorded on photographic film for each experiment. If we measure the intensity of the light on the film we get the results as indicated in Fig. 1.6. When slit-1 or slit-2 alone is opened there is no interference. When both the slits are opened we observe an interference pattern with maxima and minima, but we expect only two bright lines aligned with the slits.

The condition for the constructive interference patterns (bright bands) at the photographic film in the case of lights were in phase when they were passing through the slits is

$$d \sin \theta = m\lambda, \quad m = \pm 1, \pm 2, \dots . \tag{1.68a}$$

The condition for destructive interference patterns (dark bands) is

$$d \sin \theta = \left(m + \frac{1}{2} \right) \lambda, \quad m = \pm 1, \pm 2 \dots . \tag{1.68b}$$

These dark and bright regions are called *interference fringes*. The spacing between successive dark and bright fringes is $\Delta = \lambda L / d$. In optics, the wave amplitude is generally a complex number, ϕ. The quantity, $\phi^* \phi = |\phi|^2$ is the intensity I of the light. Now, if ϕ_1 and ϕ_2 are the amplitudes of the waves through slits 1 and 2, respectively, then

$$I_1 = |\phi_1|^2, \quad I_2 = |\phi_2|^2, \quad I_3 = |\phi_1 + \phi_2|^2 = I_1 + I_2 + 2\mathrm{Re}\left(\phi_1^* \phi_2\right). \tag{1.69}$$

---

[10]Young's double-slit experiment has been rated as the most beautiful experiment in physics by the readers of *Physics World* in 2002 [21].

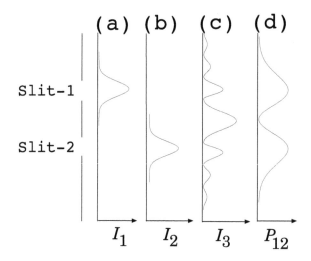

FIGURE 1.6
Intensity patterns in the double slit experiment. (a) Photon intensity I_1 on the screen with slit-1 only opened. (b) Photon intensity I_2 with slit-2 only opened. (c) Interference pattern when both the slits were opened. (d) Probability distribution P_{12} of photons when both the slits were opened and when the film was replaced by an array of photon detectors.

$I_3 \neq I_1 + I_2$ is the characteristic of interference pattern. $2\mathrm{Re}(\phi_1^* \phi_2)$ describes the interference of the waves passing through the two slits.

In the second experiment, assume that the source releases the photons slowly so that photons hit the screen without interaction. We do not expect interference pattern. Surprisingly, we get the interference pattern as in Fig. 1.6. This means that a single particle is also interfering with itself. *What will happen if we replace the photographic film by particle detectors?*

Suppose we modify the set-up to determine through which slit each particle goes through. We place one particle detector just behind slit-1 and another particle detector just behind slit-2 [24]. The screen is also present. The detectors are such that they do not absorb entire energy of the particles but only a fraction of energy and let the particles go toward the screen. A few billion particles (photons or microscopic particles) are allowed to pass through the slits. We close slit-2 first and let the microscopic particles pass through slit-1. We get the pattern similar to the one shown in Fig. 1.6a. We repeat the experiment by opening slit-2 and closing the other. We record the result and is as in Fig. 1.6b. Finally, we open both slits. The resulting pattern is as in Fig. 1.6d. The interference pattern is not realized. The result is that the detecting act forced each particle goes through one of the slits and not through both the slits simultaneously. The above experiment was performed with photons and atoms [25,26].

In the first experiment, interference occurred because the same light wave went through the two slits. Interference took place in the second experiment also, where the photons (particles) were sent one by one so that they did not interact. Here the photons behaved as a wave. These same individuals behaved as particles when a detector was used to sense the slit through which each of them passed through. The above experiments clearly demonstrate the dual nature of microscopic particles.

An interesting observation is that we have interference in the absence of observers and no interference pattern when observers are present. In other words, when we carry out a measurement, it disturbs the trajectories of photons and the results of the experiment satisfy with the classical physics – two bright lines on the film. But when we cease the measurement, the pattern become multiple lines with brightness and darkness. More precisely, particle cannot behave as a particle and a wave simultaneously. If an experiment is performed for particle character, then particle property is observed. Similarly, if an experiment looks to observe wave nature, then wave property is realized.

To explain the wave-particle duality Bohr assumed that we are able to measure a wave property of light or particle property of light but not both simultaneously. This is called *complementarity principle* of light. In 1992 Indian physicists Partha Ghose, Dipankar Home and Girish S. Agarwal [27] proposed an experimental set-up to observe both wave and particle properties of photon simultaneously. The experiment was then performed by the Japanese scientists Y. Mizobuchi and Y. Ohtake [28] using a pulsed laser. The experiment caught the wave and particle behaviours of photons simultaneously.

1.13 Concluding Remarks

In this chapter, we considered some of the experimental observations which were not adequately explained by classical theory and outlined how they have been properly interpreted by the quantum theory. In the quantum theory, certain dynamical variables take on only discrete or quantized values whereas in classical theory they take on a continuum of values. The discreteness property gave rise to the name quantum mechanics or quantum theory.

After Planck's idea of discrete energy values, initially, in the development of a new theory (quantum theory), it was thought that classical mechanics was correct but with the assumption that certain observables like energy and angular momentum take only discret values. This is often referred to as *old quantum theory*. For example, in 1905 in his work on the photoelectric effect Einstein postulated that the energy of a beam of light is quantized. In 1913 Bohr applied the idea of energy quantization to atoms and explained the various colours emitted by hydrogen glowing in a discharge tube. In 1911 Arnold Sommerfeld analyzed the influence of energy quantization on photon and momentum. We wish to stress that *in all the above work the concept of wave equation and wave function did not appear!*

Very soon after de Broglie's proposal of matter waves, Erwin Schrödinger, Werner Karl Heisenberg and Paul Adrian Maurice Dirac developed quantum theory with three different approaches. The approach of Heisenberg involved matrix calculus. Dirac introduced operator theory. In 1926 Schrödinger starting with the idea of matter waves, set up a wave equation known as Schrödinger wave equation and then the theory has seen tremendous fundamental developments. In this book we present various developments and essential features of the quantum theory.

Quantum theory is an amazingly successful theory giving predictions in very good agreement with experimental observations in minute detail. The range of phenomena to which the quantum theory has been applied is enormous. It covers phenomena from elementary particle level to the early universe. Many modern technologies would be impossible without quantum physics. Lasers, semiconductors and superconductors are just few among the various applications of the most radical predictions of quantum mechanics. Many ordinary scale physical phenomena are also explained by quantum theory. These include the existence of solid bodies, physical properties of materials, phenomena of freezing and boiling, etc.

1.14   Bibliography

[1]   M. Planck, *Annalen der Physik* 4:553, 1901.

[2]   R.L. Bobst and E. A. Karlow, *Am. J. Phys.* 53:911, 1985.

[3]   J.D. Barnett and H. T. Stokes, *Am. J. Phys.* 56:86, 1988.

[4]   R.E. Crandall and J. F. Delord, *Am. J. Phys.* 51:90, 1983.

[5]   J. Dryzek and K. Ruebenbauer, *Am. J. Phys.* 60:251, 1992.

[6]   A. Einstein, *Annalen der Physik* 17:132, 1905.

[7]   G.N. Lewis, *Nature* 118:874, 1926.

[8]   N. Bohr, *Phil. Mag.* 27:506, 1914.

[9]   J. Franck and G. Hertz, *Verhandl. deut. Physik. Ges.* 16:457, 512, 1914.

[10] D.W. Preston and E.R. Dietz, *The Art of Experimental Physics.* John Wiley, New York, 1991.

[11] G.F. Hanne, *Am. J. Phys.* 56:696, 1988.

[12] A. Sommerfeld, *Annalen der Physik* 51:1, 1916.

[13] W. Gerlach and O. Stern, *Z. Phys.* 8:110, 1922.

[14] B.L. Van der Waerden, *Sources of Quantum Mechanics.* Dover, New York, 1967.

[15] R.L. Liboff, *Physics Today* February pp. 50, 1984.

[16] A.H. Compton, *Phys. Rev.* 21:483, 1923; 22:409, 1923.

[17] A. Einstein, *Annalen der Physik* 22:180, 1907.

[18] L. de Broglie, *Phil. Mag.* 47:446, 1924; *Annals d. Physique* 3:22, 1925.

[19] C.J. Davisson and L.H. Germer, *Nature* 119:558, 1927; *Phys. Rev.* 30:705, 1927.

[20] G.P. Thomson, *Nature* 120:802, 1927; *Proc. Roy. Soc. London A* 117:600, 1928.

[21] P. Rogers, *Physics World* September pp.19, 2002.

[22] N.D. Mermin, *Physics Today* 46:9, 1993.

[23] G. Venkataraman, *The Quantum Revolution-I: The Breakthrough.* University Press, Hyderabad, 1994.

[24] R. Rolleigh, *The Double Slit Experiment and Quantum Mechanics.* 2010. https://www.hendrix.edu/uploadedFiles/Departments_and_ Programs/Physics-/Faculty/DS5Quantum.pdf.

[25] S. Durr, T. Nonn and G. Rempe, *Phys. Rev. Lett.* 81:5705, 1998.

[26] Y.H. Kim, R. Yu, S.P. Kulik, Y. Shih and M.O. Scully, *Phys. Rev. Lett.* 84:1, 2000.

[27] P. Ghose, D. Home and G.S. Agarwal, *Phys. Lett. A* 168:95, 1992.

[28] Y. Mizobuchi and Y. Ohtake, *Phys. Lett. A* 168:1, 1992.

1.15   Exercises

1.1 Calculate the energy of a photon of 253.6 nm light. State the energy in eV (1 eV = 1.6×10^{-19} J).

1.2 Calculate the wavelength of a quantum of electromagnetic radiation with energy 4000 eV.

1.3 The Planck's radiation formula for an ideal black body is

$$E(\nu)d\nu = \frac{8\pi\nu^2 d\nu}{c^3}\,\frac{h\nu}{e^{h\nu/k_B T}-1},$$

where $E(\nu)d\nu$ is the energy density in the frequency interval between ν and $\nu+d\nu$. Determine $\lambda_{\max}$.

1.4 Consider the exercise 1.3. Examine the spectral shift according to Wien's displacement law $\lambda_{\max}T = $ constant if the value of h is doubled.

1.5 A light of wavelength 1000 nm produces photoelectrons from a metal. The threshold potential V_0 is 0.5 V. Calculate the work function ϕ_0.

1.6 Consider the previous exercise. Calculate the threshold wavelength λ_c.

1.7 Calculate the frequency of light required to produce electrons of kinetic energy 5 eV from illumination of a material whose work function is 0.5 eV.

1.8 Light of wavelength 500 nm is incident on a metal with work function (ϕ_0) of 0.75 eV. What is the maximum kinetic energy ($K_{\max}$) of the ejected photoelectrons?

1.9 Suppose you are standing in front of a 40 W incandescent light bulb 5 m away. If the diameter of your pupils is about 2 mm, about how many photons, n, enter your eye every second?

1.10 Calculate the minimum voltage required for an electron to give all of its energy in a collision with a target and produce an electromagnetic radiation of wavelength 0.05 nm.

1.11 Show that the spacing between two minima in the Franck–Hertz curve increases linearly with the order n.

1.12 Find the wavelength of photons emitted by mercury atoms in Franck–Hertz experiment if an electron of energy 6 eV passed through mercury vapour with an energy 1 eV.

1.13 Assume that the resolving power of a microscope is equal to the wavelength of the light used. If electrons are used as the light source calculate the kinetic energy of electrons needed to have the resolving power as 10^{-11} m.

1.14 Suppose a laser provides 100 TW of power in 1ns pulses at a wavelength of 0.26 μm. How many photons are contained in a single pulse?

1.15 Express the change in wavelength in terms of Compton wavelength of electron when light is scattered through an angle $\pi/4$ by a proton.

1.16 A beam of X-rays of wavelength $\lambda = 0.738$ Å is Compton scattered by electrons through an angle $\pi/2$. Calculate the wavelength and the energy of the scattered radiation.

1.17   Show that the de Broglie wavelength λ_d associated with photons is equal to the electromagnetic wavelength λ.

1.18   A 5.78 MeV α-particle (assume that its velocity is $\ll c$) is emitted in the decay of radium. If the diameter of the radium nucleus is 2×10^{-14} m, how many α-particle de Broglie wavelengths fit inside the nucleus?

1.19   A diatomic gas molecule consisting of two atoms of mass m separated by a fixed distance d is rotating about a central axis perpendicular to the line joining the two atoms. Assuming that its angular momentum is quantized as in the Bohr atom, determine the possible angular velocities (ω) and the possible quantized rotational energies (E).

1.20   In a particle diffraction experiment a beam of 54 eV electrons are perpendicularly diffracted to a single crystal having spacing of the planes is $a = 0.91$ Å. The angle of incidence and scattering relative to the family of Bragg planes is $\theta = 65°$. Determine the wavelength of the diffraction electron from $n\lambda = 2a \sin\theta$ with $n = 1$ and $\lambda = h/(mv)$. What do you observe?

2

Schrödinger Equation and Wave Function

2.1 Introduction

Till 1925 the idea of what the quantum theory would be was unknown. During 1925 three different but equivalent versions of quantum theory were proposed – Schrödinger proposed wave mechanics; Heisenberg developed matrix mechanics and Dirac-introduced operator theory. Considering the de Broglie's matter waves Erwin Schrödinger, an Austrian physicist, argued that if a particle like an electron behaves as a wave then the equation of wave motion could be successfully applied to it. He postulated a function varying in both space and time in a wave-like manner (hence called *wave function* and denoted it as ψ). This function is generally complex and assumed to contain information about a system. Schrödinger set up a linear and time-dependent wave-like equation, called *Schrödinger wave equation*, to describe the wave aspect of a particle taking account of de Broglie's relation for wavelength. Physically, $|\psi(\mathbf{X}, t)|^2$, where $\psi(\mathbf{X}, t)$ is the solution of the Schrödinger equation, is interpreted as position probability density. That is, $|\psi(\mathbf{X}, t)|^2$ is the probability density of observing a particle at position $\mathbf{X}$ at time t. ψ does not give exact outcomes of observations but helps us to *know all possible events and their probabilities*. Further, the probability interpretation allows us to find the average or expected result of a set of measurements on a quantum system.

There are several interesting features of the wave function ψ. The probability interpretation of $|\psi|^2$ imposes certain conditions on meaningful ψ. Further, the $|\psi|^2$ satisfies a conservation law, an equation analogous to the continuity equation of flow in hydrodynamics. To set the total probability unity the ψ must satisfy the normalization condition $\int_{-\infty}^{\infty} |\psi|^2 d\tau = 1$. Knowing ψ we can compute expectation values of variables such as position, momentum, etc. In quantum mechanics the experimentally measurable variables are no longer dynamical variables but they become operators. The outcomes of experiments are the eigenvalues of the operators of the observables. An interesting result is that the equations of the time evolution of expectation values of position and momentum operators (called *Ehrenfest's theorem*) obey the same equation of motion of these variables in classical mechanics. On the other hand, the equation for the phase of the ψ function takes the form of Hamilton–Jacobi equation of classical mechanics. In this chapter, we set up the Schrödinger equation and discuss the above mentioned general properties of the solution of the Schrödinger equation.

2.2 Construction of Schrödinger Equation

In this section, we construct the wave equation for a particle moving in a potential $V(\mathbf{X}, t)$. We start with a free particle then generalize the result for a particle in a potential $V(\mathbf{X}, t)$.

DOI: 10.1201/9781003172178-2

2.2.1 Schrödinger Equation for a Free Particle

Suppose a particle is free and moving along a one-dimensional line. Assume that it behaves like a wave and its wave function is of the form

$$\psi(x,t) = a\mathrm{e}^{\mathrm{i}(kx-\omega t)}\,. \tag{2.1}$$

It is the simplest plane monochromatic wave. ψ refers to the case of wave propagating along x-direction whereas the disturbance is in a plane perpendicular to the x-direction. ψ given by Eq. (2.1) is called a *plane wave* since its amplitude $|\psi| = \sqrt{\psi^*\psi} = \sqrt{a^*a} = |a|$ is a constant. Schrödinger combined the particle and wave properties in constructing the wave equation. The angular frequency ω of the above wave should correspond to the energy E of the particle while the wave number k should correspond to the momentum $p_x = mv$. According to Planck and Einstein the relation between E of a particle and $\omega = 2\pi\nu$ of the wave is

$$E = \hbar\omega\,. \tag{2.2}$$

According to de Broglie

$$p_x = \frac{h}{\lambda} = \frac{h}{2\pi}\frac{2\pi}{\lambda} = \hbar k\,. \tag{2.3}$$

The wave equation satisfying the solution Eq. (2.1) and the relations (2.2)-(2.3) is constructed as follows.

Differentiating Eq. (2.1) with respect to t we obtain

$$-\mathrm{i}\frac{\partial\psi}{\partial t} = -\omega\psi\,. \tag{2.4}$$

Differentiation of Eq. (2.1) twice with respect to x gives

$$\frac{\partial^2\psi}{\partial x^2} = -k^2\psi\,. \tag{2.5}$$

Replacing k by $p_x/\hbar$ (Eq. (2.3)) and p_x in terms of E using $E = p_x^2/2$ (free particle energy) and then E by $\hbar\omega$ (Eq. (2.2)) we get

$$\frac{\hbar}{2m}\frac{\partial^2\psi}{\partial x^2} = -\omega\psi\,. \tag{2.6}$$

Comparison of Eqs. (2.4) and (2.6) leads to the equation

$$\mathrm{i}\hbar\frac{\partial\psi}{\partial t} = -\frac{\hbar^2}{2m}\frac{\partial^2\psi}{\partial x^2}\,. \tag{2.7}$$

Equation (2.7) is called a *Schrödinger wave equation* or ψ-equation for a free particle in one-dimension. For a free particle in three-dimensions the wave equation is

$$\mathrm{i}\hbar\frac{\partial\psi}{\partial t} = -\frac{\hbar^2}{2m}\left(\frac{\partial^2}{\partial x^2} + \frac{\partial^2}{\partial y^2} + \frac{\partial^2}{\partial z^2}\right)\psi\,. \tag{2.8}$$

Multiplying the equation $E = \mathbf{p}^2/(2m)$ by ψ on both sides we get

$$E\psi = \frac{1}{2m}\mathbf{p}^2\psi\,. \tag{2.9}$$

Comparison of this equation with Eq. (2.8) gives

$$E = \mathrm{i}\hbar\frac{\partial}{\partial t}\,, \quad \mathbf{p} = -\mathrm{i}\hbar\nabla\,, \quad p_x = -\mathrm{i}\hbar\frac{\partial}{\partial x}\,, \tag{2.10}$$

where $\nabla = \mathbf{i}\partial/\partial x + \mathbf{j}\partial/\partial y + \mathbf{k}\partial/\partial z$.

2.2.2 Schrödinger Equation for a Particle in a Potential

Since $E - V = p_x^2/(2m)$, the wave equation for $V(x) \neq 0$ can be obtained from Eq. (2.7) by simply replacing $i\hbar\partial/\partial t$ by $i\hbar\partial/\partial t - V(x)$:

$$i\hbar\frac{\partial\psi}{\partial t} = -\frac{\hbar^2}{2m}\frac{\partial^2\psi}{\partial x^2} + V(x)\psi\,. \tag{2.11}$$

For a three-dimensional system the Schrödinger equation is[1]

$$i\hbar\frac{\partial\psi}{\partial t} = -\frac{\hbar^2}{2m}\nabla^2\psi + V(\mathbf{X})\psi\,, \quad \nabla^2 = \left(\frac{\partial^2}{\partial x^2} + \frac{\partial^2}{\partial y^2} + \frac{\partial^2}{\partial z^2}\right)\,. \tag{2.12}$$

Equation (2.12) describes the evolution of $\psi(\mathbf{X}, t)$ in the presence of a potential function $V(\mathbf{X})$. Whenever we make a measurement on a quantum system, the results are dictated by ψ. The Schrödinger equation is used and accepted because it leads to results in agreement with observation wherever it is applied properly.

2.2.3 Schrödinger Equation in Operator Form

Equation (2.12) shows that

$$-\frac{\hbar^2}{2m}\nabla^2 + V(\mathbf{X}) = i\hbar\frac{\partial}{\partial t}\,. \tag{2.13}$$

This operator is called the *Hamiltonian operator*. In classical mechanics, Hamiltonian is a basic dynamical variable from which we obtain Hamilton's equation of motion describing the evolution of state of a system. In quantum mechanics the Hamiltonian operator of the system essentially determines the time variation of ψ. Note that to set up the wave equation for a quantum mechanical system we require classical Hamiltonian. Then we obtain the wave equation by replacing the classical dynamical variables $\mathbf{X}$, $\mathbf{p}$, $V(\mathbf{X})$ and E by $\mathbf{X}$, $-i\hbar\nabla$, $V(\mathbf{X})$ and $i\hbar\partial/\partial t$, respectively, and operating both sides of the equation $E = H(\mathbf{X}, \mathbf{p}, t)$ on ψ.

Replacing dynamical variables by operators is called *first quantization*. Replacing ψ by an operator is termed as *second quantization* and is the case in quantum field theory.

2.3 Solution of Time-Dependent Equation

Let us assume that the potential V is independent of time and search the solution of Eq. (2.12) of the form

$$\psi(\mathbf{X}, t) = \phi(\mathbf{X})f(t)\,. \tag{2.14}$$

Substitution of the above solution in Eq. (2.12) gives

$$i\hbar\frac{\dot{f}}{f} = -\frac{\hbar^2}{2m}\frac{1}{\phi}\nabla^2\phi + V\,. \tag{2.15}$$

The left-side of Eq. (2.15) is function of t only while the right-side is function of space variables only. Therefore, both sides of Eq. (2.15) must be equal to a same constant. Comparing

---

[1]In 1987 the Schrödinger equation appeared on the first-day postmark of the Austrian stamp commemorating Schrödinger's 100th anniversary [1].

the right-side of Eq. (2.15) with $H\phi = E\phi$ we write

$$-\frac{\hbar^2}{2m}\nabla^2\phi + V\phi = E\phi\,. \tag{2.16}$$

Rewrite the above equation as

$$\nabla^2\phi(\mathbf{X}) + \frac{2m}{\hbar^2}\left[E - V(\mathbf{X})\right]\phi(\mathbf{X}) = 0\,. \tag{2.17}$$

Equation (2.17) is called *time-independent Schrödinger equation* or *stationary state Schrödinger equation*.

Further, the solution of $i\hbar\dot{f}/f = E$ is

$$f(t) = a\,e^{-iEt/\hbar}\,, \tag{2.18a}$$

where a is arbitrary. Different values of E give different solutions. We write

$$f_n(t) = a_n\,e^{-iE_nt/\hbar}\,. \tag{2.18b}$$

Then ψ is given by

$$\psi(\mathbf{X},t) = \sum_{n=0}^{\infty} a_n\phi_n(\mathbf{X})e^{-iE_nt/\hbar} + \int a_E e^{-iEt/\hbar}\phi(\mathbf{X})\,dE\,, \tag{2.19}$$

where the first term in the right-side corresponds to discrete values of E while the second term is for the continuous range of values of E. In Eq. (2.19) ϕ_n and ϕ are solutions of Eq. (2.17).

2.4   Physical Interpretation of $\psi^*\psi$

In connection with the wave function ψ we ask:

1. *Is ψ an experimentally measurable quantity?*
2. *How is ψ related to the physical properties of the quantum mechanical system?*
3. *How do we determine the properties of the system from ψ?*

Schrödinger thought that ψ described a particle; that is, ψ was in some sense regarded as matter waves like sound waves and water waves. But this interpretation had several difficulties. First, when we try to locate the particles using, say, detectors we find them concentrated in a small region – it is *localized*. But the wave function $\psi(\mathbf{X},t)$ at a given time spreads out like a wave – it is *nonlocalized*. Second, suppose we interpret ψ as a wave in physical space and time of one particle. In the case of n particles, the wave has to be represented in a $3n$-dimensional space. The third point is that ψ is generally complex. We can give interpretation to real waves only. *How do we account for complex values of ψ?* Therefore, a new interpretation is essential.

ψ is in some sense related to the presence of a particle. Therefore, we do not expect a particle in the regions of space where $\psi = 0$. On the other hand, ψ must be nonzero in the region where there is a particle.

In the year 1926, Max Born (Noble Prize in 1954) considered the collision of two particles. Far before and far after the collision the two particles behave as free particles. Using a perturbation theory he obtained

$$\psi_{\text{scat}}(r \to \infty) \approx \sum_f c_{fi}\phi_f\,, \quad c_{fi} = \int \phi_f^* V \phi_i \, d\tau\,. \tag{2.20}$$

Here ϕ_f is the solution of the stationary state Schrödinger equation and represents the state of the particle far after scattering (called final state) with momentum $\mathbf{K}_f$. ϕ_i is the initial state solution (state far before collision) with the momentum $\mathbf{K}_i$ and V is the potential. In Eq. (2.20) suppose $|c_{fi}|^2$ is related to the number of particles scattered with final state ϕ_f and initial state ϕ_i. That is, $|c_{1i}|^2$, $|c_{2i}|^2$, ... represent, far after scattering, the number of particles (or percentage of incident particles) in the states ϕ_1, ϕ_2,..., respectively, with all their state far before interaction is ϕ_i. In this case ψ would be describing the particles themselves. Assume a barrier potential having the property that it would reflect 60% of the incident particles and scatter 40% of them. There is no difficulty with the above meaning of c_{fi} and ψ. Suppose only one particle incident on the barrier. Then according to the above 60% of it is to be scattered and 40% is to be reflected. Born realized that since a single particle cannot divide itself, ψ could not describe the particle itself. Therefore, only one interpretation is possible for the coefficient c_{fi} – *it is the probability amplitude that the incident particle with momentum $\mathbf{K}_i$ to be scattered with momentum $\mathbf{K}_f$.* We have no answer to the question *What is the state after the collision?* but we have the answer to the question *What is the probability of a given effect of the collision?*

Considering the above Max Born and Ernst Pascual Jordan interpreted $\psi^*\psi$ as position probability density. They interpreted $|\psi|^2$ as

$$P\,dxdydz = \psi^*\psi dxdydz = |\psi|^2 \, dxdydz\,, \tag{2.21}$$

where $Pdxdydz$ is the probability that the system will be in the volume element $dxdydz$ at the position $\mathbf{X}$. The basis for Eq. (2.21) is that in the classical theory of light, the intensity of light at each point is given by the square of the amplitude of a wave. This indicates that, in quantum mechanics a wave function can be introduced playing the role of a probability. Note that, ψ itself cannot be a probability. This is because ψ may be complex and negative but probability is always real and positive. ψ has no direct physical significance and is not experimentally measurable. It is not a physical entity existing in space such as a cricket ball or electric field. Its purpose is to calculate $|\psi|^2$. That is, it is a mathematical tool for computing the outcomes of observations.

According to probability interpretation of $\psi^*\psi$, we can never say that quantum particle exists or not at a certain place[2]. Born, Bohr and his group in Copenhagen analyzed the probability interpretation of $\psi^*\psi$ in great depth. For this reason probabilistic interpretation of $\psi^*\psi$ is also referred as the *Copenhagen interpretation*. Other interpretations have also been proposed. For example, *sum-over histories* [3,4] by Richard Feynman and *many worlds interpretation* [5,6] by Huge Everett and John Cramer's *transactional interpretation* [7,8] have also received considerable interest. The point is that a given initial conditions can give a range of final outcomes, so quantum theory makes the future open and hinder our ability to know the past.

---

[2]In connection with the above Robert Oppenheimer made a remark that [2] *we must say 'no' for whether the position of the electron remains the same or changes with time or the electron is at rest or it is in motion.*

2.5 Conditions on Allowed Wave Functions

The probability interpretation of $\psi^*\psi$ leads to the following conditions on ψ.

1. Integration of $\psi^*\psi$ with respect to space variables over all space should be unity:

$$\int_{-\infty}^{\infty} \psi^*\psi \, \mathrm{d}\tau = 1 \,. \tag{2.22}$$

2. $|\psi| \to 0$ as $|\mathbf{X}| \to \infty$.
3. ψ must be single-valued at each and every point.
4. ψ must not have the value of infinite at any point.
5. ψ, ψ_x, ψ_y and ψ_z must be continuous functions of $\mathbf{X}$.

The integral in Eq. (2.22) represents total probability. Since total probability must be one for any event ψ must satisfy the condition Eq. (2.22). Such a wave function is said to be *normalized*. The condition Eq. (2.22) is called *normalization condition*. If

$$\int_{-\infty}^{\infty} \psi^*\psi \, \mathrm{d}\tau = N^2 \,, \quad N^2 = \text{constant} \neq 1, \neq \infty \tag{2.23}$$

then ψ is said to be *square-integrable* and the constant N^2 is called *norm of ψ*. Such a wave function can always be normalized by multiplying it by the constant $1/N$. This is possible because the Schrödinger equation is linear and therefore if ψ is a solution of it then $\alpha\psi$, where α is a constant, is also a solution. The finiteness requirement of the norm imposes the condition (2). This means that $\psi^*\psi \to 0$ as $|\mathbf{X}| \to \infty$ and thus the particle is confined to a finite region of space. Such a state is said to be a *bound state*. It is called a bound state because the corresponding classical particle motion is bounded. If $\psi(\mathbf{X}, t)$ does not tend to 0 as $|\mathbf{X}| \to \infty$ then it is *nonnormalizable*. Such a state is called a *scattering state*. The motion of the corresponding classical particle is unbounded.

Note that an overall phase $e^{i\alpha}$ in a wave function is unobservable. ψ and $e^{i\alpha}\psi$ both describe the same physical state. But if a constant phase appears as a relative phase between two ψ's in a linear superposition as in $\psi = \psi_1 + e^{i\alpha}\psi_2$ then it is observable; it gives interference effects and a shift in α introduces a shift in interference fringes.

Is $\psi(\mathbf{X}, t)$ a dimensionless quantity? If not, what is the dimension of it? The probability of finding a particle in a given volume $\mathrm{d}\tau = \mathrm{d}x\mathrm{d}y\mathrm{d}z$ at a point $\mathbf{X}$ at a time t is $|\psi(\mathbf{X}, t)|^2\mathrm{d}x\mathrm{d}y\mathrm{d}z$. Probability is a dimensionless quantity. Therefore, $|\psi|^2\mathrm{d}x\mathrm{d}y\mathrm{d}z$ should also be dimensionless. $\mathrm{d}x$, $\mathrm{d}y$ and $\mathrm{d}z$ have the dimension of length. Therefore, the dimension of $\psi(\mathbf{X}, t)$ is $(\text{length})^{-3/2}$. This is for a single particle in three-dimensional space. For N-particles in a d-dimensional space the dimension of the wave function is $(\text{length})^{-Nd/2}$. As the dimensions of x and p_x are L and MLT^{-1}, respectively, the dimension of $\hbar$ is dimension of $x\times$ dimension of p_x and is $\mathrm{ML}^2\mathrm{T}^{-1}$.

Solved Problem 1:

If $\psi = \mathrm{sech}\,ax$, calculate the probability of finding the particle right of the point $x = 0$.

First, the wave function must be normalized. For this purpose we write $\psi = N\,\mathrm{sech}\,ax$. The normalization condition becomes

$$1 = N^2 \int_{-\infty}^{\infty} \mathrm{sech}^2(ax) \, \mathrm{d}x = \frac{N^2}{a}\tanh(ax)|_{-\infty}^{\infty} = \frac{2N^2}{a} \,. \tag{2.24}$$

Thus, $N = \sqrt{a/2}$ and $\psi = \sqrt{a/2}\,\mathrm{sech}\,ax$. Now, the probability of finding the particle right of the point $x = 0$ is

$$P = \int_0^\infty \psi^*\psi\,\mathrm{d}x = \frac{a}{2}\int_0^\infty \mathrm{sech}^2(ax)\,\mathrm{d}x = \frac{1}{2}\,. \tag{2.25}$$

Solved Problem 2:

If the normalized wave function of a particle is $\psi = \dfrac{1}{\sqrt{a}\pi^{1/4}}\mathrm{e}^{-x^2/(2a^2)}$ find the region of space, where the particle is most likely to be found.

The probability density is given by $|\psi|^2 = (1/(a\sqrt{\pi}))\mathrm{e}^{-x^2/a^2}$. This is a simple function. We use the graph of $|\psi|^2$ versus x to find the maximum value of $|\psi|^2$. It is maximum at $x = 0$ and decreases exponentially for $|x| > 0$. Thus, the particle is most likely to be found in a region of width a on either side of $x = 0$. For the complicated form of $|\psi|^2 = f(x)$ the conditions $\mathrm{d}f/\mathrm{d}x = 0$ and $\mathrm{d}^2 f/\mathrm{d}x^2 < 0$ are to be used to determine a maximum of $f(x)$.

2.6 Box Normalization

Normalization of a given wave function is possible only if the norm, the value of $\int_{-\infty}^{\infty}\psi^*\psi\,\mathrm{d}\tau$, is finite. But for many physically interesting wave functions the value of this integral becomes infinite. For example, for the wave function given by Eq. (2.1) the value of the integral is infinite. Hence, it is nonnormalizable. However, such wave functions are widespread in quantum mechanics, specifically in scattering problems, and they have considerable mathematical advantages.

An approach of handling a nonnormalizable ψ is the following. Assume that the particle is confined within a large box and that the $\psi^*\psi$ is taken over the interior of the box only. Discard the infinite region of space outside the box. Now, the norm is finite; it can be made unity by multiplying a suitable constant. For any calculation use this normalized wave function. After completion of calculation, impose the limit of infinite volume for the box. This procedure is called *box normalization*. In this way the nonnormalizable ψ is visualized as a limit of realizable and normalizable wave function.

Let us perform the above scheme to the wave function, Eq. (2.1). Suppose the box is a cube with edges of length L parallel to the x, y, z axes and its centre is the origin. Then the normalization gives

$$1 = \iiint_{-L/2}^{L/2}\psi^*\psi\,\mathrm{d}x\mathrm{d}y\mathrm{d}z = |a|^2\iiint_{-L/2}^{L/2}\mathrm{d}x\mathrm{d}y\mathrm{d}z = L^3|a|^2\,. \tag{2.26}$$

Hence, $a = L^{-3/2}$ and the normalized wave function is

$$\psi = L^{-3/2}\,\mathrm{e}^{-\mathrm{i}(\mathbf{k}\cdot\mathbf{X}-\omega t)}\,. \tag{2.27}$$

Note that the condition $\psi(\mathbf{X}, t) \to 0$ as $|\mathbf{X}| \to \infty$ is not applicable to the above ψ. Impose that ψ is periodic with respect to L of the box. Then $\mathrm{e}^{\mathrm{i}k_x L} = \mathrm{e}^{\mathrm{i}k_y L} = \mathrm{e}^{\mathrm{i}k_z L} = 1$ and hence k_x, k_y and k_z must be integer multiples of $2\pi/L$: $\mathbf{k} = \frac{2\pi}{L}\mathbf{n}$, n_x, n_y, $n_z = 0, \pm 1, \ldots$. Thus, the momentum vectors are discrete or quantized for the particle in a box. The box normalization discretizes k.

When the values of k are continuous the so-called *Dirac-delta normalization* is useful. Consider the integral

$$\int_{-\infty}^{\infty} \psi_k^*(\mathbf{X})\psi_l(\mathbf{X})\,d\tau = c^*(k)c(l)\iiint_{-\infty}^{\infty} e^{i(\mathbf{k}-\mathbf{l})\cdot\mathbf{X}}\,dx\,dy\,dz\,. \tag{2.28}$$

We have

$$\int_{-\infty}^{\infty} e^{i(k_x-l_x)x}\,dx = \lim_{L\to\infty}\int_{-L}^{L} e^{i(k_x-l_x)x}\,dx = 2\pi\delta\left(k_x - l_x\right)\,. \tag{2.29}$$

Therefore,

$$\int \psi_k^*\psi_l\,d\tau = c^*(k)c(l)(2\pi)^3\delta(\mathbf{k}-\mathbf{l})\,. \tag{2.30}$$

When $\mathbf{l}=\mathbf{k}$ we have $c(k)=1/(2\pi)^{3/2}$.

2.7   A Special Feature of Occurrence of i in the Schrödinger Equation

Note that the imaginary number i appears in the fundamental equation, Schrödinger equation. *What is a special feature of explicit occurrence of* i *in the Schrödinger equation?* Let us point out this [9,10].

Consider the familiar classical wave equation

$$u_{tt} = c^2 u_{xx}. \tag{2.31}$$

Suppose u is complex-valued and define it as $u(x,t) = A(x,t) + iB(x,t)$. Substitution of this solution in Eq. (2.31) leads to

$$A_{tt} = c^2 A_{xx}, \quad B_{tt} = c^2 B_{xx}. \tag{2.32}$$

That is,, A and B are independent of each other and hence one can choose them freely, however, they are solutions of Eqs. (2.32). Usually, oscillating travelling waves are chosen as $u(x,t) = ae^{i(kx-\omega t)}$ and physical meaning is attached to the real part of u.

Next, $\psi(x,t) = A(x,t) + iB(x,t)$ in the Schrödinger equation $i\hbar\partial\psi/\partial t = -(\hbar^2/(2m))\partial^2\psi/\partial x^2$ gives

$$i\hbar(A_t + iB_t) = -\frac{\hbar^2}{2m}(A_{xx} + iB_{xx}). \tag{2.33}$$

The real and imaginary parts of this equation give

$$B_t = \frac{\hbar}{2m}A_{xx}, \quad A_t = -\frac{\hbar}{2m}B_{xx}. \tag{2.34}$$

A and B are not independent to each other but they are dependent to one another. Both A and B are needed to describe the physical situation. Further, note that the probability density is now $P = \psi^*\psi = A^2+B^2$. *What do you get if you start with $\psi(x,t) = \psi(x)\cos(Et/\hbar)$?*

Why do we have i *in Eq. (2.7)?* Note that i occurs in Eq. (2.4) which is obtained by differentiating Eq. (2.1) with respect to t. That is, i appeared explicitly due to the fact that the time periodic wave function ψ given by Eq. (2.1) is represented by complex exponential.

2.8   Conservation of Probability

The probability interpretation of $\psi^*\psi$ can be made consistently, provided the conservation of probability is guaranteed. If the probability of finding a particle in a region of space decreases as time increases then the probability of finding it outside this region must increase by the same amount so that total probability is always unity. Therefore, we expect $\rho = \psi^*\psi$ to satisfy the conservation equation (also called *continuity equation*)

$$\frac{\partial \rho}{\partial t} + \nabla \cdot \mathbf{J} = 0 \,. \tag{2.35}$$

The continuity equation is necessary to prove that particles do not disappear locally. In hydrodynamics ρ in Eq. (2.35) is the fluid density and $\mathbf{J}$ represents the fluid current density. In electrodynamics ρ and $\mathbf{J}$ are the charge density and current density, respectively. Let us derive a probability conservation law for ψ from the Schrödinger equation.

Define $\rho = \psi^*\psi$ as the position probability density. The probability P of the particle in a volume V is $\int_V \rho \, d\tau$. The rate of change of P is

$$\frac{\partial P}{\partial t} = \frac{\partial}{\partial t} \int \rho \, d\tau = \int \left(\psi^* \frac{\partial \psi}{\partial t} + \psi \frac{\partial \psi^*}{\partial t} \right) d\tau \,. \tag{2.36}$$

Replace the time derivative terms in Eq. (2.36) by equivalent space derivatives using Schrödinger equation. The Schrödinger equation for ψ^* is obtained as

$$-i\hbar \frac{\partial \psi^*}{\partial t} = -\frac{\hbar^2}{2m} \nabla^2 \psi^* + V\psi^* \,. \tag{2.37}$$

Multiplications of the Schrödinger Eq. (2.12) by ψ^* and Eq. (2.37) by ψ and subtraction of one from another gives

$$\begin{aligned}
\psi^* \frac{\partial \psi}{\partial t} + \psi \frac{\partial \psi^*}{\partial t} &= -\frac{\hbar}{2mi} \left[\psi^* \nabla^2 \psi - \psi \nabla^2 \psi^* \right] \\
&= -\frac{\hbar}{2mi} \nabla \cdot \left[\psi^* \nabla \psi - \psi \nabla \psi^* \right] \\
&= -\nabla \cdot \mathbf{J} \,, \tag{2.38}
\end{aligned}$$

where

$$\mathbf{J} = \frac{\hbar}{2mi} \left[\psi^* \nabla \psi - \psi \nabla \psi^* \right] \,. \tag{2.39}$$

$\mathbf{J}$ can also be expressed as

$$\mathbf{J} = \frac{\hbar}{m} \mathrm{Re} \left(-i\psi^* \nabla \psi \right) \quad \text{or} \quad \mathbf{J} = \frac{\hbar}{m} \mathrm{Im} \left(\psi^* \nabla \psi \right) \,. \tag{2.40}$$

Using Eq. (2.38) in Eq. (2.36) we obtain

$$\frac{\partial P}{\partial t} = -\int_V \nabla \cdot \mathbf{J} \, d\tau \,. \tag{2.41}$$

Substituting $P = \int_V \rho \, d\tau$ in Eq. (2.41) we have

$$\frac{\partial}{\partial t} \int_V \rho \, d\tau + \int_V \nabla \cdot \mathbf{J} \, d\tau = 0 \,. \tag{2.42}$$

Therefore,

$$\frac{\partial \rho}{\partial t} + \nabla \cdot \mathbf{J} = 0.\tag{2.43}$$

The interpretation of $\psi^* \psi$ as a probability density leads to the satisfaction of continuity equation.

$\mathbf{J}$ is defined as a *probability current density* or *probability flux vector*. $\mathbf{J}$ at $\mathbf{X}$ is the amount of probability crossing at that point in unit time. It corresponds to the Poynting vector in classical electromagnetic theory. The increase (decrease) of the probability inside a finite volume V is due to the influx (outward flux) of the probability current through the surface A bounding this volume. *What is the significance of $\nabla \cdot \mathbf{J} = 0$?*

Solved Problem 3:

Find the current density carried by the wave function Ae^{ikx} of a free particle and verify that it satisfies the equation of continuity.

In one-dimension the probability current density J is given by

$$J = \frac{\hbar}{m} \text{Im} \left(\psi^* \frac{\partial \psi}{\partial x} \right).\tag{2.44}$$

For $\psi = Ae^{ikx}$ we get

$$J = \frac{\hbar}{m} \text{Im} \left(A^* e^{-ikx} A i k e^{ikx} \right) = \frac{\hbar k}{m} |A|^2 = v|A|^2.\tag{2.45}$$

We obtain $\partial \rho/\partial t = 0$ and $\partial J/\partial x = 0$ and hence $\partial \rho/\partial t + \partial J/\partial x = 0$.

2.9   Expectation Value

Suppose we measure the position of a quantum mechanical particle a large number of times. Even if the wave function of the particle is same before every observation we do not get same result. In this case we are interested to know the average or expected value of the set of measurements of position denoted as $\langle x \rangle$. Knowing ψ, *how do we compute $\langle x \rangle$?* To obtain the expression for $\langle x \rangle$, instead of a quantum system, consider a classical system, for example a die with six sides numbered as 1, 2, 3, 4, 5 and 6 or say $a_1, a_2, \ldots, a_6$, respectively. Throw the die N (very large) number of times and record the outcome in each throw. Let N_i denote the number of times the event a_i occurred. Then the average or expectation value of throw of the die is

$$\langle \text{outcome} \rangle = \frac{1}{N} (N_1 a_1 + \ldots + N_6 a_6) = \sum_i a_i P(a_i) = \sum_a a P(a),\tag{2.46}$$

where summation is over all possible values of a. Here the possible values of a are assumed as discrete. When the outcome of an experiment takes any value in an interval, say, $[-\infty, \infty]$ then

$$\langle \text{outcome} \rangle = \int_{-\infty}^{\infty} a P(a) \, \mathrm{d}a.\tag{2.47}$$

In quantum theory $P(\mathbf{X}, t)\mathrm{d}\tau = \psi^* \psi \mathrm{d}\tau$, where $\mathrm{d}\tau = \mathrm{d}x \mathrm{d}y \mathrm{d}z$, gives the probability of finding a particle in a volume $\mathrm{d}\tau$ centred at $\mathbf{X}$ and hence for position measurements Eq. (2.47) becomes $\langle \mathbf{X} \rangle = \int_{-\infty}^{\infty} \mathbf{X} P(\mathbf{X}, t)\, \mathrm{d}\tau$, that is,

$$\langle \mathbf{X} \rangle = \int_{-\infty}^{\infty} \mathbf{X} \psi^* \psi\, \mathrm{d}\tau = \int_{-\infty}^{\infty} \psi^*(\mathbf{X}, t)\mathbf{X}_{\mathrm{op}}\psi(\mathbf{X}, t)\, \mathrm{d}\tau\,, \tag{2.48}$$

where $\psi(\mathbf{X}, t)$ is assumed to be normalized. For simplicity drop the subscript 'op' in the operator. Equation (2.48) gives $\langle \mathbf{X} \rangle$ of a quantum mechanical particle. The expectation value of $\mathbf{X}^n$ is given by in terms of ψ or ϕ

$$\langle \mathbf{X}^n \rangle = \int_{-\infty}^{\infty} \psi^*(\mathbf{X}, t)\mathbf{X}^n\psi(\mathbf{X}, t)\, \mathrm{d}\tau\,. \tag{2.49}$$

In a similar way, the expectation value of any quantity, which is a function of position (and time) only can be defined. For a function $f(\mathbf{X}, t)$

$$\langle f(\mathbf{X}, t) \rangle = \int_{-\infty}^{\infty} \psi^* f \psi\, \mathrm{d}\tau\,, \tag{2.50}$$

where f in the integral is the operator form of the function f.

How do we define expectation values of momentum and functions of momentum? First, consider $\langle \mathbf{p} \rangle$. The first-order time derivative of $\langle \mathbf{X} \rangle$ is shown to be $\langle \mathbf{p} \rangle/m$ (see the next section for its proof). That is,

$$\frac{\mathrm{d}}{\mathrm{d}t}\langle \mathbf{X} \rangle = \frac{1}{m}\langle \mathbf{p} \rangle\,. \tag{2.51}$$

We start with the above equation and obtain

$$\begin{aligned}
\langle \mathbf{p} \rangle &= m\frac{\mathrm{d}}{\mathrm{d}t}\int_{-\infty}^{\infty} \psi^*\mathbf{X}\psi\, \mathrm{d}\tau \\
&= m\int\left(\frac{\partial \psi^*}{\partial t}\mathbf{X}\psi + \psi^*\mathbf{X}\frac{\partial \psi}{\partial t}\right)\mathrm{d}\tau \\
&= -\frac{i\hbar}{2}\int\left[(\nabla^2\psi^*)\,\mathbf{X}\psi - \psi^*\mathbf{X}\nabla^2\psi\right]\mathrm{d}\tau \\
&= -\frac{i\hbar}{2}\left[(\nabla\psi^*\mathbf{X}\psi - \psi^*\psi - \psi^*\mathbf{X}\nabla\psi)\Big|_{-\infty}^{\infty} + 2\int\psi^*\nabla\psi\, \mathrm{d}\tau\right]\,.
\end{aligned}$$

Since $\psi \to 0$ as $|\mathbf{X}| \to \infty$, we get

$$\langle \mathbf{p} \rangle = \int_{-\infty}^{\infty} \psi^*\, \mathbf{p}\, \psi\, \mathrm{d}\tau\,, \tag{2.52}$$

where $\mathbf{p}$ in the integral is operator form of momentum.

Next, find the expression for $\langle \mathbf{p}^2 \rangle$. Since $E = \mathbf{p}^2/(2m) + V(\mathbf{X})$ we write

$$\begin{aligned}
\langle \mathbf{p}^2 \rangle &= 2m\left[\int_{-\infty}^{\infty} \psi^* E\psi\, \mathrm{d}\tau - \int_{-\infty}^{\infty} \psi^*V\psi\, \mathrm{d}\tau\right] \\
&= 2m\left[-i\hbar\int\psi^*\frac{\partial \psi}{\partial t}\, \mathrm{d}\tau - \int\psi^*V\psi\, \mathrm{d}\tau\right] \\
&= -\hbar^2\int\psi^*\nabla^2\psi\, \mathrm{d}\tau \\
&= \int_{-\infty}^{\infty} \psi^*\, \mathbf{p}^2\, \psi\, \mathrm{d}\tau\,. \tag{2.53}
\end{aligned}$$

Equations (2.52) and (2.53) suggest that

$$\langle \mathbf{p}^n \rangle = \int_{-\infty}^{\infty} \psi^* \mathbf{p}^n \psi \, \mathrm{d}\tau \,. \tag{2.54}$$

Then

$$\langle f(\mathbf{p}) \rangle = \int_{-\infty}^{\infty} \psi^* f(\mathbf{p}) \psi \, \mathrm{d}\tau \,. \, (\text{Why?}). \tag{2.55}$$

In view of Eqs. (2.50) and (2.55) the expectation value of an arbitrary dynamical variable $A(\mathbf{X}, \mathbf{p}, t)$ is

$$\langle A \rangle = \int_{-\infty}^{\infty} \psi^* A \psi \, \mathrm{d}\tau \,. \tag{2.56}$$

The expectation values are functions of time only, since the space coordinates have to be integrated out. The expectation value may be real or complex depending on the nature of A. However, in an experiment we can measure only real value. Thus, only the dynamical variables whose expectation values are real to be considered directly *measurable* or *observable*. Therefore, if A is an observable then $\langle A \rangle = \langle A \rangle^*$. That is,

$$\int_{-\infty}^{\infty} \psi^* A \psi \, \mathrm{d}\tau = \left(\int_{-\infty}^{\infty} \psi^* A \psi \, \mathrm{d}\tau \right)^* = \int_{-\infty}^{\infty} (A\psi)^* \psi \, \mathrm{d}\tau \,. \tag{2.57}$$

To ensure the above requirement the operator A must have a certain property called *self-adjointness* which will be discussed in the next chapter.

Equation (2.56) is defined for a noninteracting particle. For the more general case, such as a particle interacting with a heat bath or the environment, the expected value is given by [11] $\langle A(t) \rangle = \mathrm{Tr}[\rho(t)A]$, where ρ is the *density matrix* and is the solution of the von Neumann equation

$$i\hbar \frac{\partial \rho}{\partial t} = [H, \rho] \tag{2.58}$$

with H now the Hamiltonian for this entire system and $[H, \rho] = H\rho - \rho H$ is called *commutator* of H and ρ (which will be discussed in the next chapter).

Solved Problem 4:

Calculate $\langle x \rangle$, $\langle x^2 \rangle$, $\langle p_x \rangle$, $\langle p_x^2 \rangle$ if $\psi = \mathrm{e}^{-kx}$ for $x > 0$ and 0 for $x < 0$.

To normalize the wave function ψ assume that $\psi = N\mathrm{e}^{-kx}$ for $x > 0$ and 0 for $x < 0$. The normalization condition gives

$$1 = N^2 \int_0^{\infty} \mathrm{e}^{-2kx} \, \mathrm{d}x = -\frac{N^2}{2k} \mathrm{e}^{-2kx} \Big|_0^{\infty} = \frac{N^2}{2k} \,. \tag{2.59}$$

Therefore, $N = \sqrt{2k}$. Now,

$$\langle x \rangle = 2k \int_0^{\infty} x\,\mathrm{e}^{-2kx} \, \mathrm{d}x = \frac{1}{2k} \,. \tag{2.60}$$

$$\langle x^2 \rangle = 2k \int_0^{\infty} x^2\,\mathrm{e}^{-2kx} \, \mathrm{d}x = \frac{1}{2k^2} \,. \tag{2.61}$$

$$\langle p_x \rangle = 2k \int_0^{\infty} \mathrm{e}^{-kx} \left(-i\hbar \frac{\mathrm{d}}{\mathrm{d}x} \right) \mathrm{e}^{-kx} \, \mathrm{d}x = i\hbar k \,. \tag{2.62}$$

$$\langle p_x^2 \rangle = 2k \int_0^{\infty} \mathrm{e}^{-kx} (-i\hbar)^2 \frac{\mathrm{d}^2}{\mathrm{d}x^2} \mathrm{e}^{-kx} \, \mathrm{d}x = \hbar^2 k^2 \,. \tag{2.63}$$

p_x is a Hermitian operator and hence its expectation value must be real. Here we get $\langle p_x \rangle$ an imaginary quantity. It is imaginary as the given function has a discontinuous jump at $x = 0$. The point is that the given function is not a physically acceptable wave function.

Solved Problem 5:

The ground state (lowest energy state) wave function (in spherical polar coordinate) of a hydrogen atom is given by $\psi = e^{-r/a_0}$, where a_0 is a constant. Calculate the average value of r.

Writing $\psi = N e^{-r/a_0}$ the normalization condition gives

$$1 = \int \psi^* \psi \, d\tau = N^2 \int_0^\infty \int_0^\pi \int_0^{2\pi} e^{-2r/a_0} r^2 \sin\theta \, dr \, d\theta \, d\phi, \qquad (2.64)$$

where $r^2 \sin\theta dr d\theta d\phi$ is the volume element $d\tau$ in the spherical polar coordinate system. Integrating with respect to ϕ and θ we get

$$1 = 4\pi N^2 \int_0^\infty r^2 e^{-2r/a_0} \, dr. \qquad (2.65)$$

Integrating by parts we obtain $N = 1/(\pi a_0^3)^{1/2}$. Then $\langle r \rangle$ is

$$\langle r \rangle = \int_0^\infty \int_0^\pi \int_0^{2\pi} \psi^* r \psi r^2 \sin\theta \, dr \, d\theta \, d\phi = \frac{4}{a_0^3} \int_0^\infty r^3 e^{-2r/a_0} \, dr = \frac{3}{2} a_0. \qquad (2.66)$$

Solved Problem 6:

Consider the following two wave functions $\psi_1(x) = N_1 e^{-\alpha x^2}$ and $\psi_2(x) = N_2 e^{-\alpha^* x^2}$ with α being complex with $\mathrm{Re}\,\alpha \geq 0$. Determine (a) N_1 and N_2, (b) $\psi_1^* \psi_1$ and $\psi_2^* \psi_2$ and (c) $\mathcal{F}_{12} = |\langle \psi_1 | \psi_2 \rangle|^2$. (d) What do we observe from the results of (a), (b) and (c)?

(a) The normalization condition $\int_{-\infty}^\infty \psi_1^* \psi_1 dx = 1$ gives

$$1 = |N_1|^2 \int_{-\infty}^\infty e^{-(\alpha+\alpha^*)x^2} dx = \frac{2|N_1|^2}{\sqrt{\alpha+\alpha^*}} \int_0^\infty e^{-y^2} dy = \frac{|N_1|^2 \sqrt{\pi}}{\sqrt{\alpha+\alpha^*}}. \qquad (2.67)$$

So $N_1 = ((\alpha + \alpha^*)/\pi)^{1/4}$. Similarly, we find $N_2 = ((\alpha + \alpha^*)/\pi)^{1/4} = N_1$.

(b) We obtain

$$\psi_1^* \psi_1 = |N_1|^2 e^{-(\alpha+\alpha^*)x^2} \quad \text{and} \quad \psi_2^* \psi_2 = |N_1|^2 e^{-(\alpha+\alpha^*)x^2}. \qquad (2.68)$$

That is, $\psi_1^* \psi_1 = \psi_2^* \psi_2$.

(c) We obtain

$$\langle \psi_1 | \psi_2 \rangle = \int_{-\infty}^\infty \psi_1^* \psi_2 \, dx = |N_1|^2 \int_{-\infty}^\infty e^{-2\alpha^* x^2} dx = \sqrt{\frac{\alpha + \alpha^*}{2\alpha^*}}. \qquad (2.69)$$

Then $\mathcal{F}_{12} = |\langle \psi_1 | \psi_2 \rangle|^2 = (\alpha + \alpha^*)/(2\sqrt{\alpha\alpha^*})$.

(d) We notice that ψ_1 and ψ_2 have same normalization constant and same probability densities. If ψ_1 and ψ_2 are the same then $\mathcal{F}_{12} = 1$. Since $\mathcal{F}_{12} \neq 1$ we conclude that ψ_1 and ψ_2 are different even though $\psi_1^* \psi_1 = \psi_2^* \psi_2$.

2.10   Ehrenfest's Theorem

In classical mechanics $\mathbf{X}$ and $\mathbf{p}$ satisfy the equations

$$\frac{\mathrm{d}\mathbf{X}}{\mathrm{d}t} = \frac{1}{m}\mathbf{p}, \quad \frac{\mathrm{d}\mathbf{p}}{\mathrm{d}t} = \mathbf{F} = -\nabla V. \tag{2.70}$$

What are the forms of these equations in quantum mechanics? In quantum mechanics rate of change of $\langle \mathbf{X}(t)\rangle$ and $\langle \mathbf{p}(t)\rangle$ are calculated using the Schrödinger equation as shown below.

2.10.1   Equations of Motion of $\langle \mathbf{X}\rangle$ and $\langle \mathbf{p}\rangle$

We write

$$\frac{\mathrm{d}}{\mathrm{d}t}\langle \mathbf{X}\rangle = \frac{\mathrm{d}}{\mathrm{d}t}\int_{-\infty}^{\infty} \psi^* \mathbf{X}\psi \,\mathrm{d}\tau = \int_{-\infty}^{\infty} \psi^* \mathbf{X}\frac{\partial \psi}{\partial t}\,\mathrm{d}\tau + \int_{-\infty}^{\infty} \frac{\partial \psi^*}{\partial t}\mathbf{X}\psi \,\mathrm{d}\tau. \tag{2.71}$$

Using the Schrödinger equations for ψ and ψ^* replace the time derivative terms $\partial\psi/\partial t$ and $\partial\psi^*/\partial t$ by space derivative terms. We obtain

$$\frac{\mathrm{d}}{\mathrm{d}t}\langle \mathbf{X}\rangle = \frac{i\hbar}{2m}\int_{-\infty}^{\infty} \left[\psi^* \mathbf{X}\nabla^2\psi - \left(\nabla^2\psi^*\right)\mathbf{X}\psi\right]\mathrm{d}\tau. \tag{2.72}$$

Integration of second integral in Eq. (2.72) by parts gives

$$\begin{aligned}
\int_{-\infty}^{\infty} \left(\nabla^2\psi^*\right)\mathbf{X}\psi \,\mathrm{d}\tau &= \left.\mathbf{X}\psi\nabla\psi^*\right|_{-\infty}^{\infty} - \int_{-\infty}^{\infty} \left(\nabla\psi^*\right)\nabla\cdot\left(\mathbf{X}\psi\right)\mathrm{d}\tau \\
&= -\int_{-\infty}^{\infty} \left(\nabla\psi^*\right)\nabla\cdot\left(\mathbf{X}\psi\right)\mathrm{d}\tau \\
&= \left.-\psi^*\nabla\cdot\left(\mathbf{X}\psi\right)\right|_{-\infty}^{\infty} + \int_{-\infty}^{\infty} \psi^*\nabla^2\left(\mathbf{X}\psi\right)\mathrm{d}\tau \\
&= \int_{-\infty}^{\infty} \psi^*\left(2\nabla\psi + \mathbf{X}\nabla^2\psi\right)\mathrm{d}\tau.
\end{aligned} \tag{2.73}$$

Now,

$$\begin{aligned}
\frac{\mathrm{d}}{\mathrm{d}t}\langle \mathbf{X}\rangle &= \frac{i\hbar}{2m}\int_{-\infty}^{\infty} \left[\psi^* \mathbf{X}\nabla^2\psi - 2\psi^*\nabla\psi - \psi^*\mathbf{X}\nabla^2\psi\right]\mathrm{d}\tau \\
&= -\frac{i\hbar}{m}\int_{-\infty}^{\infty} \psi^*\nabla\psi \,\mathrm{d}\tau \\
&= \frac{1}{m}\langle \mathbf{p}\rangle.
\end{aligned} \tag{2.74a}$$

In a similar manner, we calculate the time rate of change of $\langle \mathbf{p}\rangle$ as

$$\begin{aligned}
\frac{\mathrm{d}}{\mathrm{d}t}\langle \mathbf{p}\rangle &= -i\hbar\frac{\mathrm{d}}{\mathrm{d}t}\int_{-\infty}^{\infty} \psi^*\nabla\psi \,\mathrm{d}\tau \\
&= -i\hbar\left[\int_{-\infty}^{\infty} \psi^*\nabla\left(\frac{\partial \psi}{\partial t}\right)\mathrm{d}\tau + \int_{-\infty}^{\infty} \frac{\partial \psi^*}{\partial t}\nabla\psi \,\mathrm{d}\tau\right] \\
&= -\int_{-\infty}^{\infty} \psi^*\left(\nabla(V\psi) - V\nabla\psi\right)\mathrm{d}\tau \\
&= -\int_{-\infty}^{\infty} \psi^*(\nabla V)\psi \,\mathrm{d}\tau \\
&= \langle -\nabla V\rangle.
\end{aligned} \tag{2.74b}$$

Equations (2.74) are analogous to the classical equations of motion, Eq. (2.70). Thus, the quantum mechanical $\langle \mathbf{X} \rangle$ and $\langle \mathbf{p} \rangle$ obey the classical equation of motion. This is known as *Ehrenfest's theorem*. The theorem states that the *Schrödinger equation leads to the satisfaction of Newton's second-law of motion on the average*. The Eqs. (2.74) are crucial for understanding the emergence of classical mechanics from quantum mechanics.

2.10.2 Verification of Ehrenfest's Theorem

In this section, we verify the Ehrenfest's theorem in certain interesting systems. First, we solve the Eqs. (2.74) for linear harmonic oscillator and show that their solutions are identical to the solutions of classical equation of motion. Next, we consider a particle in an infinite square-well potential. We give the solutions of classical and quantum cases. We discuss the classical limit of quantum solutions.

2.10.2.1 Linear Harmonic Oscillator

For the linear harmonic oscillator with potential $V(x) = \omega^2 x^2/(2m)$ the classical equation of motion is

$$\frac{\mathrm{d}x}{\mathrm{d}t} = \frac{1}{m} p_x , \quad \frac{\mathrm{d}p_x}{\mathrm{d}t} = -m\omega^2 x . \tag{2.75}$$

Its solution is

$$x = x_0 \cos \omega t + \frac{p_{x0}}{m\omega} \sin \omega t , \tag{2.76}$$

where x_0 and p_{x0} are integration constants. The quantum equations for are

$$\frac{\mathrm{d}}{\mathrm{d}t} \langle x \rangle = \frac{1}{m} \langle p_x \rangle , \quad \frac{\mathrm{d}}{\mathrm{d}t} \langle p_x \rangle = -m\omega^2 \langle x \rangle \ \text{ or } \ \frac{\mathrm{d}^2}{\mathrm{d}t^2} \langle x \rangle + \omega^2 \langle x \rangle = 0 . \tag{2.77}$$

The solution of it is

$$\langle x \rangle = \langle x \rangle_0 \cos \omega t + \frac{\langle p_x \rangle_0}{m\omega} \sin \omega t \tag{2.78}$$

which is same as the Eq. (2.76).

2.10.2.2 Particle in an Infinite Square-Well Potential

Let us consider a particle of mass m confined to one-dimensional infinite height potential with $V(x) = 0$ for $0 < x < a$ and ∞ otherwise.

Classical Case

The particle moves between the walls $x = 0$ and a with a uniform velocity v. Its velocity gets reversed when it hits the walls. This to-and-fro motion is given by [12]

$$x = \frac{1}{m} p_x t = \begin{cases} 2at/T, & 0 < t < T/2 \\ 2a - 2at/T, & T/2 < t < T, \end{cases} \tag{2.79}$$

where $p_x = mv = ma/(T/2)$ and $T/2$ is the time taken by the particle to go from one side of the wall to the other. A Fourier series of this solution is

$$x_c = \frac{a}{2} - \frac{4a}{\pi^2} \sum_{r=0}^{\infty} \frac{\cos(2r+1)\omega t}{(2r+1)^2} , \quad \omega = 2\pi/T . \tag{2.80}$$

Quantum Solution

The normalized eigenfunctions and energies are given by (refer to sec. 4.6)

$$\psi_n(x,t) = \phi_n(x)e^{-iE_nt/\hbar} = \sqrt{2/a}\,\sin{(k_nx)}\,e^{-iE_nt/\hbar}\,, \tag{2.81}$$

where

$$k_n = \frac{n\pi}{a}\,, \quad E_n = \frac{n^2\hbar^2\pi^2}{2ma^2}\,. \tag{2.82}$$

We write the wave function as $\xi(x,t) = \sum_n c_n\psi_n$. Then

$$\langle x\rangle = \int \xi^*x\xi\,dx = \frac{a}{2} - \frac{2a}{\pi^2}\left[I(t) + I^*(t)\right]\,, \tag{2.83a}$$

where

$$I(t) = \sum_{s=1}^{\infty}\sum_{r=0}^{\infty}c_s^*c_{s+2r+1}e^{-i(2r+1)G}\left(\frac{1}{(2r+1)^2} - \frac{1}{(2s+2r+1)^2}\right) \tag{2.83b}$$

and $G = (2s+2r+1)\pi^2\hbar t/(2ma^2)$.

Classical Limit

The de Broglie wavelength of a particle of momentum p_x is $\lambda = h/p_x$. In the potential well we approximately fit N such half-wavelengths, where

$$N = \frac{a}{\lambda/2} = \frac{2p_xa}{h} = \frac{ma^2\omega}{\pi^2\hbar} \gg 1\,. \tag{2.84}$$

The energy E_n for this large quantum number N will correspond to the classical kinetic energy $E_c = p_x^2/(2m)$. c_n's are significant only over a narrow range $N - \delta' \le n \le N + \delta''$ of energy states so that $\sum_{n=N-\delta'}^{N+\delta''}|c_n|^2 = 1$ with δ', $\delta'' \ll N$. In Eq. (2.83b) terms for which $s \approx N$ and $r \ll N$ alone contribute. G goes over to $N\pi^2\hbar t/(ma^2)$, that is to ωt. Neglect $1/(2s+2r+1)^2$ comparing to $1/(2r+1)^2$ because $s \approx N$ is very large. Thus

$$I(t) \approx \sum_{r=0}^{\infty}\sum_{s=N-\delta'}^{N+\delta''}\frac{c_s^*}{(2r+1)^2}\,e^{-i(2r+1)\omega t}\,. \tag{2.85}$$

The sum over s goes to unity. Now, (2.83a) becomes

$$\langle x\rangle \approx \frac{a}{2} - \frac{4a}{\pi^2}\sum_{r=0}^{\infty}\frac{\cos(2r+1)\omega t}{(2r+1)^2} \tag{2.86}$$

which agrees with the classical result Eq. (2.80). Thus, we conclude that as $N \to \infty$, the quantum $\langle x\rangle$ goes over to the corresponding classical trajectory.

Some other examples on the significance of Ehrenfest's theorem are given in refs.[13–15]. Ehrenfest's theorem is neither a necessary nor a sufficient condition for a quantum system to display classical behaviour. The harmonic oscillator satisfies Eqs. (2.74). But a quantum oscillator has discrete energy eigenvalues contradictory to the classical system. The system obeys Ehrenfest's theorem but does not behave classically. This system is an example of *lack of sufficiency*. For an example of lack of necessity – that a system may behave classically even when Ehrenfest's theorem does not apply – see ref. [14].

2.11   Basic Postulates

In classical mechanics we work with the position, the momentum, the total energy, etc. These quantities are called *dynamical variables*. In quantum mechanics the dynamical variables are no longer variables but operators. An operator is operated on the wave function to get various outcomes of the corresponding physical variable. Based on our discussion so far, we now list important postulates in quantum mechanics.

Postulate 1:

For every dynamical system there exists a wave function (and its spatial derivatives). It is continuous, finite, single-valued function of the parameters of the system, space coordinate and time. From the wave function, predictions of physical properties of the system to be made. This postulate contains requirements that all physical waves such as sound waves, electromagnetic waves and water waves meet.

Postulate 2:

The Hamiltonian operator of a quantum mechanical system determines the evolution of the system. The classical $E = H = \mathbf{p}^2/(2m) + V(\mathbf{X})$ is converted into the wave Eq. (2.12) by the substitution of operators for dynamical variables: $x \to \hat{x}$, $\mathbf{X} \to \hat{\mathbf{X}}$, $p_x \to -i\hbar\partial/\partial x$, $\mathbf{p} \to -i\hbar\nabla$ and $E \to i\hbar\partial/\partial t$.

Postulate 3:

$\psi^*\psi$ is interpreted as position probability density and an admissible ψ must satisfy the normalization condition (2.22) to make total probability unity.

Postulate 4:

For every observable physical quantity, there exists a corresponding operator with real eigenvalues. The only measurable values of an observable are the possible eigenvalues of the corresponding operator.

Postulate 5:

Quantum mechanics does not say with certainty the outcome of an experiment. But it allows computing of probabilities for the various possible outcomes. Consequently, we are interested in the average value of a dynamical variable A and this is given by Eq. (2.56).

2.12   Time Evolution of Stationary States

Consider the time-independent Schrödinger equation. As stated earlier the state of a quantum mechanical system is represented by a wave function ψ. $|\psi|^2$ is assumed to give the position probability density. From the solution $\phi(\mathbf{X})$ of time-independent Schrödinger Eq. (2.17), the solution of the time-dependent Schrödinger equation is written as

$$\psi_n(\mathbf{X}, t) = \phi_n(\mathbf{X})e^{-iE_n t/\hbar} . \tag{2.87}$$

We note that $|\psi_n(\mathbf{X}, t)|^2$ is independent of time. Further, the time dependence of $\psi_n(\mathbf{X}, t)$ is periodic with angular frequency $\omega = E_n/\hbar$. Such states are called *stationary states*. In quantum mechanics a stationary state does not mean that the system is strictly in a rest state or in a classical equilibrium point. In this state, expectation values of all dynamical variables

$$\langle A \rangle = \int_{-\infty}^{\infty} \psi_n^*(\mathbf{X}, t) A \psi_n(\mathbf{X}, t) \mathrm{d}\tau = \int_{-\infty}^{\infty} \phi_n^*(\mathbf{X}) A \phi_n(\mathbf{X}) \mathrm{d}\tau \qquad (2.88)$$

are independent of time. It means that though ψ may vary with time, $\psi^*\psi$ at any point in the configuration space is independent of time.

The solution $\phi_n(\mathbf{X})$ of the time-independent Schrödinger Eq. (2.17) for a fixed value of E_n is obviously time-independent and thus $\phi_n^*\phi_n$ is independent of time. Therefore, we call Eq. (2.17) the *stationary state Schrödinger equation* and its solutions as *stationary state solutions* or *eigenfunctions*. This is an energy eigenvalue equation, $H\phi = E\phi$. Consequently, when the state of the particle is the solution of Eq. (2.17) then the energy of it has a definite value given by the eigenvalue E.

Suppose a particle is initially in a stationary state represented by a $\psi_n(\mathbf{X}, 0) = \phi_n(\mathbf{X})$. Then from Eq. (2.87) we write

$$\psi_n(\mathbf{X}, t) = \psi_n(X, 0) \mathrm{e}^{-iE_n t/\hbar} \qquad (2.89)$$

and $|\psi_n(\mathbf{X}, t)|^2 = |\psi_n(\mathbf{X}, 0)|^2$ is independent of time. That is, *if a system is initially in a stationary state then it remains in that state forever.* Though $\psi_n(\mathbf{X}, t)$ changes with time the energy and expectation values of observables remain the same.

We note that the wave function given by Eq. (2.89) is a particular solution of the time-dependent Schrödinger equation. A combination of such particular solutions is given by the Eq. (2.19). Now we ask: *Can the wave function given by Eq. (2.19) represent a stationary state?* The answer is no, as shown below.

Let us calculate $\psi^*\psi$ of ψ given by Eq. (2.19). For simplicity assume that the energy values are discrete. We write ψ^* as $\psi^* = \sum_m a_m^* \phi_m^* \mathrm{e}^{iE_m t/\hbar}$. Then

$$\psi^*\psi = \sum_n |a_n|^2 |\phi_n|^2 + \sum_{m \neq n} \sum_n a_m^* a_n \phi_m^* \phi_n \mathrm{e}^{-i(E_n - E_m)t/\hbar}. \qquad (2.90)$$

Since the second term in the above equation contains t, $\psi^*\psi$ is dependent on time. Therefore, solution ψ given by the Eq. (2.19) does not represent a stationary state solution.

From our above discussion we say that a stationary state solution of a quantum system is obtained by finding the solution of the time-independent Schrödinger Eq. (2.17) and then multiply it by $\mathrm{e}^{-iEt/\hbar}$, provided the potential is time-independent. Now, an interesting question is: *What are the allowed values of E? Can E be arbitrary? Does there exist a (bounded) solution of Eq. (2.17) for all real values of E?* The answer is no. Solutions exist only for certain specific values of E. This is because the solution of Eq. (2.17) satisfies the admissible conditions only for certain values of E. The allowed energy values are called the *eigenvalue spectrum* or *energy levels* of the system. *Can you recall some of the equations that are in the form of $A\phi = \lambda\phi$ where A is a linear differential operator for which bounded solution exist only for certain discrete real values of λ? What are their eigenvalues and eigenfunctions?*

2.13   Conditions for Allowed Transitions

Suppose we wish to know whether a particle is able to make a transition from a given state to another state. Also, we ask: *Can a particle jump to an arbitrary state? If not, to which states can it jump? What does happen when the particle goes from one energy state to another?* We will address these questions in this section.

Bohr postulated that when a particle makes transition from a higher energy level E_m to a lower energy level E_n then the frequency of the radiation emitted is $\nu_{mn} = (E_m - E_n)/\hbar$. *Can a particle emit radiation when it is in a stationary state?* The answer is no. To prove this consider a quantum system of a particle moving only in the x-direction and is in a state $\phi_n(x)$ with energy E_n. The time-dependent eigenfunction associated with this state is

$$\psi_n(x,t) = \phi_n(x) e^{-iE_n t/\hbar}. \tag{2.91}$$

Now, the expectation value of the position of the particle is

$$\langle x \rangle = \int_{-\infty}^{\infty} x \psi_n^*(x,t) \psi_n(x,t)\, dx = \int_{-\infty}^{\infty} x \phi_n^*(x) \phi_n(x)\, dx \tag{2.92}$$

which is independent of time. Thus, the $\langle x \rangle$ of the particle does not oscillate and no radiation occurs.

Next, we calculate $\langle x \rangle$ for the transition from E_m to E_n. The ψ of the particle is $a\phi_m + b\phi_n$, where a^*a and b^*b are the probabilities for the particle to be in ϕ_m and ϕ_n, respectively. We obtain

$$\begin{aligned}
\langle x \rangle = \; & a^*a \int_{-\infty}^{\infty} x \phi_m^* \phi_m\, dx + b^*b \int_{-\infty}^{\infty} x \phi_n^* \phi_n\, dx \\
& + \cos(\nu_{mn} t) \int_{-\infty}^{\infty} x \left[a^*b \phi_m^* \phi_n + ab^* \phi_m \phi_n^* \right] dx \\
& + i \sin(\nu_{mn} t) \int_{-\infty}^{\infty} x \left[a^*b \phi_m^* \phi_n - ab^* \phi_m \phi_n^* \right] dx,
\end{aligned} \tag{2.93}$$

where $\nu_{mn} = (E_m - E_n)/\hbar$. If the particle is in a specific state E_m or E_n then $\nu = 0$ and $\langle x \rangle = $ constant, independent of time. When the particle makes a transition from the state E_m to the state E_n then $\langle x \rangle$ oscillates with time and the frequency of radiation emitted is ν_{mn}. From Eq. (2.93) we obtain the condition for $\langle x \rangle \neq$ constant as

$$\int_{-\infty}^{\infty} \phi_m^* \, x \, \phi_n\, dx \neq 0. \tag{2.94}$$

If the value of the above integral is nonzero then the transition from the level E_m to E_n is *allowed* otherwise the transition is *forbidden*. The conditions on m and n for which the integral in Eq. (2.94) is nonzero are called *selection rules*. Since the electric dipole operator is given as ex, the transition based on the above selection rule is called *electric dipole transition*.

2.14   Orthogonality of Two States

A wave function belonging to a specific energy (eigen)value E_n is commonly called an *energy eigenfunction* ϕ_n. An important property of eigenfunctions ϕ_n of stationary state Schrödinger equation is that two eigenfunctions belonging to two different energy values E_m and E_n satisfy the integral equation

$$\int_{-\infty}^{\infty} \phi_m^* \phi_n \, d\tau = 0. \tag{2.95}$$

In vector analysis two vectors $\mathbf{u}$ and $\mathbf{v}$ for which $\mathbf{u} \cdot \mathbf{v} = \sum_i u_i v_i = 0$ are called *orthogonal*. (If the number of the components of $\mathbf{u}$ and $\mathbf{v}$ are very large and the increment is infinitesimally small then we replace the summation in the above by integral). In analogy with this ϕ_m and ϕ_n satisfying the condition (2.95) are called *orthogonal functions*. Orthogonal functions refer to states that are independent of one another. The different possible energy states of a particle are all orthogonal to one another.

Let ϕ_m and ϕ_n are two eigenfunctions with energies E_m and E_n, respectively. These functions satisfy the equations (refer to Eq. (2.17))

$$\nabla^2 \phi_m + \frac{2m}{\hbar^2} (E_m - V) \phi_m = 0, \qquad (2.96a)$$

$$\nabla^2 \phi_n + \frac{2m}{\hbar^2} (E_n - V) \phi_n = 0. \qquad (2.96b)$$

The equation for ϕ_m^* is

$$\nabla^2 \phi_m^* + \frac{2m}{\hbar^2} (E_m - V) \phi_m^* = 0. \qquad (2.96c)$$

Multiplying Eq. (2.96b) by ϕ_m^* and (2.96c) by ϕ_n and subtracting one from another we obtain

$$\phi_m^* \nabla^2 \phi_n - \phi_n \nabla^2 \phi_m^* + \frac{2m}{\hbar^2} (E_n - E_m) \phi_m^* \phi_n = 0. \qquad (2.97)$$

Integrating Eq. (2.97) over the space variables we get

$$\int_{-\infty}^{\infty} \left(\phi_m^* \nabla^2 \phi_n - \phi_n \nabla^2 \phi_m^* \right) d\tau + \frac{2m}{\hbar^2} (E_n - E_m) \int_{-\infty}^{\infty} \phi_m^* \phi_n \, d\tau = 0. \qquad (2.98)$$

Using *Green's second identity*

$$\int_V \left(f \nabla^2 g - g \nabla^2 f \right) dV = \int_S \left(f \nabla g - g \nabla f \right) dS \qquad (2.99)$$

Eq. (2.98) becomes

$$\int_S \left(\phi_m^* \nabla \phi_n - \phi_n \nabla \phi_m^* \right) dS + \frac{2m}{\hbar^2} (E_n - E_m) \int_{-\infty}^{\infty} \phi_m^* \phi_n \, d\tau = 0. \qquad (2.100)$$

Since ϕ_m, $\phi_n \to 0$ as $|\mathbf{X}| \to \infty$ the surface integral vanishes and we have

$$(E_n - E_m) \int_{-\infty}^{\infty} \phi_m^* \phi_n \, d\tau = 0. \qquad (2.101)$$

Because $E_m \neq E_n$ we have condition Eq. (2.95). When $E_m = E_n$ and ϕ_n is linearly independent of ϕ_m then the two states are said to be *degenerate states* having the same eigenvalue but different eigenfunctions. In such cases $\int_{-\infty}^{\infty} \phi_m^* \phi_n d\tau$ need not be zero. But it is possible to choose a set of orthogonal functions by using Schmidt's orthogonalization procedure. The integral for $m = n$ is $\int_{-\infty}^{\infty} \phi_n^* \phi_n \, d\tau =$ finite and it can be normalized to one by normalization condition. Hence, if each ϕ is normalized then

$$\int_{-\infty}^{\infty} \phi_m^* \phi_n \, d\tau = \delta_{mn}. \qquad (2.102)$$

Functions satisfying Eq. (2.102) are called *orthonormal* since they are orthogonal to each other and also normalized.

Solved Problem 7:

The normalized state of a system is $\psi = c_1\phi_1 + c_2\phi_2 + c_3\phi_3$, where c_i's are constants. If $\int_{-\infty}^{\infty} \phi_i^*\phi_j \, d\tau = \delta_{ij}$, $c_1 = 2/3$ and $c_2 = \sqrt{2}/3$ find the value of c_3.

Using the normalization condition $\int_{-\infty}^{\infty} \psi^*\psi \, d\tau = 1$ we obtain

$$
\begin{aligned}
1 &= \int_{-\infty}^{\infty} \left(c_1^*\phi_1^* + c_2^*\phi_2^* + c_3^*\phi_3^*\right)\left(c_1\phi_1 + c_2\phi_2 + c_3\phi_3\right) d\tau \\
&= |c_1|^2 + |c_2|^2 + |c_3'|^2 \\
&= 2/3 + |c_3|^2 \,.
\end{aligned}
\tag{2.103}
$$

Thus, $c_3 = 1/\sqrt{3}$.

2.15 Phase of the Wave Function

In 1926 Gregor Wentzel, Hendrik Anthony Kramers and Leon Nicolas Brillouin showed the possibility of getting the classical Hamilton–Jacobi equation from the Schrödinger equation in Planck's limit $h \to 0$. This is an another example for Planck's correspondence principle. In sec. 2.10 we have seen that in quantum mechanics the expectation values of $\mathbf{X}$ and $\mathbf{p}$ operators obey the same equation of motion of the $\mathbf{X}$ and $\mathbf{p}$ variables in classical mechanics. Now, we show that in quantum mechanics the equation for phase of the complex ψ takes the form of the Hamilton–Jacobi equation of classical physics [16].

Assume the solution of a time-dependent Schrödinger equation in the form

$$
\psi(\mathbf{X}, t) = A(\mathbf{X}, t)\, e^{iS(\mathbf{X},t)/\hbar} \,.
\tag{2.104}
$$

Substituting Eq. (2.104) in the Schrödinger Eq. (2.12) and separating real and imaginary parts we get

$$
A\frac{\partial S}{\partial t} - \frac{\hbar^2}{2m}\nabla^2 A + \frac{1}{2m}A(\nabla S)^2 + VA = 0 \,,
\tag{2.105a}
$$

$$
\frac{\partial A}{\partial t} + \frac{1}{2m}\left[A\nabla^2 S + 2\nabla A \cdot \nabla S\right] = 0 \,.
\tag{2.105b}
$$

Defining $\rho = |\psi|^2 = A^2$ Eq. (2.105b) takes the form, with the substitution $\partial A = \partial\rho/(2\sqrt{\rho})$ and $\nabla A = \nabla\rho/(2\sqrt{\rho})$,

$$
\frac{\partial\rho}{\partial t} + \frac{1}{m}\nabla \cdot (\rho\nabla S) = 0 \,.
\tag{2.106}
$$

This is the continuity equation for conservation of probability with the probability flux $\mathbf{J} = \rho\nabla S/m$. Equation (2.105a) is rewritten as

$$
\frac{\partial S}{\partial t} + \frac{1}{2m}(\nabla S)^2 + V + V_q = 0 \,, \quad V_q = -\frac{\hbar^2}{2m}\frac{\nabla^2 A}{A} \,.
\tag{2.107}
$$

Since V and V_q appears in the equation in the same way, V_q is called *quantum potential*. V_q is state dependent. Equation (2.107) is of the form of classical Hamilton–Jacobi equation. It is called *quantum Hamilton–Jacobi equation*. The observation is that the phase of ψ obeys the same equation as the Hamilton principle function in classical mechanics in the limit $V_q \to 0$.

Introducing $\mathbf{v}(\mathbf{X}, t)$ as $\mathbf{J}/\rho = \nabla S/m$ and taking the gradient of Eq. (2.107) gives

$$m\frac{\partial \mathbf{v}}{\partial t} + m\left(\mathbf{v} \cdot \nabla\right)\mathbf{v} + \nabla\left(V + V_q\right) = 0. \tag{2.108}$$

The first two terms in the above equation is $m\mathrm{d}\mathbf{v}/\mathrm{d}t$. The system thus moves with the acceleration

$$m\frac{\mathrm{d}\mathbf{v}}{\mathrm{d}t} = -\nabla\left(V + V_q\right). \tag{2.109}$$

The force is given not only by the gradient of V but also by ∇V_q. If $V_q \to 0$ as $\hbar \to 0$ then the particle trajectories will obey Newton's equation of motion.

2.16 Classical Limit of Quantum Mechanics

Classical mechanics has been verified in a wide variety of phenomena. It is thus natural to know the limit with which quantum mechanics agrees with classical mechanics. We cannot simply define $\hbar \to 0$ as the classical limit. Special relativity reduces to classical Galilean relativity in the limit $c \to \infty$ or $v/c \ll 1$, where v is the speed of a particle. In this limit relativistic mechanics includes Newtonian mechanics as a special case. But quantum mechanics is formulated in terms of probabilities. It does not predict the individual observed possibility but predicts only the probabilities of the possibilities.

If quantum mechanics has to give an individual trajectory in the classical limit, it is necessary for the probability distribution to become arbitrarily narrow as $\hbar \to 0$. Certain special case of states behave like this. But many realistic states do not. Therefore, we cannot expect to get an individual classical trajectory when we let the classical limit of quantum mechanics. Instead we may expect the probability distribution of quantum mechanics to become equivalent to the probability distribution of an ensemble of classical trajectories.

Consider an ensemble of identical particles with an initial position x_0 as $F(x_0)$ and with the identical initial momenta p_{x0} [17]. In classical mechanics, the position probability density function of an ensemble is given by [18]

$$P(x, t) = \frac{\partial^2 S}{\partial p_x \partial x}\left|F\left(\frac{\partial S}{\partial p_x}\right)\right|^2, \tag{2.110}$$

where S is the *Hamilton principle function*. An approximate solution of the time-dependent Schrödinger equation, up to terms linear in $\hbar$ obtained using WKB approximation method is [18]

$$\psi(x, t) = e^{iS/\hbar}\left(\frac{\partial^2 S}{\partial p_x \partial x}\right)^{1/2} F\left(\frac{\partial S}{\partial p_x}\right). \tag{2.111}$$

The $|\psi(x, t)|^2$ obtained from Eq. (2.111) agrees with $P(x, t)$ given by Eq. (2.110). Thus, if one restricts to the effects of the first-order in $\hbar$ then in the classical limit the time evolution of the quantum mechanical ensemble is similar to that of the classical ensemble.

To get deeper insight of the classical limit of quantum mechanics let us introduce the following statement and call it as the classical limit postulate.

In the classical limit the time development of a quantum mechanical ensemble is identical with that of an ensemble of a classical system whose initial distribution of position and momentum are same as that of the quantum mechanical ensemble at the initial instant.

To test the above postulate, we take a classical limit in the construction of $\psi(x,t)$ by suitably combining the eigenfunctions belonging to the continuum limit so that the time-dependent wave equation is exactly solvable. The evolution of the classical ensemble can be computed using the equations of motion applied to the phase space density function. According to the above postulate the evolution of two such ensembles should be the same if the initial position and momentum probability distribution functions are the same. In the following we verify this connection for a free particle. For the case of harmonic oscillator and a particle in a homogeneous potential field the reader may refer to ref. [17].

Let us choose $\psi(x,0)$ of an ensemble of free particles as the *Gaussian function* centred at the origin

$$\psi(x,0) = \left(\frac{1}{\sqrt{\pi}\sigma_x}\right)^{1/2} \exp\left[\frac{ik_0 x - x^2}{2\sigma_x^2}\right],\qquad (2.112)$$

where $p_{x0} = \hbar k_0$. The position distribution function is

$$P(x,0) = |\psi(x,0)|^2 = \frac{1}{\sqrt{\pi}\,\sigma_x}\,\mathrm{e}^{-x^2/\sigma_x^2}.\qquad (2.113)$$

Then $\psi(x,t) = \iint \psi_E^*(\xi)\psi_E(x)\psi(\xi,0)\mathrm{e}^{-iEt/\hbar}\mathrm{d}E\mathrm{d}\xi$, where $\psi_E(x)$ is the box normalized eigenfunction of a free particle. Evaluating the integrals we obtain

$$P(x,t) = \left[\pi\sigma_x^2 + \frac{\hbar^2 t^2}{m^2\sigma_x^2}\right]^{-1/2} \exp\left[-\frac{(x - (\hbar k_0 t/m))^2}{\sigma_x^2} + \frac{\hbar^2 t^2}{m^2\sigma_x^2}\right].\qquad (2.114)$$

Next, consider the time evolution of a classical ensemble of a free particle with P given by Eq. (2.113). The classical phase space density function is $D(x,k,0) = |\psi(x,0)|^2 |\phi(k,0)|^2$, where $k = p_x/\hbar$ and $\phi(k,0)$ is the Fourier transform of $\psi(x,0)$ given by Eq. (2.112). Evaluating the Fourier transform

$$D(x,k,0) = \frac{1}{\pi\hbar} \exp\left[-\frac{x^2}{\sigma_x^2} - \frac{(k - k_0)^2}{\sigma_k^2}\right],\qquad (2.115)$$

where $\sigma_x \sigma_k = 1$. Applying classical Liouville's theorem we find

$$D(x,k,t) = \frac{1}{\pi\hbar} \exp\left[-\frac{(x - (\hbar k t/m))^2}{\sigma_x^2} - \frac{(k - k_0)^2}{\sigma_k^2}\right].\qquad (2.116)$$

Then $P(x,t) = \hbar \int D(x,k,t)\mathrm{d}k$. It is easy to verify that the resulting expression for $P(x,t)$ is same as that given by Eq. (2.114). When the classical probability density for a particle of particular energy coincides with $P_{\mathrm{QM}} = \psi^*\psi$ it is called *configuration correspondence*.

2.17 Concluding Remarks

Note that the time evolution of ψ is described by the Schrödinger equation which is deterministic, but the observable quantities associated with this ψ are stated in terms of probability only. The probability concept is also used in classical mechanics in some occasions. For most of the traditional experiments of quantum physics, it is not possible to set up an individual quantum object, for example, an atom and probe it. Instead, we can probe an ensemble of identically prepared objects then the ensemble average is treated as the quantum average. In an electron-atom collision experiment, thousands of different electrons knock electrons out of different atoms and for each individual case, the momenta of target electrons before the collision are determined by momentum conservation. The result is interpreted as the probability distribution of finding a given momentum in a single atom. Thus, even if the fundamental mechanics of atoms and molecules were classical, we still need to describe most experiments with atoms using probabilities.

The spectacular advances in science in 20th century could not have occurred without the use of quantum theory and the Schrödinger equation. Though the quantum theory gives the probability interpretation of the wave function, there is no other theory which gives a comparable quantitative description of the atom, molecule, nucleus, fields and even macroscopic matters like solids, liquids, conductors and semiconductors. Without quantum theory, phenomena like superconductivity, superfluidity, Bose–Einstein condensation cannot be explained. Every advanced technology makes use of the tools provided by quantum mechanics.

2.18   Bibliography

[1]   J.P. McEvoy and O. Zarate, *Introducing Quantum Theory*. Icon Books, Cambridge, 1996.

[2]   F. Capra, *The Tao of Physics*. Bantam Books, New York, 1984.

[3]   R.P. Feynman and A.R. Hibbs, *Quantum Mechanics and Path Integrals*. McGraw–Hill, New York, 1965.

[4]   J.Z. Justin, *Path Integrals in Quantum Mechanics*. Oxford University Press, Oxford, 2005.

[5]   H. Everett, *Rev. Mod. Phys.* 29:454, 1957.

[6]   M.A. Rubin, *Found. Phys. Lett.* 14:301, 2001.

[7]   J.G. Cramer, *Rev. Mod. Phys.* 58:647, 1986.

[8]   J.G. Cramer, *Inter. J. Theor. Phys.* 27:227, 1988.

[9]   D. Bohm, *Quantum Theory*. Prentice Hall, New York, 1951.

[10] R. Karam, *Am. J. Phys.* 88:39, 2020.

[11] V. Fano, *Rev. Mod. Phys.* 29:74, 1957.

[12] J. Nag, V.J. Menon and S.N. Mukherjee, *Am. J. Phys.* 55:802, 1987.

[13] G. Manfredi, S. Mola and M.R. Feix, *Eur. J. Phys.* 14:101, 1993.

[14] L.E. Ballentine, *Quantum Mechanics: A Modern Development*. World Scientific, Singapore, 1998.

[15] D. Sen, S.K. Das, A.N. Basu and S. Sengupta, *Current Science* 80:536, 2001.

[16] R.A. Leacock and M.J. Pedgett, *Phys. Rev. D* 28:2491, 1983.

[17] D. Home and S. Sengupta, *Am. J. Phys.* 51:265, 1983.

[18] L.S. Brown, *Am. J. Phys.* 40:371, 1972.

[19] S.B. Haley, *Am. J. Phys.* 65:237, 1997.

2.19   Exercises

2.1   Write the Schrödinger equation for a particle in a potential $\sin x$.

2.2   For a particle carrying a charge q in a uniform time-dependent electric field $\mathbf{E}(t)$ the force experienced by it is $\mathbf{F} = -\nabla(-q\mathbf{E}(t)\cdot\mathbf{X})$. Write the Schrödinger equation of the system.

2.3   Starting from the equation for ψ, obtain the equation for ψ^*.

2.4   What is the Schrödinger equation of a system of two particles of masses m_1 and m_2 carrying charges q_1 and q_2, respectively, with $H = p_1^2/(2m_1) + p_2^2/(2m_2) + q_1q_2/r_{12}$ and $r_{12} = |\mathbf{r}_1 - \mathbf{r}_2|$?

2.5   Write the Schrödinger equation for a charged particle in an electromagnetic field with $H = \frac{1}{2m}\left(\mathbf{p} - \frac{q}{c}\mathbf{A}(\mathbf{X},t)\right)^2 + q\phi(\mathbf{X},t)$. Rewrite the Schrödinger equation for $\mathbf{A} = (Bz, 0, 0)$ and $\phi = -Ez$.

2.6   If E_1 and E_2 are the eigenvalues and ϕ_1 and ϕ_2 are the corresponding eigenfunctions, respectively, of a Hamiltonian then find whether the energy corresponding to the superposition state $\phi_1 + \phi_2$ is equal to $E_1 + E_2$ or not.

2.7   Express $\left[-(\hbar^2/2m)\nabla^2 + V(x,y,z)\right]\psi(x,y,z) = E\psi(x,y,z)$ in the spherical polar coordinates defined by $x = r\sin\theta\cos\phi$, $y = r\sin\theta\sin\phi$, $z = r\cos\theta$ and in the parabolic coordinates defined by $\xi = r(1 - \cos\theta)$, $\eta = r(1 + \cos\theta)$, $\phi = \phi\cos\theta$.

2.8   Find the condition under which both $\psi(\mathbf{X},t)$ and $\psi^*(\mathbf{X},-t)$ will be the solutions of the same time-dependent Schrödinger equation.

2.9   What is the major difference between a real and complex ψ?

2.10   What is the difference between the wave function $\psi_1 = e^{i(kx-\omega t)}$ and $\psi_2 = e^{i(\mathbf{k}\cdot\mathbf{X}-\omega t)}$?

2.11   Write the Hamiltonian of a photon.

2.12   What is the physical meaning of $\langle x \rangle = 0$?

2.13   Write the operators of kinetic energy and angular momentum $\mathbf{L} = \mathbf{r} \times \mathbf{p}$.

2.14   Consider the one-dimensional Schrödinger equation $\psi_{xx} + (\lambda - g(x))\psi = 0$, $\lambda = 2mE/\hbar^2$ and $g = 2mV/\hbar^2$. Obtain the Schrödinger equation under the change of variable $u(x) = -\frac{d}{dx}\ln\psi(x)$ [19].

2.15   Find the conditions to be satisfied by the functions f and g such that under the transformation $\psi = f(x)F(g(x))$ the equation $\psi_{xx} + (E - V)\psi = 0$ can be written as $F_{gg} + Q(g)F_g + R(g)F(g) = 0$. Determine $E - V$.

2.16   What are the effects of addition of a constant to a potential on the time-independent Schrödinger equation and the energy levels?

2.17   Which of the following wave functions are admissible in quantum mechanics? State the reasons.

(a) e^{-x^2}.   (b) $\text{sech}x$.   (c) e^{-x}.   (d) $\tanh x$.   (e) $\sin x$, $0 < x < 2\pi$.   (f) $\sin x$, $-\infty < x < \infty$.   (g) $\sqrt{e^{-x^2}}$.   (h) $\tan x$.   (i) $\sec x$.   (j) xe^{-x^2}.

2.18   Normalize the wave function $\psi = Ne^{ikx - x^2/(2\sigma^2)}$.

2.19   Find the value of N for which the wave function $\psi(x) = N$ for $|x| < a$ and 0 for $|x| > a$ is normalized.

2.20   Normalize the wave function $\psi = e^{-|x|}\sin\alpha x$. It is given that $\int_0^\infty e^{-x}\sin\alpha x\, dx$
       $= \alpha^2/(1+\alpha^2)$.

2.21   A particle of mass m moves in a one-dimensional box of length L with origin as
       the centre. If the wave function of this particle is $\psi(x) = Nx(1-x^2)$ for $|x| < L/2$
       and 0 otherwise find the factor N.

2.22   A quantum mechanical particle moving in one-dimension has the wave function
       $\psi(x) = cxe^{-|x|/b}$, $-\infty < x < \infty$, where c and b are constants ($b > 0$). Find the
       probability that the position of the particle lies in the region $-\infty < x \leq b$.

2.23   What is the probability current density corresponding to $\psi(x) = Ae^{-\alpha x}$, where
       A and α are constants?

2.24   A free particle in one-dimension is in a state described by $\psi = A\,e^{i(p_x x - Et)/\hbar} +$
       $B\,e^{-i(p_x x + Et)/\hbar}$, where p_x and E are constants. Find the probability current den-
       sity.

2.25   If $\psi(x,0) = Ne^{-ikx-(x^2/2a^2)}$ for a particle then calculate the probability density
       and current density.

2.26   For the Gaussian wave function $\psi(x) = [1/(\sigma\sqrt{\pi})]^{1/2}\,e^{-x^2/2\sigma^2}$ calculate the prob-
       ability current density.

2.27   Verify whether the wave function $\psi = Ne^{ikx-x^2/(2a^2)}$ satisfies the continuity
       equation or not.

2.28   The wave function of a linear harmonic oscillator with potential $V = m\omega^2 x^2/2$
       is $\psi = xe^{-m\omega x^2/(2\hbar)}$. Find its energy.

2.29   If $\psi = \sqrt{1/L}\cos(\pi x/(2L))$ for the system confined to the potential $V(x) = 0$ for
       $|x| < L$ and ∞ for $|x| > L$ then calculate E.

2.30   The wave function of a particle confined to a box of length L is $\sqrt{2/L}\,\sin(\pi x/L)$
       in the region $0 < x < L$ and 0 otherwise. Calculate the probability of finding the
       particle in the region $0 < x \leq L/2$.

2.31   Write the law of conservation of energy $H = T+V$ in terms of expectation values.

2.32   Are the wave functions
       $$\psi_1 = \left(\frac{1}{\pi a_0^3}\right)^{1/2} e^{-r/a_0} \text{ and } \psi_2 = \left(\frac{1}{32\pi a_0^3}\right)^{1/2}\left(2-\frac{r}{a_0}\right)e^{-r/(2a_0)}$$
       of the electron in hydrogen atom orthogonal?

2.33   Find the potential corresponding to the following wave functions: (a) $\psi =$
       $(1/\pi^{1/4})e^{-x^2/2}$, $E = \hbar\omega/2$. (b) $\psi = A\sin(\pi x/(2L))$ for $|x| < L$ and 0 for $|x| > L$,
       $E = \hbar^2\pi^2/(8mL^2)$.

2.34   A particle of mass m is confined in the potential $V = 0$ for $0 < x \leq L$ and
       $V = \infty$ otherwise. It has the normalized stationary state $\phi_n(x)$ and eigenvalues
       $E_n = n^2\hbar^2\pi^2/(2mL^2)$. Its wave function at time $t = 0$ is given by $\psi(x,0) =$
       $(\phi_1(x) + \phi_2(x))/\sqrt{2}$. What is the smallest positive time τ for which $\psi(x,t)$ will
       be orthogonal to $\psi(x,0)$?

2.35   Show that $\langle xp_x + p_x x \rangle$ is real.

2.36   Show that $\langle A^n \rangle = \langle A \rangle^n$ in its eigenstates.

2.37   Calculate $\langle p_x^2 \rangle$ for $\psi = \sqrt{k}e^{-k|x|}$.

2.38   A particle of mass m in the potential $V(x) = 0$ for $0 \leq x \leq L$ and ∞ otherwise
       is in a state with $\psi(x) = Nx(L - x)$, where N is the normalization constant.
       Determine $\langle E \rangle$ in this state.

2.39 Show that $\dfrac{d}{dt}\langle x^2 \rangle = \dfrac{2}{m}\langle xp_x \rangle - \dfrac{i\hbar}{m}$.

2.40 Show that $\dfrac{d^2}{dt^2}\langle x^2 \rangle = \dfrac{2}{m^2}\langle p_x^2 \rangle + \dfrac{2}{m}\langle xF \rangle$.

2.41 Show that $\dfrac{d}{dt}\langle p_x^2 \rangle = \langle 2Fp_x + p_xF \rangle$.

2.42 Show that $\dfrac{d}{dt}\langle p_x x \rangle = \dfrac{1}{m}\langle p_x^2 \rangle + \langle xF \rangle$.

2.43 For a free particle determine (a) $\dfrac{d}{dt}\langle p_x^2 \rangle$ and (b) $\dfrac{d}{dt}\langle xp_x + p_xx \rangle$.

2.44 ϕ_1 and ϕ_2 are the only eigenfunctions of a system belonging to the energy eigenvalues E_0 and $-E_0$, respectively. If $\langle E \rangle$ is found to be $E_0/2$. Find the wave function of the system.

2.45 The H of a charged particle in uniform electric and magnetic fields is given by $H = (\mathbf{p} - e\mathbf{A})^2/(2m) - eE_0\mathbf{k}.\mathbf{k}$, where $\mathbf{A} = B_0(-y, x, 0)/2$. Both $\mathbf{E}$ and $\mathbf{B}$ are applied along the z-direction. Applying Ehrenfest's theorem find $d\langle \mathbf{r} \rangle/dt$ and $d\langle \mathbf{p} - e\mathbf{A} \rangle/dt$.

2.46 A particle of mass m enclosed in a one-dimensional box of length L such that $0 < x < L$, has $\phi_n(x) = A\sin(n\pi x/L)$, $n = 1, 2, \ldots$ and $E_n = n^2\pi^2\hbar^2/(2mL^2)$. At $t = 0$, the particle has $\psi(x) = B\sin(2\pi x/L)\cos(\pi x/L)$, where B is a constant. (a) If the energy of the particle is measured at $t = 0$, what are the possible results of the measurement? (b) What is the expectation value of energy?

2.47 Given the normalized wave function of hydrogen atom $\psi_{100} = 1/(\pi a_0^3)^{1/2}\,e^{-r/a_0}$ find the expectation value of its z-coordinate.

2.48 If $H\phi_n(x) = E_n\phi_n(x)$ find the expectation value of the Hamiltonian H in the superposition state $\psi(x, t) = \sum_{n=0}^{\infty} C_n\phi_n(x)\,e^{-iE_nt/\hbar}$.

2.49 Consider a system in a state $\psi = (\phi_1 + \phi_2)/\sqrt{2}$, where ϕ_1 and ϕ_2 are orthonormal eigenfunctions with the eigenvalues E_1 and E_2, respectively. What is the probability of finding the system in the energy E_1? What is $\langle E \rangle$?

2.50 Consider a spherically symmetric potential energy function given by $V(r) = 0$, for $0 < r < a$ and ∞ for $r > a$. Given the solution $\psi(r) = A(\sin kr)/r + B(\cos kr)/r$, where $k = (2mE/\hbar^2)^{1/2}$ satisfying the Schrödinger equation, obtain the corresponding eigenvalues by applying proper boundary conditions, seen from the frame (x, t).

3

Operators, Eigenvalues and Eigenfunctions

3.1 Introduction

A basic quantity in quantum mechanics is the wave function ψ. Given a quantum mechanical system we write the Schrödinger equation for it. Its solution is the wave function. When introducing the wave function we stated that the maximum information about the state of a system can be obtained only from ψ. How do we extract observable properties of the system from the wave function? In quantum mechanics for every classical dynamical variable there exists an operator. For example, the operator of momentum $\mathbf{p}$ is $-i\hbar\nabla$ while the operator of energy variable $H = \mathbf{p}^2/(2m) + V(\mathbf{X})$ is $-(\hbar^2/2m)\nabla^2 + V(\mathbf{X})$. Operation of an operator on ψ gives another function. If the new function is $\lambda\psi$, then λ is the eigenvalue of the operator. For example, the energy of a particle in a particular state represented by ψ can be obtained by operating H on ψ. That is, we have $H\psi = E\psi$, where E is the energy eigenvalue. The expectation value of $A(\mathbf{X}, \mathbf{p})$ can be obtained from the formula

$$\langle A \rangle = \int_{-\infty}^{\infty} \psi^* A \psi \, \mathrm{d}\tau \,, \tag{3.1}$$

where ψ is the solution of the equation $H\psi = E\psi$.

Operators in quantum mechanics which represent observables are of a special type: They are *linear* and *Hermitian*. Such operators not only form computation tools, but also form the most effective language in terms of which the theory can be formulated. The fascinating point is that using linear operators we can understand the quantum mechanics. Therefore, it is important to study the basic theory of linear operators. In this chapter, we present important properties of linear operators and their eigenfunctions. We discuss the features of various operators which are necessary for studying quantum mechanical systems. This chapter will show how rich and beautiful the mathematical structure of quantum mechanics is.

3.2 Linear Operators

An *operator* A_{op} is a mathematical operation which, when it acts on a function $f(x)$ gives another function $g(x)$:

$$A_{\mathrm{op}} f(x) = g(x) \,. \tag{3.2}$$

For simplicity we drop the subscript 'op' in A_{op}. Mathematical operations such as addition, subtraction, differentiation, integration, etc. can be represented by appropriate operators. An example of operator is $\mathrm{d}/\mathrm{d}x$.

Usually an operator is specified by its action. However, an operator essentially consists of its action and its domain. By *action* we mean what the operator does to the functions. On

DOI: 10.1201/9781003172178-3

the other hand, the specified set of functions on which the operator acts are referred as the *domain* of the operator. Why should one define the domain of an operator? We will come to this question when discussing about self-adjoint and Hermitian operators in sec. 3.4.

3.2.1   Definition

An operator A is said to be a *linear operator* if it satisfies the rules

$$A\left(\phi_1 + \phi_2\right) = A\phi_1 + A\phi_2 \,, \tag{3.3a}$$

$$A\left(c\phi_1\right) = cA\phi_1 \,, \quad A\left(c\phi_2\right) = cA\phi_2 \,, \tag{3.3b}$$

where ϕ_1 and ϕ_2 are given functions and c is any arbitrary complex constant. The operator $\mathrm{d}/\mathrm{d}x$ is linear because

$$\frac{\mathrm{d}}{\mathrm{d}x}\left(\phi_1 + \phi_2\right) = \frac{\mathrm{d}}{\mathrm{d}x}\phi_1 + \frac{\mathrm{d}}{\mathrm{d}x}\phi_2 \,, \tag{3.4a}$$

$$\frac{\mathrm{d}}{\mathrm{d}x}c_1\phi_1 = c_1\frac{\mathrm{d}}{\mathrm{d}x}\phi_1 \,, \quad \frac{\mathrm{d}}{\mathrm{d}x}c_2\phi_2 = c_2\frac{\mathrm{d}}{\mathrm{d}x}\phi_2 \,. \tag{3.4b}$$

Operators do not generally have the property given by Eq. (3.3a). Suppose an operator A is defined by $A\phi = \sin\phi$. Then left-side of Eq. (3.3a) is

$$A\left(\phi_1 + \phi_2\right) = \sin\left(\phi_1 + \phi_2\right) \,. \tag{3.5a}$$

The right-side of Eq. (3.3a) is

$$A\phi_1 + A\phi_2 = \sin\phi_1 + \sin\phi_2 \,. \tag{3.5b}$$

Right-sides of Eqs. (3.5) are not same. That is, the operator A defined by $A\phi = \sin\phi$ is not a linear operator.

3.2.2   Eigenvalues and Eigenfunctions of Linear Operators

Suppose we assume that

$$A\phi = \lambda\phi \,, \tag{3.6}$$

where λ is a number. Equation (3.6) is called the *eigenvalue equation* of A. ϕ is called an *eigenfunction* of A. λ is called the *eigenvalue* of the operator A corresponding to the eigenfunction ϕ. Given an eigenvalue equation we ask: What are the values of λ for which nontrivial solutions which are finite everywhere even at $x = \pm\infty$ occur? In general, nontrivial solutions exist for *certain* values of λ only. The certain values are called *eigenvalues*[1]. The eigenvalues are the important values of λ because only for these specific set of values Eq. (3.6) has nontrivial solutions. These solutions, called *eigenfunctions*, are the important solutions because they are the only possible meaningful solutions. The set of eigenvalues is called a *spectrum*.

Suppose A and B are two linear operators. The linear combination of them is $C = \alpha A + \beta B$. Is C a linear operator? The answer is yes (see the exercise 5 at the end of this chapter). That is, a linear combination of the two or more linear operators is also a linear operator. It is easy to show that product of two or more linear operators is also a linear operator. If $A\phi = 0$ then A is called a *singular operator*. When $AB = BA = I$ then $A^{-1} = B$ and $B^{-1} = A$. Now, the operators A and B are called *nonsingular operators*.

---

[1] *Eigen* is a German word which means *certain*.

Some of the properties of linear operators are summarized below:

$$A + B = B + A, \tag{3.7a}$$
$$A + B + C = (A + B) + C = A + (B + C), \tag{3.7b}$$
$$A(BC) = (AB)C + ABC, \tag{3.7c}$$
$$(A + B)C = AC + BC, \tag{3.7d}$$
$$A(B + C) = AB + AC, \tag{3.7e}$$
$$A^2 = AA. \tag{3.7f}$$

In quantum mechanics, the operators associated with the observables such as position, linear momentum, angular momentum and energy all satisfy the properties of linear operators.

Example 1:

Let us see how eigenfunctions and eigenvalues of an operator can be calculated. For illustrative purpose we consider the operator $x + d/dx$. The eigenvalue equation for the given operator is

$$\left(x + \frac{d}{dx}\right)\phi = \lambda\phi. \tag{3.8}$$

Rewriting it as

$$\frac{d\phi}{\phi} = (\lambda - x)\,dx \tag{3.9}$$

and integrating on both sides we obtain

$$\phi(x) = c\,e^{\lambda x - x^2/2}, \tag{3.10}$$

where c is a constant. The eigenfunction ϕ is continuous and well defined for any value of λ. Thus, the eigenvalues of the given operator form a continuous spectrum. However, in quantum mechanics a physically acceptable wave function should satisfy certain boundary conditions. When these boundary conditions are applied, then the eigenvalues may not take continuous values. Only for certain values of λ, the wave function may be physically acceptable. Under such situations, λ values will take discrete values. In the case of the Schrödinger equation $H\psi = E\psi$, the boundary conditions on ψ lead to discrete energy eigenvalues and quantum numbers.

A linear operator that when acted on all the normalizable eigenfunctions of a system gives normalizable eigenfunctions is termed as a *bounded opertor*. The operator which leads to nonnormalizable eigenfunctions is an *unbounded operator*. The momentum operator $-i\hbar\nabla$ and the operator multiplication by x are unbounded operators.

Example 2:

There are a number of operators for which the eigenvalues are discrete. For example, consider the operator

$$L = \left(1 - x^2\right)\frac{d^2}{dx^2} - 2x\frac{d}{dx}. \tag{3.11}$$

It is a linear operator. Its eigenvalue equation is $Ly = \lambda y$. With $-\lambda = l(l + 1)$ we can identify the equation $Ly = \lambda y$ as the Legendre differential equation. It has bounded solutions only for integer values of l and the associated solutions (called *Legendre polynomials*) are defined for $x \in [-1, 1]$ [1,2]. The eigenvalues of L are certain integers and the corresponding eigenfunctions are Legendre polynomials. If λ is not chosen as $-l(l+1)$ with l taking integer values, then we will get a diverging infinite series solution, which is not an acceptable physical wave function. So, the condition $\lambda = -l(l+1)$ with l taking 0, 1, 2, ... gives a finite polynomial solutions which are physically acceptable.

An important property of eigenfunctions of linear operators is that a linear combination of two eigenfunctions belonging to two distinct eigenvalues of a linear operator A is not an eigenfunction of it. To prove this assume that ϕ_a and ϕ_b are two eigenfunctions of an operator A with distinct eigenvalues a and b, respectively. The eigenvalue equations for ϕ_a and ϕ_b are

$$A\phi_a = a\phi_a , \quad A\phi_b = b\phi_b . \tag{3.12}$$

Suppose the linear combination $\psi = \alpha\phi_a + \beta\phi_b$ is an eigenfunction of A. Then we must have $A\psi = \lambda\psi$, that is

$$A\left(\alpha\phi_a + \beta\phi_b\right) = \lambda\left(\alpha\phi_a + \beta\phi_b\right) , \tag{3.13}$$

where λ is the eigenvalue of A corresponding to ψ. Using Eq. (3.12) in Eq. (3.13) we obtain

$$\alpha(a - \lambda)\phi_a + \beta(b - \lambda)\phi_b = 0 . \tag{3.14}$$

This equation is true only if $\lambda = a$ and $\lambda = b$. Since a and b are distinct eigenvalues the Eq. (3.14) is not true. Therefore, $\psi = \alpha\phi_a + \beta\phi_b$ cannot be an eigenfunction of A. Equation (3.14) implies that if ϕ_1 and ϕ_2 are two linearly independent eigenfunctions of a linear operator belonging to a same eigenvalue λ then a linear combination of ϕ_1 and ϕ_2 will also be an eigenfunction of A with same eigenvalue λ.

One can easily verify that if ϕ is an eigenfunction of an operator A with the eigenvalue a then ϕ is also an eigenfunction of the operators A^2, A^n and e^A. *What are the corresponding eigenvalues?* (See the exercise 11 at the end of this chapter.)

Solved Problem 1:

Obtain the eigenvalues and eigenfunctions of the operator $A = \dfrac{1}{r^2}\dfrac{d}{dr}\left(r^2\dfrac{d}{dr}\right)$.

The eigenvalue equation of A is $A\psi = \lambda\psi$ which becomes

$$\frac{d^2\psi}{dr^2} + \frac{2}{r}\frac{d\psi}{dr} = \lambda\psi . \tag{3.15}$$

Introducing the change of variable $\psi = y/r$ we can eliminate the second term in the above equation and obtain

$$y'' + \alpha^2 y = 0, \quad \lambda = -\alpha^2 . \tag{3.16}$$

Its particular solutions are $\sin\alpha r$ and $\cos\alpha r$. Then $\psi = N(\sin\alpha r)/r$ and $N(\cos\alpha r)/r$. In the limit $r \to 0$, $(\sin\alpha r)/r = \alpha$ while $(\cos\alpha r)/r \to \infty$. Therefore, we discard the solution $N\cos\alpha r/r$ and choose the eigenfunctions as $N\sin\alpha r/r$, where N is the normalization constant. The eigenvalues are $-\alpha^2$ and form a continuous spectrum.

3.3   Commuting and Noncommuting Operators

In classical mechanics if we know some set of physical quantities such as $(\mathbf{r}, \mathbf{p})$ then certain other quantities of interest can be determined. For example, given $\mathbf{r}$ and $\mathbf{p}$ one can determine orbital angular momentum $\mathbf{L} = \mathbf{r} \times \mathbf{p}$. From the set of all dynamical variables many such compatible subsets can be chosen. Their member is restricted by the Poisson brackets. If the coordinates of a particle are denoted by x_1, x_2, x_3 and momenta by p_1, p_2, p_3 then the Poisson bracket [3–5] of two functions f and g (which are functions of x_i's and p_i's) is

$$\{f, g\} = \sum_{i=1}^{3} \left(\frac{\partial f}{\partial x_i} \frac{\partial g}{\partial p_i} - \frac{\partial f}{\partial p_i} \frac{\partial g}{\partial x_i} \right). \tag{3.17}$$

In quantum mechanics the Poisson bracket conditions are replaced by commutator brackets.

3.3.1 Definition, Identities and Examples

The operator $AB - BA$ is called the *commutator* of A and B and is denoted $[A, B]$. The properties of commutator brackets are similar to those of Poisson brackets. If $[A, B] = 0$, that is, $AB = BA$, then the two operators A and B are said to be *commuting operators*; otherwise *noncommuting operators*. For commuting operators their order of occurrence can be freely changed. When $AB = -BA$ the operators A and B are said to be *anticommuting operators*.

The following identities can be easily verified:

$$[A, B] = -[B, A], \tag{3.18a}$$
$$[AB, C] = [A, C] B + A [B, C], \tag{3.18b}$$
$$[A, BC] = [A, B] C + B [A, C], \tag{3.18c}$$
$$[\alpha_1 A + \alpha_2 B, C] = \alpha_1 [A, C] + \alpha_2 [B, C], \tag{3.18d}$$
$$[A, \alpha_1 B + \alpha_2 C] = \alpha_1 [A, B] + \alpha_2 [A, C], \tag{3.18e}$$
$$[A, A^n] = 0, \quad e^A = \sum_{n=0}^{\infty} \frac{1}{n!} A^n. \tag{3.18f}$$

The rule for the calculation of area of a rectangle of length l and breadth b is commutative; $l \times b = b \times l$. However, there are certain mathematical operations which are noncommutative. Let us consider the following two operations:

1. Move two steps in the forward direction. Call this operation as S.
2. Rotate by $90°$ to right-side. Call this operation as R.

Suppose we stand in a room and perform the operations 1 and 2. Note down the final position and it be P_{12}. Next, we go to the starting position and perform the operations 2 and 1. The position of us is say P_{21}. Obviously, P_{21} is different from P_{12}. That is, symbolically $SR \neq RS$ or $SR - RS \neq 0$. The operators S and R are noncommutative.

For the coordinate and momentum operators we have

$$\begin{aligned} [x, p_x] \phi &= (xp_x) \phi - (p_x x) \phi \\ &= -i\hbar \left(x \frac{\partial \phi}{\partial x} - \frac{\partial}{\partial x} (x\phi) \right) \\ &= i\hbar\phi. \end{aligned} \tag{3.19}$$

Thus, $[x, p_x] = i\hbar$ which implies that the order of these operators cannot be changed. *What is the consequence of this relation?* A consequence of it is that $x(p_x\phi) \neq p_x(x\phi)$, where ϕ is a state of the system. That is, a position measurement followed by a momentum measurement might give different result from a momentum measurement followed by a position measurement.

Similarly, we obtain

$$[y, p_y] = [z, p_z] = i\,\hbar\,, \quad [p_x, x] = [p_y, y] = [p_z, z] = -i\,\hbar\,. \tag{3.20}$$

Note that the conjugate pairs (x, p_x), (y, p_y) and (z, p_z) do not commute. But for x, p_y we have $[x, p_y] = 0$ so that the order of these operators can be changed without affecting the result. Further,

$$[x, p_z] = 0\,, \quad [y, p_x] = 0\,, \quad [y, p_z] = 0\,, \quad [z, p_x] = 0\,, \quad [z, p_y] = 0\,. \tag{3.21}$$

3.3.2   Features of Commuting and Noncommuting Operators

There are three important features associated with commuting and noncommuting quantum mechanical operators. They are given below:

1. If the operators A and B of two observables do not commute then they cannot be measured simultaneously with arbitrary accuracy (for details see sec. 8.3).

2. If A and B are two commuting operators then the eigenfunctions of A (or B) are also the eigenfunctions of B (A). (For a proof see sec. 3.3.3).

3. The expectation value of an operator commuting with the Hamiltonian H is a constant of motion or time-independent. Later in sec. 7.2 we show that the time-independency of expectation value of an operator A is governed by the equation

$$\frac{\mathrm{d}}{\mathrm{d}t}\langle A \rangle = \left\langle \frac{\partial A}{\partial t} \right\rangle + \frac{i}{\hbar}\langle [H, A] \rangle\,. \tag{3.22}$$

If A has no explicit time-dependence then

$$\frac{\mathrm{d}}{\mathrm{d}t}\langle A \rangle = \frac{i}{\hbar}\langle [H, A] \rangle\,. \tag{3.23}$$

When $[H, A] = 0$ we have $\mathrm{d}\langle A \rangle/\mathrm{d}t = 0$. Hence, $\langle A \rangle$ of an operator commuting with the Hamiltonian is a constant of motion. Further, for any time-independent Hamiltonian operator $\mathrm{d}\langle H \rangle/\mathrm{d}t = 0$ because of $[H, H] = 0$. Therefore, for any system with time-independent Hamiltonian operator, energy is conserved.

In classical mechanics with respect to generalized position and momentum coordinates q and p, the equation of motion is [3–5]

$$\frac{\mathrm{d}}{\mathrm{d}t}f(q, p, t) = \frac{\partial f}{\partial t} - \{H, f\}\,. \tag{3.24}$$

Comparison of Eqs. (3.22) and (3.24) shows that the Poisson bracket of classical mechanics corresponds to the commutator multiplied by $-i/\hbar$.

Solved Problem 2:

Compute $[H, p_x]$ for the linear harmonic oscillator with

$$H = -\frac{\hbar^2}{2m}\frac{\mathrm{d}^2}{\mathrm{d}x^2} + \frac{1}{2}kx^2\,, \quad k = m\omega^2\,. \tag{3.25}$$

We obtain

$$
\begin{aligned}
[H, p_x]\phi &= H(p_x\phi) - p_x(H\phi) \\
&= \left(-\frac{\hbar^2}{2m}\frac{d^2}{dx^2} + \frac{1}{2}kx^2\right)\left(-i\hbar\frac{d}{dx}\phi\right) \\
&\quad - \left(-i\hbar\frac{d}{dx}\right)\left(-\frac{\hbar^2}{2m}\frac{d^2}{dx^2} + \frac{1}{2}kx^2\right)\phi \\
&= i\hbar kx\phi. \\
&= i\hbar m\omega^2 x\phi.
\end{aligned}
\tag{3.26}
$$

Thus, $[H, p_x] = i\hbar m\omega^2 x$ for the linear harmonic oscillator. Since $[H, p_x] \neq 0$, $\langle p_x \rangle$ is not a constant of motion. For a free particle $k = 0$ and hence $[H, p_x] = 0$, that is, $\langle p_x \rangle$ is a constant of motion.

3.3.3 Simultaneous Eigenfunctions

Suppose A and B are two operators and ϕ is a function such that

$$
A\phi = a\phi, \quad B\phi = b\phi
\tag{3.27}
$$

then ϕ is called a *simultaneous eigenfunction* of A and B belonging to the eigenvalues a and b, respectively. The equation $A\phi = a\phi$ implies

$$
BA\phi = aB\phi = ab\phi.
\tag{3.28}
$$

Similarly, from $B\phi = b\phi$ we get

$$
AB\phi = bA\phi = ba\phi.
\tag{3.29}
$$

From Eqs. (3.28) and (3.29) we obtain

$$
(AB - BA)\phi = [A, B]\phi = 0.
\tag{3.30}
$$

Therefore, if ϕ is a simultaneous eigenfunction of the operators A and B then $[A, B] = 0$. If two operators A and B are commuting operators then they admit simultaneous eigenfunction. $[A, B] = 0$ is the necessary condition for ϕ to be a simultaneous eigenfunction of A and B. In fact, it is also to be the sufficient condition. For a free particle $[H, p_x] = 0$. Hence, the operators $H = -(\hbar^2/2m)d^2/dx^2$ and $p_x = -i\hbar d/dx$ have same set of eigenfunctions. (Verify this with the solutions of the time-independent Schrödinger equation of the free particle).

Suppose A and B are commuting operators and $\phi_a^{(1)}$ and $\phi_a^{(2)}$ are two linearly independent eigenfunctions of A with the same eigenvalue a. Then there is a state with two different eigenfunctions but with same eigenvalue. This state is said to be *doubly degenerate*. We write $A\phi_a^{(1)} = a\phi_a^{(1)}$, $A\phi_a^{(2)} = a\phi_a^{(2)}$ and $B\phi_a^{(1)} = b_1\phi_a^{(1)}$. We have $BA\phi_a^{(1)} = ab_1\phi_a^{(1)}$ and $AB\phi_a^{(1)} = b_1a\phi_a^{(1)}$. Then $(AB - BA)\phi_a^{(1)} = 0$. That is,

$$
AB\phi_a^{(1)} = BA\phi_a^{(1)} = aB\phi_a^{(1)}.
\tag{3.31}
$$

$B\phi_a^{(1)}$ is an eigenfunction of A with eigenvalue a. As $\phi_a^{(1)}$ and $\phi_a^{(2)}$ are eigenfunctions of A with eigenvalue a we write $B\phi_a^{(1)} \propto \phi_a^{(1)}$ and $B\phi_a^{(1)} \propto \phi_a^{(2)}$. $\phi_a^{(1)}$ is not an eigenfunction of B. That is we cannot claim that $B\phi_a^{(1)}$ is $\propto \phi_a^{(1)}$. Though $\phi_a^{(1)}$ and $\phi_a^{(2)}$ may be degenerate eigenfunctions of the operator A, they need not be degenerate eigenfunctions of the operator B which commutes with A.

Next, consider $\psi_+ = \alpha\phi_a^{(1)} + \beta\phi_a^{(2)}$ which is an eigenfunction of B with eigenvalue b. ψ_+ is also an eigenfunction of A with the eigenvalue a. Then $A\psi_+ = a\psi_+$ and $B\psi_+ = b\psi_+$. Then $(AB - BA)\psi_+ = 0$ gives

$$AB\psi_+ = BA\psi_+ = aB\psi_+ . \tag{3.32}$$

That is, $B\psi_+$ is an eigenfunction of A with eigenvalue a and $B\psi_+ \propto \psi_+$. Similarly, $B\psi_- \propto \psi_-$. Note that ψ_+ and ψ_- of B may correspond to different eigenvalues. Therefore, if $\phi_a^{(1)}$, $\phi_a^{(2)}$, ..., $\phi_a^{(m)}$ are m-fold degenerate eigenfunctions of an operator A with an eigenvalue a and $[A, B] = 0$ then linear combinations of m-fold eigenfunctions form simultaneous eigenfunctions of A and B.

3.4   Self-Adjoint and Hermitian Operators

In this section, we introduce the adjoint, self-adjoint and Hermitian operators.

3.4.1   Definition

Consider the integral

$$(\phi_1, A\phi_2) = \int_{-\infty}^{\infty} \phi_1^* A\phi_2 \, d\tau \tag{3.33}$$

which involves two different functions, ϕ_1 and ϕ_2, and A is an operator. For an operator A one can always find another operator $A^\dagger$ (A-dagger) called the *adjoint* of A such that

$$(\phi_1, A\phi_2) = \left(A^\dagger\phi_1, \phi_2\right) , \tag{3.34}$$

that is, $\int_{-\infty}^{\infty} \phi_1^* A\phi_2 \, d\tau = \int_{-\infty}^{\infty} \left(A^\dagger\phi_1\right)^* \phi_2 \, d\tau$. As far as the value of the integral is concerned it does not matter whether A acts on ϕ_2 or its adjoint $A^\dagger$ acts on other function ϕ_1. When $A^\dagger = A$ the operator A is said to be *self-adjoint*. A self-adjoint operator with certain boundary conditions on the eigenfunctions is called *Hermitian*[2].

When an operator A is self-adjoint then

$$(\phi_1, A\phi_2) = (A\phi_1, \phi_2) . \tag{3.35}$$

An interesting feature of a self-adjoint (Hermitian) operator is that the expectation value of the corresponding dynamical variable is real as shown below. When $A^\dagger = A$ and $\phi_1 = \phi_2 = \phi$ using Eqs. (3.34) and (3.35) we get

$$\langle A \rangle = \int_{-\infty}^{\infty} \phi^* A\phi \, d\tau = \int_{-\infty}^{\infty} \left(A^\dagger\phi\right)^* \phi \, d\tau = \left[\int_{-\infty}^{\infty} \phi^*(A\phi) \, d\tau\right]^* = \langle A \rangle^*. \tag{3.36}$$

$\langle A \rangle$ is thus real. If the operator of a dynamical variable is self-adjoint then the possible values of the variable are only real. In an experiment, the result of any single measurement of a variable is necessarily a real number. Therefore, *self-adjoint operators are suitable for representing observable dynamical variables.*

If $(\phi_1, A\phi_2) = -(A\phi_1, \phi_2)$ then A is said to be *anti-self-adjoint* (or *anti-Hermitian*). In this case $\langle A \rangle = -\langle A \rangle^*$ – the expectation value of an anti-self-adjoint operator is always a pure imaginary.

---

[2]Usually, self-adjoint operators are also called Hermitian. However, there is a distinction between these two operators. The difference lies in the domain of the operators. For details see appendix D.

3.4.2 Examples

In an experiment position and momentum of a particle can be measured and hence their operators must be self-adjoint. The position operator is obviously self-adjoint. For the momentum operator we have

$$
\begin{aligned}
(\phi_1, p_x \phi_2) &= \int_{-\infty}^{\infty} \phi_1^* \left(p_x \phi_2 \right) \mathrm{d}x \\
&= -\mathrm{i}\hbar \int_{-\infty}^{\infty} \phi_1^* \frac{\mathrm{d}\phi_2}{\mathrm{d}x} \, \mathrm{d}x \\
&= -\mathrm{i}\hbar \phi_1^* \phi_2 \big|_{-\infty}^{\infty} + \mathrm{i}\hbar \int_{-\infty}^{\infty} \frac{\mathrm{d}\phi_1^*}{\mathrm{d}x} \phi_2 \, \mathrm{d}x \,.
\end{aligned} \tag{3.37}
$$

Since ϕ_1, $\phi_2 \to 0$ as $|x| \to \infty$ the above equation becomes

$$
(\phi_1, p_x \phi_2) = \int_{-\infty}^{\infty} \left(-\mathrm{i}\hbar \frac{\mathrm{d}\phi_1}{\mathrm{d}x} \right)^* \phi_2 \, \mathrm{d}x = \int_{-\infty}^{\infty} (p_x \phi_1)^* \phi_2 \, \mathrm{d}x = (p_x \phi_1, \phi_2) \,. \tag{3.38}
$$

The adjoint of p_x is defined through the equation

$$
(\phi_1, p_x \phi_2) = \left(p_x^\dagger \phi_1, \phi_2 \right) \,. \tag{3.39}
$$

Comparison of Eqs. (3.38) and (3.39) gives $p_x^\dagger = p_x$. Thus, $p_x = -\mathrm{i}\hbar\,\mathrm{d}/\mathrm{d}x$ is a self-adjoint operator. But the operator $A = \mathrm{d}/\mathrm{d}x$ is not a self-adjoint. For this operator we have

$$
\int_{-\infty}^{\infty} \phi_1^* \frac{\mathrm{d}\phi_2}{\mathrm{d}x} \mathrm{d}x = \phi_1^* \phi_2 \big|_{-\infty}^{\infty} - \int_{-\infty}^{\infty} \frac{\mathrm{d}\phi_1^*}{\mathrm{d}x} \phi_2 \, \mathrm{d}x = \int_{-\infty}^{\infty} \left(-\frac{\mathrm{d}\phi_1}{\mathrm{d}x} \right)^* \phi_2 \, \mathrm{d}x. \tag{3.40}
$$

Thus, the adjoint of $\mathrm{d}/\mathrm{d}x$ is $-\mathrm{d}/\mathrm{d}x$ and it is not a self-adjoint.

An eigenfunction of the form $\mathrm{e}^{\mathrm{i}kx}$ corresponds to a momentum $\hbar k$ along the x-direction. It is not normalizable if x extends from $-\infty$ to $+\infty$ as $\int_{-\infty}^{\infty} |\mathrm{e}^{\mathrm{i}kx}|^2 \mathrm{d}x = \infty$. Under this situation, $p_x = -\mathrm{i}\hbar\,\mathrm{d}/\mathrm{d}x$ is not Hermitian as $\mathrm{i}\hbar \phi_1^* \phi_2 \big|_{-\infty}^{\infty}$ in Eq. (3.37) is not zero. In order to make p_x Hermitian, we use a box normalization by confining the system in an arbitrarily large box of length L and imposing the periodic boundary condition $\phi(x) = \phi(x+L)$. Under this boundary condition $\phi_k(x) = \mathrm{e}^{\mathrm{i}kx} = \mathrm{e}^{\mathrm{i}k(x+L)}$. Then k will take discrete values $2\pi n/L$, $n = 0, \pm 1, \ldots$. Then $\mathrm{i}\hbar \phi_1^* \phi_2 \big|_{-\infty}^{\infty}$ (in Eq. (3.37)) $\to \mathrm{i}\hbar \phi_1^* \phi_2 \big|_0^L = 0$. Hence, p_x is a Hermitian.

3.4.3 Properties of Self-Adjoint Operators

Some of the properties of self-adjoint operators are given below with necessary proofs.

1. *The eigenvalues of a self-adjoint operator are real.*

   Proof:

   Let A be a self-adjoint operator and ϕ_a and $\phi_{a'}$ be two eigenfunctions of it so that

$$
A\phi_a = a\phi_a \,, \quad A\phi_{a'} = a'\phi_{a'} \,. \tag{3.41}
$$

   The self-adjoint condition with ϕ_a and $\phi_{a'}$ is $(\phi_a, A\phi_{a'}) = (A\phi_a, \phi_{a'})$, that is,

$$
\int_{-\infty}^{\infty} \phi_a^* A\phi_{a'} \, \mathrm{d}x = \int_{-\infty}^{\infty} (A\phi_a)^* \phi_{a'} \, \mathrm{d}x \,. \tag{3.42}
$$

Using Eq. (3.41) in Eq. (3.42) we obtain

$$(a' - a^*) \int_{-\infty}^{\infty} \phi_a^* \phi_{a'} \, \mathrm{d}x = 0 \,. \tag{3.43}$$

When $\phi_{a'} = \phi_a$ Eq. (3.43) becomes

$$(a - a^*) \int_{-\infty}^{\infty} \phi_a^* \phi_a \, \mathrm{d}x = 0 \,. \tag{3.44}$$

The integral in the above equation is the norm of ϕ_a and it should be nonzero and finite. Therefore, we have $a^* = a$, a is real.

2. *Any two eigenfunctions belonging to distinct eigenvalues of a self-adjoint operator are mutually orthogonal.*

    Proof:

    Because of $a^* = a$, Eq. (3.43) can be rewritten as

$$(a' - a) \int_{-\infty}^{\infty} \phi_a^* \phi_{a'} \, \mathrm{d}x = 0 \,. \tag{3.45}$$

If $a' \neq a$ then $\int_{-\infty}^{\infty} \phi_a^* \phi_{a'} \mathrm{d}x = 0$ which completes the proof.

3. *The product of the self-adjoint operators is self-adjoint if and only if they commute.*

    Proof:

    Assume that A and B are two self-adjoint operators. If the product operator AB is a self-adjoint then $(\phi, AB\phi_2) = (AB\phi_1, \phi_2)$:

$$\int_{-\infty}^{\infty} \phi_1^* AB\phi_2 \, \mathrm{d}x = \int_{-\infty}^{\infty} (AB\phi_1)^* \phi_2 \, \mathrm{d}x \,. \tag{3.46}$$

Now, consider $\int_{-\infty}^{\infty} \phi_1^* AB\phi_2 \mathrm{d}x$. As A and B are self-adjoint we write

$$\int_{-\infty}^{\infty} \phi_1^* AB\phi_2 \, \mathrm{d}x = \int_{-\infty}^{\infty} (A\phi_1)^* B\phi_2 \, \mathrm{d}x = \int_{-\infty}^{\infty} (BA\phi_1)^* \phi_2 \, \mathrm{d}x \,. \tag{3.47}$$

From Eqs. (3.46) and (3.47) we find $AB = BA$. That is, $[A, B] = 0$.

4. *Because of $[A, A^n] = 0$ and the property (3) any power of a self-adjoint operator is also a self-adjoint operator.*

5. *A linear combination of two or more self-adjoint operators is also a self-adjoint operator.* The proof of it is left as an exercise to the reader.

6. *Self-adjoint operators can be used to generate a unitary transformation,* $\mathrm{e}^{\mathrm{i}A\theta}$. *This* will be discussed in secs. 6.4 and 11.2.

Operators of observables such as position, linear momentum, angular momentum and energy—all are self-adjoint and they satisfy the above properties. Real eigenvalues are not restricted to self-adjoint operators alone. There are nonself-adjoint operators with all eigenvalues being real. For example, it has been shown that the eigenvalues of the non-Hermitian operator

$$A = -\frac{\mathrm{d}^2}{\mathrm{d}x^2} - (\mathrm{i}x)^{2M} - \alpha(\mathrm{i}x)^{M-1} + l(l+1)\frac{1}{x^2} \,, \tag{3.48}$$

where $M > 1$, α and l are real, all are real [6]. For discussion on non-Hermitian operators one may refer to refs. [6–11] and also sec. 3.8.

Solved Problem 3:

For a Hermitian operator A prove that $\langle AA^\dagger \rangle \geq 0$.

We obtain

$$
\begin{aligned}
\langle AA^\dagger \rangle &= \int_{-\infty}^{\infty} \phi^* A A^\dagger \phi \, \mathrm{d}\tau \\
&= \int \left(A^\dagger \phi \right)^* A^\dagger \phi \, \mathrm{d}\tau \\
&= \int_{-\infty}^{\infty} (A\phi)^* A\phi \, \mathrm{d}\tau \\
&= \int_{-\infty}^{\infty} \psi^* \psi \, \mathrm{d}\tau \,, \quad (3.49)
\end{aligned}
$$

where $\psi = A\phi$. The above integral is norm of ψ which is positive. Therefore, $\langle AA^\dagger \rangle \geq 0$.

3.5 Discrete and Continuous Eigenvalues

In order to treat a dynamical variable A as an observable the operator representing it must be Hermitian so that $\langle A \rangle$ = real. Another requirement is that the eigenfunction of the operator should form a complete set. The eigenfunctions $\{\phi_a\}$ of the self-adjoint operator A are said to form a *complete set* if any arbitrary wave function ψ of the system can be expressed into a linear combination

$$
\psi = \sum_{a \in \mathrm{D}} c_a \phi_a + \int_{a \in \mathrm{C}} c_a \phi_a \, \mathrm{d}a \,. \quad (3.50)
$$

of the members of the set[3]. In Eq. (3.50) the summation is over the discrete part of the eigenvalue spectrum and integration is over the continuous part. Any dynamical variable represented by a Hermitian operator having a complete set of eigenfunctions is called an *observable*.

Let us normalize the wave function ψ assuming that all the eigenvalues of the operator A are discrete. We have

$$
1 = \int_{-\infty}^{\infty} \psi^* \psi \, \mathrm{d}\tau = \sum_a \sum_{a'} c_a^* c_{a'} \int_{-\infty}^{\infty} \phi_a^* \phi_{a'} \, \mathrm{d}\tau = \sum_a |c_a|^2 \,, \quad (3.51)
$$

---

[3]In the language of linear vector space, this complete set of vectors ϕ_a span the infinite-dimensional vector space known as *Hilbert space*. In this space any arbitrary vector ψ can be represented as the linear combination of these basis vector with the expansion coefficients being the coordinates of the vectors. These coordinates form column matrix which represents the vector ψ. Then any operation on ψ is represented as an infinite-dimensional matrix. Diagonalization of this matrix gives the eigenvalues and eigenvectors which are the basis of Heisenberg's matrix mechanics.

where $\int_{-\infty}^{\infty} \phi_a^* \phi_{a'} = \delta_{aa'}$ with $\delta_{aa'}$ as Kronecker-delta function, which is defined as $\delta_{aa'} = 1$ for $a = a'$ otherwise 0. By suitably choosing the values of c_a we can make $\sum |c_a|^2$ equal to 1. Thus, ψ is normalizable. When $a' = a$

$$\int_{-\infty}^{\infty} \phi_a^* \phi_a \, \mathrm{d}\tau = 1 \, . \tag{3.52}$$

That is, ϕ_a's are normalized. Thus, *the eigenfunctions belonging to discrete eigenvalues are normalizable.*

When all the eigenvalues form a continuous spectrum then the normalization condition becomes

$$1 = \int_{-\infty}^{\infty} c_a^* \mathrm{d}a \int_{-\infty}^{\infty} c_{a'} \mathrm{d}a' \int_{-\infty}^{\infty} \phi_a^* \phi_{a'} \, \mathrm{d}\tau \, . \tag{3.53}$$

Since a is continuous we write

$$\int_{-\infty}^{\infty} \phi_a^* \phi_{a'} \, \mathrm{d}\tau = \delta(a, a') = \delta(a - a') \, , \tag{3.54}$$

where $\delta(a, a')$ is Dirac-delta function given by

$$\delta(a, a') \;=\; \infty \quad \text{for} \quad a' = a \quad \text{otherwise } 0, \tag{3.55a}$$

$$\int_{-\infty}^{\infty} \delta(a, a') \, \mathrm{d}a' \;=\; 1 \, . \tag{3.55b}$$

Using the result

$$\int_{-\infty}^{\infty} F(x') \, \delta(x, x') \, \mathrm{d}x' = F(x) \tag{3.56}$$

Eq. (3.53) becomes

$$1 = \int_{-\infty}^{\infty} |c_a|^2 \, \mathrm{d}a \, . \tag{3.57}$$

Therefore, ψ is normalizable. However, using Eq. (3.55) in Eq. (3.54) we get

$$\int_{-\infty}^{\infty} \phi_a^* \phi_a \, \mathrm{d}\tau = \infty \, . \tag{3.58}$$

The eigenfunctions are nonnormalizable. Thus, *the eigenfunctions belonging to continuous eigenvalues are nonnormalizable.*

When we look for normalizable eigenfunctions we would end up with discrete eigenvalues. If the eigenvalues are continuous then the eigenfunctions are nonnormalizable. If the eigenvalue spectrum has both discrete and continuous parts then the normalization condition gives

$$\sum_{a \in \mathrm{D}} |c_a|^2 + \int_{a \in \mathrm{C}} |c_a|^2 \, \mathrm{d}a = 1 \, . \tag{3.59}$$

In connection with the eigenvalue problems, the following questions may be raised. If we assign an arbitrary real value to a in the eigenvalue equation $A\phi_a = a\phi_a$ does there exist a corresponding eigenfunction ϕ_a? The answer is negative. This is because the solution of the eigenvalue equation has to satisfy the physical admissibility conditions. Therefore, only for specific values of a eigenfunctions exist. The next question to be discussed is whether there can be more than one eigenfunction for a particular eigenvalue. It is quite possible for a set of eigenfunctions to have same eigenvalue. If there exists only one eigenfunction belonging to a given eigenvalue, the eigenvalue is said to be *nondegenerate*; otherwise, it is *degenerate*.

If ϕ_a and χ_a belong to the same eigenvalue a, for the operator A, then the linear combination $\psi = c_1\phi_a + c_2\chi_a$ is also an eigenfunction of the operator A with the same eigenvalue a. We can construct an infinite number of eigenfunctions corresponding to any degenerate eigenvalue. If there are r-linearly independent eigenfunctions corresponding to a particular degenerate eigenvalue, we say that the eigenvalue is r-fold degenerate.

3.6   Meaning of Eigenvalues and Eigenfunctions

In connection with the eigenvalues (a), eigenfunctions (ϕ_a) and the normalized wave function (ψ) we ask the following questions: *What are actually a, ϕ_a, c_a and ψ? What do they represent in a quantum mechanical system? Can we measure them in an experiment?* The aim of this section is to give precise answers to these questions.

Recall the experiment of throwing a die of six sides considered in sec. 2.9. Let it be thrown a very large number of times, say N times. The average or mean or expectation value of an outcome of a throw of the die is given by

$$\langle \text{outcome} \rangle = \sum P(a)a, \quad a = 1, 2, \ldots, 6, \tag{3.60}$$

where $P(a)$ is the probability of getting a as an outcome. The formula, Eq. (3.60), is not only for the throw of a die but also for outcome of any experiment. In quantum mechanics for the position measurement we write

$$\langle x \rangle = \sum P(x)x = \int_{-\infty}^{\infty} \psi^* x \psi \, dx. \tag{3.61}$$

For an observable A we can write

$$\langle A \rangle = \int_{-\infty}^{\infty} \psi^* A \psi \, dx. \tag{3.62}$$

Substituting $\psi = \sum c_a \phi_a$ in the above equation we get

$$
\begin{aligned}
\langle A \rangle &= \int_{-\infty}^{\infty} \sum_a c_a^* \phi_a^* A \sum_{a'} c_{a'} \phi_{a'} \, dx \\
&= \int_{-\infty}^{\infty} \sum_a c_a^* \phi_a^* \sum_{a'} a' c_{a'} \phi_{a'} \, dx \\
&= \sum_a \sum_{a'} c_a^* c_{a'} a' \delta_{aa'} \\
&= \sum_a |c_a|^2 a. \tag{3.63}
\end{aligned}
$$

Comparing the Eqs. (3.60) and (3.63) we infer that a in the eigenvalue equation $A\phi_a = a\phi_a$ is a possible outcome of an experiment. Suppose we measure the energy of a system. We get a number. This number is one of the possible energy eigenvalues. In the experiment, we say that the outcome of energy measurement is the energy value of the system. In the theory, we call a possible energy value as an energy eigenvalue. *The physical significance of the eigenvalues of an operator of an observable is that they are the possible results of measurement of an observable.* Eigenvalues can be measured in an experiment.

Comparing Eqs. (3.60) and (3.63) we find that $|c_a|^2 = P(a)$. The absolute square of the coefficient c_a in the expansion of ψ in terms of eigenfunctions of the operator A is the probability that a measurement of A will yield the eigenvalue a. c_a is the probability amplitude.

We note that quantum theory does not give exact value of an observable of a particle at a time. Rather, it gives various possible values of the observable and their corresponding probabilities. Each eigenfunction is associated with each possible value of the observable. That is, quantum states are described by the mathematical functions called eigenfunctions. Operating the corresponding operator A of the observable on an eigenfunction ϕ_a we get the corresponding eigenvalue a. If the value of the observable is a' then we say that the system is in an eigenstate $\phi_{a'}$. In quantum theory all possible (eigen)states are combined together with complex number c_a weighings as

$$\psi = \sum_a c_a \phi_a, \tag{3.64}$$

where the eigenvalues are assumed to be discrete. This collection of all possible states with complex number weighings is called the *wave function* and it describes arbitrary quantum state of the particle. Generally, ϕ_a is a particular state of the system and ψ is the arbitrary state of the system. Further, ϕ_a is a physically realizable state. In an experiment a system can be found in the stationary eigenstate ϕ_a. However, ψ is not a physically realizable stationary state. In fact, ψ given by Eq. (3.50) or (3.64) is not a solution of the eigenvalue equation $A\psi = \lambda\psi$ (why?). Then, what does ψ represent? ψ *states the various possible eigenstates and their corresponding probabilities*. What do you infer if ψ of a system is $c_1\phi_1 + c_2\phi_2$ while ψ of another system is $c_1\phi_1 + c_2\phi_2 + c_3\phi_3$, where ϕ_i's are stationary state eigenfunctions?

c_a can be called as the *A-space wave function* just $\psi(x)$ is called the *coordinate space* or *configuration space wave function*. c_a may be used instead of ψ for describing the state since ψ can be obtained from c_a through Eq. (3.64). That is, c_a and ψ are different representations of the same state.

3.7   Parity Operator

Suppose P is an operator such that for any function $\phi(x)$, $P\phi(x) = \phi(-x)$. Then

$$P^2\phi = PP\phi(x) = P\phi(-x) = \phi(x). \tag{3.65}$$

The operator P is called a *parity operator*. We find that $P^2 = 1$. The operator P reverses x to $-x$ and p into $-p$ etc. When $P\phi(\mathbf{X}) = \phi(\mathbf{X})$ then P reverses $\mathbf{X}$ to $-\mathbf{X}$ and is equivalent to a reflection in $x - y$ plane followed by a rotation of π about the z-axis.

3.7.1   Properties of the Parity Operator

The operator P has the following properties:

1. Eigenvalues of P are ± 1.
2. For a symmetric potential $[P, H] = 0$.
3. Expectation value of P is a constant of motion.
4. P is Hermitian and linear.
5. $Px_{\mathrm{op}}P = -x_{\mathrm{op}}$, $Pp_{\mathrm{op}}P = -p_{\mathrm{op}}$.

We now prove the above properties. To find the eigenvalues of P assume that ϕ_λ is an eigenfunction of P corresponding to the eigenvalue λ. Then

$$P\phi_\lambda = \lambda\phi_\lambda\,, \quad P^2\phi_\lambda = P\lambda\phi_\lambda = \lambda^2\phi_\lambda\,. \tag{3.66}$$

But according to Eq. (3.65) $P^2\phi_\lambda$ must be ϕ_λ. Therefore, $\lambda^2 = 1$ or $\lambda = \pm 1$. When $\lambda = 1$ we have $P\phi_\lambda = \phi_\lambda$ and the operator P is called an *even parity operator*. When $\lambda = -1$ we have $P\phi_\lambda = -\phi_\lambda$ and the operator P is called an *odd parity operator*. Any function which is even, that is $\phi(x) = \phi(-x)$, is an eigenfunction of parity operator with eigenvalue $+1$. Similarly, any odd function of x, $\phi(x) = -\phi(-x)$, is an eigenfunction of parity operator with eigenvalue -1. The spherical harmonics $Y_l^m(\theta, \phi)$ are examples of eigenfunctions of a parity operator.

For any symmetric potential $V(x, y, z) = V(-x, -y, -z)$ the parity operator commutes with H. It is a constant of motion. Further, P has even or odd parity. Therefore, a system with symmetric potential will be described by eigenfunctions with even or odd parity. In central force field problems $V(r) = V(\sqrt{x^2 + y^2 + z^2})$. The potential is symmetric with respect to parity operator. An eigenfunction for a central force problem in spherical polar coordinates will contain the radial part $R_{nl}(r)$ and the spherical part $Y_l^m(\theta, \phi)$ (see sec. 12.2). Since R_{nl} has even parity, the total parity of the state is given by the spherical harmonics $Y_l^m(\theta, \phi)$.

In spherical polar coordinates, the parity operator causes $r \to r$, $\theta \to \pi - \theta$, $\phi \to \pi + \phi$. We can easily show that $PY_l^m = (-1)^l Y_l^m$. Eigenfunctions with $\phi(\mathbf{r}) = \phi(-\mathbf{r})$ and $\phi(\mathbf{r}) = -\phi(-\mathbf{r})$ are said to have even parity and odd parity, respectively. The significance of even parity is that the waves travelling to the left and right described by $\psi(\mathbf{r})$ have the same energy. One can easily verify that P commutes with L_z and $\mathbf{L}^2$ given by Eqs. (11.27c) and (11.28), respectively. Since $Y_l^m(\theta, \phi)$ are the eigenfunctions of $\mathbf{L}^2$ and L_z (see sec. 11.3.2), they are also the eigenfunctions of P.

We notice that parity is quantum mechanical and has no classical analogue as there is no classical dynamical variable from which it can be obtained by the operator correspondence $\mathbf{r} \to \mathbf{r}_{op}$ and $\mathbf{p} \to -i\hbar\nabla$.

3.7.2 Commutator of P and H

For a reflection symmetric potential $V(x)$, $PV(x) = V(x)$. Further, $PH\phi(x) = H\phi(-x) = HP\phi(x)$ which implies $[P, H] = 0$. Thus, for symmetric potentials P and H are commuting operators and they have simultaneous eigenfunctions. Further, $\mathrm{d}\langle P\rangle/\mathrm{d}t = 0$ because $[P, H] = 0$ (see, Eq. (3.23)). Let us verify the relation $[P, H] = 0$ for the linear harmonic oscillator potential $V(x) = kx^2/2$. We obtain

$$
\begin{aligned}
PH\phi(x) &= P\left(-\frac{\hbar^2}{2m}\frac{\mathrm{d}^2}{\mathrm{d}x^2} + \frac{1}{2}kx^2\right)\phi(x) \\
&= \left(-\frac{\hbar^2}{2m}\frac{\mathrm{d}^2}{\mathrm{d}(-x)^2} + \frac{1}{2}k(-x)^2\right)\phi(-x) \\
&= H\phi(-x) \\
&= HP\phi(x) \tag{3.67}
\end{aligned}
$$

which implies $[P, H] = 0$.

There is an another parity operator called *time reversal parity* which reverses time coordinate and is denoted as T. It does not satisfy the eigenvalue equation $T\phi(t) = \alpha\phi(t) = \phi(-t)$ but $T\phi(t) = \phi^*(t)$. T transforms p_x into $-p_x$ and x into x and $i \to -i$. It preserves the relation $[x, p_x] = i\hbar$.

Solved Problem 4:

Show that P is Hermitian.

Consider the integral $\int_{-\infty}^{\infty} \phi_1^*(x) P^\dagger \phi_2(x) dx$. We obtain

$$
\begin{aligned}
\int_{-\infty}^{\infty} \phi_1^*(x) P^\dagger \phi_2(x)\, dx &= \int_{-\infty}^{\infty} \left[\left(P^\dagger\right)^\dagger \phi_1(x) \right]^* \phi_2(x) dx \\
&= \int \left[P\phi_1(x) \right]^* \phi_2(x)\, dx \\
&= \int \phi_1^*(-x)\phi_2(x)\, dx \\
&= \int \phi_1^*(x)\phi_2(-x)\, dx \\
&= \int_{-\infty}^{\infty} \phi_1^*(x) P \phi_2(x)\, dx \,.
\end{aligned} \tag{3.68}
$$

Therefore, $P^\dagger = P$ and P is a Hermitian.

Solved Problem 5:

How do the parity operators act on the position and momentum operators?

Suppose we operate $P x_{\mathrm{op}} P$ on $\phi(x)$. We get

$$
\left(P x_{\mathrm{op}} P\right) \phi(x) = \left(P x_{\mathrm{op}}\right) \phi(-x) = P(x\phi(-x)) = -x\phi(x) = -x_{\mathrm{op}}\phi(x) \,. \tag{3.69}
$$

Therefore, $P x_{\mathrm{op}} P = -x_{\mathrm{op}}$. Similarly, we can show that $P p_{x\mathrm{op}} P = -p_{x\mathrm{op}}$.

3.7.3 All Hermitian Hamiltonians Have Parity

Let us consider a Hermitian Hamiltonian, for example,

$$
H = \frac{1}{2}p_x^2 + \frac{1}{3}x^3 + \frac{1}{4}x^4 \,. \tag{3.70}
$$

This Hamiltonian is symmetric under the operation of time reversal operator T. A parity operator for a Hermitian Hamiltonian can be constructed as follows [12].

Given a Hamiltonian like that in Eq. (3.70) we can in principle solve the time-independent Schrödinger equation $H\phi_n(x) = E_n\phi_n(x)$. Now, we consider a new operator $\mathcal{P}(x, y)$ given by

$$
\mathcal{P}(x, y) = \sum_{n=0}^{\infty} (-1)^n \phi_n(x)\phi_n^*(y) \,. \tag{3.71}
$$

The operator $\mathcal{P}$ has the following four properties:

1. $\mathcal{P}$ is linear and Hermitian.
2. $\mathcal{P}$ commutes with H and hence it is conserved.
3. $\mathcal{P}^2 = 1$, that is, in coordinate space

$$
\int_{-\infty}^{\infty} \mathcal{P}(x, z)\mathcal{P}(z, y)\, dz = \delta(x - y) \,. \tag{3.72}
$$

4. By virtue of orthonormality ϕ_n is an eigenfunction of $\mathcal{P}$ with the eigenvalue $(-1)^n$, that is,

$$\int_{-\infty}^{\infty} \mathcal{P}(x, z)\phi_n(z)\,dz = (-1)^n \phi_n(x). \tag{3.73}$$

Because of the properties 3 and 4 the operator $\mathcal{P}$ exhibits the characteristics of the parity operator even though the Hamiltonian may not be symmetric under space reflection. If the potential $V(x)$ of the Hamiltonian $H = p_x^2/(2m) + V(x)$ is invariant under the transformation $x \to -x$ then $\mathcal{P}(x, y)$ is the usual parity operator.

We can follow the above procedure for constructing many differential operators that commute with H. For example, we can construct a tri-parity operator $Q(x, y)$

$$Q(x, y) = \sum_{n=0}^{\infty} \omega^n \phi_n(x)\phi_n^*(y), \tag{3.74}$$

where $\omega = e^{\pm i2\pi/3}$ and $\omega^3 = 1$. However, $Q(x, y)$ is not an observable. *Why?*

3.8 Some Other Useful Operators

In addition to Hermitian and parity operators there are other operators which play important role in quantum theory. We define some of these operators in this section.

3.8.1 Identity and Unitary Opeators

An operator I is said to be *identity* if $I\phi = \phi$. The effect of I on ϕ is to leave it as such. That is, the operator I does not change anything. The square of the parity operator, P^2, is an identity operator. Suppose A and B are two operators such that $A\phi_1 = \phi_2$ and $B\phi_2 = \phi_1$, where ϕ_1 and ϕ_2 are two functions of x. We obtain

$$BA\phi_1 = B\phi_2 = \phi_1, \quad AB\phi_2 = A\phi_1 = \phi_2. \tag{3.75}$$

The effect of AB on ϕ_2 is to leave it as such and so $AB = I$. Since we can write $AA^{-1} = A^{-1}A = I$ and hence $B = A^{-1}$: the operator B is the *inverse* of the operator A and vice-versa. The operator $(A - zI)^{-1}$, where I is the unit operator and z is a complex number called the *resolvent operator* of the operator A [13].

An operator U with the property $U^\dagger = U^{-1}$ is called a *unitary operator*. Since $UU^{-1} = 1$ we have $UU^\dagger = 1$ or $U^\dagger U = 1$. In this case the operator U behaves like a transformation. For any two eigenfunctions we get $(\phi_1, \phi_2) = (\phi_1, U^\dagger U \phi_2) = (U\phi_1, U\phi_2)$ which shows that the action of U on eigenfunctions preserve their norm. That is, $U\phi$ and ϕ both have the same norm which means the net probability is unchanged by the unitary transformation U. Since $P^2\phi = \phi$ we write $P^2 = 1$ or $P = P^{-1}$. Because of $UU^\dagger = 1$ where U is a unitary operator we note that $P^\dagger = P$, hence $P^{-1} = P^\dagger$ and P is a unitary.

3.8.2 Parity-Time Reversal Symmetric Hamiltonian

There exists a class of eigenvalue problems, the so-called *parity-time reversal* $(\mathcal{PT})$-*symmetric non-Hermitian Hamiltonian* problems. They have arisen in recent years in a number of physical contexts such as localization transitions, vortex pinning and population

biology [14–16]. Such non-Hermitian Hamiltonians can be used to describe certain experimental phenomena. They have been used for Bose system of hard sphere [17], complex
Toda lattice [18] and field theory models [19–22]. $\mathcal{PT}$-symmetry has been noticed in certain
optical systems, such as optical couplers, microwave billiard and large-scale temporal lattices [23]. An optical system with periodic complex refractive index is essentially equivalent
to a time-periodic non-Hermitian quantum system. A necessary condition for an optical
structure to be $\mathcal{PT}$-symmetric is that its complex refractive index distribution becomes
$n(x) = n^*(-x)$. In this case the real part of the index profile is symmetric in space while
the imaginary component (gain-loss) is antisymmetric [24]. It has been shown that induc-
tively coupled two LRC circuits exhibit all the salient features of non-Hermitian systems
which commute with $\mathcal{PT}$ operator [25,26].

A $\mathcal{PT}$-symmetric Hamiltonian is a Hamiltonian, which is invariant under the product
of the parity operator, that is, $\mathcal{P}(: x \rightarrow -x)$, and the time reversal operation, that is,
$\mathcal{T}(: \mathrm{i} \rightarrow -\mathrm{i})$. For example, the operator

$$H = -\frac{\mathrm{d}^2}{\mathrm{d}x^2} + \alpha\left(x^2\right) - (\mathrm{i}x)^{2n+1} , \tag{3.76}$$

where $\alpha(x^2)$ is a polynomial of degree $n \geq 1$ with positive real coefficients, is $\mathcal{PT}$-symmetric.
A simple example is $H = p_x^2/(2m) + \mathrm{i}x^3$. The parity operator $\mathcal{P}$ is linear and has the effect
of $p_x \rightarrow -p_x$, $x \rightarrow -x$. The time-reversal operator $\mathcal{T}$ is anti-linear and has the effect
$p_x \rightarrow -p_x$, $x \rightarrow x$, $\mathrm{i} \rightarrow -\mathrm{i}$.

An example of an operator which is not $\mathcal{PT}$-symmetric is

$$H = -\frac{\mathrm{d}^2}{\mathrm{d}x^2} + x - (\mathrm{i}x)^3 . \tag{3.77}$$

If $H = -\mathrm{d}^2/\mathrm{d}x^2 + V(x)$ is $\mathcal{PT}$-symmetric then $V^*(-x) = V(x)$ and so $\mathrm{Re}V(x)$ is
an even function and $\mathrm{Im}V(x)$ is an odd function. Hence, if $V(x)$ is a polynomial then
$V(x) = P(x^2) + \mathrm{i}xQ(x^2)$ for some real polynomials P and Q. It has been shown that the
$\mathcal{PT}$-symmetric Hamiltonian $H = -\mathrm{d}^2/\mathrm{d}x^2 + x^2 - g(\mathrm{i}x)^3$ has a discrete spectrum for g real
and the eigenvalues are real if g is small enough [27–28]. There are some $\mathcal{PT}$-symmetric
Hamiltonians that have no eigenvalues [29] or no real eigenvalues [30]. For more details
about $\mathcal{PT}$-symmetric Hamiltonians one may read refs. [14–33]. Some of the applications
of non-Hermitian $\mathcal{PT}$-symmetric Hamiltonian have been reported in refs. [17, 18, 34]. It
is noteworthy to mention that $\mathcal{PT}$-symmetric materials can exhibit power oscillations and
nonreciprocity of light propagation [35], absorption enhanced transmission and unidirec-
tional invisibility.

3.8.3 Time Operator

We note that in Schrödinger quantum mechanics time is considered as a parameter and
it can be chosen with any desired precision while the position variation is an operator.
$\psi^*(x,t)\psi(x,t)$ gives the probability of finding a particle between x and $x + \mathrm{d}x$ at time t.

Operators for time have been discussed in different contexts [36–41]. Consider the time
arrival of a particle in a detector localized spatially. In this case one can define an operator
T for time so that $[T, H] = \mathrm{i}\hbar$. Denoting $\psi(t)$ as the wave function of T we can write
$T\psi(t) = t\psi(t)$ in an analogy with $x_{\mathrm{op}}\psi(x) = x\psi(x)$, where $\psi(x)$ is the wave function of x_{op}.
Observe that x_{op} is Hermitian while T is non-Hermitian (verify). We have $\langle\psi(t)|T|\psi(t)\rangle = t$
and $\langle\psi(t)|T^2|\psi(t)\rangle = t^2$ leading to the *uncertainty* $\Delta T = \langle T^2\rangle - \langle T\rangle^2 = 0$. That is, temporal
uncertainty becomes zero and time appears as a parameter that can be chosen arbitrary.

3.9 Concluding Remarks

In this chapter, we have seen that the time-independent Schrödinger equation is an eigenvalue equation. We have seen that the operators representing an observable must be a Hermitian operator and discussed a few important properties of Hermitian operators. *How do we obtain an anti-Hermitian operator from a Hermitian operator?* Multiplying it by i. We note that the operator representing an observable should be (i) linear, (ii) Hermitian and (iii) its eigenfunctions must form a complete set. It is easy to verify that the operators **L** and H are Hermitian if the potential $V(x, y, z)$ is real. Further, the parity operator P commutes with L_z and $\mathbf{L}^2$. In chapters 4 and 5 we will consider some simple and physically interesting quantum mechanical systems and solve the eigenvalue equations associated with them. We will find that in some systems normalizable solutions occur with discrete eigenvalues while in certain systems the solutions are nonnormalizable with continuous eigenvalues. Classical mechanics describes the dynamical variables as functions in phase space. We have seen that in quantum mechanics, they are Hermitian operators. As all the operators in quantum mechanics are linear operators, the state of a system is represented by a superposition of eigenfunctions.

3.10 Bibliography

[1]   A. Jeffrey, *Advanced Engineering Mathematics*. Academic Press, San Diego, 2003.

[2]   E. Kreyszig, *Advanced Engineering Mathematics*. John Wiley, New York, 2001.

[3]   H. Goldstein, *Classical Mechanics*. Narosa, New Delhi, 1985.

[4]   V.B. Bhatia, *Classical Mechanics*. Narosa, New Delhi, 1997.

[5]   E. Corinaldesi, *Classical Mechanics for Physics Graduate Students*. World Scientific, Singapore, 1998.

[6]   P. Dorey, C. Dunning and R. Tateo, *J. Phys. A* 34:L391, 2001.

[7]   K.C. Shin, *J. Math. Phys.* 42:2513, 2001.

[8]   C.M. Bender, S. Boettcher and P.N. Meisinger, *J. Math. Phys.* 40:2201, 1999.

[9]   C.M. Bender, F. Cooper, P.N. Meisinger and V.M. Savage, *Phys. Lett. A* 259:224, 1999.

[10] C.M. Bender and G.V. Dunne, *J. Math. Phys.* 40:4616, 1999.

[11] C.M. Bender, D.C. Brody and H.F. Jones, *Am. J. Phys.* 71:1095, 2003.

[12] C.M. Bender, P.N. Meisinger and Q. Wang, *J. Phys. A* 36:1029, 2003.

[13] M. Amaku, F.A.B. Coutinho and F.M. Toyama, *Am. J. Phys.* 85:692, 2017.

[14] N. Hatano and D.R. Nelson, *Phys. Rev. Lett.* 77:570, 1996.

[15] N. Hatano and D.R. Nelson, *Phys. Rev. B* 56:8651, 1997.

[16] D.R. Nelson and N.M. Shnerb, *Phys. Rev. E* 58:1383, 1998.

[17] T.T. Wu, *Phys. Rev.* 115:1390, 1959.

[18] T. Hollowood, *Nucl. Phys. B* 384:523, 1992.

[19] R.C. Brower, M.A. Furman and M. Moshe, *Phys. Lett. B* 76:213, 1978.

[20] C.M. Bender and K.A. Milton, *Phys. Rev. D* 57:3595, 1998; *J. Phys. A* 32:L87, 1999.

[21] C.M. Bender, K.A. Milton and V.M. Savage, *Phys. Rev. D* 62:85001, 2000.

[22] C.M. Bender, P.N. Meisinger and H. Yang, *Phys. Rev. D* 63:045001, 2001.

[23] J. Luo, J. Huang, H. Zhong, X. Qin, Q. Xie, Y.S. Kivshar and C. Lee, *Phys. Rev. Lett.* 110:243902, 2013.

[24] A. Regensburger, M. Ali Miri, C. Bersch, J. Nager, G. Onishchukov, D.N. Christodoulides and U. Peschel, *Phys. Rev. Lett.* 110:223902, 2013.

[25] J. Schindler, Z. Liu, J.M. Lee, H. Ramezani, F.M. Ellis and T. Kottos, *J. Phys. A: Math. Theor.* 45:444029, 2012.

[26] J. Schindler, A. Li, M.C. Zheng, F.M. Ellis and T. Kottos, *Phys. Rev. A* 84:040101, 2011.

[27] E. Caliceti, *J. Phys. A* 33:3753, 2000.

[28] E. Caliceti, S. Graffi and M. Maioli, *Commun. Math. Phys.* 75:51, 1980.

[29] G.A. Mezincescu, *J. Phys. A* 33:4911, 2000.

[30] E. Delabaere and F. Pham, *Phys. Lett. A* 250:25, 1998.

[31] P. Dorey, C. Dunning and R. Tateo, *J. Phys. A* 34:5679, 2001.

[32] C.M. Bender, S. Boettcher and P.N. Meisinger, *J. Math. Phys.* 40:2201, 1999.

[33] C.M. Bender, J. Brod, A. Refig and M.E. Reuter, *J. Phys. A* 37:10139, 2004.

[34] C.M. Bender, G.V. Dunne and P.N. Meisinger, *Phys. Lett. A* 252:272, 1999.

[35] M.C. Zheng, D.N. Christodoulides, R. Fleischmann and T. Kottos, *Phys. Rev. A* 82:010103, 2010.

[36] E.P. Wigner, *Phys. Rev.* 98:145, 1955.

[37] J. Kijowski, *Re. Math. Phys.* 6:361, 1974.

[38] N. Kumar, *Pramana J. Phys.* 75:363, 1985.

[39] R. Werner, *J. Math. Phys.* 27:793, 1986.

[40] G. Torres-Vega, *Phys. Rev. A* 75:032112, 2007.

[41] E.O. Dias and F. Parisio, arXiv 1507.02899 v1 quant-ph 10 July 2015.

3.11 Exercises

3.1 Show that the following operators are linear.

(a) x^2, (b) $x\dfrac{d}{dx}$, (c) $\left(x + \dfrac{d}{dx}\right)^2$ and (d) $\dfrac{d}{dx}\left(e^{-x^2}\dfrac{d}{dx}\right)$.

3.2 If $Af = f^2$ evaluate $A(x^3 + 5x)$.

3.3 Distinguish between $A\phi = 5\phi$ and $A\phi$ not directly proportional to ϕ.

3.4 If A and B are two linear operators prove that AB and BA are also linear operators.

3.5 Show that linear combination of two linear operators A and B is also a linear operator. Also prove that $A+B = B+A$, $A+B+C = (A+B)+C = A+(B+C)$, $A(BC) = (AB)C = ABC$, $(A + B)C = AC + BC$, $A(B + C) = AB + AC$ and $A^2 = AA$.

3.6 Find whether the operators defined through the equations $A\phi = \phi^2$ and $B\phi = e^\phi$ are linear or nonlinear.

3.7 Let the operator Ω be defined by $\Omega\psi(x) = \int_{-\infty}^{\infty} U(x - x')\psi(x') \, dx'$, where the function $U(x)$ is the unit step-function $U(x) = \begin{cases} 0, & x < 0 \\ 1, & x > 0. \end{cases}$ If the class of functions ψ is the class for which ψ and $d\psi/dx$ are integrable for $|x| \to \infty$, show that $\Omega = (d/dx)^{-1}$.

3.8 What is the effect of $e^{-iH\tau/\hbar}$ on $\psi(t)$?

3.9 Show that $e^{iap_x/\hbar} f(x) = f(x + a)$.

3.10 If ϕ is an eigenfunction of a linear operator A with an eigenvalue λ find the eigenfunctions and the eigenvalues of A^2 and A^3. What do you infer from your result?

3.11 Verify that if ϕ is an eigenfunction of A then it is also an eigenfunction of A^n, e^A and A^{-n}.

3.12 Find the eigenvalues (λ) and the eigenfunctions (y) of the operator d^2/dx^2. Assume that the eigenfunctions are defined in the interval $0 \leq x \leq \pi$ and they vanish at $x = 0, \pi, \ldots$.

3.13 For the class of functions satisfying proper boundary conditions show that the Hermitian condition $(\psi, A\psi) = (A\psi, \psi)$ implies $(\phi_1, A\phi_2) = (A\phi_1, \phi_2)$.

3.14 Write down the eigenvalue equations for position and momentum operators and solve them. Express the normalization conditions for the eigenfunctions of position and momentum operators.

3.15 Show that the standard deviation of the measurement of a dynamical variable in its eigenstate is zero.

3.16 Find the eigenfunctions and eigenvalues of a particle of mass m in the potential $V(x) = V_0$, (a constant), $V_0 > 0$.

3.17 Show that product of two unitary operators is also a unitary operator.

3.18 Given $\hat{d}f(x) = \partial f/\partial x$ and $\hat{I}f(x) = \int_0^x f(x)dx$ find (a) $[\hat{d}, \hat{I}]f(x)$ and (b) $[x, \hat{I}]$.

3.19 If X and Y are two operators and $Z = e^{\alpha X} Y e^{-\alpha X}$ show that $dZ/d\alpha = [X, Z]$. Hence, show that the Taylor expansion of Z (in powers of α) is

$$e^{\alpha X} Y e^{-\alpha X} = Y + \alpha[X, Y] + \frac{\alpha^2}{2!}[X, [X, Y]] + \frac{\alpha^3}{3!}[X, [X, [X, Y]]] + \ldots .$$

3.20 Show that if A and B are two linear operators then $[A, B] = -[B, A]$ and $[A, B^{-1}] = -B^{-1}[A, B]B^{-1}$.

3.21 Is $\sin x$ a Hermitian operator?

3.22 If A and B are two linear operators with $[A, B] = a$, where a is a constant, by induction, show that $[A^n, B] = naA^{n-1}$ and $[A, B^n] = naB^{n-1}$. Then deduce the value of $[x^n, p_x]$ and $[x, p_x^n]$.

3.23 Show that $[H, t] = i\hbar$.

3.24   Suppose the operators A and B commute with their commutator. Show that (a) $[A, B^n] = nB^{n-1}[A, B]$ and (b) $[A^n, B] = nA^{n-1}[A, B]$.

3.25   If $[A, B] = 0$ and $[B, C] = 0$, are A and C commuting operators? Give examples.

3.26   What are the eigenvalues of the operators (a) $[x, p_x]$ and (b) $[x, p_y]$?

3.27   Suppose the operators A and B commute with their commutator, that is $[B, [A, B]] = 0$. Prove that (a) for any analytic function $F(x)$, $[A, F(B)] = [A, B]F'(B)$. (b) $e^A e^B = e^{A+B} e^{[A,B]/2}$.

3.28   Prove that $[A, e^{-\alpha B}] = -\alpha e^{-\alpha B}[A, B]$, if A and B commutes with their commutation.

3.29   Find $[X, P]$ for $X = \sinh z$, $P = -\dfrac{1}{2}i\hbar \left[\cosh z \dfrac{d}{dz} + \dfrac{d}{dz}\cosh z\right]$.

3.30   Consider a particle in a box potential $V(x) = 0$ for $|x| < a$ and ∞ elsewhere. Write the Hamiltonian of the system for $|x| < a$. Compute $[H, x]$ and $[H, p]$.

3.31   Prove the Jacobi identity $[[A, B], C] + [[B, C], A] + [[C, A], B] = 0$.

3.32   If A and B are operators such that $[A, B] = 1$ and if $Z = \alpha A + \beta B$, where α and β are numbers, verify that

   (a) $[A, Z^n] = nBZ^{n-1}$. Hence, show that

   (b) $(\partial/\partial\alpha)Z^n = nAZ^{n-1} - n(n-1)\beta Z^{n-2}/2$ and

   (c) $(\partial/\partial\alpha)e^Z = (A - \beta/2)e^Z$.

   (d) By integrating the differential equation for e^Z, with due attention to the order of operator factors, show that $e^{\alpha A + \beta B} = e^{\alpha A} e^{\beta B} e^{-\alpha\beta/2}$.

3.33   Prove the adjoint operators relations: (a) adj $d^2/dx^2 = d^2/dx^2$, (b) adj $d^3/dx^3 = -d^3/dx^3$ and (c) adj $d^n/dx^n = (-1)^n d^n/dx^n$.

3.34   What is the value of $\langle[A, B]\rangle$, where A and B are linear operators, associated with any eigenstate of A or B?

3.35   Find the adjoint $a^\dagger$ of the operator $a = (x + d/dx)/\sqrt{2}$. Is a a self-adjoint operator? Verify that $a^\dagger a$ is self-adjoint.

3.36   Suppose A, B, C are three linear operators such that $C = A + B$. If A and C are Hermitian, is B necessarily Hermitian?

3.37   Is the operator $A = [x, p_x]$ Hermitian? Why?

3.38 A and B are two linear operators. For the product AB to become a Hermitian, are A and B necessarily Hermitian?

3.39   Is $H = -(\hbar^2/2m)d^2/dx^2 + V(x)$, where $V(x)$ is a polynomial in x, a Hermitian operator? Why?

3.40   Are the operators xd/dx and $x^2 d^2/dx^2$ Hermitian? Why?

3.41   Show that the operators $\mathbf{L}$ and H are Hermitian if the potential $V(x, y, z)$ is real.

3.42   Show that if H_1 and H_2 are Hermitian then $i[H_1, H_2]$ is also a Hermitian.

3.43   Show that the anticommutator of two Hermitian operators is also a Hermitian.

3.44   Prove whether or not the following are eigenfunctions of the parity operator P. What are their corresponding eigenvalues?

   (a) $x^4 + 6x^2$. (b) x^5. (c) $\sin x$. (d) $\cos x$. (e) $\sin^2 x$. (f) $x^3 + 5x$. (g) $x^3 + x^2 + x$. (h) e^{-x}. (i) e^{-x^2}.

3.45 Show that

(a) eigenfunctions of the parity operator belonging to different eigenvalues are orthogonal and

(b) every function can be expressed as a linear combination of the eigenfunctions of the parity operator.

3.46 What is the effect of time reversal on any time-independent operator A representing a dynamical variable?

3.47 Show that the Hamiltonian $H = -\mathrm{d}^2/\mathrm{d}x^2 - (\mathrm{i}x)^3$ is not a Hermitian.

3.48 Show that (a) $H = p_x^2 + \mathrm{i}\sin x$ is not a Hermitian but $\mathcal{PT}$-symmetric and (b) $H = p_x^2 + x^2 + 2x$ is not $\mathcal{PT}$-symmetric.

3.49 Show that the time reversal operator changes any number into its complex conjugate and hence it is not a linear operator.

3.50 If $H = AB + \epsilon$ where $A = \dfrac{\mathrm{d}}{\mathrm{d}x} + \beta(x)$ and $B = -\dfrac{\mathrm{d}}{\mathrm{d}x} + \beta(x)$ with $\beta(x)$ and ϵ being real show that H is self-adjoint.

4

Exactly Solvable Systems I: Bound States

4.1 Introduction

Analytical solutions to the Schrödinger equation serve a variety of purposes. They help us

1. to identify the features of potentials and state functions,
2. to construct appropriate potentials from the experimental data and
3. to get analytical insight of the system considered.

The problem of solving a time-dependent Schrödinger equation for a given time-dependent potential is difficult. However, for some simple and physically interesting potentials the time-independent Schrödinger equation is solvable exactly and the solutions are obtainable in terms of well-known functions. Useful informations about a given system are extracted from the solutions of time-independent Schrödinger equation. The time-independent Schrödinger equation gives stationary state solutions which are independent of time. For a particle in a one-dimensional potential $V(x)$, the time-independent Schrödinger equation with $H = p_x^2/(2m) + V(x)$ is given by

$$\psi_{xx} + \frac{2m}{\hbar^2}(E - V(x))\psi = 0, \tag{4.1}$$

where the subscript x in ψ denotes differentiation with respect to x.

The eigenvalues of energy eigenstates can be discrete or continuous. The energy eigenstates with discrete eigenvalues are called *bound states* because they correspond to the classical bound orbits confined to a finite region in space. The bound state eigenfunctions are normalizable. On the other hand, the energy eigenstates with continuous eigenvalues are called *unbound states* or *scattering states* and they correspond to the classical unbound orbits in which the particle approaches from and recedes to infinite distance. The eigenfunctions of scattering states are nonnormalizable.

The present chapter is primarily concerned with the study of bound states of some simple one-dimensional systems. Study of bound systems and their bound states is very important because atoms, molecules and other fundamental matters are bound systems and their characteristics are understood by quantum mechanics. We consider a one-dimensional linear harmonic oscillator, an anharmonic oscillator, a particle in a box, Morse oscillator, Pöschl–Teller potentials and pendulum system. Since free particle is the simplest quantum mechanical system and form the starting point of the study of quantum systems we include the study of it in this chapter even though its solutions are unbound. Determination of eigenspectra of damped and forced linear harmonic oscillator is presented. In two-dimension the linear harmonic oscillator and a particle in a box are discussed. The quantum mechanical treatment of three-dimensional version of the above mentioned systems are easy to analyze and are given as problems at the end of this chapter. However, we will consider the important systems like rigid rotator and hydrogen atom. We solve the rigid rotator in this chapter while the hydrogen atom will be explored in chapter 12. Potentials giving scattering states will be analyzed in the next chapter.

DOI: 10.1201/9781003172178-4

The time evolution of stationary state eigenfunctions is easy to obtain. We illustrate this with reference to linear harmonic oscillator. When the Schrödinger equation of a potential is not exactly solvable we can study the solution by means of numerical techniques. In chapter 23 we solve the time-independent and time-dependent Schrödinger equations of certain systems numerically.

4.2   Classical Probability Distribution

How do we establish the connection between the Schrödinger wave function and the classical motion of a particle for stationary state solutions? An approach is to compare the probability density $P_{\mathrm{QM}} = |\psi_n(x)|^2$ with a classical probability distribution $P_{\mathrm{CM}}(x)$ [1]. Then show that these probabilities approach each other in a locally averaged manner, in the limit of large quantum number n. Let us describe the calculation of $P_{\mathrm{CM}}(x)$ and $P_{\mathrm{CM}}(p)$ for a classical particle exhibiting periodic motion.

Consider a particle exhibiting a bounded periodic motion (with period τ) with $E = H = p_x^2/(2m) + V(x)$. We have $p_x^2/(2m) = E - V(x)$. Since $p_x^2/(2m)$ must be greater than or equal to zero, the classical motion is confined to the regions, where $E - V(x) \geq 0$. The points at which $E - V(x) = 0$, (or $E = V(x)$), are called classical *turning points*. At these points the velocity of the particle vanishes and it turns back. Let us denote these points as a and b. Then the particle will move back and forth between a and b. The time taken by the particle for one traversal from a to b is half the period, $\tau/2$. The time $\mathrm{d}t$ spend by the particle in the small region, $\mathrm{d}x$, near the point x is $\mathrm{d}t = \mathrm{d}x/v(x)$, where $v(x)$ is the speed. The probability of finding the particle in $\mathrm{d}x$ region is ratio of $\mathrm{d}t$ and the time taken for one traversal:

$$P_{\mathrm{CM}}(x)\,\mathrm{d}x = \frac{\mathrm{d}t}{\tau/2} = \frac{2}{\tau}\frac{\mathrm{d}x}{v} \quad \text{or} \quad P_{\mathrm{CM}}(x) = \frac{2}{\tau v}. \tag{4.2}$$

The particle will spend more time in regions where its speed is low. Further, P_{CM} given by Eq. (4.2) is normalized ($\int_a^b P_{\mathrm{CM}}(x)\mathrm{d}x = 1$) . Using $E = mv^2/2 + V(x)$ for v in Eq. (4.2) we obtain [1]

$$P_{\mathrm{CM}}(x) = \frac{2}{\tau}\sqrt{\frac{m}{2(E - V(x))}}. \tag{4.3}$$

Solved Problem 1:

Find an expression for $P_{\mathrm{CM}}(p_x)$.

The probability of measuring the classical particle with momentum p_x is

$$P_{\mathrm{CM}}(p_x)\,\mathrm{d}p_x = \frac{\mathrm{d}t}{\tau/2} = \frac{2}{\tau}\frac{\mathrm{d}p_x}{|\mathrm{d}p_x/\mathrm{d}t|}. \tag{4.4}$$

Since $\mathrm{d}p_x/\mathrm{d}t = F(x) = -\mathrm{d}V/\mathrm{d}x$ we have $P_{\mathrm{CM}}(p_x) = 2/(\tau|F|)$.

4.3   Free Particle

For a free particle the force is zero and hence the potential is a constant which can be taken as 0 since a constant potential does not influence the dynamics of the system. A free particle has significance in problems related to collision and metallic state. The Schrödinger equation for the free particle is

$$\psi_{xx} + k^2\psi = 0, \quad k^2 = 2mE/\hbar^2. \tag{4.5}$$

Its solution is

$$\psi = Ae^{\pm ikx}, \tag{4.6}$$

where A is a constant. This $\psi(x)$ is always single-valued, finite and continuous for all values of k. Hence, there is no restriction on the values of k and E. They can take continuous values. But the ψ given by Eq. (4.6) is nonnormalizable! *Why is this? What does it imply?*

For the free particle the energy is always positive and the expectation value of kinetic energy is E. *What is the expectation value of momentum?*

Let us calculate the flux with $\psi = Ae^{ikx}$. We have

$$J(x) = -\frac{i\hbar}{2m}\left[\psi^*\frac{\partial\psi}{\partial x} - \psi\frac{\partial\psi^*}{\partial x}\right] = \frac{\hbar k|A|^2}{m} = \frac{|A|^2}{m}p_x. \tag{4.7}$$

Classically, if $\rho(x)$ is the spatial charge density of particles and v is the speed of the particles then $J(x) = v\rho(x) = \rho p_x/m$. Thus, $\psi = Ae^{ikx}$ physically represents flow of particles.

In Eq. (4.6) for a given k or E there are two independent eigenfunctions: e^{ikx} and e^{-ikx}. The energy eigenvalue corresponding to these eigenfunctions is thus degenerate. However, for $k \neq 0$:

$$\int_{-\infty}^{\infty}\left(e^{-ikx}\right)^* e^{ikx}dx = \int_{-\infty}^{\infty} e^{2ikx}dx \propto \delta(k) = 0. \tag{4.8}$$

The eigenfunctions e^{ikx} and e^{-ikx} are orthogonal. *How do we distinguish these degenerate eigenfunctions?* Operating $\hat{p}_x$ on them we have

$$\hat{p}_x e^{\pm ikx} = \hat{p}_x e^{ip_x x/\hbar} = \pm p_x e^{\pm ip_x x/\hbar}. \tag{4.9}$$

The two eigenfunctions have the same energy but different momenta.

Instead of $e^{\pm ikx}$, suppose consider the eigenfunctions $\sin kx$ and $\cos kx$. *Does $\hat{p}_x$ distinguish them?* The answer is no. Because $\hat{p}_x \sin kx = -i\hbar k\cos kx$, $\sin kx$ is not a momentum eigenfunction. Therefore, $\hat{p}_x$ is not useful to distinguish them. However, as $\sin kx$ and $\cos kx$ are odd and even functions of x, respectively, we can distinguish them by the parity operator.

Even though $\psi = Ae^{\pm ikx}$ is not normalizable we can find the constant A. The orthogonality condition for eigenfunctions with continuous eigenvalues is

$$\int_{-\infty}^{\infty}\psi_\lambda^*\psi_{\lambda'}dx = \delta(\lambda' - \lambda), \tag{4.10}$$

where δ is Dirac-delta function. Hence, for the free particle eigenfunctions the above condition is

$$\int_{-\infty}^{\infty}\psi_k^*(x)\psi_{k'}(x)dx = \delta(k' - k). \tag{4.11}$$

Substituting for ψ_k and $\psi_{k'}$ we obtain

$$|A|^2\int_{-\infty}^{\infty} e^{i(k'-k)x}dx = \delta(k' - k). \tag{4.12}$$

That is, $2\pi|A|^2\delta(k'-k) = \delta(k'-k)$ (see the exercise 3 at the end of this chapter). Integrating this with respect to k' we get $2\pi|A|^2 = 1$ or $A = 1/\sqrt{2\pi}$. Hence,

$$\psi_k(x) = \frac{1}{\sqrt{2\pi}}e^{\pm ikx}. \tag{4.13}$$

4.4   Harmonic Oscillator

Consider the linear harmonic oscillator, a particle of mass m bounded to the origin by a restoring force $F = -kx$. Here k is the restoring force constant and x is the displacement from equilibrium. The sign of the force is opposite to the sign of x. The force is always directed towards the equilibrium point.

The potential of the system is $V = kx^2/2$. The Hamiltonian is

$$H = \frac{1}{2m}p_x^2 + \frac{1}{2}kx^2 \ \text{ or } \ H = \frac{1}{2m}p_x^2 + \frac{1}{2}m\omega^2 x^2, \ \ \omega = \sqrt{k/m}\,. \tag{4.14}$$

The classical equation of motion is

$$\ddot{x} + \omega^2 x = 0 \ \text{ or } \ \dot{x} = \frac{1}{m}p_x, \ \ \dot{p}_x = -m\omega^2 x\,. \tag{4.15}$$

The Schrödinger equation of the quantum mechanical linear harmonic oscillator is

$$\psi_{xx} + \frac{2m}{\hbar^2}\left(E - \frac{1}{2}kx^2\right)\psi = 0\,, \tag{4.16}$$

where E is the energy of the particle with respect to the bottom of the potential well.

4.4.1   Physical Systems Modelled by the Linear Harmonic Oscillator

The linear harmonic oscillator (LHO) is a fundamental system solved exactly in both classical and quantum mechanics. It is used in the modelling of subnuclear quarks to cosmic systems. Any system with small amplitude of vibration about its equilibrium point is approximately described by LHO. Vibration of diatomic molecules is approximately a LHO. Electrons restricted in three-dimensional electromagnetic traps are found to behave as quantized harmonic oscillators. The LHO description is used in the study of probability distribution of hadronic matter. The model is found to give a simple explanation for a wide range of hadronic phenomena noticed in high energy physics experiments.

For a very general potential $V_c(x)$ and for small oscillations of a system about a stable state $x^* = a$ we consider Taylor series expansion of $V_c(x)$ as

$$V_c(x) = V(a) + V'(a)(x-a) + \frac{1}{2}V''(a)(x-a)^2 + \dots\,. \tag{4.17}$$

Because $V'(a) = 0$ and $V''(a) > 0$ the system behaves approximately as a harmonic oscillator with $k/2 = V'(a)$.

An oscillating electromagnetic field is similar to a LHO [2]. An electromagnetic wave at a point in space can be considered as oscillating electric and magnetic fields or a collection of photons. Consider a standing electromagnetic wave in a one-dimensional enclosure without a source. An example is a resonant cavity of a laser. Suppose the enclosure is kept along the x-axis. Its volume is, say V. The electric field is plane polarized parallel to the y-axis and its angular frequency is ω. The strength of electric and magnetic fields are given by

$$E_y(x,t) = \sqrt{\frac{2N\omega^2}{\epsilon_0 V}}\,\mathcal{E}(t)\sin\left(\frac{\omega x}{c}\right)\,, \ \ B_z(x,t) = \sqrt{\frac{2\mu_0}{NV}}\,\mathcal{B}(t)\cos\left(\frac{\omega x}{c}\right)\,, \tag{4.18}$$

where N is the normalization factor. $\mathcal{E}$ and $\mathcal{B}$ are given by the Maxwell's equations

$$\frac{d\mathcal{E}}{dt} = \frac{1}{N}\mathcal{B}, \ \ \frac{d\mathcal{B}}{dt} = -N\omega^2\mathcal{E}\,. \tag{4.19}$$

The total energy obtained integrating the energy density over the enclosure is

$$U = \frac{1}{2N}\mathcal{B}^2 + \frac{1}{2}N\omega^2\mathcal{E}^2. \tag{4.20}$$

Equations (4.19) and (4.20) are similar to the Eqs. (4.15) and (4.14), respectively.

There are two standard approaches to determine the eigenvalues and eigenfunctions of quantum LHO – one is differential and another is algebraic. First we take up the former.

4.4.2 Transformation of Schrödinger Equation to Hermite Equation

Equation (4.16) is a variable coefficient and linear ordinary differential equation. We do not have a systematic method to construct exact analytical solutions of such equations. However, the Schrödinger Eq. (4.16) can be transformed into the Hermite differential equation by introducing a suitable change of variable and parameters. Writing $\xi = \alpha x$, we have $\psi_x = \psi_\xi d\xi/dx = \alpha\psi_\xi$ and $\psi_{xx} = \alpha^2\psi_{\xi\xi}$, where $\alpha = d\xi/dx$. Then Eq. (4.16) becomes

$$\psi_{\xi\xi} + \left(\frac{2mE}{\hbar^2\alpha^2} - \frac{mk}{\hbar^2\alpha^4}\xi^2\right)\psi = 0. \tag{4.21}$$

Defining

$$\alpha^4 = \frac{mk}{\hbar^2} = \frac{m^2\omega^2}{\hbar^2}, \quad \lambda = \frac{2mE}{\hbar^2\alpha^2} = \frac{2E}{\hbar}\sqrt{\frac{m}{k}} = \frac{2E}{\hbar\omega} \tag{4.22}$$

we rewrite the Eq. (4.21) as

$$\psi_{\xi\xi} + \left(\lambda - \xi^2\right)\psi = 0. \tag{4.23}$$

With the change of variable

$$\psi = H(\xi)\,e^{-\xi^2/2} \tag{4.24}$$

Eq. (4.23) takes the form

$$H_{\xi\xi} - 2\xi H_\xi + (\lambda - 1)H = 0. \tag{4.25}$$

Redefining $\lambda = 2n + 1$ the above equation becomes the well-known Hermite differential equation

$$H_{\xi\xi} - 2\xi H_\xi + 2nH = 0. \tag{4.26}$$

Equation (4.26) is not an exactly solvable equation. However, it admits a power series solution. We are seeking bound state solutions of (4.16): $\psi(x) \to 0$ as $x \to \pm\infty$. The particular solutions which are finite in the interval $-\infty < \xi < \infty$ and decay to zero as $\xi \to \pm\infty$ exist for integer values of n and are called *Hermite polynomials* [3–5].

4.4.3 Eigenfunctions and Eigenvalues

The Hermite polynomial solutions of Eq. (4.26) are given by [3–5]

$$H_n(\xi) = (-1)^n\,e^{\xi^2}\,\frac{\partial^n}{\partial\xi^n}\,e^{-\xi^2}. \tag{4.27}$$

The first few Hermite polynomials are

$$H_0 = 1, \quad H_1 = 2\xi, \quad H_2 = 4\xi^2 - 2, \quad H_3 = 8\xi^3 - 12\xi. \tag{4.28}$$

From Eq. (4.24) the eigenfunctions are obtained as

$$\psi_n(\xi) = N_n H_n(\xi)\,e^{-\xi^2/2}, \tag{4.29}$$

where N_n's are normalization constants. The substitution $\xi = \alpha x$ in Eq. (4.29) gives

$$\psi_n(\alpha x) = \psi_n(x) = N_n H_n(\alpha x) \, e^{-\alpha^2 x^2/2} \, . \tag{4.30}$$

Thus, the eigenfunctions of LHO are Hermite polynomials times a Gaussian function and are called *Weber functions*. The normalization condition gives

$$1 = \int_{-\infty}^{\infty} \psi_n^*(x)\psi_n(x)\,\mathrm{d}x = \left|N_n^2\right| \int_{-\infty}^{\infty} H_n^2(\alpha x)\, e^{-\alpha^2 x^2}\,\mathrm{d}x = |N_n|^2 \sqrt{\pi}\, 2^n n!/\alpha \, . \tag{4.31}$$

Thus,

$$N_n = \left(\frac{\alpha}{\sqrt{\pi}\, 2^n n!}\right)^{1/2} . \tag{4.32}$$

The eigenfunction ψ_n with a particular n is called a *quantum state n*. The number n is called *quantum number*.

Because of $\lambda = 2n + 1$, Eq. (4.22) gives

$$E_n = \left(n + \frac{1}{2}\right)\hbar\omega\,, \quad n = 0, 1, 2, \ldots \tag{4.33}$$

where $\omega = \sqrt{k/m}$ is the classical angular frequency of the oscillator. E_n's are the only stationary state energy levels of the oscillator.

Some of the features of a LHO are summarized below.

1. E_n's are discrete because n is an integer and E_n depend on n. Hence, the quantum LHO has a discrete energy spectrum. This is not true for the classical LHO since an arbitrary amount of energy can be imparted to the classical system. We note that the energy spectrum of a free particle is continuous.

2. The states represented by the eigenfunctions ψ_n are normalized and hence they are *bound states*.

3. The energy difference between two successive energy levels is $\hbar\omega$, a fixed value, and thus the energy levels are equally spaced. This is due to the potential energy function $V(x) = kx^2/2$. This potential expands in such a manner that, as the amplitude increases with energy the levels remain equally spaced.

4. The ground state energy obtained by substituting $n = 0$ is $\hbar\omega/2$ and is nonzero. This is known as *zero-point energy*. The classical minimum energy is zero. In contrast, quantum mechanically it is $\hbar\omega/2$. This is because $\psi_0(x)$ does not minimize $V(x)$ alone but rather the sum of kinetic and potential energies.

5. *What would happen if $E_n = 0$?* If $E_n = 0$ then

$$E_n = 0 = \int_{-\infty}^{\infty} \psi_n^* H \psi_n \,\mathrm{d}x = \frac{1}{2m}\langle p_x^2\rangle + \frac{k}{2}\langle x^2\rangle$$

would imply $\langle x^2\rangle = \langle p_x^2\rangle = 0$. For a statistical variable x, which is not always zero, $\langle x^2\rangle$ should be greater than zero.

6. Figure 4.1 shows first few eigenfunctions of LHO. $\psi_n(-x) = \psi_n(x)$ for even values of n while $\psi_n(-x) = -\psi_n(x)$ for odd integer of n. Thus, the parity associated with ψ_n is $P = (-1)^n$.

7. ψ_n given by Eq. (4.30) are the eigenfunctions of the Hamiltonian operator H given by Eq. (4.14) with the eigenvalues E_n. H is a Hermitian operator (verify). Earlier, in sec. 3.4 we have shown that for a Hermitian operator eigenfunctions belonging to distinct eigenvalues are mutually orthogonal. Therefore, the eigenfunctions of LHO are orthogonal. We can verify the above using the ψ_n's given by Eq. (4.30).

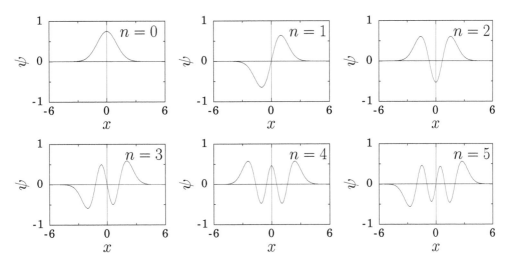

FIGURE 4.1
First few eigenfunctions of LHO where $k = 1$.

4.4.4 Classical and Quantum Mechanical Probabilities

The solution of the classical equation of motion (4.15) is $x(t) = A \sin(\omega t + \phi)$, where A and ϕ are integration constants. The solution is periodic with period $\tau = 2\pi/\omega$. The maximum amplitude of the solution is A and the turning points are $+A$ and $-A$. The classical motion is bounded between $x = -A$ and $+A$, that is, $-A \leq x(t) \leq A$. Now, for the LHO Eq. (4.3) becomes

$$P_{\text{CM}}(x) = \frac{2\omega}{2\pi} \sqrt{\frac{m}{2\left(E - \frac{1}{2}kx^2\right)}} = \frac{1}{\pi\sqrt{\frac{2E}{k} - x^2}}. \qquad (4.34)$$

We express E in terms of the turning point A. At $x(t) = \pm A$ the velocity dx/dt is zero and $E = V(x = \pm A) = kA^2/2$ or $2E/k = A^2$. Then

$$P_{\text{CM}}(x) = \frac{1}{\pi\sqrt{A^2 - x^2}}, \qquad (4.35a)$$

where

$$A^2 = \frac{2E_n}{k} = \frac{2}{k}\left(n + \frac{1}{2}\right)\hbar\omega = \frac{2}{\alpha^2}\left(n + \frac{1}{2}\right). \qquad (4.35b)$$

Figure 4.2 depicts $P_{\text{QM}} = |\psi_n(x)|^2$ for $n = 0$, 1, 2, 3 and 4. In this plot $|\psi|^2$ is scaled into $|\psi|^2 + E$ and $k = 1$. Figure 4.3 shows $P_{\text{CM}}(x)$ and $P_{\text{QM}}(x)$ for $n = 10$ with $k = 1$. The vertical dashed lines indicate classical turning points. Region within the vertical dashed lines is classically allowed region and outside region is classically forbidden. From the Figs. 4.2 and 4.3 we note that quantum mechanically there is a small probability in the classical forbidden region. It is clear from these figures that the position probability densities $|\psi_n|^2$ associated with the stationary state eigenfunctions have little resemblance to the corresponding densities of the classical oscillation. The agreement between classical and quantum probability densities improves with increase in n. The agreement is fairly good, on the average for large n, the principal discrepancy being the rapid oscillation of $|\psi_n|^2$.

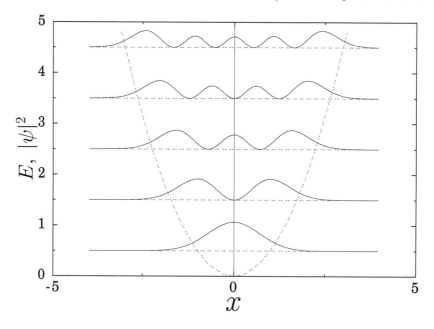

FIGURE 4.2

$|\psi|^2$ versus n for LHO. From the bottom curve to top the values of n are 0, 1, 2, 3 and 4. k is chosen as 1. E is in units of $\hbar\omega$. The potential is also shown.

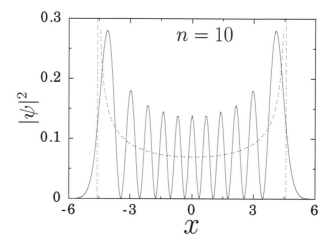

FIGURE 4.3

P_{CM} (dashed curve) and $P_{\mathrm{QM}}(x)$ (continuous curve) versus x for $n = 10$, $k = 1$ of LHO. The two vertical dashed lines represent classical turning points.

Solved Problem 2:

What are the probabilities of the LHO system in the ground state to be inside and outside the classical regions?

The classically allowed region is obtained from Eq. (4.35b) with $n = 0$. It is $(-A, A) \rightarrow (-1/\alpha, 1/\alpha)$. The P_{QM} in this region is

$$P_{\mathrm{QM}} = \int_{-1/\alpha}^{1/\alpha} |\psi_0|^2 \, \mathrm{d}x = \frac{2\alpha}{\sqrt{\pi}} \int_0^{1/\alpha} \mathrm{e}^{-\alpha^2 x^2} \, \mathrm{d}x = \frac{2}{\sqrt{\pi}} \int_0^1 \mathrm{e}^{-y^2} \, \mathrm{d}y = 0.83. \qquad (4.36)$$

The probability of finding the particle in the classically allowed region is 0.83 (obtained numerically computing the value of the integral in Eq. (4.36)). The probability in the outside region is 0.17.

4.4.5 Classical and Quantum Moments of Linear Harmonic Oscillator

Let us calculate the first and second moments of classical and quantum x and p_x for LHO [6].

Classical Mechanics

Using Eq. (4.35a) for $|x| \le A$ and 0 otherwise, $X = x/A$, $P = p_x/(m\omega A)$ and $E = m\omega^2 A^2/2$ we obtain

$$\langle X \rangle_{\mathrm{CM}} = \frac{1}{A} \int_{-A}^{A} P_{\mathrm{CM}}(x) x \, \mathrm{d}x = \frac{1}{A\pi} \int_{-A}^{A} \frac{x}{\sqrt{A^2 - x^2}} \mathrm{d}x = 0 \qquad (4.37a)$$

and

$$\langle X^2 \rangle_{\mathrm{CM}} = \frac{1}{A^2} \int_{-A}^{A} P_{\mathrm{CM}}(x) x^2 \, \mathrm{d}x = \frac{1}{A^2\pi} \int_{-A}^{A} \frac{x^2}{\sqrt{A^2 - x^2}} \mathrm{d}x = \frac{1}{2}. \qquad (4.37b)$$

For an arbitrary function $F(x, p_x)$ one can show that

$$\begin{aligned} \langle F(x, p_x) \rangle_{\mathrm{CM}} &= \frac{1}{2} \int P_{\mathrm{CM}}(x) \left[F\left(x, -\sqrt{2m(E - V)}\right) \right. \\ &\quad \left. + F\left(x, \sqrt{2m(E - V)}\right) \right] \mathrm{d}x. \end{aligned} \qquad (4.38)$$

Then

$$\langle P \rangle_{\mathrm{CM}} = \frac{1}{2m\omega A} \int_{-A}^{A} P_{\mathrm{CM}}(x) \left[-\sqrt{2m(E - V)} + \sqrt{2m(E - V)} \right] \mathrm{d}x = 0 \qquad (4.39a)$$

and

$$\begin{aligned} \langle P^2 \rangle_{\mathrm{CM}} &= \frac{1}{2m^2\omega^2 A^2} \int_{-A}^{A} P_{\mathrm{CM}}(x) \left[2m(E - V) + 2m(E - V) \right] \mathrm{d}x \\ &= \frac{2}{m\omega^2 A^2} \int_{-A}^{A} P_{\mathrm{CM}}(x)(E - V) \mathrm{d}x \\ &= \frac{2}{m\omega^2 A^2 \pi} \int_{-A}^{A} \frac{1}{\sqrt{A^2 - x^2}} \left(\frac{1}{2}m\omega^2 A^2 - \frac{1}{2}m\omega^2 x^2 \right) \mathrm{d}x \\ &= \frac{1}{A^2\pi} \int_{-A}^{A} \sqrt{A^2 - x^2} \, \mathrm{d}x \\ &= \frac{1}{2}. \end{aligned} \qquad (4.39b)$$

For the quantum LHO with $A_n = \sqrt{2E_n/(m\omega^2)} = \sqrt{(2n+1)\hbar/(m\omega)}$, $X = x/A_n = x\sqrt{m\omega/((2n+1)\hbar)}$, $P = p_x/\sqrt{2mE_n} = p_x/\sqrt{(2n+1)\hbar m\omega}$ it is easy to obtain

$$\langle X \rangle_{\mathrm{QM}} = \sqrt{\frac{m\omega}{(2n+1)\hbar}} \int_{-\infty}^{\infty} |\psi_n|^2 x \, \mathrm{d}x = 0, \tag{4.40a}$$

$$\langle X^2 \rangle_{\mathrm{QM}} = \frac{m\omega}{(2n+1)\hbar} \int_{-\infty}^{\infty} |\psi_n|^2 x^2 \, \mathrm{d}x = \frac{1}{2}, \tag{4.40b}$$

$$\langle P \rangle_{\mathrm{QM}} = -\mathrm{i}\sqrt{\frac{\hbar}{(2n+1)m\omega}} \int_{-\infty}^{\infty} \psi_n^* \frac{\mathrm{d}\psi_n}{\mathrm{d}x} \, \mathrm{d}x = 0, \tag{4.40c}$$

$$\langle P^2 \rangle_{\mathrm{QM}} = -\frac{\hbar}{(2n+1)m\omega} \int_{-\infty}^{\infty} \psi_n^* \frac{\mathrm{d}^2\psi_n}{\mathrm{d}x^2} \, \mathrm{d}x = \frac{1}{2}. \tag{4.40d}$$

From Eqs. (4.39) and (4.40) we notice that the quantum expectation values match with the classical expectation values.

4.4.6 Selection Rules

It is possible to find the selection rules for the allowed transitions for LHO. We know from classical electrodynamics, any oscillatory electric or magnetic multipole emits or absorbs radiation. We can show that the presence of any electric or magnetic multipole matrix elements between any two states results in induced absorption or emission of the electromagnetic wave. The electric dipole transition is the predominant one. Therefore, normally selection rules are worked-out by considering the matrix element of electric dipole operation $e\mathbf{r}$ between the states involved in the transition. As shown in sec. 2.13, a transition between two states characterized by the quantum numbers m and n is possible only if the condition (2.94) is satisfied. Selection rules for allowed transitions are obtained starting from the recurrence relation of the Hermite polynomials

$$H_{n+1} - 2\xi H_n + 2n H_{n-1} = 0. \tag{4.41}$$

Using Eq. (4.29) the above equation is expressed in terms of ψ_n as

$$\sqrt{n+1}\,\psi_{n+1} - \sqrt{2}\,\xi\psi_n + \sqrt{n}\,\psi_{n-1} = 0. \tag{4.42}$$

Multiplication of Eq. (4.42) by ψ_n^* and integrating with respect to ξ we get

$$\sqrt{n+1} \int_{-\infty}^{\infty} \psi_n^* \psi_{n+1} \, \mathrm{d}\xi - \sqrt{2} \int_{-\infty}^{\infty} \psi_n^* \, \xi \, \psi_n \, \mathrm{d}\xi + \sqrt{n} \int_{-\infty}^{\infty} \psi_n^* \psi_{n-1} \, \mathrm{d}\xi = 0. \tag{4.43}$$

The first and last integrals in Eq. (4.43) become zero because of the orthogonality of the eigenfunctions. Then

$$\int_{-\infty}^{\infty} \psi_n^* \, \xi \, \psi_n \, \mathrm{d}\xi = 0. \tag{4.44}$$

Similarly, multiplication of Eq. (4.42) with ψ_{n+1}^* and integrating with respect to ξ give

$$\int_{-\infty}^{\infty} \psi_{n+1}^* \, \xi \, \psi_n \, \mathrm{d}\xi = \sqrt{(n+1)/2}. \tag{4.45}$$

Replacing n by $n-1$ in Eq. (4.45) we have

$$\int_{-\infty}^{\infty} \psi_n^* \, \xi \, \psi_{n-1} \, \mathrm{d}\xi = \sqrt{n/2}. \tag{4.46}$$

From Eqs. (4.44) and (4.46) we write

$$\int_{-\infty}^{\infty} \psi_n^* \xi \psi_m \, d\xi \neq 0 \quad \text{for} \quad n = m + 1 . \tag{4.47a}$$

Interchanging n and m we get

$$\int_{-\infty}^{\infty} \psi_m^* \xi \psi_n \, d\xi \neq 0 \quad \text{for} \quad m = n + 1 . \tag{4.47b}$$

Therefore, $m - n = \pm 1$ or $\Delta_{mn} = \pm 1$. Hence, the electric dipole matrix element $\int_{-\infty}^{\infty} \psi_m^* e x \psi_n \, dx$ is not equal to zero if $m - n = \pm 1$. Thus, transition can occur between adjacent energy levels. Transitions between energy levels involve emission or absorption of quanta of energy by the oscillator. When transition from level n to level $n - 1$ occurs then $\Delta E = E_{n-1} - E_n = -\hbar\omega$ which is negative and hence energy is emitted by the oscillator. For the transition from level n to level $n + 1$ we have $\Delta E = \hbar\omega$ and therefore absorption of energy by the oscillator takes place. A quantum mechanical system in its ground state cannot radiate energy because there is no lower energy state for it to go.

4.4.7 Abstract Operator Method to Linear Harmonic Oscillator

Let us solve the LHO problem by the operator method.

First, we express H of LHO as $H = BA + \epsilon$, where A and B are two new operators functions of x and p_x and ϵ is a real constant. To determine them we assume that

$$A = c(W(x) + ip_x), \quad B = c(W(x) - ip_x), \tag{4.48}$$

where c and $W(x)$ to be determined. Then we obtain

$$\begin{aligned} BA + \epsilon &= c^2(W(x) - ip_x)(W(x) + ip_x) + \epsilon \\ &= c^2\left(W^2 + iWp_x - ip_xW + p_x^2\right) + \epsilon \\ &= c^2\left(p_x^2 - \hbar W' + W^2\right) + \epsilon \end{aligned} \tag{4.49}$$

Equating $BA + \epsilon$ to H gives

$$\frac{1}{2m}p_x^2 + \frac{1}{2}kx^2 = c^2 p_x^2 - c^2\hbar W' + c^2 W^2 + \epsilon. \tag{4.50}$$

With the choice $c = 1/\sqrt{2m}$ the above equation becomes

$$\frac{1}{2}kx^2 = -\frac{1}{2m}\hbar W' + \frac{1}{2m}W^2 + \epsilon. \tag{4.51}$$

The choice $W(x) = \sqrt{km}\,x = m\omega x$ eliminates x^2 terms. Then Eq. (4.51) gives $\epsilon = \hbar\omega/2$.

Now, we write

$$H = BA + \epsilon = BA + \frac{1}{2}\hbar\omega = \hbar\omega\left(\frac{1}{\sqrt{\hbar\omega}}B\frac{1}{\sqrt{\hbar\omega}}A + \frac{1}{2}\right). \tag{4.52}$$

We rewrite the above Eq. (4.52) as

$$H = \hbar\omega\left(a^\dagger a + \frac{1}{2}\right), \tag{4.53}$$

where

$$a = \frac{1}{\sqrt{\hbar\omega}}A = \frac{1}{\sqrt{2m\omega\hbar}}(m\omega x + ip_x) = \frac{1}{\sqrt{2}}\left(\alpha x + \frac{1}{\alpha}\frac{d}{dx}\right), \quad (4.54a)$$

$$a^\dagger = \frac{1}{\sqrt{\hbar\omega}}B = \frac{1}{\sqrt{2m\omega\hbar}}(m\omega x - ip_x) = \frac{1}{\sqrt{2}}\left(\alpha x - \frac{1}{\alpha}\frac{d}{dx}\right) \quad (4.54b)$$

with $\alpha = \sqrt{m\omega/\hbar}$. (Express $\bar{H} = \hbar\omega\left(aa^\dagger + \epsilon\right)$ in terms of x and p_x. Compare H and $\bar{H}$. What do you infer?)

From Eqs. (4.54) x and p_x are expressed in terms of a and $a^\dagger$ as

$$x = \frac{1}{\sqrt{2}\alpha}\left(a + a^\dagger\right), \quad p_x = -\frac{i\hbar\alpha}{\sqrt{2}}\left(a - a^\dagger\right). \quad (4.55)$$

What is the commutation relation of a, $a^\dagger$? To find $[a, a^\dagger]$ consider $[x, p_x]$. We obtain $[x, p_x] = xp_x - p_x x = i\hbar\left[a, a^\dagger\right]$. Since $[x, p_x] = i\hbar$ we get

$$\left[a, a^\dagger\right] = aa^\dagger - a^\dagger a = 1. \quad (4.56)$$

We expressed the H of LHO in terms of a and $a^\dagger$ as in Eq. (4.53). Thus, the problem of finding the eigenvalues of H is reduced to the problem of finding the eigenvalues of $a^\dagger a$. Note that the two quadratic operators x^2 and p_x^2 in the Hamiltonian operator of the LHO is converted into one squared form of the operators $a^\dagger$ and a and plus a constant ($\hbar\omega/2$). The constant $\hbar\omega/2$ does not affect the determination of the energy spectrum.

4.4.7.1   Eigenvalues

We define an operator $\hat{n} = a^\dagger a$ and denote ψ_n as the eigenfunction of $\hat{n}$ with the eigenvalue n:

$$\hat{n}\psi_n = n\psi_n. \quad (4.57)$$

We find $[\hat{n}, a^\dagger] = [a^\dagger a, a^\dagger] = a^\dagger [a, a^\dagger] + [a^\dagger, a^\dagger] a = a^\dagger$. From this we have

$$\hat{n}a^\dagger - a^\dagger\hat{n} = a^\dagger. \quad (4.58)$$

From Eq. (4.58) we obtain

$$\hat{n}a^\dagger\psi_n = \left(a^\dagger + a^\dagger\hat{n}\right)\psi_n = (n+1)a^\dagger\psi_n. \quad (4.59)$$

Next,

$$\hat{n}\left(a^\dagger\right)^2\psi_n = \left(a^\dagger + a^\dagger\hat{n}\right)a^\dagger\psi_n = (n+2)\left(a^\dagger\right)^2\psi_n. \quad (4.60)$$

In general

$$\hat{n}\left(a^\dagger\right)^m\psi_n = (n+m)\left(a^\dagger\right)^m\psi_n. \quad (4.61)$$

Similarly,

$$\hat{n}a^m\psi_n = (n-m)a^m\psi_n. \quad (4.62)$$

The news from (4.61) is that if ψ_n is the eigenfunction of $\hat{n}$ with the eigenvalue n then $a^\dagger\psi_n$, $(a^\dagger)^2\psi_n$, ... are the eigenfunctions of $\hat{n}$ belonging to the eigenvalues $(n+1)$, $(n+2)$, ..., respectively. On the other hand, Eq. (4.62) implies that, $a\psi_n$, $a^2\psi_n$, ... are the eigenfunctions of $\hat{n}$ corresponding to the eigenvalues $(n-1)$, $(n-2)$, ..., respectively. The effects of $a^\dagger$ and a are the raising and lowering, respectively, of the eigenvalue by one. $a^\dagger$ essentially adds one quantum of energy while a removes one quantum of energy. For this reason $a^\dagger$ and a are called *raising and lowering operators* or simply *ladder operators*. They are also called *creation* and *annihilation* operators, respectively. These operators appear particularly in fields of applied physics, where quantum mechanics is of primary importance. The areas include lasers, nanometer scales devices and studies of electrical conduction in a variety of devices.

Now, we find the range of eigenvalues of $\hat{n}$. Note that $a^\dagger a$ is Hermitian irrespective of whether a is Hermitian or not. Further,

$$\langle \hat{n} \rangle = \langle a^\dagger a \rangle = \int_{-\infty}^{\infty} \psi^* a^\dagger a \psi \, \mathrm{d}x = \int_{-\infty}^{\infty} (a\psi)^* a\psi \, \mathrm{d}x \geq 0. \tag{4.63a}$$

because the value of the last integral is the norm of $a\psi$ which should be > 0. If the system is in the state ψ_λ then the eigenvalues of $\hat{n} = a^\dagger a$ is λ. This gives

$$\langle \hat{n} \rangle = \int_{-\infty}^{\infty} \psi_\lambda^* \hat{n} \psi_\lambda \, \mathrm{d}x = \lambda \int_{-\infty}^{\infty} \psi_\lambda^* \psi_\lambda \mathrm{d}x = \lambda \,. \tag{4.63b}$$

Comparison of (4.63a) and (4.63b) gives $\lambda > 0$. Thus, the eigenvalues of $\hat{n}$ must be non-negative. Therefore, $(n-1)$, $(n-2)$, ... must be nonnegative and $a\psi_n$, $a^2\psi_n$, ... must terminate. Suppose ψ_0 is the last eigenfunction in the above set. Since the effect of a is to decrease an eigenvalue by one, we have $a\psi_0 = 0$ and

$$a^\dagger a \psi_0 = 0 \quad \text{or} \quad \hat{n}\psi_0 = 0 \,. \tag{4.64}$$

ψ_0 is the eigenstate of $a^\dagger a$ or $\hat{n}$ with the eigenvalue 0. The other eigenfunctions are constructed by the repeated application of $a^\dagger$ on ψ_0. Thus, ψ_0, $a^\dagger\psi_0$, $(a^\dagger)^2\psi_0$, ... are the eigenfunctions belonging to the eigenvalues 0, 1, 2, ..., respectively. The values of n are 0, 1, 2, ... Then the eigenvalues of H are $E_n = \left(n + \frac{1}{2}\right)\hbar\omega$.

4.4.7.2 Eigenfunctions

The norm $(a^\dagger\psi_n, a^\dagger\psi_n)$ is written as

$$\left(a^\dagger\psi_n, a^\dagger\psi_n\right) = \left(\psi_n, aa^\dagger\psi_n\right) = (n+1)\left(\psi_n, \psi_n\right) \,. \tag{4.65}$$

If ψ_n and ψ_{n+1} are normalized to unity then $(\psi_n, \psi_n) = (\psi_{n+1}, \psi_{n+1})$ and

$$\begin{aligned}
\left(a^\dagger\psi_n, a^\dagger\psi_n\right) &= (n+1)\left(\psi_{n+1}, \psi_{n+1}\right) \\
&= \left(\sqrt{n+1}\,\psi_{n+1}, \sqrt{n+1}\,\psi_{n+1}\right)
\end{aligned} \tag{4.66}$$

which implies

$$a^\dagger\psi_n = \sqrt{n+1}\,\psi_{n+1} \,. \tag{4.67}$$

$a^\dagger$ on the nth state gives $\sqrt{n+1}$ times the next higher state. Similarly, it is easy to show that a on the nth state gives $\sqrt{n}$ times the next lower state, except when $n = 0$, that is, $a\psi_n = \sqrt{n}\psi_{n-1}$, $n \neq 0$. Replacement of n by $n-1$ in Eq. (4.67) gives

$$\psi_n = \frac{a^\dagger}{\sqrt{n}}\psi_{n-1}. \tag{4.68}$$

Repeated use of Eq. (4.67) in Eq. (4.68) leads to

$$\psi_n = \frac{a^\dagger}{\sqrt{n}}\frac{a^\dagger}{\sqrt{n-1}}\psi_{n-2} = \ldots = \frac{(a^\dagger)^n}{\sqrt{n!}}\psi_0 \,, \tag{4.69}$$

where ψ_0 is the solution of $a\psi_0 = 0$. Integrating the equation

$$a\psi_0 = \frac{1}{\sqrt{2}}\left(\alpha x + \frac{1}{\alpha}\frac{d}{dx}\right)\psi_0 = 0 \qquad (4.70)$$

we get $\psi_0(x) = Ae^{-\alpha^2 x^2/2}$. Normalization gives $A = [\alpha/(2^n n!\sqrt{\pi})]^{1/2}$. Then

$$\begin{aligned}
\psi_n(x) &= \left(\frac{\alpha}{2^n n!\sqrt{\pi}}\right)^{1/2}\left(a^\dagger\right)^n e^{-\alpha^2 x^2/2} \\
&= \left(\frac{\alpha}{2^n n!\sqrt{\pi}}\right)^{1/2} e^{-\alpha^2 x^2/2} H_n(\alpha x)\,.
\end{aligned} \qquad (4.71)$$

We note that the Schrödinger Eq. (4.21) is second-order while the differential Eq. (4.70) used in the operator algebra approach to determine the ground state eigenfunction is a first-order equation. That is, the order of the equation to be solved to determine the energy eigenfunction is reduced in the operator algebra method. Further, once ψ_0 is known then ψ_n is obtained by operating $(a^\dagger)^n$ on ψ_0, that is, by means of differentiation.

From Eq. (4.53) we write $\hbar\omega a^\dagger a = H + E_0$, where $E_0 = \hbar\omega/2$ is the ground state energy. $a^\dagger a$ measures the number of energy quanta (photons) present above the ground state and hence is called *number operator*. The energy eigenstates are also eigenstates of this operator and are thus often called *number states*. There is a straightforward interpretation of the above formalism for the radiation field. Each quantum of energy, $\hbar\omega$, represents a photon. The eigenstates ψ_n represents a system with n photons. The ground state is the vacuum state (no photons) with nonzero energy $\hbar\omega/2$. The operators $a^\dagger$ and a represent the physical processes by which photons are absorbed or emitted. Therefore, $a^\dagger$ and a are called *creation* and *annihilation operators*, respectively. *Does an operator of an observable of a conservative system contain a or $a^\dagger$ alone? Why?*

4.5 Particle in the Potential $V(x) = x^{2k}$, $k = 1, 2, \ldots$

We consider a particle of mass m, moving along the x-axis with the potential energy function $V(x) = Ax^n$, where A is a constant [7]. As n increases $V(x)$ becomes more and more flat-bottomed and the sides become steeper and steeper near the turning points. For very large n the $V(x)$ approximates a square-well or box potential. It is of interest to know the change in the energy levels due to deformation of $V(x)$. For convenience, we rewrite the potential $V(x)$ as

$$V(x) = \alpha\left(\frac{x}{a}\right)^{2k}, \qquad k = 1, 2, \ldots \qquad (4.72)$$

where α, a and p are parameters. Defining

$$\theta^2 = \frac{2ma^2}{\hbar^2}\alpha, \qquad \lambda = \theta^{2k/(k+1)}\frac{E}{\alpha}, \qquad u = \theta^{1/(k+1)}\frac{x}{a} \qquad (4.73)$$

the Schrödinger equation for the problem takes the form

$$\psi_{uu} + \left(\lambda - u^{2k}\right)\psi = 0\,. \qquad (4.74)$$

The boundary conditions are $\psi(0) = $ finite and $\psi \to 0$ as $|u| \to \infty$. When $k = 1$, Eq. (4.74) becomes the LHO Eq. (4.23)

For $k \to \infty$, the boundary conditions become $\psi(u) = 0$ for $|u| \geq 1$ and Eq. (4.74) reduces to

$$\psi_{uu} + \lambda \psi = 0, \quad |u| \leq 1 \tag{4.75}$$

Its solution subjected to the boundary conditions $\psi(\pm 1) = 0$, $\psi(0) = 1$ is

$$\psi_0(u) = \begin{cases} \cos \sqrt{\lambda_0}\, u, & \text{if } |u| < 1 \\ 0, & \text{if } u \geq 1, \end{cases} \tag{4.76}$$

where $\sqrt{\lambda_0} = \pi/2$. Though the solution of Eq. (4.74) for arbitrary p is not known, we get approximate solution for large k. For this purpose $w = u^{k+1}/(k+1)$ or $u = [(k+1)w]^{1/(k+1)}$. We assume solution of the form

$$\psi(u) = e^{-w} f(u). \tag{4.77}$$

Substitution of Eq. (4.77) in Eq. (4.74) gives

$$f_{uu} - 2u^k f_u + \left(\lambda - k u^{k-1}\right) f = 0. \tag{4.78}$$

For large p and $|u| < 1$, the terms containing u^k and u^{k-1} are small. In this case $f_{uu} + \lambda f \approx 0$ whose solution is given by Eq. (4.76). Thus,

$$\psi(u) = e^{-w} \cos \sqrt{\lambda}\, u \tag{4.79}$$

is an approximate solution of Eq. (4.74). For large k, e^{-w} is almost constant at 1 until u is close to 1, so the variation of ψ is described by the cosine function. For $u \gg 1$, e^{-w} in Eq. (4.79) is dominant and the eigenfunction very rapidly decays to zero. Near $u \approx 1$ we obtain

$$\psi(w) = e^{-w} \cos \left[\sqrt{\lambda}\,(k+1)^{1/(k+1)} w^{1/(k+1)} \right]. \tag{4.80}$$

For $k \gg 1$, $w^{1/(k+1)}$ is ≈ 1. Therefore, $\psi(\omega)$ will vanish if

$$\sqrt{\lambda}\,(k+1)^{1/(k+1)} = \pi/2. \tag{4.81}$$

Therefore, the eigenvalue condition is

$$\lambda_1 = \left(\frac{\pi}{2}\right)^2 \frac{1}{(k+1)^{2/(k+1)}}, \tag{4.82}$$

where λ_1 is the ground state eigenvalue of Eq. (4.74).

4.6 Particle in a Box

Another interesting exactly solvable quantum mechanical system is a single particle of mass m confined within a potential well V given by

$$V(x) = \begin{cases} 0, & |x| < a \\ \infty, & |x| > a \end{cases} \tag{4.83}$$

and is depicted in Fig. 4.4. Realization of this kind of potentials arises for molecules of a dilute gas in a long thin tube with a rigid stoppers at the ends. The vibration of a string clamped at two points is modelled by the above potential. The thermodynamic equilibrium

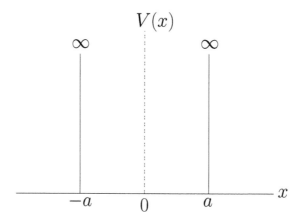

FIGURE 4.4
Sketch of an infinite height potential, Eq. (4.83).

of a cold white dwarf star is modelled by balancing the dynamics of electrons, represented as particles in a box, with the gravitational forces. This model has predicted the stellar mass threshold for final collapse of a white dwarf to a black hole. The electrons in metals are also modelled by particles in a three-dimensional box potential. This model explains the motion of conduction electrons and their contribution to magnetism and specific heat. It is now possible to manufacture a box trapping electrons in a few nanometers-wide region. The box potential is a good description for one-dimensional bound-state problems, when the confining potential is much steeper compared to the parabolically curved LHO potential.

4.6.1   Classical Result

The particle is confined between the two walls of the potential and bounces elastically when it incidents on the walls. The problem is a bound state problem. The total energy is $E = mv^2/2$. The classical equation of motion is $\ddot{x} = 0$, $|x| < a$. Its solution is $x(t) = At + B$. The choice $x(0) = 0$ gives $x(t) = At$ and A is the velocity v of the particle. The particle starting from $x = 0$ moves towards $x = a$ with velocity v. Its trajectory is a straight-line as shown in the subplot of Fig. 4.5. When the particle reaches $x = a$, it turns back with the velocity $-v$ and proceeds towards $-a$. At $x = -a$, it turns back towards $+a$ with $+v$ and so on. The turning points are $-a$ and a. The distance the particle traverses to come back to the origin with velocity v is $4a$. Therefore, the period of motion is $\tau = 4a/|v|$. The fundamental frequency f (in Hertz) is $1/\tau$ (and $\omega = 2\pi f$, where ω is in rad/sec).

The solution is written as [1]

$$x(t) = -a + \frac{4a}{\tau}|t|, \quad \text{for} \ \ -\tau/2 < t < \tau/2 \tag{4.84}$$

or

$$x(t) = -\frac{4a}{\pi^2} \sum_{\substack{n=1 \\ \text{odd}}}^{\infty} \frac{1}{n^2}\cos(2\pi n t/\tau). \tag{4.85}$$

Figure 4.5 shows classical trajectory and velocity. Substitution of $V(x) = 0$, $E = mv^2/2$, $\tau = 4a/|v|$ in Eq. (4.3) gives $P_{\text{CM}}(x) = 1/(2a)$.

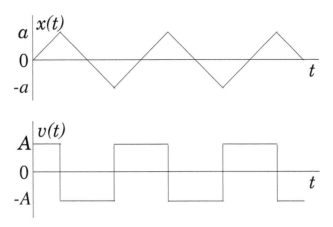

FIGURE 4.5
Classical trajectory and velocity of a particle in the infinite height potential, Eq. (4.83).

4.6.2 Quantum Mechanical Result

The Schrödinger equation for $|x| \leq a$ is given by

$$\psi_{xx} + k^2\psi = 0, \quad k^2 = 2mE/\hbar^2. \tag{4.86}$$

For $|x| > a$, $\psi(x) = 0$. The solution of Eq. (4.86) is

$$\psi(x) = A\sin kx + B\cos kx. \tag{4.87}$$

This solution must satisfy the boundary conditions $\psi(a) = 0$, $\psi(-a) = 0$. Applying the boundary conditions the following equations are obtained:

$$A\sin ka + B\cos ka = 0, \quad -A\sin ka + B\cos ka = 0. \tag{4.88}$$

For nontrivial solutions ($A \neq 0$ or $B \neq 0$) the condition is the determinant of the coefficient matrix must be zero:

$$\begin{vmatrix} \sin ka & \cos ka \\ -\sin ka & \cos ka \end{vmatrix} = 0. \tag{4.89}$$

This gives $\sin 2ka = 0$. In other words

$$k = \frac{n\pi}{2a}, \quad n = 1, 2, \ldots. \tag{4.90}$$

Thus, the allowed values of ka are discrete. Since k is related to E in Eq. (4.86) E becomes discrete. Since $\hbar k$ is the momentum we write $p_{xn} = \hbar k_n = n\pi\hbar/(2a)$. The news is that p_x is discrete. For a free particle p_x is continuous. From the Eqs. (4.88) with $k_n = n\pi/(2a)$ we obtain

$$2A\sin(n\pi/2) = 0, \quad 2B\cos(n\pi/2) = 0. \tag{4.91}$$

When n is an odd integer ($n = 1, 3, \ldots$) $A = 0$ while B is arbitrary. For even integer values of n, $n = 2, 4, \ldots$, A is arbitrary and $B = 0$. Thus, Eqs. (4.88) lead to the following two cases.

Case 1:

$$A = 0, \ B \neq 0, \ ka = \frac{n\pi}{2}, \ n = 1, 3, \ldots \text{ (odd } n). \tag{4.92}$$

Case 2:

$$A \neq 0, \quad B = 0, \quad ka = \frac{n\pi}{2}, \quad n = 2, 4, \ldots \text{ (even } n\text{)}. \tag{4.93}$$

($n = 0$ is not acceptable. *Why?*)

The discreteness of n has an immediate significance. For example, for a string fixed at the ends $x = \pm a$ we have $2\pi/\lambda = n\pi/(2a)$, giving $\lambda = 4a/n$, $n = 1, 2, \ldots$. The result is certain wavelengths alone are allowed. In other words, only certain frequencies of vibrations are allowed. This is the *quantization of frequency*. The waves oscillate more rapidly as n increases. For a particle in a box, the de Broglie wavelengths are $\lambda_n = 4a/n$, $n = 1, 2, \ldots$. The lowest energy of the particle in the box potential is not zero. If it is zero $v = 0$ and hence $\lambda = h/mv = \infty$. But there is no way to reconcile an infinite wavelength to the particle in a box.

The features of a particle in a box are listed in the following:

1. From Eqs. (4.86) and (4.90) the energy levels are obtained as

$$E_n = \frac{n^2\hbar^2\pi^2}{8ma^2}, \quad n = 1, 2, \ldots. \tag{4.94}$$

   There are infinite number of energy levels. However, the energy levels are discrete or quantized. The lowest energy is $E_1 = \hbar^2\pi^2/(8ma^2)$ which is nonzero. The particle will make to and fro motion in the box in the region $-a$ to a even at absolute zero temperature. E_1 is the zero-point energy.

2. The energy difference between the successive levels is

$$\triangle E_n = E_{n+1} - E_n = \frac{(2n+1)\hbar^2\pi^2}{8ma^2}. \tag{4.95}$$

   The energy levels are not equally spaced. E_n are proportional to n^2 so they are close at lower energy levels than at higher levels.

3. The momentum of the particle is $p_x = \hbar\pi n/(2a)$ and $E_n \propto 1/a^2$. Therefore, squeezing a particle increases its momentum and the total energy.

4. Because of discreteness of energy spectrum we discretize ψ, k, A and B and rewrite Eq. (4.87) as

$$\psi_n(x) = \begin{cases} A_n \sin k_n x, & k_n = n\pi/(2a), \quad n = 2, 4, \ldots \\ B_n \cos k_n x, & k_n = n\pi/(2a), \quad n = 1, 3, \ldots. \end{cases} \tag{4.96}$$

   These eigenfunctions are normalizable. Applying the normalization condition for ψ_n with even n we get $1 = \int_{-a}^{a} A_n^2 \sin^2 k_n x \, dx = A_n^2 a$. Thus, $A_n = 1/\sqrt{a}$. Similarly, for odd n $B_n = 1/\sqrt{a}$.

5. For $|x| < a$ the values of x at which ψ vanishes is called a *node*. For odd n the nodes are $x = \pm a/m$, $m = 1, 3, \ldots, n$. For even n the nodes are $x = 0$, $\pm 2a/m$, $m = 2, 4, \ldots, n$. For $n = 1$ the nodes are $x = \pm a$ and for $n = 2$ they are 0, $\pm a$. For a given state the number of nodes is $n + 1$.

6. Because $\cos(-x) = \cos x$, the eigenfunction for an odd n is symmetric with respect to $x = 0$. Since $\sin(-x) = -\sin x$, ψ is anti-symmetric about $x = 0$ for even n. Adjacent states have opposite symmetry. For even n, $P\psi_n(x) = \psi_n(x)$. Therefore, $\sin(n\pi x/2a)$ and $\cos(n\pi x/2a)$ are also the eigenfunctions of the parity operator.

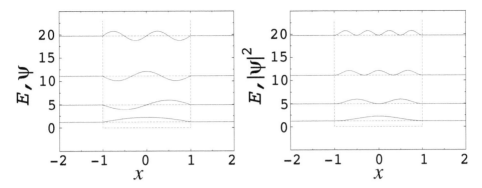

FIGURE 4.6
First few energy levels and the eigenfunctions of infinite height potential (left subplot). Here $a = 1$ and $\hbar^2$ and m are set to unity. $\psi_n(x)$ is scaled into $\psi_n(x) + E_n$. From bottom to top the values of n are 1, 2, 3 and 4. $|\psi_n|^2$ versus x (right subplot) for the eigenfunctions shown in the left subplot.

7. Figure 4.6 shows first few energy levels and eigenfunctions. The corresponding probability distribution

$$P_{\text{QM}}(x) = |\psi_n(x)|^2 = \frac{1}{a} \cos^2 \left[\left(n - \frac{1}{2} \right) \frac{\pi x}{a} \right] \tag{4.97}$$

is depicted in Fig. 4.6. For large value of n, the $\cos^2$ term in Eq. (4.97) averages to $1/2$ showing that P_{QM} approaches the classical limit $1/2a$.

8. A transition between a pair of states with the quantum numbers n and n' is allowed if one of the numbers is even and the other is odd, that is $\triangle n = \pm 1$. Transitions between two odd or even quantum numbers states is forbidden.

9. Since $V(x) = V(-x)$ we have $[P, H] = 0$, where P is the parity operator. The parity of the eigenfunctions of a particle in a box will not change in time.

10. Let us describe the time dynamics in the infinite height potential with an analogy to the classical wave propagation on a stretched string of length L. Assume that $\psi(x) = \sum c_n \psi_n(x)$ be the initial profile of a stretched string which then propagate without dispersion at speed $v = \sqrt{2\bar{E}_n/m}$, where $\bar{E}_n$ is the mean energy level. The elastic collisions at the infinitely high potential barriers are analogous to perfect reflections from the fixed ends of the string. When a wave form propagating on a string is reflected at the string ends, it travels in the opposite direction with a change of phase π in its amplitude. Its analogous behaviour is found in the box potential [8].

Consider $\psi_1^* \psi_1$ in Fig. 4.6. A classical interpretation of the ground state function suggests that because there is no possibility of finding the particle next to a wall, the particle travels at an infinite speed immediately before and after hitting the wall, but travels comparatively slowly at the centre of the box. So although there are no forces inside the box the particle appears to speed up and slow down. The first excited state is even more difficult to understand classically because at the centre of the box there is no possibility of finding the particle. Thus, classically the particle moves through this region at an infinite speed.

The eigenstates given by the Eqs. (4.94) and (4.96) are the stationary states of the system with energy eigenvalue being constant, independent of the space variable x. A nonstationary but normalizable eigenfunction of the system is

$$\psi(x) = \begin{cases} N\left(x^2 - a^2\right), & |x| < a \\ 0, & |x| > a, \end{cases} \tag{4.98}$$

where N is the normalization constant. But the energy of this eigenfunction is space dependent.

In the domain of our everyday experience, we do not feel the quantum effects, since on macroscopic scale they are imperceptible. Suppose a ball of mass 10 g is rolling on a smooth floor of 10cm wide box with a speed of 3.3 cm/sec. The permitted energies for the ball according to Eq. (4.94) are $E_n = 5.49 \times 10^{-57} n^2$erg. The kinetic energy of the ball is 54.4 ergs. This corresponds to the energy level of quantum number $n = 10^{29}$. This is very large. Therefore, it is practically impossible to find whether the ball is in the energy states $n = 10^{29}$ or $10^{29} + 1$, $10^{29} + 2$, ..., etc. (discrete). Therefore, we say that continuous range of energy is possible.

For the case of $V(x) = 0$ for $0 \leq x \leq a$ and ∞ otherwise the energy eigenvalues and the eigenfunctions are given by

$$E_n = \frac{n^2 \pi^2 \hbar^2}{2ma^2}, \quad \psi_n(x) = \sqrt{\frac{2}{a}} \sin \frac{n\pi x}{a}, \quad n = 1, 2, \dots. \tag{4.99}$$

Solved Problem 3:

The size of a nucleus is say, 10^{-14} m. Assuming it as a one-dimensional box calculate the ground state energy of an electron in the nucleus. Show why the electron does not exist in the nucleus.

The lowest energy is

$$E_1 = \frac{\pi^2 \hbar^2}{2ma^2} = \frac{\pi^2 \times \left(1.054 \times 10^{-34}\right)^2}{2 \times 9.11 \times 10^{-31} \times \left(10^{-14}\right)^2} \,\text{Js} = 6.018 \times 10^{-10} \,\text{Js}. \tag{4.100}$$

The Coulombic attraction energy is calculated as

$$\begin{aligned} E_c &= -\frac{1}{4\pi\epsilon_0} \frac{e^2}{r} = -\frac{(1.602 \times 10^{-19})^2}{4\pi \times 8.85418 \times 10^{-12} \times 10^{-14}} \,\text{Js} \\ &= -2.3 \times 10^{-14} \,\text{Js}. \end{aligned} \tag{4.101}$$

Since E_c is very weak compared to E_1 the nucleus cannot hold an electron.

4.6.3   Classical and Quantum Mechanical $\langle x \rangle$ and $\langle x^2 \rangle$

The classical and quantum mechanical $\langle x \rangle$ and $\langle x^2 \rangle$ are obtained as

$$\langle x \rangle_{\text{CM}} = \frac{1}{2a} \int_{-a}^{a} x \, dx = 0, \tag{4.102a}$$

$$\langle x^2 \rangle_{\text{CM}} = \frac{1}{2a} \int_{-a}^{a} x^2 \, dx = \frac{a^2}{3} \tag{4.102b}$$

and for both odd and even values of n

$$\langle x \rangle_{\text{QM}} = \frac{1}{a} \int_{-a}^{a} \psi_n^* x \psi_n \mathrm{d}x = 0, \tag{4.103a}$$

$$\langle x^2 \rangle_{\text{QM}} = \frac{1}{a} \int_{-a}^{a} \psi_n^* x^2 \psi_n \mathrm{d}x = \frac{a^2 (n^2 \pi^2 - 6)}{3 n^2 \pi^2}. \tag{4.103b}$$

The $\langle x \rangle_{\text{QM}}$ has the same value as the classical case but the value of $\langle x^2 \rangle_{\text{QM}}$ differs from the classical case. However, for large n, $\langle x^2 \rangle_{\text{QM}} = \langle x^2 \rangle_{\text{CM}}$. That is, the classical solution is the short-wavelength high energy limit of the quantum solution.

4.6.4 Abstract Operator Method to a Particle in a Box

In sec. 4.4 we obtained eigenspectra for LHO by operator method. The problem of a particle in a box is also solved in this approach [9].

We define the number operator $\hat{n}$ and its inverse $\hat{n}^{-1}$ as

$$\hat{n}\psi_n = n\psi_n, \quad \hat{n}^{-1}\psi_n = n^{-1}\psi_n. \tag{4.104}$$

Suppose $a^\dagger$ and a are the creation and annihilation operators such that

$$a^\dagger \psi_n = n\psi_{n+1}, \quad a\psi_n = (n-1)\psi_{n-1}. \tag{4.105}$$

For convenience we write the width of the box as $2L$ and $z = \pi x/(2L)$ so that the Hamiltonian is

$$H_z = -\frac{1}{2m} \left(\frac{\hbar \pi}{2L} \right)^2 \frac{\mathrm{d}^2}{\mathrm{d}z^2}. \tag{4.106}$$

The choice [9]

$$a^\dagger = (\cos z)\hat{n} + \sin z \frac{\mathrm{d}}{\mathrm{d}z}, \quad a = \left[(\cos z)\hat{n} - \sin z \frac{\mathrm{d}}{\mathrm{d}z} \right] \hat{n}^{-1}(\hat{n} - 1) \tag{4.107}$$

with

$$\psi_n = \frac{1}{\sqrt{L}} \sin nz, \quad z = \frac{\pi x}{2L} \tag{4.108}$$

gives

$$a^\dagger \psi_n = n\psi_{n+1}, \quad a\psi_n = (n-1)\psi_{n-1}. \tag{4.109}$$

Equations (4.109) are simply the Eqs. (4.105). That is, $a^\dagger$ and a given by the Eqs. (4.107) are the creation and annihilation operators of the eigenstates of the Hamiltonian (4.106). Determination of energy eigenvalues using the ladder operators is given as an exercise (see the exercise 57 at the end of this chapter).

4.7 Morse Oscillator

The Morse oscillator was introduced as a useful model for the interatomic potential and fitting the vibrational spectra of diatomic molecules. It is also used to describe the photodissociation of molecules, multi-photon excitation of the diatomic molecules in a dense medium or in a gaseous cell under high-pressure and pumping of a local mode of a polyatomic molecule by an infrared laser [10–13].

The potential of the Morse oscillator is

$$V(x) = \frac{1}{2}\beta e^{-x}\left(e^{-x} - 2\right).\tag{4.110}$$

The potential is nonpolynomial and $V(x) \to \infty$ as $x \to -\infty$ while it becomes 0 in the limit $x \to \infty$. The Schrödinger equation for the Morse oscillator is

$$-\frac{\hbar^2}{2m}\phi_{xx} + \frac{1}{2}\beta e^{-x}\left(e^{-x} - 2\right)\phi = E\phi.\tag{4.111}$$

To eliminate the exponential terms introduce the change of variable $z = 2\lambda e^{-x}$ with $\lambda^2 = m\beta/\hbar^2$ and $z > 0$. We have $\phi_{xx} = z\phi_z + z^2\phi_{zz}$. Then Eq. (4.111) becomes

$$\phi_{zz} + \frac{1}{z}\phi_z + \frac{1}{4}\left(\frac{4\lambda}{z} - 1\right)\phi = \frac{\mathcal{E}}{z^2}\phi, \quad \mathcal{E} = -\frac{2m}{\hbar^2}E.\tag{4.112}$$

For large z we can write the Eq. (4.112) as $\phi_{zz} - \frac{1}{4}\phi = 0$ which has the solution $\phi = e^{-z/2}$. We seek the solution

$$\phi(z) = e^{-z/2}u(z).\tag{4.113}$$

Now, Eq. (4.112) takes the form

$$u'' + \left(\frac{1}{z} - 1\right)u' + \left(\frac{2\lambda - 1}{2z} - \frac{\mathcal{E}}{z^2}\right)u = 0.\tag{4.114}$$

The generalized Laguerre differential equation is

$$zy'' + (k + 1 - z)y' + ny = 0.\tag{4.115}$$

The particular bounded solutions of Eq. (4.115) are called *generalized Laguerre polynomials* and are given by $(k > 0)$

$$L_n^{(k)}(z) = \frac{z^{-k}e^z}{n!}\frac{d^n}{dz^n}\left(e^{-z}z^{n+k}\right), \quad n = 0, 1, 2, \ldots.\tag{4.116}$$

We can bring Eq. (4.114) into the from (4.115) by introducing

$$u(z) = z^\alpha v(z).\tag{4.117}$$

The result is

$$zv'' + (2\alpha + 1 - z)v' + \left[\frac{\alpha^2 - \mathcal{E}}{z} + \left(\lambda - \alpha - \frac{1}{2}\right)\right]v = 0.\tag{4.118}$$

Comparing Eq. (4.118) with Eq. (4.115) we choose $\alpha^2 = \mathcal{E}$, $k = 2\alpha$ and $\lambda - \alpha - \frac{1}{2} = n$, that is, $\alpha = \lambda - n - \frac{1}{2}$. We obtain $k = 2\lambda - 2n - 1$. As $k > 0$ the expression $k = 2\lambda - 2n - 1$ gives $n < \lambda - \frac{1}{2}$. Then as n is discrete the energy eigenvalues are

$$E_n = -\frac{\hbar^2}{2m}\mathcal{E}_n = -\frac{\hbar^2}{2m}\left(\lambda - n - \frac{1}{2}\right)^2, \quad n = 0, 1, \ldots \text{ and } n < \lambda - \frac{1}{2}.\tag{4.119}$$

$E_n < 0$ for bound states. The eigenfunctions are given by

$$\phi_n(x) = N_n z^{k/2}e^{-z/2}L_n^{(k)}(z), \quad z = 2\lambda e^{-x},\tag{4.120}$$

where $\lambda^2 = m\beta/\hbar^2$, $k = 2\lambda - 2n - 1$ and $L_n^{(k)}$ are given by Eq. (4.116). The normalization condition

$$|N_n|^2 \int_0^\infty z^k e^{-z}\left(L_n^{(k)}\right)^2 dz = 1\tag{4.121}$$

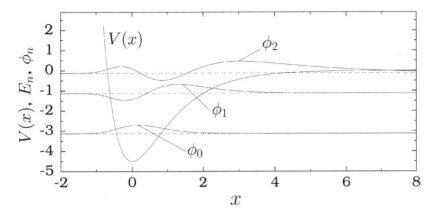

FIGURE 4.7
Bound state energy eigenvalues and eigenfunctions of the Morse oscillator for $\beta = 9$. The dashed lines from bottom to top represent the values of E_0, E_1 and E_2. The potential $V(x)$ is also shown.

with the known result

$$\int_0^\infty z^k e^{-z} \left(L_n^{(k)} \right)^2 dz = \frac{(n+k)!}{n!} \qquad (4.122)$$

gives $N_n = (n!/(n+k)!)^{1/2}$.

As $n < \lambda - \frac{1}{2}$ the Morse oscillator has a finite number of bound states and the number of bound states can be controlled by the parameter β. Setting the values of $\hbar$ and m as unity for convenience and $\beta = 9$ we obtain $n < 5/2$ and $E_0 = -3.125$, $E_1 = -1.125$ and $E_2 = -0.125$. There are only three bound states. Figure 4.7 shows the energy eigenvalues and the eigenfuntions for $\beta = 9$.

4.8 Pöschl–Teller Potentials

A class of potentials which are exactly solvable and possessing bound states are the Pöschl–Teller potentials given by [14–16]

$$V(x) = \frac{1}{2} V_0 \left[\frac{\lambda(\lambda - 1)}{\cos^2(x/2a)} + \frac{\kappa(\kappa - 1)}{\sin^2(x/2a)} \right], \qquad 0 \le x \le \pi a \qquad (4.123)$$

where λ, $\kappa > 1$ and V_0 is a constant. These potentials were first introduced in a molecular physics context. Figure 4.8 shows the potentials $V(x)$ for $a = 1/\pi$, $V_0 = 1$ and $(\lambda, \kappa) = (2, 2)$, $(2, 32)$, $(2, 64)$. When $\lambda = \kappa = 1$ the potential $V(x)$ becomes the box potential.

At the turning points of the classical motion $E = V(x)$. Solving this equation for the turning points we obtain

$$x_\pm = a \cos^{-1} \left[\frac{\alpha - \beta}{2} \pm \sqrt{\triangle} \right], \quad \triangle = 1 - (\alpha + \beta) + \frac{1}{4}(\alpha - \beta)^2, \qquad (4.124a)$$

where

$$\alpha = \lambda(\lambda - 1)V_0/E, \quad \beta = \kappa(\kappa - 1)V_0/E. \qquad (4.124b)$$

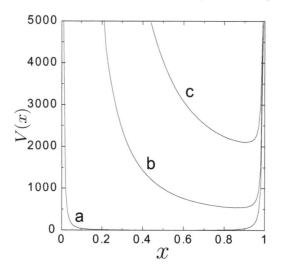

FIGURE 4.8
Pöschl–Teller potentials, Eq. (4.123), for $(\lambda, \kappa) = (2, 2)$(a), $(2, 32)$(b) and $(2, 64)$(c).

Therefore, the classical motion is possible only if

$$E > \frac{1}{2}V_0 \left(\sqrt{\lambda(\lambda - 1)} + \sqrt{\kappa(\kappa - 1)} \right)^2 . \tag{4.125}$$

The solution $x(t)$ is given by [16]

$$x(t) = a \cos^{-1} \left[\frac{(\alpha - \beta)}{2} + \sqrt{\Delta} \cos \omega t \right] , \tag{4.126}$$

where $\omega = \sqrt{2E/m}/a$, $x(0) = x_-$ and $\dot{x}(0) = 0$. The period of the solution is $\tau = 2\pi/\omega = 2\pi a\sqrt{m/(2E)}$. Note that τ is independent of V_0, λ and κ.

For the quantum mechanical problem, we impose $\psi(x) = 0$ for $x \geq \pi a$ and $x \leq 0$, $\psi(0) = \psi(\pi a) = 0$. The stationary state Schrödinger equation is

$$\left[\frac{d^2}{dx^2} + \frac{2m}{\hbar^2} \left\{ E - V(x) + \frac{\hbar^2(\lambda + \kappa)^2}{8ma^2} \right\} \right] \psi(x) = 0 , \quad 0 \leq x \leq \pi a , \tag{4.127}$$

where for convenience the Hamiltonian of the system is shifted by the amount $-\hbar^2(\lambda + \kappa)^2/(8ma^2)$. Without loss of generality choose $V_0 = \hbar^2/(4ma^2)$ in $V(x)$. The normalized eigenfunctions and eigenvalues are worked out as [16]

$$\psi_n(x) = N \left(\cos \frac{x}{2a} \right)^\lambda \left(\sin \frac{x}{2a} \right)^\kappa {}_2F_1 \left(-n, n + \lambda + \kappa; \kappa + \frac{1}{2}; \sin^2 \frac{x}{2a} \right) , \tag{4.128}$$

where N is a normalization factor, ${}_2F_1$ is a hypergeometric function and

$$E_n = \frac{\hbar^2}{2ma^2} n(n + \lambda + \kappa) , \quad \lambda, \kappa > 1 , \; n = 0, 1, \ldots . \tag{4.129}$$

4.9 Quantum Pendulum

A rigid pendulum is a point particle restricted to a circular radius and subjected to a uniform gravitational or electric field. Classical and quantum versions of this system are interesting. At low energies the potential energy of a quantum pendulum is approximated by that of a harmonic oscillator and for high energies it is thought of as the free rotor [17]. The dynamics of the quantum pendulum are closely related to molecular motions known as *hindered rotations*. The system is used to describe almost free rotations in ethene (C_2H_2) and low energy torsional vibrations in K_2PtCl_6 (potassium hexachloroplatinate) [18].

4.9.1 Classical Result

Consider a point particle of mass m restricted to a circle of radius l. It has a rotational inertia $I = ml^2$. The system is subjected to a uniform gravitational force (mg) (or electric force QE) in the vertical direction. The corresponding potential energy function is $V(\theta) = -V_0 \cos\theta$ where $V_0 = mgl$ (or QEl). The classical Hamiltonian is

$$H = E = \frac{I}{2}\dot{\theta}^2 - V_0 \cos\theta. \tag{4.130}$$

The classical equation of motion is

$$\ddot{\theta} + \frac{g}{l}\sin\theta = 0. \tag{4.131}$$

Assuming $\dot{\theta}(0) = 0$ when $\theta(0) = \theta_{max}$ where $\theta_{max} = \cos^{-1}(-E/V_0)$ the solution of Eq. (4.131) is given by [19]

$$\theta(t) = 2\sin^{-1}\left[K\,\mathrm{sn}\left(t\sqrt{g/l},K\right)\right], \quad K = \sin(\theta_{max}/2), \tag{4.132}$$

where sn is the Jacobian elliptic function. From Eq. (4.130) we write

$$dt = \sqrt{\frac{ml^2}{2}}\frac{d\theta}{\sqrt{E + V_0\cos\theta}}. \tag{4.133}$$

For $-V_0 < E < V_0$ the motion will be periodic with $\tilde{\theta} = \pm\cos^{-1}(-E/V_0)$ being the turning points. Then the classical period is

$$\tau(E) = 2\sqrt{\frac{ml^2}{2}}\int_{-\tilde{\theta}}^{\tilde{\theta}}\frac{d\theta}{\sqrt{E + V_0\cos\theta}}. \tag{4.134}$$

For low energies (small oscillations), with $-V_0 \le E \ll V_0$ or $\tilde{\theta} \ll 1$, the motion is oscillator-like. In this case the time period is $\tau = 2\pi/\sqrt{V_0/ml^2}$, which is independent of energy. For the case of unbounded motion for which $E > V_0$ one period corresponds to one revolution and is given by

$$\tau(E) = \sqrt{\frac{ml^2}{2}}\int_{-\pi}^{\pi}\frac{d\theta}{\sqrt{E + V_0\cos\theta}}. \tag{4.135}$$

In the higher-energy (rotor-like) limit, where $E = I\omega^2/2 \gg V_0$ we have

$$\tau \to \sqrt{\frac{ml^2}{2}}\frac{2\pi}{\sqrt{E}} = \frac{2\pi}{\omega}. \tag{4.136}$$

4.9.2   Quantum Result

The Schrödinger equation for the quantum pendulum is given by

$$-\frac{\hbar}{2I}\psi_{\theta\theta} - V_0\cos\theta\,\psi = E\psi\,. \tag{4.137}$$

Introducing the change of variable $\theta = 2x$ and defining $a = 8IE/\hbar^2$ and $q = -4IV_0/\hbar^2$ Eq. (4.137) becomes

$$\psi_{xx} + (a - 2q\cos x)\psi = 0\,. \tag{4.138}$$

Equation (4.138) is identified as angular Mathieu equation. Since $\theta = 2x$ the boundary condition is $\psi(2x) = \psi(2(x+\pi))$. The eigenfunction has to be periodic in x with period π.

Because of the intrinsic parity of the potential, there are two types of solutions. Even solutions $ce_r(x,q)$ (cosine-elliptic) for $r = 0, 2, 4, \ldots$ and odd solutions, $se_r(x,q)$ (sine-elliptic) for $r = 1, 3, \ldots$ [4]. ce_r and se_r are expanded in terms of Fourier series. They are given below [20]:

$$ce_{2m}(x,q) = \sum_{k=0}^{\infty} A_{2k}(q)\cos 2kx\,, \tag{4.139a}$$

$$ce_{2m+1}(x,q) = \sum_{k=0}^{\infty} A_{2k+1}(q)\cos(2k+1)x\,, \tag{4.139b}$$

$$se_{2m+2}(x,q) = \sum_{k=0}^{\infty} B_{2k+2}(q)\sin(2k+2)x\,, \tag{4.139c}$$

$$se_{2m+1}(x,q) = \sum_{k=0}^{\infty} B_{2k+1}(q)\sin(2k+1)x\,, \tag{4.139d}$$

where $m = 0,1,2,\ldots$. The recurrence relations between the coefficients are found by substituting the above series in Eq. (4.138). For example, for ce_{2m} the recurrence relations are [20]

$$aA_0 = qA_2\,, \quad (a-4)A_2 = q(2A_0 + A_4)\,, \tag{4.140a}$$

$$\left[a - (2k)^2\right]A_{2k} = q(A_{2k-2} + A_{2k+2})\,, \tag{4.140b}$$

where $k \geq 2$. When $q = 0$ the solutions (after normalization) are

$$ce_0 = 1/\sqrt{2}, \quad ce_r = \cos rx, \ r \geq 1, \quad se_r = \sin rx, \ r \geq 1. \tag{4.141}$$

For fixed values of q, periodic solutions with period π or 2π exist only for specific values (characteristic values) of a and are denoted by a_r for the even solutions. For odd solutions the characteristic values are, say, b_r. Mathematical packages, such as Mathematica, are useful to evaluate the values of a and b. It is quite difficult to compute Mathieu functions. For discussion on computation of Mathieu functions one may refer to refs. [21–23].

Figure 4.9 shows the plot of first few eigenvalues (a) as a function of q [18]. The eigenvalues are doubly-degenerate when $q = 0$ and partially small values of q. Observe the similarity of the probability density functions of the first three energy levels [18] shown in Fig. 4.9 with the harmonic oscillator.

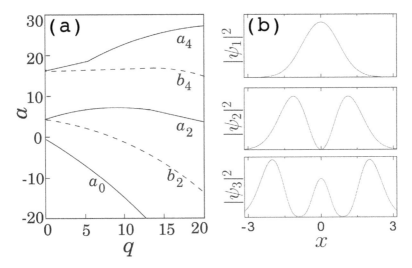

FIGURE 4.9
Eigenvalues a as a function of q and probability density function for the first three energy levels of the Mathieu's equation.

4.10 Criteria for the Existence of a Bound State

Is it possible to know whether a given potential admits at least one bound state? The answer is not positive for arbitrary potential. However, for certain potentials the conditions for existence of a bound state have been obtained [24]. In this section, we consider the one-dimensional potential not vanishing identically but vanishing everywhere outside some interval:

$$V(x) \neq 0 \quad \text{and} \quad V(x) = 0 \quad \text{for} \quad |x| > a \,. \tag{4.142}$$

For this type of potential the conditions for the existence of a bound state are obtained [24].
For a bound state

$$\langle H \rangle = \int_{-\infty}^{\infty} \psi^* H \psi \, \mathrm{d}x \Big/ \int_{-\infty}^{\infty} \psi^* \psi \, \mathrm{d}x \; < \; 0 \,, \tag{4.143}$$

where ψ is unnormalized and integrations are performed over the interval $-\infty < x < \infty$. Denoting the above quotient by N/D and since D is positive, the sufficient condition (4.143) becomes

$$D\langle H \rangle = N = \int_{-\infty}^{\infty} \left(-\frac{1}{2} \psi \psi_{xx} + V \psi^2 \right) \mathrm{d}x < 0 \,, \tag{4.144}$$

where we have set $\hbar = m = 1$ for simplicity. Integration of the first term in the integral by parts gives

$$D\langle H \rangle \; = \; N = \int_{-\infty}^{\infty} \left(\psi_x^2 + V \psi^2 \right) \mathrm{d}x < 0 \,. \tag{4.145}$$

Let us choose the normalizable trial eigenfunction as

$$\psi(x) = P(x) + \lambda V(x) \,, \tag{4.146}$$

where λ is a real parameter, $-\infty < \lambda < \infty$, and $P(x)$ is the Pyramid function

$$P(x) = \begin{cases} 1, & \text{if } 0 \le |x| \le a \\ 1 - (|x| - a)/L, & \text{if } a < |x| \le L + a \\ 0, & \text{if } L + a < |x| < \infty. \end{cases} \tag{4.147}$$

ψ given by Eq. (4.146) is normalizable for $L > 0$. Substituting Eq. (4.146) in Eq. (4.145) we obtain

$$D\langle H \rangle = A\lambda^2 + B\lambda + C + \frac{2}{L}, \tag{4.148a}$$

where

$$A = 2 \int_{-\infty}^{\infty} \left(V'^2 + V^3 \right) \, \mathrm{d}x, \quad B = 2 \int_{-\infty}^{\infty} V^2 \, \mathrm{d}x > 0, \quad C = \int_{-\infty}^{\infty} V \, \mathrm{d}x \tag{4.148b}$$

and $V' = \mathrm{d}V/\mathrm{d}x$. Since A, B and C are independent of L, for large L, the condition (4.148a) is written as

$$D\langle H \rangle = A\lambda^2 + B\lambda + C < 0. \tag{4.149}$$

Therefore, a bound state must exist if the above condition is satisfied for some values of λ. From Eqs. (4.148) and (4.149) we infer the following:

1. When $\lambda = 0$, ψ becomes $P(x)$ and the sufficiency condition is [25,26]

$$\int_{-\infty}^{\infty} V(x) \, \mathrm{d}x < 0. \tag{4.150}$$

2. If $A \le 0$, then Eq. (4.149) is satisfied for some λ of sufficiently large magnitude.
3. If $A > 0$ then Eq. (4.149) is satisfied if $C < B^2/(4A)$.

Combining the above results, the Hamiltonian H will have a bound state if one of the following conditions is realized:

$$(1) \quad \int_{-\infty}^{\infty} \left(V'^2 + V^3 \right) \, \mathrm{d}x \le 0, \quad A < 0. \tag{4.151a}$$

$$(2) \quad \int_{-\infty}^{\infty} V \, \mathrm{d}x \le \frac{\left(\int_{-\infty}^{\infty} V^2 \mathrm{d}x \right)^2}{\int_{-\infty}^{\infty} \left(V'^2 + V^3 \right) \mathrm{d}x}, \quad C < \frac{B^2}{4A}, \quad A > 0. \tag{4.151b}$$

$$(3) \quad \int_{-\infty}^{\infty} V \, \mathrm{d}x \le 0. \tag{4.151c}$$

The third condition is due to the fact that either (4.151a) is true or the right-side of (4.151b) is positive. Further, it includes the case of equality. Thus, for any one-dimensional potential, $V(x) \ne 0$, which vanishes outside some interval, the system must have at least one bound state if $\int_{-\infty}^{\infty} V(x) \, \mathrm{d}x = 0$, (the potential is zero on the average). The condition (4.151c) is weaker.

It has been pointed out that a weak attractive potential in two-dimensions can generate a bound state [27]. The problem of the existence of bound states in one- and two-dimensions for N-particle systems has also been examined [28]. The mathematical and physical properties for the occurrence and nonoccurrence of bound states in a square-well potential have been discussed [29].

4.11 Time-Dependent Harmonic Oscillator

In sec. 4.4 we studied the time-independent Schrödinger equation of LHO. Now, we solve the time-dependent Schrödinger equation of it given by

$$i\hbar\frac{\partial\psi}{\partial t} = -\frac{\hbar^2}{2m}\frac{\partial^2\psi}{\partial x^2} + \frac{1}{2}kx^2\psi. \tag{4.152}$$

Let us write the solution as

$$\psi(x,t) = \sum_{n=0}^{\infty} a_n\phi_n(x)\,e^{-iE_n t/\hbar} = e^{-i\omega t/2}\sum a_n\phi_n(x)\,e^{-in\omega t}, \tag{4.153}$$

where $\phi_n(x)$ is the solution of the time-independent Schrödinger equation and a_n are the constants to be determined. Assume that at $t = 0$, the solution is

$$\psi(x,0) = \sum a_n\phi_n(x) = \left(\frac{\alpha}{\sqrt{\pi}}\right)^{1/2} e^{-\alpha^2(x-b)^2/2}, \tag{4.154}$$

where the centre of the wave packet is shifted by an amount b, $b > 0$. To calculate a_m, multiply Eq. (4.154) by ϕ_m^* and integrate over x. The result is

$$a_m = \left(\frac{\alpha}{\sqrt{\pi}}\right)^{1/2} N_m^* \int_{-\infty}^{\infty} e^{-(\alpha x-\alpha b)^2/2}\,H_m(\alpha x)\,e^{-\alpha^2 x^2/2}\,dx, \tag{4.155}$$

where we substituted the expression for $\phi_m(x)$. Introducing the change of variable $\alpha x = \xi$ and $\alpha b = 2\xi_0$ Eq. (4.155) becomes

$$a_m = \left(\frac{1}{\alpha\sqrt{\pi}}\right)^{1/2} N_m^* e^{-\xi_0^2} \int_{-\infty}^{\infty} H_m(\xi)\,e^{-(\xi-\xi_0)^2}\,d\xi. \tag{4.156}$$

The value of the integral is $2^m\sqrt{\pi}\xi_0^m$. Then

$$a_m = \left(\frac{2^m}{m!}\right)^{1/2} \xi_0^m e^{-\xi_0^2}. \tag{4.157}$$

Next,

$$\begin{aligned}
\psi(x,t) &= e^{-i\omega t/2}\sum a_n\phi_n(x)e^{-in\omega t} \\
&= \left(\frac{\alpha}{\sqrt{\pi}}\right)^{1/2} e^{-\frac{1}{2}\xi^2-\xi_0^2-\frac{1}{2}i\omega t}\sum\frac{H_n(\xi)}{n!}\left(\xi_0 e^{-i\omega t}\right)^n.
\end{aligned} \tag{4.158}$$

Using

$$\sum_{n=0}^{\infty}\frac{H_n(\xi)s^n}{n!} = e^{-s^2+2s\xi} \tag{4.159}$$

we obtain

$$\psi(x,t) = \left(\frac{\alpha}{\sqrt{\pi}}\right)^{1/2}\exp\left[-\frac{1}{2}\xi^2 - \xi_0^2 - \frac{1}{2}i\omega t - \xi_0^2 e^{-2i\omega t} + 2\xi_0\xi e^{-i\omega t}\right]. \tag{4.160}$$

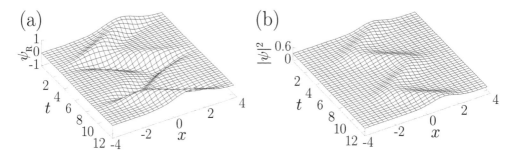

FIGURE 4.10
Plots of ψ_R and $|\psi|^2$ as a function of x and t for $\alpha = b = \omega = 1$.

Simplifying the expression for $\psi(x, t)$ in the above equation results in

$$
\psi(x, t) = \left(\frac{\alpha}{\sqrt{\pi}}\right)^{1/2} \exp\left[-\frac{1}{2}\left(\xi - 2\xi_0 \cos \omega t\right)^2\right.
$$
$$
\left. -i\left(\frac{\omega t}{2} + 2\xi_0\xi \sin \omega t - \xi_0^2 \sin 2\omega t\right)\right]. \tag{4.161}
$$

Now,

$$
P(x, t) = |\psi|^2 = \frac{\alpha}{\sqrt{\pi}} e^{-\alpha^2 (x - b \cos \omega t)^2}. \tag{4.162}
$$

Figure 4.10 shows the variation of ψ_R (real part of ψ) and $|\psi|^2$ as a function of x and t. We find that both ψ_R and $|\psi|^2$ are oscillating with time. $|\psi|^2$ oscillates about $x = b$ without change of shape.

4.12 Damped and Forced Linear Harmonic Oscillator

To study nonconservative and damped systems we need to introduce time-dependent Hamiltonian describing damped oscillations. Analysis of damped oscillators, particularly, linearly damped harmonic oscillators, started from 1931 [30]. Since then a great deal of interest has been paid to damped quantum mechanical systems [31–39].

Consider the classical Hamiltonian of a particle of mass m in one-dimension as

$$
H(x, p, t) = \frac{1}{2m}e^{-dt}p^2 + \frac{1}{2}m\omega_0^2 e^{dt}x^2 - mf(t)e^{dt}x, \tag{4.163}
$$

where $d > 0$ and $\omega_0^2 > 0$ are constant parameters and $f(t)$ is a function of time. The classical equation of motion is

$$
\dot{x} = \frac{1}{m}e^{-dt}p, \quad \dot{p} = -\omega_0^2 m e^{dt}x + mf(t)e^{dt}. \tag{4.164}
$$

Equation (4.164) can be rewritten as

$$
\ddot{x} + d\dot{x} + \omega_0^2 x = f(t). \tag{4.165}
$$

Equation (4.165) is the classical damped and forced LHO. We can obtain useful information of the unforced system by identifying its equilibrium point and its stability [40]. For $d \neq 0$ and $f(t) = 0$ the system (4.165) has only one equilibrium point and is $(x^*, \dot{x}^*) = (0,0)$. This equilibirum point is *always stable*. The trajectories started in the neighbourhood of this equilibrium point approach it in the limit of $t \to \infty$. However, the nature of trajectories near the equilibrium point depends on the sign of the quantity $d^2 - 4w_0^2$.

In the under-damped case $(d^2 - 4w_0^2 < 0)$, that is, $0 < d < 2w_0$ the trajectories in the $(x - \dot{x})$ phase space wind around the equilibrium point a number of times before reaching it asymptotically. The equilibrium point is called a *stable focus*. When $d^2 - 4w_0^2 > 0$, that is, $d > 2w_0$ (over-damped case) the trajectories in the phase space approach the equilibrium along parabolic paths. Such an equilibrium point is termed as a *stable node*. For the choice $d = 2w_0$ corresponding to the critically damping the trajectories approach the equilibirum point along straight-line paths. This type of equilibrium point is called a *stable focus*. Consequently, we have the following three case:

1. Under-damped system: $d^2 - 4w_0^2 < 0$.
2. Over-damped system: $d^2 - 4w_0^2 > 0$.
3. Critical-damped system: $d^2 = 4w_0^2$.

4.12.1 Solution of the Classical System

The solution of the system (4.165) for the above mentioned three cases can be obtained by Green's function approach [34]. For the under-damped case the solution of Eq. (4.165) is

$$x(t) = e^{-dt/2} \left[C_1 e^{i\omega t} + C_2 e^{-i\omega t} \right] + x_\mathrm{p}(t), \tag{4.166a}$$

where $\omega^2 = -\dfrac{d^2}{4} + w_0^2$ and the particular solution x_p, which is independent of initial conditions is obtained from Green's function approach as

$$x_\mathrm{p}(t) = \int_{t_0}^t G(t,t')f(t')\mathrm{d}t'. \tag{4.166b}$$

The Green's function G is given by

$$G(t,t') = \frac{1}{\omega} e^{-d(t-t')/2} \sin \omega(t - t'). \tag{4.166c}$$

In the case of over-damped case with $\omega^2 = \dfrac{d^2}{4} - w_0^2$, the solution is

$$x(t) = e^{-dt/2} \left[C_1 e^{\omega t} + C_2 e^{-\omega t} \right] + x_\mathrm{p}(t), \tag{4.167a}$$

where

$$x_\mathrm{p}(t) = \int_{t_0}^t G(t,t')f(t')\mathrm{d}t', \quad G(t,t') = \frac{1}{\omega} e^{-d(t-t')/2} \sinh \omega(t - t'). \tag{4.167b}$$

The solution for the critical-damped system is

$$x(t) = e^{-dt/2} \left[C_1 + C_2 t \right] + x_\mathrm{p}(t), \tag{4.168a}$$

where

$$x_\mathrm{p}(t) = \int_{t_0}^t G(t,t')f(t')\mathrm{d}t', \quad G(t,t') = (t - t')e^{-d(t-t')/2}. \tag{4.168b}$$

In all the three cases t_0 is the initial time and C_1 and C_2 are to be determined from the given initial conditions. Construction of x_p for $f(t) = \cos \Omega t$ is left as an exercise for readers.

4.12.2 Eigenfunctions of Quantum System

Eigenspectra and wave functions for the quantum mechanical damped and forced LHO for the under-damped, over-damped and critical-damped cases can be obtained separately. In the following we consider the under-damped case [34]. For the other cases see the ref.[34].

For a classical Hamiltonian system if an *invariant quantity* (constant of motion) $I(x, p, t)$ other than the Hamiltonian exists then it can be determined. Further,

$$\frac{dI}{dt} = \frac{\partial I}{\partial t} + \{I, H\} = \frac{\partial I}{\partial t} + \frac{\partial I}{\partial x}\frac{\partial H}{\partial p} - \frac{\partial I}{\partial p}\frac{\partial H}{\partial x} = 0. \tag{4.169}$$

One can assume $I = a(t)x + b(t)p + c(t)$ and using Eq. (4.169) the quantities a, b and c can be determined. The above I is linear in x and p. We can also assume that I is quadrative in x and p.

For a quantum mechanical system a quantum invariant operator satisfies the Eq. (7.12). The eigenvalue equation for an I is

$$I(x, p, t)\phi(x, t) = \lambda\phi(x, t). \tag{4.170}$$

The general solutions of the Schrödinger equation of a system with $H(x, p, t)$ for discrete eigenvalues are related to ϕ_n as

$$\psi(x, t) = \sum_n c_n e^{\gamma_n(t)}\phi_n(x, t), \tag{4.171}$$

where c_n and $\gamma_n(t)$ are to be determined. There are innumerable number of I satisfying Eq. (4.170).

The quadratic invariant operator obtained for the quantum mechanical damped and forced LHO for the under-damped case is [33,34]

$$\begin{aligned} I &= \frac{1}{2}\left\{\frac{1}{m}e^{-dt}(p - p_{\mathrm{p}})^2 + me^{dt}\omega_0^2(x - x_{\mathrm{p}})^2 \right. \\ &\quad \left. + \frac{d}{2}\left[(x - x_{\mathrm{p}})(p - p_{\mathrm{p}}) + (p - p_{\mathrm{p}})(x - x_{\mathrm{p}})\right]\right\}, \end{aligned} \tag{4.172}$$

where x_{p} is the particular solution of the classical equation of motion and p_{p} is the corresponding canonical momentum.

We can express the I given by Eq. (4.172) in terms of appropriate ladder operators a and $a^\dagger$ [33,34]. We can attempt to factorize I given by Eq. (4.172) as

$$\begin{aligned} I &= \frac{1}{\sqrt{2m}}e^{-dt/2}\left(i(p - p_{\mathrm{p}}) + me^{dt}\left(\omega + i\frac{d}{2}\right)(x - x_{\mathrm{p}})\right) \\ &\quad \times \frac{1}{\sqrt{2m}}e^{-dt/2}\left(-i(p - p_{\mathrm{p}}) + me^{dt}\left(\omega - i\frac{d}{2}\right)(x - x_{\mathrm{p}})\right) + \epsilon_0. \end{aligned} \tag{4.173}$$

Compare the Eqs. (4.172) and (4.173) and find the value of ϵ_0. Equation (4.173) suggests that as in the case of undamped LHO considered in sec. 4.4.7 we can define

$$a = \frac{e^{-dt/2}}{\sqrt{2m\hbar\omega}}\left[i(p - p_{\mathrm{p}}) + me^{dt}\left(\omega + i\frac{d}{2}\right)(x - x_{\mathrm{p}})\right], \tag{4.174a}$$

$$a^\dagger = \frac{e^{-dt/2}}{\sqrt{2m\hbar\omega}}\left[-i(p - p_{\mathrm{p}}) + me^{dt}\left(\omega - i\frac{d}{2}\right)(x - x_{\mathrm{p}})\right]. \tag{4.174b}$$

For the above a and $a^\dagger$ we can find $[a, a^\dagger]$ as 1. Starting from $a^\dagger a$ we can arrive that $a^\dagger a = \dfrac{I}{\hbar\omega} - \dfrac{1}{2}$ (see the exercise 53 at the end of this chapter). That is,

$$I = \hbar\omega \left(a^\dagger a + \frac{1}{2} \right). \tag{4.175}$$

Refering to secs. 4.4.7.1 and 4.4.7.2 the ground state eigenfunction is obtained from $a\phi_0(x, t) = 0$ and the other eigenfunctions are given by (4.68) or (4.69). We obtain

$$\phi_0 = FH_0(X), \quad \phi_n = \left(\frac{1}{2^n n!} \right)^{1/2} FH_n(X), \tag{4.176a}$$

where

$$
\begin{aligned}
F &= \left(\frac{\alpha}{\sqrt{\pi}} \right)^{1/2} e^{dt/4} \exp\left[-\frac{\alpha^2}{2} e^{dt} (x - x_{\mathrm{p}})^2 \right] \\
&\quad \times \exp\left[-\frac{i}{\hbar} \frac{md}{4} e^{dt} (x - x_{\mathrm{p}})^2 \right] e^{ip_{\mathrm{p}}x/\hbar},
\end{aligned} \tag{4.176b}
$$

$$X = \alpha e^{dt/2}(x - x_{\mathrm{p}}), \quad \alpha = \sqrt{m\omega/\hbar}. \tag{4.176c}$$

Now, we need to find $\gamma_n(t)$ in Eq. (4.171). Substituting ψ given by Eqs. (4.171) and using Eq. (4.176a) in the time-dependent Schrödinger equation and solving the resulting equation for $\gamma_n(t)$ we obtain

$$\gamma_n(t) = -i\omega \left(n + \frac{1}{2} \right)(t - t_0) - \frac{i}{\hbar} \int_{t_0}^{t} e^{dt} \left(\frac{1}{2}m\dot{x}_{\mathrm{p}}^2 - \frac{1}{2}m\omega_0^2 x_{\mathrm{p}}^2 \right) dt, \tag{4.177}$$

where t_0 is the initial time at which the external driving force is turned-on.

4.12.3   Energy Eigenvalues

In terms of a and $a^\dagger$ we can express x and p as

$$x = \frac{1}{\sqrt{2}\alpha} e^{-dt/2}(a^\dagger + a) + x_{\mathrm{p}}, \tag{4.178a}$$

$$p = -\frac{i}{\sqrt{2}} e^{dt/2}\alpha\hbar \left[\left(1 - \frac{id}{2\omega} \right) a - \left(1 + \frac{id}{2\omega} \right) a^\dagger \right] + p_{\mathrm{p}}. \tag{4.178b}$$

We have

$$\langle x^2 \rangle = \frac{\hbar e^{-dt}}{m\omega} \left(n + \frac{1}{2} \right) + x_{\mathrm{p}}^2, \tag{4.179a}$$

$$\langle p^2 \rangle = \hbar m\omega e^{dt} \left(1 + \frac{d^2}{4\omega^2} \right) \left(n + \frac{1}{2} \right) + p_{\mathrm{p}}^2. \tag{4.179b}$$

From the first subequation of Eq. (4.164) we write $p = e^{dt}m\dot{x}$. The energy is

$$E = \frac{1}{2}m\dot{x}^2 + \frac{1}{2}kx^2 = \frac{1}{2m}e^{-2dt}p^2 + \frac{1}{2}m\omega_0^2 x^2. \tag{4.180}$$

Then the expectation value of the energy operator over the nth eigenstate is given by

$$E_n = \frac{1}{2m}e^{-2dt}\langle p^2 \rangle + \frac{1}{2}m\omega_0^2 \langle x^2 \rangle. \tag{4.181}$$

We obtain

$$
\begin{aligned}
E_n &= \frac{e^{-dt}\hbar}{8\omega}\left(4\omega^2 + 4\omega_0^2 + d^2\right)\left(n + \frac{1}{2}\right) + \frac{e^{-2dt}}{2m}p_{\mathrm{p}}^2 + \frac{1}{2}m\omega_0^2 x_{\mathrm{p}}^2 \\
&= \frac{e^{-dt}\hbar\omega_0^2}{\omega}\left(n + \frac{1}{2}\right) + \frac{e^{-2dt}}{2m}p_{\mathrm{p}}^2 + \frac{1}{2}m\omega_0^2 x_{\mathrm{p}}^2.
\end{aligned} \tag{4.182}
$$

For $t \gg 1/d$ the first term in E_n can be neglected and then E_n is independent of n. In the limit of $t \to \infty$ we have $E_n = E = \frac{1}{2}m\omega_0^2 x_{\mathrm{p}}^2$. This is the classical energy due to the particular solution and is conserved.

4.13   Two-Dimensional Systems

Having studied certain one-dimensional quantum systems, next, we take up two-dimensional systems. One novel feature of multi-dimensional systems is the realization of *degeneracy* in energy levels. That is, different eigenfunctions correspond to same energy eigenvalue. The source of degeneracy is an existence of symmetry in the potential.

We study the quantum mechanics of LHO and a particle in a box in two-dimension Cartesian coordinates. Also, we treat the LHO in two-dimension polar coordinates.

4.13.1   Linear Harmonic Oscillator in Two-Dimension Cartesian Coordinates

The potential of the LHO in two-dimension is $V(x,y) = \frac{1}{2}m\omega^2(x^2 + y^2)$, $\omega = \sqrt{k/m}$. The Schrödinger equation in two-dimension is

$$
\left(\frac{\partial^2}{\partial x^2} + \frac{\partial^2}{\partial y^2}\right)\psi + \frac{2m}{\hbar^2}\left(E - \frac{1}{2}k\left(x^2 + y^2\right)\right)\psi = 0. \tag{4.183}
$$

To solve this equation we use the method of separation of variables. Writing $\psi(x,y) = \psi_1(x)\psi_2(y)$, Eq. (4.183) becomes

$$
\psi_{1xx}\psi_2 + \psi_1\psi_{2yy} + \frac{2m}{\hbar^2}\left(E - \frac{1}{2}k\left(x^2 + y^2\right)\right)\psi_1\psi_2 = 0. \tag{4.184}
$$

Dividing Eq. (4.184) by $1/(\psi_1\psi_2)$ gives

$$
\frac{1}{\psi_1}\psi_{1xx} - \frac{mk}{\hbar^2}x^2 + \frac{1}{\psi_2}\psi_{2yy} - \frac{mk}{\hbar^2}y^2 + \frac{2mE}{\hbar^2} = 0. \tag{4.185}
$$

Assume that $E = E_x + E_y$. As ψ_1 is function of x only and ψ_2 is function of y only, from Eq. (4.185), we write

$$
\psi_{1xx} + \frac{2m}{\hbar^2}\left(E_x - \frac{1}{2}kx^2\right)\psi_1 = 0, \tag{4.186a}
$$

$$
\psi_{2yy} + \frac{2m}{\hbar^2}\left(E_y - \frac{1}{2}ky^2\right)\psi_2 = 0. \tag{4.186b}
$$

TABLE 4.1
Energy eigenvalues of the two-dimensional LHO in units of $\hbar\omega$ for first few values of n_x and n_y.

n_x	n_y	E	n_x	n_y	E	n_x	n_y	E	n_x	n_y	E
0	0	1	1	0	2	2	0	3	3	0	4
0	1	2	1	1	3	2	1	4	3	1	5
0	2	3	1	2	4	2	2	5	3	2	6
0	3	4	1	3	5	2	3	6	3	3	7

Equations (4.186a) and (4.186b) are the Schrödinger equations of the one-dimensional LHO. Referring to the sec. 4.4 we have

$$E_{n_x} = \left(n_x + \frac{1}{2}\right)\hbar\omega, \quad E_{n_y} = \left(n_y + \frac{1}{2}\right)\hbar\omega, \quad n_x, n_y = 0, 1, 2, \ldots . \qquad (4.187)$$

Therefore,

$$E_{n_x, n_y} = E_x + E_y = (n_x + n_y + 1)\hbar\omega, \quad n_x, n_y = 0, 1, 2, \ldots . \qquad (4.188)$$

Table 4.1 lists the energy eigenvalues for $n_x, n_y = 0, 1, 2, 3$. $E_{0,0} = \hbar\omega$ is nondegenerate. As $E_{1,1} = E_{1,0} = 2\hbar\omega$ the first excited state with $E = 2\hbar\omega$ is doubly degenerate. Other higher energy levels are also degenerate. nth excited level is $(n + 1)$-fold degenerate.

4.13.2 Harmonic Oscillator in Polar Coordinates

We now consider the LHO with mass μ in polar coordinate system [41]. With $x = r\cos\phi$, $y = r\sin\phi$ the Laplacian is

$$\nabla^2 = \frac{1}{r}\frac{\partial}{\partial r}\left(r\frac{\partial}{\partial r}\right) + \frac{1}{r^2}\frac{\partial^2}{\partial\phi^2}. \qquad (4.189)$$

Then the Schrödinger Eq. (4.183) becomes

$$\frac{1}{r}\frac{\partial}{\partial r}\left(r\frac{\partial\psi}{\partial r}\right) + \frac{1}{r^2}\frac{\partial^2\psi}{\partial\phi^2} + \frac{2\mu}{\hbar^2}\left[E - \frac{1}{2}kr^2\right]\psi = 0. \qquad (4.190)$$

Seeking the solution of the form $\psi(r, \phi) = f(r)g(\phi)$ gives

$$\frac{r^2}{f}f'' + \frac{r}{f}f' + \frac{2\mu r^2}{\hbar^2}\left[E - \frac{1}{2}kr^2\right] = -\frac{g''}{g}. \qquad (4.191)$$

In Eq. (4.191) left-side is function of r only and right-side is function of ϕ only. Therefore, both sides must be equal to same constant and let it be m^2. Then Eq. (4.191) gives

$$g'' + m^2 g = 0, \qquad (4.192a)$$

$$f'' + \frac{1}{r}f' + \left[\frac{2\mu E}{\hbar^2} - \frac{\mu^2\omega^2}{\hbar^2}r^2 - \frac{m^2}{r^2}\right]f = 0. \qquad (4.192b)$$

The solution of (4.192a) is $g = Ae^{im\phi} + Be^{-im\phi}$. The condition $g(0) = g(2\pi)$ gives $m = 0, \pm 1, \ldots$ and hence $g(\phi) = Ae^{im\phi}$.

Next, consider the asymptotic limits of Eq. (4.192b). In the limit of $r \to 0$, Eq. (4.192b) becomes

$$f'' + \frac{1}{r}f' - \frac{m^2}{r^2}f = 0. \tag{4.193}$$

Seeking its solution as $f \sim r^\alpha$ gives $r^{\alpha-2}(\alpha^2 - m^2) = 0$. That is, $\alpha = \pm m$. The choice $\alpha = m$ gives physically meaningful solution near $r = 0$. So, near $r = 0$ we write

$$f(r \to 0) \sim r^{|m|}. \tag{4.194}$$

In the other limit $r \to \infty$, Eq. (4.192b) takes the form

$$f'' - \frac{r^2}{\xi^2}f = 0, \quad \xi^2 = \frac{\hbar}{\mu\omega}. \tag{4.195}$$

For this equation in the limit of $r \to \infty$

$$f \sim e^{-r^2/(2\xi^2)}. \tag{4.196}$$

Introducing the change of variable $x = r/\xi$ and using $f' = f_x/\xi$, $f'' = f_{xx}/\xi^2$, $m = |m|$ Eq. (4.192b) becomes

$$f_{xx} + \frac{1}{x}f_x + \left(\frac{2E}{\hbar\omega} - x^2 - \frac{|m|^2}{x^2}\right)f = 0. \tag{4.197}$$

Referring to the solutions given by Eqs. (4.194) and (4.196) we assume the solution of (4.197) as

$$f(r) = x^{|m|}e^{-x^2/2}F(x). \tag{4.198}$$

Substitution of (4.198) in (4.197) gives

$$F'' + \left[\frac{2|m|+1}{x} - 2x\right]F' + \left[\frac{2E}{\hbar\omega} - 2(|m|+1)\right]F = 0. \tag{4.199}$$

Introduction of the change of variable $x^2 = y$ gives

$$2yF_{yy} + 2\left[|m| + \frac{1}{2} - y^2\right]F_y + \left[\frac{E}{\hbar\omega} - (|m|+1)\right]F = 0. \tag{4.200}$$

Power series solution can be constructed for Eq. (4.200). Assuming

$$F(y) = \sum_{n=0}^{\infty} a_n y^n \tag{4.201}$$

we obtain the recursion relation as

$$a_{n+1} = \frac{2n + |m| + 1 - (E/(\hbar\omega))}{2(n+1)(2n + |m| + 1)}a_n. \tag{4.202}$$

The finiteness of the wave function requires terminating series, that is, $a_{n+1} = 0$ for some integer value of n. This gives

$$E_{n,m} = (2n + |m| + 1)\hbar\omega. \tag{4.203}$$

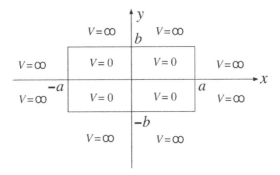

FIGURE 4.11
Two-dimensional box potential.

Define $n_1 = 2n$ and $n_2 = |m|$. Then

$$E_{n_1, n_2} = (n_1 + n_2 + 1)\hbar\omega. \tag{4.204}$$

In terms of n and m the choice $n = 0$, $m = 0$ corresponds to ground state. The ground state energy is $E_{0,0} = \hbar\omega$. When $n_1 = 0$, $n_2 = 1$ we have $E_{0,1} = 2\hbar\omega$. In this case $(n, m) = (0, \pm 1)$. Therefore, the first excited state is doubly degenerate. The second excited state is three-fold degenerate.

Another interesting two-dimensional system that can be solved in polar coordinates system is a charged particle of charge q and mass μ confined to a two-dimensional plane subjected to a magnetic field $B\mathbf{k}$ perpendicular to the plane. The Schrödinger equation of the system is

$$\frac{1}{2\mu}\left(\mathbf{p} - \frac{q}{c}\mathbf{A}\right)^2 \psi = E\psi, \quad A = (B/2)(-y, x, 0). \tag{4.205}$$

The bound state energy eigenvalues are [41]

$$E_{n,m} = (2n + |m| - m + 1)\frac{\hbar\omega_c}{2}, \quad \omega_c = \frac{qB}{\mu c}. \tag{4.206}$$

These energy levels are known as *Landau levels*.

4.13.3 Particle in a Two-Dimensional Box

We consider an electron in a two-dimensional box potential (Fig. 4.11). The potential V is zero inside the box and becomes ∞ outside the box. The Schrödinger equation for $|x| < a$ and $|y| < b$ is

$$\left(\frac{\partial^2}{\partial x^2} + \frac{\partial^2}{\partial y^2}\right)\psi(x, y) = -\frac{2mE}{\hbar^2}\psi(x, y) \tag{4.207}$$

and $\psi = 0$ for $|x| \geq a$ and $|y| \geq b$. The relevant boundary conditions are

$$\psi(x = -a, y) = \psi(x = a, y) = 0, \quad \psi(x, y = -b) = \psi(x, y = b) = 0. \tag{4.208}$$

Substitution of $\psi(x, y) = \psi_1(x)\psi_2(y)$ in Eq. (4.207) and dividing it by $\psi_1(x)\psi_2(y)$ lead to the equation

$$\frac{1}{\psi_1}\psi_{1xx} + \frac{1}{\psi_2}\psi_{2yy} = -\frac{2mE}{\hbar^2}. \tag{4.209}$$

Writing $E = E_x + E_y$, $k_x^2 = 2mE_x/\hbar^2$, $k_y^2 = 2mE_y/\hbar^2$ and as ψ_1 is a function of x only and ψ_2 is function of y only we write

$$\psi_{1xx} + k_x^2 \psi_1 = 0, \quad \psi_{2yy} + k_y^2 \psi_2 = 0. \tag{4.210}$$

The solutions of Eqs. (4.210) are

$$\psi_1(x) = A_x \sin k_x x + B_x \cos k_x x, \quad \psi_2(y) = A_y \sin k_y y + B_y \cos k_y y. \tag{4.211}$$

Then

$$\psi(x,y) = (A_x \sin k_x x + B_x \cos k_x x)(A_y \sin k_y y + B_y \cos k_y y). \tag{4.212}$$

This ψ should satisfy the boundary conditions given by Eq. (4.208). Applying these boundary conditions we obtain

$$A_x \sin k_x a + B_x \cos k_x a = 0, \quad -A_x \sin k_x a + B_x \cos k_x a = 0, \tag{4.213}$$
$$A_y \sin k_y b + B_y \cos k_y b = 0, \quad -A_y \sin k_y b + B_y \cos k_y b = 0. \tag{4.214}$$

For nontrivial solution $A_x \neq 0$ or $B_x \neq 0$ and $A_y \neq 0$ or $B_y \neq 0$. These give

$$k_x = \frac{n_x \pi}{2a}, \quad n_x = 1,2,\ldots, \quad k_y = \frac{n_y \pi}{2b}, \quad n_y = 1,2,\ldots. \tag{4.215}$$

As k_x and k_y are discrete E is also discrete. When $k_x = n_x\pi/(2a)$ from Eqs. (4.213) and (4.214) we obtain $A_x \sin(n_x\pi/2) = 0$ and $B_x \cos(n_x\pi/2) = 0$. When n_x is an odd integer ($n_x = 1,3,\ldots$) $A_x = 0$ while B_x is arbitrary. For even integer values of n_x ($2,4,\ldots$) $B_x = 0$ while A_x is arbitrary. Similarly, for $n_y = 1,3,\ldots$ we have $A_y = 0$ while B_y is arbitrary and for $n_y = 2,4,\ldots$ A_y is arbitrary and $B_y = 0$. Therefore, we have the following four cases:

Case 1:

$$A_x = 0, B_x \neq 0, A_y = 0, B_y \neq 0, \quad k_x = n_x\pi/(2a), k_y = n_y\pi/(2b),$$
$$n_x = 1,3,\ldots, \quad n_y = 1,3,\ldots.$$

Case 2:

$$A_x = 0, B_x \neq 0, A_y \neq 0, B_y = 0, \quad k_x = n_x\pi/(2a), k_y = n_y\pi/(2b),$$
$$n_x = 1,3,\ldots, \quad n_y = 2,4,\ldots.$$

Case 3:

$$A_x \neq 0, B_x = 0, A_y = 0, B_y \neq 0 \quad k_x = n_x\pi/(2a), k_y = n_y\pi/(2b),$$
$$n_x = 2,4,\ldots, \quad n_y = 1,3,\ldots.$$

Case 4:

$$A_x \neq 0, B_x = 0, A_y \neq 0, B_y = 0 \quad k_x = n_x\pi/(2a), k_y = n_y\pi/(2b),$$
$$n_x = 2,4,\ldots, \quad n_y = 2,4,\ldots.$$

We obtain

$$E = E_x + E_y = \frac{\hbar^2\pi^2}{8ma^2}\left(n_x^2 + \alpha^2 n_y^2\right), \quad \alpha^2 = \frac{a^2}{b^2}. \tag{4.216}$$

Table 4.2 displays the energy eigenvalues for first few values of n_x and n_y for $\alpha^2 = 0.5$ and 1. Note the degeneracy of certain energy values.

TABLE 4.2
Presents the values of E of a particle in a two-dimensional box potential for $n_x = 1, 2, 3, 4$ and $n_y = 1, 2, 3, 4$ for $\alpha^2 = 0.5$ and 1 in units of $\hbar^2\pi^2/(8ma^2)$. Certain energy levels are degenerate.

n_x	n_y	E for		n_x	n_y	E for	
		$\alpha^2 = 0.5$	$\alpha^2 = 1$			$\alpha^2 = 0.5$	$\alpha^2 = 1$
1	1	1.5	2	3	1	9.0	10
1	2	3.0	5	3	2	11.0	13
1	3	5.5	10	3	3	13.5	18
1	4	9.0	17	3	4	17.0	25
2	1	4.5	5	4	1	16.5	17
2	2	6.0	8	4	2	18.0	20
2	3	8.5	13	4	3	20.5	25
2	4	12.0	20	4	4	24.0	32

4.14   Rigid Rotator

In this section, we focus on an interesting three-dimensional system. Consider two particles of masses m_1 and m_2 joined about an axis passing through a point O. The particle of mass m_1 is at a distance r_1 from O and the other particle is at a distance r_2 from O and $r_1 + r_2 = R$ (Fig. 4.12). Such a system is called a *rigid rotator* because the distances r_1 and r_2 are fixed. The magnitude of the vibration of a diatomic molecule is generally small compared with the equilibrium bond length so the rigid rotator is a good first approximation.

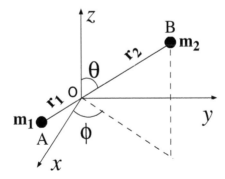

FIGURE 4.12
Parameters of the rigid rotator AB.

4.14.1   Hamiltonian and Schrödinger Equation

Let us denote the coordinates of the two particles with respect to the origin O as (x_1, y_1, z_1) and (x_2, y_2, z_2), respectively. The classical equation of motion for the rigid rotator is

$$\text{Kinetic Energy (K.E)} = \frac{1}{2}m_1 v_1^2 + \frac{1}{2}m_2 v_2^2 = \frac{1}{2}\left(m_1 r_1^2 + m_2 r_2^2\right)\omega^2 , \qquad (4.217)$$

where $v_1 = r_1\omega$, $v_2 = r_2\omega$ and ω is the angular frequency of rotation. Thus, K.E. $= I\omega^2/2$ where $I = m_1 r^2 + m_2 r_2^2$ is the moment of inertia. The potential of the system is zero because there is no external force on the system and hence $H = I\omega^2/2$.

The system is spherically symmetric. Therefore, it is convenient to deal it in spherical polar coordinates. In this coordinates system the coordinates of the particles are (a, θ, ϕ) and $(b, \pi - \theta, \pi + \phi)$, respectively. We have

$$\begin{aligned}
x_1 &= a\sin\theta\cos\phi, \quad x_2 = b\sin\theta\cos\phi, \quad y_1 = a\sin\theta\sin\phi, & (4.218\text{a})\\
y_2 &= b\sin\theta\sin\phi, \quad z_1 = a\cos\theta, \quad z_2 = -b\cos\theta. & (4.218\text{b})
\end{aligned}$$

The component of velocity of the particles are

$$v_{1x} = \frac{dx_1}{dt} = a\left(\cos\theta\cos\phi\frac{d\theta}{dt} - \sin\theta\sin\phi\frac{d\phi}{dt}\right), \qquad (4.219\text{a})$$

$$v_{1y} = \frac{dy_1}{dt} = a\left(\cos\theta\sin\phi\frac{d\theta}{dt} + \sin\theta\cos\phi\frac{d\phi}{dt}\right), \qquad (4.219\text{b})$$

$$v_{1z} = \frac{dz_1}{dt} = -a\sin\theta\frac{d\theta}{dt}. \qquad (4.219\text{c})$$

Then

$$v_1^2 = v_{1x}^2 + v_{1y}^2 + v_{1z}^2 = r_1^2\left(\dot{\theta}^2 + \sin^2\theta\,\dot{\phi}^2\right). \qquad (4.220)$$

Similarly $v_2^2 = r_2^2\left(\dot{\theta}^2 + \sin^2\theta\,\dot{\phi}^2\right)$. If we choose the centre of mass coordinate and the relative coordinate $\mathbf{r} = \mathbf{r}_2 - \mathbf{r}_1 = (r\sin\theta\cos\phi, r\sin\theta\sin\phi, r\cos\theta)$ then we can get

$$\text{K.E.} = \frac{1}{2}M\dot{R}^2 + \frac{1}{2}\mu r^2\left(\dot{\theta}^2 + \sin^2\dot{\phi}^2\right), \qquad (4.221)$$

where $M = m_1 + m_2$ and $\mu = m_1 m_2/(m_1 + m_2)$.

The centre of mass Hamiltonian has the solution $\psi = Ae^{i\mathbf{k}\cdot\mathbf{R}}$ with the translational energy $E_T = \hbar^2 k^2/(2M)$. As $\mu r^2 = I$, the moment of inertia, the kinetic energy due to the relative motion of the two masses is given by

$$E_R = \frac{1}{2}I\left(\dot{\theta}^2 + \sin^2\dot{\phi}^2\right) = -\frac{\hbar^2}{2\mu}\nabla^2. \qquad (4.222)$$

The Laplacian ∇^2 in spherical polar coordinates is

$$\nabla^2 = \frac{1}{r^2}\frac{\partial}{\partial r}\left(r^2\frac{\partial}{\partial r}\right) + \frac{1}{r^2\sin\theta}\frac{\partial}{\partial\theta}\left(\sin\theta\frac{\partial}{\partial\theta}\right) + \frac{1}{r^2\sin^2\theta}\frac{\partial^2}{\partial\phi^2}. \qquad (4.223)$$

For the rigid rotator r is constant so

$$\nabla^2 = \frac{1}{\sin\theta}\frac{\partial}{\partial\theta}\left(\sin\theta\frac{\partial}{\partial\theta}\right) + \frac{1}{\sin^2\theta}\frac{\partial^2}{\partial\phi^2}. \qquad (4.224)$$

The Hamiltonian is

$$H = -\frac{\hbar^2}{2I} \left[\frac{1}{\sin\theta} \frac{\partial}{\partial\theta} \left(\sin\theta \frac{\partial}{\partial\theta} \right) + \frac{1}{\sin^2\theta} \frac{\partial^2}{\partial\phi^2} \right]. \tag{4.225}$$

The orientation of the rigid rotator depends only on θ and ϕ. Therefore, its ψ depends only on θ and ϕ and is independent of r. The Schrödinger equation $H\psi = E\psi$ takes the form

$$-\frac{\hbar^2}{2I} \left[\frac{1}{\sin\theta} \frac{\partial}{\partial\theta} \left(\sin\theta \frac{\partial}{\partial\theta} \right) + \frac{1}{\sin^2\theta} \frac{\partial^2}{\partial\phi^2} \right] \psi(\theta,\phi) = E\psi(\theta,\phi) \tag{4.226}$$

or

$$\frac{1}{\sin\theta} \frac{\partial}{\partial\theta} \left(\sin\theta \frac{\partial\psi}{\partial\theta} \right) + \frac{1}{\sin^2\theta} \frac{\partial^2\psi}{\partial\phi^2} + \frac{2I}{\hbar^2} E\psi = 0. \tag{4.227}$$

The operator in the left-side of Eq. (4.226) is identified as $\mathbf{L}^2/2I$, where $\mathbf{L}^2$ is the square of the orbital angular momentum (see chapter 11). Equation (4.226) implies that the eigenvalues E of the rigid rotator are $(1/2I)$ times the eigenvalues of the operator $\mathbf{L}^2$.

4.14.2 Eigenfunctions and Eigenvalues

We show that the solution of Eq. (4.227) is simply the spherical harmonics. We obtain the solution by variable separable method. Assume the solution of the form

$$\psi(\theta,\phi) = \Theta(\theta)\Phi(\phi). \tag{4.228}$$

Substituting Eq. (4.228) in Eq. (4.227) and multiplying by $\sin^2\theta/(\Theta\Phi)$ we get

$$\frac{\sin\theta}{\Theta} \frac{\mathrm{d}}{\mathrm{d}\theta} \left(\sin\theta \frac{\mathrm{d}\Theta}{\mathrm{d}\theta} \right) + \frac{2IE}{\hbar^2} \sin^2\theta = -\frac{1}{\Phi} \frac{\mathrm{d}^2\Phi}{\mathrm{d}\phi^2}. \tag{4.229}$$

Since θ and ϕ are two independent variables the above equation is true only if both sides are equal to the same constant. We choose the constant as m^2. Then

$$-\frac{1}{\Phi} \frac{\mathrm{d}^2\Phi^2}{\mathrm{d}\phi^2} = m^2. \tag{4.230}$$

The solution of it is

$$\Phi_m = Ae^{\pm im\phi}. \tag{4.231}$$

A is determined from the normalization condition $\int_0^{2\pi} \Phi^*\Phi \,\mathrm{d}\phi = 1$. A is obtained as $1/\sqrt{2\pi}$. Equation (4.229) is now written as

$$-\frac{1}{\sin\theta} \frac{\mathrm{d}}{\mathrm{d}\theta} \left(\frac{\sin^2\Theta}{\sin\theta} \frac{\mathrm{d}\Theta}{\mathrm{d}\theta} \right) - \left(\frac{2IE}{\hbar^2} - \frac{m^2}{\sin^2\theta} \right) \Theta = 0. \tag{4.232}$$

Substituting

$$\cos\theta = x, \quad 2IE/\hbar^2 = l(l+1) \tag{4.233}$$

in Eq. (4.232) we have

$$(1 - x^2)\,\Theta'' - 2x\Theta' + \left[l(l+1) - \frac{m^2}{1 - x^2} \right] \Theta = 0 \tag{4.234}$$

which is the associated Legendre differential equation.

Let us find a solution of Eq. (4.234) of the form

$$\Theta(x) = \left(1 - x^2\right)^{m/2} X(x). \tag{4.235}$$

Now, Eq. (4.234) becomes

$$\left(1 - x^2\right) \frac{d^2 X}{dx^2} - 2(m+1)x \frac{dX}{dx} + \left[l(l+1) - m(m+1)\right] X = 0. \tag{4.236}$$

Differentiating the Legendre differential equation

$$\left(1 - x^2\right) \frac{d^2 P_l}{dx^2} - 2x \frac{dP_l}{dx} + l(l+1)P_l = 0 \tag{4.237}$$

m times we get

$$\left(1 - x^2\right) \frac{d^2}{dx^2} \left(\frac{d^m P_l}{dx^m}\right) - 2(m+1)x \frac{d}{dx} \left(\frac{d^m P_l}{dx^m}\right)$$

$$+ \left[l(l+1) - m(m+1)\right] \frac{d^m P_l}{dx^m} = 0. \tag{4.238}$$

Comparing the Eq. (4.238) with Eq. (4.236) we write

$$X(x) = \frac{d^m}{dx^m} P_l(x) \tag{4.239}$$

provided m assumes only positive integers. Then the solution $\Theta_l^{(m)}$ is

$$\Theta_l^{(m)}(x) = \left(1 - x^2\right)^{|m|/2} \frac{d^{|m|}}{dx^{|m|}} P_l(x) = P_l^{|m|}. \tag{4.240}$$

$P_l^{|m|}$ are called *associated Legendre polynomials*. We note that since Eq. (4.234) does not depend on sign of m, because $P_l(x)$ are polynomials of degree l, we have $d^{|m|} P_l/dx^{|m|} = 0$ if $|m| > l$. Therefore, $|m| \leq l$. The orthogonality relation for the associated Legendre polynomials $\Theta_l^{|m|}$ is

$$\int_{-1}^{1} \Theta_l^{|m|} \Theta_{l'}^{|m|} \, dx = \frac{(l - |m|)!}{(l + |m|)!} \frac{2}{2l + 1} \delta_{ll'} \tag{4.241}$$

which gives the normalization factor for $\Theta_l^{|m|}$. Now

$$\Theta_l^{|m|} \Phi_m = \left[\frac{(2l+1)(l - |m|)!}{4\pi(l + |m|)!}\right]^{1/2} P_l^{|m|}(\cos \theta) \, e^{im\phi}. \tag{4.242}$$

The right-side of Eq. (4.242) are the so-called *spherical harmonics* and are the eigenfunctions of the rigid rotator.

What are the allowed energy eigenvalues of the system? From Eq. (4.233)

$$E_l = \frac{\hbar^2}{2I} l(l+1), \quad l = 0, 1, \ldots. \tag{4.243}$$

The energy levels are thus discrete. The separation between two successive energy levels are given by

$$\triangle E_l = E_{l+1} - E_l = \frac{\hbar^2}{2I}[l(l+1) - l(l-1)] = \frac{\hbar^2}{I}(l+1). \tag{4.244}$$

The distance between successive energy levels increases with an increase in l. For $l = 0$ only one value of m is possible: $m = 0$. For other values of l there are $(2l + 1)$ allowed values of m: $-l, -l+1, \ldots, 0, \ldots, l-1, l$. Therefore, each energy level is $(2l + 1)$-fold degenerate. It can be shown that the allowed transitions between rotational states obey $\triangle l = \pm 1$ and $\triangle m = 0, \pm 1$.

Solved Problem 4:

A rigid body freely rotates in the $x - y$ plane. At $t = 0$ the wave function is assumed as $\psi(0) = C \sin^2 \phi$, where ϕ is the angle between the x-axis and the rotator axis. Express $\psi(0)$ in terms of the eigenfunctions $\Phi_m = \dfrac{1}{\sqrt{2\pi}} e^{im\phi}$, $m = 0, \pm 1, \pm 2, \ldots$.

In terms of Φ_m we can write an arbitrary wave function as

$$\psi(0) = \sum_m C_m \Phi_m = C_0 \Phi_0 + C_1 \Phi_1 + C_{-1} \Phi_{-1} + C_2 \Phi_2 + C_{-2} \Phi_{-2} + \ldots \ . \tag{4.245}$$

We can rewrite the given $\psi(0)$ as

$$
\begin{aligned}
\psi(0) &= C \sin^2 \phi \\
&= \frac{C}{2}(1 - \cos 2\phi) \\
&= \frac{C}{2} - \frac{C}{4}\left(e^{i2\phi} + e^{-i2\phi}\right) \\
&= \sqrt{2\pi}\left(\frac{C}{2}\Phi_0 - \frac{C}{4}\Phi_2 - \frac{C}{4}\Phi_{-2}\right).
\end{aligned}
\tag{4.246}
$$

Comparison of the above two equations for $\psi(0)$ gives

$$C_0 = C\sqrt{\frac{\pi}{2}}, \ \ C_1 = 0, \ \ C_{-1} = 0, \ \ C_2 = -\frac{C\sqrt{\pi}}{2\sqrt{2}}, \ \ C_{-2} = -\frac{C\sqrt{\pi}}{2\sqrt{2}} \tag{4.247}$$

and other C_i's are zero.

For the quantum mechanical treatment of hydrogen atom in three-dimension see chapter 12.

4.15   Concluding Remarks

Apart from the potentials considered in this chapter many other physically interesting potential problems have also been solved exactly. Solving some of them are given as exercises in the problem section. For example, a particle in a symmetric linear potential and the case of a constant acceleration in one-dimension with $V(x) = Fx$, $F > 0$ are solvable exactly and the solutions of the corresponding Schrödinger equation are expressed in terms of Airy function [1]. Another interesting potential is given by $V(x) = \infty$ for $x < 0$ and mgx for $x > 0$ and is equivalent to a bouncing ball. The solutions to this problem are the odd states of the linear potential [1,42]. Particle in a box with a delta-function potential [43] and in the presence of a static electric field [44] are solved exactly. The quantum states for a particle in the potential $V(x) = -\alpha/\epsilon^2$ if $0 < x < \epsilon$ and $-\alpha/x^2$ if $\epsilon < x < \infty$ [45], $V(x) = \frac{1}{2}kx^2 + \beta/x^2$ [46] and in the asymmetric Manning–Rosen Type potential [47] are obtained.

Ladder operators for Pöschl–Teller potentials [48] have been constructed. The stationary one-dimensional Schrödinger equation with a real polynomial potential $V(q)$ of degree > 2 on the real axis is studied [49,50]. There exists a class of potentials called *quasi-exactly solvable* [51–53] which depends on a parameter J. For these systems it is possible to find J number of eigenstates.

A method of constructing analytical solutions of the Schrödinger equation for $V(x) = \sum_m V_m f^m$, where V_m are constants and $f(x)$ is a function satisfying certain conditions are proposed [54]. The dynamical group structure of the one-dimensional quantum system with the infinitely deep square-well potential and its creation and annihilation operators are studied [55]. The influence of distant boundaries on quantum mechanical energy levels is also investigated. For a simple and explicit formula for the ground state energy change with application to LHO and hydrogen atom the readers may refer the ref. [56]. There exists some potentials tending to zero at infinity and yet possessing eigenfunctions with positive energy. For details about such potentials and a method to construct them see for example ref. [1]. There has been a considerable interest in recent years in obtaining exact and quasi-exact solutions of the position-dependent mass Schrödinger equation for various potentials and mass functions by using various methods [57]. A Green's function procedure to construct ground state and excited state solutions has been illustrated in refs. [58,59]. A variational method has been developed for obtaining energy eigenvalues of N-dimensional nonpolynomial potential systems [60]. In ref. [61] an approximate formulas for bound state energy eigenvalues of particle in a box potential and finite square-well potential are obtained without solving the Schrödinger equation.

The connection between the quantized energy eigenvalue spectrum of a bound state and the classical motions of the corresponding classical system has become increasingly interesting as the ability to experimentally probe the quantum-classical interface has dramatically improved. Methods such as periodic orbit theory [62,63] are very useful to explore this. *Can degenerate bound states occur in one-dimensional quantum systems?* It has been pointed out that bound states with degeneracy in energy but differing in parity can occur even if the potential is nonsingular in an finite domain [64].

4.16   Bibliography

[1]   R.W. Robinett, *Am. J. Phys.* 63:823, 1995.

[2]   I.D. Johnston, *Am. J. Phys.* 64:245, 1996.

[3]   I.N. Sneddon, *Special Functions of Mathematical Physics and Chemistry*. Longman, London, 1980.

[4]   M. Abramowitz and I.A. Stegun, *Hand Book of Mathematical Functions with Formulas, Graphs and Mathematical Tables*. Dover, New York, 1966.

[5]   G.B. Arfken and H.J. Weber, *Mathematical Methods for Physicists*. Academic Press, San Diego, 2005.

[6]   A.R. Usha Devi and H.S. Karthik, *Am. J. Phys.* 80:708, 2012.

[7]   L.S. Salter, *Am. J. Phys.* 58:961, 1990.

[8]   D.L. Aronstein and C.R. Stroud Jr., *Phys. Rev. A* 55:4526, 1997.

[9]   S.H. Dong and Z.Q. Ma, *Am. J. Phys.* 70:520, 2002.

[10]   J.R. Ackerhalt and P.W. Milonni, *Phys. Rev. A* 34:1211, 1986.

[11] M.E. Goggin and P.W. Milonni, *Phys. Rev. A* 37:796, 1988.

[12] D. Beigie and S. Wiggins, *Phys. Rev. A* 45:4803, 1992.

[13] A. Memboeuf and S. Aubry, *Physica D* 207:1, 2005.

[14] S. Flügge, *Practical Quantum Mechanics-I.* Springer, Berlin, 1971.

[15] G. Pöschl and E. Teller, *Z. Phys.* 83:143, 1933.

[16] J.P. Antoine, J.P. Gazeau, P. Monceau, J.R. Klauder and K.A. Penson, *J. Math. Phys.* 42:2349, 2001.

[17] M.A. Doncheski and R.W. Robinett, *Ann. Phys.* 308:578, 2003.

[18] G.L. Baker, J.A. Blackburn and H.J.T. Smith, *Am. J. Phys.* 70:525, 2002.

[19] T. Davis, *Introduction to Nonlinear Differential and Integral Equations.* Dover, New York, 1962.

[20] J.C.G. Vega, R.M.R. Dagnino, M.A.M. Nava and S.C. Cerda, *Am. J. Phys.* 71:233, 2003.

[21] A. Lindner and H. Freese, *J. Phys. A* 27:5565, 1994.

[22] D. Frenkel and R. Portugal, *J. Phys. A: Math. Gen.* 34:3541, 2001.

[23] F.A. Alhargan, *ACM Trans. Math. Software* 26:390, 2001.

[24] K.R. Brownstein, *Am. J. Phys.* 68:160, 2000.

[25] W.F. Buell and B.A. Shadwick, *Am. J. Phys.* 63:256, 1995.

[26] C.A. Kocher, *Am. J. Phys.* 45:71, 1977.

[27] M.L. Goldberger and B. Simon, *Ann. Phys.* 108:69, 1977.

[28] F.A.B. Coutinho, C.P. Malta and J.F. Perez, *Phys. Lett. A* 97:242, 1983.

[29] B. Sahu, M.Z. Rahmankhan, C.S. Shastry, B. Dey and S.C. Phatak, *Am. J. Phys.* 57:886, 1989.

[30] H. Bateman, *Phys. Rev.* 38:815, 1931.

[31] V.V. Dodonov and V.I. Man'ko, *Phys. Rev. A* 20:550, 1979.

[32] K.H. Yeon, K.K. Lec, C.I. Um, T.F. George and L.N. Pandey, *Phys. Rev. A* 48:2716, 1993.

[33] K.H. Yeon, D.H. Kim, C.I. Um, T.F. George and L.N. Pandey, *Phys. Rev. A* 55:4023, 1997.

[34] K.H. Yeon, S.S. Kim, Y.M. Moon, S.K. Hong, C.I. Um and T.F. George, *J. Phys. A: Math. Gen.* 34:7719, 2001.

[35] C.I. Um, K.H. Yeon, *J. Korean Phys. Soc.* 41:594, 2002.

[36] T.J. Li, *Cent. Eur. J. Phys.* 6:891, 2008.

[37] A.N. Ikot, L.E. Akpabio, I.O. Akpan, M.I. Umo and E.E. Ituen, *Int. J. Opt.* 2010:275910, 2010.

[38] T.G. Philbin, *New. J. Phys.* 14:083043, 2012.

[39] F. Kheirandish, *Eur. Phys. J. Plus* 135:243, 2020.

[40] M. Lakshmanan and S. Rajasekar, *Nonlinear Dynamics: Integrability, Chaos and Patterns.* Springer, Berlin, 2002.

[41] D. Banerjee and J.K. Bhattacharjee, *Eur. J. Phys.* 34:435, 2013.

[42] I. Gea-Banacloclin, *Am. J. Phys.* 67:776, 1999.

[43] I.R. Lapidus, *Am. J. Phys.* 55:172, 1987.

[44] D. Nguyen and T. Odagaki, *Am. J. Phys.* 55:466, 1987.

[45] T.X. Nguyen and F. Marsiglio, *Am. J. Phys.* 88:746, 2020.

[46] G. Palma and V. Raff, *Am. J. Phys.* 71:247, 2003

[47] A. Tas, S. Alpdogan and A. Havare, *Adv. High Energy Phys.* 2014:619241, 2014.

[48] S.H. Dong and R. Lemus, *Int. J. Quan. Chem.* 86:265, 2002.

[49] A. Voros, *J. Phys. A* 32:5993, 1999.

[50] P. Dorey and R. Tateo, *J. Phys. A* 32:L419, 1999 and references therein.

[51] C.M. Bender, G.V. Dunne, *J. Math. Phys.* 37:6, 1996.

[52] A. Turbiner, *Sov. Phys. JETP* 67:230, 1988.

[53] A.G. Ushveridze, *Quasi-exactly Solvable Models in Quantum Mechanics*. Institute of Physics, Briston, 1993.

[54] L. Skala, J. Cizek, J. Dvorak and V. Spirko, *Phys. Rev. A* 53:2009, 1996.

[55] S.H. Dong and Z.Q. Ma, *Am. J. Phys.* 70:520, 2002.

[56] G. Barton, A.J. Bray and A.J. Mckane, *Am. J. Phys.* 58:751, 1990.

[57] B. Midya and B. Roy, *J. Phys. A: Math. Gen.* 42:285301, 2009 and references therein.

[58] R.G. Winter, *Am. J. Phys.* 45:569, 1977.

[59] H.T. Williams, *Am. J. Phys.* 70:532, 2002.

[60] N. Saad, R.L. Hall and H. Ciftci, *J. Phys. A: Math. Gen.* 39:7745, 2006.

[61] V. Barsan, *Eur. J. Phys.* 36:065009, 2015.

[62] M. Gutzwiller, *The Interplay Between Classical and Quantum Mechanics*. Am. Assoc. Phys. Teachers College Park, Maryland, 2001.

[63] M. Brack and R.K. Bhaduri, *Semiclassical Physics*. Addison–Wesley, Reading, Massachusetts, 1997.

[64] S. Kar and R.R. Parwani, *Europhys. Lett.* 80:30004, 2007.

4.17   Exercises

4.1 Why does not a quantum system in its stationary energy states radiate energy?

4.2 Find the wave function and the energy eigenvalue for a free particle with potential $V = a$.

4.3 Using the representation of the δ-function $\delta(x) = \lim\limits_{a \to \infty} (\sin ax)/(\pi x)$ show that $\int_{-\infty}^{\infty} e^{i(k'-k)x}\,dx = 2\pi\delta(k'-k)$.

4.4 Write the time-independent Schrödinger equation for a free particle in three-dimensions. Solve it by the variable separable method.

4.5 If $\psi = (1/\hbar\pi)^{1/4}\,e^{-x^2/2\hbar}$ is an eigenfunction of the one-dimensional linear harmonic oscillator with potential $V(x) = x^2/2$ then obtain the expression for corresponding energy. Assume the mass of the particle as unity.

4.6 For a linear harmonic oscillator ground state eigenfunction compute $\langle x \rangle$ and $\langle p \rangle$.

4.7 For the linear harmonic oscillator with Hamiltonian $H = p_x^2/(2m) + m\omega^2 x^2/2$ calculate $\langle x \rangle$, $\langle x^2 \rangle$, $\langle p_x \rangle$ and $\langle p_x^2 \rangle$ in its nth state.

4.8 Find the expectation value of energy E when the state of the linear harmonic oscillator is described by the wave function $\psi = [\phi_0(x,t) + \phi_1(x,t)]/\sqrt{2}$, where ϕ_0 and ϕ_1 are the eigenfunctions of the ground state and first excited state, respectively.

4.9 For a linear harmonic oscillator using the expectation values of x^2 and p^2 show that $E \geq \hbar\omega/2$.

4.10 Calculate the probability of finding a linear harmonic oscillator with first excited state eigenfunction in a classically allowed region. Assume that $\int_0^{\sqrt{3}} y^2 e^{-y^2}\, dy \approx 0.39366$.

4.11 Show that the Hamiltonian $H = m\dot{q}^2/2 + m\omega^2 q^2/2$, by means of the transformation $q = \sqrt{\hbar/2m\omega}\left(ae^{-i\omega t} + a^\dagger e^{i\omega t}\right)$ and $[a, a^\dagger] = 1$, can be brought into the form $H = \hbar\omega\left(aa^\dagger + 1/2\right)$.

4.12 Show that the eigenfunctions ϕ_1 and ϕ_2 of linear harmonic oscillator are orthogonal.

4.13 If $u(x)$ is such that $\psi' + u\psi = 0$ then using the Schrödinger equation of linear harmonic oscillator show that u is the solution of the Riccati equation $u' - u^2 - \lambda + g(x) = 0$, $\lambda = 2mE/\hbar^2$ and $g = m^2\omega^2 x^2/\hbar^2$. Express ψ in terms of u. If $u = m\omega x/\hbar$ find E and ψ of the harmonic oscillator.

4.14 Consider the one-dimensional linear harmonic oscillator in an applied electric field. The Hamiltonian of the system is $H_e = p_x^2/(2m) + m\omega^2 x^2/2 - eEx$. Obtain the eigenfunction and energy eigenvalues of the system.

4.15 Show that a linear harmonic oscillator with the same energy and same amplitude as a particle in a box has frequency that is $2/\pi$ times the frequency of the particle in the box.

4.16 Show that the eigenvalues of $\hat{n} = a^\dagger a$ of the linear harmonic oscillator are 0 and 1 if $\{a, a\} = 0$, where $\{\,\}$ denotes anticommutator bracket.

4.17 The ground state energy and eigenfunction of a one-dimensional particle of unit mass in a potential $V(x)$ are $\hbar/2$ and $\phi_0 = (1/\hbar\pi)^{1/4}e^{-x^2/2\hbar}$, respectively. Calculate the potential.

4.18 The eigenfunction of a linear harmonic oscillator is of the form $\phi = N(a + x^2)e^{-bx^2/2}$. Determine a, b, N and the energy of the system.

4.19 Show by induction $(a^\dagger)^n g(x)$, where $a^\dagger$ is the raising operator of the linear harmonic oscillator, is proportional to

$$e^{\alpha^2 x^2/2}\frac{d^n}{dx^n}\left(e^{-\alpha^2 x^2/2}g(x)\right)$$

for any arbitrary function $g(x)$. Then establish the relation

$$\psi_n(x) = \left(\frac{\alpha}{n!2^n\sqrt{\pi}}\right)^{1/2} e^{-\alpha^2 x^2/2} H_n(\alpha x).$$

4.20 The time-dependent wave function of a linear harmonic oscillator at $t = 0$ is given by $\psi(0) = 1/\sqrt{2s}\sum_{n=N-s}^{N+s}\phi_n$, $N \gg s \gg 1$.

(a) Show that for large n $\langle x(t)\rangle = [2\hbar N/(m\omega)]^{1/2}\cos\omega t$.

(b) Compare the classical amplitude of the oscillation with the quantum mechanical amplitude.

4.21 Obtain the energy levels of three-dimensional linear harmonic oscillator described by the Hamiltonian $H = \mathbf{p}^2/(2m) + m(\omega_1^2 x^2 + \omega_2^2 y^2 + \omega_3^2 z^2)/2$. Show that the energy levels are degenerate when $\omega_1 = \omega_2 = \omega_3$ and the degree of degeneracy is $(n+1)(n+2)/2$, where $n = n_1 + n_2 + n_3$.

4.22 An electron of mass $m = 9.109 \times 10^{-31}$ kg is confined within a box with perfectly rigid walls. The width of the box is 2 nm$(= 2 \times 10^{-9}$ m$)$. If the electron is in the ground state, compute the energy.

4.23 Calculate the lowest energy gap for an electron of mass $m = 9.109 \times 10^{-31}$ kg moving in a box of length $2a$. What would be the total energy of the system for the energy spacing between two adjacent levels to be 1 J?

4.24 Obtain the electric dipole selection rule for allowed transitions for a particle in a box with the potential $V(x) = 0$ for $|x| < a$ and ∞ otherwise.

4.25 For a particle in the box potential $V(x) = 0$ for $|x| < a$ and ∞ otherwise calculate $\langle H \rangle$.

4.26 For a particle in a box potential $V(x) = 0$ for $|x| < a$ and ∞ otherwise calculate $\langle x \rangle$, $\langle x^2 \rangle$, $\langle p_x \rangle$ and $\langle p_x^2 \rangle$.

4.27 The normalized wave function of a particle in a box potential $V(x) = 0$ for $|x| < a$ and ∞ otherwise at time $t = 0$ is

$$\psi(x,0) = \frac{1}{\sqrt{10a}} \cos\left(\frac{\pi x}{2a}\right) \left[1 + 6\sin\left(\frac{\pi x}{2a}\right)\right], \quad |x| < a$$

and zero otherwise. Find the probability of finding the particle in the region $0 \le x \le a$.

4.28 Consider the wave function given in the previous problem.

(a) Obtain the wave function at $t = t_0 \ne 0$.
(b) Calculate the average energy of the particle at $t = 0$.

4.29 The wave function of a particle in an infinite square-well potential of width $2a$ in the interval $|x| < a$ is given by $\psi(x) = C\left[\cos(\pi x/2a) + \sin(\pi x/a)/4\right]$.

(a) Determine the coefficient C.
(b) What are the possible energy eigenvalues obtained when energy is measured?

4.30 The even eigenfunctions of a particle in a box with sides at $x = -a$ and $x = a$ are $\phi_n(x) = \frac{1}{\sqrt{a}} \cos\left[(2n-1)\frac{\pi x}{2a}\right]$, $n = 1, 2, \ldots$. An even normalized wave function is given by $\psi(x) = \left(\frac{15}{16a^5}\right)^{1/2} (a^2 - x^2)$ for $|x| \le a$ and $\psi(x) = 0$ for $|x| > a$. Expand $\psi(x)$ on the complete basis of eigenfunction. Also compute $\langle E \rangle$.

4.31 Show that the eigenfunctions $\phi_1 = \frac{1}{\sqrt{a}} \cos\frac{\pi x}{2a}$ and $\phi_3 = \frac{1}{\sqrt{a}} \cos\frac{3\pi x}{2a}$ of a particle in the box potential $V(x) = 0$ for $|x| < a$ and ∞ otherwise, are orthogonal.

4.32 For the infinite height potential $V(x) = 0$, $|x| < a$ and ∞ otherwise obtain the probability $P_{\mathrm{QM}}(b < x < b+c)$ in the limit of the quantum number $n \to \infty$.

4.33 The wave function of a particle in a box is $\psi(x) = N(a^2 - x^2)$ for $|x| < a$ and 0 for $|x| > a$. Normalize the wave function. Also, calculate the probability that a measurement of energy yields the eigenvalue E_n, $n = 1, 3, \ldots$.

4.34 An electron of mass $m = 9.109 \times 10^{-31}$ kg is moving in a one-dimensional box of size 1 Å. Determine the probability of finding it in the interval 0 to 0.5 Å for the states with the quantum numbers $n = 1$ and $n = 2$.

4.35 An electron in a box potential $V(x) = 0$ for $|x| < a = 0.5$ nm and ∞ otherwise absorbs the incident radiation and excited to the energy level $n = 2$ from $n = 1$. Calculate the wave length of the incident radiation.

4.36 Consider a gas molecule of molecular weight 50 units in a box of size 1 cm. Show that there is no quantization of energy.

4.37 The wave function of a quantum mechanical particle of mass m in a square-well potential $V(x) = -V_0$ for $0 < x < a$ and otherwise 0 is given by $\psi = \sin(\pi x/a)$. Calculate $\langle H \rangle$ at $t = 0$.

4.38 A particle in a box is found to be in the ground state. Suddenly the walls of the box at $x = -a$ and a are shifted to $x = -2a$ and $x = 2a$, respectively.

    (a) What is the probability that the particle to be found in the ground state of the expanded box?

    (b) What would happen if the shifting of the walls is adiabatic?

4.39 Consider the system in the previous problem.

    (a) Show that the probability for finding the particle in the first excited state $(n = 2)$ of the expanded box is zero.

    (b) Show that the probability for the transition to the second excited state of the expanded box is $64/(25\pi^2)$.

4.40 Find the energy eigenvalues and eigenfunctions of a particle in a three-dimensional box-potential.

4.41 Assuming the solution of the form $\psi(x, t) = e^{-\alpha(t)x^2 - \beta(t)}$ for the time-dependent linear harmonic oscillator equation determine $\alpha(t)$ and $\beta(t)$.

4.42 Consider the solved problem 4 in sec. 4.14. Find $\psi(t)$.

4.43 The Schrödinger equation for a rigid body that is constrained to rotate about a fixed axis with the moment of inertia I is $i\hbar \partial\psi/\partial t = -(\hbar^2/2I)\partial^2\psi/\partial\phi^2$, where $\psi(\phi, t)$ is a function of time and the angle of rotation about the axis. What boundary condition must be applied to the solution of this equation? Find the normalized energy eigenfunctions and eigenvalues. Is there any degeneracy?

4.44 Find the selection rules for electric dipole transition between the rotational levels of a rigid rotator.

4.45 An electron of mass m moves in a potential $V(x) = -a/x$, $x > 0$, $a = $ constant, $V(x) = \infty$, $x \le 0$. Construct the ground state eigenfunction and calculate ground state energy.

4.46 Consider the harmonic oscillator with the potential $V_-(x) = -kx^2/2$, $k > 0$. The difference between the linear harmonic oscillator potential $V(x) = kx^2/2$, $k > 0$ and the potential $V_-(x)$ is that the former has a stable equilibrium point while the latter has an unstable one. In other words, the motion of the particle is stable for $V(x)$ while it is unstable for $V_-(x)$. Solve the classical equation of motion and the Schrödinger equation. Also, obtain P_{CM} and P_{QM}.

4.47 The one-dimensional linear harmonic oscillator potential in the presence of a dipole-like interaction is written as [46] $V(x) = (m\omega^2/2)x^2 + (2m\alpha/\hbar^2)(1/x^2)$ with $\alpha > 0$. Determine the energy eigenfunctions and eigenvalues.

4.48 Solve the Schrödinger equation with the potential $V(x) = Fx$, $F > 0$ and then show that $P_{CM}(x) \propto 1/\sqrt{x}$ and $P_{QM}(x) \propto 1/\sqrt{x}$.

4.49 Find the unnormalized wave function of the Schrödinger equation with the potential $V(x) = F|x|$. Also, obtain the energy eigenvalues.

4.50 Consider a charged particle of charge q and mass μ confined to a two-dimensional plane subjected to a magnetic field $B\mathbf{k}$ perpendicular to the plane. The Schrödinger equation of the system is

$$\frac{1}{2\mu}\left(\mathbf{p} - \frac{q}{c}\mathbf{A}\right)^2 \psi = E\psi, \quad A = (B/2)(-y, x, 0).$$

Determine the energy eigenvalues of the system [41].

4.51 Consider the equation

$$f'' + \frac{1}{r}f' + \left[\frac{2\mu E}{\hbar^2} - \frac{m^2}{r^2} - \frac{\mu^2\omega^2}{\hbar^2}r^2\right]f = 0$$

which is realized in the study of harmonic oscillator in polar coordinates. Introducing the change of variable $f = u(r)/\sqrt{r}$ obtain the equation for u and identify the potential associated with the equation for u and the energy eigenvalues of the potential.

4.52 The simplified Hamiltonian of a free particle of charge e in a constant homogeneous magnetic field pointing in positive z-direction without any other potential is $H = H_0 - \omega L_z$, $H_0 = \frac{1}{2}\left(p_x^2 + \omega^2 x^2 + p_y^2 + \omega^2 y^2\right)$. Express H in terms of appropriate creation and annihilation operators.

4.53 For the damped and forced linear harmonic oscillator starting from $a^\dagger a$ with a and $a^\dagger$ given by Eq. (4.174) arrive the result $I = \hbar\omega\left(a^\dagger a + \frac{1}{2}\right)$.

4.54 For the classical Hamiltonian $H = \frac{1}{2}e^{-dt}p^2 + \frac{1}{2}\omega_0^2 e^{-dt}x^2$ construct the equation of motion and assuming $x = \eta(t)e^{i\gamma(t)}$ obtain the equations for η and γ.

4.55 For the classical Hamiltonian $H = \frac{1}{2}e^{-dt}p^2 + \frac{1}{2}\omega_0^2 e^{dt}x^2$ assuming an invariant as $I = \frac{1}{2}\alpha(t)p^2 + \beta(t)xp + \frac{1}{2}\delta(t)x^2$ determine α, β and δ.

4.56 For the a and $a^\dagger$ given by Eq. (4.174) of the damped and forced harmonic oscillator show that $\left[a, a^\dagger\right] = 1$.

4.57 Determine the energy eigenvalues and ground state eigenfunction of a particle in a box potential using ladder operators.

4.58 For the time-dependent harmonic oscillator starting from Eq. (4.160) obtain Eq. (4.161).

4.59 The wave function of a particle in a box is $\psi(x) = \sqrt{15/(16a^5)}(a^2 - x^2)$ for $|x| < a$ and 0 for $|x| > a$. Calculate the probability that a measurement of energy yields the eigenvalue E_n, $n = 2, 4, \ldots$.

4.60 An electron moves in a one-dimensional potential $V(x) = 0$ for $0 \le x \le a$ and ∞ otherwise. At $t = 0$ the normalized wave function is $\psi(x, 0) = \sqrt{30/a^5}\left(ax - x^2\right)$ for $0 \le x \le a$ and 0 otherwise. Express $\psi(x, t)$ in terms of the basis eigenfunctions $\sqrt{\frac{2}{a}}\sin\frac{n\pi x}{a}$. Assume that $E_n = n^2\pi^2\hbar^2/(2ma^2)$.

5

Exactly Solvable Systems II: Scattering States

5.1 Introduction

In the previous chapter, we focused our discussion on certain exactly solvable quantum mechanical systems possessing bound states. In the present chapter, we wish to concentrate on some fascinating and interesting one-dimensional systems, where the quantum states are essentially scattering states. For scattering states energy eigenvalues are continuous and energy eigenfunctions are nonnormalizable. An interesting feature of many scattering potentials is the occurrence of tunnelling. It is the most striking qualitative difference between classical mechanics and quantum mechanics. In tunnelling a quantum mechanical particle passes through a region where the potential energy exceeds the total energy of the particle. This phenomenon is impossible in classical mechanics. Classically, a particle is completely reflected from a barrier. Quantum tunnelling has many technological applications.

In the present chapter, we consider the following potentials and discuss the features of scattering states of them:

1. Barrier potential.
2. Square-well potential.
3. Step potential.
4. Locally periodic potentials.
5. Reflectionless potentials – sech^2 potentials.

We will construct the reflected and transmitted waves for the above potentials subjected to appropriate boundary conditions. We obtain the expressions for reflection and transmission amplitudes and analyze their dependence on the energy of the incident particle and the parameters of the potentials. The square-well and sech^2 potentials admit bound states also. We construct the bound state eigenfunctions and the energy eigenvalues for these potentials. Finally, we point out a new kind of tunnelling called *dynamical tunnelling* which refers to a motion allowed by quantum mechanics but forbidden by classical mechanics.

5.2 Potential Barrier: Tunnel Effect

As a first intriguing example of a scattering system, we consider a particle in the presence of a square-barrier potential (Fig. 5.1)

$$V(x) = \begin{cases} V_0, & \text{for } |x| < a \\ 0, & \text{for } |x| > a . \end{cases} \tag{5.1}$$

DOI: 10.1201/9781003172178-5

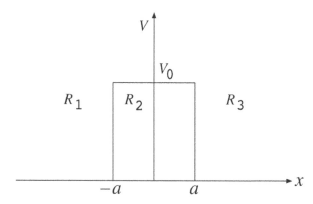

FIGURE 5.1
One-dimensional barrier potential.

5.2.1   Classical Solution

A classical particle moving towards the barrier will be completely reflected from it if $E < V_0$. For the particle in region R_1 moving towards the barrier from a position x_0 with velocity v, the position $x(t)$ is given by

$$x(t) = \begin{cases} x_0 + vt, & 0 < t < t_1 = (|x_0| - a)/v \\ -a - |v|(t - t_1), & t > t_1 . \end{cases} \tag{5.2}$$

Similar solution can also be written for the region R_3. The motion of the particle in the regions R_1 and R_3 is not confined to a finite region of space. It does not exhibit periodic oscillation. The motion is classically forbidden in the barrier region $|x| < a$. The problem thus turns out to be scattering of the particle by the potential barrier.

5.2.2   Solution of the Wave Equation for $E < V_0$

The Schrödinger equations in the regions R_1, R_2 and R_3 are given by

$$\begin{align}
\psi_{1xx} + k_0^2 \psi_1 &= 0, & x < -a & \tag{5.3a} \\
\psi_{2xx} - k^2 \psi_2 &= 0, & |x| < a & \tag{5.3b} \\
\psi_{3xx} + k_0^2 \psi_3 &= 0, & x > a & \tag{5.3c}
\end{align}$$

where $k_0^2 = 2mE/\hbar^2$, $k^2 = 2m(V_0 - E)/\hbar^2$. We treat the quantum cases with $E < V_0$ and $E > V_0$ separately.

The solutions of Eqs. (5.3) are

$$\psi(x) = \begin{cases} \psi_1 = Ae^{ik_0x} + Be^{-ik_0x} \\ \psi_2 = Fe^{-kx} + Ge^{kx} \\ \psi_3 = Ce^{ik_0x} + De^{-ik_0x}. \end{cases} \tag{5.4}$$

The functions e^{ik_0x} and e^{-ik_0x} in ψ_1 represent waves moving towards the barrier (from $x < -a$) and moving away from the barrier (towards $x = -\infty$), respectively. They are the incident and reflected waves. In the region R_3, the functions e^{ik_0x} and e^{-ik_0x} are the waves travelling away from the barrier towards $x = \infty$ and moving towards the barrier. The unknowns in Eqs. (5.4) are determined by applying the appropriate boundary conditions. Note that the wave functions given by Eqs. (5.4) are nonnormalizable.

The boundary conditions are set by requiring that ψ's and their first derivative are continuous at $x = a$ and $-a$. The boundary conditions are

$$\text{at } x = -a: \qquad \psi_1 = \psi_2 \text{ and } \psi_{1x} = \psi_{2x}, \tag{5.5a}$$

$$\text{at } x = a \quad: \qquad \psi_2 = \psi_3 \text{ and } \psi_{2x} = \psi_{3x}. \tag{5.5b}$$

These boundary conditions give four linear simultaneous equations for the six unknowns A, B, C, D, F and G. We need to fix the values of two unknowns. For a particle incident from left $D = 0$. As A represents the amplitude of the incident wave we can freely set the value of it in an experiment. Therefore, the value of A is treated as known. Then there are four equations for the four unknowns B, C, F and G.

The boundary conditions (5.5) lead to the following set of equations:

$$Ae^{-ik_0a} + Be^{ik_0a} = Fe^{ka} + Ge^{-ka}, \tag{5.6a}$$

$$Ae^{-ik_0a} - Be^{ik_0a} = iF\alpha e^{ka} - iG\alpha e^{-ka}, \quad \alpha = k/k_0, \tag{5.6b}$$

$$Ce^{ik_0a} = Fe^{-ka} + Ge^{ka}, \tag{5.6c}$$

$$Ce^{ik_0a} = iF\alpha e^{-ka} - iG\alpha e^{ka}. \tag{5.6d}$$

These equations can be solved by Gauss elimination method [1].

If J_{inc}, J_{ref} and J_{tra} represent incident, reflected and transmitted currents, respectively, then the reflected and transmitted coefficients are $R = J_{\text{ref}}/J_{\text{inc}}$ and $T = J_{\text{tra}}/J_{\text{inc}}$, where $J = (\hbar/m)\text{Im}(\psi^*\partial\psi/\partial x)$. With

$$\psi_{\text{inc}} = Ae^{ik_0x}, \quad \psi_{\text{ref}} = Be^{-ik_0x}, \quad \psi_{\text{tra}} = Ce^{ik_0x} \tag{5.7}$$

we find

$$J_{\text{inc}} = \frac{\hbar k_0}{m}|A|^2, \quad J_{\text{ref}} = -\frac{\hbar k_0}{m}|B|^2, \quad J_{\text{tra}} = \frac{\hbar k_0}{m}|C|^2, \tag{5.8}$$

where the negative sign in J_{ref} indicates the direction of flow of J is in the direction opposite to the incident direction and the sign can be dropped. Then the probability for the particle to be transmission, called *transmission amplitude*, is given by $T = J_{\text{tra}}/J_{\text{inc}} = |C|^2/|A|^2$. The reflection amplitude R is $R = J_{\text{ref}}/J_{\text{inc}} = |B|^2/|A|^2$.

C and B are obtained as (for the details of solving the Eqs. (5.6) see the solved problem 1 at the end of the subsection 5.2.4)

$$\begin{aligned} C &= \frac{2\alpha Ae^{-2ik_0a}}{2\alpha\cosh 2ka + i(\alpha^2 - 1)\sinh 2ka} \\ &= \frac{Ae^{-2ik_0a}}{\cosh 2ka + i(\epsilon/2)\sinh 2ka}, \quad \epsilon = \frac{k^2 - k_0^2}{kk_0} \end{aligned} \tag{5.9}$$

and

$$\begin{aligned} B &= -\frac{i(1 + \alpha^2)Ae^{-2ik_0a}\sinh 2ka}{2\alpha\cosh 2ka + i(\alpha^2 - 1)\sinh 2ka} \\ &= -\frac{iC\eta}{2}\sinh 2ka, \quad \eta = \frac{k^2 + k_0^2}{kk_0}. \end{aligned} \tag{5.10}$$

Then with the choice $A = 1$ the quantities T and R are worked out as

$$T = |C|^2 = \left[1 + \left(1 + \frac{\epsilon^2}{4}\right)\sinh^2 2ka\right]^{-1}, \tag{5.11a}$$

$$R = |B|^2 = \left[\frac{1}{4}\eta^2\sinh^2 2ka\right]T. \tag{5.11b}$$

Since the total probability must be 1 we require that $T + R = 1$. This can be easily verified from Eqs. (5.11). T is called *tunnelling probability*.

5.2.3 Solution of the Wave Equation for $E > V_0$

For $E > V_0$, $k^2 = 2m(V_0 - E)/\hbar^2 < 0$ while for $E < V_0$, $k^2 > 0$. Therefore, the results for $E > V_0$ can be obtained from the results of $E < V_0$ by replacing k by ik, sinh2ka by isin2ka. Then T and R are obtained as

$$T = \left[1 - \left(1 + \frac{\epsilon^2}{4}\right)\sin^2 2ka\right]^{-1}, \quad \epsilon = \frac{(-k^2 - k_0^2)}{ikk_0}, \tag{5.12a}$$

$$R = 1 - T. \tag{5.12b}$$

For a particle with energy $E > V_0$, Eq. (5.12c) predicts complete transmission when $\sin^2 2ka = 0$. In this case, $T_{\max} = 1$ and $2ka = n\pi$, $n = 1, 2, \ldots$. Such *resonances* in transmission correspond to the situation in which the scattered waves originating from $x = -a$ and $x = a$ interfere and exactly cancel any reflection from the potential barrier.

5.2.4 Quantum Tunnelling

A classical particle would be reflected from the barrier for $E < V_0$. In contrast, in quantum theory there is a finite transmission probability. That is, if a microscopic particle is incident on the barrier with energy relatively lower than the height of the potential barrier then it will not necessarily be reflected by the barrier. But there is always a finite nonzero probability for the particle to cross the barrier and continue its motion in the classically forbidden region R_3. This phenomenon of crossing the barrier is called *tunnel effect*. (When a metal surface is subjected to a large electric field the potential at the surface changes and some of the free electrons in the metal tunnel through the resulting barrier and are emitted. This is known as *field emission*.)

The tunnelling effect is one of the phenomena for which the difference between classical and quantum physics is most striking. It describes the propagation of a particle in a classically forbidden domain. (*What are the kinetic energy and the velocity of the particle in this zone?*) Without tunnelling, tunnel diode would not exist. The probability of tunnelling depends in general on (i) the energy E of the particle, (ii) the height V_0 of the barrier and (iii) the width $2a$ of the barrier.

Figure 5.2 illustrates the Re ψ for a few values of E, where $V_0 = 1$ and $a = 6$. The behaviour of Re ψ inside the barrier is not always an exponential decay. For $E = 0.6$, Re ψ exhibits exponential decay with x; for $E = 0.9$ it displays a power-law type variation; for $E = 1.0001$ (that is $\approx V_0$) it varies linearly; for $E = 1.1$ and 2 it shows oscillation.

Figure 5.3 shows T versus E/V_0 for two fixed values of $V_0 a^2$. For $E < V_0$ ($E/V_0 < 1$), T increases with increase in E. For $E > V_0$, an interesting feature is that T (as well as R) varies in an oscillatory fashion. It shows a series of *resonances* peaked at specific E/V_0 values where $T = 1$, the maximum allowed value. The resonances come from the multiple scattering effects within the barrier. (*What will happen if we consider a double-barrier or a finite number of square-barrier? For some details see sec. 5.5.*) For $2ka \gg 1$ Eq. (5.11a) becomes

$$T \approx \frac{16}{4 + \epsilon^2} e^{-4ka}. \tag{5.13}$$

The exponential factor comes from the exponential decay of the wave function from the incident side to the transmitted side a distance $2a$ away. The factors in front of the exponential term come from the matching conditions. T decreases exponentially with a and

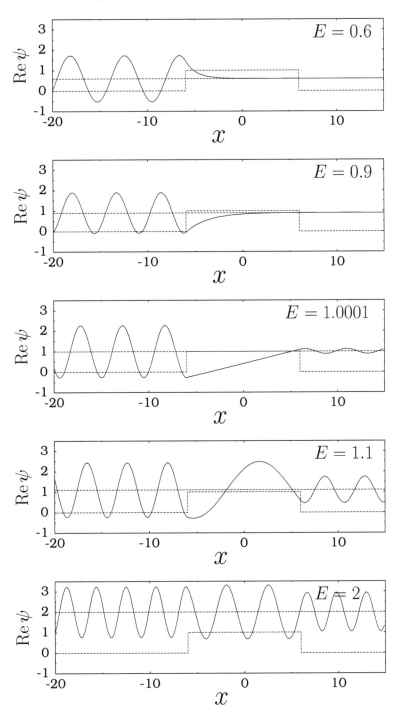

FIGURE 5.2

Re ψ is plotted for 5 values of E for the barrier potential (5.1) with $V_0 = 1$ and $a = 6$. For illustration Re ψ is shifted as Re $\psi + E$.

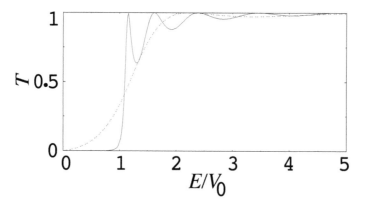

FIGURE 5.3
Variation of transmission amplitude T as a function of E/V_0 for $V_0 a^2 = 1$ (dashed curve)
and $V_0 a^2 = 8$ (continuous curve) for the barrier potential (5.1). For $E/V_0 < 1$, Eq. (5.11a)
is used to compute T. For $E/V_0 > 1$, T is calculated using Eq. (5.12a).

quantum tunnelling is extremely sensitive to small changes in a. This dependence of T on
a has been experimentally confirmed in the phenomenon of vacuum tunnelling [2].

We wish to know whether the oscillatory behaviour of T is peculiar to the rectangular
shape or characteristic of other potential barriers also. Oscillatory behaviour of T and R
has been observed in the truncated Gaussian potential [3]

$$V(x) = \begin{cases} V_0\, e^{-\beta^2 x^2}, & |x| < a \\ 0, & |x| > a \end{cases} \tag{5.14}$$

and truncated parabolic potential

$$V(x) = \begin{cases} V_0 \left(1 - \dfrac{x^2}{b^2}\right), & |x| < a \\ 0, & |x| > a, \end{cases} \tag{5.15}$$

where V_0, β and a are constant parameters. When the width of the potential barrier is
increased the magnitude of oscillations in T for $E < V_0$ is found to decrease. The results of
the potentials given by Eqs. (5.1), (5.14) and (5.15) suggest that the discontinuous change
in the potential and its slope essentially determine the relatively large variations in T that
occur for $E > V_0$. We expect from these results that the smaller the abrupt change in a
potential barrier, the smaller will be the variations in T for $E > V_0$.

Solved Problem 1:

Solve the Eqs. (5.6) and obtain the expressions for C and B.

We use the Gauss elimination method to solve Eqs. (5.6). First, we eliminate F in
Eqs. (5.6b)–(5.6d) using Eq. (5.6a). Equation (5.6b)−Eq. (5.6a)$\times i\alpha$, Eq. (5.6c)$\times e^{ka}$−
Eq. (5.6a)$\times e^{-ika}$ and Eq. (5.6d)$\times e^{ka}$−Eq. (5.6a)$\times i\alpha e^{-ika}$ give

$$A(1 - i\alpha)e^{-ik_0 a} - B(1 + i\alpha)e^{ik_0 a} = -2i\alpha G e^{-ka}, \tag{5.16a}$$

$$Ae^{-ik_0 a}e^{-ka} + Be^{ik_0 a}e^{-ka} - Ce^{ik_0 a}e^{ka} = -2G\sinh 2ka, \tag{5.16b}$$

$$Ai\alpha e^{-ik_0 a}e^{-ka} + Bi\alpha e^{ik_0 a}e^{-ka} - Ce^{ik_0 a}e^{ka} = 2Gi\alpha\cosh 2ka. \tag{5.16c}$$

Next, to eliminate G in Eqs. (5.16b) and (5.16c) consider Eq. (5.16b)$\times i\alpha e^{-ka}$ $-$Eq. (5.16a)$\times \sinh 2ka$ and Eq. (5.16c)$\times e^{-ka}$ $+$Eq. (5.16a)$\times \cosh 2ka$. The results are

$$Ae^{-2ik_0a}\left[(1-i\alpha)\sinh 2ka - i\alpha e^{-2ka}\right]$$
$$-B\left[(1+i\alpha)\sinh 2ka + i\alpha e^{-2ka}\right] + Ci\alpha = 0, \qquad (5.17a)$$

and

$$Ae^{-2ik_0a}\left[(1-i\alpha)\cosh 2ka + i\alpha e^{-2ka}\right]$$
$$-B\left[(1+i\alpha)\cosh 2ka - i\alpha e^{-2ka}\right] - C = 0. \qquad (5.17b)$$

Then (5.17b)$\times \left[(1+i\alpha)\sinh 2ka + i\alpha e^{-2ka}\right] - $(5.17a)$\times \left[(1+i\alpha)\cosh 2ka - i\alpha e^{-2ka}\right]$ gives after some simple mathematics

$$C = \frac{2\alpha A e^{-2ik_0a}}{2\alpha\cosh2ka + i(\alpha^2-1)\sinh2ka}. \qquad (5.18)$$

On the other hand, Eq.(5.17a)$+$Eq.(5.17b)$\times i\alpha$ gives

$$B = -\frac{i(1+\alpha^2)Ae^{-2ik_0a}\sinh2ka}{2\alpha\cosh2ka + i(\alpha^2-1)\sinh2ka}. \qquad (5.19)$$

Solved Problem 2:

A beam of electrons incident on the square-barrier potential, Eq. (5.1), from left. Find the fraction of electrons reflected and transmitted if the energy of the incident electrons is $V_0/2$ and $2a\sqrt{mV_0}/\hbar = 3/2$.

From Eq. (5.11a) the transmission amplitude T is given by

$$\frac{1}{T} = 1 + \left(1 + \frac{\epsilon^2}{4}\right)\sinh^2 2ka, \qquad (5.20a)$$

where

$$\epsilon = \frac{k^2 - k_0^2}{kk_0} = \frac{V_0 - 2E}{\sqrt{E(V_0-E)}}. \qquad (5.20b)$$

Substituting $E = V_0/2$ we get $\epsilon = 0$. Further,

$$2ka = \frac{2a}{\hbar}\sqrt{2m\left(V_0 - \frac{V_0}{2}\right)} = \frac{3}{2}. \qquad (5.21)$$

Writing $\sinh x = (e^x - e^{-x})/2$ we get

$$\sinh^2 2ka = \sinh^2 \frac{3}{2} = \left(\frac{e^{3/2} - e^{-3/2}}{2}\right)^2 = 4.534. \qquad (5.22)$$

Then $1/T = 1 + 4.534 = 5.534$. Therefore, $T = 0.181$ and $R = 0.819$.

5.2.5   Physical Systems Modelled by the Square-Barrier Potential

In 1838 Heinrich Friedrich Emil Lenz observed the freezing of water into ice on a bismuth-antimony junction when an electric current passed through the junction in one direction. When the direction of the current was reversed the melting of the ice into water was found. Consequently, a thermo-electric power generator works as a refrigerator. The fundamental quantum process of an electron moving across a potential barrier plays the vital role in thermionic power generation and refrigeration.

Consider two parallel metal electrodes with a vacuum gap between them acting as a barrier. The charge accumulation only affects the magnitude of the thermionic current. Because of the nature of the quantum transmission probability and the temperature behaviour of the Fermi distribution under certain conditions the thermionic current reverses its direction. This effect occurs in a system of two identical metallic conducting wires separated by a layer of insulating material [4]. There is no bias applied across the wires. The wires are connected to two thermal reservoirs kept at a higher temperature T_h and a lower temperature T_c. The difference in chemical potentials μ_h and μ_c is due to the different temperatures. Such a system can be realized with the semiconductor hetero-structures fabrication technology.

The barrier potential is also useful to model metal-oxide-semiconductor (MOS) and field-effect-transistors (FETs). A barrier can arise if a metal is heated to give rise to thermionic emission. Lothar Wolfgang Nordheim and Ralph Howard Fowler (1920s) realized that the barrier penetration could be used to understand the emission of electrons from a metal in a strong electric field.

5.2.6   Importance of Quantum Tunnelling

There are many phenomena in physics where tunnelling plays a central role. Five Nobel Prizes in physics were awarded for research concerned with tunnelling in semiconductor and for the invention of scanning tunnelling microscopy. Tunnelling is crucial for nucleo-synthesis in stars and in the evolution of early universe. Molecular spectra, electron emission from metals, cold emission of electrons from metals, current-voltage characteristics of transistors, magnetization in certain metals, microwave absorption of ammonia and alpha decay are explained through the tunnelling phenomenon. Some of the examples and applications of tunnelling are given below:

1. In tunnel diodes, the tunnelling effect is a source for a region in the current-voltage characteristic where the differential resistance is negative. α-decay occurs due to the tunnel effect.

2. In 1973, Brain David Josephson was awarded the Nobel Prize together with Leo Esaki and Ivar Giaever for an invention of a junction called *Josephson junction*. In a Josephson junction transition regions between two superconducting materials are separated by a thin insulating layer. In these devices pairs of electrons known as cooper pairs tunnel from one superconducting region to another through thin insulating layer. This effect gives rise to currents flowing without an applied voltage.

3. A scanning tunnelling microscope (STM) working by quantum tunnelling between the surface of an electrical conductor and a fine conducting probe was invented by Gerd Binning and Heinrich Rohrer. They received the Nobel Prize in 1986 for this invention. In a STM a thin needle moves across the surface of a sample. If the needle and the sample are close to each other then electrons can tunnel from the needle to the sample. In this way the picture of the surface of the sample is obtained.

4. Friedrich Hund (1927) was the first to use the quantum mechanical barrier penetration in the theory of molecular spectra. Internal structure of complex molecules and the rotationally induced dissociation of a diatomic molecule from an excited state can be studied by the tunnelling phenomenon.

5. A series of nuclear reactions in the sun produces light energy. In this scenario, two protons come very close and form a bound consisting of a proton, a neutron, a positron and a neutrino. First, two protons come closer. Since they have same electric charge they repel each other. According to classical mechanics this reaction would take place very slowly and no light would come out from the sun. But quantum mechanics predicts that one proton tunnels through the barrier of repulsion separating the two protons and allows the reaction to continue [5].

5.3 Finite Square-Well Potential

Consider a potential which is zero within a certain interval and V_0 elsewhere. Such a potential is called a *square-well potential* and is represented as

$$V(x) = \begin{cases} 0, & \text{for } |x| < a \\ V_0, & \text{for } |x| > a. \end{cases} \qquad (5.23)$$

The potential is schematically represented in Fig. 5.4. An interesting remarkable feature of this potential is that it admits both bound and scattering states. Further, *the bound states of the square-well potential are finite in number*.

5.3.1 Physical Systems

Certain materials crystallize in an almost similar manner and possess exactly the same lattice constants. Therefore, they may be grown together as a single crystal. A junction in a single crystal between two materials whose lattices match but are otherwise different is known as *heterojunction*. In a double heterojunctions, two junctions are used in order to sandwich a layer of a narrow band gap material between wide band gap materials. In this way, two heterostructure interfaces form a quantum well. Such double-heterostructures

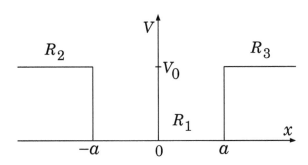

FIGURE 5.4
Finite square-well potential.

are very common for light-emitting diodes and lasers used for optical communication. This potential is used to model emission of electrons in metals and for short-range forces.

5.3.2   Solution of the Wave Equation for $E < V_0$

For a classical particle with kinetic energy less than the potential energy, its motion is bounded within the region R_1. Its dynamics are similar to that of infinite height potential discussed in the sec. 4.6. The particle in R_1 cannot enter into the region R_2 or R_3. When the kinetic energy is greater than the potential energy the particle is able to overcome the potential barrier and enter into the region R_2 or R_3. Then the motion is not bounded.

Now, we focus on quantum dynamics of the particle. The Schrödinger equation for $|x| < a$ and $|x| > a$ are given by

$$\psi_{xx} + k^2\psi = 0, \quad |x| < a \tag{5.24a}$$
$$\psi_{xx} - \alpha^2\psi = 0, \quad |x| > a \tag{5.24b}$$

where $k^2 = 2mE/\hbar^2$, $\alpha^2 = 2m(V_0 - E)/\hbar^2$. Evidently, Eq. (5.24a) is a free particle equation. α^2 is positive for $0 < E < V_0$ and negative for $E > V_0$. The nature of the solution of Eq. (5.24b) is different for $\alpha^2 < 0$ and $\alpha^2 > 0$. Therefore, these two cases must be studied separately.

5.3.2.1   Bound States: $0 < E < V_0$, $\alpha^2 > 0$

The solution of Eq. (5.24a) (for the region R_1) is

$$\psi_1 = A\sin kx + B\cos kx, \quad |x| < a. \tag{5.25a}$$

The solution of Eq. (5.24b) for the regions R_2 and R_3 is $\psi = Ce^{-\alpha x} + De^{\alpha x}$. In the region R_2 when $x < -a$, as $x \to -\infty$ the first term in the above solution tends to ∞. To avoid this, set $C = 0$ so that

$$\psi_2 = De^{\alpha x}, \quad x < -a. \tag{5.25b}$$

In the region R_3 as $x \to \infty$ the second term in ψ tends to ∞. Therefore, we choose $D = 0$. Now, the solution is

$$\psi_3 = Ce^{-\alpha x}, \quad x > a. \tag{5.25c}$$

The solutions ψ_1, ψ_2 and ψ_3 must satisfy certain appropriate boundary conditions.

5.3.2.2   Boundary Conditions

The boundary conditions arise due to the continuity of the solutions and their derivatives at $x = \pm a$. Matching the solutions (5.25) and their derivatives at $x = \pm a$ give a system of four equations for the four unknowns A, B, C and D. The compatibility condition for these equations leads to an equation for the energy E_n.

The boundary conditions are

$$\text{at } x = a, \quad \psi_1 = \psi_3, \ \psi_{1x} = \psi_{3x}, \tag{5.26a}$$
$$\text{at } x = -a, \quad \psi_1 = \psi_2, \ \psi_{1x} = \psi_{2x}. \tag{5.26b}$$

These conditions lead to the following set of equations:

$$A\sin ka + B\cos ka = Ce^{-\alpha a}, \tag{5.27a}$$
$$Ak\cos ka - Bk\sin ka = -\alpha Ce^{-\alpha a}, \tag{5.27b}$$
$$-A\sin ka + B\cos ka = De^{-\alpha a}, \tag{5.27c}$$
$$-Ak\cos ka - Bk\sin ka = -\alpha De^{-\alpha a}. \tag{5.27d}$$

Adding Eq. (5.27a) with (5.27c) and (5.27b) with (5.27d) we get

$$2B\cos ka = (C+D)e^{-\alpha a}, \tag{5.28a}$$
$$2Bk\sin ka = \alpha(C+D)e^{-\alpha a}. \tag{5.28b}$$

Eliminating $(C+D)e^{-\alpha a}$ in Eq. (5.28a) using Eq. (5.28b) we have

$$2B\left(1 - \frac{k}{\alpha}\tan ka\right) = 0. \tag{5.29}$$

Equation (5.29) is satisfied if the following two conditions are satisfied:
(i) $B = 0$, k is arbitrary.
(ii) $B \neq 0$, $\tan ka = \alpha/k$.

Next, subtracting Eq. (5.27c) from Eq. (5.27a) and Eq. (5.27d) from Eq. (5.27b) we obtain

$$2A\sin ka = (C-D)e^{-\alpha a}, \tag{5.30a}$$
$$2Ak\cos ka = -\alpha(C-D)e^{-\alpha a}. \tag{5.30b}$$

Elimination of $(C-D)e^{-\alpha a}$ in Eq. (5.30a) gives

$$2A\left(\frac{k}{\alpha} + \tan ka\right) = 0 \tag{5.31}$$

which implies either
(iii) $A = 0$, k is arbitrary or
(iv) $A \neq 0$, $\tan ka = -k/\alpha$.

Because $k^2 = 2mE/\hbar^2$, arbitrariness of k means E is arbitrary. From the results (i)-(iv) note that E is arbitrary only if $A = B = 0$. In this case the wave function ψ_1 is zero which we do not want. For both $A \neq 0$, $B \neq 0$ the condition is

$$\tan^2 ka = \frac{\alpha}{k}\left(-\frac{k}{\alpha}\right) = -1. \tag{5.32}$$

For real ka this is not possible and hence the conditions (iii) and (iv) are not satisfied simultaneously. The news is, $A \neq 0$, $B \neq 0$ is not possible. Therefore, either $A = 0$ or $B = 0$. Now, we have the following two cases:

Case 1: $A = 0$, $B \neq 0$, $\tan ka = \alpha/k$.

Case 2: $A \neq 0$, $B = 0$, $\tan ka = -k/\alpha$.

Since k is not arbitrary, E is not arbitrary and it can take only specific sets of values. This leads to discreteness of energy levels. For the cases 1 and 2 we

$$C = D = Be^{\alpha a}\cos ka, \quad C = -D = Ae^{\alpha a}\sin ka. \tag{5.33}$$

Further, from $k^2 = 2mE/\hbar^2$ and $\alpha^2 = 2m(E-V_0)/\hbar^2$ we obtain

$$k^2 + \alpha^2 = \frac{2mV_0}{\hbar^2} = \text{constant}. \tag{5.34}$$

5.3.2.3   Calculation of E_n and ψ_n

Case 1:

For the case 1, as $\tan(k/a) > 0$, consider only the values of ka with

$$2r\frac{\pi}{2} \le ka \le (2r+1)\frac{\pi}{2}, \quad r = 0, 1, \ldots . \tag{5.35a}$$

Then Eq. (5.34) gives

$$\frac{2mV_0a^2}{\hbar^2} = k^2a^2 \sec^2 ka . \tag{5.35b}$$

Similarly, for the case 2 we need to consider

$$(2r-1)\frac{\pi}{2} \le ka \le 2r\frac{\pi}{2}, \quad r = 1, 2, \ldots \tag{5.36a}$$

and consequently

$$\frac{2mV_0a^2}{\hbar^2} = k^2a^2 \mathrm{cosec}^2 ka . \tag{5.36b}$$

Equations (5.35) and (5.36) are satisfied only for certain discrete values of k. Denote these values as k_n. Then the energy values from $k^2 = 2mE/\hbar^2$ are

$$E_n = \frac{\hbar^2}{2m}k_n^2, \quad n = 0, 1, 2, \ldots . \tag{5.37}$$

The energy levels are discrete. The energy levels with $n = 0, 2, 4, \ldots$ correspond to the case 1. The eigenfunctions are

$$\psi = \begin{cases} Be^{\alpha_n a} \cos k_n a \, e^{\alpha_n x}, & x < -a \\ B\cos k_n x, & -a < x < a \\ Be^{\alpha_n a} \cos k_n a \, e^{-\alpha_n x}, & x > a . \end{cases} \tag{5.38}$$

Case 2:

The energy levels with $n = 1, 3, \ldots$ are for the case 2. In this case

$$\psi = \begin{cases} -Ae^{\alpha_n a} \sin k_n a \, e^{\alpha_n x}, & x < -a \\ A\sin k_n x, & -a < x < a \\ Ae^{\alpha_n a} \sin k_n a \, e^{-\alpha_n x}, & x > a . \end{cases} \tag{5.39}$$

The constants A and B are to be determined by the normalization condition.

Next, we determine the eigenfunctions for specific values of V_0 and a. Let us fix $V_0 = 35$, $a = 1$ and $\hbar^2 = m = 1$. Using the Newton–Raphson method the roots of Eq. (5.35b) are found to be $k = 1.4024, 1.7859, 4.1881, 5.4166$ and 6.8870. Among these 5 values, the values of k satisfying the condition (5.35a) are $k_0 = 1.4024$, $k_2 = 4.1881$ and $k_4 = 6.8870$. The notable result is that there are only three discrete energy levels for the case 1. The corresponding unnormalized eigenfunctions are depicted in Fig. 5.5. For the case 2 we obtain $k_1 = 2.8003$, $k_3 = 5.5568$ and $k_5 = 8.1048$. In this case also there are only three discrete energy levels. The unnormalized eigenfunctions of the discrete energy levels are shown in Fig. 5.6.

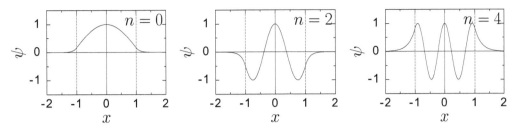

FIGURE 5.5
Three eigenfunctions for the case 1 of the finite square-well potential with $V_0 = 35$, $a = 1$ and $\hbar^2 = m = 1$. The eigenfunctions are unnormalized ($B = 1$ in Eq. (5.38)).

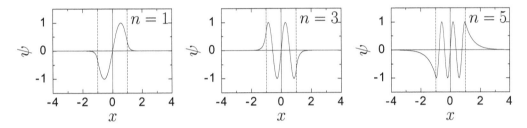

FIGURE 5.6
Three eigenfunctions for the case 2 of the finite square-well potential with $V_0 = 35$, $a = 1$ and $\hbar^2 = m = 1$. The eigenfunctions are unnormalized ($A = 1$ in Eq. (5.39)).

The energy difference between V_0 and the top-most bound state energy is called the *work function* ϕ_0. An electron bound in a metal would be dislodged only if the incident photon energy exceeds ϕ_0. The ϕ_0 of the metals Ni, Ta, Cd and Ag are 4.2, 4.5, 4.6 and 4.8eV, respectively.

In the regions R_2 and R_3, the total energy E is less than V_0. Classically, a particle in R_1 cannot enter the regions R_2 and R_3. When it reaches $x = \pm a$, it turns back and hence the motion is confined within $|x| < a$. In contrast to this, in quantum mechanics $|\psi_2|^2$ and $|\psi_3|^2$ are nonzero in R_2 and R_3 and hence the probability for the presence of particle in the classically forbidden regions is finite and nonzero.

Let us summarize the significant features of the square-well potential.

1. The potential admits only a few discrete energy eigenvalues and normalizable eigenfunctions. All the allowed discrete energy values are less than the depth of the well.

2. The number of discrete states increases with increase in the depth of the well. However, since α increases with increase in V_0, the extent of the tail of the eigenfunctions in the classically forbidden region decreases (refer to Eq. (5.39)).

3. From Eq. (5.39) it is clear that the eigenfunctions are oscillatory in the classically allowed region ($-a < x < a$) and exponentially decaying in the classically forbidden regions ($|x| > a$).

4. From Figs. 5.5 and 5.6 we note that ψ_n has n nodes.

5. As a consequence of the symmetry in V, $V(x) = V(-x)$, the eigenfunctions have alternating even (Eq. (5.38)) and odd (Eq. (5.39)) parity.

Solved Problem 3:

Determine the expression for the transmission probability T for the case of $E > V_0$ with the particle moing from left to right in the presence of the finite square-well potential.

For $E > V_0$ defining $\alpha^2 = 2m(E - V_0)/\hbar^2 > 0$ Eq. (5.24b) becomes

$$\psi_{xx} + \alpha^2 \psi = 0, \quad |x| > a. \tag{5.40}$$

The solutions of the Schrödinger Eqs. (5.24a) and (5.40) are

$$\psi_1 = A \sin kx + B \cos kx, \quad |x| < a, \tag{5.41a}$$

$$\psi_2 = Ce^{i\alpha x} + De^{-i\alpha x}, \quad x < -a, \tag{5.41b}$$

$$\psi_3 = Fe^{i\alpha x} + Ge^{-i\alpha x}, \quad x > a. \tag{5.41c}$$

For $E > V_0$ the wave function in the interval $x \in [-\infty, \infty]$ is nonnormalizable and the problem is essentially scattering problem. For a particle incident from left $G = 0$. The boundary conditions at $x = -a$ and $x = a$ give

$$A \sin ka - B \cos ka = -Ce^{-i\alpha a} - De^{i\alpha a}, \tag{5.42a}$$

$$A \cos ka + B \sin ka = Ci\beta e^{-i\alpha a} - Di\beta e^{i\alpha a}, \quad \beta = \alpha/k, \tag{5.42b}$$

$$A \sin ka + B \cos ka = Fe^{i\alpha a}, \tag{5.42c}$$

$$A \cos ka - B \sin ka = Fi\beta e^{i\alpha a}. \tag{5.42d}$$

Treat C, the amplitude of the incident wave, as a known quantity. Then for the four unknowns A, B, D and F there are four equations given by Eqs. (5.42). As done in the case of square-barrier potential the set of Eqs. (5.42) can be solved by the Gauss elimination method.

Solving of Eqs. (5.42) give

$$F = \frac{iC\beta e^{-2i\alpha a}}{(\sin ka + i\beta \cos ka)(\cos ka - i\beta \sin ka)}. \tag{5.43}$$

The transmission amplitude $T = FF^*$ is obtained as (with $C = 1$)

$$T = \frac{1}{1 + \epsilon \sin^2 2ka}, \quad \epsilon = \frac{(k^2 - \alpha^2)^2}{4k^2\alpha^2} = \frac{1}{4(E/V_0)((E/V_0) - 1)}. \tag{5.44}$$

Figure 5.7 shows the plot of variation of T with E/V_0. As E increases from V_0 the value of ϵ decreases and for sufficiently large E/V_0 the value of $T \approx 1$. Similar to the barrier potential a series of resonances occur.

5.4   Potential Step

As an example of another interesting scattering state system we consider the problem of a homogeneous beam of free particles moving along a straight-line crossing one region to another of a different potential. The two regions are separated by a plane surface. Suppose the potential of the above system is of the form

$$V(x) = \begin{cases} 0, & x < 0 \\ V_0, & x > 0 \end{cases} \tag{5.45}$$

which is depicted in Fig. 5.8. The potential is zero for $x < 0$ and V_0 for $x > 0$. The potential given by Eq. (5.45) is called a *single-step potential* barrier.

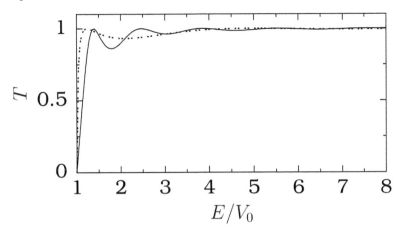

FIGURE 5.7
Variation of transmission coefficient T as a function of E/V_0 for $V_0 a^2 = 1$ (dashed curve) and $V_0 a^2 = 8$ (continuous curve) for the square-well potential with $m = 1$ units, $\hbar = 1$ units and $2ka = (8ma^2 V_0 (E/V_0)/\hbar^2)^{1/2}$.

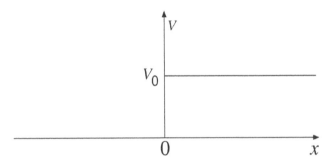

FIGURE 5.8
Step potential.

5.4.1 Physical Systems with Step Potential

Heterojunctions are commonly grown with semiconductors, which may be undoped or doped n or p type. Consider two conductors A and B with conduction and valence band edges E_c and E_v, respectively, and with different magnitudes of forbidden gap. An electron at conduction band edge in A will have lower energy than in material B. When junction is made between A and B, a potential barrier is developed at the interface with region $x < 0$ and $x > 0$ representing the materials A and B, respectively. This restricts the motion of electrons incident from the A (left) side. This type of situation is realized in GaAs and $Al_x Ga_{1-x} As$ junctions. The electron in the GaAs encounters a barrier to enter in AlGaAs part of the structure. It is reflected back by the potential.

5.4.2 Particle Energy Above the Potential Step: $E > V_0$

Let us first mention the dynamics of the classical particle. The total energy is greater than the potential energy. Further, the kinetic energy is positive. Therefore, a classical particle moving towards $x = 0$ from $x < 0$ will not be reflected at $x = 0$. In the quantum case, the Schrödinger equation corresponding to the potential, Eq. (5.45), is

$$\psi_{1xx} + k_0^2 \psi_1 = 0, \quad k_0^2 = \frac{2mE}{\hbar^2}, \quad x < 0 \tag{5.46a}$$

$$\psi_{2xx} + k^2 \psi_2 = 0, \quad k^2 = \frac{2m(E - V_0)}{\hbar^2}, \quad x > 0. \tag{5.46b}$$

When $E > V_0$, k_0^2 and k^2 are positive. k^2 is negative for $E < V_0$. We analyze these two cases separately.

The solutions of the Eqs. (5.46) are

$$\psi_1 = A e^{ik_0 x} + B e^{-ik_0 x}, \quad x < 0 \tag{5.47a}$$

$$\psi_2 = C e^{ikx} + D e^{-ikx}, \quad x > 0 \tag{5.47b}$$

where A, B, C and D are constants to be determined. In Eq. (5.47a) $e^{ik_0 x}$ represents the incident wave moving towards positive x-direction and $e^{-ik_0 x}$ represents the reflected wave moving towards negative x-direction. In Eq. (5.47b) e^{ikx} represents the transmitted wave moving in positive x-direction and e^{-ikx} represents the reflected wave moving towards negative x-direction. Since discontinuity occurs only at $x = 0$ there is no reflected wave for $x > 0$ and so D becomes zero. Further, we set the amplitude of the incident wave as unity. We write the Eqs. (5.47) as

$$x < 0: \quad \psi_1 = \psi_{\text{inc}} + \psi_{\text{ref}}, \tag{5.48a}$$

$$x > 0: \quad \psi_2 = \psi_{\text{tra}}, \tag{5.48b}$$

where $\psi_{\text{inc}} = e^{ik_0 x}$, $\psi_{\text{ref}} = B e^{-ik_0 x}$ and $\psi_{\text{tra}} = C e^{ikx}$. Here

$$\psi_{\text{inc}}^* \psi_{\text{inc}} = 1, \quad \psi_{\text{ref}}^* \psi_{\text{ref}} = |B|^2, \quad \psi_{\text{tra}}^* \psi_{\text{tra}} = |C|^2. \tag{5.48c}$$

The boundary conditions $\psi_1(0) = \psi_2(0)$ and $\psi_{1x}(0) = \psi_{2x}(0)$ lead to $1 + B = C$ and $k_0(1 - B) = kC$. Solving the above set of equations for B and C gives

$$B = \frac{1 - (k/k_0)}{1 + (k/k_0)}, \quad C = \frac{2}{1 + (k/k_0)}. \tag{5.49}$$

We find $R = J_{\text{ref}}/J_{\text{inc}} = |B|^2$ and $T = J_{\text{tra}}/J_{\text{inc}} = (k/k_0)|C|^2$. Then

$$R = |B|^2 = \left[\frac{1 - (k/k_0)}{1 + (k/k_0)} \right]^2, \quad T = \frac{k}{k_0}|C|^2 = \frac{4k/k_0}{(1 + k/k_0)^2}. \tag{5.50}$$

Here $T + R = 1$ and $k/k_0 = \sqrt{1 - (V_0/E)}$.

Since $V_0 > 0$ we have $k < k_0$, or $k/k_0 < 1$. As a result the probability of finding the particle anywhere in the region $x > 0$ is greater than the probability of finding it anywhere in the region $x < 0$. We account for this from the classical argument that the probability of finding a particle is more in the region where its velocity is less. Since the velocity of the particle in the region $x > 0$ is less than its velocity in $x < 0$, the probability is more in the region $x > 0$. Figure 5.9 shows the reflection (R) and transmission (T) probabilities as a function of k/k_0 for the step potential. Continuous and dashed curves represent R and T, respectively.

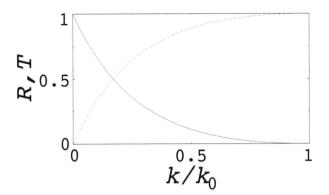

FIGURE 5.9
Variation of R (continuous curve) and T (dashed curve) of the step potential.

5.4.3 Particle Energy Below the Potential Step: $0 < E < V_0$

Next, consider the case of the energy E of the particle is smaller than the height V_0 of
the potential step. The classical particle in the left region will remain in the left region.
The particle moving towards the right will be completely reflected at $x = 0$. In classical
mechanics finding the particle in $x > 0$ is impossible because in this region the total energy
is less than the potential energy, so that the kinetic energy $p_x^2/(2m) < 0$ and the momentum
p_x is imaginary.

Defining $k^2 = 2m(V_0 - E)/\hbar^2$, the Schrödinger equation takes the form

$$\psi_{1xx} + k_0^2\psi_1 \;=\; 0, \quad x < 0 \tag{5.51a}$$
$$\psi_{2xx} - k^2\psi_2 \;=\; 0, \quad x > 0. \tag{5.51b}$$

The solutions of Eqs. (5.51) are

$$\psi_1 \;=\; e^{ik_0x} + Be^{-ik_0x}, \quad x < 0 \tag{5.52a}$$
$$\psi_2 \;=\; Ce^{kx} + De^{-kx}, \quad x > 0. \tag{5.52b}$$

In the limit $x \to \infty$, we have $e^{kx} \to \infty$, therefore, choose $C = 0$. The probability of finding
the particle in the region $x > 0$ decreases rapidly with x as e^{-2kx}. But there is a finite
nonzero probability for finding the particle in the region $x > 0$. The penetration into the
classically forbidden region is a striking prediction of quantum mechanics. The penetrating
distance is $\Delta x = \hbar/(2\sqrt{2m(V_0 - E)})$.

As done earlier for $E > V_0$, we determine B and D using the boundary conditions at
$x = 0$. We obtain

$$B = \frac{k_0 - ik}{k_0 + ik}, \quad D = \frac{2k_0}{k_0 + ik}. \tag{5.53}$$

Further, $R = |B|^2 = 1$ while $T = 0$. Even though T is zero in the region $x > 0$, the wave
function is not zero!

For the linear harmonic oscillator and the particle in a box, the stationary state eigenfunctions are normalizable and the energy eigenvalues are discrete. The eigenfunctions for
the cases of the free particles, square-barrier and step potentials are nonnormalizable and
their energy eigenvalues are continuous.

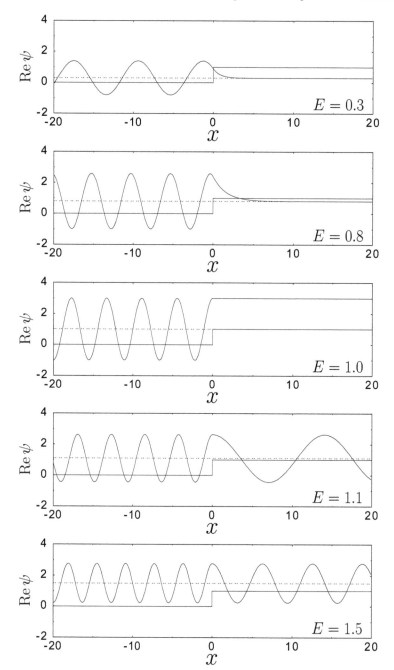

FIGURE 5.10
$\mathrm{Re}\,\psi$ is plotted for a few values of E for the step potential with $V_0 = 1$, $\hbar^2 = 1$ and $m = 1$. For illustration $\mathrm{Re}\,\psi$ is shifted as $\mathrm{Re}\,\psi + E$.

Figure 5.10 shows $\mathrm{Re}\,\psi$ (actually $\mathrm{Re}\,\psi + E$) for a few chosen values of E for the step potential with $V_0 = 1$. For $E < V_0$, the eigenfunction is decaying for $x > 0$. For $E > V_0$, it is oscillatory for $x > 0$. The eigenfunction does not fall abruptly to zero at $x = 0$ but diminishes exponentially with x. The probability of finding the particle in $x > 0$ decreases exponentially with increasing x.

Solved Problem 4:

A beam of electrons incident on an energy barrier of height V_0eV of the step potential. Find the fraction of electrons reflected and transmitted at the barrier if the E of the incident electrons is (a) $3V_0/2$eV, (b) V_0 and (c) $V_0/4$eV.

Case (a): $E = (3V_0/2)\,$eV

T and R can be calculated from Eq. (5.50). In this case

$$E = \frac{3}{2}V_0\,\text{eV}\,,\quad k_0^2 = \frac{3mV_0}{\hbar^2}\,,\quad k^2 = \frac{2m(E-V_0)}{\hbar^2} = \frac{mV_0}{\hbar^2}\,. \tag{5.54}$$

Using these in Eq. (5.50) we obtain

$$R = \left[\frac{\sqrt{3mV_0} - \sqrt{mV_0}}{\sqrt{3mV_0} + \sqrt{mV_0}}\right]^2 = \left(\frac{\sqrt{3}-1}{\sqrt{3}+1}\right)^2 \approx 0.0718\,. \tag{5.55}$$

Then $T = 1 - R \approx 0.9282$.

Case (b): $E = V_0\,$eV

When $E = V_0$, k is 0 and hence $R = 1$ and $T = 0$. That is, all the incident electrons are reflected back.

Case (c): $E = (V_0/4)\,$eV

From Eq. (5.53) we have $T = 0$ and $R = 1$.

5.5 Locally Periodic Potential

In this section, we consider scattering by a potential consisting of N-identical cells. The potential is locally periodic, that is, it is periodic on a finite interval and zero elsewhere. It has been shown that the N-cell transmission and reflection probabilities can be expressed in terms of single cell probabilities [6]. The N-delta function and square-barrier potentials exhibit band structure, a striking special character of transmission probability [6–9]. This feature is observed even for quite small number of unit cell.

First, we define the transfer matrix M [6] associated with the scattering by a single cell. Next, we show the determination of the transmission probability for multiple scattering using M. Then we obtain a simple closed form expression for transmission probability for arbitrary Dirac-delta and rectangular barrier cells.

5.5.1 M Matrix

Let us consider a locally periodic potential consisting of three regions:

$$\begin{array}{llll}
\text{Region}-I & -\infty < x < a & : & V = 0 \\
\text{Region}-II & a \le x \le b & : & V = V(x) \\
\text{Region}-III & b < x < \infty & : & V = 0\,.
\end{array} \tag{5.56}$$

In the region-II, the potential V is not always zero and denote $\psi_{II}(x) = \psi_{ab}(x)$. The general solutions in the regions I and II are written as

$$\psi_I(x) = Ae^{ikx} + Be^{-ikx}, \tag{5.57a}$$

$$\psi_{II}(x) = Ce^{ikx} + De^{-ikx}, \quad k = \sqrt{2mE}/\hbar. \tag{5.57b}$$

If particles are incident from left then $D = 0$ and $T = |C|^2/|A|^2$. To determine T find the solution in the region-II and apply the boundary conditions at $x = a$ and b. The result is a set of equations of the form

$$\begin{pmatrix} A \\ B \end{pmatrix} = M \begin{pmatrix} C \\ D \end{pmatrix}, \quad M = \begin{pmatrix} \alpha_{11} & \alpha_{12} \\ \alpha_{21} & \alpha_{22} \end{pmatrix}, \tag{5.58}$$

where the matrix M is called the *transfer matrix* [6].

If $\psi(x,t)$ is a solution of the Schrödinger equation then $\psi^*(x,-t)$ is also a solution. ψ^* is given by

$$\psi^*(x) = \begin{cases} A^*e^{-ikx} + B^*e^{ikx}, & \text{for region} - I \\ \psi_{ab}^*(x), & \text{for region} - II \\ C^*e^{-ikx} + D^*e^{ikx} & \text{for region} - III. \end{cases} \tag{5.59}$$

ψ^* interchanges the incoming and outgoing waves. In this case

$$\begin{pmatrix} A \\ B \end{pmatrix} = \begin{pmatrix} \alpha_{22}^* & \alpha_{21}^* \\ \alpha_{12}^* & \alpha_{11}^* \end{pmatrix} \begin{pmatrix} C \\ D \end{pmatrix}. \tag{5.60}$$

Comparison of Eqs. (5.58) and (5.60) gives $\alpha_{11} = \alpha_{22}^*$, $\alpha_{21} = \alpha_{12}^*$. Next, make use of conservation of probability current J given by

$$J = \frac{\hbar}{2mi} \left(\psi^* \frac{d\psi}{dx} - \frac{d\psi^*}{dx} \psi \right) \tag{5.61}$$

and is independent of x. The condition $J|_{x<a} = J|_{x>b}$ together with $\alpha_{11} = \alpha_{22}^*$, $\alpha_{21} = \alpha_{12}^*$ gives $\det M = 1$. Now, the transfer matrix M becomes

$$M = \begin{pmatrix} \alpha_{11} & \alpha_{12} \\ \alpha_{12}^* & \alpha_{11}^* \end{pmatrix}, \quad |\alpha_{11}|^2 - |\alpha_{12}|^2 = 1. \tag{5.62}$$

5.5.2   Multiple Cells

Suppose the basic unit cell is replicated N times at regular intervals. Then the goal is to construct the matrix M for the full array making use of the given M of a single cell. For simplicity, assume the potential of the unit cell as $V = V(x)$ for $-a < x < a$ and zero for $|x| > a$. Now, we study the scattering from a Dirac comb [8].

Interestingly, the scattering of particles by a Dirac-delta potential $V = V_0\delta(x - x_0)$ is exactly solvable. The M matrix for the single cell with $x_0 = 0$ is obtained as (see the exercise 19 at the end of this chapter)

$$M = \begin{pmatrix} 1 + i\beta & i\beta \\ -i\beta & 1 - i\beta \end{pmatrix}, \quad \beta = \frac{mV_0}{\hbar^2 k}. \tag{5.63}$$

Assume that the cells are separated by a distance $s = 2a$. Then the wave function between the cells is

$$\psi_n(x) = A_n e^{ik(x-ns)} + B_n e^{-ik(x-ns)}, \quad (n-1)s + a < x < ns - a \tag{5.64}$$

where $0 < n < N$. ψ_0 and ψ_N are given by

$$\psi_0(x) = A_0 e^{ikx} + B_0 e^{-ikx}, \quad x < -a \tag{5.65a}$$
$$\psi_N(x) = A_N e^{ik(x-Ns)} + B_N e^{-ik(x-Ns)}, \quad x > (N-1)s + a. \tag{5.65b}$$

For the nth cell, using Eqs. (5.58), (5.62) and (5.64) we obtain

$$\begin{pmatrix} A_n \\ B_n \end{pmatrix} = P \begin{pmatrix} A_{n+1} \\ B_{n+1} \end{pmatrix}, \quad P = M \begin{pmatrix} e^{-iks} & 0 \\ 0 & e^{iks} \end{pmatrix}. \tag{5.66}$$

From Eq. (5.66) we have

$$\begin{pmatrix} A_0 \\ B_0 \end{pmatrix} = P^N \begin{pmatrix} A_N \\ B_N \end{pmatrix}, \quad \det P = 1. \tag{5.67}$$

We need to find P^N. Using Cayley–Hamilton theorem [10,11] we find a formula for P^N. The characteristic equation for P is $\det(P - IP) = 0$, that is,

$$P^2 - P\mathrm{Tr}(P) + \det P = P^2 - 2\xi P + I = 0, \tag{5.68}$$

where $\xi = \mathrm{Tr}(P)/2$. According to the Cayley–Hamilton theorem any matrix satisfies its own characteristic equation

$$P^2 - 2P\xi + I = 0. \tag{5.69}$$

The point is that any higher power of P can be reduced to a linear combination of P and I:

$$P^N = PU_{N-1}(\xi) - IU_{N-2}(\xi), \tag{5.70}$$

where $U_N(\xi)$ is a polynomial of degree N in ξ. We can easily find U's. Multiplying Eq. (5.70) by P and using Eq. (5.69) for P^2 we obtain

$$P^{N+1} = (2P\xi - I)U_{N-1}(\xi) - PU_{N-2}(\xi). \tag{5.71}$$

On the other hand, replacing N by $N+1$ in Eq. (5.70) we get

$$P^{N+1} = PU_N(\xi) - IU_{N-1}(\xi). \tag{5.72}$$

Comparison of Eqs. (5.71) and (5.72) gives

$$U_{N+2}(\xi) - 2\xi U_{N+1}(\xi) + U_N(\xi) = 0. \tag{5.73}$$

Equation (5.73) is the recursion relation for Chebychev polynomials. For $N = 2$, Eq. (5.70) becomes

$$P^2 = PU_1(\xi) - IU_0(\xi). \tag{5.74}$$

Equating this expression with Eq. (5.69) we have $U_0 = 1$ and $U_1 = 2\xi$. Using these U_2 can be computed from Eq. (5.73). Similarly, other U's can be computed. Now,

$$M_N = P^N \begin{pmatrix} e^{ikNs} & 0 \\ 0 & e^{-ikNs} \end{pmatrix} = \begin{pmatrix} W_N & Z_N \\ Z_N^* & W_N^* \end{pmatrix} \tag{5.75}$$

where

$$W_N = \left[We^{-iks}U_{N-1} - U_{N-2} \right] e^{ikNs}, \quad Z_n = ZU_{N-1}e^{-ik(N-1)s} \tag{5.76}$$

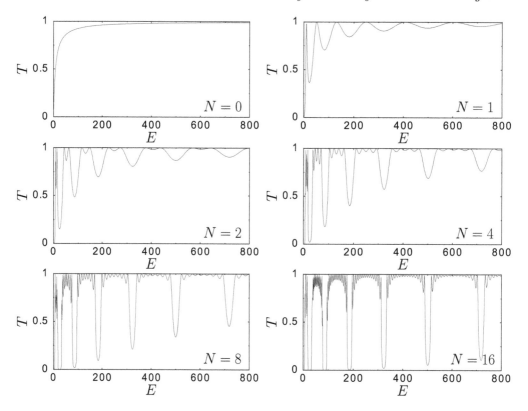

FIGURE 5.11
Transmission probability as a function of energy [7,8]. Here $s = 0.5$, $V_0 = 4$ and $m = \hbar^2 = 1$.
$N = 0$ corresponds to one barrier.

with $|W_N|^2 - |Z_N|^2 = 1$. For a single cell (with $D = 0$), from Eq. (5.58) C/A is found to be
$1/\alpha_{11}$. Therefore,

$$T = \left|\frac{C}{A}\right|^2 = \frac{1}{|\alpha_{11}|^2} = \frac{1}{|W_1|^2}. \tag{5.77}$$

Hence,

$$T_N = \frac{1}{|W_N|^2} = \frac{1}{1 + |Z_N|^2} = \frac{1}{1 + |Z|^2 (U_{N-1}(\xi))^2}, \tag{5.78}$$

where

$$\xi = \frac{1}{2}\mathrm{Tr}P = \frac{1}{2}\left(We^{-iks} + W^*e^{iks}\right) = \mathrm{Re}(W)\cos ks + \mathrm{Im}(W)\sin ks. \tag{5.79}$$

From (5.63) we find $W = 1 + i\beta$, $Z = i\beta$ and thus $\xi = \cos ks + \beta \sin ks$. Then

$$T_N = \frac{1}{1 + \beta^2 \left(U_{N-1}(\xi)\right)^2}. \tag{5.80}$$

In the limit $E \to \infty$, k is also $\to \infty$, $\beta \to 0$ and hence $T \to 1$. Figure 5.11 shows the
plots of T against E for $N = 0$ (one barrier), $N = 1, 2, 4, 8, 16$ with $V_0 = 4$ and $s = 0.5$.

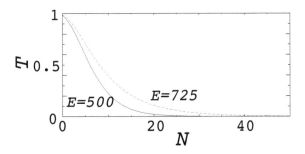

FIGURE 5.12
Transmission probability T versus N for two values of E.

Some of the special features of T versus E are summarized below:

1. For more than one barrier the graph of T has number of energy bands in which $T(N) \approx 1$.

2. The bands are separated by valleys where $T(N)$ approaches zero for large N. They can be regarded as the precursors of the band structure of periodic potentials. The bands occur even for small numbers of barriers.

3. Each band contains N ripples.

4. As N increases the valleys become more and more narrow. The number of valleys that are flatter and deeper increases with an increase in N.

5. For a fixed energy value, T changes significantly as the number of barriers is increased. This is shown in Fig. 5.12 where T is plotted as a function of N for $E = 500$ and 725.

For a detailed discussion on multiple rectangular barrier one may refer to ref. [8]. For discussion on quantum mechanical tunnelling in neural networks refer to refs. [12,13].

5.6   Reflectionless Potentials

In this present chapter so far we studied scattering of particles by some simple potentials. In the case of square-barrier oscillating T is observed. For very large E greater than V_0 we have $T \approx 1$. For the Dirac-delta function potential T is found to increase with E and approach to the limiting value 1. Now, we ask an interesting question: Are there potentials for which $T = 1$ for arbitrary energy? In other words, do we have a potential which perfectly transmits incoming waves with any energy? The answer is yes. There are infinite number of potentials with the above property. Such potentials are called *reflectionless potentials* [14–17] and they can be constructed using inverse scattering transform [14].

A class of one-dimensional reflectionless potentials with their bound and scattering states expressed in terms of elementary functions are the sech^2 potentials, called *modified Pöschl–Teller potentials*, given by [15–17]

$$V_l(x) = -\frac{\hbar^2}{2m}V_0^2 l(l+1)\mathrm{sech}^2\beta x, \quad l = 0, 1, \ldots . \tag{5.81}$$

These potentials essentially have two remarkable properties: (i) Bound states at zero energy and (ii) reflectionless scattering. Introducing the change of variables $x' = \beta x$, $k^2 = 2mE/(\hbar^2 V_0^2)$ and dropping the prime the Schrödinger equation of the given problem takes the form

$$\frac{d^2\psi}{dx^2} + \left[k^2 + l(l+1)\text{sech}^2 x\right]\psi = 0. \tag{5.82}$$

Making a new change of variable $y = \cosh^2 x$ and writing $\psi = y^{(l+1)/2}v(y)$ the Eq. (5.82) becomes

$$y(1-y)v'' + \left[\left(l+\frac{3}{2}\right) - (l+2)y\right]v' - \frac{1}{4}\left[(l+1)^2 + k^2\right]v = 0. \tag{5.83}$$

Equation (5.83) is the hypergeometric differential equation and its solution is [18,19]

$$v(y) = A\,_2F_1\left(a, b, \frac{1}{2}; 1-y\right) + B\sqrt{1-y}\,_2F_1\left(a+\frac{1}{2}, b+\frac{1}{2}, \frac{3}{2}; 1-y\right), \tag{5.84}$$

where $a = (l+1+ik)/2$, $b = (\lambda+1-ik)/2$ and $_2F_1(a, b, c; x)$ is the hypergeometric function with the usual notation.

We look for a fundamental system of two real solutions with one is odd and the other is even in x. The even solution is obtained with $A = 1$, $B = 0$:

$$\psi_e = \cosh^{(l+1)} x\,_2F_1\left(a, b, \frac{1}{2}; -\sinh^2 x\right). \tag{5.85}$$

The odd solution is obtained with $A = 0$, $B = -i$:

$$\psi_o = \cosh^{(l+1)} x \sinh^{(l+1)} x\,_2F_1\left(a+\frac{1}{2}, b+\frac{1}{2}, \frac{3}{2}; -\sinh^2 x\right). \tag{5.86}$$

When $l = 0$, the Hamiltonian is that of the free particle. For $l > 0$, $p_x = 0$ at the turning points of classically bounded solutions. When $p_x = 0$ the Hamiltonian becomes $H_l = -l(l+1)\text{sech}^2 x^*$, where x^* is a turning point. Since $\text{sech}^2 x^*$ is an even function of x^* we have $E < 0$. Therefore, $E < 0$ corresponds to bound states and $E > 0$ corresponds to scattering states. We now consider the cases $E < 0$ and $E > 0$ separately.

Case 1: Bound States, $E < 0$

We write $k = i\epsilon$, $\epsilon > 0$ so that $E_l = -\epsilon^2$. By performing an asymptotic analysis of (5.85), we obtain the eigenvalues for even eigenstates from the relation $\epsilon = l - 2n$, $n = 0, 1, \ldots$. For the odd states $\epsilon = l - 1 - 2n$, $n = 0, 1, \ldots$. Then the energy eigenvalues are given by

$$E_l^{(n)} = (l-n)^2, \quad n \leq l \tag{5.87}$$

where $n = 0, 1, \ldots$. Since $n \leq l$ for a given l, there are only a finite number of bound states. From Eq. (5.87) we note that $E_l = 0$ is an energy eigenvalue for any integer value of l.

Case 2: Scattering States

From the asymptotic form of ψ_o and ψ_e we infer the following. When $E > 0$, a and b are complex conjugates and the general solution for $|x| \to \infty$ is written as

$$\psi = \frac{\tilde{A}C_e}{2}\left[e^{i\phi_e}e^{-ikx} + e^{-i\phi_e}e^{ikx}\right] - \frac{\tilde{B}C_o}{2}\left[e^{i\phi_o}e^{-ikx} + e^{-i\phi_o}e^{ikx}\right], \quad x \to -\infty \tag{5.88a}$$

and

$$\psi = \frac{\tilde{A}C_e}{2}\left[e^{i\phi_e}e^{ikx} + e^{-i\phi_e}e^{-ikx}\right] + \frac{\tilde{B}C_o}{2}\left[e^{i\phi_o}e^{ikx} + e^{-i\phi_o}e^{-ikx}\right], \quad x \to \infty \tag{5.88b}$$

where the constants C_e, C_o, ϕ_e and ϕ_o are to be determined by making an asymptotic analysis of the solutions at $x \to \pm\infty$. The constants $\tilde{A}$ and $\tilde{B}$ are found to be arbitrary while ϕ_o and ϕ_e are determined as [16]

$$\phi_e = \arg\frac{\Gamma(ik)e^{-ik\log 2}}{\Gamma[(l+1+ik)/2]\Gamma[(-l+ik)/2]}, \qquad (5.88c)$$

$$\phi_o = \arg\frac{\Gamma(ik)e^{-ik\log 2}}{\Gamma[(l+ik)/2]\Gamma[(-l+ik)/2]}. \qquad (5.88d)$$

Now, we look for the specific asymptotic form ($|x| \to \infty$) of the solution

$$\psi = \begin{cases} e^{ikx} + Re^{-ikx}, & x \to -\infty \\ Te^{ikx}, & x \to \infty, \end{cases} \qquad (5.89)$$

where R and T are reflection and transmission coefficients. Comparison of Eqs. (5.88) and (5.89) results in

$$R = \frac{1}{2}\left[(\cos 2\phi_e + \cos 2\phi_o) + i(\sin 2\phi_e + \sin 2\phi_o)\right], \qquad (5.90a)$$

$$T = \frac{1}{2}\left[(\cos 2\phi_e - \cos 2\phi_o) + i(\sin 2\phi_e - \sin 2\phi_o)\right]. \qquad (5.90b)$$

Then

$$|R|^2 = \frac{1}{1+s^2}, \quad |T|^2 = \frac{s^2}{1+s^2}, \qquad (5.91)$$

where $s = \sinh \pi k/\sin \pi l$. When l is an integer $|s| = \infty$ and hence $|R|^2 = 0$ and $|T|^2 = 1$. That is, a particle incident on the potential is transmitted with unit probability. As a result the potentials given by Eq. (5.81) are known as reflectionless potentials. The reflectionless potential systems can also be studied by means of creation and annihilation operators [20].

5.7 Dynamical Tunnelling

The tunnelling phenomenon encountered in the study of potential barrier is a process permitted by quantum theory but disallowed by classical mechanics on energy consideration. A new kind of tunnelling called *dynamical tunnelling* was found. It refers to a behaviour allowed by quantum mechanics but forbidden by classical mechanics. It involves no potential barrier. However, a constant of motion (other than energy) forbids classically the quantum allowed phenomenon. Dynamical tunnelling has been observed in

1. a periodically driven undamped pendulum,
2. ultracold atoms forming a Bose–Einstein condensate in an amplitude-modulated optical standing wave and
3. a ball bouncing between two semicircular mirrors.

In the third example, it is easy to notice that classically the ball cannot escape from the region between the mirrors. Even though escape is possible energetically, and potential barrier would not prevent it, but still the classical motion of the ball disallows it to escape. Here escape is prevented not by energy considerations. Note that escape is not ruled out by the quantum formalism. The ball would be able to escape by dynamical tunnelling slowly. Another example, is the formaldehyde molecule H_2CO. In classical picture, oxygen in H_2CO molecule spinning around the C-O axis always points out in one direction. In quantum mechanics it can oscillate between pointing up and down without violating the conservation of energy or angular momentum. For details on dynamical tunnelling see refs. [21–28].

5.8 Concluding Remarks

In the systems we studied in this chapter, the bound states and continuum states, (scattering states) are well separated in energy. However, there are quantum systems with bound states embedded in continuum (BSECs) [29–35]. Such states are regarded as *exotic quantum states* and they do not have classical analogues. BSECs have been observed experimentally [30]. Scattering by a double square-well potential [36], double delta-function potential [37, 38], a Coulomb field in two-dimensions [39], asymmetrical potential [40], a finite periodic potential [41], double-barrier potential [42], perturbed finite square-well potential [43], Hermitian and non-Hermitian potentials [44] and two-dimensional potentials [45] have also been analyzed. Quantum mechanical transmission probability has been computer for periodic, quasiperiodic and random finite lattices [46], rectangular double-barrier potential and parabolic double-barrier potential [47].

There are number of modern opto-electronic devices involving low-dimensional hetero-structure. The unbound states of semiconductor quantum wells and quantum dots play significant role in their optical properties. Quantum mechanical transmission and reflection of electrons at potential energy steps can be applied to the design of nonequilibrium transistors.

5.9 Bibliography

[1]   J.H. Mathews, *Numerical Methods for Mathematics, Science and Engineering*. Prentice–Hall, New Delhi, 1998.

[2]   G. Binnig, H. Rohrer, Ch. Gerber and E. Weibel, *Physica B+C* 109 & 110:2075, 1982.

[3]   J.D. Chalk, *Am. J. Phys.* 56:29, 1988.

[4]   M. Larsson, V.B. Antonyuk, A.G. Mal'shukov, Z. Ma and K.A. Chao, *J. Phys. A* 35:L531, 2002.

[5]   D.F. Styer, *The Strange World of Quantum Mechanics*. Cambridge University Press, Cambridge, 2000.

[6]   D.W.L. Sprung, H. Wu and J. Martorell, *Am. J. Phys.* 61:1118, 1993.

[7]   D.J. Griffiths and N.F. Taussig, *Am. J. Phys.* 60:883, 1992.

[8]   D.J. Griffiths and C.A. Steinke, *Am. J. Phys.* 69:137, 2001.

[9]   P.R. Berman, *Am. J. Phys.* 81:190, 2013.

[10] L.A. Pipes and L.R. Harvill, *Applied Mathematics for Physicists*. McGraw–Hill, Singapore, 1984.

[11] T.L. Chow, *Mathematical Methods for Physicists*. Cambridge University Press, Cambridge, 2000.

[12] M. Dugic and D. Rakovic, *Eur. J. Phys. B* 13:781, 2000.

[13] D. Rakovic, *Informatica* 21:507, 1997.

[14] S. Novikov, S.V. Manakov, L.P. Pitaevskii and V.E. Zakharov, *Theory of Solitons: The Inverse Scattering Method*. Consultants Bureau, New York, 1984.

[15] J. Lekner, *Am. J. Phys.* 75:1151, 2007.

[16] S. Flügge, *Practical Quantum Mechanics–I*. Springer, Berlin, 1971.

[17] N. Kiriushcheva and S. Kuzmin, *Am. J. Phys.* 66:867, 1998.

[18] J. Mathews and R.L. Walker, *Mathematical Methods of Physics*. Pearson Education, New Delhi, 2004.

[19] M. Abramowitz and I.A. Stegun, *Handbook of Mathematical Functions*. Applied Mathematics Series No.55, 1964.

[20] R.L. Jaffe, *An algebraic approach to reflectionless potentials in one-dimension*. Unpublished, 2009.

[21] M.J. Davis and E.J. Heller, *J. Chem. Phys.* 75:246, 1981.

[22] E.J. Heller, *Nature* 412:33, 2001.

[23] D.A. Steck, W.H. Oskay and M.G. Raizen, *Science* 293:274, 2001.

[24] W.K. Hensinger, H. Haffner, A. Browaeys, N.R. Heckenberg, K. Helmerson, C. McKenzie, G.J. Milburn, W.D. Phillips, S.L. Rolston, H. Rubinsztein-Dunlop and B. Upcroft, *Nature* 412:52, 2001.

[25] A. Ramamoorthy, R. Akis, J.P. Bird, T. Maemoto, D.K. Ferry and M. Inoue, *Phys. Rev. E* 68:026221, 2003.

[26] M. Henseler, T. Dittrich and K. Richter, *Europhys. Lett.* 49:289, 2000.

[27] G. Hackenbroich and J.U. Nöckel, *Europhys. Lett.* 39:371, 1997.

[28] P. Jangid, A.K. Chauhan and S. Wuster, *Phys. Rev. A* 102:043513, 2020.

[29] J. von Neumann and E. Wigner, *Phys. Z.* 30:467, 1929.

[30] F. Capasso, C. Sirtori, J. Faist, D.L. Sivco, S.N.G. Chu and A.Y. Cho, *Nature* 358:565, 1992.

[31] B.A. Arbuzov, S.A. Shichanin and V.I. Savrin, *Phys. Lett. B* 275:144 , 1992.

[32] J.R. Spence and J.P. Vary, *Phys. Lett. B* 271:27, 1991.

[33] J. Pappademos, U. Sukhatme and A. Pagnamenta, *Phys. Rev. A* 48:3525, 1993.

[34] T.A. Weber and D.L. Pursey, *Phys. Rev. A* 50:4478, 1994.

[35] A.A. Stahlhofen, *J. Phys. A* 27:8279, 1994.

[36] B. Stec and C. Jedrzejek, *Eur. J. Phys.* 11:75, 1990.

[37] I.R. Lapidus, *Am. J. Phys.* 50:663, 1982.

[38] P. Senn, *Am. J. Phys.* 56:916, 1988.

[39] Q.G. Lin, *Am. J. Phys.* 65:1007, 1997.

[40] Y. Nogami and C.K. Ross, *Am. J. Phys.* 64:923, 1996.

[41] D.W.L. Sprung, H. Wu and J. Martorell, *Am. J. Phys.* 61:1118, 1993.

[42] M.L. Strekalov, *J. Math. Chem.* 56:890, 2018.

[43] D.C. Colladay and J.E. Goff, *Am. J. Phys.* 88:711, 2020.

[44] R.L. Chai, Q.T. Xie and X.L. Liu, *Chin. Phys. B* 29:090301, 2020.

[45] F. Loran and A. Mostafazadeh, *Phys. Rev. A* 96:063837, 2017.

[46] B. Gutierrez-Medina, *Am. J. Phys.* 81:104, 2013.

[47] P. Boonserm, T. Ngampitipan and K. Sansuk, *J. Phys. Conf. Series* 1366:012035, 2019.

5.10 Exercises

5.1 Express the transmission probability for the potential barrier as a function of energy for $E < V_0$ and $E > V_0$.

5.2 For a symmetrical barrier potential of energy V_0 and thickness $2a$ find the minimum transition probability for $E > V_0$.

5.3 For the potential barrier

$$ T = \left[1 + \left(1 + \frac{\epsilon^2}{4} \right) \sinh^2 2ka \right]^{-1} , \quad R = \frac{T\eta^2}{4} \sinh^2 2ka. $$

Using the above equations prove that $T = 0$ and $R = 1$ if $V_0 = \infty$.

5.4 A beam of electrons is incident on the square-barrier potential $V(x) = V_0$ for $|x| < a$ and 0 otherwise from left. Find the fraction of electrons reflected and transmitted if the energy of the electrons is $3V_0/4$eV. Assume that $\sqrt{2mV_0}\, a/\hbar = 3/2$.

5.5 For the square-barrier potential find the probability current, J_x, through the barrier for the incident particles.

5.6 An object of mass $10\,\text{kg}$ is moving towards a $120\,\text{J}$ potential barrier of 2m wide with a speed $2\,\text{m/s}$. Compute the probability for it to tunnel through the barrier.

5.7 For the barrier potential $V(x) = V_0$ for $0 \le x \le a$ and 0 otherwise, write down the expression for transmission probability (T) for $E < V_0$ and $E > V_0$.

5.8 For the barrier potential show that $T + R = 1$.

5.9 For a particle in the presence of the square-well potential solve the equations obtained from the boundary conditions at $x = -a$ and $x = a$ and obtain an expression for F.

5.10 For the square-well potential $V(x) = 0$ for $|x| < a$ and V_0 for $|x| > a$ with particles incident from left with energy $E > V_0$ find the incident and tranmission probability current densities.

5.11 A beam of 10^5 electrons incident on the square-well potential $V(x) = 0$ for $|x| < a$ and V_0 for $|x| > a$ from left. Compute the number of electrons transmitted if the energy of the incident electrons is $5V_0/4$ and $a\sqrt{mV_0}/\hbar = 1/\sqrt{10}$.

5.12 Express the transmission probability T of the square-well potential in terms of E and V_0

5.13 Show that the energy eigenvalues and eigenfunctions of a particle in a box can be obtained from the expression for the finite square-well potential in the limit $V_0 \to \infty$.

5.14 Find the relation between the energy of the electrons incident on the step potential barrier and the strength of the potential V_0 in order to have 75% of the incident particles to be transmitted.

5.15 A beam of electrons is incident on an energy barrier of height V_0 eV of the step potential. Compute the energy of the incident electrons in order to have $T = R = 0.5$.

5.16 In the treatment of step potential, the mass is assumed as constant in both the regions $x < 0$ and $x > 0$. Assuming that the masses are different (which is the case in LEDs and laser diodes) in the two regions determine the reflection and transmission coefficients. Also, find the condition for complete transmission.

5.17 Estimate the penetration distance Δx for (a) a small dust particle of mass 4×10^{-14} kg moving at a very low velocity $v = 10^{-2}$ m/sec and (b) an electron moving at a velocity 10^7 m/sec if they are incident on a potential step of height equal to twice their kinetic energy.

5.18 For the reflectionless potential system with the $H_l = (p_x^2/\hbar^2) - l(l+1)\mathrm{sech}^2 x$ introducing $a_l = (p_x/\hbar) - il\tanh x$ and $a_l^\dagger = (p_x/\hbar) + il\tanh x$ determine the ground state eigenfunctions and eigenvalues for $l = 0$ and $l = 1$. Also, construct the continuum eigenstates for $l = 1$ and 2 and obtain the reflection and transmission amplitudes.

5.19 Consider the delta-function potential $V(x) = V_0\delta(x)$. Obtain the bound state eigenvalues and eigenfunctions for $V_0 < 0$ and reflection and transmission amplitudes for scattering states for $V_0 > 0$.

5.20 For the Dirac-delta potential of the previous problem compute the required energy of the incident particles to make $T = R = 1/2$.

6

Matrix Mechanics

6.1 Introduction

The quantum theoretical description of microscopic systems discussed so far is based on differential operators and the solution of the underlying eigenvalue equations. This approach is called *Schrödinger approach or wave mechanics*. Heisenberg developed another equivalent approach called *matrix mechanics*. Because the trajectory of an electron is unobservable and moreover line spectra implied discrete allowed states, Heisenberg discarded the idea of electron orbits and formulated a theory for line spectra of excited systems based on observables. When Heisenberg arranged the allowed set of transitions in a matrix form, to his surprise, each element of the matrix was found to be related to the probability of a particular transition. This gave the birth to matrix mechanics. Then Heisenberg, Born and Jordan worked out the rules of these matrices.

In the Heisenberg formulation, the wave function ψ becomes a vector while the operators are represented by matrices [1,2]. The transformations from one vector into another is described by matrices. It is thus possible to represent the operators as matrices. The differential equations of the Schrödinger's wave mechanics became matrix equations. For example, consider the differential equation $A\phi = \phi'$, where ϕ and ϕ' are two functions and A is a linear differential operator. In matrix method the functions ϕ and ϕ' are column (or row) vectors and A is a square matrix. The wave mechanical approach and matrix approach, although mathematically different, they are found to give equivalent solutions to the same problems[1] [3,4].

In matrix mechanics all observables are represented by matrices. All possible eigenvalues of a matrix representing an observable are the possible outcomes of experimental measurement on that observable. Since observables take only real values and because the eigenvalues of Hermitian matrices are real they are the most suitable matrices to represent observables. More precisely, in matrix mechanics the properties of the matrices are of interest and not the space on which these matrices operate. In contrast to this, in wave mechanics we study the properties of the elements of this space.

In this chapter first we give the preliminary concepts relative to the use of the matrix method. Next, we present the various elements of matrix mechanics. Then we discuss the matrix representation of operators of certain systems. We introduce Dirac's notation of state vectors, Hilbert space, projection operator, displacement operators and the generators of translations.

---

[1]D.F. Styer described the equivalence of the two theories as *very much as the process of adding algebraic numerals is very different from the process of adding roman numerals but the two processes nevertheless always give the same result* [5].

DOI: 10.1201/9781003172178-6

6.2   Linear Vector Space and Tensor Products

The quantum description of a state of a system is in terms of a vector. To understand the implications of this kind of representation of a system, in this section, let us first introduce the notion of linear vector space.

6.2.1   Linear Vector Space

A *linear vector space* is defined as a set of vectors **A**, **B**, **C**, ... satisfying the following:

1. For every pair of vectors **A** and **B** there exists a **C**: $\mathbf{C} = \mathbf{A} + \mathbf{B}$.

2. Vector addition is commutative and associative:

$$\mathbf{A} + \mathbf{B} = \mathbf{B} + \mathbf{A}, \quad (\mathbf{A} + \mathbf{B}) + \mathbf{C} = \mathbf{A} + (\mathbf{B} + \mathbf{C}) = \mathbf{A} + \mathbf{B} + \mathbf{C}. \tag{6.1}$$

3. The origin of the vector space is the zero vector, **0**.

4. Multiplication is distributive and associative:

$$a(\mathbf{A} + \mathbf{B}) = a\mathbf{A} + a\mathbf{B}, \quad (a + b)\mathbf{A} = a\mathbf{A} + b\mathbf{A}, \quad a(b\mathbf{A}) = (ab)\mathbf{A}, \tag{6.2}$$

where a and b are complex numbers.

5. For every vector **A** there exists $-\mathbf{A}$ such that $\mathbf{A} + (-\mathbf{A}) = 0$.

If a set of vectors $\mathbf{A}_1$, $\mathbf{A}_2$, ..., $\mathbf{A}_N$ obey the relation

$$c_1 \mathbf{A}_1 + c_2 \mathbf{A}_2 + \ldots + c_N \mathbf{A}_N = 0, \tag{6.3}$$

where c_1, c_2, ..., c_N are not all zero then **A**'s are said to be *linearly dependent*. When no such c's exist then the **A**'s are *linearly independent*. The maximum number of linearly independent vectors in a vector space is the *dimension* of the vector space. The dimension can be finite or infinite.

A linearly independent set of vectors $\mathbf{A}_n$ satisfying

$$\psi = \sum_{n=1}^{N} c_n \mathbf{A}_n \tag{6.4}$$

for every vector ψ in the linear vector space of dimension N is called a *coordinate system* or *bases vectors* or simply *bases* for the vector space. Here any arbitrary vector in the space may be written as a linear combination of the basis vectors in one and only one way. c_n in Eq. (6.4) are the components of ψ. The N-tuples of complex numbers $(c_1, c_2, \ldots, c_N)$ are called *vectors* and are represented in matrix form as

$$\begin{bmatrix} c_1 \\ c_2 \\ \vdots \\ c_N \end{bmatrix} \quad \text{or} \quad \begin{bmatrix} c_1 & c_2 & \ldots & c_N \end{bmatrix}. \tag{6.5}$$

Thus, a vector can be expressed as either a row or a column matrix. $\mathbf{A}_1$, $\mathbf{A}_2$, ..., $\mathbf{A}_N$ are called *orthonormal* if

$$(\mathbf{A}_n, \mathbf{A}_m) = \mathbf{A}_n \cdot \mathbf{A}_m^\dagger = \delta_{nm}, \quad n, m = 1, 2, \ldots, N. \tag{6.6}$$

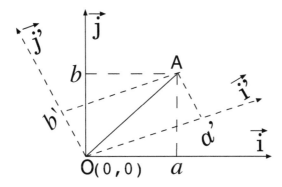

FIGURE 6.1
Illustration of representation of a vector **OA** in two different base vectors $(\mathbf{A}_1, \mathbf{A}_2)$ and $(\mathbf{A}_1', \mathbf{A}_2')$.

From a set of N linearly independent vectors we can construct an orthogonal set employing the Schmidt orthogonalization procedure [6,7].

If the basis set $\mathbf{A}_n$ has N vectors then the space is N-dimensional. We can choose any set of N linearly independent vectors which can span the same vector space. Any vector represented in two different bases will have different components and the column matrices will be different. We can write the basis vectors of one representation as a linear combination of the basis vectors of a second representation. Hence, any two representations of the same linear vector space will be related by a transformation and the transformation matrix will be unitary if the norm of the vector is same in any representation.

Consider a vector **OA** shown in Fig. 6.1. Using the base vectors $(\mathbf{A}_1, \mathbf{A}_2)$ the vector **OA** is expressed as $\mathbf{OA} = a\mathbf{A}_1 + b\mathbf{A}_2$. On the other hand, in terms of the base vectors $(\mathbf{A}_1', \mathbf{A}_2')$ we write $\mathbf{OA} = a'\mathbf{A}_1' + b'\mathbf{A}_2'$. The change from one basis set to another is called *transformation of the base vectors*. Suppose a problem is defined in $(\mathbf{A}_1, \mathbf{A}_2)$. Assume that solving it in terms of $(\mathbf{A}_1', \mathbf{A}_2')$ is much simpler than with $(\mathbf{A}_1, \mathbf{A}_2)$. Then we can go from the set $(\mathbf{A}_1, \mathbf{A}_2)$ to $(\mathbf{A}_1', \mathbf{A}_2')$. This is similar to $(q, p) \to (Q, P)$ in classical mechanics.

Suppose a set of N orthonormal vectors $\{\phi_n\}$ forms a basis for a N-dimensional vector space. Then we express any vector in the space as

$$\psi = \sum_{n=1}^{N} c_n \phi_n, \quad c_n = (\phi_n, \psi), \quad \phi_1 = [1\ 0\ \ldots], \quad \phi_2 = [0\ 1\ \ldots] \tag{6.7}$$

and so on. The inner product of two vectors ψ_1 and ψ_2 is given by

$$\langle \psi_1, \psi_2 \rangle = \left\langle \sum_n c_n \phi_n, \sum_m d_m \phi_m \right\rangle \tag{6.8}$$

which is a number. We have $\langle \psi_1, \psi_2 \rangle = \langle \psi_2, \psi_1 \rangle^*$, where * denotes the complex conjugate. If $\langle \psi_1, \psi_2 \rangle = 0$ then ψ_1 and ψ_2 are said to be *orthogonal*. If the basis vectors $\{\phi_n\}$ are orthonormal then $\langle \phi_n, \phi_m \rangle = \delta_{nm}$. Then in an orthonormal basis we have

$$\langle \psi_1, \psi_2 \rangle = \sum_{n,m} c_n^* d_m \langle \phi_n, \phi_m \rangle = \sum_{n,m} c_n^* d_m \delta_{nm} = \sum_n c_n^* d_n \,. \tag{6.9}$$

The square of norm of ψ_1 is $\sum_n |c_n|^2$. If ψ_1 is normalized then $\langle \psi_1, \psi_1 \rangle = 1$.

Solved Problem 1:

Determine whether the following set of vectors are linearly independent or dependent. $\psi_1 = 3\mathbf{i} + 2\mathbf{j} - \mathbf{k}$, $\psi_2 = 3\mathbf{i} + 6\mathbf{j} - 3\mathbf{k}$, $\psi_3 = \mathbf{i} + 2\mathbf{j} - \mathbf{k}$.

Three vectors A_1, A_2 and A_3 are linearly dependent if at least one C_i, $i = 1, 2, 3$ is nonzero in $C_1\mathbf{A}_1 + C_2\mathbf{A}_2 + C_3\mathbf{A}_3 = 0$. For the given vectors we write

$$
\begin{aligned}
C_1\psi_1 + C_2\psi_2 + C_3\psi_3 &= (3C_1 + 3C_2 + C_3)\mathbf{i} + (2C_1 + 6C_2 + 2C_3)\mathbf{j} \\
&\quad -(C_1 + 3C_2 + C_3)\mathbf{k} = 0 .
\end{aligned}
\tag{6.10}
$$

From the above we have

$$
3C_1 + 3C_2 + C_3 = 0 , \quad 2C_1 + 6C_2 + 2C_3 = 0 , \quad C_1 + 3C_2 + C_3 = 0 .
\tag{6.11}
$$

Solving the above three equations we get $C_1 = 0$, $C_2 = 1$, $C_3 = -3$. Since not all the C_i's are zero the three vectors are linearly dependent.

6.2.2   Tensor Products

The tensor product is a way of making vector spaces together to form larger vector spaces. The construction is crucial for understanding the quantum mechanics of multiparticle system. Suppose V and W are vector spaces of dimensions m and n, respectively. The $V \otimes W$ (read V tensor W) is an mn-dimensional vector space. The elements of $V \otimes W$ are linear combination of tensor products $|v\rangle \otimes |w\rangle$ of elements of $|v\rangle$ of V and $|w\rangle$ of W.

What sorts of linear operators act on the space $V \otimes W$? Suppose $|v\rangle$ and $|w\rangle$ are vectors in V and W and A and B are linear operators in V and W, respectively. Then we define a linear operator $A \otimes B$ in $V \otimes W$ by the equation

$$
(A \otimes B(|v\rangle \otimes |w\rangle)) = A|v\rangle \otimes B|w\rangle .
\tag{6.12}
$$

Suppose A is an $m \times n$ matrix and B is a $p \times q$ matrix. Then we have the matrix representation

$$
A \otimes B = \begin{bmatrix}
A_{11}B & A_{12}B & \ldots & A_{1n}B \\
A_{21}B & A_{22}B & \ldots & A_{2n}B \\
& & \vdots & \\
A_{m1}B & A_{m2}B & \ldots & A_{mn}B
\end{bmatrix} .
\tag{6.13}
$$

For example, the tensor product of $\sigma_x = \begin{pmatrix} 0 & 1 \\ 1 & 0 \end{pmatrix}$ and $\sigma_y = \begin{pmatrix} 0 & -i \\ i & 0 \end{pmatrix}$ is

$$
\sigma_x \otimes \sigma_y = \begin{pmatrix} 0 \times \sigma_y & 1 \times \sigma_y \\ 1 \times \sigma_y & 0 \times \sigma_y \end{pmatrix} = \begin{pmatrix} 0 & 0 & 0 & -i \\ 0 & 0 & i & 0 \\ 0 & -i & 0 & 0 \\ i & 0 & 0 & 0 \end{pmatrix} .
\tag{6.14}
$$

6.3   Matrix Representation of Operators and Wave Function

An operator A operating on a vector ψ_1 in a linear vector space changes it into another vector ψ_2 in the same space. Rewrite $\psi_2 = A\psi_1$ as

$$
\sum_n d_n\phi_n = A \sum_m c_m\phi_m = \sum_m c_m A\phi_m .
\tag{6.15}
$$

Then $d_n = \sum_m \langle \phi_n, A\phi_m \rangle c_m$. Defining

$$A_{nm} = \langle \phi_n, A\phi_m \rangle = \int_{-\infty}^{\infty} \phi_n^* A\phi_m \, \mathrm{d}\tau \tag{6.16}$$

we have

$$d_n = \sum_m A_{nm} c_m, \quad n = 1, 2, \ldots, N. \tag{6.17}$$

The numbers A_{nm} form a square-array and set up a matrix. The equation $\psi_2 = A\psi_1$ in a matrix form is

$$\psi_1 = \begin{bmatrix} c_1 \\ c_2 \\ \vdots \\ c_N \end{bmatrix}, \quad \psi_2 = \begin{bmatrix} d_1 \\ d_2 \\ \vdots \\ d_N \end{bmatrix}, \quad [A_{mn}] = \begin{bmatrix} A_{11} & \cdots & A_{1N} \\ A_{21} & \cdots & A_{2N} \\ & \vdots & \\ A_{N1} & \cdots & A_{NN} \end{bmatrix}. \tag{6.18}$$

Note that any valid arithmetical relation between two or more operators is also applicable between the corresponding matrices. The eigenvalues of an operator A are also the eigenvalues of the matrix representation of the same operator. Similar result holds for eigenfunctions also.

Hermitian operators are represented by Hermitian matrices. A matrix A is *Hermitian* if $A = A^\dagger$, where $(A_{mn})^\dagger = (A_{nm})^*$. When $A = -A^\dagger$ it is *anti-Hermitian*. The following properties are easy to verify:

1. The eigenvalues of a Hermitian matrices are real.

2. The eigenvectors of a Hermitian matrix corresponding to different eigenvalues are orthogonal.

3. The product of Hermitian matrices is a Hermitian if and only if they commute.

4. Commuting Hermitian matrices have simultaneous eigenvectors.

There exists an orthonormal basis $\phi_1, \phi_2, \ldots, \phi_N$ for a Hermitian operator A in a linear vector space of N-dimensions. In relative to ϕ's the operator A is represented by a diagonal matrix

$$A = \begin{pmatrix} \alpha_1 & 0 & \cdots & 0 \\ 0 & \alpha_2 & \cdots & 0 \\ & & \vdots & \\ 0 & 0 & \cdots & \alpha_N \end{pmatrix}, \tag{6.19}$$

where α's are real numbers. ϕ's and α's are solutions of $A\phi = \alpha\phi$. Then

$$A_{nm} = \int_{-\infty}^{\infty} \phi_n^* \alpha_m \phi_m \, \mathrm{d}\tau = \alpha_m \delta_{nm}. \tag{6.20}$$

Notice that any operator in its own eigenvector basis is represented by a diagonal matrix. The diagonal elements of A are the eigenvalues. Any vector ψ can be represented as a linear combination of the eigenvectors of A, as

$$\psi = \sum_n c_n \phi_n, \quad A\phi_n = \alpha_n \phi_n. \tag{6.21}$$

6.4   Unitary Transformation

The representations in different basis system are related by unitary transformation as shown below.

Suppose the set of functions $\{\phi_n(x)\}$ are orthonormal and A is an operator. Then the matrix representation of A in the basis $\{\phi_n(x)\}$ is defined as $A_{nm} = (\phi_n, A\phi_m)$. Any arbitrary state $\psi(x)$ is represented as

$$\psi(x) = \sum_n c_n \phi_n(x), \quad c_n = (\phi_n, \psi). \tag{6.22}$$

Now, consider a second orthonormal basis $\{\phi_n'(x)\}$. The operator A and $\psi(x)$ in this new basis system are

$$A_{nm}' = (\phi_n', A\phi_m'), \quad \psi(x) = \sum_n c_n' \phi_n'(x), \quad c_n' = (\phi_n', \psi). \tag{6.23}$$

What is the relationship between the representations $\phi_n(x)$ and $\phi_n'(x)$? Let us expand $\phi_n'(x)$ in terms of $\{\phi_n(x)\}$ as

$$\phi_n'(x) = \sum_m U_{mn} \phi_m(x), \quad U_{mn} = (\phi_m, \phi_n') = \int_{-\infty}^{\infty} \phi_m^*(x) \phi_n'(x)\, dx. \tag{6.24}$$

The quantity $\sum_n U_{mn} U_{m'n}^*$ is calculated as

$$\begin{aligned}
\sum_n U_{mn} U_{m'n}^* &= \sum_n \int_{-\infty}^{\infty} dx \int_{-\infty}^{\infty} dy\, \phi_m^*(x)\phi_n'(x)\phi_{m'}(y)\phi_n'^*(y)\\
&= \int_{-\infty}^{\infty} dx \int_{-\infty}^{\infty} dy\, \delta(x-y)\phi_{m'}(y)\phi_m^*(x)\\
&= \delta_{mm'}.
\end{aligned} \tag{6.25}$$

Further, $\sum_n U_{nm}^* U_{nm'} = \delta_{mm'}$. Thus, $UU^\dagger = U^\dagger U = I$, where U is unitary. Hence, $\det(U^\dagger U) = \det U^2 = 1$. Therefore, $\det U = \pm 1$. Now, we write

$$c_n' = \sum_m (U^\dagger)_{nm} c_m, \quad c_m = \sum_n U_{mn} c_n'. \tag{6.26}$$

In matrix form Eqs. (6.26) take the form

$$\begin{pmatrix} c_1' \\ c_2' \\ \vdots \end{pmatrix} = U^\dagger \begin{pmatrix} c_1 \\ c_2 \\ \vdots \end{pmatrix}, \quad \begin{pmatrix} c_1 \\ c_2 \\ \vdots \end{pmatrix} = U \begin{pmatrix} c_1' \\ c_2' \\ \vdots \end{pmatrix}. \tag{6.27}$$

The proof of Eq. (6.27) is left as an exercise to the reader. For the matrix form of the operator we have

$$A_{nm}' = \sum_{l,k} U_{ln}^* A_{lk} U_{km} \quad \text{or} \quad A' = U^\dagger A U. \tag{6.28}$$

Thus, the representations in different basis systems are related by unitary transformations. We have $\text{Tr}AB = \text{Tr}BA$, $\det AB = (\det A)(\det B)$ and

$$\text{Tr}A' = \text{Tr}(U^\dagger A U) = \text{Tr}(U^\dagger U)A = \text{Tr}A. \tag{6.29}$$

Therefore, the trace and determinant of any operator are invariant under unitary transformations. Hence, no operator has unique matrix representation. Any operator will have different matrix representation in different basis. But all these matrices will have same eigenvalues. Any matrix representation of an operator on an arbitrary basis can be brought into a diagonal form by means of a unitary transformation of the arbitrary basis to the eigenvector basis.

There are many ways of expressing a unitary operator U in terms of a Hermitian operator. The most common one is

$$U = e^{iQ} = \sum_{n=0}^{\infty} \frac{(iQ)^n}{n!}, \tag{6.30}$$

where Q is Hermitian so that $U^\dagger = e^{-iQ^\dagger} = e^{-iQ}$ and $UU^\dagger = U^\dagger U = I$. Another useful form is $U = (1 + ik)/(1 - ik)$, where k is Hermitian.

Solved Problem 2:

What will happen to a unitary operator under a unitary transformation?

Let $A' = UAU^\dagger$. Consider $A'^\dagger A'$. We get

$$
\begin{aligned}
A'^\dagger A' &= (UAU^\dagger)^\dagger (UAU^\dagger) \\
&= (UA^\dagger U^\dagger)(UAU^\dagger) \\
&= UA^\dagger(U^\dagger U)AU^\dagger \\
&= U(A^\dagger A)U^\dagger \\
&= UU^\dagger \\
&= I.
\end{aligned} \tag{6.31}
$$

Similarly, we can show that $A'A'^\dagger = I$. Thus, $A'A'^\dagger = A'^\dagger A' = I$. Thus, a unitary operator remains unitary under a unitary transformation.

6.5 Schrödinger Equation and Other Quantities in Matrix Form

Suppose the energy eigenfunctions of a Hamiltonian operator are $\phi_1, \phi_2, \ldots, \phi_n$ and the corresponding energy eigenvalues are $E_1, E_2, \ldots, E_n$ so that $H\phi_n = E_n\phi_n$. Now,

$$H_{nm} = \langle \phi_n, H\phi_m \rangle = E_m \langle \phi_n, \phi_m \rangle = E_m \delta_{nm}, \tag{6.32}$$

where ϕ_n's are an orthonormal set of basis functions. H is a diagonal square matrix with the diagonal elements E_m.

Let us write the time-dependent Schrödinger equation in matrix form. We write

$$\psi(\mathbf{X}, t) = \sum_i c_i(t)\phi_i(\mathbf{X}). \tag{6.33}$$

Then $H\psi = E\psi$ is given by

$$\sum_i Hc_i(t)\phi_i(\mathbf{X}) = i\hbar \sum_i \frac{d}{dt}c_i(t)\phi_i(\mathbf{X}). \tag{6.34}$$

Multiplying Eq. (6.34) by $\phi_j^*(\mathbf{X})$ and integrating over all space we obtain

$$\sum_i c_i(t) \int_{-\infty}^{\infty} \phi_j^* H \phi_i \, \mathrm{d}\tau = \mathrm{i}\hbar \sum_i \frac{\mathrm{d}}{\mathrm{d}t} c_i(t) \int_{-\infty}^{\infty} \phi_j^* \phi_i \, \mathrm{d}\tau \qquad (6.35)$$

or

$$\sum_i c_i(t) H_{ji} = \mathrm{i}\hbar \sum_i \frac{\mathrm{d}}{\mathrm{d}t} c_i(t) \delta_{ji} = \mathrm{i}\hbar \frac{\mathrm{d}}{\mathrm{d}t} c_j(t) \,. \qquad (6.36)$$

Interchanging i and j gives

$$\sum_j c_j(t) H_{ij} = \mathrm{i}\hbar \frac{\mathrm{d}}{\mathrm{d}t} c_i(t) \,. \qquad (6.37)$$

In matrix form the normalization condition $1 = \int_{-\infty}^{\infty} \psi^* \psi \, \mathrm{d}\tau = \sum_n |c_n|^2$ is

$$\int_{-\infty}^{\infty} \psi^* \psi \, \mathrm{d}\tau = [c_1^* \ c_2^* \ \cdots \ c_N^*] \begin{bmatrix} c_1 \\ c_2 \\ \vdots \\ c_N \end{bmatrix} = c^\dagger c = I \,. \qquad (6.38)$$

The orthogonality condition $\int_{-\infty}^{\infty} \psi_a^* \psi_b \, \mathrm{d}\tau = 0$ in matrix form is

$$\begin{aligned} 0 &= \int_{-\infty}^{\infty} \sum_n \sum_m c_{an}^* \phi_n^* c_{bm} \phi_m \, \mathrm{d}\tau \\ &= \sum_n c_{an}^* c_{bn} \\ &= [c_{a1}^* \ c_{a2}^* \ \cdots \ c_{aN}^*] \begin{bmatrix} c_{b1} \\ c_{b2} \\ \vdots \\ c_{bN} \end{bmatrix} \\ &= c_a^\dagger c_b \,. \end{aligned} \qquad (6.39)$$

In matrix form the expectation value of an observable A is

$$\langle A \rangle = \int_{-\infty}^{\infty} \sum_n c_n^* \phi_n^* A \sum_m c_m \phi_m \, \mathrm{d}\tau = \sum_n \sum_m c_n^* A_{nm} c_m \,. \qquad (6.40)$$

What is $\langle H \rangle$ if ψ is any one of the eigenfunctions, say, ϕ_k?

6.6   Application to Certain Systems

In this section, we work out the matrix forms of operators of certain quantum mechanical systems. We start with the linear harmonic oscillator.

6.6.1 Linear Harmonic Oscillator

The matrix elements of x are determined from $x_{nm} = \int_{-\infty}^{\infty} \phi_n^* x \phi_m \, dx$. The eigenfunctions of linear harmonic oscillator are given by Eq. (4.30). Substituting the expressions for ϕ_n^* and ϕ_m in the above integral and introducing the change of variable $x = y/\alpha$ we get

$$x_{nm} = \frac{N_n N_m}{\alpha^2} \int_{-\infty}^{\infty} H_n(y) y H_m(y) e^{-y^2} \, dy \,. \tag{6.41}$$

The value of the integral in the above equation is given by

$$
\begin{aligned}
\int_{-\infty}^{\infty} H_n(y) y H_m(y) e^{-y^2} \, dy &= \sqrt{\pi} \left[2^{n-1} n! \delta_{m,n-1} + 2^n (n+1)! \delta_{m,n+1} \right] \\
&= \begin{cases} \sqrt{\pi}\, 2^{n-1} n!, & \text{for } m = n - 1 \\ \sqrt{\pi}\, 2^n (n+1)!, & \text{for } m = n + 1 \\ 0, & \text{for } m \neq n \pm 1 . \end{cases}
\end{aligned} \tag{6.42}
$$

When $m = n + 1$ we have $x_{nm} = \sqrt{(n+1)/2}/\alpha$. For $m = n - 1$ we get $x_{nm} = \sqrt{n/2}/\alpha$. Then

$$x_{nm} = \begin{cases} \sqrt{n/2}/\alpha\,, & \text{for } m = n - 1 \\ \sqrt{(n+1)/2}/\alpha\,, & \text{for } m = n + 1 \\ 0, & \text{for } m \neq n \pm 1 \end{cases} \tag{6.43}$$

where $n, m = 0, 1, \ldots$. The matrix form of the operator x is

$$
x = \begin{pmatrix} x_{00} & x_{01} & \cdots \\ x_{10} & x_{11} & \cdots \\ \vdots & \vdots & \vdots \end{pmatrix} = \sqrt{\frac{\hbar}{2m\omega}} \begin{pmatrix} 0 & \sqrt{1} & 0 & 0 & \cdots \\ \sqrt{1} & 0 & \sqrt{2} & 0 & \cdots \\ 0 & \sqrt{2} & 0 & \sqrt{3} & \cdots \\ 0 & 0 & \sqrt{3} & 0 & \cdots \\ \vdots & \vdots & \vdots & \vdots & \vdots \end{pmatrix}. \tag{6.44}
$$

The matrix elements of the momentum operator $p_x = -i\hbar d/dx$ are computed as follows. We obtain

$$
\begin{aligned}
p_{xnm} &= \int_{-\infty}^{\infty} \phi_n^*(\alpha x) \left(-i\hbar \frac{d}{dx} \right) \phi_m(\alpha x) \, dx \\
&= -i\hbar N_n N_m \int_{-\infty}^{\infty} H_n(\alpha x) e^{-\alpha^2 x^2/2} \frac{d}{dx} H_m(\alpha x) e^{-\alpha^2 x^2/2} \, dx \\
&= i\hbar N_n N_m \left[\alpha^2 \int_{-\infty}^{\infty} H_n H_m x e^{-\alpha^2 x^2} \, dx - \int_{-\infty}^{\infty} H_n \frac{dH_m}{dx} e^{-\alpha^2 x^2} \, dx \right] \\
&= i\hbar N_n N_m \left[\int_{-\infty}^{\infty} H_n H_m y\, e^{-y^2} \, dy - \int_{-\infty}^{\infty} H_n H_m' y\, e^{-y^2} \, dy \right]. \tag{6.45}
\end{aligned}
$$

The value of the first integral in Eq. (6.45) is given by Eq. (6.42) while that of the second integral is

$$\int_{-\infty}^{\infty} H_n(y) H_m'(y) e^{-y^2} \, dy = 2m \int_{-\infty}^{\infty} H_n(y) H_{m-1}(y) e^{-y^2} \, dy = 2m\sqrt{\pi} 2^n n! \delta_{n,m-1}, \tag{6.46}$$

where we have used the relation $H_m'(y) = 2m H_{m-1}(y)$. Now Eq. (6.45) becomes

$$p_{xnm} = i\hbar N_n N_m \sqrt{\pi} \left[2^{n-1} n! \delta_{m,n-1} + 2^n (n+1)! \delta_{m,n+1} - 2m 2^n n! \delta_{n,m-1} \right]. \tag{6.47}$$

Or

$$p_{xnm} = i\hbar \begin{cases} N_n N_{n+1}\sqrt{\pi}\,[2^n(n+1)! - 2(n+1)2^n n!], & m = n+1 \\ N_n N_{n-1}\sqrt{\pi}\,2^{n-1}n!, & m = n-1 \\ 0, & m \neq n \pm 1. \end{cases} \tag{6.48}$$

Using the values of N_{n+1}, N_n and N_{n-1} we get

$$N_n N_{n+1}\sqrt{\pi}\,[2^n(n+1)! - 2(n+1)2^n n!] = -\alpha\sqrt{\frac{n+1}{2}}, \tag{6.49}$$

and $p_{nm} = -i\hbar\alpha\sqrt{(n+1)/2}$. For $m = n-1$ we obtain

$$N_n N_{n-1}\sqrt{\pi}\,2^{n-1}n! = \alpha\sqrt{n/2}. \tag{6.50}$$

Then

$$p_{xnm} = i\hbar \begin{cases} \alpha\sqrt{n/2}, & \text{for } m = n-1 \\ -\alpha\sqrt{(n+1)/2}, & \text{for } m = n+1 \\ 0, & \text{for } m \neq n \pm 1. \end{cases} \tag{6.51}$$

Comparison of Eqs. (6.43) and (6.51) indicates that the matrix of p_x is analogous to the matrix of x with $1/\alpha$ in Eq. (6.43) is replaced by $-i\hbar\alpha$. The matrix form of p_x is

$$p_x = i\sqrt{\frac{m\hbar\omega}{2}} \begin{pmatrix} 0 & -\sqrt{1} & 0 & 0 & 0 & \cdots \\ \sqrt{1} & 0 & -\sqrt{2} & 0 & 0 & \cdots \\ 0 & \sqrt{2} & 0 & -\sqrt{3} & 0 & \cdots \\ 0 & 0 & \sqrt{3} & 0 & -\sqrt{4} & \cdots \\ \vdots & \vdots & \vdots & \vdots & \vdots & \vdots \end{pmatrix}. \tag{6.52}$$

The operators a and $a^\dagger$ given by Eq. (4.54) are obtained as

$$a = \frac{1}{\sqrt{2m\omega\hbar}}(m\omega x + ip_x) = \begin{pmatrix} 0 & \sqrt{1} & 0 & 0 & 0 & \cdots \\ 0 & 0 & \sqrt{2} & 0 & 0 & \cdots \\ 0 & 0 & 0 & \sqrt{3} & 0 & \cdots \\ \vdots & \vdots & \vdots & \vdots & \vdots & \vdots \end{pmatrix} \tag{6.53}$$

and

$$a^\dagger = \frac{1}{\sqrt{2m\omega\hbar}}(m\omega x - ip_x) = \begin{pmatrix} 0 & 0 & 0 & 0 & 0 & \cdots \\ \sqrt{1} & 0 & 0 & 0 & 0 & \cdots \\ 0 & \sqrt{2} & 0 & 0 & 0 & \cdots \\ 0 & 0 & \sqrt{3} & 0 & 0 & \cdots \\ \vdots & \vdots & \vdots & \vdots & \vdots & \vdots \end{pmatrix}. \tag{6.54}$$

The Hamiltonian operator $H = \hbar\omega(a^\dagger a + aa^\dagger)/2$ is then worked out to be

$$H = \hbar\omega \begin{pmatrix} \dfrac{1}{2} & 0 & 0 & 0 & 0 & \cdots \\ 0 & \dfrac{3}{2} & 0 & 0 & 0 & \cdots \\ 0 & 0 & \dfrac{5}{2} & 0 & 0 & \cdots \\ \vdots & \vdots & \vdots & \vdots & \vdots & \vdots \end{pmatrix}. \tag{6.55}$$

Solved Problem 3:

For the linear harmonic oscillator using the matrix forms of the operators x and p_x verify that $xp_x - p_x x = i\hbar I$.

We obtain

$$
xp_x - p_x x = \sqrt{\frac{\hbar}{2m\omega}}\, i\sqrt{\frac{m\hbar\omega}{2}} \left[\begin{pmatrix} 0 & \sqrt{1} & 0 & 0 & 0 & \cdots \\ \sqrt{1} & 0 & \sqrt{2} & 0 & 0 & \cdots \\ 0 & \sqrt{2} & 0 & \sqrt{3} & 0 & \cdots \\ 0 & 0 & \sqrt{3} & 0 & \sqrt{4} & \cdots \\ 0 & 0 & 0 & \sqrt{4} & 0 & \cdots \\ \vdots & \vdots & \vdots & \vdots & \vdots & \vdots \end{pmatrix} \right.
$$

$$
\times \begin{pmatrix} 0 & -\sqrt{1} & 0 & 0 & 0 & \cdots \\ \sqrt{1} & 0 & -\sqrt{2} & 0 & 0 & \cdots \\ 0 & \sqrt{2} & 0 & -\sqrt{3} & 0 & \cdots \\ 0 & 0 & \sqrt{3} & 0 & -\sqrt{4} & \cdots \\ 0 & 0 & 0 & \sqrt{4} & 0 & \cdots \\ \vdots & \vdots & \vdots & \vdots & \vdots & \vdots \end{pmatrix}
$$

$$
- \begin{pmatrix} 0 & -\sqrt{1} & 0 & 0 & 0 & \cdots \\ \sqrt{1} & 0 & -\sqrt{2} & 0 & 0 & \cdots \\ 0 & \sqrt{2} & 0 & -\sqrt{3} & 0 & \cdots \\ 0 & 0 & \sqrt{3} & 0 & -\sqrt{4} & \cdots \\ 0 & 0 & 0 & \sqrt{4} & 0 & \cdots \\ \vdots & \vdots & \vdots & \vdots & \vdots & \vdots \end{pmatrix}
$$

$$
\times \left. \begin{pmatrix} 0 & -\sqrt{1} & 0 & 0 & 0 & \cdots \\ \sqrt{1} & 0 & -\sqrt{2} & 0 & 0 & \cdots \\ 0 & \sqrt{2} & 0 & -\sqrt{3} & 0 & \cdots \\ 0 & 0 & \sqrt{3} & 0 & -\sqrt{4} & \cdots \\ 0 & 0 & 0 & \sqrt{4} & 0 & \cdots \\ \vdots & \vdots & \vdots & \vdots & \vdots & \vdots \end{pmatrix} \right]
$$

$$
= \frac{i\hbar}{2} \left[\begin{pmatrix} 1 & 0 & -\sqrt{2} & 0 & \cdots \\ 0 & 1 & 0 & -\sqrt{6} & \cdots \\ \sqrt{2} & 0 & 1 & 0 & \cdots \\ 0 & \sqrt{6} & 0 & 1 & \cdots \\ \vdots & \vdots & \vdots & \vdots & \vdots \end{pmatrix} \right.
$$

$$
\left. - \begin{pmatrix} -1 & 0 & -\sqrt{2} & 0 & \cdots \\ 0 & -1 & 0 & -\sqrt{6} & \cdots \\ \sqrt{2} & 0 & -1 & 0 & \cdots \\ 0 & \sqrt{6} & 0 & -1 & \cdots \\ \vdots & \vdots & \vdots & \vdots & \vdots \end{pmatrix} \right]
$$

The result is

$$xp_x - p_x x = i\hbar \begin{pmatrix} 1 & 0 & 0 & 0 & \cdots \\ 0 & 1 & 0 & 0 & \cdots \\ 0 & 0 & 1 & 0 & \cdots \\ 0 & 0 & 0 & 1 & \cdots \\ \vdots & \vdots & \vdots & \vdots & \vdots \end{pmatrix} = i\hbar I. \tag{6.56}$$

Similarly, one can verify that $Hp_x - p_x H = i\hbar m\omega^2 x$.

6.6.2 Particle in a Box

We consider the infinite height potential $V(x) = 0$ for $0 < x < a$ and ∞ elsewhere. The normalized eigenfunctions are $\phi_n(x) = \sqrt{2/a}\sin(n\pi x/a)$, $n = 1, 2, \ldots$.

The matrix elements of x_{nm} are obtained as

$$x_{nm} = \frac{2}{a}\int_0^a \sin\frac{n\pi x}{a} \, x \, \sin\frac{m\pi x}{a}\,\mathrm{d}x = \frac{2a}{\pi^2}\int_0^\pi y\sin ny\sin my\,\mathrm{d}y \,.$$

The value of the integral in the above equation is obtained as

$$\begin{aligned}
x_{nm} &= \frac{a}{\pi^2}\left[\int_0^\pi y\cos(n-m)y\,\mathrm{d}y - \int_0^\pi y\cos(n+m)y\,\mathrm{d}y\right] \\
&= \frac{a}{\pi^2}\left[\frac{1}{(n-m)^2}\left\{(-1)^{n-m} - 1\right\} - \frac{1}{(n+m)^2}\left\{(-1)^{n+m} - 1\right\}\right] \\
&= \frac{4anm}{\pi^2(n^2 - m^2)^2}\left[(-1)^{n-m} - 1\right], \quad n \neq m, \ n, m = 1, 2, \ldots.
\end{aligned} \tag{6.57}$$

When $n = m$ we find

$$x_{nm} = \frac{a}{\pi^2}\left[\int_0^\pi y\,\mathrm{d}y - \int_0^\pi y\cos 2ny\,\mathrm{d}y\right] = \frac{a}{2}\,. \tag{6.58}$$

Combining Eqs. (6.57) and (6.58) we write

$$x_{nm} = \begin{cases} -8anm/(\pi^2(n^2 - m^2)^2), & \text{for } m \neq n, \ n - m = \text{odd} \\ a/2, & \text{for } m = n \\ 0, & \text{for } m \neq n, \ n - m = \text{even}. \end{cases} \tag{6.59}$$

The matrix form of x is

$$x = \frac{a}{2\pi^2}\begin{pmatrix} \pi^2 & -32/9 & 0 & -64/225 & \cdots \\ -32/9 & \pi^2 & -96/25 & 0 & \cdots \\ 0 & -96/25 & \pi^2 & -192/49 & \cdots \\ \vdots & \vdots & \vdots & \vdots & \vdots \end{pmatrix}. \tag{6.60}$$

Next, we evaluate the matrix elements of the momentum operator p_x. We have

$$\begin{aligned}
p_{xnm} &= -\frac{2i\hbar m\pi}{a^2}\int_0^a \sin\frac{n\pi x}{a}\cos\frac{m\pi x}{a}\,\mathrm{d}x \\
&= \frac{2i\hbar nm}{a(n^2 - m^2)}\left[(-1)^{n+m} - 1\right], \quad n \neq m.
\end{aligned} \tag{6.61}$$

When $m = n$

$$p_{xnn} = -\frac{2i\hbar n}{a} \int_0^\pi \sin ny \cos ny \, dy = 0. \tag{6.62}$$

Combining the Eqs. (6.61) and (6.62) we write

$$p_{xnm} = \begin{cases} -4inm\hbar/(a(n^2 - m^2)), & \text{for } n + m = \text{odd} \\ 0, & \text{otherwise}. \end{cases} \tag{6.63}$$

The matrix form of p_x is

$$p_x = \frac{i\hbar}{a} \begin{pmatrix} 0 & 8/3 & 0 & 16/15 & \cdots \\ -8/3 & 0 & 24/5 & 0 & \cdots \\ 0 & -24/5 & 0 & 48/7 & \cdots \\ \vdots & \vdots & \vdots & \vdots & \vdots \end{pmatrix}. \tag{6.64}$$

Since the energy eigenvalues are given by $E_n = \pi^2\hbar^2 n^2/(2ma^2)$, $n = 1, 2, \ldots$ the matrix form of the Hamiltonian is

$$H = \frac{\pi^2\hbar^2}{2ma^2} \begin{pmatrix} 1 & 0 & 0 & 0 & \cdots \\ 0 & 4 & 0 & 0 & \cdots \\ 0 & 0 & 9 & 0 & \cdots \\ \vdots & \vdots & \vdots & \vdots & \vdots \end{pmatrix}. \tag{6.65}$$

6.7 Dirac's Bra and Ket Notations

As mentioned earlier the quantum state of a system can be represented in terms of a vector. The values of the wave functions $\psi(\mathbf{X})$ at various points $\mathbf{X}$ are the components of a vector. The components depend on the choice of coordinate system or basis in the vector space. Instead of specifying one or another set of infinitely many components, Dirac introduced a vector notation for the state vector [8]. A state vector is represented by a symbol $|\rangle$ called a *ket*. For example, $\psi(\mathbf{X})$ is written as $|\psi\rangle$. Corresponding to every vector $|\psi\rangle$ there is a vector $\langle\psi|$ called *bra* vector, the complex conjugate of $|\psi\rangle$.

6.7.1 Basis Vectors

Recall from our discussion in earlier sections of this chapter that any vector can be written in terms of a basis vector. *What are the basis of a quantum system?* A basis is essentially constituted by the set of vectors with each possible outcome of a measurement of a particular observable of the system. In other words, the eigenvectors of the operator of an observable can always be used as the basis for the vector space.

In a three-dimensional vector (in physical space) $\mathbf{v} = a\mathbf{i} + b\mathbf{j} + c\mathbf{k}$, where $\mathbf{i}$, $\mathbf{j}$ and $\mathbf{k}$ are unit vectors along the x, y and z-axis, respectively, and a, b and c are the components of $\mathbf{v}$ along the x, y and z-axes, respectively. In the same way, a quantum state $|\psi\rangle$ is written as

$$|\psi\rangle = c_1|\phi_1\rangle + c_2|\phi_2\rangle + \ldots, \tag{6.66}$$

where the unit vectors $|\phi_1\rangle$, $|\phi_2\rangle$, $\ldots$, etc are like the unit vectors $\mathbf{i}$, $\mathbf{j}$, $\mathbf{k}, \ldots$ and the coefficients $c_1, c_2, \ldots$ are like a, b, $c, \ldots$. However, c_i's are generally complex. We write Eq. (6.66) as

$$|\psi\rangle = \sum_i c_i|\phi_i\rangle \ \text{ or } \ \sum_i c_i|i\rangle \,, \tag{6.67}$$

where $|i\rangle$ denotes a unit vector or a *base vector* $|\phi_i\rangle$. Then the bra state is $\langle\psi| = \sum_i c_i^*\langle i|$. The number of base vectors required in quantum mechanics depends on the situation. In matrix form

$$|\psi\rangle = \begin{bmatrix} c_1 \\ c_2 \\ \vdots \end{bmatrix} \,, \quad \langle\psi| = [c_1^* \ c_2^* \ \dots] \,. \tag{6.68}$$

6.7.2 Examples of Basis in Quantum Theory

In this section, we give a few examples of basis in quantum mechanics [9].

1. Energy representation

Let the discrete energy eigenvalues of a particle are E_1, E_2, ... A measurement of energy of the particle would yield one of E_i's. The set of basis vectors in this case is $|E_1\rangle$, $|E_2\rangle$, ... The number of basis vectors is infinite.

2. Position representation

Consider a one-dimensional particle. A measurement of its position would give a real number, say q. Denote $|q\rangle$ the state of the particle for which the outcome of position measurement is q. The basis states are taken as the set $\{|q\rangle\}$, where $-\infty < q < \infty$. As the possible values of q are infinite, the number of basis vectors is infinite.

3. Momentum representation

Instead of position values we may measure the momentum p of the particle. In this case the set $\{|p\rangle\}$, where $-\infty < p < \infty$ forms the basis states. In this representation also the number of basis vectors is infinite.

4. Spin representation

Consider a particle whose spin measurement in any direction gives $\pm\hbar/2$. If the spin measurement in a direction $\mathbf{e}$ yields $\hbar/2$ then the corresponding state is, for example, labeled as $|1/2, \mathbf{e}\rangle$. If the outcome is $-\hbar/2$ then the state is $|-1/2, \mathbf{e}\rangle$. The state of the spin component in any direction is then represented in terms of the states $|1/2, \mathbf{e}\rangle$ and $|-1/2, \mathbf{e}\rangle$. Consequently, the number of basis vectors is only two.

The basis vectors in energy, position and momentum representations are

$$\psi_n(x) \ \rightarrow \ |\psi_n\rangle \rightarrow |n\rangle \text{ or } |i\rangle \,, \ \psi_E(x) \rightarrow |E\rangle \,, \tag{6.69a}$$

$$\psi_q(x) \ \rightarrow \ |q\rangle \,, \quad \psi_p(x) \rightarrow |p\rangle \,, \tag{6.69b}$$

respectively. In terms of the basis vectors the state vector in these three representations in ket notation are

$$\psi(x) \ \rightarrow \ |\psi\rangle = \sum c_n|n\rangle \text{ or } \sum c_i|i\rangle \,, \tag{6.70a}$$

$$\psi(x) \ \rightarrow \ |\psi\rangle = \int_{-\infty}^{\infty} c_q|q\rangle \, \mathrm{d}q \,, \tag{6.70b}$$

$$\psi(x) \ \rightarrow \ |\psi\rangle = \int_{-\infty}^{\infty} c_p|p\rangle \, \mathrm{d}p \,. \tag{6.70c}$$

To determine the complex coefficients c_n in the linear vector space, define inner product between ψ_m and ψ_n as

$$(\psi_n, \psi_m) = \int_{-\infty}^{\infty} \psi_n^* \psi_m \, d\tau = \langle \psi_n | \psi_m \rangle \text{ or } \langle n | m \rangle \tag{6.71}$$

and is a number δ_{nm}. Observe that when the bra is on the left of a ket, that is, $((\langle \psi_n |)(|\psi_m \rangle))$, then they contract to give a number. $\langle \psi_n |$ and $|\psi_m \rangle$ are said to be *orthogonal* if $\langle \psi_n | \psi_m \rangle = 0$. The basis vectors are orthonormal if $\langle \phi_i | \phi_j \rangle = \delta_{ij}$. On taking the inner product of Eq. (6.67) successively with $|\phi_i \rangle$ and on using $\langle \phi_i | \phi_j \rangle = \delta_{ij}$ we obtain $c_i = \langle \phi_i | \psi \rangle$ which gives the value of the component of $|\psi \rangle$ along $|\phi_i \rangle$.

The c's in Eq. (6.70) are given by

$$c_n = \langle n | \psi \rangle, \quad c_q = \langle q | \psi \rangle, \quad c_p = \langle p | \psi \rangle. \tag{6.72}$$

If λ_i is the eigenvalue corresponding to the eigenvector $|\phi_i \rangle$ of an operator Λ then the probability of getting λ_i as an eigenvalue in a measurement is

$$P(\lambda_i) = |c_i|^2 = |\langle \phi_i | \psi \rangle|^2. \tag{6.73}$$

Since the total probability is 1 we have $\sum |c_i|^2 = 1$. If a system is in the state $|\phi_i \rangle$ then the state $|\psi \rangle$ no longer exists. If $|\psi \rangle$ is normalized to unity then

$$\langle \psi | \psi \rangle = \sum \langle \psi | \phi_i \rangle \langle \phi_i | \psi \rangle = \sum |\langle \phi_i | \psi \rangle|^2 = \sum |c_i|^2 = 1. \tag{6.74}$$

6.7.3 Properties of Ket and Bra Vectors

An operator by definition acts on one ket to produce another ket. In Dirac's notation there is a useful way of representing linear operators, known as the outer product representation. For example, the inner product $\langle v | w \rangle$ is a complex number whereas $|v \rangle \langle w|$ called the *outer product* is an operator. If $|w' \rangle$ is some ket then the action of $|v \rangle \langle w|$ on $|w' \rangle$ is defined by $|v \rangle \langle w | w' \rangle = \langle w | w' \rangle | v \rangle$.

What is the rule to obtain Hermitian conjugate of an expression? The rule contains

1. reversing the order of the terms in the expression and
2. changing
    (a) the numbers into their complex conjugates
    (b) the operators into their adjoints,
    (c) the kets into bras and bras into kets.

The Hermitian conjugate of $\lambda | \psi_2 \rangle \langle \psi_1 | AB^\dagger C$ is $\lambda^* C^\dagger BA^\dagger | \psi_1 \rangle \langle \psi_2 |$.

Some of the properties of the ket and bra vectors are summarized below, where A and B are Hermitian operators:

$$A|\psi \rangle = |\psi' \rangle, \quad \langle \psi | A = \langle \psi' |, \tag{6.75a}$$
$$\langle A \rangle = \langle \psi | A | \psi \rangle, \quad A_{nm} = \langle \psi_n | A | \psi_m \rangle, \tag{6.75b}$$
$$\langle \psi_m | \psi_n \rangle = \langle \psi_n | \psi_m \rangle^*, \quad \langle \psi_n | \psi_m \rangle = \delta_{nm}, \tag{6.75c}$$
$$(A + B)|\psi \rangle = A|\psi \rangle + B|\psi \rangle, \tag{6.75d}$$
$$\langle q | q' \rangle = \delta(q - q'), \quad \langle p | p' \rangle = \delta(p - p'). \tag{6.75e}$$

The time-dependent Schrödinger equation is written as

$$i\hbar\frac{\partial}{\partial t}|\psi\rangle = H|\psi\rangle.\tag{6.76}$$

The time-independent Schrödinger equation for $|\psi_n\rangle$ is $H|\psi_n\rangle = E_n|\psi_n\rangle$. Then

$$|\psi_n(t)\rangle = e^{-iE_nt/\hbar}|\psi_n(0)\rangle.\tag{6.77}$$

The arbitrary state $|\psi\rangle$ is expressed as

$$|\psi\rangle = \sum_n c_n e^{-iE_nt/\hbar}|\psi_n\rangle.\tag{6.78}$$

Solved Problem 4:

Compare two-dimensional ordinary vectors and quantum state vectors.

The following table gives a comparison of ordinary and quantum state vectors.

S.No.	Ordinary Vectors	Quantum State Vectors
1.	$\mathbf{X} = a\mathbf{i} + b\mathbf{j}$	$|\psi\rangle = \sum c_n|n\rangle$
2.	$\mathbf{i}\cdot\mathbf{j} = 0$ $\mathbf{i}\cdot\mathbf{i} = \mathbf{j}\cdot\mathbf{j} = 1$	$\langle n|m\rangle = 0$ $\langle n|n\rangle = \langle m|m\rangle = 1$
3.	$\mathbf{A}\cdot\mathbf{B} = \mathbf{B}\cdot\mathbf{A}$	$\langle x|\psi\rangle = \langle\psi|x\rangle^*$
4.	$\mathbf{A} = a_1\mathbf{i} + a_2\mathbf{j}$ $\mathbf{B} = b_1\mathbf{i} + b_2\mathbf{j}$ $\mathbf{A}\cdot\mathbf{B} = a_1b_1 + a_2b_2$	$|\psi\rangle = \sum c_n|n\rangle$ $|\chi\rangle = \sum d_n|n\rangle$ $\langle\chi|\psi\rangle = \sum c_n d_n$
5.	$\mathbf{X}\cdot\mathbf{i} = a$, $\mathbf{X}\cdot\mathbf{j} = b$, $\mathbf{X} = \mathbf{i}(\mathbf{X}\cdot\mathbf{i}) + \mathbf{j}(\mathbf{X}\cdot\mathbf{j})$	$|\psi\rangle = \sum|n\rangle\langle n|\psi\rangle$

6.8 Dimensions of Kets and Bras

In sec. 2.5 we noticed that the dimension of $\psi(\mathbf{X},t)$ for a single particle in three-dimension is (length)$^{-3/2}$, that is, $L^{-3/2}$ and that of $\psi(x,t)$ in one-dimension is $L^{-1/2}$. For the momentum wave function $\Phi(p,t)$ the normalization condition $\int_{-\infty}^{\infty}|\Phi(p,t)|^2 dp = 1$ gives $[\Phi(p,t)] = (MLT^{-1})^{-1/2}$. *Do kets and bras have dimensions?* In this section, we determine the dimensions of kets and bras [10].

Denote the dimension of a quantity Q as $[Q]$. If Q is dimensionless then $[Q] = 1$. The dimensions of x and its conjugate momentum p are L and MLT^{-1}, respectively. For the kets $|\alpha\rangle$ and $|\beta\rangle$ we write $[\langle\alpha|\beta\rangle] = [\langle\alpha|][|\beta\rangle]$ and as the complex conjugation does not alter the dimension we write $[\langle\alpha|\beta\rangle] = [\langle\beta|\alpha\rangle]$ and $[|\alpha\rangle] = [\langle\alpha|]$. For the physical state $|\psi\rangle$, the continuous position basis $\{|x\rangle\}$ with $\hat{x}|x\rangle = x|x\rangle$, where $\hat{x}$ is the position operator and the continuous momentum basis $\{|p\rangle\}$ with $\hat{p}|p\rangle = p|p\rangle$, where $\hat{p}$ is the momentum opterator we denote

$$[|\psi\rangle] = \mathrm{K}, \quad [\langle\psi|] = \mathrm{B}, \quad [|x\rangle] = \mathrm{X_K}, \quad [\langle x|] = \mathrm{X_B}, \quad [|p\rangle] = \mathrm{P_K}, \quad [\langle p|] = \mathrm{P_B},\tag{6.79}$$

where K, B, X_B, X_K, P_K and P_B are dimensional formulas. $[\langle\psi|\psi\rangle] = 1$ gives $[\langle\psi|][|\psi\rangle] = BK = 1$. That is, $B = 1/K$.

The position space wave function $\Psi(x,t)$ is defined through the equation

$$\langle x|\Psi\rangle = \langle\Psi|x\rangle^* = \Psi(x,t) \tag{6.80}$$

and this equation gives

$$X_B K = B X_K = L^{-1/2}. \tag{6.81}$$

Substitution of B=1/K in the above equation gives

$$X_B K = X_K K^{-1} = L^{-1/2}. \tag{6.82}$$

Similarly, for $\Phi(p,t)$

$$\langle p|\Psi\rangle = \langle\Psi|p\rangle^* = \Phi(p,t). \tag{6.83}$$

This equation gives

$$P_B K = P_K K^{-1} = \left(MLT^{-1}\right)^{-1/2}. \tag{6.84}$$

We can expand $|\Psi\rangle$ in a discrete orthonormal basis $\{|n\rangle\}$ as

$$|\Psi\rangle = \sum_{n=1}^{\infty} C_n(t)|n\rangle, \quad C_n(t) = \langle n|\Psi\rangle, \quad \langle n|m\rangle = \delta_{nm}. \tag{6.85}$$

The normalization condition is

$$\langle\Psi|\Psi\rangle = \sum_{m,n=1}^{\infty} C_m^* C_n \langle m|n\rangle = \sum_{n=1}^{\infty} |C_n|^2 \langle n|n\rangle = 1. \tag{6.86}$$

Equation (6.86) implies $[C_n]^2[\langle n|n\rangle] = 1$. This gives $[C_n] = 1$. With $[|n\rangle] = N_K$ and $[\langle n|] = N_B$ the Eq. (6.85) and its bra case lead to the result

$$K = N_K, \quad B = N_B = K^{-1}. \tag{6.87}$$

From the above we note that we cannot able to uniquely find the dimensions of the various kets and bras. Therefore, for simplicity we decide that the dimension of the physical state $|\psi\rangle$ is K = 1.

Then considering

$$|\psi\rangle = \int_{-\infty}^{\infty} \Psi(y,t)|y\rangle dy, \quad \langle x|y\rangle = \delta(x-y) \tag{6.88}$$

and

$$\langle\psi| = \int_{-\infty}^{\infty} \Psi(y,t)^* \langle y| dy \tag{6.89}$$

we can write $[|\psi\rangle] = [\langle\psi|] = 1$. Further, $[|\alpha\rangle] = [\langle\alpha|] = K = 1$. Now, from the Eqs. (6.79), (6.82), (6.84) and (6.87) we write

$$K = B = N_K = N_B = 1, \quad X_B = X_K = L^{-1/2}, \quad P_B = P_K = \left(MLT^{-1}\right)^{-1/2}. \tag{6.90a}$$

That is,

$$[|\psi\rangle] = [\langle\psi|] = 1, \quad [|x\rangle] = [\langle x|] = L^{-1/2}, \tag{6.91a}$$

$$[|p\rangle] = [\langle p|] = \left(MLT^{-1}\right)^{-1/2}, \quad [|n\rangle] = [\langle n|] = 1. \tag{6.91b}$$

These results can be easily generalized to a *N*-particle system.

6.9   Hilbert Space

Suppose a physical system is described by a set of linearly independent finite norm states denoted by $|\phi_1\rangle$, $|\phi_2\rangle$, ... The set $\{|\phi_n\rangle\}$ represents all the possible outcomes of measurements of an observable A. Assume that

1. there is no linear combination of these vectors vanishing except when all the coefficients are zero and

2. the set is complete so that $|\psi\rangle = \sum c_n|\phi_n\rangle$.

Then the space of the ket vectors $|\psi\rangle$ is called a *Hilbert space* [11]. The basis of the Hilbert space are the set $|\phi_n\rangle$. The number of c_n's are the component of $|\psi\rangle$ and gives a representation of $|\psi\rangle$. Thus, we write the c_n's in a column (or row) vector such that a state function is represented by a column (or a row) vector in the Hilbert space. In this space we imagine an axis for each $\phi_n(x)$ and treat c_n the component of the vector in the direction of that axis. The quantum states in Hilbert space are called *state vectors*.

Suppose $|\phi_1\rangle$ and $|\phi_2\rangle$ are two states in Hilbert space. The linear superposition of them is $|\psi\rangle = |\phi_1\rangle + |\phi_2\rangle$. We cannot form a new state by superposition of a state with itself. The state $c_1|\phi_1\rangle + c_2|\phi_1\rangle$ is not a new state. This corresponds to the same state that $|\phi_1\rangle$ does. That is, in the Hilbert space, we can multiply a state vector by a nonzero complex number, however, the physical state is unchanged. The vectors $|\psi\rangle$, $5|\psi\rangle$, $-|\psi\rangle$, $-\sqrt{2}|\psi\rangle$, $\mathrm{i}|\psi\rangle$ and $(2+\mathrm{i}3)|\psi\rangle$ all represent the same physical state. Their magnitudes or lengths are physically unimportant. In this sense state vectors differ from conventional vectors.

A single point of Hilbert space represents the quantum state of the system. The origin does not represent a state. A state vector is represented by a line passing through the origin. Since $|\psi\rangle$ and $5|\psi\rangle$ are equivalent, a line passing through the origin represents all possible multiples of a state vector.

Each dimension of the Hilbert space corresponds to one of the rows of the one-column matrix describing the state. That is, each dimension corresponds to one of the various independent physical states of a quantum state. Essentially, the quantum description of a system with observable assuming continuous values requires a continuous states. In this case the dimension of the Hilbert space is infinite. Thus, the dimension of the Hilbert space for energy, position and momentum representations is infinite. However, there are many systems with discrete and finite sets of values of observables so that the dimension of the associate Hilbert space is finite. In spin representation, for $s = 1/2$ the dimension is two. Another example of an observable with finite values is angular momentum $J = \sqrt{j(j+1)}\hbar$. The corresponding dimension of the Hilbert space is $2j+1$. Some mathematical features of finite-dimensional Hilbert space for the position and momentum of a particle on a one-dimensional lattice is discussed in ref. [12]. An interesting result associated with the finite-dimensional Hilbert space is that one can choose an arbitrary normalized state as a member of an orthonormal basis of the space [32] (see the exercise 6.22). The formulation of quantum mechanics in a finite-dimensional Hilbert space has found applications in quantum cryptography, quantum computers and quantum optics.

Let us see how position states and momentum states are represented in the same Hilbert space. Consider a Hilbert space with each coordinate axis corresponds to a possible position state of a single particle. The momentum state can also be represented in the same Hilbert space. Momentum states are expressed as combination of position states. Therefore, each momentum state corresponds to an axis going off-diagonally tilted with respect to the position-space axes. The set of all momentum states forms a new set of axes. Transformation from the position-space axes to the momentum-space axes involves a rotation in the Hilbert space.

Solved Problem 5:

Compare the Hilbert space with the N-dimensional vector space.

The following table gives a comparative study of N-dimensional vector space and Hilbert space.

S.No.	Vector Space	Hilbert Space
1.	The basis vectors of a vector space are unit vectors.	The basis vectors of the Hilbert space are the functions $\phi_n(x)$, $n = 1, 2, \ldots$ or $\phi_k(x)$, $-\infty < k < \infty$.
2.	The orthogonality condition for the basis vectors is $\mathbf{v}_i \cdot \mathbf{v}_j = \delta_{ij}$, $i, j = 1, 2, \ldots, N$, where δ is the Kronecker-delta function.	The orthogonality condition is $\int_{-\infty}^{\infty} \phi_m^*(x)\phi_n(x)\,\mathrm{d}x = \delta(m - n)$, where $\delta(m - n)$ is the Dirac-delta function.
3.	The completeness property for a vector $\mathbf{F}$ is given by $\mathbf{F} = \sum_{i=1}^{N} f_i \mathbf{v}_i$, where f_i's are the N components of $\mathbf{F}$.	For a function $F(x)$ the completeness property is $F(x) = \int_{-\infty}^{\infty} f(k)\phi_k(x)\,\mathrm{d}k$ if the basis is continuous. For discrete basis $F(x) = \sum_n f_n\phi_n(x)$.
4.	The projection formula is $f_i = \mathbf{v}_i \cdot \mathbf{F}$	The projection formula is $f(k) = \int_{-\infty}^{\infty} \phi_k(x)F(x)\,\mathrm{d}x$.

6.10 Symmetry Operators in Hilbert Space

A pure state of a quantum mechnical system is usually described by a certain unit norm vector ψ of a Hilbert space. If one prepares the system in a pure state ψ then the probability of observing it to be in another pure state $|\phi\rangle$ is given by

$$P(|\psi\rangle \to |\phi\rangle) = |\langle\phi|\psi\rangle|^2. \tag{6.92}$$

If $|\psi\rangle$ and $|\phi\rangle$ states have phase factors as $|\psi\rangle_\mathrm{R} = e^{i\alpha}|\psi\rangle$ and $|\phi\rangle_\mathrm{R} = e^{i\beta}|\phi\rangle$, where α and β are all real then the probability $|\langle\phi|\psi\rangle|^2$ is not changed. Therefore, a pure state is really described by a ray $|\psi\rangle_\mathrm{R} = \{e^{i\alpha}|\psi\rangle;\ \alpha \in \mathbb{R}\}$, that is, the equivalence class of all vectors that is obtained by multiplying $|\psi\rangle$ by all possible phases. A vector in $|\psi\rangle_\mathrm{R}$ is called a *representative* of the ray $|\psi\rangle_\mathrm{R}$.

One of the most powerful techniques of extracting consequences about a physical system is to identify the various symmetries exhibited by the system. A *symmetry* is a transformation of a physical system that does not change the results of possible experiments. Eugene Paul Wigner [13,14] was the first who introduced the definition of symmetry operators in quantum mechanics. *Wigner symmetry* is defined as the transformation of pure states described by the rays $|\psi\rangle_\mathrm{R}$ and $|\phi\rangle_\mathrm{R}$ into pure states described by the rays $|\psi\rangle'_\mathrm{R}$ and $|\phi\rangle'_\mathrm{R}$ such that all probabilities are preserved. That is,

$$|_\mathrm{R}\langle\phi|\psi\rangle_\mathrm{R}|^2 = |'_\mathrm{R}\langle\phi|\psi\rangle'_\mathrm{R}|^2. \tag{6.93}$$

Hence, a symmetry transformation is a ray transformation $T|\psi\rangle_R = |\psi\rangle'_R$ satisfying Eq. (6.93). The nature of the symmetry transformation T is given by *Wigner's theorem* which states that:

Any symmetry transformation can be represented on the Hilbert space of physical states by an operator that is either linear and unitary or anti-linear and anti-unitary.

In fact, all unitary operators are linear and all anti-unitary operators are anti-linear. The transformation defined by $T|\psi\rangle_R = |\psi\rangle'_R$ is *unitary* if

$$'_R\langle\psi|\phi\rangle'_R = {}_R\langle\psi|\phi\rangle_R \tag{6.94}$$

and *anti-unitary* if

$$'_R\langle\psi|\phi\rangle'_R = {}_R\langle\psi|\phi\rangle^*_R = {}_R\langle\phi|\psi\rangle_R. \tag{6.95}$$

It can be seen that Eqs. (6.94) and (6.95) satisfy (6.93). A *unitary operator* U is such that

$$_R\langle U\psi|U\phi\rangle_R = {}_R\langle\psi|U^\dagger U|\phi\rangle_R = {}_R\langle\psi|\phi\rangle_R \tag{6.96}$$

for all states $|\phi\rangle_R$ and $|\psi\rangle_R$ so that $U^\dagger U = UU^\dagger = I$ and is linear so that

$$U\left(C_1|\phi\rangle_R + C_2|\psi\rangle_R\right) = C_1 U|\phi\rangle_R + C_2 U|\psi\rangle_R, \tag{6.97}$$

where C_1 and C_2 are arbitrary complex numbers. An *anti-unitary operator* is such that

$$_R\langle A\psi|A\phi\rangle_R = {}_R\langle\psi|\phi\rangle^*_R = {}_R\langle\phi|\psi\rangle_R \tag{6.98}$$

and is *anti-linear* so that

$$A\left(C_1|\phi\rangle_R + C_2|\psi\rangle_R\right) = C_1^* A|\phi\rangle_R + C_2^* A|\psi\rangle_R \tag{6.99}$$

for arbitrary complex constants C_1 and C_2.

As the symmetry operators form a group Wigner laid the foundations both for the application of group theory to quantum mechanics and for the role of symmetry principles in quantum mechanics. These symmetry groups can be classified into either continuous groups or discrete groups. Continuous symmetries are characterized by invariance following a continuous changes in some parameters of the system. Space-time symmetries such as rotation, boost (together forms proper orthochronous Lorentz transformations) and transition belong to this group. It can be shown that continuous symmetry transformations are represented by unitary operators. Discrete symmetries are characterized by invariance following a non-continuous change in the system space-time symmetries. The space-time symmetries such as time reversal (T) and parity (P), which together form improper Lorentz transformations, belong to this class. It can be shown that parity symmetry can be represented by unitary operators. But the time reversal symmetry T can be represented only by anti-unitary operators.

Apart from the space-time symmetries other symmetry principles used in the fundametnal physical theories are:

1. Charge conjugate symmetry (discret symmetry existing between particles and the anti-particles).

2. The permutation symmetry existing between the identical particles (which leads to two different statistics, namely, Bose-Einstein statistics and Fermi–Dirac statistics).

3. Unitary internal symmetry (gauge symmetries) described by the unitary groups SU(N).

4. Supersymmetry (the symmetry between fermions and bosons).

Physical symmetries play a crucial role in the progress of theoretical physics. They occupy a leading place in the eloboration and testing of fundamental theories and they offer a basis for proceeding towards theoretical unification.

6.11 Projection and Displacement Operators

Let the eigenkets $|i\rangle$ of an operator Ω have eigenvalues ω_i. Since the set of kets $|i\rangle$ form a complete set, any arbitrary ket $|\alpha\rangle$ is expanded as $|\alpha\rangle = \sum_i a_i |i\rangle$, where $a_i = \langle i|\alpha\rangle$ is a number. We write

$$|\alpha\rangle = \sum_i |i\rangle\langle i|\alpha\rangle = \sum_i P_i |\alpha\rangle, \qquad (6.100)$$

where $P_i = |i\rangle\langle i|$ is called the *projection operator* [15] for the ith state. In

$$P_i |\alpha\rangle = |i\rangle\langle i|\alpha\rangle \qquad (6.101)$$

the operator P_i projects only the particular eigenstate $|i\rangle$ of Ω among $|\alpha\rangle$. Hence, it is called a *projection operator*. From (6.100) and (6.101) we find that

$$\sum_i P_i |\alpha\rangle = \sum_i |i\rangle\langle i|\alpha\rangle = |\alpha\rangle. \qquad (6.102)$$

The result is $\sum P_i = I$. *What does it state?* We have

$$P_i^2 |\alpha\rangle = |i\rangle\langle i|i\rangle\langle i|\alpha\rangle = |i\rangle\langle i|\alpha\rangle = P_i |\alpha\rangle. \qquad (6.103)$$

This gives $P_i^2 = P_i$. Further, $P_j P_i |\alpha\rangle = |j\rangle\langle j|i\rangle\langle i|\alpha\rangle = 0$. Therefore, $P_j P_i = 0$. Combining $P_i^2 = P_i$ and $P_j P_i = 0$ we get $P_i P_j = P_j P_i = \delta_{ij} P_i$.

Any operator A can be written as $A = IAI$. Therefore, the completeness relation is

$$A = \sum_i |i\rangle\langle i|A \sum_j |j\rangle\langle j| = \sum_i \sum_j |i\rangle\langle i|A|j\rangle\langle j| = \sum_i \sum_j |i\rangle\langle j|A_{ij}, \qquad (6.104)$$

where A_{ij} is the matrix element $\langle i|A|j\rangle$. Thus, A is written as a linear combination (with coefficients A_{ij}) of the operator $|i\rangle\langle j|$. If $A = \Omega$, then $\langle i|\Omega|j\rangle = \omega_j\langle i|j\rangle = \omega_j \delta_{ij}$ and

$$\Omega = \sum_i \sum_j |i\rangle\langle j|\omega_j \delta_{ij} = \sum_i P_i \omega_i. \qquad (6.105)$$

Thus, any observable can be written as a weighted sum of projection operators to their own eigenstates. The weighting factors are the eigenvalues. More generally, for any function of Ω, we have $f(\Omega) = \sum_i f(\omega_i)|i\rangle\langle i| = \sum f(\omega_i) P_i$.

When the eigenstates of an observable are used as bases for representations then the observable is defined in relation to a coordinate system, say, $S : oxyz$ in ordinary space. A change in the observable arises if there is a change in the coordinate system. This results in a transformation of the basis in the Hilbert space. Suppose the basis transformation is made by a translation (shifting the origin without altering the orientation of the axes). Let us investigate the effect of translation on a wave function of a system.

Let us consider that a one-dimensional system is displaced through a distance along the x-axis. The final wave function is unknown because it can be multiplied by a phase factor. Since superposition relations between eigenstates are unchanged under the displacements we write

$$\psi_{\mathrm{UD}} = \phi_{\mathrm{UD}}^a + \phi_{\mathrm{UD}}^b \,, \tag{6.106}$$

where ψ_{UD} is the wave function of the undisplaced system. Here assume that the system has only two eigenstates ϕ_{UD}^a and ϕ_{UD}^b. Equation (6.106) for the displaced system is $\psi_{\mathrm{D}} = \phi_{\mathrm{D}}^a + \phi_{\mathrm{D}}^b$. Now, define an operator called *displacement operator* [16], the effect of it on ψ_{UD} is to give ψ_{D}: $D\psi_{\mathrm{UD}} = \psi_{\mathrm{D}}$. An example is $D\psi(x) = \psi(x + \delta)$. The displacement operator is widely used in quantum computation.

Let us calculate $\langle D^\dagger D\rangle$. We obtain

$$\langle D^\dagger D\rangle = \int_{-\infty}^{\infty} \psi_{\mathrm{UD}}^* D^\dagger D\psi_{\mathrm{UD}} \, \mathrm{d}x = \int_{-\infty}^{\infty} \psi_{\mathrm{D}}^* \psi_{\mathrm{D}} \, \mathrm{d}x = 1 \,. \tag{6.107}$$

Since this is valid for any eigenfunction we have $DD^\dagger = 1$, D is a unitary.

Suppose γ is an operator such that $\gamma\phi_{\mathrm{UD}}^a = \phi_{\mathrm{UD}}^b$ and $\gamma_{\mathrm{d}}\phi_{\mathrm{D}}^a = \phi_{\mathrm{D}}^b$, where γ_{d} is the *displaced operator* [16]. We have

$$\gamma_{\mathrm{d}}\phi_{\mathrm{D}}^a = D\phi_{\mathrm{UD}}^b = D\gamma\phi_{\mathrm{UD}}^a = D\gamma D^\dagger \phi_{\mathrm{D}}^a \,. \tag{6.108}$$

Thus, $\gamma_{\mathrm{d}} = D\gamma D^\dagger$. For an infinitesimal displacement δ of the system we write

$$\lim_{\delta \to 0} \frac{(\phi_{\mathrm{D}}^a - \phi_{\mathrm{UD}}^a)}{\delta} = \lim_{\delta \to 0} \frac{D\phi_{\mathrm{UD}}^a - \phi_{\mathrm{UD}}^a}{\delta} = \lim_{\delta \to 0} \frac{(D-1)\phi_{\mathrm{UD}}^a}{\delta} \,. \tag{6.109}$$

Defining

$$d = \lim_{\delta \to 0} \frac{D-1}{\delta} \tag{6.110}$$

and since D can be replaced by $De^{\mathrm{i}\theta}$, where θ is a phase angle we write

$$\lim_{\delta \to 0} \frac{De^{\mathrm{i}\theta} - 1}{\delta} = \lim_{\delta \to 0} \frac{D - 1 + \mathrm{i}\theta}{\delta} = d + \mathrm{i}\gamma \,, \tag{6.111}$$

where $\gamma = \lim_{\delta \to 0}(\theta/\delta)$.

Is the operator d Hermitian? To answer this, for small δ we write $D = 1 + \delta d$. Then $DD^\dagger = 1$ becomes $(1 + \delta d^\dagger)(1 + \delta d) = 1$. Neglecting δ^2 we obtain $d^\dagger + d = 0$ or $d^\dagger = -d$. Thus, the displacement operator d is *anti-Hermitian*.

Examples:

Assume that the system is displaced along x-direction by an amount, say, δ_x. The old and new wave functions are $\psi(x)$ and $\psi(x - \delta_x)$, respectively. Then $xd = x - \delta_x$. Replacing x by γ we have $\gamma d = \gamma - \delta_x$ or $(\gamma_{\mathrm{d}} - \gamma)/\delta_x = -1$. Now, the above equation becomes $[d, \gamma] = [d_x, x] = -1$. That is, $xd_x - d_x x = 1$. Multiplication on both sides by $\mathrm{i}\hbar$ gives

$$x\mathrm{i}\hbar d_x - \mathrm{i}\hbar d_x x = \mathrm{i}\hbar[x, d_x] = \mathrm{i}\hbar \,. \tag{6.112}$$

Comparison of this equation with $xp_x - p_x x = \mathrm{i}\hbar$ gives $d_x = -\mathrm{d}/\mathrm{d}x$. Thus, the result is that the displacement operator in the x-direction is proportional to the momentum conjugate to x. Hence, p_x is the generator of translations along the x-axis. Since any finite translation along the x-axis can be constructed from a series of many infinitesimal translations we write

$$D(\delta_x) = \lim_{N \to \infty} \left(1 - \frac{\mathrm{i}\delta_x p_x}{N\hbar}\right)^N = e^{-\mathrm{i}p_x \delta_x/\hbar} \,. \tag{6.113}$$

Alternatively, we write $\psi(x + \delta_x)$ as a Taylor series

$$
\begin{aligned}
\psi(x + \delta_x) &= \psi(x) + \delta_x \psi'(x) + \frac{1}{2!}\delta_x^2 \psi''(x) + \dots \\
&= \left[\sum_{n=0}^{\infty} \frac{1}{n!} \left(\frac{\mathrm{i}p_x \delta_x}{\hbar}\right)^n\right] \psi(x) \\
&= D(\delta_x)\psi(x).
\end{aligned}
\tag{6.114}
$$

In general, for a three-dimensional wave function we write

$$
D(\delta_j) = \mathrm{e}^{-\mathrm{i}p_j \delta_j/\hbar}, \quad D(\delta_j)\psi(x_j) = \psi(x_j + \delta_j), \quad j = x, y \text{ or } z.
\tag{6.115}
$$

Next, consider the change of representation induced by a rotation of coordinate axes, say, from $S : oxyz$ to $S' : ox'y'z'$ about oz through an angle θ. We have

$$
x' = x\cos\theta + y\sin\theta, \quad y' = -x\sin\theta + y\cos\theta, \quad z' = z.
\tag{6.116}
$$

Since the wave function is independent of the orientation of the coordinate axes we write $\psi'(\mathbf{X}') = \psi(\mathbf{X})$. Substituting Eq. (6.116) and assuming θ to be infinitesimally small so that $\cos\theta \approx 1$ and $\sin\theta \approx \theta$ we obtain

$$
\psi'(x + y\theta, y - x\theta, z) = \psi(x, y, z).
\tag{6.117}
$$

Replacing x by $x - y\theta$ and y by $y + x\theta$ in Eq. (6.117) we get

$$
\begin{aligned}
\psi'(x, y, z) &= \psi(x - y\theta, y + x\theta, z) \\
&= \psi(x, y, z) + \theta\left[x\frac{\partial}{\partial y} - y\frac{\partial}{\partial x}\right]\psi(x, y, z),
\end{aligned}
\tag{6.118}
$$

where higher-order terms in θ are neglected. The terms in the square-bracket of Eq. (6.118) is the z-component of orbital angular momentum without $-\mathrm{i}\hbar$. Then

$$
\psi'(\mathbf{X}) = \left(1 + \frac{\mathrm{i}\theta}{\hbar}L_z\right)\psi(\mathbf{X}).
\tag{6.119}
$$

Thus, the operator $L_z/\hbar$ plays the role of the generators of infinitesimal rotations. For a rotation about an arbitrary axis along the unit vector $\mathbf{n}$

$$
\psi'(\mathbf{X}) = \left(1 + \frac{\mathrm{i}\theta}{\hbar}\mathbf{n} \cdot \mathbf{L}\right)\psi(\mathbf{X}).
\tag{6.120}
$$

6.12 Quaternionic Quantum Mechanics

In the year 1843, William Rowan Hamilton discovered a new mathematical object called *quaternion* [17]. A complex number c is defined as $c = a + b\mathrm{i}$, where $(a, b) \in \mathbb{R}^2$ and $\mathrm{i} = \sqrt{-1}$. A *quaternion* q is given as $q = a + b\mathrm{i} + c\mathrm{j} + d\mathrm{k}$ with 1, i, j, k are the directions of q. The components of q form the set $\mathbf{H}(a, b, c, d) \in \mathbb{R}^4$, where $\mathbb{R}^4$ is a four-dimensional vector space over the real numbers. $\mathbf{H}$ has three operations: addition, scalar multiplication and quaternion multiplication. The sum of two elements of $\mathbf{H}$, say, $q_1 = a_1 + b_1\mathrm{i} + c_1\mathrm{j} + d_1\mathrm{k}$ and $q_2 = a_2 + b_2\mathrm{i} + c_2\mathrm{j} + d_2\mathrm{k}$ is

$$
q_1 + q_2 = (a_1 + a_2) + (b_1 + b_2)\mathrm{i} + (c_1 + c_2)\mathrm{j} + (d_1 + d_2)\mathrm{k}.
\tag{6.121}
$$

The multiplication rules for i, j and k are: $i^2 = j^2 = k^2 = ijk = -1$. The multiplication of two quaternions q_1 and q_2 is given by

$$
\begin{aligned}
q_1 * q_2 &= (a_1 + b_1 i + c_1 j + d_1 k)(a_2 + b_2 i + c_2 j + d_2 k) \\
&= (a_1 a_2 - b_1 b_2 - c_1 c_2 - d_1 d_2) + (a_1 b_2 + b_1 a_2 + c_1 d_2 - d_1 c_2) i \\
&\quad + (a_1 c_2 - b_1 d_2 + c_1 a_2 + d_1 b_2) j + (a_1 d_2 + b_1 c_2 - c_1 b_2 + d_1 a_2) k. \quad (6.122)
\end{aligned}
$$

The theory of quaternions has found applications in classical mechanics, quantum mechanics and relativity. It has been used in aerospace applications, flight simulators, computer graphics, animation, visualization, fractals and virtual reality.

6.12.1 Operators and States

As the quaternions are noncommutative the left and right action of the quaternionic imaginary units i, j and k are different. The left $\mathbf{L} = (L_i, L_j, L_k)$ and right $\mathbf{R} = (R_i, R_j, R_k)$ operators are introduced. They act on the quaternionic function ψ_H as $L_q \psi_H = q \psi_H$ and $R_p \psi_H = \psi_H p$, $p, q = $ i, j, k. These quaternionic operators satisfy the relations $L_q L_p = L_{qp}$, $R_q R_p = R_{pq}$ and $[L_q, R_p] = 0$.

Jordan, von Neumann and Wigner [18] proposed a possible generalization of quantum theory with respect to the background field way back in 1934. Birkhoff and von Neumann [19] proposed the use of quaternions to develop a new quantum mechanics which was developed by Finkelstein [20,21] and Adler [22]. We give below some basic concepts of quaternionic quantum mechanics.

Similar to the complex quantum mechanics the states of the quaternionic quantum mechanics is also described by wave functions belonging to an abstract vector space, namely, the *quaternionic Hilbert space*. Quaternionic quantum mechanics is a modification of the complex quantum theory in which the wave functions belong to a Hilbert space defined over the quaternion field. The quantum state of particle is defined at a given instant by quaternionic wave function interpreted as a probability amplitude given by

$$
\psi_H(\mathbf{r}) = \psi_{R,0}(\mathbf{r}) + i \psi_{R,1}(\mathbf{r}) + j \psi_{R,2}(\mathbf{r}) + k \psi_{R,3}(\mathbf{r}), \quad (6.123)
$$

where $\psi_{R,0}$, $\psi_{R,1}$, $\psi_{R,2}$ and $\psi_{R,3}$ are real functions of real variables $\mathbf{r}$. The probability interpretation of ψ_H requires that it belongs to the Hilbert space of square-integral functions.

The same ψ_H can be represented by several distinct sets of components in the same vector space corresponding to different choice of basis. Let ψ_H and ϕ_H belong to the same vector space. Then the quaternionic scalar product is given by $(\psi_H, \phi_H) = \int \bar{\psi}_H \phi_H d^3 r$, where $\bar{\psi}_H = \psi_{R,0} - i \psi_{R,1} - j \psi_{R,2} - k \psi_{R,3}$ represents the quaternionic conjugate of ψ_H.

Due to the noncommutative nature of quaternionic multiplication, the right and left multiplications by quaternionic scalars will not be the same. For the quaternionic quantum mechanics we have to consider the Hilbert space of linearity of right multiplication [22].

6.12.2 Schrödinger Equation

In the standard formulation of quantum mechanics, the Schrödinger equation is given by

$$
i\hbar \frac{\partial}{\partial t} \psi(\mathbf{r}, t) = H \psi(\mathbf{r}, t) = \left(-\frac{\hbar^2}{2m} \nabla^2 + V(\mathbf{r}, t) \right) \psi(\mathbf{r}, t), \quad (6.124)
$$

where the Hamiltonian operator H is Hermitian. In quaternionic quantum mechanics, in the quaternionic Hilbert space, the time development of $\psi_H(\mathbf{r}, t)$ is governed by the Schrödinger equation [23,24]

$$\frac{\partial \psi_{\mathbb{H}}}{\partial t} = H_{\mathbb{H}} \psi_{\mathbb{H}}, \tag{6.125}$$

where $H_{\mathbb{H}}$ is the anti-self-adjoint operator and is given by

$$H_{\mathbb{H}} = -\frac{i}{\hbar} H = \frac{1}{\hbar} \left[i\frac{\hbar^2}{2m} \nabla^2 - V_{\mathbb{H}} \right], \quad V_{\mathbb{H}} = iV_{\mathbb{R},1} + jV_{\mathbb{R},2} + kV_{\mathbb{R},3}, \tag{6.126}$$

where $V_{\mathbb{H}}$ is the quaternionic potential. We can seek the solution of the form $\psi_{\mathbb{H}}(\mathbf{r}, t) = \phi_{\mathbb{H}}(\mathbf{r}) f_{\mathbb{H}}(t)$. We obtain

$$\hbar \dot{f}_{\mathbb{H}} f_{\mathbb{H}}^{-1} = \phi_{\mathbb{H}}^{-1} \left[i\frac{\hbar^2}{2m} \nabla^2 - V_{\mathbb{H}} \right] \phi_{\mathbb{H}}. \tag{6.127}$$

This equation is true only if both sides are equal to the same constant. We can set this constant as E. The left-side of Eq. (6.127) gives $f_{\mathbb{H}}(t) = e^{-iEt/\hbar}$. The right-side of Eq. (6.127) then gives

$$\left[i\frac{\hbar^2}{2m} \nabla^2 - V_{\mathbb{H}} \right] \phi_{\mathbb{H}} = \phi_{\mathbb{H}} E. \tag{6.128}$$

That is, $\hbar H_{\mathbb{H}} \phi_{\mathbb{H}} = \phi_{\mathbb{H}} E$ or

$$H_{\mathbb{H}} \phi_{\mathbb{H}} = \phi_{\mathbb{H}} E/\hbar. \tag{6.129}$$

Equation (6.129) is the stationary state eigenvalue problem.

As $H_{\mathbb{H}}$ is anti-Hermiticity, that is, $(\phi_{\mathbb{H}}, H_{\mathbb{H}} \psi_{\mathbb{H}}) = -(H_{\mathbb{H}} \phi_{\mathbb{H}}, \psi_{\mathbb{H}})$ we have $E = -\bar{E} = -(iE_1 + jE_2 + kE_3)$. Then the quaternionic stationary state eigenvalue equation is

$$H_{\mathbb{H}} \phi_{\mathbb{H}} = -\phi_{\mathbb{H}}(iE_1 + jE_2 + kE_3)/\hbar. \tag{6.130}$$

Now, consider $\psi_{\mathbb{H}} = \phi_{\mathbb{H}} e^{-iEt/\hbar}$. Then Eq. (6.125) gives

$$H_{\mathbb{H}} \psi_{\mathbb{H}} = -\psi_{\mathbb{H}} iE/\hbar. \tag{6.131}$$

Thus, the stationary state quaternionic Schrödinger equation becomes

$$\left[i\frac{\hbar^2}{2m} \nabla^2 - V_{\mathbb{H}} \right] \psi_{\mathbb{H}} = -\psi_{\mathbb{H}} iE. \tag{6.132}$$

In this equation due to the noncommutative nature of quaternionic multiplication the order of multiplication is important.

As like in the standard quantum mechanics one can define the current density $\mathbf{J}$ as

$$\mathbf{J} = \frac{\hbar}{2m} \left[(\nabla \bar{\psi}_{\mathbb{H}}) i\psi_{\mathbb{H}} - \bar{\psi}_{\mathbb{H}} i\nabla \psi_{\mathbb{H}} \right] \tag{6.133}$$

and the probability density $\rho = \bar{\psi}_{\mathbb{H}} \psi_{\mathbb{H}}$. They satisfy the continuity equation $\partial \rho/\partial t + \nabla \cdot \mathbf{J} = 0$. Hence, the probability is conserved.

Quaternionic bound states and scattering states of certain systems were reported [23–27]. An experiment was proposed [28] to identify quaternionic deviations from standard quantum mechanics. The noncommutativity of quaternionic phases could be observable experimentally in Bragg scattering by a crystal made of three different atoms in neutron interferometry. The experiment results were reported [29]. In the experiment, phase-shifts commute to better than one in 3×10^4. In order to explain the observed null shift it has been argued that quaternionic potentials act only for certain fundamental forces [30].

Quaternionic quantum mechanics is still in early stage of development. It is quite difficult physically interpret quaternionic potentials and the quaternionic solutions. Consequently, no conclusive experimental proof has been found for quaternionic quantum mechanics. Still it is generally, believed that quaternionic Hilbert space dynamics may play a role as an underlying dynamics of the standard model.

6.13 Concluding Remarks

In this chapter, we outlined the matrix mechanics for quantum mechanical systems. In this theory the behaviour of a quantum mechanical system is determined by the matrices of **p** and **X**. The Hamiltonian is a diagonal matrix having diagonal elements E_n, the energy eigenvalues. To solve $H\psi = E\psi$ as a matrix problem, we need to find an appropriate linear vector space. We note that instead of characterizing a vector ψ by its components with respect to a particular coordinate system, it is convenient to use the coordinate independent vector. We can characterize a vector on some basis $\{\phi_n\}$. In this case the vector ψ itself does not change, only its components with respect to the coordinate axes change under the transformation from one basis to another basis, for example, say from $\{\phi_n\}$ to $\{\phi_{n'}\}$. We have shown that unitary transformation relates the representations in different basis systems. A unitary transformation from one representation of $|\psi\rangle$ to another corresponds to a rotation of axis in the Hilbert space without change in the state vectors. For example, the energy representation corresponds to choosing axes in such a way that a state vector oriented along one of these axes is an eigenstate of the Hamiltonian. Though the wave mechanics and matrix mechanics are logically distinct, they are equivalent in all respects.

Schrödinger emphasized that wave mechanics and matrix mechanics have very intimate inner connections [3] and established the mathematical identity of the two theories based on the coordination of a matrix with a well-ordered function symbols [3,31]. von Neumann did this by considering F_z and F_Ω are isomorphic [19] with F_z and F_Ω denoting the space of square-summable sequences and space of square-integrable functions, respectively. For a discussion of this inner connection with reference to a particle in a box potential one may refer to the ref. [31].

6.14 Bibliography

[1]   W. Heisenberg, *Z. Phys.* 33:879, 1925.

[2]   M. Born, W. Heisenberg and P. Jordan, *Z. Phys.* 35:557, 1926.

[3]   E. Schrödinger, *Annalen der Physik* 79:734, 1926.

[4]   C. Eckart, *Phys. Rev.* 28:711, 1926.

[5]   D.F. Styer, *The Strange World of Quantum Mechanics.* Cambridge University Press, Cambridge, 2000 pp.127.

[6]   G.B. Arfken, *Mathematical Methods for Physicists.* Academic Press, Florida, 1985.

[7] G.H. Golub and C.F. Van Loan, *Matrix Computations*. Johns Hopkins, Baltimore, 1996.

[8] P.A.M. Dirac, *The Principles of Quantum Mechanics*. Oxford, New York, 1958.

[9] R.R. Puri, *Physics News* January–March 2002, pp.26-35.

[10] C. Semay and C.T. Willemyns, *Eur. J. Phys.* 42:025404, 2021.

[11] R. Courant and D. Hilbert, *Methods of Mathematical Physics*. Volume V. Interscience, New York, 1953.

[12] A.C. de la Torre and D. Goyeneche, *Am. J. Phys.* 71:49, 2003.

[13] E.P. Wigner, *Gruppentheorie*. Vieweg, Braunschweig, 1931.

[14] E.P. Wigner, *Group Theory and its Application to the Quantum Mechanics of Atomic Spectra*. Academic Press, New York, 1959.

[15] L.I. Schiff, *Quantum Mechanics*. McGraw-Hill, Singapore, 1968.

[16] R. Fitzpatrick, *Quantum Mechanics*. Unpublished, 2010.

[17] B.L. van der Waerden, *Mathematics Magazine* 49:227, 1976.

[18] F. Jordan, J. von Neumann and E.P. Wigner, *Ann. Maths.* 35:29, 1934.

[19] J. von Neumann, *Mathematical Foundations of Quantum Mechanics*. Princeton University Press, Princeton, New Jersey, 1955.

[20] D. Finkelstein, J.M. Jauch, S. Schiminovich and D. Speiser, *J. Math. Phys.* 3:207, 1962.

[21] D. Finkelstein, J.M. Jauch, S. Schiminovich and D. Speiser, *J. Math. Phys.* 4:788, 1963.

[22] S.L. Adler, *Quaternion Quantum Mechanics and Quantum Fields*. Oxford University Press, New York, 1995.

[23] S.D. Leo and G.C. Ducate, *J. Math. Phys.* 44:2224, 2003.

[24] S.D. Leo and G.C. Ducati, *J. Phs. A: Math. Gen.* 38:3443, 2005.

[25] A.J. Davies and B.H.J. McKellar, *Phys. Rev. A* 40:4209, 1989.

[26] A.J. Davies and B.H.J. McKellar, *Phys. Rev. A* 46:3671, 1992.

[27] S.D. Leo, G.C. Ducate and T.M. Madureira, *J. Math. Phys.* 47:082106, 2006.

[28] A. Peres, *Phys. Rev. Lett.* 42:683, 1979.

[29] H. Kaiser, E.A. George and S.A. Werner, *Phys. Rev. A* 29:2276, 1984.

[30] A.G. Klein, *Physica B & C* 151:44, 1988.

[31] J. Perentis and B. Ty, *Am. J. Phys.* 82:583, 2014.

[32] I. Sargolzahi and E. Anjidani, *Phys. Edu.* 33 (2):1 (article number 4), 2017.

6.15 Exercises

6.1 Find out whether the set of vectors $\psi_1 = 3\mathbf{i} - \mathbf{j} + 2\mathbf{k}$, $\psi_2 = -2\mathbf{i} + \mathbf{j}$, $\psi_3 = \mathbf{j} + 2\mathbf{k}$ are linearly independent.

6.2 If $|\phi\rangle = \sum_i C_i|i\rangle$, $|\chi\rangle = \sum_i D_i|i\rangle$ and A is an operator write down $\langle\chi|A|\phi\rangle$.

6.3 If $A|\phi_i\rangle = \lambda_i|\phi_i\rangle$, $i = 1, 2, \ldots, n$ write the expression for $\langle\psi|A|\psi\rangle$, where $|\psi\rangle$ is a general state.

6.4 The wave function of a system is given by $|\psi(t)\rangle = a_1(t)|A\rangle + a_2(t)|B\rangle$, $t \geq t_1$, where $a_1(t) = e^{-i\alpha(t-t_1)}\cos[\omega(t-t_1)/2]$, $a_2(t) = -i\,e^{-i\alpha(t-t_1)}\sin[\omega(t-t_1)/2]$ with α and ω being two constants. Check whether the wave function is normalized.

6.5 Suppose V is a vector space with basis vectors $|v_1\rangle = \begin{bmatrix} 1 \\ 0 \end{bmatrix}$ and $|v_2\rangle = \begin{bmatrix} 0 \\ 1 \end{bmatrix}$.

If A is a linear operator such that $A|v_1\rangle = |v_2\rangle$ and $A|v_2\rangle = |v_1\rangle$ find the matrix representation of A.

6.6 (a) Show that a two-dimensional vector space can be spanned by the two set of basis $|v_1\rangle = \begin{bmatrix} 1 \\ 0 \end{bmatrix}$, $|v_2\rangle = \begin{bmatrix} 0 \\ 1 \end{bmatrix}$ and $|v_1'\rangle = \frac{1}{\sqrt{2}}\begin{bmatrix} 1 \\ 1 \end{bmatrix}$, $|v_2'\rangle = \frac{1}{\sqrt{2}}\begin{bmatrix} 1 \\ -1 \end{bmatrix}$.

(b) Find the transformation matrix between the two basis.

(c) Express the vector $|v\rangle = \begin{bmatrix} 2 \\ -3 \end{bmatrix}$ in both the basis.

6.7 Show that the basis $(1, x, x^2, \ldots)$ in the domain $-1 \leq x \leq 1$ is not orthogonal. Find an orthogonal basis using Gram–Schmidt procedure.

6.8 Express $i\hbar\partial|\psi\rangle/\partial t = H|\psi\rangle$ in terms of H and U.

6.9 If $|n\rangle$ and $|m\rangle$ are two arbitrary state vectors and $|i\rangle$ are the complete set of basis vectors show that $\langle n|m\rangle = \sum_i \langle n|i\rangle\langle i|m\rangle$.

6.10 For a set of operators A, B and C if $AB = C$ then what is $[A][B]$?

6.11 Show that if A is an anti-Hermitian matrix then e^A is unitary.

6.12 State the condition for a Hermitian matrix to become unitary.

6.13 Express the definition of Hermitian operator in Dirac notation.

6.14 Write the Hermitian conjugate of $\lambda A|\psi_1\rangle\langle\psi_2|B^\dagger$.

6.15 Prove that $\langle A^m\rangle = \sum_n a_n^m |c_n|^2$, where a's are eigenvalues of A.

6.16 If λ_i is the eigenvalue of an operator Λ corresponding to the eigenvector $|\phi_i\rangle$ and if $|\psi\rangle = \sum_i c_i|\phi_i\rangle$ then find the probability of getting λ_i as an eigenvalue in a measurement.

6.17 Show that the eigenvalues of an operator represented in two different basis connected by a transformation matrix U are identical.

6.18 Using $f(\Omega) = \sum_i f(\omega_i)|i\rangle\langle i|$ find $e^{\theta\sigma_z}$, where σ_z is the Pauli matrix $\begin{pmatrix} 1 & 0 \\ 0 & -1 \end{pmatrix}$.

6.19 Find the eigenkets of the Pauli's spin matrices

$$\sigma_x = \begin{pmatrix} 0 & 1 \\ 1 & 0 \end{pmatrix}, \sigma_y = \begin{pmatrix} 0 & -i \\ i & 0 \end{pmatrix} \text{ and } \sigma_z = \begin{pmatrix} 1 & 0 \\ 0 & -1 \end{pmatrix}$$

for spin$-1/2$ systems. Find the unitary transformation matrices from eigenket basis of (a) σ_z to σ_x, (b) σ_z to σ_y and (c) σ_x to σ_y.

6.20 With $a = \dfrac{1}{\sqrt{2m\omega\hbar}}(m\omega x + ip_x)$, $a^\dagger = \dfrac{1}{\sqrt{2m\omega\hbar}}(m\omega x - ip_x)$ being the lowering and raising operators, respectively, of the linear harmonic oscillator, show that the matrix elements $\langle m|A|n\rangle$ vanish unless $r - s = m - n$ if A is an operator which contains r factors of $a^\dagger$ and s factors of a in any order.

6.21 Consider the previous exercise. Calculate $\langle n|x^4|n\rangle$.

6.22 Show that in a two-dimensional Hilbert space each normalized state is a member of an orthonormal basis [32].

6.23 Show that a unitary operator is linear.

6.24 Show that an anti-unitary operator is anti-linear.

6.25 Show that the product of two anti-unitary transformations is a unitary transformation and hence the continuous groups can be represented only by unitary transformations.

6.26 Show that an orthonormal and complete set of basis vectors remains so after Wigner symmetry transformation.

6.27 Show that $L_i^2 = L_j^2 = L_k^2 = R_i^2 = R_j^2 = R_k^2 = L_i L_j L_k = R_k R_j R_i = -1$.

6.28 Find the value of $[L_q, R_p]$.

6.29 Show that the basis vector $[1, i\sigma_x, i\sigma_y, i\sigma_z]$ with $i = \sqrt{-1}$, σ_x, σ_y and σ_z − Pauli matrices constitute a representation of the quaternions basis.

6.30 Set up the radial quaternionic Schrödinger equation for the spherically symmetric potentials [24].

7

Various Pictures and Density Matrix

7.1 Introduction

In classical mechanics the state of a system is defined by the numerical values of a set of dynamical variables such as the position coordinates and the velocities. In quantum mechanics the state of a system at an instant is represented by a unit vector in a Hilbert space, where a set of axes are the eigenvectors of a complete set of observables. Here the observable are not the state vectors but the expectation values of a set of self-adjoint or Hermitian operators corresponding to the dynamical variables. The variation of $\langle A \rangle$ with time is seen as arising in any one of the following ways:

1. The state vector changes with time, but the operator A remains unchanged.

2. The operator changes with time, whereas the state vector remains unchanged.

3. Both the operator A and the state vector change with time.

These three cases are called *Schrödinger*, *Heisenberg* and *interaction pictures*, respectively. The time evolution of ensembles is described by means of a density matrix. We discuss the above mentioned pictures and density matrix in this chapter.

7.2 Schrödinger Picture

The description of quantum dynamics where the wave function is time-dependent and the operators describing the observables are time-independent is known as the Schrödinger picture.

7.2.1 Matrix Form of Schrödinger Equation

The time-dependent Schrödinger equation in Dirac's notation is given by Eq. (6.76). Its Hermitian adjoint is

$$-i\hbar \frac{\partial}{\partial t} \langle \psi | = \langle \psi | H \, . \tag{7.1}$$

Since the dependence of $|\psi\rangle$ on coordinate or other variables is not explicit we replace $\partial/\partial t$ by d/dt. If H is independent of time t then the solution of Eq. (7.1) is

$$|\psi(t)\rangle = e^{-iHt/\hbar}|\psi(0)\rangle \ \text{ and } \ \langle \psi(t)| = \langle \psi(0)|e^{iHt/\hbar} \, . \tag{7.2}$$

DOI: 10.1201/9781003172178-7

$\mathrm{e}^{-\mathrm{i}Ht/\hbar}$ is also an operator (because H is an operator) and write it as

$$
\begin{aligned}
U(t) &= \mathrm{e}^{-\mathrm{i}Ht/\hbar} = 1 + \frac{(-\mathrm{i}Ht)}{\hbar} + \frac{1}{2}\frac{(-\mathrm{i}Ht)}{\hbar}\frac{(-\mathrm{i}Ht)}{\hbar} + \cdots \\
&= \sum_{n=0}^{\infty} \frac{1}{n}\left(-\mathrm{i}Ht/\hbar\right)^{n}.
\end{aligned}
\tag{7.3}
$$

$U(t)$ can also be defined as

$$
U(t) = \mathrm{e}^{-\mathrm{i}Ht/\hbar} = \lim_{n\to\infty}\left(1 + \frac{\mathrm{i}t}{\hbar n}H\right)^{-n}.
\tag{7.4}
$$

It is possible to represent $U(t)$ as a square matrix. *What is the effect of $U(t) = \mathrm{e}^{-\mathrm{i}Ht/\hbar}$ on $|\psi(0)\rangle$?* Its significant effect is to change $|\psi(0)\rangle$ into $|\psi(t)\rangle$. Therefore, the operator $U(t)$ is termed as the *evolution operator*. It is also called the *propagator*. We note that if H is Hermitian then $U^{\dagger}U = \mathrm{e}^{\mathrm{i}(H^{\dagger}-H)t/\hbar} = \mathrm{e}^{0} = 1$ implying that U is *unitary*. As $|\psi(t)\rangle = U(t)|\psi(0)\rangle$ one can say that $U(t)$ governs the intrinsic features of quantum dynamics. Suppose the states are specified by the eigenectors of the position operator. Then the transition amplitude to go from a point x' at time $t = 0$ to the point x at time t is determined by $\langle x|U(t)|x'\rangle$. It is also called the *Green's function* [1,2].

Equation (7.2) shows that the time evolution is represented by a unitary transformation. Because the origin of time has no significance U depends on time differences only. From Eq. (7.2) for $t = t_1$ we write

$$
|\psi(t_1)\rangle = \mathrm{e}^{-\mathrm{i}Ht_1/\hbar}|\psi(0)\rangle
\tag{7.5}
$$

and

$$
|\psi(0)\rangle = \mathrm{e}^{\mathrm{i}Ht_1/\hbar}|\psi(t_1)\rangle.
\tag{7.6}
$$

Then

$$
|\psi(t)\rangle = \mathrm{e}^{-\mathrm{i}H(t-t_1)/\hbar}|\psi(t_1)\rangle.
\tag{7.7}
$$

In Eqs. (6.76) and (7.2) the operator is time-independent, whereas the wave function is time-dependent. This representation of a quantum mechanical system is referred as a *Schrödinger picture*. In this picture, all operators are written as functions of (the time-independent operators) x_S and p_{xS} where the subscript S is used to denote the Schrödinger representation.

7.2.2 Equation of Motion in the Schrödinger Picture

We proceed to obtain the evolution equation of an operator. The expectation value of an observable A_S is

$$
\langle A_S \rangle = \langle \psi_S(t)|A_S|\psi_S(t)\rangle.
\tag{7.8}
$$

The time-dependent kets are given by Eq. (7.2). Differentiation of Eq. (7.8) with respect to time gives

$$
\frac{\mathrm{d}}{\mathrm{d}t}\langle A_S \rangle = \left\langle \frac{\mathrm{d}\psi_S}{\mathrm{d}t}\middle| A_S \middle| \psi_S \right\rangle + \left\langle \psi_S \middle| \frac{\partial A_S}{\partial t} \middle| \psi_S \right\rangle + \left\langle \psi_S \middle| A_S \middle| \frac{\mathrm{d}\psi_S}{\mathrm{d}t} \right\rangle.
\tag{7.9}
$$

Using Eqs. (6.76) and (7.1) in the above equation we get

$$
\begin{aligned}
\frac{\mathrm{d}}{\mathrm{d}t}\langle A_{\mathrm{S}}\rangle &= \frac{1}{\mathrm{i}\hbar}\left[-\langle\psi_{\mathrm{S}}|HA_{\mathrm{S}}|\psi_{\mathrm{S}}\rangle + \langle\psi_{\mathrm{S}}|A_{\mathrm{S}}H|\psi_{\mathrm{S}}\rangle\right] + \left\langle\frac{\partial A_{\mathrm{S}}}{\partial t}\right\rangle \\
&= \frac{1}{\mathrm{i}\hbar}\langle\psi_{\mathrm{S}}|A_{\mathrm{S}}H - HA_{\mathrm{S}}|\psi_{\mathrm{S}}\rangle + \left\langle\frac{\partial A_{\mathrm{S}}}{\partial t}\right\rangle \\
&= \frac{1}{\mathrm{i}\hbar}\langle\psi_{\mathrm{S}}|\left[A_{\mathrm{S}}, H\right]|\psi_{\mathrm{S}}\rangle + \left\langle\frac{\partial A_{\mathrm{S}}}{\partial t}\right\rangle \\
&= \left\langle\frac{\partial A_{\mathrm{S}}}{\partial t}\right\rangle + \frac{1}{\mathrm{i}\hbar}\langle\left[A_{\mathrm{S}}, H\right]\rangle,
\end{aligned}
\tag{7.10}
$$

where $[A_{\mathrm{S}}, H]$ is the commutator of A_{S} and H. This is the equation of motion of the expectation value of a dynamical variable A_{S}. The first term in the right-side of this equation arises from the explicit time-dependence in A_{S} and the second term is from the time development of the states. If A_{S} has no explicit time-dependence then

$$
\frac{\mathrm{d}}{\mathrm{d}t}\langle A_{\mathrm{S}}\rangle = \frac{1}{\mathrm{i}\hbar}\langle\left[A_{\mathrm{S}}, H\right]\rangle \quad \text{or} \quad \frac{\mathrm{i}}{\hbar}\langle\left[H, A_{\mathrm{S}}\right]\rangle.
\tag{7.11}
$$

This is the equation of motion of the expectation value of a dynamical variable A_{S} in the Schrödinger representation and is called the *Schrödinger equation of motion*. *What do we infer if* $[A_{\mathrm{S}}, H] = 0$? In this case $\mathrm{d}\langle A_{\mathrm{S}}\rangle/\mathrm{d}t = 0$ and the operator is said to be an *invariant*. *What is the significance of an invariant property of an operator?* We deduce the following two major results from Eq. (7.11).

1. The expectation value of an observable whose operator commutes with the Hamiltonian is time-independent.

2. If the system is initially in an energy eigenstate then $\langle A\rangle$ will become time-independent. The system will remain in that eigenstate forever.

For convenience define the total time derivative of any operator as

$$
\frac{\mathrm{d}A_{\mathrm{S}}}{\mathrm{d}t} = \frac{\partial A_{\mathrm{S}}}{\partial t} + \frac{1}{\mathrm{i}\hbar}[A_{\mathrm{S}}, H].
\tag{7.12}
$$

It is easy to prove that

$$
\frac{\mathrm{d}}{\mathrm{d}t}(AB) = \frac{\mathrm{d}A}{\mathrm{d}t}B + A\frac{\mathrm{d}B}{\mathrm{d}t}, \quad \frac{\mathrm{d}}{\mathrm{d}t}(f(t)A) = \frac{\mathrm{d}f}{\mathrm{d}t}A + f\frac{\mathrm{d}A}{\mathrm{d}t}
\tag{7.13}
$$

These imply that the total derivative satisfies the same algebraic rules as ordinary derivatives. Because of this the total time derivative is not defined as a rate of change of anything. *Does it matter?* It is not necessary to think it as differentiation in the usual sense; it is enough to know that it obeys the same algebraic rules. The fact that it is not defined to be a rate of change of anything does not matter because the operator relations are never the end result, they are means to find eigenfunctions, expectation values of observables and eigenvalues.

The use of total time derivative of operator that depends on the time evolution of the wave function as well as on any intrinsic time-dependence in the operators simplifies the development of quantum mechanics and allows the development to more closely follow the corresponding development of classical mechanics.

We illustrate below the use of the total time derivative for a free particle and linear harmonic oscillator. For a few other interesting systems the reader may refer to ref. [3].

Solved Problem 1:

Obtain the evolution equations for $\langle x_S \rangle$ and $\langle p_{xS} \rangle$ and their solutions for a free particle with $H = p_x^2/(2m)$.

Making use of the relations

$$[x_S, P(p_{xS})] = i\hbar \frac{\partial P}{\partial p_{xS}} , \quad [p_{xS}, X(x_S)] = -i\hbar \frac{\partial X}{\partial x_S} \tag{7.14}$$

from Eq. (7.11) we write

$$\frac{d}{dt} \langle x_S \rangle = \frac{1}{i\hbar} \langle [x_S, H(p_{xS})] \rangle = \frac{1}{i\hbar} \left\langle i\hbar \frac{\partial}{\partial p_{xS}} \frac{p_{xS}^2}{2m} \right\rangle = \frac{1}{m} \langle p_{xS} \rangle . \tag{7.15a}$$

Further,

$$\frac{d}{dt} \langle p_{xS} \rangle = \frac{1}{i\hbar} \langle [p_{xS}, H] \rangle = \frac{1}{i\hbar} \left\langle -i\hbar \frac{\partial H}{\partial x_S} \right\rangle = 0 . \tag{7.15b}$$

Thus, $\langle p_{xS} \rangle$ is independent of time. The solution of Eqs. (7.15) is

$$\langle x_S(t) \rangle = \frac{1}{m} \langle p_{xS}(0) \rangle t + \langle x_S(0) \rangle , \quad \langle p_{xS}(t) \rangle = \langle p_{xS}(0) \rangle . \tag{7.16}$$

This solution is similar to the classical solution $x_S(t) = x_0 + v_0 t$, $v(t) = v(0)$.

Solved Problem 2:

For the linear harmonic oscillator with $H = p_{xS}^2/(2m) + k x_S^2/2$ find the equations of motion for $\langle x_S \rangle$ and $\langle p_{xS} \rangle$.

The equations of motion for $\langle x_S \rangle$ and $\langle p_{xS} \rangle$ are obtained as

$$\frac{d}{dt} \langle x_S \rangle = \frac{1}{i\hbar} \langle [x_S, H] \rangle = \frac{1}{i\hbar} \left\langle i\hbar \frac{\partial H}{\partial p_{xS}} \right\rangle = \frac{1}{m} \langle p_{xS} \rangle , \tag{7.17a}$$

$$\frac{d}{dt} \langle p_{xS} \rangle = \frac{1}{i\hbar} \langle [p_{xS}, H] \rangle = \frac{1}{i\hbar} \left\langle -i\hbar \frac{\partial H}{\partial x_S} \right\rangle = -m\omega^2 \langle x_S \rangle , \tag{7.17b}$$

where $\omega = \sqrt{k/m}$. We rewrite the above equations in terms of $\langle x_S \rangle$ alone as

$$\frac{d^2}{dt^2} \langle x_S \rangle + \omega^2 \langle x_S \rangle = 0 . \tag{7.18}$$

Its solution is

$$\langle x_S(t) \rangle = \langle x_S(0) \rangle \cos \omega t + \frac{1}{m\omega} \langle p_{xS}(0) \rangle \sin \omega t . \tag{7.19}$$

7.2.3 Calculation of Action of the Evolution Operator

Let us present the calculation of the effect of $\left(1 + \dfrac{it}{\hbar n} H \right)^{-n}$ on an arbitrary $|\psi(0)\rangle$. We write

$$\frac{1}{\left(1 + \frac{it}{\hbar n} H \right)^n} |\psi^{(0)}\rangle = |\psi^{(n)}\rangle . \tag{7.20}$$

We can solve Eq. (7.20) iteratively [2]. Defining $\alpha = it/(\hbar n)$ and for $n = 1$ Eq. (7.20) gives

$$\frac{1}{(1 + \alpha H)}|\psi^{(0)}\rangle = |\psi^{(1)}\rangle \tag{7.21}$$

or

$$|\psi^{(0)}\rangle = (1 + \alpha H)|\psi^{(1)}\rangle. \tag{7.22}$$

Solving Eq. (7.22) $|\psi^{(1)}\rangle$ can be determined. Then $|\psi^{(2)}\rangle$ can be found by solving

$$|\psi^{(1)}\rangle = (1 + \alpha H)|\psi^{(2)}\rangle. \tag{7.23}$$

In this manner by successive iteration we can find $|\psi^{(n)}\rangle$.

7.3 Heisenberg Picture

Consider the circumstances where an operator is time-dependent and the wave function is time-independent.

7.3.1 Operators and Kets in a Heisenberg Picture

Consider the expectation value of the operator A given by

$$\langle A_\mathrm{S} \rangle = \langle \psi_\mathrm{S}(t)|A_\mathrm{S}|\psi_\mathrm{S}(t)\rangle. \tag{7.24}$$

Using Eq. (7.2) we have

$$\langle A_\mathrm{S} \rangle = \langle \psi_\mathrm{S}(0)|e^{iHt/\hbar} A_\mathrm{S}\, e^{-iHt/\hbar}|\psi_\mathrm{S}(0)\rangle = \langle \psi_\mathrm{S}(0)|A_\mathrm{H}(t)|\psi_\mathrm{S}(0)\rangle, \tag{7.25a}$$

where

$$A_\mathrm{H}(t) = e^{iHt/\hbar} A_\mathrm{S}\, e^{-iHt/\hbar}. \tag{7.25b}$$

Suppose we define $|\psi_\mathrm{H}(t)\rangle$ such that

$$\langle A_\mathrm{H}(t) \rangle = \langle \psi_\mathrm{H}(t)|A_\mathrm{H}|\psi_\mathrm{H}(t)\rangle. \tag{7.25c}$$

Comparison of Eqs. (7.25a) and (7.25c) gives $\langle A_\mathrm{H}(t) \rangle = \langle A_\mathrm{S} \rangle$ provided

$$|\psi_\mathrm{H}(t)\rangle = |\psi_\mathrm{H}(0)\rangle, \quad A_\mathrm{H}(t) = e^{iHt/\hbar} A_\mathrm{S}\, e^{-iHt/\hbar}. \tag{7.26}$$

In this case the ket $|\psi_\mathrm{H}(t)\rangle$ is time-independent while A_H is time-dependent. This representation is called *Heisenberg representation*. Here

$$\langle A_\mathrm{S} \rangle = |A_\mathrm{H}(t)\rangle = \langle \psi_\mathrm{S}(0)|A_\mathrm{H}(t)|\psi_\mathrm{S}(0)\rangle. \tag{7.27}$$

The result is that the expectation value of a time-independent operator A_S in a time-dependent state is same as the expectation value of a time-dependent operator $A_\mathrm{H}(t)$ in the time-independent state $|\psi_\mathrm{S}(0)\rangle$. Though, the operator in Eq. (7.25b) is written as $A_\mathrm{H}(t)$, it is actually function of $x(0)$, $p(0)$ and t. It should be noted from Eq. (7.25b) that H and any operator that commutes with H are same in both Schrödinger and Heisenberg pictures. This is because $A_\mathrm{H} = e^{iHt/\hbar} A_\mathrm{S} e^{-iHt/\hbar} = e^{iHt/\hbar}e^{-iHt/\hbar}A_\mathrm{S} = A_\mathrm{S}$.

In the Heisenberg picture, the kets and operators are labeled with the subscript H. Because the kets are time-independent we simply write $|\psi_{\mathrm{H}}\rangle$. The kets are given by $|\psi_{\mathrm{H}}(t)\rangle = |\psi_{\mathrm{H}}(0)\rangle = |\psi_{\mathrm{S}}(0)\rangle$. The basic set of observables for the Heisenberg picture is generally defined in such a way that no explicit time-dependence appears in the state vectors. From this and from Eq. (7.2) it is clear that the kets in the Heisenberg picture are simply the kets in the Schrödinger picture at $t = 0$. *What about the operators in these two representations? Do they coincide at all times t or only at $t = 0$?*

7.3.2   Equations of Motion

Differentiation of $A_{\mathrm{H}} = \mathrm{e}^{\mathrm{i}Ht/\hbar} A_{\mathrm{S}} \mathrm{e}^{-\mathrm{i}Ht/\hbar}$ with respect to t gives

$$
\begin{aligned}
\frac{\mathrm{d}}{\mathrm{d}t} A_{\mathrm{H}} &= \frac{\mathrm{i}}{\hbar} H \mathrm{e}^{\mathrm{i}Ht/\hbar} A_{\mathrm{S}} \mathrm{e}^{-\mathrm{i}Ht/\hbar} + \mathrm{e}^{\mathrm{i}Ht/\hbar} \frac{\partial A_{\mathrm{S}}}{\partial t} \mathrm{e}^{-\mathrm{i}Ht/\hbar} - \frac{\mathrm{i}}{\hbar} \mathrm{e}^{\mathrm{i}Ht/\hbar} A_{\mathrm{S}} H \mathrm{e}^{-\mathrm{i}Ht/\hbar} \\
&= \mathrm{e}^{\mathrm{i}Ht/\hbar} \frac{\partial A_{\mathrm{S}}}{\partial t} \mathrm{e}^{-\mathrm{i}Ht/\hbar} + \frac{\mathrm{i}}{\hbar} \mathrm{e}^{\mathrm{i}Ht/\hbar} \left(H A_{\mathrm{S}} - A_{\mathrm{S}} H \right) \mathrm{e}^{-\mathrm{i}Ht/\hbar}.
\end{aligned} \tag{7.28}
$$

Frequently A_{S} is function of x and p_x only. If it does not contain t explicitly then $\partial A_{\mathrm{S}}/\partial t = 0$ so that

$$
\begin{aligned}
\frac{\mathrm{d}}{\mathrm{d}t} A_{\mathrm{H}} &= \frac{1}{\mathrm{i}\hbar} \mathrm{e}^{\mathrm{i}Ht/\hbar} \left(A_{\mathrm{S}} H - H A_{\mathrm{S}} \right) \mathrm{e}^{-\mathrm{i}Ht/\hbar} \\
&= \frac{1}{\mathrm{i}\hbar} \left(\mathrm{e}^{\mathrm{i}Ht/\hbar} A_{\mathrm{S}} \, \mathrm{e}^{-\mathrm{i}Ht/\hbar} H - H \mathrm{e}^{\mathrm{i}Ht/\hbar} A_{\mathrm{S}} \, \mathrm{e}^{-\mathrm{i}Ht/\hbar} \right) \\
&= \frac{1}{\mathrm{i}\hbar} \left(A_{\mathrm{H}} H - H A_{\mathrm{H}} \right) \\
&= \frac{1}{\mathrm{i}\hbar} \left[A_{\mathrm{H}}, H \right].
\end{aligned} \tag{7.29}
$$

If A_{S} has explicit time dependence then

$$
\frac{\mathrm{d}}{\mathrm{d}t} A_{\mathrm{H}} = \frac{\partial A_{\mathrm{H}}}{\partial t} + \frac{1}{\mathrm{i}\hbar} \left[A_{\mathrm{H}}, H \right]. \tag{7.30}
$$

Equation (7.30) governs time evolution of an operator in Heisenberg representation and is called the *Heisenberg equation of motion*.

What is the connection between the total time derivative of operators in Schrödinger and Heisenberg pictures? We obtain

$$
\frac{\mathrm{d}A_{\mathrm{S}}}{\mathrm{d}t} = \mathrm{e}^{\mathrm{i}Ht/\hbar} \frac{\mathrm{d}A_{\mathrm{H}}}{\mathrm{d}t} \mathrm{e}^{-\mathrm{i}Ht/\hbar}, \qquad \frac{\mathrm{d}A_{\mathrm{H}}}{\mathrm{d}t} = \mathrm{e}^{-\mathrm{i}Ht/\hbar} \frac{\mathrm{d}A_{\mathrm{S}}}{\mathrm{d}t} \mathrm{e}^{\mathrm{i}Ht/\hbar}. \tag{7.31}
$$

Thus, the total time derivative of an operator is the Schrödinger picture version of the time derivative of its Heisenberg picture version.

Because the canonical operators q_i and p_i satisfy the rules $[q_i, p_j] = \mathrm{i}\hbar \delta_{ij}$, we can prove that

$$
[q, G(p)] = \mathrm{i}\hbar \frac{\partial G}{\partial p}, \qquad [p, F(q)] = -\mathrm{i}\hbar \frac{\partial F}{\partial q}. \tag{7.32}
$$

Equation (7.29) gives

$$
\frac{\mathrm{d}q_i}{\mathrm{d}t} = \frac{1}{\mathrm{i}\hbar} [q_i, H] = \frac{\partial H}{\partial p_i}, \tag{7.33a}
$$

$$
\frac{\mathrm{d}p_i}{\mathrm{d}t} = \frac{1}{\mathrm{i}\hbar} [p_i, H] = -\frac{\partial H}{\partial q_i}. \tag{7.33b}
$$

Equations (7.33) have the same form of Hamilton's classical canonical equation of motion. If the potential V has only a constant or linear or quadratic term corresponding to the cases of no force, constant force and harmonic motion, respectively, then it is easy to establish that $\langle q_i \rangle$ and $\langle p_i \rangle$ in any state evolve exactly like those of classical mechanics.

Solved Problem 3:

For the linear harmonic oscillator obtain the evolutions equations of x_{H} and $p_{x\mathrm{H}}$.

The operators x_{H} and $p_{x\mathrm{H}}$ are given by

$$x_{\mathrm{H}}(t) = e^{iHt/\hbar} x_{\mathrm{S}} e^{-iHt/\hbar}, \quad p_{x\mathrm{H}}(t) = e^{iHt/\hbar} p_{x\mathrm{S}} e^{-iHt/\hbar}. \tag{7.34}$$

The Hamiltonian in the Heisenberg picture is $H = p_{x\mathrm{H}}^2/(2m) + kx_{\mathrm{H}}^2/2$, where $\omega = \sqrt{k/m}$. The evolution equations of x_{H} and $p_{x\mathrm{H}}$ are obtained as follows:

$$
\begin{aligned}
i\hbar \frac{dx_{\mathrm{H}}}{dt} &= [x_{\mathrm{H}}, H] \\
&= e^{iHt/\hbar} x_{\mathrm{S}} e^{-iHt/\hbar} H - H e^{iHt/\hbar} x_{\mathrm{S}} e^{-iHt/\hbar} \\
&= e^{iHt/\hbar} [x_{\mathrm{S}}, H] e^{-iHt/\hbar}.
\end{aligned}
$$

Then

$$
\begin{aligned}
\frac{dx_{\mathrm{H}}}{dt} &= -\frac{i}{2m\hbar} e^{iHt/\hbar} \left[p_{x\mathrm{S}}^2, x_{\mathrm{S}} \right] e^{-iHt/\hbar} \\
&= \frac{1}{m} e^{iHt/\hbar} p_{x\mathrm{S}} e^{-iHt/\hbar} \\
&= \frac{1}{m} p_{x\mathrm{H}}(t).
\end{aligned}
\tag{7.35a}
$$

Next,

$$
\begin{aligned}
\frac{dp_{x\mathrm{H}}}{dt} &= \frac{1}{i\hbar} [p_{x\mathrm{H}}, H] \\
&= \frac{1}{i\hbar} \left(e^{iHt/\hbar} p_{x\mathrm{S}} e^{-iHt/\hbar} H - H e^{iHt/\hbar} p_{x\mathrm{S}} e^{-iHt/\hbar} \right) \\
&= \frac{1}{i\hbar} \left(e^{iHt/\hbar} \left[p_{x\mathrm{S}}, p_{x\mathrm{S}}^2/2m \right] e^{-iHt/\hbar} + e^{iHt/\hbar} \left[p_{x\mathrm{S}}, kx_{\mathrm{S}}^2/2 \right] e^{-iHt/\hbar} \right) \\
&= \frac{1}{i\hbar} e^{iHt/\hbar} \left[p_{\mathrm{S}}, kx_{\mathrm{S}}^2/2 \right] e^{-iHt/\hbar} \\
&= -\frac{1}{2} k e^{iHt/\hbar} \frac{\partial}{\partial x_{\mathrm{S}}} x_{\mathrm{S}}^2 e^{-iHt/\hbar} \\
&= -k^2 x_{\mathrm{H}}.
\end{aligned}
\tag{7.35b}
$$

Equations (7.35) are analogous in structure to the classical ones.

7.4   Interaction Picture

Now, we take up the case of both the operators and the wave function varying with time.

7.4.1   Kets and Operators

Suppose the Hamiltonian of a system is split into two parts H_0 and H', where H_0 is independent of time and H' is time-dependent. Such could be the case if a system is subjected to an external time-dependent field. In this case, the Schrödinger equation is

$$i\hbar \frac{\partial}{\partial t}|\psi_{\mathrm{S}}\rangle = (H_0 + H')|\psi_{\mathrm{S}}\rangle \,. \tag{7.36}$$

Let us introduce the interaction representation as

$$|\psi_{\mathrm{int}}\rangle = e^{iH_0 t/\hbar}|\psi_{\mathrm{S}}\rangle \,. \tag{7.37}$$

Replacing $|\psi_{\mathrm{S}}\rangle$ using Eq. (7.37) in Eq. (7.36) we get

$$i\hbar e^{-iH_0 t/\hbar}\frac{\partial}{\partial t}|\psi_{\mathrm{int}}\rangle = H' e^{-iH_0 t/\hbar}|\psi_{\mathrm{int}}\rangle \,. \tag{7.38}$$

Premultiplying the above equation by $e^{iH_0 t/\hbar}$ gives

$$i\hbar \frac{\partial}{\partial t}|\psi_{\mathrm{int}}\rangle = e^{iH_0 t/\hbar} H' e^{-iH_0 t/\hbar}|\psi_{\mathrm{int}}\rangle = H'_{\mathrm{int}}|\psi_{\mathrm{int}}\rangle \,, \tag{7.39a}$$

where

$$H'_{\mathrm{int}} = e^{iH_0 t/\hbar} H' e^{-iH_0 t/\hbar} \,. \tag{7.39b}$$

For all operators we write

$$A_{\mathrm{int}} = e^{iH_0 t/\hbar} A_{\mathrm{S}} e^{-iH_0 t/\hbar} \,. \tag{7.40}$$

The time evolution of ket is governed by the part H'_{int} while that of the operator is by the unperturbed part H_0. All operators are treated as functions of $x_{\mathrm{S}}(0)$, $p_{x\mathrm{S}}(0)$ and t. If A_{int} commute with H_0 then $A_{\mathrm{int}} = A_{\mathrm{S}}$. That is, it is identical to the operator in the Schrödinger and Heisenberg pictures. From Eqs. (7.37) and (7.40) we see that the time dependence is associated with $|\psi_{\mathrm{int}}\rangle$ and also with A_{int}. In this way interaction picture is a sort of intermediate between Schrödinger and Heisenberg pictures.

7.4.2   Equation of Motion

Differentiation of Eq. (7.40) with respect to t leads to

$$\frac{\mathrm{d}A_{\mathrm{int}}}{\mathrm{d}t} = \frac{\partial A_{\mathrm{int}}}{\partial t} + \frac{1}{i\hbar}\left[A_{\mathrm{int}}, H_{0,\mathrm{int}}\right] \,, \quad H_{0,\mathrm{int}} = e^{iH_0 t/\hbar} H_0 e^{-iH_0 t/\hbar} \,. \tag{7.41}$$

The evolution equations of x_{int} and $p_{x\mathrm{int}}$ for a free particle and linear harmonic oscillator can be obtained easily (given as problems at the end of this chapter).

7.5   Comparison of Three Representations

The difference between the Schrödinger and Heisenberg pictures is best understood as the difference in the basic set of observables. The basic set of observables for Heisenberg picture is defined so that no explicit time dependence appears in the state vectors. In this picture all operators are written as functions of x_S, p_{xS} and t. In the Schrödinger picture, all operators are functions of x_S and p_{xS}. An operator A which has no explicit time dependence in the Schrödinger picture, appears with an explicit time dependence in the Heisenberg picture. Table 7.1 gives the comparison of the three pictures.

7.6   Density Matrix for a Single System

Generally, the many-body wave function $\psi(q_1, \ldots, q_{3N}, t)$ is too large to calculate for a macroscopic system. In this case we may leave the information about individual systems and wish to know the average properties of an ensemble of identical and noninteracting systems. That is, we aim to use the concept of ensembles to derive the expectation value of a dynamical variable A without directly finding the wave function. In this case we may speak of *density operator* and its matrix representation called *density matrix*. This formalism was introduced by John von Neumann and independently by Lev Landau and Felix Bloch.

Consider a single member of an ensemble and let it be in the state $|\psi\rangle$. We expand $|\psi\rangle$ as

$$|\psi\rangle = \sum_n c_n |\phi_n\rangle = \sum_n c_n |n\rangle, \quad \sum |c_n|^2 = 1, \tag{7.42}$$

where $|\phi_n\rangle$ are orthonormal set of vectors. We define the trace of an operator A in matrix representation with the basis set $\{|\phi_n\rangle\}$ as

$$\mathrm{tr} A = \sum_n \langle \phi_n | A | \phi_n \rangle = \sum_n \langle n | A | n \rangle. \tag{7.43}$$

Then $\langle A \rangle$ is given by

$$\langle A \rangle = \langle \psi | A | \psi \rangle = \sum_m \sum_n c_m c_n^* \langle \phi_n | A | \phi_m \rangle. \tag{7.44}$$

We define a matrix ρ with elements ρ_{mn} given by

$$\rho_{mn} = c_m c_n^*. \tag{7.45}$$

Writing $A_{nm} = \langle \phi_n | A | \phi_m \rangle$ we have

$$\langle A \rangle = \sum_m \sum_n \rho_{mn} A_{nm} = \sum_n (\rho A)_{nn} = \mathrm{Tr}(\rho A). \tag{7.46}$$

The matrix ρ with the elements given by Eq. (7.45) is called a *density matrix* of the single system. There exists an operator called a *density operator* corresponding to the density matrix and is defined as

$$\rho = |\psi\rangle\langle\psi|. \tag{7.47}$$

TABLE 7.1
Comparison of Schrödinger, Heisenberg and interaction pictures.

No.	Quantity	Schrödinger picture	Heisenberg picture	Interaction picture
1.	Hamiltonian	H	H	$H_{\text{int}} = H_0 + H'$
2.	State	Time-dependent $$i\hbar\frac{\partial}{\partial t}|\psi_S(t)\rangle = H|\psi_S(t)\rangle$$ $$|\psi_S(t)\rangle = e^{-iHt/\hbar}|\psi_S(0)\rangle$$	Time-independent $$\frac{\partial}{\partial t}|\psi_H(t)\rangle = 0$$ $$|\psi_H(t)\rangle = |\psi_S(0)\rangle$$	Time-dependent $$i\hbar\frac{\partial}{\partial t}|\psi_{\text{int}}(t)\rangle = H_{\text{int}}|\psi_{\text{int}}(t)\rangle$$ $$|\psi_{\text{int}}(t)\rangle = e^{iH_0t/\hbar}|\psi_S(0)\rangle$$
3.	Operators	All operators are written as functions of x_S and p_{xS}. $$A_S = A_S(x_S, p_{xS})$$	All operators are written as functions of t and position and momentum operators at $t = 0$. $$A_H(t) = e^{iHt/\hbar}A_S e^{-iHt/\hbar}$$	All operators are written as functions of t and position and momentum operators at $t = 0$. $$A_{\text{int}}(t) = e^{iH_0t/\hbar}A_S e^{-iH_0t/\hbar}$$
4.	Equation of motion	$$\frac{\text{d}}{\text{d}t}A_S = \frac{1}{i\hbar}[A_S, H]$$	$$\frac{\text{d}}{\text{d}t}A_H = \frac{1}{i\hbar}[A_H, H]$$	$$\frac{\text{d}}{\text{d}t}A_{\text{int}} = \frac{1}{i\hbar}[A_{\text{int}}, H_{0,\text{int}}]$$
5.	$\langle A_S\rangle$	$\langle A_S\rangle = \langle\psi_S(t)|A_S|\psi_S(t)\rangle$	$\langle A_H\rangle = \langle\psi_S(0)|A_H(t)|\psi_S(0)\rangle$	$\langle A_{\text{int}}\rangle = \langle\psi_{\text{int}}(t)|A_{\text{int}}|\psi_{\text{int}}(t)\rangle$

The density operator is the quantum analog of the classical density function. It is easy to verify that the elements of ρ defined by Eq. (7.47) are simply given by Eq. (7.45). Considering $\{|\phi_n\rangle\}$, ρ_{mn} are given by $\langle\phi_m|\rho|\phi_n\rangle$. We have

$$\langle\phi_m|\rho|\phi_n\rangle = c_m c_n^* \langle\phi_m|\phi_m\rangle\langle\phi_n|\phi_n\rangle = c_m c_n^* = \rho_{mn}. \tag{7.48}$$

When $m = n$ we obtain $\rho_{nn} = |c_n|^2$ which implies that the diagonal matrix elements of the density operator represent the probabilities of the system in the base states.

Solved Problem 4:

For the density operator ρ determine (i) $\text{Tr}(\rho A)$, (ii) $\text{Tr}(\rho) = 1$, (iii) $\text{Tr}(\rho^2)$ and (iv) whether ρ is Hermitian.

(i) Starting from $\text{Tr}(\rho A)$ we get

$$\text{Tr}(\rho A) = \sum_n \langle n|\rho A|n\rangle = \sum_n \langle n|\psi\rangle\langle\psi|A|n\rangle = \langle\psi|A|\psi\rangle = \langle A\rangle. \tag{7.49}$$

(ii) We obtain

$$\text{Tr}(\rho) = \text{Tr}(\rho 1) = \langle\psi|1|\psi\rangle = 1. \tag{7.50}$$

(iii) First, we find ρ^2. Using the definition of ρ we get

$$\rho^2 = |\psi\rangle\langle\psi|\psi\rangle\langle\psi| = |\psi\rangle\langle\psi| = \rho. \tag{7.51}$$

Since $\rho^2 = \rho$ we have $\text{Tr}(\rho^2) = \text{Tr}(\rho) = 1$.

(iv) From Eq. (7.45) we have

$$\rho_{mn} = c_m c_n^* = (c_m^* c_n)^* = (c_n c_m^*)^* = \rho_{nm}^*. \tag{7.52}$$

Hence, ρ is a Hermitian.

7.7 Density Matrix for an Ensemble

Suppose we consider an ensemble of systems with N members. If all the members are in the same state $|\psi\rangle$ then the ensemble is said to be a *pure ensemble* and the state is a *pure state*. In a *mixed ensemble* at least one member must be in a state different from $|\psi\rangle$. Let N_n be the number of identical members out of N members. If a_n is the eigenvalue of an operator of N_n identical members with $|\psi\rangle = \sum_n c_n|\phi_n\rangle$ then the probability $|c_n|^2$ is given by

$$|c_n|^2 = \lim_{N\to\infty} \frac{N_n}{N}. \tag{7.53}$$

The expectation value $\langle\overline{A}\rangle$ of an observable taken over the ensemble is

$$\langle\overline{A}\rangle = \sum_n |c_n|^2 a_n = \lim_{N\to\infty} \frac{1}{N}\sum N_n a_n. \tag{7.54}$$

Instead of identical members, an ensemble may have members with distinct states. When there are N_i members in the ith state $|\psi_i\rangle$ then the probability of finding a member in the ith state $|\psi_i\rangle$ is given by $p_i = N_i/N$ and $\sum p_i = 1$. Then $\langle \overline{A} \rangle$ is

$$\langle \overline{A} \rangle = \frac{1}{N} \sum_i N_i a_i = \sum_i p_i a_i = \sum_i p_i \langle \psi_i | A | \psi_i \rangle. \tag{7.55}$$

$\langle \overline{A} \rangle$ can be expressed in terms $\overline{\rho}$ of the ensemble where $\overline{\rho} = \sum_i p_i |\psi_i\rangle\langle\psi_i|$.

Let us enumerate the properties of the density operator $\overline{\rho}$.

1. $\langle \overline{A} \rangle = \text{Tr}(\overline{\rho}\overline{A})$.

    $\text{Tr}(\overline{\rho}\overline{A}) = \sum_n \langle n | \overline{\rho}\overline{A} | n \rangle$ becomes

    $$\text{Tr}(\overline{\rho}\overline{A}) = \sum_n \sum_i p_i \langle n | \psi_i \rangle \langle \psi_i | \overline{A} | n \rangle = \sum_n \sum_i p_i \langle n | \psi_i \rangle \langle \psi_i | A | n \rangle. \tag{7.56}$$

    Since $\langle n | \psi_i \rangle$ and $\langle \psi_i | A | n \rangle$ are numbers we rewrite Eq. (7.56) as

    $$\text{Tr}(\overline{\rho}\overline{A}) = \sum_n \sum_i p_i \langle \psi_i | A | n \rangle \langle n | \psi_i \rangle = \sum_i p_i \langle \psi_i | A | \psi_i \rangle = \langle \overline{A} \rangle. \tag{7.57}$$

    Thus, $\langle \overline{A} \rangle$ depends only on $\overline{\rho}$. Once $\overline{\rho}$ is known we need not consider the individual systems to calculate ensemble average.

2. $\overline{\rho}$ is a Hermitian.

    Defining $\rho^{(i)}$ is the density matrix of a member in the ith state

    $$\overline{\rho}_{mn} = \sum_i p_i \rho_{mn}^{(i)} = \sum_i p_i \left(c_m^{(i)*} c_n^{(i)} \right)^* = \overline{\rho}_{nm}. \tag{7.58}$$

3. $\text{Tr}(\overline{\rho}) = 1$.

    When $\overline{A} = 1$ from Eq. (7.57) we obtain $\text{Tr}(\overline{\rho}) = 1$.

4. $\overline{\rho}^2 \neq \overline{\rho}$.

    Using the definition of $\overline{\rho}$ we obtain

    $$\begin{aligned}
    \overline{\rho}^2 &= \sum_i p_i \Big| \psi_i \Big\rangle \Big\langle \psi_i \Big| \sum_j p_j \Big| \psi_j \Big\rangle \Big\langle \psi_j \Big| \\
    &= \sum_i \sum_j p_i p_j | \psi_i \rangle \delta_{ij} \langle \psi_j | \\
    &= \sum_i p_i^2 | \psi_i \rangle \langle \psi_i |. \tag{7.59}
    \end{aligned}$$

If all p_i's are zero except one, say p_a, then $p_a = 1$ and $\overline{\rho}^2 = |\psi_i\rangle\langle\psi_i| = \rho$. On the other hand, if more than one p_i's are nonzero then $\overline{\rho}^2 \neq \overline{\rho}$.

5. $\text{Tr}\left(\bar{\rho}^2\right) < 1$.

We obtain

$$
\begin{aligned}
\text{Tr}\left(\bar{\rho}^2\right) &= \sum_n \langle n|\bar{\rho}^2|n\rangle \\
&= \sum_n \sum_i \sum_j p_i p_j \langle n|\psi_i\rangle\langle\psi_i|\psi_j\rangle\langle\psi_j|n\rangle \\
&= \sum_i \sum_j p_i p_j \left(\delta_{ij}\right)^2 \\
&= \sum_i p_i^2 \, .
\end{aligned}
\tag{7.60}
$$

Since $0 < p_i < 1$ and $\sum_i p_i = 1$ for a mixed state we have $\text{Tr}\left(\bar{\rho}^2\right) < 1$.

Solved Problem 5:

Show that the eigenvalues of $\langle\bar{\rho}\rangle$ are real and nonnegative.

For each $|\psi_i\rangle$ we have

$$
\langle\bar{\rho}\rangle = \langle\psi_i|\,\bar{\rho}\,|\psi_i\rangle = \sum_i p_i \langle\psi_i|\psi_i\rangle\langle\psi_i|\psi_i\rangle = \sum_i p_i |\langle\psi_i|\psi_i\rangle|^2 \, .
\tag{7.61}
$$

Because p_i and $|\langle\psi_i|\psi_i\rangle|^2$ are positive we have $\langle\bar{\rho}\rangle \geq 0$. Defining a_i as the eigenvalue of $\bar{\rho}$ associated with $|\psi_i\rangle$ we write $\langle\bar{\rho}\rangle = \sum_i p_i a_i$. Comparison of this with the Eq. (7.61) indicates that $a_i = |\langle\psi_i|\psi_i\rangle|^2$ and hence the eigenvalues of $\langle\bar{\rho}\rangle$ are nonnegative and the matrix $\bar{\rho}$ is positive.

7.8　Time Evolution of Density Operator

The time evolution of the density operator $\bar{\rho}$ can be predicted making use of the Schrödinger equations

$$
i\hbar\frac{\partial}{\partial t}|\psi_i\rangle = H|\psi_i\rangle \, , \quad -i\hbar\frac{\partial}{\partial t}\langle\psi_i| = \langle\psi_i|H \, .
\tag{7.62}
$$

The time derivative of $\bar{\rho}$ is given by

$$
\frac{\partial\bar{\rho}}{\partial t} = \frac{\partial}{\partial t}\sum_i p_i|\psi_i\rangle\langle\psi_i| = \frac{1}{i\hbar}\left[H,\bar{\rho}\right] \, .
\tag{7.63}
$$

Equation (7.63) is called a *von Neumann equation* and is the quantum mechanical analog of the Liouville equation of classical statistical mechanics. Equation (7.63) is valid for time-dependent Hamiltonian also.

Let us consider the Schrödinger equation with $H(t)$. We write

$$
i\hbar\frac{\partial}{\partial t}|\psi(t)\rangle = H(t)|\psi(t)\rangle \, .
\tag{7.64}
$$

Its solution can be represented as

$$
|\psi(t)\rangle = U(t,0)|\psi(0)\rangle,
\tag{7.65}
$$

where U expresses the time development of $|\psi(0)\rangle$. Substituting Eq. (7.65) in Eq. (7.64) we get $i\hbar\partial U/\partial t = HU$. Assuming $U(0,0) = 1$ the solution of the above equation is $U(t,0) = e^{-iHt/\hbar}$. In the Heisenberg picture

$$\bar{\rho}_H(t) = e^{-iHt/\hbar}\,\bar{\rho}(0)\,e^{iHt/\hbar} = U(t,0)\,\bar{\rho}(0)\,U^\dagger. \tag{7.66}$$

The equation of motion of $\langle\overline{A}\rangle$ is obtained as

$$i\hbar\frac{d}{dt}\langle\overline{A}\rangle = i\hbar\frac{d}{dt}\mathrm{Tr}\left(\bar{\rho}\,\overline{A}\right) = \mathrm{Tr}\left[\bar{\rho}\left([\overline{A},H] + i\hbar\frac{\partial\overline{A}}{\partial t}\right)\right]. \tag{7.67}$$

7.9   Concluding Remarks

In this chapter, we presented the description of quantum dynamics in three different pictures. We have shown the connection between the total time derivative of operators in the various pictures. We have compared the three representations. The Heisenberg picture is better suited to bring out fundamental features, such as symmetries and conservation laws, and it is indispensable in systems with many degrees of freedom, like those dealt within quantum field theory and statistical physics. No matter which picture is applied to explain the dynamics of a quantum system, the laws of motion of the micro-system are all identical. As the three pictures are equivalent and connected by unitary transformations, we must choose the picture in which the physical properties of a system are more evident and the calculations are simple. We discussed the features of density matrix. For more details on density matrix formalism and its applications refer to refs. [4–6]. Quantun dynamics of a driven damped harmonic oscillator in Heisenberg picture is reported [7,8].

7.10   Bibliography

[1]   J. Saha, *Am. J. Phys.* 84:770, 2016.

[2]   M. Amaku, F.A.B. Coutinho and F.M. Toyama, *Am. J. Phys.* 85:692, 2017.

[3]   M. Andrews, *Am. J. Phys.* 71:326, 2003.

[4]   K. Blum, *Density Matrix Theory and Applications.* Springer, Berlin, 1996.

[5]   J. Vanier, *Basic Theory of Lasers and Masers: A Density Matrix Approach.* Taylor & Francis, New York, 1971.

[6]   D.A. Mazziotti, *Reduced Density Matrix Mechanics with Applications to Many-Electron Atoms and Molecules.* Wiley-Interscience, New Jersey, 2007.

[7]   T.G. Philbin, *New J. Phys.* 14:083043, 2012.

[8]   F. Kheirandish, *Eur. Phys. J. Plus* 135:243, 2020.

7.11   Exercises

7.1 (a) Show that if a Hamiltonian in Schrödinger picture is independent of time then it is independent of time in Heisenberg picture also. (b) Show that $e^{-iHt/\hbar}|\psi(0)\rangle = \lim\limits_{n\to\infty}\left(1 + \dfrac{it}{\hbar n}H\right)^{-n}|\psi(0)\rangle$ satisfies the time-dependent Schrödinger equation.

7.2 Show that $\dfrac{d}{dt}(AB) = \dot{A}B + A\dot{B}$ where A and B are operators. Then find $\dfrac{d}{dt}(f(t)A)$.

7.3 If T and V are kinetic energy and potential energy operators and the time average of the time derivative of the quantity $\mathbf{r}\cdot\mathbf{p}$ is 0, then establish the virial theorem $2\langle T\rangle = \langle\mathbf{r}\cdot\nabla V\rangle$.

7.4 Find (a) $\dfrac{d}{dt}\langle x^2\rangle$ and (b) $\dfrac{d}{dt}\langle xp_x\rangle$.

7.5 Show that for the Hamiltonian $H = p_x^2/2 + (p_x x + xp_x)/2 + x^2/2$ the total time derivative of x and p_x are $\dot{x} = p_x + x$, $\dot{p}_x = -p_x - x$.

7.6 In the previous exercise suppose for any wave packet we define $X = x - \langle x\rangle$ and $P_x = p_x - \langle p_x\rangle$. Then find (a) $\dot{X}$ and (b) $\dot{P}_x$.

7.7 Prove the following:

(a) The expectation value of a dynamical variable whose operator commutes with the Hamiltonian is time-independent regardless of initial state.

(b) If the system is initially in an energy eigenstate then the expectation value of any dynamical variable will be time-independent.

(c) If the system is initially in a superposition of energy eigenstates (nonstationary state) then the expectation value consists of oscillatory terms with Bohr's frequencies.

7.8 For $H = p_x^2/2 + f(t)p_x + g(t)x$ obtain $\dot{x}$ and $\dot{p}_x$.

7.9 $A(x, p_x)$ and $B(x, p_x)$ can be expressed as power series in x and p_x and $[x, p_x] = i\hbar$. Show by purely matrix method that $\lim_{\hbar\to 0}(1/i\hbar)[A, B] = \{A, B\}$, where the right-side is the Poisson bracket calculated as if x and p_x were classical variables.

7.10 Starting from the equation $dA_H/dt = (1/i\hbar)[A_H, H] + \partial A_H/\partial t$ find dA_S/dt.

7.11 For a free particle the Hamiltonian in the Schrödinger picture is $H = p_x^2/(2m)$. Can you write $i\hbar\dot{x}_H = [x_H, H_H]$, where H_H is the Hamiltonian in a Heisenberg picture?

7.12 Determine the evolution equation for x_{int} and $p_{x\text{int}}$ in the interaction picture for a free particle.

7.13 Obtain the evolution equations for x_{int} and $p_{x\text{int}}$ in the interaction picture for the harmonic oscillator Hamiltonian $H = p_x^2/(2m) + kx^2/2$.

7.14 In terms of the ladder operators a and $a^\dagger$ the Hamiltonian of the harmonic oscillator can be written as $H = \hbar\omega aa^\dagger + \hbar\omega/2$. Obtain the equation of motions for a and $a^\dagger$ and solutions of them.

7.15 A particle is in a one-dimensional harmonic oscillator potential. Determine the conditions for the expectation value of an operator to depend on time, if

(a) the particle is initially in a momentum eigenstate and

(b) the particle is initially in an energy eigenstate.

7.16 If the Hamiltonian of a particle of a charge q and mass m subjected to a uniform electrostatic field $\mathbf{E}$ is $H = \mathbf{p}^2/(2m) - q\mathbf{E}\cdot\mathbf{r}$ then find (a) dx/dt, (b) dp_x/dt and (c) $md^2\mathbf{r}/dt^2$.

7.17 The Hamiltonian operator of a charged particle subjected to a magnetic field is $H = [\mathbf{p} - (e/c)\mathbf{A}(r)]^2 /(2m)$. Show that in the Heisenberg picture $m\dot{\mathbf{r}} = \mathbf{p} - (e/c)\mathbf{A}$, $m\ddot{\mathbf{r}} = (e/2c)(\dot{\mathbf{r}} \times \mathbf{B} - \mathbf{B} \times \dot{\mathbf{r}})$.

7.18 Prove the Baker–Hausdorff lemma

$$e^{iA\lambda} B e^{-iA\lambda} = B + i\lambda[A, B] + \frac{(i\lambda)^2}{2!} [A, [A, B]] + \dots.$$

Then, for a linear harmonic oscillator find $x(t)$.

7.19 Using the equation $\rho(t) = U(t, 0)\rho(0)U^\dagger$ show that $\mathrm{Tr}\rho^2(t)$ is independent of time.

7.20 Consider two quantum systems–the first one in the state $|0\rangle$ with probability $3/4$ and in the state $|1\rangle$ with probability $1/4$ and the second one must be prepared in the state $|a\rangle$ with probability $1/2$ and in the state $|b\rangle$ with probability $1/2$, where $|a\rangle = \sqrt{3/4}\,|0\rangle + \sqrt{1/4}\,|1\rangle$, $|b\rangle = \sqrt{3/4}\,|0\rangle - \sqrt{1/4}\,|1\rangle$. Determine the density matrices of the above two different ensembles of quantum states.

8

Heisenberg Uncertainty Principle

8.1 Introduction

One of the most striking deviations of quantum mechanics from classical mechanics is that any measurement of a microscopic system involves the interaction with the measuring instruments. The interaction cannot be neglected and in fact leads to uncontrollable disturbance on the system. Let a measurement of an observable A yield the value a and be the eigenvalue of the operator A belonging to the eigenfunction ϕ_a. The individual measurements will deviate from the average value ($\langle a \rangle$) computed over large number (N) of measurements. Denote the deviation of a from $\langle a \rangle$ as $\tilde{a}$. *What would be the value of average deviation?* We obtain

$$\langle \tilde{a} \rangle = \frac{1}{N} \sum (a - \langle a \rangle) = \langle a \rangle - \langle a \rangle = 0 \,. \tag{8.1}$$

This is because some of the values of a are below $\langle a \rangle$ while the others are above $\langle a \rangle$. Therefore, $\langle \tilde{a} \rangle$ is not useful for describing how much an individual value of an observable deviates from the expected or mean value. The mean-square deviation or variance (σ^2), that is, the average of the squares of the deviations would be nonzero. The variance is given by

$$\sigma^2 = (\Delta a)^2 = \frac{1}{N} \sum (a - \langle a \rangle)^2 = \langle a^2 \rangle - \langle a \rangle^2 \,. \tag{8.2}$$

The square-root of mean-square deviation is called *standard deviation* or *dispersion*. This quantity could be taken as a measure of the spread in the measured values. When most of the values a are near $\langle a \rangle$ then the distribution of a peaks about $\langle a \rangle$ and in this case σ is small. If most of the values of a are found over a wide range about $\langle a \rangle$ then the distribution spreads over a wide range of a with a large value of σ. This spread, denoted as ΔA, in the measured values of A, is called the *uncertainty* in the measurement of A.

In 1927, Werner Karl Heisenberg in his celebrated paper [1] (Nobel Prize in 1932) established the quantitative expression for unavoidable momentum disturbance caused by any position measurement. Essentially, any measurement of the position x of a mass with Δx would cause a disturbance Δp_x on the momentum (p_x) measurement and Δx and Δp_x should always obey the relation $\Delta x \Delta p_x \geq \hbar/2$. Heisenberg explained the physical intuition underlying the above expression through the gamma-ray microscope thought experiment.

In this chapter, first we derive the classical uncertainty relation. Next, we give a proof of the Heisenberg uncertainty relation. Then we present a modified form of it.

DOI: 10.1201/9781003172178-8

8.2   The Classical Uncertainty Relation

Let A be a classical statistical variable with mean $\langle A \rangle$. The uncertainty in A, ΔA, is defined by

$$(\Delta A)^2 = \sigma_A^2 = \frac{1}{N} \sum (A - \langle A \rangle)^2 = \langle (A - \langle A \rangle)^2 \rangle \qquad (8.3)$$

The above can be rewritten as

$$(\Delta A)^2 = \langle A^2 \rangle - \langle A \rangle^2. \qquad (8.4)$$

This is the variance of A. The variance between two variables A and B is given by [2,3]

$$\sigma_{AB} = \langle (A - \langle A \rangle)(B - \langle B \rangle) \rangle = \langle AB \rangle - \langle A \rangle \langle B \rangle. \qquad (8.5)$$

Let us define

$$\tilde{A} = A - \langle A \rangle, \quad \tilde{B} = B - \langle B \rangle \qquad (8.6)$$

which implies $\langle \tilde{A} \rangle = \langle \tilde{B} \rangle = 0$ and

$$(\Delta A)^2 = \langle \tilde{A}^2 \rangle, \quad (\Delta B)^2 = \langle \tilde{B}^2 \rangle, \quad \sigma_{AB} = \langle \tilde{A} \tilde{B} \rangle. \qquad (8.7)$$

For a statistical variable x that is not always 0 we have $\langle x^2 \rangle \geq 0$. When $x = \tilde{A} + \lambda \tilde{B}$ we find

$$\langle x^2 \rangle = \langle \tilde{A}^2 \rangle + \lambda^2 \langle \tilde{B}^2 \rangle + 2\lambda \langle \tilde{A} \tilde{B} \rangle \geq 0. \qquad (8.8)$$

An equality sign will be realized if and only if $\tilde{A} + \lambda \tilde{B} = 0$. The nature of the roots of λ is determined from the quantity

$$\alpha = 4\langle \tilde{A} \tilde{B} \rangle^2 - 4\langle \tilde{A}^2 \rangle \langle \tilde{B}^2 \rangle. \qquad (8.9)$$

When α is 0 the roots of λ are real and identical. That is, there is one real root of λ. Roots are complex for $\alpha < 0$ while they are real and distinct if $\alpha > 0$. Thus, the condition for at most one real root of λ is (*What would happen if there are two real roots of λ?*)

$$\langle \tilde{A} \tilde{B} \rangle^2 - \langle \tilde{A}^2 \rangle \langle \tilde{B}^2 \rangle \leq 0. \qquad (8.10)$$

When the Eqs. (8.7) are used the above equation becomes $\sigma_{AB}^2 - (\Delta A)^2 (\Delta B)^2 \leq 0$ or

$$(\Delta A)(\Delta B) \geq |\sigma_{AB}|. \qquad (8.11)$$

This is the uncertainty principle for classical statistics [2,3].

8.3   Heisenberg Uncertainty Relation

Let us consider two sets of complex numbers: $A_1, A_2, \ldots, A_n$ and $B_1, B_2, \ldots, B_n$. We have

$$\sum_i \sum_j |A_i B_j - A_j B_i|^2 \geq 0. \qquad (8.12)$$

An equality sign holds when these sets are identical. Expanding the left-side of Eq. (8.12) we get

$$
\begin{aligned}
0 &\leq \sum_i \sum_j (A_i B_j - A_j B_i)^* (A_i B_j - A_j B_i) \\
&\leq \sum_i \sum_j \left(|A_i|^2 |B_j|^2 - A_i^* B_j^* A_j B_i - A_j^* B_i^* A_i B_j + |A_j|^2 |B_i|^2 \right) \\
&\leq 2 \left[\left(\sum |A_i|^2 \right) \left(\sum |B_i|^2 \right) - \left(\sum A_i^* B_i \right) \left(\sum A_i B_i^* \right) \right]
\end{aligned}
$$

or

$$
\left(\sum |A_i|^2 \right) \left(\sum |B_i|^2 \right) \geq \left| \sum A_i^* B_i \right|^2 . \tag{8.13}
$$

If A and B are continuous functions of x then the Eq. (8.13) takes the form

$$
\int_{-\infty}^{\infty} A^* A \, dx \int_{-\infty}^{\infty} B^* B \, dx \geq \int_{-\infty}^{\infty} A^* B \, dx . \tag{8.14}
$$

In a quantum state ψ, the uncertainty in an observable A is given by

$$
(\Delta A)^2 = \langle (A - \langle A \rangle)^2 \rangle = \langle A^2 \rangle - \langle A \rangle^2 . \tag{8.15}
$$

Then with $\tilde{A} = A - \langle A \rangle$

$$
\Delta A = \left[\langle \tilde{A}^2 \rangle \right]^{1/2} = \left[\int_{-\infty}^{\infty} \psi^* \tilde{A}^2 \psi \, d\tau \right]^{1/2} = \left[\int_{-\infty}^{\infty} \psi^* \tilde{A}(\tilde{A}\psi) \, d\tau \right]^{1/2} . \tag{8.16}
$$

Since $\tilde{A}$ is Hermitian we have

$$
\Delta A = \left[\int_{-\infty}^{\infty} (\tilde{A}\psi)^* (\tilde{A}\psi) \, d\tau \right]^{1/2} = \left[\int_{-\infty}^{\infty} \psi_1^* \psi_1 \, d\tau \right]^{1/2} = ||\psi_1|| , \tag{8.17}
$$

where

$$
\psi_1 = \tilde{A}\psi = (A - \langle A \rangle)\psi . \tag{8.18}
$$

Similarly,

$$
\psi_2 = \tilde{B}\psi = (B - \langle B \rangle)\psi . \tag{8.19}
$$

If ψ_1 and ψ_2 are two states then from Eq. (8.14) we write

$$
||\psi_1|| \cdot ||\psi_2|| \geq \left| \int_{-\infty}^{\infty} \psi_1^* \psi_2 \, d\tau \right| . \tag{8.20}
$$

This is known as *Schwarz's inequality*. The proof of this inequality is presented in appendix E. We rewrite the Eq. (8.20) as

$$
||\psi_1|| \cdot ||\psi_2|| \geq \left| \mathrm{Im} \int_{-\infty}^{\infty} \psi_1^* \psi_2 \, d\tau \right| . \tag{8.21}
$$

Suppose we write

$$
\int_{-\infty}^{\infty} \psi_1^* \psi_2 \, d\tau = R + \mathrm{i}I , \qquad \int_{-\infty}^{\infty} \psi_2^* \psi_1 \, d\tau = R - \mathrm{i}I , \tag{8.22}
$$

where R and I are real numbers. Then I is given by

$$
I = \mathrm{Im}(R + \mathrm{i}I) = \frac{1}{2\mathrm{i}} \left[(R + \mathrm{i}I) - (R + \mathrm{i}I)^* \right] . \tag{8.23}
$$

In terms of the integrals in Eq. (8.22) the Eq. (8.23) is written as

$$\text{Im} \int_{-\infty}^{\infty} \psi_1^* \psi_2 \, d\tau = \frac{1}{2i} \left[\int_{-\infty}^{\infty} \psi_1^* \psi_2 \, d\tau - \int_{-\infty}^{\infty} \psi_2^* \psi_1 \, d\tau \right]. \tag{8.24}$$

Then

$$||\psi_1|| \cdot ||\psi_2|| \geq \left| \frac{1}{2i} \left[\int_{-\infty}^{\infty} \psi_1^* \psi_2 \, d\tau - \int_{-\infty}^{\infty} \psi_2^* \psi_1 \, d\tau \right] \right|. \tag{8.25}$$

That is,

$$\begin{aligned}
\Delta A \Delta B \quad &\geq \quad \frac{1}{2} \left| \frac{1}{i} \left[\int_{-\infty}^{\infty} (\tilde{A}\psi)^* (\tilde{B}\psi) \, d\tau - \int_{-\infty}^{\infty} (\tilde{B}\psi)^* (\tilde{A}\psi) \, d\tau \right] \right| \\
&\geq \quad \frac{1}{2} \left| \frac{1}{i} \left[\int_{-\infty}^{\infty} \psi^* \tilde{A}\tilde{B}\psi \, d\tau - \int_{-\infty}^{\infty} \psi^* \tilde{B}\tilde{A}\psi \, d\tau \right] \right| \\
&\geq \quad \frac{1}{2} \left| \frac{1}{i} \left\langle \left[\tilde{A}, \tilde{B} \right] \right\rangle \right|.
\end{aligned} \tag{8.26}$$

The commutator $[\tilde{A}, \tilde{B}]$ is shown to be

$$\begin{aligned}
\left[\tilde{A}, \tilde{B} \right] &= \tilde{A}\tilde{B} - \tilde{B}\tilde{A} \\
&= (A - \langle A \rangle)(B - \langle B \rangle) - (B - \langle B \rangle)(A - \langle A \rangle) \\
&= [A, B].
\end{aligned} \tag{8.27}$$

Thus,

$$\Delta A \Delta B \geq \frac{1}{2} \left| \frac{1}{i} \langle [A, B] \rangle \right| \tag{8.28}$$

or

$$(\Delta A)^2 (\Delta B)^2 \geq -\frac{1}{4} \langle [A, B] \rangle^2. \tag{8.29}$$

Equation (8.29) is known as *Heisenberg uncertainty principle*[1]. It is the uncertainty principle for any pair of observables A and B. This equation with equality sign is commonly referred as the *minimum uncertainty product*. In Eq. (8.29)

$$(\Delta A)^2 = \langle A^2 \rangle - \langle A \rangle^2 \tag{8.30}$$

is variance in A. Similarly $(\Delta B)^2$ is defined.

For a canonically conjugate pair of operators A and B we have $[A, B] = i\hbar$. Hence,

$$(\Delta A)^2 (\Delta B)^2 \geq \hbar^2/4 \tag{8.31}$$

or

$$(\Delta A)(\Delta B) \geq \hbar/2. \tag{8.32}$$

Thus, the statement of the uncertainty principle is the following:

The value of both the observables of a canonically conjugate pairs of a quantum mechanical system cannot be specified precisely and simultaneously.

The crucial word here is *simultaneously*. The uncertainty relation gives the lower bound on the product of the uncertainties of two pairs of variables. But it says nothing about the

---

[1]This principle was enunciated by Heisenberg when he was just 25 years old.

accuracy with which individual observables can be measured. An important point is that the uncertainty relation has nothing to do with the precision of the measuring instrument or technique. Rather, it is an inherent characteristic of the quantum mechanical systems. The uncertainty relation is not obtained based on experimental data. Without the uncertainty principle, quantum mechanics becomes a deterministic theory.

Because $[x, p_x] = i\hbar$, the position-momentum uncertainty principle is

$$(\Delta x)(\Delta p_x) \geq \hbar/2 \,, \tag{8.33}$$

where

$$(\Delta x)^2 = \langle x^2 \rangle - \langle x \rangle^2 \,, \quad (\Delta p_x)^2 = \langle p_x^2 \rangle - \langle p_x \rangle^2 \,. \tag{8.34}$$

$(\Delta x)^2$ and $(\Delta p_x)^2$ are the variance of position and momentum of a particle, respectively. Note that only the mean values of x and x^2 are needed to compute Δx. The uncertainty principle always has members of a conjugate pair. The position-momentum uncertainty relation was first called as *indetermination principle* by Arthur Edward Ruark [4] and then as *uncertainty principle* by Edward Uhler Condon [5]. The Heisenberg relation was given in the inequality form by Earle Hesse Kennard [6].

In momentum space $p = \hbar k$ and hence the uncertainty relation is

$$(\Delta x)(\Delta k) \geq 1/2 \,. \tag{8.35}$$

From the equality sign we have $\Delta x = 1/(2\Delta k)$. This implies that if Δx is small then Δk must be large and vice-versa. This accounts for the wave-particle paradox. We cannot make Δx and Δk small simultaneously (beyond the restriction $\Delta x \Delta k \geq 1/2$), that is, both a wave and a particle. A similar condition we can have for the energy and time measurements:

$$(\Delta E)(\Delta t) \geq \hbar/2 \,. \tag{8.36}$$

For example, suppose we wish to measure the time t at which a particle arrives a particular position with an uncertainty Δt. As per the uncertainty principle, ΔE of the measured energy of the particle at that position cannot be smaller than $\hbar/(2\Delta t)$. Another result of (8.36) is that a system, which is not in a steady state but changing with time, cannot have a well-defined energy. The physical meaning of energy-time uncertainty is different from that of $x - p_x$ uncertainty. Position and momentum are observables to which Hermitian operators are assigned and which are measured at a particular time. But time appears only as a parameter in quantum theory as stated in sec. 3.8.3.

Solved Problem 1:

With an example show that the uncertainty principle is insignificant to larger objects or particles and significant to microscopic particles.

Suppose a cricket ball of mass 300 gm is moving with a speed of 20 m per second. The error in momentum measurement is, say, 0.1%. Then the uncertainty in the position measurement is

$$\Delta x \geq \frac{\hbar}{2\Delta p_x} = \frac{1.0544 \times 10^{-34}}{2 \times 0.3 \times 20 \times \dfrac{0.1}{100}} = 8.78667 \times 10^{-33} \, \text{m} \,. \tag{8.37}$$

Δx is of the order of 10^{-33} m. As this quantity is very small we cannot realize the uncertainty principle. This is true for heavier bodies.

Next, consider an electron with velocity 2×10^7m. Let the uncertainty in its velocity measurement be 0.1%. We obtain

$$\Delta x = \frac{\hbar}{2m\Delta v}$$

$$= \frac{1.0544 \times 10^{-34}}{2 \times 9.109 \times 10^{-31} \times \frac{0.1}{100} \times 2 \times 10^7} \, \text{m}$$

$$= 2.89384 \times 10^{-9} \, \text{m} \,. \tag{8.38}$$

The size of the nucleus in any atom is $\approx 10^{-14}$ m. The calculated Δx is very much greater than the size of the nucleus and is thus matter in atomic scales.

Solved Problem 2:

Calculate the ground state energy of a harmonic oscillator ($H = p_x^2/(2m) + kx^2/2$) using the Heisenberg's uncertainty principle assuming $\langle x \rangle = \langle p_x \rangle = 0$.

For the linear harmonic oscillator

$$(\Delta x)^2 = \langle x^2 \rangle - \langle x \rangle^2 = \langle x^2 \rangle \,, \tag{8.39}$$

$$(\Delta p_x)^2 = \langle p_x^2 \rangle - \langle p_x \rangle^2 = \langle p_x^2 \rangle \,. \tag{8.40}$$

From the relation $(\Delta x)^2(\Delta p_x)^2 \geq \hbar^2/4$ we write

$$(\Delta p_x)^2 \geq \frac{\hbar^2}{4(\Delta x)^2} = \frac{\hbar^2}{4\langle x^2 \rangle} \tag{8.41}$$

For the linear harmonic oscillator $\langle H \rangle$ is

$$\langle H \rangle = E = \frac{1}{2m}\langle p_x^2 \rangle + \frac{1}{2}m\omega^2\langle x^2 \rangle$$

$$= \frac{1}{2m}(\Delta p_x)^2 + \frac{1}{2}m\omega^2\langle x^2 \rangle$$

$$\geq \frac{\hbar^2}{8m\langle x^2 \rangle} + \frac{1}{2}m\omega^2\langle x^2 \rangle \,. \tag{8.42}$$

The ground state energy is the minimum of E. The conditions for minimization of E are $dE/d\langle x^2 \rangle = 0$ and $d^2E/d\langle x^2 \rangle^2 > 0$. The first condition gives $\langle x^2 \rangle = \hbar/(2m\omega)$. With this $\langle x^2 \rangle$ the equality sign in the equation for $\langle H \rangle$ gives $E_{\min} = \hbar\omega/2$.

In sec. 4.4.5 we compared the classical and quantum moments of linear harmonic oscillator. From Eqs. (4.37) and (4.39) we find

$$(\Delta X)_{\text{CL}}^2 = \langle X^2 \rangle_{\text{CL}} - \langle X \rangle_{\text{CL}}^2 = \frac{1}{2}, \tag{8.43a}$$

$$(\Delta P)_{\text{CL}}^2 = \langle P^2 \rangle_{\text{CL}} - \langle P \rangle_{\text{CL}}^2 = \frac{1}{2} \tag{8.43b}$$

and hence $(\Delta X)_{\text{CL}}^2 (\Delta P)_{\text{CL}}^2 = 1/4$. Similarly, we find from Eqs. (4.40) $(\Delta X)_{\text{QM}}^2 (\Delta P)_{\text{QM}}^2 = 1/4$. For the damped harmonic oscillator discussed in sec. 4.12 the uncertainty relation is found to be $(\Delta x)(\Delta p_x) = \frac{\hbar\omega_0}{\omega}\left(n + \frac{1}{2}\right)$.

8.4 Condition for Minimum Uncertainty Product

For minimum uncertainty product, in the Schwarz inequality, the equal sign must hold. We have the Schwarz inequality (refer Eq. (8.20))

$$(\tilde{A}\psi,\ \tilde{A}\psi)(\tilde{B}\psi,\ \tilde{B}\psi) \geq |(\tilde{A}\psi,\ \tilde{B}\psi)|^2\,. \tag{8.44}$$

When $\tilde{A}\psi = z\tilde{B}\psi$ we get

$$(z\tilde{B}\psi,\ z\tilde{B}\psi)(\tilde{B}\psi,\ \tilde{B}\psi) \geq |(z\tilde{B}\psi,\ \tilde{B}\psi)|^2\,. \tag{8.45}$$

That is,

$$|z|^2|(\tilde{B}\psi,\ \tilde{B}\psi)|^2 \geq |z|^2|(\tilde{B}\psi,\ \tilde{B}\psi)|^2\,. \tag{8.46}$$

The equality sign holds for $\tilde{A}\psi = z\tilde{B}\psi$. We have

$$(\Delta A)^2 = (\psi,\ \tilde{A}^2\psi)\,,\quad (\Delta B)^2 = (\psi,\ \tilde{B}^2\psi)\,. \tag{8.47}$$

For minimum product of uncertainties we have

$$(\Delta A)^2(\Delta B)^2 = |(\tilde{B}\psi,\ \tilde{A}\psi)|^2 = |(\psi,\ \tilde{B}\tilde{A}\psi)|^2\,. \tag{8.48}$$

Then from

$$\Delta A\Delta B = \frac{1}{2}\left|\frac{1}{\mathrm{i}}\langle[\tilde{A},\ \tilde{B}]\rangle\right| \tag{8.49}$$

we get

$$|(\psi,\ \tilde{B}\tilde{A}\psi)|^2 = \frac{1}{4}|(\psi,\ [\tilde{B},\ \tilde{A}]\psi)|^2\,. \tag{8.50}$$

We have

$$\tilde{B}\tilde{A} = \frac{1}{2}\left\{\tilde{B},\ \tilde{A}\right\} + \frac{1}{2}\left[\tilde{B},\ \tilde{A}\right]\,, \tag{8.51}$$

where $\left\{\tilde{B},\tilde{A}\right\}$ is the anticommutation. Comparing with the above expression for $|(\psi,\ \tilde{B}\tilde{A}\psi)|^2$ we find that the condition for minimum uncertainties as $\left\langle\left\{\tilde{B},\tilde{A}\right\}/2\right\rangle = 0$. That is,

$$(\psi,\tilde{B}\tilde{A}\psi) + (\psi,\tilde{A}\tilde{B}\psi) = 0 \quad\text{or}\quad (\tilde{B}\psi,\tilde{A}\psi) + (\tilde{A}\psi,\tilde{B}\psi) = 0\,. \tag{8.52}$$

Substituting $\tilde{A}\psi = z\tilde{B}\psi$ we get $(z+z^*)(\tilde{B}\psi,\tilde{B}\psi) = 0$. Since $(\tilde{B}\psi,\tilde{B}\psi) \neq 0$ we get $z+z^* = 0$. Therefore, z is pure imaginary. Suppose $z = \mathrm{i}\lambda$, λ is a real. The condition for $\Delta A\Delta B$ to be minimum is $\tilde{A}\psi = \mathrm{i}\lambda\tilde{B}\psi$. For $B = x$ and $A = -\mathrm{i}\hbar\mathrm{d}/\mathrm{d}x$ the condition is

$$\left(-\mathrm{i}\hbar\frac{\mathrm{d}}{\mathrm{d}x} - \langle p_x\rangle\right)\psi = \mathrm{i}\lambda(x - \langle x\rangle)\psi\,. \tag{8.53}$$

We rewrite it as

$$\frac{\mathrm{d}\psi}{\psi} = -\frac{\lambda}{\hbar}(x - \langle x\rangle)\,\mathrm{d}x + \frac{\mathrm{i}}{\hbar}\langle p_x\rangle\,\mathrm{d}x\,. \tag{8.54}$$

Integration of the above equation gives

$$\psi(x) = N\exp\left[-\frac{\lambda}{2\hbar}(x - \langle x\rangle)^2 + \mathrm{i}\frac{\langle p_x\rangle}{\hbar}x\right]\,, \tag{8.55}$$

where N is an arbitrary constant.

8.5   Implications of Uncertainty Relation

The uncertainty relation has a wide-ranging significance[2]. In the following we point out some of them.

1. Suppose we wish to find the position and momentum of a moving particle. For this purpose we allow light to fall on the particle. The particle will scatter part of the incident light which will give details about the position of it. Note that the position cannot be found more accurately than the wavelength of the incident light. However, we can improve the accuracy using short wavelength wave. But according to Planck's hypothesis we need to use at least one quantum. This will affect the particle and alter its momentum in a way that cannot be determined. Further, to measure the position more precisely we have to use shorter wavelength light. In such a case, energy of a single quantum (hc/λ) is higher and hence the disturbance to the velocity of the particle will be large. Consequently, the more accurately we attempt to measure the position, the less accurately we can measure the speed and vice versa.

2. Consider the measurement of radius r_0 of the electron orbit in a ground state of the hydrogen atom. Theoretically predicted value of r_0 is 5.3×10^{-11} m. According to quantum mechanics this value is the most probable radius. This means in experiment, most trials will give a different value, larger or smaller than 5.3×10^{-11} m.

3. An important implication of the uncertainty principle is that quantum systems do not follow classical trajectories which conflict with our everyday experience.

4. Another consequence is that quantum mechanical systems in their lowest energy state (ground state) could not be at rest. For example, the electron of an atom in its ground state could not be at rest. *What will happen if it is at rest?* In this case its velocity is exactly known to be zero and its position would be known precisely. This would violate the Heisenberg uncertainty relation.

5. Suppose $\psi(x)$ is a delta-function at $x = x_0$. The position of the particle is determined exactly. *What about the corresponding momentum wave function?* The momentum wave function $C(k)$ is a sinusoidal function extending over all space. Thus, the momentum uncertainty is infinity. Similarly, if $\psi(x)$ is sinusoidal then the position uncertainty is infinity whereas because the corresponding momentum wave function is delta-function the uncertainty in momentum is zero.

6. It gives a resolution limit for the determination of the spectrum of a signal in terms of the measurement [7]. It also states resolution limits for imaging systems in terms of their aperture size. It serves as the basis of one of the standard measure of laser beam quality [8]. The uncertainty principle indicates the existence of quantum fluctuations in physical states that cannot be controlled completely. Particularly, the fluctuations of energy and momentum go arbitrarily large when the space-time interval becomes shorter and shorter.

7. The interaction between two particles occurs by means of exchange of a quantum of the interacting field. For example, Yukawa's strong nuclear force between

---

[2]Hans Bethe once said: *Without the uncertainty principle there could not exist any atoms and there could not be any certainty in the behaviour of matter whatever. It is really the certainty principle* [I. Goodwin, Physics Today, June 1995, pp.39].

nucleons is due to the exchange of pions. The mass of the exchanged quantum is found to have an inverse relation with the range of the interaction. This can be accounted only by the uncertainty relation. If R is the range and m is the mass of the exchanged quantum then $\Delta t \approx R/c$ and $\Delta E \approx mc^2$. The energy-time uncertainty relation gives $m \approx \hbar/(2Rc)$. If $R \approx 10^{-15}$ m then $m \approx 200m_e$, which is roughly the mass of pions.

8. *What prevents an electron from falling into nucleus?* Suppose the electron was assumed to be in the nucleus. In this case we exactly know its position. Then according to uncertainty principle the momentum would be very large and hence its energy would be very large. This large energy would then able to break away the electron from the nucleus.

9. An important application of Heisenberg uncertainty pointed out by John Gribbin is that in nuclear fusion reactions the quantum uncertainty allows atomic nuclei that are not close enough to touch one another to overlap with one another to combine. Some of such nuclear reactions keep the stars hot. That is, without uncertainty principle the sun would not shine the way it does.

8.6   Illustration of Uncertainty Relation

Let us explain the uncertainty relation with two examples.

8.6.1   Gamma-Ray Microscope

We can measure the x-coordinate of an electron using a gamma-ray microscope by scattering a photon off it. Here, to measure a position as accurate as possible, we need to use the light of short wavelength. The accuracy in the measurement is given by

$$\Delta x = \lambda/(2\sin\theta)\,, \tag{8.56}$$

where θ is the half-angle of the cone formed in Fig. 8.1 and λ is the wavelength of the gamma-ray used.

In order to see the electron through the microscope the gamma-rays scattered by the electron has to enter the objective of the microscope. In this process, the Compton effect occurs, namely, a decrease in energy or increase in the wavelength of a photon happens. Let λ represent the wavelength of the photon before interaction, and λ' is the wavelength of it after interaction. In the process the electron undergoes a recoil and it can be set minimum by using minimum gamma-ray energy. However, at least one gamma-ray photon need to be used. In this case the scattered photon with momentum h/λ' may enter the objective. If its path is along PA then its momentum component along the direction of the incident ray will be $(h/\lambda')\sin\theta$. The momentum that is transferred to the electron is

$$p_1 = \frac{h}{\lambda} - \frac{h}{\lambda'}\sin\theta\,. \tag{8.57}$$

In the case of the photon scattered along PB, the momentum given to the electron will be

$$p_2 = \frac{h}{\lambda} + \frac{h}{\lambda'}\sin\theta\,. \tag{8.58}$$

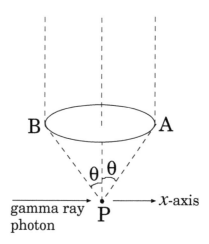

FIGURE 8.1
Position measurement of an electron.

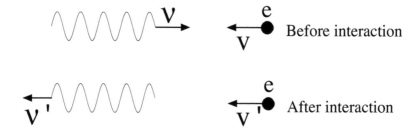

FIGURE 8.2
Doppler effect.

Thus, the value of the momentum transferred to the electron will be between p_1 and p_2. Then

$$\Delta p_x = p_2 - p_1 = \frac{2h}{\lambda'} \sin\theta. \tag{8.59}$$

For a small aperture we can assume that $\lambda' \approx \lambda$. Hence,

$$\Delta x \Delta p_x = h = 2\pi\hbar. \tag{8.60}$$

8.6.2 Doppler Effect

The Doppler effect refers to the change in the frequency of waves noticed by a detector or receiver due to the motion of the source of waves and the detector/observer. Let us consider the problem of finding momentum of an electron using Doppler effect (Fig. 8.2). We have

$$\nu' = \nu \left(\frac{1 + v/c}{1 - v/c} \right). \tag{8.61}$$

When $v \ll c$,

$$\nu' \approx \nu \left(1 + \frac{v}{c}\right) \left(1 + \frac{v}{c}\right) \approx \nu \left(1 + \frac{2v}{c}\right) \quad \text{or} \quad \frac{v}{c} = \frac{\nu' - \nu}{2\nu}. \tag{8.62}$$

The electron momentum is

$$p = mv = \frac{mc(\nu' - \nu)}{2\nu}. \tag{8.63}$$

Then

$$\frac{\Delta p}{p} = \frac{\Delta(\nu' - \nu)}{\nu' - \nu} \tag{8.64}$$

gives the error in p. The accuracy of the measurement of $\nu' - \nu$ given by $\Delta(\nu' - \nu)$ determines the limitation on p. The details of the particle's position will be lost due to the time interval Δt needed to measure $\nu' - \nu$.

In the above considered process at least one photon must be scattered. After the scattering the velocity of the electron is changed. We do not know at what exact instant during Δt, the scattering took place. This results in an uncertainty in the velocity of the electron and hence in its position.

Solved Problem 3:

Determine the uncertainty product for an electron passing through a diffraction grating of infinite length.

Diffraction grating is a device used to produce spectra by diffraction. The well known grating formula is $d \sin \theta = n\lambda$, where d is the spacing between grating lines, θ is the diffraction angle of the diffracted maximum, n is the order of the spectral lines and λ is the wavelength of the light. The above equation is the condition for diffraction to take place.

Let us consider the uncertainty in x-direction. After the diffraction we have $d \sin \theta = 2n\pi/k_x$. We do not know where the entrance of electron took place. Therefore, if the grating has m slits each of width d then the uncertainty in the position coordinate along x-direction is $\Delta x = md$.

Initially the electrons are moving normal to x-axis and hence no component of the momentum k_x exists. When they are diffracted they have momentum component along k_x-direction. Then the uncertainty in k_x component of momentum is $\Delta k_x = (2\pi/d)\Delta n$. Taking $\Delta n = 1/m$ we get $\Delta k_x = 2\pi/\Delta x$. That is, $\Delta x \Delta k_x = 2\pi$. Multiplying both sides by $\hbar$ we obtain $\Delta x \Delta p_x = 2\pi\hbar > \hbar/2$.

Solved Problem 4:

An electron of energy $300\,\text{eV}$ is allowed to move through a circular aperture of radius $10^{-4}\,\text{cm}$. Calculate the uncertainty in the angle of divergence.

The uncertainty in the angle of divergence is $\Delta\theta \approx \Delta p_x/p_x$. We calculate p_x and Δp_x in eV as ($1\,\text{eV} = 1.602 \times 10^{-19}\,\text{J}$)

$$
\begin{aligned}
p_x &= \sqrt{2mE} \\
&= \left(2 \times 9.109 \times 10^{-31} \times 300 \times 1.602 \times 10^{-19}\right)^{1/2} \text{kg/sec} \\
&= 9.3571207 \times 10^{-24} \text{kg/sec} \tag{8.65} \\
\Delta p_x &\approx \hbar/\Delta x = \frac{1.0544 \times 10^{-34}}{2 \times 10^{-6}} \text{kg/sec} = 5.272 \times 10^{-29} \text{kg/sec}. \tag{8.66}
\end{aligned}
$$

Then

$$\Delta\theta \approx \frac{5.272 \times 10^{-29}}{9.3571207 \times 10^{-24}} = 5.6342 \times 10^{-6} \text{ radians}. \tag{8.67}$$

8.7 Some Extensions of Uncertainty Relation

An extension of the uncertainty relation (8.29) is [9]

$$(\Delta A)^2 (\Delta B)^2 \geq \left| \frac{1}{2} \langle [A, B] \rangle \right|^2 + \left| \frac{1}{2} \langle \{A, B\} \rangle - \langle A \rangle \langle B \rangle \right|^2, \tag{8.68}$$

where $\{A, B\} = AB + BA$. The uncertainty relations based on the sum of variances are given by [10]

$$(\Delta A)^2 + (\Delta B)^2 \geq \frac{1}{2} \left| \langle \psi_{A+B}^{\perp} | A + B | \psi \rangle \right|^2, \tag{8.69}$$

$$(\Delta A)^2 + (\Delta B)^2 \geq \pm i |\langle [A, B] \rangle| + |\langle \psi | A \pm i B | \psi^{\perp} \rangle|^2, \tag{8.70}$$

where $\psi^{\perp}$ is an arbitrary state and $\psi_{A+B}^{\perp} \propto (A + B - \langle A + B \rangle)\psi$ and both are orthogonal to the state ψ of the system. The sign is to be suitably fixed in order to have $\pm i \langle [A, B] \rangle$ positive.

For the observables incompatible on ψ it has been obtained that [11]

$$\begin{aligned} (\Delta A)^2 + (\Delta B)^2 \geq{} & |\langle [A, B] \rangle + \langle \{A, B\} \rangle - 2\langle A \rangle \langle B \rangle| \\ & + |\langle \psi | A - e^{i\alpha} B | \psi^{\perp} \rangle|^2, \end{aligned} \tag{8.71}$$

where α is a real constant and

$$(\Delta A)^2 (\Delta B)^2 \geq \frac{\frac{1}{2} \langle [A, B] \rangle|^2 + |\frac{1}{2} \langle \{A, B\} \rangle - \langle A \rangle \langle B \rangle|^2}{\left(1 - \frac{1}{2} |\langle \psi | (A/\Delta A) - e^{i\alpha} (B/\Delta B) | \psi^{\perp} \rangle|^2 \right)^2}. \tag{8.72}$$

Two more uncertainty inequalities have been constructed and are [12]

$$\begin{aligned} (\Delta A)^2 + (\Delta B)^2 ={} & |\langle [A, B] \rangle + \langle \{A, B\} \rangle - 2\langle A \rangle \langle B \rangle| \\ & + \sum_{n=1}^{d-1} |\langle \psi | A - e^{i\alpha} B | \psi_n^{\perp} \rangle|^2, \end{aligned} \tag{8.73}$$

$$(\Delta A)^2 (\Delta B)^2 = \frac{|\frac{1}{2} \langle [A, B] \rangle|^2 + |\frac{1}{2} \langle \{A, B\} \rangle - \langle A \rangle \langle B \rangle|^2}{\left(1 - \frac{1}{2} \sum_{n=1}^{d-1} |\langle \psi | (A/\Delta A) - e^{i\alpha} (B/\Delta B) | \psi_n^{\perp} \rangle|^2 \right)^2}, \tag{8.74}$$

where $\left\{ |\psi_n^{\perp}\rangle_{n=1}^{d-1} \right\}$ forms an orthonormal complete basis in the d-dimensional Hilbert space.

Uncertainty relations for two non-Hermitian operators corresponding to the two non-commuting observables for weak measurements [12] and in the presence of quantum memory have been developed [13–15].

8.8 The Modified Heisenberg Relation

First, we show the need to modify the standard uncertainty relation, Eq. (8.28). In order to hold the Heisenberg inequality, the state of a quantum system must possess certain criteria. Before pointing out these restrictions let us verify the Heisenberg uncertainty relation for the canonical pair L_z and ϕ.

8.8.1 What is $(\Delta\phi)(\Delta L_z)$?

The components of orbital angular momentum L are (see sec. 11.3)

$$L_x = yp_z - zp_y = -i\hbar\left(y\frac{\partial}{\partial z} - z\frac{\partial}{\partial y}\right), \tag{8.75a}$$

$$L_y = zp_x - xp_z = -i\hbar\left(z\frac{\partial}{\partial x} - x\frac{\partial}{\partial z}\right), \tag{8.75b}$$

$$L_z = xp_y - yp_x = -i\hbar\left(x\frac{\partial}{\partial y} - y\frac{\partial}{\partial x}\right). \tag{8.75c}$$

In spherical polar coordinates we have (see sec. 11.3 for details) $L_z = -i\hbar\partial/\partial\phi$. L_z and the azimuthal angle ϕ form a canonical pair. It is easy to verify that $[\phi, L_z] = i\hbar$. From this relation we expect $\Delta\phi\Delta L_z \geq \hbar/2$. Let us find $\Delta\phi\Delta L_z$ for the eigenstate

$$\psi(\phi) = \frac{1}{\sqrt{2\pi}}e^{im\phi} \tag{8.76}$$

of L_z. We obtain

$$\langle\phi\rangle = \int_0^{2\pi}\psi^*\phi\psi\,d\phi = \pi\,, \quad \langle\phi^2\rangle = \int_0^{2\pi}\psi^*\phi^2\psi\,d\phi = \frac{4}{3}\pi^2, \tag{8.77a}$$

$$\langle L_z\rangle = \int_0^{2\pi}\psi^*L_z\psi d\phi = m\hbar, \quad \langle L_z^2\rangle = \int_0^{2\pi}\psi^*L_z^2\psi d\phi = m^2\hbar^2 \tag{8.77b}$$

and therefore

$$(\Delta\phi)^2 = \langle\phi^2\rangle - \langle\phi\rangle^2 = \frac{\pi^2}{3}, \quad (\Delta L_z)^2 = \langle L_z^2\rangle - \langle L_z\rangle^2 = 0\,. \tag{8.78}$$

Thus,

$$(\Delta\phi)(\Delta L_z) = 0 < \frac{\hbar}{2}\,. \tag{8.79}$$

What went wrong? The reason is that eigenstates of L_z do not satisfy the criteria required for the Heisenberg principle. Specifically, ϕ is not a proper dynamical variable and hence does not represent a quantum mechanical operator. This is because ϕ converts an acceptable eigenfunction ψ to an inadmissible one when it operates on ψ. When ϕ is limited to the domain 0 to 2π then after it has increased to 2π, it jumps *discontinuously* down to zero. Hence, $\phi\psi$ will also have a discontinuous jump after ϕ taking 2π. As $\phi\psi$ is discontinuous, it is not physically acceptable. Therefore, ϕ is not a good operator and the uncertainty relation between ϕ and L_z does not exist. In the following we describe these criteria and show why ψ of L_z do not satisfy them.

8.8.2   Restrictions on States to Apply Standard Uncertainty Relation

For operators their domains are vital. The *domain* of an operator A is the set of all vectors ψ_n in the Hilbert space (see sec. 6.9 for vector representation of ψ and the definition and significance of Hilbert space) of the system so that $A\psi$ is also a well defined member of the Hilbert space. We denote the domain of A as $D(A)$. To apply the uncertainty relation (8.28) $A\psi$ and $B\psi$ must be defined in Hilbert space. There are reasons for a given ψ not in $D(A)$ [3]:

1. For the given ψ the prescription of the operator A is not defined.

2. $A\psi$ is not in the Hilbert space. Consider $A = p_x = -i\hbar d/dx$ and $\psi(x) = \sqrt{2|x|}\,e^{-|x|}$. We have

$$A\psi = -i\hbar \frac{x}{|x|} \frac{e^{-|x|}}{\sqrt{2|x|}} \left(1 - 2|x|\right). \qquad (8.80)$$

    Observe that $A\psi$ is undefined at $x = 0$ and so it is not in Hilbert space. In other words, ψ is not in $D(A)$.

3. In order to make A Hermitian its domain is restricted. Note that L_z is Hermitian only if its $D(L_z)$ is restricted to a set of functions ψ for which $\psi(2\pi) = \psi(0)$ holds. In fact, this is the source of the problem with the uncertainty relation between ϕ and L_z [3].

8.8.3   Derivation of Modified Uncertainty Relation

Now, we present a brief derivation of modified uncertainty relation [3]. Let us define $(\Delta A)^2$ as

$$(\Delta A)^2 = \langle (A - \langle A \rangle)\psi, (A - \langle A \rangle)\psi \rangle = ||A - \langle A \rangle \psi||^2. \qquad (8.81)$$

This is for all the states in $D(A)$. But the Eq. (8.28) is only for the states in $D(A^2)$. For two Hermitian operators A and B we have

$$
\begin{aligned}
\sigma_{AB} &= \frac{1}{2}\langle (A - \langle A \rangle)\psi, (B - \langle B \rangle)\psi \rangle + \frac{1}{2}\langle (B - \langle B \rangle)\psi, (A - \langle A \rangle)\psi \rangle \\
&= \mathrm{Re}\langle (A - \langle A \rangle)\psi, (B - \langle B \rangle)\psi \rangle. \qquad (8.82)
\end{aligned}
$$

This is satisfied for larger set of states in both $D(A)$ and $D(B)$. With $\tilde{A} = A - \langle A \rangle$, Eqs. (8.81) and (8.82) become

$$\Delta A = ||\tilde{A}\psi||, \quad \sigma_{AB} = \mathrm{Re}\langle \tilde{A}\psi, \tilde{B}\psi \rangle. \qquad (8.83)$$

The Schwarz inequality, Eq. (8.20), becomes

$$
\begin{aligned}
(\Delta A)(\Delta B) &= ||\tilde{A}\psi||\,||\tilde{B}\psi|| \\
&\geq |\langle \tilde{A}\psi, \tilde{B}\psi \rangle| \\
&= \sqrt{(\mathrm{Re}\langle \tilde{A}\psi, \tilde{B}\psi \rangle)^2 + (\mathrm{Im}\langle \tilde{A}\psi, \tilde{B}\psi \rangle)^2} \\
&= \sqrt{\sigma_{AB}^2 + (\mathrm{Im}\langle \tilde{A}\psi, \tilde{B}\psi \rangle)^2} \\
&= \sqrt{\sigma_{AB}^2 + (\mathrm{Im}\langle A\psi, B\psi \rangle)^2}. \qquad (8.84)
\end{aligned}
$$

This is the *modified Heisenberg uncertainty relation* and is guaranteed to valid in all cases of well defined ΔA and σ_{AB}.

Interestingly, we recover (8.29) from Eq. (8.84). If ψ is in both $D(AB)$ and $D(BA)$ then

$$
\begin{aligned}
\mathrm{Im}\langle A\psi,\, B\psi\rangle &= -\frac{i}{2}\langle A\psi,\, B\psi\rangle + \frac{i}{2}\langle B\psi,\, A\psi\rangle \\
&= -\frac{i}{2}\langle \psi,\, AB\psi\rangle + \frac{i}{2}\langle \psi,\, BA\psi\rangle \\
&= -\frac{i}{2}\langle [A, B]\rangle\,.
\end{aligned}
\tag{8.85}
$$

Thus,

$$
(\Delta A)(\Delta B) \geq \sqrt{\sigma_{AB}^2 - \frac{1}{4}\langle [A, B]\rangle^2}
\tag{8.86}
$$

implying the standard Heisenberg inequality.

8.8.4 Modified Uncertainty Relation for the Pair (ϕ, L_z)

Note that L_z is Hermitian only if $\psi(2\pi) = \psi(0)$. Observe that $\phi\psi(\phi) = (1/\sqrt{2\pi})\phi e^{im\phi}$ is zero at $\phi = 0$ but nonzero at $\phi = 2\pi$. Hence, $(L_z\phi)\psi$ does not exist and therefore the commutator inequality is not applicable. However, Eq. (8.84) is applicable. Let us find $\mathrm{Im}\langle \phi\psi,\, L_z\psi\rangle$. We obtain

$$
\begin{aligned}
\mathrm{Im}\langle \phi\psi,\, L_z\psi\rangle &= -\frac{i}{2}\langle \phi\psi,\, L_z\psi\rangle + \frac{i}{2}\langle L_z\psi,\, \phi\psi\rangle \\
&= -\frac{\hbar}{2}\int_0^{2\pi}\phi\psi^*\frac{\mathrm{d}\psi}{\mathrm{d}\phi}\,\mathrm{d}\phi - \frac{\hbar}{2}\int_0^{2\pi}\frac{\mathrm{d}\psi^*}{\mathrm{d}\phi}\phi\psi\,\mathrm{d}\phi \\
&= -\frac{\hbar}{2}\left[\phi\psi^*\psi\right]_0^{2\pi} + \frac{\hbar}{2}\int_0^{2\pi}\psi^*\psi\,\mathrm{d}\phi \\
&= \frac{\hbar}{2}\left(1 - 2\pi|\psi(2\pi)|^2\right)\,.
\end{aligned}
\tag{8.87}
$$

For the ψ given by Eq. (8.76), $|\psi(2\pi)|^2$ is $1/(2\pi)$. Substituting this in the above equation yields $\mathrm{Im}\langle \phi\psi,\, L_z\psi\rangle = 0$. Next, we ask: *What is $\sigma_{\phi L_z}$ for the ψ given by Eq. (8.76)?* It is easy to check that

$$
\sigma_{\phi L_z} = \mathrm{Re}\langle \phi\psi,\, L_z\psi\rangle - \langle \phi\rangle\langle L_z\rangle = \hbar m\pi - \hbar m\pi = 0\,.
\tag{8.88}
$$

This is due to the fact that ψ is an eigenstate of L_z. With this the modified Heisenberg uncertainty relation, Eq. (8.84), for ϕ and L_z is $(\Delta\phi)(\Delta L_z) \geq 0$.

8.8.5 Modified Uncertainty Relation for (x, p_x)

For x and p_x we find $\mathrm{Im}\langle x\psi,\, p_x\psi\rangle = \frac{\hbar}{2}\left(1 - [x\psi^*\psi]_{-\infty}^{\infty}\right)$ The usual Heisenberg inequality holds if ψ falls-off faster than $1/\sqrt{|x|}$ as $|x| \to \infty$, which is ensured by the normalization condition.

8.9 Concluding Remarks

Generally, in the case of macroscopic systems the uncertainties associated with the outcomes of measurements are small compared with the outcomes. As a result the perturbation can be reduced as desired in considering joint measurements. In microsystems measurements may lead to a large perturbation. Therefore, in the case of simultaneous measurements of microscopic systems the perturbation may be very relevant. Essentially, one may assume that in classical mechanics the observables are compatible and can be measured simultaneously. For microscopic systems there are certain incompatible observables and with the same experimental set up they cannot be measured simultaneously.

The equation of motion of classical mechanics is a second-order differential equation for each dependent variable. Determining each dynamical variable of a system needs two integrations to be done and so two integration constants to be determined using, say, the coordinates and conjugate momenta at $t = t_0$ as initial conditions. Since the uncertainty relation says that both of them cannot be determined simultaneously at $t = t_0$, the trajectory of the particles cannot be determined at least for microscopic particles for which $\hbar$ is significant.

From this chapter note that quantum theory establishes a set of following rules:

1. It is impossible to make a measurement without disturbing the system.

2. Simultaneous measurement of a canonically conjugate pair of observables with arbitrarily high accuracy is not possible.

3. We cannot draw pictures of individual quantum processes.

4. We cannot duplicate an unknown quantum state.

The above negative view points of quantum mechanics have recently been turned positive and quantum cryptography is one of the best illustrations of this.

8.10   Bibliography

[1] W. Heisenberg, *Z. Phys.* 43:172, 1927.

[2] J. Peslak, *Am. J. Phys.* 47:39, 1979.

[3] E.D. Chisolm, *Am. J. Phys.* 69:368, 2001.

[4] A.E. Ruark, *Phys. Rev.* 31:311, 1928.

[5] E.U. Condon, *Science* 69:573, 1929.

[6] E.H. Kennard, *Z. Physik* 44:326, 1927.

[7] G.W. Forbes and M.A. Alonso, *Am. J. Phys.* 69:340, 2001.

[8] T.F. Johnston Jr., *Laser Focus World* 26:173, 1990.

[9] E. Schrödinger, *Phys. Math. Klasse* 14:296, 1930.

[10] L. Maccone and A.K. Pati, *Phys. Rev. Lett.* 113:260401, 2014.

[11] Qiu-Cheng Song and Cong-Feng Qiao, *Phys. Lett. A* 380:2925, 2016.

[12] Q.C. Song and C.F. Qiao, arXiv 1505.02233 quant-ph 9 May 2015.

[13] M. Berta, M. Christandl, R. Colbeck, J.M. Renes and R. Renner, *Nat. Phys.* 6:659, 2010.

[14] P.J. Coles and M. Piani, *Phys. Rev. A* 89:022112, 2014.

[15] Y. Xiao, N. Jing, S.M. Fei and X. Li-Jost, *J. Phys. A: Math. Theor.* 49:49LT01, 2016.

8.11 Exercises

8.1 (a) Starting from $(\Delta x)^2 = \langle (x - \langle x \rangle)^2 \rangle$, show that $(\Delta x)^2 = \langle x^2 \rangle - \langle x \rangle^2$.

8.2 An electron is moving with velocity 2×10^7 m/sec. Determine the smallest uncertainty in its position.

8.3 Calculate the smallest possible uncertainty in the momentum of a particle whose uncertainty in position measurement is 0.001 cm.

8.4 The uncertainty in the measurement of velocity of an electron is 0.02%. If the measured velocity is 50 m/sec find the smallest uncertainty with which the position of the electron can be determined.

8.5 The uncertainty in momentum corresponding to a one-dimensional problem is 10^{-22} kg/sec. What is the minimum uncertainty in x?

8.6 The uncertainty in the location of a particle is twice its de Broglie wavelength. Find the uncertainty in its velocity.

8.7 A beam of atoms of mass 1.836×10^{-25} kg at a temperature 1000° C are made to pass through a slit of size $2d$. Determine the optimum slit size in order to have smallest spread on the screen, which is placed at a distance $D = 1$ m from the slit.

8.8 Obtain the value of $\Delta x \Delta p_x$ for diffraction of a beam of electrons by a slit.

8.9 Show that according to uncertainty principle an electron cannot exist inside the nucleus.

8.10 Determine the radius of the Bohr's first orbit applying uncertainty principle.

8.11 Show that $\Delta E \Delta t \geq \hbar/2$.

8.12 Calculate the frequency spread $\Delta \nu$ for a picosecond (10^{-12} s) pulse from a CO_2 laser in which the nominal photon energy is $h\nu = 0.112$ eV.

8.13 A beam of mono-energetic electron is used to raise atoms to an excited state in a Franck–Hertz experiment. If the excited state has a life-time of 10^{-7} sec calculate the spread in energy of the inelastically scattered electron.

8.14 The expectation value of time between the excitation of an atom and the radiation of energy is 10^{-7} sec. Determine the uncertainties in the energy in eV of the emitted photon and the frequency of the radiation.

8.15 Calculate $\Delta x \Delta p_x$ for the wave function $\psi = \sqrt{a/2} \operatorname{sech} ax$.

8.16 Calculate $(\Delta x)(\Delta p_x)$ for the eigenstate $\psi = \dfrac{1}{\sqrt{a}} \cos \dfrac{\pi x}{2a}$ of a particle in the box potential $V(x) = 0$ for $|x| < a$ and ∞ otherwise.

8.17 Show that for $\psi = e^{-m\omega x^2/\hbar}$ the value of $\Delta x \Delta p_x$ is $\hbar/2$.

8.18 The ladder operators of under-damped linear harmonic oscillator are given in sec. 4.12.2. Making use of the expressions of x and p given in Eqs. (4.178) in terms of a and $a^\dagger$ determine $(\Delta x)(\Delta p_x)$.

8.19 Show that the ground state energy E_0 of a system can be determined exactly.

8.20 Estimate the binding energy of the hydrogen atom in its ground state using position-momentum uncertainty relation.

8.21 Show that ΔA of an observable A is 0 if $\psi = \phi_a$, where ψ is a wave function and ϕ_a is the eigenfunction of A with the eigenvalue a.

8.22 The Fourier transform of $f(x)$ is $F(p_x) = 1/\sqrt{2\pi} \int_{-\infty}^{\infty} f(x) e^{-i p_x x} \, dx$. Derive the analogs of uncertainty principle for x and p_x.

8.23 Show that $\mathrm{Im}\langle A\psi, B\psi \rangle = -(i/2)\langle [A, B] \rangle$, where A and B are Hermitian operators.

8.24 Define T as a time operator with $T|\psi(t)\rangle = t|\psi(t)\rangle$ and $|\psi(t)\rangle = |t\rangle$. That is, for each instant of t the eigenvalues of T correspond to the solutions of the time-dependent Schrödinger equation. Determine the value of ΔT. What is the significance of the value of ΔT?

8.25 Compute $\{(\Delta A)^2, (\Delta B)^2\}$.

9

Momentum Representation

9.1 Introduction

In classical physics one needs to know the velocity or momentum of a particle in order to determine its evolution. In quantum mechanics the wave function ψ already contains information of possible momenta. What one has to do is to apply harmonic analysis to ψ. Sometimes it is helpful to describe quantum states in terms of wave function of momentum. This requires decomposition of ψ in terms of various momentum states and constructing a new function $C(k)$. The value of $C(k)$ of each k gives the strength of the contribution to the $k-$momentum state of ψ. The space of k is called *momentum space*.

One may ask: *Where do we use momentum wave function? What is the form of Schrödinger equation in momentum representation? What are the operator form of momentum* $\mathbf{p}$ *and coordinate* $\mathbf{X}$*?* These will be answered in the present chapter.

The position-space distribution is useful when we perform measurements of position of a particle. Photo-cells and photographic plates are useful for position measurements for photons. The momentum representation is helpful in the case of measurement of momentum of a particle. Recoil effects or Diffraction from a crystal are useful for momentum measurements. The momentum space probability distribution $P(k) = |C(k)|^2$ is of increasing relevance as related quantities are directly measurable in many fields of experimental physics. Examples include:

1. Momentum distributions obtained from neutron scattering in solids and liquids.

2. Nuclear momentum distributions in nuclear scattering experiments.

3. Quark-gluon distributions in the proton necessary for understanding the high-energy collisions in elementary particle physics.

9.2 Momentum Eigenfunctions

The eigenvalue equation of the operator $p_{x\text{op}}$ (in one-dimension) is

$$-i\hbar \frac{\mathrm{d}}{\mathrm{d}x}\phi_{p_x} = p_x\,\phi_{p_x}(x)\,, \tag{9.1}$$

where ϕ_{p_x} is the eigenfunction of $p_{x\text{op}}$. The eigenvalue of $p_{x\text{op}}$ is denoted as p_x. Here onwards in this chapter, we denote $p_{x\text{op}}$ as simply the operator p. The solution of Eq. (9.1) is

$$\phi_p(x) = C\,\mathrm{e}^{ipx/\hbar}\,, \tag{9.2}$$

where C is a constant. Since $p = \hbar k$ we write

$$\phi_k(x) = C\,\mathrm{e}^{ikx}\,. \tag{9.3}$$

DOI: 10.1201/9781003172178-9

Therefore, ϕ_p is an eigenfunction of the momentum operator p with the eigenvalue $\hbar k$. The possible momenta are $\hbar k$. It is easy to prove that the set of functions (9.3) with the continuum values of k is an orthogonal set since

$$\int_{-\infty}^{\infty} \phi_k^* \phi_{k'} \, dx = |C|^2 \int_{-\infty}^{\infty} e^{i(k'-k)x} \, dx = \delta(k'-k) \,. \tag{9.4}$$

Since $\phi_k(x)$ form a complete set an arbitrary function can be expanded in terms of $\phi_k(x)$ of the operator p. The summation is now an integral so

$$\psi(x) = \frac{1}{\sqrt{2\pi}} \int_{-\infty}^{\infty} C(k) e^{ikx} \, dk \,, \tag{9.5a}$$

where $1/\sqrt{2\pi}$ is introduced for convenience. Equation (9.5a) is identified as the Fourier inverse transform. Then $C(k)$ is written as

$$C(k) = \frac{1}{\sqrt{2\pi}} \int_{-\infty}^{\infty} \psi(x) e^{-ikx} \, dx \,. \tag{9.5b}$$

Any two variables whose wave function representations transform as in Eq. (9.5b) are said to be *complementary*.

Solved Problem 1:

Determine the momentum wave function of a free particle.

The solution of a time-independent Schrödinger equation in coordinate space of a free particle in one-dimension is given by Eq. (4.6). The superposition of these eigenfunctions is given by

$$\psi(x) = \int_{-\infty}^{\infty} C(k) e^{ikx} \, dk \,. \tag{9.6}$$

Then the momentum wave function $C(k)$ is

$$C(k) = \int_{-\infty}^{\infty} \psi(x) e^{-ikx} \, dx \,. \tag{9.7}$$

The acceptable solutions are obtained by the normalization condition. In sec. 4.3 we have noted that free particle wave function is nonnormalizable by the usual normalization condition $\int_{-\infty}^{\infty} \psi^* \psi \, dx = 1$. Normalization is possible by applying periodic boundary conditions. This leads to the solution

$$\psi(x) = (1/\sqrt{L}) e^{ikx} \,. \tag{9.8}$$

For continuous distribution of momentum eigenvalues, that is, for the box of infinite extent, the normalization condition is

$$\int_{-\infty}^{\infty} \psi_l^*(x) \psi_k(x) \, dx = \delta(k-l) \quad \text{or} \quad C^*(l) C(k) 2\pi \delta(k-l) = \delta(k-l) \tag{9.9}$$

which gives $|C(k)|^2 = 1/(2\pi)$ and $C(k) = 1/\sqrt{2\pi}$.

What does $C(k)$ represent? In Eq. (9.3), $C(k)$ is the amplitude of the eigenfunction. $|\psi|^2 dx$ gives the probability that a particle is being between x and $x + dx$. Similarly, $|C(k)|^2 dk$ gives the probability that the particle is lying between $\hbar k$ and $\hbar(k + dk)$. For this reason, $C(k)$ is termed as the *momentum wave function* or wave function in momentum space. *It is the Fourier transform of ψ.* If Eqs. (9.5) give ψ and $C(k)$ at $t = 0$ then at other times they are given by

$$\psi(x,t) = \frac{1}{\sqrt{2\pi}} \int_{-\infty}^{\infty} C(k,t)\,e^{ikx}\,dk, \tag{9.10a}$$

$$C(k,t) = \frac{1}{\sqrt{2\pi}} \int_{-\infty}^{\infty} \psi(x,t)\,e^{-ikx}\,dx. \tag{9.10b}$$

The above two equations are also written as

$$\psi(x,t) = \frac{1}{\sqrt{2\pi\hbar}} \int_{-\infty}^{\infty} \phi(p,t)e^{ipx/\hbar}\,dp, \tag{9.11a}$$

$$\phi(p,t) = \frac{1}{\sqrt{2\pi\hbar}} \int_{-\infty}^{\infty} \psi(x,t)e^{-ipx/\hbar}\,dx. \tag{9.11b}$$

In three-dimensions the relations between $\psi(\mathbf{X},t)$ and $C(\mathbf{k},t)$ are

$$\psi(\mathbf{X},t) = \frac{1}{(2\pi)^{3/2}} \int_{-\infty}^{\infty} C(\mathbf{k},t)e^{i\mathbf{k}\cdot\mathbf{X}}\,d\mathbf{k}, \tag{9.12a}$$

$$C(\mathbf{k},t) = \frac{1}{(2\pi)^{3/2}} \int_{-\infty}^{\infty} \psi(\mathbf{x},t)e^{-i\mathbf{k}\cdot\mathbf{X}}\,d\mathbf{X}, \tag{9.12b}$$

where $\int_{-\infty}^{\infty}$ is a triple-integral. Note that the knowledge of ψ is helpful in finding $C(\mathbf{k})$ and vice-versa. This fact implies that these two functions provide different representations of the same state. These two equally valid representations are called as the *coordinate representation* and the *momentum representation*.

9.3 Schrödinger Equation

To obtain the wave equation for $C(k)$, differentiate Eq. (9.12b) with respect to t and then multiply by $i\hbar$:

$$i\hbar\frac{\partial C}{\partial t} = \frac{i\hbar}{(2\pi)^{3/2}} \int_{-\infty}^{\infty} \frac{\partial \psi}{\partial t} e^{-i\mathbf{k}\cdot\mathbf{X}}\,d\mathbf{X}. \tag{9.13}$$

Use of the Schrödinger equation for $\partial\psi/\partial t$ in Eq. (9.13) gives

$$i\hbar\frac{\partial C}{\partial t} = \frac{1}{(2\pi)^{3/2}} \int_{-\infty}^{\infty} e^{-i\mathbf{k}\cdot\mathbf{X}}\left[\frac{-\hbar^2}{2m}\nabla^2 + V(\mathbf{X})\right]\psi\,d\mathbf{X}. \tag{9.14}$$

From Eq. (9.12a) $\nabla^2\psi$ is found to be

$$\nabla^2\psi = \frac{1}{(2\pi)^{3/2}} \int_{-\infty}^{\infty} C(\mathbf{k},t)\nabla^2 e^{i\mathbf{k}\cdot\mathbf{X}}\,d\mathbf{k} = -k^2\psi(\mathbf{X},t). \tag{9.15}$$

Then Eq. (9.14) becomes

$$i\hbar\frac{\partial C}{\partial t} = \frac{\hbar^2 k^2}{2m}\frac{1}{(2\pi)^{3/2}} \int_{-\infty}^{\infty} \psi e^{-i\mathbf{k}\cdot\mathbf{X}}\,d\mathbf{X} + \frac{1}{(2\pi)^{3/2}} \int_{-\infty}^{\infty} V(\mathbf{X})\psi e^{-i\mathbf{k}\cdot\mathbf{X}}\,d\mathbf{X}. \tag{9.16}$$

The last integral in the above equation is the Fourier transform of $V\psi$. Denoted it as $V(\mathbf{k})C(\mathbf{k},t)$. Then

$$i\hbar\frac{\partial C(\mathbf{k},t)}{\partial t} = \frac{\hbar^2 k^2}{2m}C(\mathbf{k},t) + V(\mathbf{k})C(\mathbf{k},t). \tag{9.17}$$

This equation is the Schrödinger equation in momentum representation.

Rewrite the Eq. (9.17) as

$$EC(\mathbf{k},t) = \frac{\hbar^2 k^2}{2m}C(\mathbf{k},t) + V(\mathbf{k})C(\mathbf{k},t). \tag{9.18}$$

Comparison of this equation with the Schrödinger equation in operator form gives the operator form of $\mathbf{p}$ in momentum space as $\hbar\mathbf{k}$. In terms of $\mathbf{p}$ Eq. (9.18) becomes

$$E\phi(\mathbf{p},t) = \frac{1}{2m}\mathbf{p}^2\phi(\mathbf{p},t) + V\phi(\mathbf{p},t), \tag{9.19}$$

where the symbol ϕ is used instead of C for the notational convenience.

9.4   Expressions for $\langle \mathbf{X} \rangle$ and $\langle \mathbf{p} \rangle$

Having derived the Schrödinger equation in momentum space now we obtain the formulas for the expectation values of position and momentum.

9.4.1   Expectation Value of Position

$\langle \mathbf{X} \rangle$ is obtained as follows:

$$\begin{aligned}
\langle \mathbf{X} \rangle &= \int_{-\infty}^{\infty} \psi^* \mathbf{X}\psi\,d\mathbf{X} \\
&= \int_{-\infty}^{\infty} \psi^* \mathbf{X}\left[\frac{1}{(2\pi)^{3/2}}\int_{-\infty}^{\infty} C(\mathbf{k})e^{i\mathbf{k}\cdot\mathbf{X}}\,d\mathbf{k}\right]d\mathbf{X} \\
&= \frac{1}{(2\pi)^{3/2}}\int_{-\infty}^{\infty} C(\mathbf{k})\left[\int_{-\infty}^{\infty}\psi^*\mathbf{X}e^{i\mathbf{k}\cdot\mathbf{X}}\,d\mathbf{X}\right]d\mathbf{k} \\
&= \frac{1}{(2\pi)^{3/2}}\int_{-\infty}^{\infty} C(\mathbf{k})\left[\int_{-\infty}^{\infty}\psi^*\frac{1}{i}\nabla_{\mathbf{k}}\left(e^{i\mathbf{k}\cdot\mathbf{X}}\right)\,d\mathbf{X}\right]d\mathbf{k}.
\end{aligned} \tag{9.20}$$

Integrating by parts (with respect to $\mathbf{k}$) gives

$$\begin{aligned}
\langle \mathbf{X} \rangle &= \frac{1}{(2\pi)^{3/2}}\int_{-\infty}^{\infty}\psi^*\frac{1}{i}\left\{\left[C(\mathbf{k})e^{i\mathbf{k}\cdot\mathbf{X}}\right]_{-\infty}^{\infty} - \int_{-\infty}^{\infty}\nabla_{\mathbf{k}}C(\mathbf{k})e^{i\mathbf{k}\cdot\mathbf{X}}\,d\mathbf{k}\right\}d\mathbf{X} \\
&= \frac{1}{(2\pi)^{3/2}}\int_{-\infty}^{\infty}\psi^*i\left[\int_{-\infty}^{\infty}\nabla_{\mathbf{k}}C(\mathbf{k})e^{i\mathbf{k}\cdot\mathbf{X}}\,d\mathbf{k}\right]d\mathbf{X} \\
&= \int_{-\infty}^{\infty}\left[\frac{1}{(2\pi)^{3/2}}\int_{-\infty}^{\infty}\psi^* e^{i\mathbf{k}\cdot\mathbf{X}}\,d\mathbf{X}\right]i\nabla_{\mathbf{k}}C(\mathbf{k})\,d\mathbf{K} \\
&= \int_{-\infty}^{\infty} C^*(\mathbf{k})i\nabla_{\mathbf{k}}C(\mathbf{k})\,d\mathbf{k}.
\end{aligned} \tag{9.21}$$

Or

$$\langle \mathbf{X} \rangle = \int_{-\infty}^{\infty} \phi^*(\mathbf{p}) i\hbar \nabla_{\mathbf{p}} \phi(\mathbf{p}) \, d\mathbf{p}, \tag{9.22}$$

where $\nabla_{\mathbf{p}} = \mathbf{i}\dfrac{\partial}{\partial p_x} + \mathbf{j}\dfrac{\partial}{\partial p_y} + \mathbf{k}\dfrac{\partial}{\partial p_z}$. The Eq. (9.22) suggests that $i\nabla_{\mathbf{k}}$ (or $i\hbar\nabla_{\mathbf{p}}$) is the operator form of $\mathbf{X}$ denoted as $\mathbf{X_p}$ in the momentum representation.

The eigenvalue equation for $\mathbf{X_p}$ is

$$\mathbf{X_p} C(\mathbf{p}) = \mathbf{X} C(\mathbf{p}) \text{ or } i\hbar\nabla_{\mathbf{p}} C = \mathbf{X}C. \tag{9.23}$$

Its solution is $C(\mathbf{p}) = A e^{-i\mathbf{X}\cdot\mathbf{p}/\hbar}$. The eigenvalues of $\mathbf{X_p}$ form a continuous spectrum.

9.4.2 Expectation Value of Momentum

The expression for $\langle \mathbf{p} \rangle$ in momentum space is obtained as follows:

$$\begin{aligned}
\langle \mathbf{p} \rangle &= -\int_{-\infty}^{\infty} \psi^* i\hbar\nabla\psi \, d\mathbf{X} \\
&= -i\hbar \int_{-\infty}^{\infty} \psi^* \left[\frac{1}{(2\pi)^{3/2}} \nabla \int_{-\infty}^{\infty} C(\mathbf{k}) e^{i\mathbf{k}\cdot\mathbf{X}} \, d\mathbf{k} \right] d\mathbf{X} \\
&= -i\hbar \int_{-\infty}^{\infty} \psi^* \left[\frac{1}{(2\pi)^{3/2}} \int_{-\infty}^{\infty} C(\mathbf{k}) i\mathbf{k} e^{i\mathbf{k}\cdot\mathbf{X}} \, d\mathbf{k} \right] d\mathbf{X} \\
&= \int_{-\infty}^{\infty} \left[\frac{1}{(2\pi)^{3/2}} \int_{-\infty}^{\infty} \psi^* e^{i\mathbf{k}\cdot\mathbf{X}} \, d\mathbf{X} \right] \hbar\mathbf{k} C(\mathbf{k}) \, d\mathbf{k} \\
&= \int_{-\infty}^{\infty} C^* \hbar\mathbf{k} C \, d\mathbf{k}. \tag{9.24}
\end{aligned}$$

Equations (9.22) and (9.24) represent the transformation of $\langle \mathbf{X} \rangle$ and $\langle \mathbf{p} \rangle$ from coordinate space to momentum space.

Solved Problem 2:

The normalization of $C(\mathbf{k})$ would ensure the consistency of the interpretation of $|C(\mathbf{k})|^2$ as momentum probability density. Interestingly, if $\psi(\mathbf{X})$ is normalized then $C(\mathbf{k})$ given by Eq. (9.12b) is automatically normalized. Verify this.

We start with the normalization condition of $\psi(\mathbf{X})$. We obtain

$$\begin{aligned}
1 &= \int_{-\infty}^{\infty} \psi^* \psi \, d\mathbf{X} \\
&= \int_{-\infty}^{\infty} \left[\frac{1}{(2\pi)^{3/2}} \int_{-\infty}^{\infty} C^*(\mathbf{k}) e^{-i\mathbf{k}\cdot\mathbf{X}} \, d\mathbf{k} \right] \\
&\quad \times \left[\frac{1}{(2\pi)^{3/2}} \int_{-\infty}^{\infty} C(\mathbf{k'}) e^{-i\mathbf{k'}\cdot\mathbf{X}} \, d\mathbf{k'} \right] d\mathbf{X} \\
&= \int_{-\infty}^{\infty} C^*(\mathbf{k}) \left\{ \int_{-\infty}^{\infty} C(\mathbf{k'}) \left[\frac{1}{(2\pi)^3} \int_{-\infty}^{\infty} e^{i(\mathbf{k'}-\mathbf{k})\cdot\mathbf{X}} \, d\mathbf{X} \right] d\mathbf{k'} \right\} d\mathbf{k} \\
&= \int_{-\infty}^{\infty} C^*(\mathbf{k}) \left[\int_{-\infty}^{\infty} C(\mathbf{k'}) \delta(\mathbf{k'}-\mathbf{k}) \, d\mathbf{k'} \right] d\mathbf{k} \\
&= \int_{-\infty}^{\infty} |C(\mathbf{k})|^2 \, d\mathbf{k}. \tag{9.25}
\end{aligned}$$

Thus, the normalization of $\psi(\mathbf{X})$ guarantees the normalization of $C(\mathbf{k})$.

9.5   Transformation Between Momentum and Coordinate Representations

We represent an electrical signal either in time domain or frequency domain. Like-wise a quantum state is represented either in coordinate representation or in the momentum representation.

Let $|x'\rangle$ represents an eigenstate of the operator $\hat{x}$. Then $\hat{x}|x'\rangle = x'|x'\rangle$. These eigenfunctions obey the orthonormality condition $\langle x'|x\rangle = \delta(x - x')$. The matrix elements of the operator $\hat{x}$ are then given by

$$\langle x|\hat{x}|x'\rangle = x'\langle x|x'\rangle = x'\delta(x - x')\,. \tag{9.26}$$

Similarly, if $|p\rangle$ represents an eigenstate of $\hat{p}$, then $\hat{p}|p'\rangle = p'|p'\rangle$. The matrix of $\hat{p}$ in the coordinate representation is given by

$$\langle x|\hat{p}|x'\rangle = -\mathrm{i}\hbar\frac{\partial}{\partial x}\langle x|x'\rangle = -\mathrm{i}\hbar\frac{\partial}{\partial x}\delta(x - x')\,. \tag{9.27}$$

This relation allows us to obtain an explicit form of the transfer matrix $\langle x|p\rangle$ between the two representations

$$
\begin{aligned}
\hat{p}\langle x|p\rangle &= \langle x|\hat{p}|p\rangle \\
&= \int_{-\infty}^{\infty} \langle x|\hat{p}|x'\rangle\langle x'|p\rangle\,\mathrm{d}x' \\
&= -\mathrm{i}\hbar \int_{-\infty}^{\infty} \frac{\partial}{\partial x}\delta(x - x')\langle x'|p\rangle\,\mathrm{d}x' \\
&= -\mathrm{i}\hbar\frac{\partial}{\partial x}\langle x|p\rangle\,.
\end{aligned}
\tag{9.28}
$$

That is,

$$\frac{\partial}{\partial x}\langle x|p\rangle = \frac{\mathrm{i}}{\hbar}p\langle x|p\rangle\,. \tag{9.29}$$

The solution of (9.29) is $\langle x|p\rangle = A\mathrm{e}^{\mathrm{i}px/\hbar}$. Since the transformation is unitary, the unitary condition

$$\int_{-\infty}^{\infty} \langle p|x\rangle^*\langle p|x'\rangle\,\mathrm{d}p = \delta(x - x') \tag{9.30}$$

gives $A = 1/\sqrt{2\pi\hbar}$. Then we get the continuum transformation matrix

$$\langle x|p\rangle = \frac{1}{\sqrt{2\pi\hbar}}\,\mathrm{e}^{\mathrm{i}px/\hbar}\,. \tag{9.31}$$

Equation (9.31) not only connects the coordinate and momentum representation of the state vectors, but also relates these for operators.

9.6   Operators in Momentum Representation

A quantum operator A acts on vectors in Hilbert space and yields other Hilbert space vectors. When A operates on $|\beta\rangle$, it gives, say, $|\alpha\rangle$ in the same Hilbert space as $|\alpha\rangle = A|\beta\rangle$.

In coordinate representation $\psi_\alpha(\mathbf{X}) = \langle \mathbf{X}|\alpha\rangle$ and $\psi_\beta(\mathbf{X}) = \langle \mathbf{X}|\beta\rangle$. $|\alpha\rangle = A|\beta\rangle$ in terms of $\langle \mathbf{X}|A|\mathbf{X}'\rangle$ takes the form

$$\psi_\alpha(\mathbf{X}) = \int_{-\infty}^{\infty} \langle \mathbf{X}|A|\mathbf{X}'\rangle \, \psi_\beta(\mathbf{X}') \, d\mathbf{X}'. \tag{9.32}$$

If A is a real operator then

$$\langle \mathbf{X}|A|\mathbf{X}'\rangle = A(\mathbf{X}')\langle \mathbf{X}|\mathbf{X}'\rangle = A(\mathbf{X}')\delta(\mathbf{X}-\mathbf{X}'). \tag{9.33}$$

Using (9.33) in (9.32) we get $\psi_\alpha(\mathbf{X}) = A(\mathbf{X})\psi_\beta(\mathbf{X})$. Since $A = \langle \mathbf{X}|A|\mathbf{X}'\rangle$ the matrix elements of A in the momentum representation is

$$\langle \mathbf{p}'|A|\mathbf{p}\rangle = \int_{-\infty}^{\infty}\int_{-\infty}^{\infty} \langle \mathbf{p}'|\mathbf{X}'\rangle\langle \mathbf{X}'|A|\mathbf{X}\rangle\langle \mathbf{X}|\mathbf{p}\rangle \, d\mathbf{X}\, d\mathbf{X}'. \tag{9.34}$$

Using

$$\langle \mathbf{X}|\mathbf{p}\rangle = \left(\frac{1}{2\pi\hbar}\right)^{3/2} e^{i\mathbf{p}\cdot\mathbf{X}/\hbar}, \quad \langle \mathbf{p}'|\mathbf{X}'\rangle = \left(\frac{1}{2\pi\hbar}\right)^{3/2} e^{-i\mathbf{p}'\cdot\mathbf{X}'/\hbar} \tag{9.35}$$

in Eq. (9.34) we have

$$\langle \mathbf{p}'|A|\mathbf{p}\rangle = \left(\frac{1}{2\pi\hbar}\right)^{3} \int_{-\infty}^{\infty}\int_{-\infty}^{\infty} e^{-i\mathbf{p}'\cdot\mathbf{X}'/\hbar}\langle \mathbf{X}'|A|\mathbf{X}\rangle e^{i\mathbf{p}\cdot\mathbf{X}/\hbar} \, d\mathbf{X}\, d\mathbf{X}'. \tag{9.36}$$

Since a quantum operator A is a local operator

$$\langle \mathbf{X}'|A|\mathbf{X}\rangle = A(\mathbf{X}')\delta(\mathbf{X}-\mathbf{X}'). \tag{9.37}$$

Using (9.37) in (9.36) we obtain

$$\langle \mathbf{p}'|A|\mathbf{p}\rangle = \left(\frac{1}{2\pi\hbar}\right)^{3} \int_{-\infty}^{\infty} e^{i(\mathbf{p}-\mathbf{p}')\cdot\mathbf{X}/\hbar} A(\mathbf{X}) \, d\mathbf{X}. \tag{9.38}$$

Unfortunately $\phi_\mathbf{k}(\mathbf{X}) = (1/2\pi\hbar)^{3/2} e^{i\mathbf{p}\cdot\mathbf{X}/\hbar}$ is not an eigenfunction in a Hilbert space as it does not vanish at infinity and has infinite norm. Therefore, evaluation of Eq. (9.38) is a difficult task. We must identify another approach like box normalization.

For a short range potential $\langle \mathbf{X}'|V|\mathbf{X}\rangle = \delta(\mathbf{X}-\mathbf{X}')V(\mathbf{X})$ its momentum representation from Eq. (9.38) is

$$\begin{aligned}\langle \mathbf{p}'|V|\mathbf{p}\rangle &= \left(\frac{1}{2\pi\hbar}\right)^{3} \int_{-\infty}^{\infty} e^{i(\mathbf{p}-\mathbf{p}')\cdot\mathbf{X}/\hbar} V(\mathbf{X}) \, d\mathbf{X}\\ &= \left(\frac{1}{2\pi\hbar}\right)^{3/2}\left[\left(\frac{1}{2\pi\hbar}\right)^{3/2}\int_{-\infty}^{\infty} e^{i(\mathbf{p}-\mathbf{p}')\cdot\mathbf{X}/\hbar} V(\mathbf{X}) \, d\mathbf{X}\right].\end{aligned} \tag{9.39}$$

Since $V(\mathbf{X})$ is of finite range (it is zero for large and finite $\mathbf{X}$) the integral in (9.39) exists. In fact, the quantity in square-bracket is the Fourier transform $\bar{V}((\mathbf{p}'-\mathbf{p})/\hbar)$ of $V(\mathbf{X})$. Hence, in the momentum representation, a local potential $V(\mathbf{X})$ goes over to its Fourier transform with arguments $\mathbf{k}-\mathbf{k}'$:

$$\langle \mathbf{k}'|V|\mathbf{k}\rangle = (1/2\pi)^{3/2}\bar{V}(\mathbf{k}-\mathbf{k}'). \tag{9.40}$$

9.7 Momentum Function of Some Systems

In this section, we obtain the momentum eigenfunctions and momentum probability distribution for some simple time-independent quantum systems.

9.7.1 Linear Harmonic Oscillator

The Schrödinger equation in momentum space for the linear harmonic oscillator is

$$E\phi(p) = \frac{1}{2m}p^2\phi(p) + V(x)\phi(p). \tag{9.41}$$

Since $x = i\hbar d/dp$, Eq. (9.41) is rewritten as [1]

$$\frac{1}{2m}p^2\phi(p) - \frac{1}{2}\hbar^2 m\omega^2\frac{d^2\phi}{dp^2} = E\phi(p). \tag{9.42}$$

This equation is similar to the Eq. (4.21). Comparing these two equations and from Eq. (4.30) we obtain

$$\phi_n(p) = \alpha_n H_n(q)e^{-q^2/2}, \tag{9.43a}$$

where

$$q = \frac{p}{\sqrt{m\omega\hbar}}, \quad \alpha_n^2 = \frac{1}{2^n n!\pi\sqrt{m\omega\hbar}}. \tag{9.43b}$$

The time development of $\phi(p)$ is $\phi(p)e^{-iEt/\hbar}$. Since $F = -kx$ and $V = kx^2/2$ we write

$$|F| = \sqrt{2kV} = \left[2k\left(E - \frac{1}{2m}p^2\right)\right]^{1/2} = \left[\frac{k}{m}\left(p_0^2 - p^2\right)\right]^{1/2}, \tag{9.44}$$

where $p_0 = \pm\sqrt{2mE}$ is the maximum momentum. Then

$$P_{\rm CM}(p) = \frac{1}{\pi\sqrt{p_0^2 - p^2}}. \tag{9.45}$$

The classical momentum distribution is similar to the form of the classical position distribution, Eq. (4.35). This is infact consistent with the complete symmetry between x and p. $P_{\rm QM}(p)$ is given by $|\phi_n(p)|^2$. Note that $P_{\rm CM}(p)$ versus $P_{\rm QM}(p)$ have same features of $P_{\rm CM}(x)$ versus $P_{\rm QM}(x)$.

9.7.2 Particle in a Box

Next, we consider a particle in a box with the even solutions given by Eq. (4.96) where $n = 2, 4, \ldots$. Rewrite them as

$$\psi_n = \frac{1}{\sqrt{a}}\cos\left[(n - 1/2)\pi x/a)\right], \quad n = 1, 2, \ldots \tag{9.46}$$

with $E_n = (2n-1)^2\hbar^2\pi^2/(8ma^2)$. We obtain $\phi_n(p)$ [1,2] as

$$
\begin{aligned}
\phi_n(p) &= \frac{1}{\sqrt{2\pi\hbar}} \int_{-a}^{a} \psi_n e^{-ipx/\hbar}\,dx \\
&= \frac{1}{\sqrt{2\pi\hbar a}} \int_{-a}^{a} \left[\cos\left(n-\frac{1}{2}\right)\frac{\pi x}{a} \cos(px/\hbar) \right. \\
&\qquad\qquad \left. -i\cos\left(n-\frac{1}{2}\right)\frac{\pi x}{a}\sin(px/\hbar) \right]dx \\
&= \frac{1}{2\sqrt{2\pi\hbar a}} \int_{-a}^{a} [\cos X_+ + \cos X_- - i\sin X_+ \\
&\qquad\qquad +i\sin X_-]\,dx, \quad X_\pm = \left(n-\frac{1}{2}\right)\frac{\pi x}{a} \pm \frac{px}{\hbar}.
\end{aligned} \tag{9.47}
$$

As $\sin(\cdot)$ is an odd function the last two integrals in Eq. (9.47) vanish. Then

$$
\begin{aligned}
\phi_n(p) &= \frac{1}{\sqrt{2\pi\hbar a}} \int_0^a [\cos X_+ + \cos X_-]\,dx \\
&= \sqrt{\frac{a}{2\pi\hbar}} \left\{ \frac{\sin\left[\left(n-\frac{1}{2}\right)\pi + \frac{pa}{\hbar}\right]}{\left[\left(n-\frac{1}{2}\right)\pi + \frac{pa}{\hbar}\right]} + \frac{\sin\left[\left(n-\frac{1}{2}\right)\pi - \frac{pa}{\hbar}\right]}{\left[\left(n-\frac{1}{2}\right)\pi - \frac{pa}{\hbar}\right]} \right\} \\
&= \sqrt{\frac{a}{2\pi\hbar}}\, u(p).
\end{aligned} \tag{9.48}
$$

Analytical determination of $P_{\mathrm{CM}}(p)$ is tedious. However, from Fig. 4.5 we infer that only two values of p are possible and are $p = \pm\sqrt{2mE}$. $P_{\mathrm{QM}}(p)$ is proportional to $|u(p)|^2$. Figure 9.1 shows the variation of $|u(p)|^2$ with $pa/\hbar$ for $n = 1, 2, 5$ and 10. $P_{\mathrm{QM}}(p)$ is nonzero for certain range of other values of classically allowed two values of $pa/\hbar$.

Defining $p_0 = \left(n-\frac{1}{2}\right)\hbar\pi/a = 2mE_n$ and $\Delta p = \hbar/a$, the momentum eigenfunction becomes

$$
\phi = \sqrt{\frac{\Delta p}{2\pi}} \left(\frac{\sin[(p_0+p)/\Delta p]}{(p_0+p)} + \frac{\sin[(p_0-p)/\Delta p]}{(p_0-p)} \right). \tag{9.49}
$$

The $P_{\mathrm{QM}}(p)$ is then worked out as

$$
\begin{aligned}
P_{\mathrm{QM}}(p) &= \frac{\Delta p}{2\pi} \left[\frac{\sin^2(p_+/\Delta p)}{p_+^2} + \frac{\sin^2(p_-/\Delta p)}{p_-^2} \right. \\
&\qquad\qquad \left. + \frac{2\sin(p_+/\Delta p)\sin(p_-/\Delta p)}{(p_0^2-p^2)} \right],
\end{aligned} \tag{9.50}
$$

where $p_\pm = p_0 \pm p$. In the limit $\Delta p = \hbar/a \to 0$ (or $a \to \infty$), the last term vanishes. Using the representation of Dirac-delta function

$$
\lim_{\epsilon \to 0} \delta_\epsilon(x) = \lim_{\epsilon \to 0} \frac{\epsilon \sin^2(x/\epsilon)}{\pi x^2} = \delta(x) \tag{9.51}
$$

we get

$$
P_{\mathrm{QM}}(p) = \lim_{\Delta p \to 0} |\phi(p)|^2 = \frac{1}{2}[\delta(p-p_0) + \delta(p+p_0)] = P_{\mathrm{CM}}(p). \tag{9.52}
$$

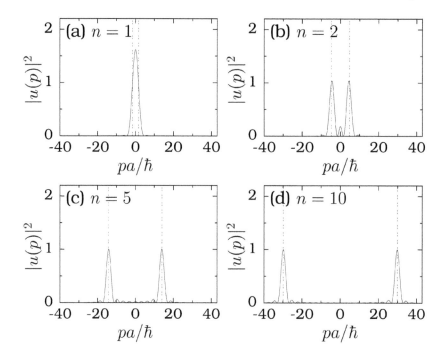

FIGURE 9.1
Plot of $|u(p)|^2$ as a function of $pa/\hbar$ for four values of n. The vertical dashed lines represent the two values of $P_{\text{CM}}(pa/\hbar)$.

9.7.3   Particle in a Linear Potential

Schrödinger equation is exactly solved in momentum space for the linear potential $V(x) = kx$ [3]. This potential in momentum representation is $V(p) = i\hbar k\partial/\partial p$. Then the Schrödinger equation of the problem is

$$\frac{1}{2m}p^2\phi + i\hbar k\frac{\partial\phi}{\partial p} = E\phi. \tag{9.53}$$

The direct integration of the above equation yields

$$\phi(p) = N\exp\left[i\left(-Ep + \frac{1}{6m}p^3\right)\Big/(\hbar k)\right], \tag{9.54}$$

where the constant N is to be determined by the normalization condition. As E forms a continuous spectrum the normalization condition is

$$\int_{-\infty}^{\infty}\phi_{E'}^*(p)\phi_E(p)\,\mathrm{d}p = \delta(E - E'). \tag{9.55}$$

Use of Eq. (9.54) in (9.55) results in

$$\begin{aligned}
|N|^2\int_{-\infty}^{\infty}e^{i(E'-E)p/\hbar k}\,\mathrm{d}p &= \delta(E'-E)\\
&= \frac{1}{2\pi}\int_{-\infty}^{\infty}e^{i(E'-E)y}\,\mathrm{d}y\\
&= \frac{1}{2\pi\hbar k}\int_{-\infty}^{\infty}e^{i(E'-E)p/\hbar k}\,\mathrm{d}p. \tag{9.56}
\end{aligned}$$

From this equation we obtain $|N|^2 = 1/(2\pi\hbar k)$. Then

$$\phi(p) = \frac{1}{(2\pi\hbar k)^{1/2}} \exp\left[i \left(\frac{1}{6m}p^3 - Ep \right) \Big/ \hbar k \right] \qquad (9.57)$$

and $P(p) = \phi^*(p)\phi(p) = 1/(2\pi\hbar k)$.

9.8 Concluding Remarks

Though we so far considered only one-dimensional cases, the wave function in momentum space for higher-dimensional systems has also been reported. For example, the wave functions for hydrogen atom in the momentum representation [4,5] have obtained. The momentum representation is completely equivalent to the coordinate representation. In spite of their complete equivalence, the momentum representation could throw much insight on certain features that may remain hidden in the coordinate representation. For examples, the $SO(4)$ symmetry of the hydrogen atom was first uncovered by Fock [6] treating the problem in momentum representation. Further, the resonant Gamow states in the momentum representation were found to be square-integrable solutions to a homogeneous Lippmann–Schwinger equation. Generally, the position representation has been useful for most bound state problems. With a few exceptions, the momentum representation has been used almost exclusively to study scattering.

A method using delta function interactions has been introduced for constructing momentum wave function for hydrogen atom and hydrogen molecule ion [7]. Momentum space is shown to be better suited for exploring electronic wave functions of quasiperiodic systems [8].

9.9 Bibliography

[1] R.W. Robinett, *Am. J. Phys.* 63:823, 1995.

[2] Y.Q. Liang, H. Zhang and Y.X. Dardenne, *J. Chem. Edu.* 72:148, 1995.

[3] F. Constantinescu and E. Magyari, *Problems in Quantum Mechanics*. Pergamon, London, 1978.

[4] J.R. Lombardi, *Phys. Rev. A* 22:797, 1980.

[5] H.N.N. Yepez, C.A. Vargas and A.L.S. Brito, *Eur. J. Phys.* 8:189, 1987.

[6] V. Fock, *Z. Phys.* 98:145, 1935.

[7] A. Binesh, H. Arabshahi and A. Sadremomtaz, *Asian J. Sci. Tech.* 2:1, 2011.

[8] S. Rolof, S. Thiem and M. Schreiber, *Eur. Phys. J. B* 86:372, 2013.

9.10   Exercises

9.1   Show that $\psi(x,t) = (1/\sqrt{2\pi}) \int_{-\infty}^{\infty} C(k) e^{i(kx - \omega_k t)} \, dk$ satisfies the time-dependent Schrödinger equation of a free particle if $\hbar \omega_k = \hbar^2 k^2/(2m)$.

9.2   What is the matrix representation of the momentum operator p in the momentum representation.

9.3   Consider the function $\phi_k = (1/\sqrt{a}) e^{ikx}$ defined over the interval $(-a/2, a/2)$. Show that these functions are all normalized to unity and maintain the normalization in the limit $a \to \infty$.

9.4   If the state functions for a free particle moving in one-dimension vanish at $x \to \pm\infty$ then find $\langle p \rangle$ for purely real states.

9.5   Show that the functions $\phi_k = (1/\sqrt{a}) e^{ikx}$ defined over the interval $(-a/2, a/2)$ are an orthogonal set in the limit $a \to \infty$.

9.6   Obtain the momentum eigenfunction for $\psi(x) = A$ for $|x| < a$ and 0 otherwise.

9.7   Can we write $\langle g(p) \rangle = \int_{-\infty}^{\infty} \phi^*(p,t) g(p) \phi(p,t) \, dp$, where g is an arbitrary function of p? Why?

9.8   Find the momentum wave function for $\psi = (1/\sqrt{2L}) e^{ik_0 x}$ for $|x| \le L$ and 0 for $|x| > L$.

9.9   If the wave function of a particle in one-dimension is $\psi = e^{-|x|}$ determine the corresponding momentum eigenfunction $\phi(p,t)$.

9.10 Calculate the position wave function if the momentum wave function is given by $\phi(\mathbf{p}) = N e^{-|\mathbf{p}|/\Delta \mathbf{p}}$.

9.11 If the wave function of a free particle in one-dimension at $t = 0$ is given by $\psi(x,0) = A e^{-x^2}$, where A is the normalization constant find $C(k,0)$, $C(k,t)$ and $\psi(x,t)$.

9.12 Obtain the momentum wave function for the Gaussian wave function $\psi = (1/(\sigma\sqrt{\pi})^{1/2} e^{-x^2/(2\sigma^2)}$.

9.13 Obtain the momentum wave function for the normalized wave function $\psi_n(x) = \sqrt{2/L} \sin(n\pi x/L)$ for $0 < x < L$ and 0 otherwise.

9.14 Given the momentum wave function $C(k) = A e^{-\alpha|k|}$ find $\psi(x)$.

9.15 Find a formula for $\langle x^n \rangle$ in terms of $\phi(p,t)$.

9.16 Show that $\langle p^n \rangle = \int_{-\infty}^{\infty} \phi^*(p,t) p^n \phi(p,t) \, dp$.

9.17 Find $[\hat{x}, \hat{p}] C(p)$ if $p_{\text{op}} = \hat{p} = p$ and $x_{\text{op}} = \hat{x} = i\hbar \dfrac{\partial}{\partial p}$.

9.18 Consider the screened Coulomb potential $V_0 e^{-\mu r}/r$. Find its momentum representation.

9.19 What is the eigenfunction of the operator $\hat{x} = i\hbar \partial/\partial p$ (in the momentum representation) corresponding to the eigenvalue x?

9.20 The matrix of $\hat{p}$ in p-representation is $\langle p|\hat{p}|p' \rangle = p' \delta(p - p')$. Using the transfer matrix $\langle x|p \rangle = A e^{ipx/\hbar}$ show that the matrix of $\hat{p}$ in the coordinate representation is $\langle x|\hat{p}|x' \rangle = -i\hbar \partial/\partial x [\delta(x - x')]$.

10

Wave Packet

10.1 Introduction

A plane wave considered earlier in chapter 2, Eq. (2.1), extends over all space. This wave has a single wave number k and is spread over all space. If we use this plane wave then the particle may be found anywhere in space with equal probability. In this case $\Delta x = \infty$ and $\Delta p_x = 0$. However, particles are localized in space. If we want to associate a wave with a particle in a consistent way, then the wave-like phenomenon should be localized in the neighbourhood of the particle. That is, we wish to construct a localized wave with nonzero amplitude in a small region of space and zero elsewhere. Such a description of a localized particle can be achieved by constructing a wave packet.

To localize a free particle, we form a superposition of plane waves with various momenta. Then Δx and Δp_x are finite. Because the particle is localized, the probability distribution is confined to a finite region as $P = \psi^*\psi \to 0$ as $|\mathbf{x}| \to \infty$. It is possible to construct a sufficiently narrow *wave packet* (superposition of number of waves of different frequencies) which moves in such a way that the average position of the particle confirms classical laws. Earlier, in chapter 4 we have noticed that the energy levels of linear harmonic oscillator are equally spaced. It has a consequence on a wave packet. If we construct a wave packet for a linear harmonic oscillator by suitably superposing its eigenstates then it will return to its identical shapes at every classical period $2\pi/\omega$. This is because each state will have changed phase by an integral number of 2π's. Such states are called *coherent states*.

The behaviour of localized quantum wave packet has lately become an object of much interest to forge a link between classical and quantum mechanics. We can consider the wave packets synthesized by the superposition of a great many bound states centred around a rather large mean value $\bar{n}$ of the quantum number n. Such a wave packet may begin by mimicking classical dynamics, but after a short while, quantum effects set in, causing, at first a spreading and, at much later times, a partial or total restoration of the original form at a point, which is not necessarily the same as that where the packet was initially localized.

In this chapter, we discuss some of the features of wave packets.

10.2 Phase and Group Velocities

Consider a plane wave $u = e^{i(kx - \omega t)}$ of frequency $\nu = \omega/(2\pi)$ and wavelength $\lambda = 2\pi/k$. $kx - \omega t$ is called *phase* of the wave. The velocity of the wave is simply the velocity of a point at which the phase is constant. Such a position is given by $x = \text{constant} + (\omega/k)t = \text{constant} + v_\text{p}t$, where $v_\text{p} = \omega/k$ is known as the *phase velocity* of the wave. Suppose we superpose a large number of waves of the form $u = e^{i(kx - \omega t)}$ and denote it as $\psi(x, t)$. It is

DOI: 10.1201/9781003172178-10

given by

$$\psi(x,t) = \frac{1}{\sqrt{2\pi}} \int_{-\infty}^{\infty} C(k) \, \mathrm{e}^{\mathrm{i}(kx-\omega t)} \, \mathrm{d}k \,. \qquad (10.1a)$$

As this equation appears as the Fourier inverse transform we write

$$C(k) = \frac{1}{\sqrt{2\pi}} \int_{-\infty}^{\infty} \psi(x,t) \, \mathrm{e}^{-\mathrm{i}(kx-\omega t)} \, \mathrm{d}x \,. \qquad (10.1b)$$

Although, t appears explicitly in the integral in Eq. (10.1b), $C(k)$ does not depend on t provided ψ is a superposition of plane waves. In this case $\partial C(k)/\partial t = 0$:

$$\int_{-\infty}^{\infty} \left(\frac{\partial \psi}{\partial t} + \mathrm{i}\omega\psi \right) \mathrm{e}^{-\mathrm{i}(kx-\omega t)} \, \mathrm{d}x = 0 \,. \qquad (10.2)$$

The implication is $\psi(x,t)$ can be replaced by $\psi(x,0)$ in Eqs. (10.1).

The wave function $\psi(x,t)$ given by Eq. (10.1a) is called a *wave group* or a *wave packet*. Its localizations in x and p_x (or k) can be adjustable and can be made very narrow. Wave packet is often taken as the best approximation of quantum theory to a classical particle. If the wave packet is regarded as a point, with zero amplitude everywhere except in a very small region, then its motion can be interpreted as the motion of a single particle. Such a wave packet is necessarily not stationary and spread (or shrink) in time.

When a wave travels without change of form then it is called a *nondispersive wave*. In this case the connection between ω and k, called *dispersion relation*, is linear and the phase velocity is same for each harmonic component of ψ. Suppose the connection between k and ω is not linear. Because of the change of relative phase of the components in time the wave packet cannot propagate without change of shape. The wave group will get dispersed and die down in due course. It is a dispersive propagation. Dispersion can be avoided by limiting the range of k.

Assume the simple case of major contribution to the integral in Eq. (10.1a) is coming from groups of waves in a small neighbourhood of $k = k_0$, $(k_0 - \Delta k) < k < (k_0 + \Delta k)$. Then the Taylor series of $\omega(k)$ about $k = k_0$ given by

$$\omega(k) = \omega(k_0) + \frac{\mathrm{d}\omega}{\mathrm{d}k}\bigg|_{k=k_0} \Delta k + \text{higher orders in } \Delta k \qquad (10.3)$$

suggests the approximate $\psi(x,t)$ as

$$
\begin{aligned}
\psi(x,t) &= \frac{C(k_0)}{\sqrt{2\pi}} \mathrm{e}^{-\mathrm{i}[\omega(k_0)t-k_0 x]} \int_{-\Delta k}^{\Delta k} \mathrm{e}^{\mathrm{i}(x-\omega' t)(k-k_0)} \, \mathrm{d}k \\
&= \frac{2C(k_0)}{\sqrt{2\pi}} \frac{\sin[\Delta k(x-\omega' t)]}{(x-\omega' t)} \mathrm{e}^{-\mathrm{i}[\omega(k_0)t-k_0 x]} \,,
\end{aligned}
\qquad (10.4)
$$

where $\omega' = \mathrm{d}\omega/\mathrm{d}k|_{k=k_0}$. Equation (10.4) represents a wave of wavelength $2\pi/k_0$ and frequency $\omega(k_0)/2\pi$. While any elementary wave of the form $u = \mathrm{e}^{\mathrm{i}(kx-\omega t)}$ moves with a phase velocity $v_\mathrm{p} = \omega/k$, *how does the wave group itself move? Does it move with the same velocity v_p or with a different velocity?* In Eq. (10.4) $x - \omega' t$ indicates that the wave packet moves with a group velocity $v_\mathrm{g} = \mathrm{d}x/\mathrm{d}t = \omega' = \mathrm{d}\omega/\mathrm{d}k$, which is in general not equal to v_p. *When does $v_\mathrm{g} = v_\mathrm{p}$?* $v_\mathrm{g} = v_\mathrm{p}$ only when $\mathrm{d}\omega/\mathrm{d}k$ is independent of k, $\omega \propto k$.

For a free particle, from $E = \hbar\omega$, we have $\omega = E/\hbar = p_x^2/(2m\hbar) = \hbar k^2/(2m)$. This is the dispersion relation for a free particle. Now,

$$v_\mathrm{g} = \frac{\mathrm{d}\omega}{\mathrm{d}k} = \frac{\hbar k}{m} = \frac{p_x}{m} = v_\mathrm{particle} \,. \qquad (10.5)$$

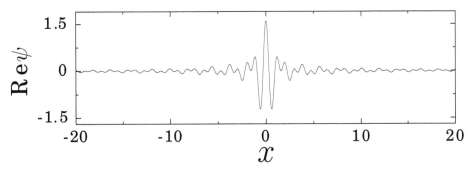

FIGURE 10.1
Real part of $\psi(x,0)$ given by Eq. (10.4) where $k_0 = 5$, $\Delta k = 2$ and $C(k_0) = 1$.

Equation (10.5) implies the following:

1. Group velocity of quantum wave function of a free particle is same as the velocity of the classical free particle.

2. The kinetic energy and momentum are carried with the group velocity of the quantum wave function.

Figure 10.1 shows the wave packet ψ at $t = 0$. In this figure real part of ψ (Eq. (10.4)) is plotted where $k_0 = 5$, $\Delta k = 2$ and $C(k_0) = 1$. The wave packet given by Eq. (10.4) is a localized wave with its amplitude zero except in a small region of space. When more and more waves are added the wave packet gets more and more narrower. In the limit of an infinite number of waves covering all wavelengths, we can localize a particle to a single point in space.

Solved Problem 1:
Show that the phase velocity and group velocity of the de Broglie wave describing a free particle with classical velocity v are c^2/v and v, respectively.

We obtain

$$v_p = \frac{\omega}{k} = \frac{\hbar\omega}{\hbar k} = \frac{E}{p}, \quad v_g = \frac{d\omega}{dk} = \frac{dE}{dp}. \tag{10.6}$$

Since $E^2 = c^2 p^2 + m^2 c^4$ we obtain $(E/p)dE/dp = c^2$. Using the expressions for v_p and v_g in $(E/p)dE/dp = c^2$ we get $v_p v_g = c^2$. Also,

$$v_g = \frac{c^2}{v_p} = \frac{c^2}{E/p}. \tag{10.7}$$

Using

$$E = \frac{mc^2}{\sqrt{1-v^2/c^2}}, \quad p = \frac{mv}{\sqrt{1-v^2/c^2}} \tag{10.8}$$

we get $v_g = c^2/(E/p) = v$. Thus, the group velocity is the classical velocity of the particle. Further, $v_p = c^2/v$.

Many one-dimensional quantum mechanical systems have been studied based on a wave packet approach. For example, transmission and reflection from a square-barrier [1–4] and linear potential steps [5], bound state wave packets in a single square-well in position space [6] and momentum space [7], double-wells [8,9] and systems relevance to solid state physics [10,11], temperature effect on wave packet spreading [12] and wave packet motion using expectation values [13] have been reported. The behaviour of Coulomb wave packets on circular [14] and elliptical orbits [15] are tested experimentally on Rydberg atom systems [16,17].

Study of wave packet is the subject of much current research in atomic, molecular, chemical and condensed matter physics. They are well suited for studying the classical limit of quantum mechanical systems. The advent of short-pulsed lasers made it possible to produce and detect superposition of electron states for many physical systems. Such superposition leads to the formation of localized electron wave packets.

Solved Problem 2:

Find the relation between the phase and group velocities of a wave packet obtained by superposing the three waves given by

$$\psi_1 = A\mathrm{e}^{\mathrm{i}(kx-\omega t)}, \ \psi_2 = \frac{A}{2}\mathrm{e}^{\mathrm{i}[(k+\mathrm{d}k)x-(\omega+\mathrm{d}\omega)t]} \text{ and } \psi_3 = \frac{A}{2}\mathrm{e}^{\mathrm{i}[(k-\mathrm{d}k)x-(\omega-\mathrm{d}\omega)t]}.$$

We obtain

$$
\begin{aligned}
\psi &= \psi_1 + \psi_2 + \psi_3 \\
&= A\mathrm{e}^{\mathrm{i}(kx-\omega t)}\left[1 + \frac{1}{2}\mathrm{e}^{\mathrm{i}(x\mathrm{d}k-\mathrm{d}\omega t)} + \frac{1}{2}\mathrm{e}^{-\mathrm{i}(x\mathrm{d}k-\mathrm{d}\omega t)}\right] \\
&= A\mathrm{e}^{\mathrm{i}(kx-\omega t)}[1 + \cos(x\mathrm{d}k - \mathrm{d}\omega t)] \\
&= A\mathrm{e}^{\mathrm{i}(kx-\omega t)}\left[1 + \cos\left(\mathrm{d}k\left(x - \frac{\mathrm{d}\omega}{\mathrm{d}k}t\right)\right)\right].
\end{aligned}
$$

The envelope moves with the velocity $\mathrm{d}\omega/\mathrm{d}k$ and is the group velocity v_g.

10.3   Wave Packets and Uncertainty Principle

In quantum mechanics, a particle is described by a wave packet $\psi(x,t)$. The wave packet surrounds the position of the classical particle. The particle may be found anywhere within the region with a probability density $|\psi(x,t)|^2$. This implies that the position of the particle is indeterminate within the domain of the wave packet. A similar argument for the momentum wave function $\phi(p,t)$ says that the momentum of the particle is also indeterminate within the limits of $\phi(p,t)$. Since $\psi(x,t)$ and $\phi(p,t)$ are related by Fourier transforms the squaring of the wave packet in one domain will elongate in the other domain as space domain and momentum domain have inverse relationship. If the wave packet is confined to a small region in space and is highly localized then the spread of the wave packet in momentum space become very large.

The uncertainties in position measurement Δx and the momentum measurement Δp_x always satisfy the relation $\Delta x \Delta p_x \geq \hbar/2$. This Heisenberg uncertainty principle is independent of the shape of the wave packet. To solve the equation of motion of a classical particle we need to completely specify the initial values of position and momentum which means $\Delta x = \Delta p_x = 0$. But for a quantum mechanical particle it has been confirmed by experiments that it is physically impossible to measure simultaneously position and momentum of a particle with a higher degree of accuracy than the uncertainty relation allows. Since the constant $\hbar$ is very small, for macroscopic objects, it becomes insignificant and hence classical mechanics is fairly accurate. But for microscopic bodies, it is highly significant and hence classical mechanics fails in describing systems like atoms, molecules, nucleus and elementary particles.

The uncertainties Δx and Δp_x are given by $\Delta x = \langle (x - \langle x \rangle)^2 \rangle^{1/2}$ and $\Delta p_x = \langle (p_x - \langle p_x \rangle)^2 \rangle^{1/2}$. We shall consider, for simplicity, a wave packet with $\langle x \rangle = \langle p_x \rangle = 0$. If the wave packet is represented by the position-space wave function $\psi_\alpha(x)$ then

$$\langle x \rangle = \int_{-\infty}^{\infty} \psi_\alpha^*(x) x \psi_\alpha(x) \, dx \,, \tag{10.9a}$$

$$\langle p_x \rangle = \int_{-\infty}^{\infty} \psi_\alpha^*(x) \left(-i\hbar \frac{d}{dx} \right) \psi_\alpha(x) \, dx \,. \tag{10.9b}$$

For the case of momentum-space wave function $\phi_\alpha(p_x)$

$$\langle x \rangle = \int_{-\infty}^{\infty} \phi_\alpha^*(p) \left(i\hbar \frac{d}{dp} \right) \phi_\alpha(p) \, dp \,, \tag{10.10a}$$

$$\langle p_x \rangle = \int_{-\infty}^{\infty} \phi_\alpha^*(p) p_x \phi_\alpha(p) \, dp \,. \tag{10.10b}$$

We obtain

$$\int_{-\infty}^{\infty} \left(i\hbar \frac{d\psi_\alpha^*}{dx} \right) x \psi_\alpha \, dx = i\hbar \psi_\alpha^* x \psi_\alpha |_{-\infty}^{\infty} - i\hbar \int_{-\infty}^{\infty} \psi_\alpha^* \frac{d}{dx}(x\psi_\alpha) \, dx$$

$$= -i\hbar \int_{-\infty}^{\infty} \psi_\alpha^* \psi_\alpha \, dx - i\hbar \int_{-\infty}^{\infty} \frac{d\psi_\alpha}{dx} x \psi_\alpha^* \, dx \,. \tag{10.11}$$

That is,

$$\int_{-\infty}^{\infty} \left(i\hbar \frac{d\psi_\alpha^*}{dx} x\psi_\alpha + i\hbar \frac{d\psi_\alpha}{dx} x\psi_\alpha^* \right) dx = -i\hbar \int_{-\infty}^{\infty} \psi_\alpha^* \psi_\alpha \, dx \,. \tag{10.12}$$

Or

$$2i\mathrm{Im} \int_{-\infty}^{\infty} i\hbar \frac{d\psi_\alpha^*}{dx} x\psi_\alpha \, dx = -i\hbar \int_{-\infty}^{\infty} |\psi_\alpha|^2 \, dx \,. \tag{10.13}$$

The square of the modulus of the integral gives

$$\hbar^2 \left| \int_{-\infty}^{\infty} |\psi_\alpha|^2 \, dx \right|^2 = 4 \left| \mathrm{Im} \int_{-\infty}^{\infty} i\hbar \frac{d\psi_\alpha^*}{dx} x \, \psi_\alpha dx \right|^2$$

$$\leq 4 \left| \int_{-\infty}^{\infty} i\hbar \frac{d\psi_\alpha^*}{dx} x\psi_\alpha \, dx \right|^2 \,. \tag{10.14}$$

By Schwarz inequality, the right-side of Eq. (10.14) is smaller than

$$4 \int_{-\infty}^{\infty} i\hbar \frac{d\psi_\alpha^*}{dx}(-i\hbar) \frac{d\psi_\alpha}{dx} \, dx \int_{-\infty}^{\infty} x\psi_\alpha^* x\psi_\alpha \, dx \,. \tag{10.15}$$

Hence,

$$\frac{\int_{-\infty}^{\infty} x^2 |\psi_\alpha|^2 \, dx}{\int_{-\infty}^{\infty} |\psi_\alpha|^2 \, dx} \frac{\int_{-\infty}^{\infty} \left| -i\hbar \frac{d\psi_\alpha}{dx} \right|^2 dx}{\int_{-\infty}^{\infty} |\psi_\alpha|^2 \, dx} \geq \frac{\hbar^2}{4} \,. \tag{10.16}$$

It is easy to show that

$$\langle p_x^2 \rangle = \frac{\displaystyle\int_{-\infty}^{\infty} p_x^2 |\phi_\alpha(p)|^2 \, \mathrm{d}p}{\displaystyle\int_{-\infty}^{\infty} |\phi_\alpha(p)|^2 \, \mathrm{d}p} = \frac{\displaystyle\int_{-\infty}^{\infty} \left| -\mathrm{i}\hbar \frac{\mathrm{d}\phi_\alpha}{\mathrm{d}x} \right|^2 \mathrm{d}x}{\displaystyle\int_{-\infty}^{\infty} |\phi_\alpha|^2 \, \mathrm{d}x}. \tag{10.17}$$

Equation (10.16) gives $\langle x^2 \rangle \langle p_x^2 \rangle \geq \hbar^2/4$. Since $\langle x \rangle = \langle p_x \rangle = 0$ we have $(\Delta x)^2 = \langle x^2 \rangle$ and $(\Delta p_x)^2 = \langle p_x^2 \rangle$. Then using $\langle x^2 \rangle \langle p_x^2 \rangle \geq \hbar^2/4$ we get $\Delta x \Delta p_x \geq \hbar/2$.

10.4   Gaussian Wave Packet

In this section, we discuss the features of Gaussian wave packet.

10.4.1   Uncertainty Relation

A Gaussian minimum uncertainty wave packet is a superposition of plane waves localized in position and momentum and with $\Delta x \Delta p_x = \hbar/2$. This wave packet at $t = 0$ is given by

$$\psi(x, 0) = \left(\frac{1}{\sqrt{2\pi}\sigma} \right)^{1/2} \mathrm{e}^{-(x-x_0)^2/(4\sigma^2)}, \tag{10.18}$$

where σ is a constant and it measures the width of the packet. It is a simple initial quantum state. This wave packet is normalized. The probability of observing the particle decays to zero very rapidly for $|x - x_0| > \sqrt{2}\,\sigma$. An interesting feature of the Gaussian wave packet is that it has $\Delta x \Delta p_x = \hbar/2$ at $t = 0$, a minimum uncertainty product. One can easily show that (see the exercise 10.4 at the end of this chapter)

$$\langle x \rangle = x_0, \quad \langle x^2 \rangle = \sigma^2 + x_0^2, \quad \langle p_x \rangle = 0, \quad \langle p_x^2 \rangle = \frac{\hbar^2}{4\sigma^2}, \tag{10.19a}$$

$$(\Delta x)^2 = \langle x^2 \rangle - \langle x \rangle^2 = \sigma^2, \quad (\Delta p_x)^2 = \langle p_x^2 \rangle - \langle p_x \rangle^2 = \frac{\hbar^2}{4\sigma^2} \tag{10.19b}$$

and $(\Delta x)(\Delta p_x) = \hbar/2$. Hence, the Gaussian wave packet is a suitable candidate to represent a particle. In this case the position and momentum can be determined simultaneously with minimum uncertainty relation. If we make the Gaussian narrower and narrower in x by reducing σ then Δp_x grows, as $1/\sigma$, in such a way that the product of Δx and Δp_x satisfies the bound $\hbar/2$. It is of interest to know how a wave packet evolves in various physically interesting quantum systems. This is explored in the rest of this section.

10.4.2   Construction of Gaussian Wave Packet for a System

To construct a wave function that is Gaussian at $t = 0$ first the time-independent Schrödinger equation $H\phi_n = E_n \phi_n$ is solved. Then consider the linear superposition of ϕ_n as

$$\psi(x, 0) = \sum_n C_n \phi_n, \quad \sum |C_n|^2 = 1. \tag{10.20}$$

If the energy eigenvalues are continuous then the summation is replaced by integral. Choosing $\psi(x,0)$ as Gaussian form we have

$$\psi(x,0) = \left(\frac{1}{\sqrt{2\pi}\,\sigma}\right)^{1/2} \mathrm{e}^{-(x-x_0)^2/(4\sigma^2)} = \sum_n C_n \phi_n . \tag{10.21}$$

We need to find C_n. Multiplying both sides of Eq. (10.21) by ϕ_m^* and integrating over space gives

$$\sum_n C_n \int_{-\infty}^{\infty} \phi_m^* \phi_n \, \mathrm{d}x = \int_{-\infty}^{\infty} \phi_m^* \psi(x,0) \, \mathrm{d}x$$

from which we write

$$C_m = \int_{-\infty}^{\infty} \phi_m^* \psi(x,0) \, \mathrm{d}x \ \ \text{or} \ \ C_n = \int_{-\infty}^{\infty} \phi_n^* \psi(x,0) \, \mathrm{d}x . \tag{10.22}$$

Thus, choosing C_n as dictated in the above equation a Gaussian wave packet can be constructed for any system. The time evolution of $\psi(x,t)$ is

$$\psi(x,t) = \sum_n C_n \phi_n(x) \mathrm{e}^{-\mathrm{i}E_n t/\hbar} . \tag{10.23}$$

10.4.3   Gaussian Wave Packet of a Free Particle

We choose the Gaussian wave packet at $t=0$ as

$$\psi(x,0) = \left(\frac{1}{\sqrt{2\pi}\,\sigma}\right)^{1/2} \mathrm{e}^{-(x-x_0)^2/(4\sigma^2)} \mathrm{e}^{\mathrm{i}k_0 x} . \tag{10.24}$$

It is a plane wave with wave number k_0 modulated by a Gaussian profile. For this wave packet $\Delta x \Delta p_x = \hbar/2$. Figure 10.2a depicts $\mathrm{Re}\psi(x,0)$ for $k_0 = 5$, $\sigma = 1$, $x_0 = 0$. Figure 10.2b shows $\psi(x,0)$ for $k_0 = 0$, $\sigma = 1$, $x_0 = 0$.

Let us compute the time evolution of (10.24). Suppose $x_0 = 0$. The coefficients $C(k)$ are evaluated as

$$C(k) = \frac{1}{\sqrt{2\pi}} \int_{-\infty}^{\infty} \psi(x,0) \mathrm{e}^{-\mathrm{i}kx} \, \mathrm{d}x = \left(\frac{2\sigma^2}{\pi}\right)^{1/4} \mathrm{e}^{-\sigma^2(k-k_0)^2} . \tag{10.25}$$

The Fourier inverse transform of $C(k)$ with $E = \hbar\omega = \hbar^2 k^2/(2m)$, that is, $\omega = \hbar k^2/(2m)$ gives

$$
\begin{aligned}
\psi(x,t) &= \frac{1}{\sqrt{2\pi}} \int_{-\infty}^{\infty} C(k) \mathrm{e}^{\mathrm{i}(kx-\omega t)} \mathrm{d}k && (10.26)\\[4pt]
&= \frac{1}{\sqrt{2\pi}} \left(\frac{2\sigma^2}{\pi}\right)^{1/4} \int_{-\infty}^{\infty} \mathrm{e}^{-\sigma^2(k-k_0)^2} \mathrm{e}^{\mathrm{i}kx} \mathrm{e}^{-\mathrm{i}\omega t} \mathrm{d}k && \\[4pt]
&= \frac{1}{\sqrt{2\pi}} \left(\frac{2\sigma^2}{\pi}\right)^{1/4} \mathrm{e}^{-\sigma^2 k_0^2} \int_{-\infty}^{\infty} \mathrm{e}^{-\alpha^2 k^2 + \beta k} \mathrm{d}k, && (10.27)
\end{aligned}
$$

where

$$\alpha^2 = \sigma^2 + \mathrm{i}\hbar t/(2m), \quad \beta = -\mathrm{i}\left(x - 2\mathrm{i}\sigma^2 k_0\right) . \tag{10.28}$$

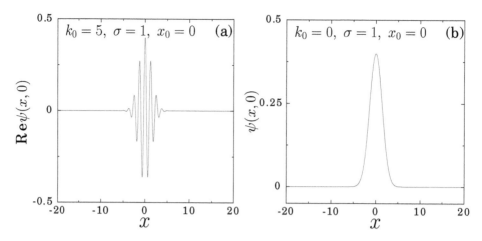

FIGURE 10.2
Gaussian wave packet solution (Eq. (10.24)) for (a) $k_0 = 5$, $\sigma = 1$, $x_0 = 0$ and (b) $k_0 = 0$, $\sigma = 1$, $x_0 = 0$.

The integral in Eq. (10.27) is evaluated as

$$
\begin{aligned}
\int_{-\infty}^{\infty} e^{-\alpha^2 k^2 + \beta k} dk &= e^{\beta^2/(4\alpha^2)} \int_{-\infty}^{\infty} e^{-\alpha^2 \left(k - \frac{\beta}{2\alpha^2}\right)^2} dk \\
&= e^{\beta^2/(4\alpha^2)} \int_{-\infty}^{\infty} e^{-\alpha^2 y^2} dy \\
&= e^{\beta^2/(4\alpha^2)} \frac{\sqrt{\pi}}{\alpha} .
\end{aligned}
\tag{10.29}
$$

Then

$$
\psi(x,t) = \left(\frac{\sigma}{\sqrt{2\pi}\alpha^2}\right)^{1/2} e^{-\left(\sigma^2 k_0^2 - \frac{\beta^2}{4\alpha^2}\right)} .
\tag{10.30}
$$

$\psi(x,t)$ for $x_0 \neq 0$ is obtained from Eq. (10.26) by replacing x by $x - x_0$. $|\psi(x,t)|^2$ is calculated as

$$
|\psi(x,t)|^2 = \frac{\sigma}{\sqrt{2\pi}} \frac{1}{\alpha\alpha^*} \exp\left[-2\sigma^2 k_0^2 + \frac{\beta^2}{4\alpha^2} + \frac{(\beta^2)^*}{4(\alpha^2)^*}\right] .
\tag{10.31}
$$

Defining

$$
\alpha\alpha^* = \sqrt{\sigma^4 + \frac{\hbar^2 t^2}{4m^2}}, \quad b = \frac{\hbar t}{2m} ,
\tag{10.32a}
$$

$$
\sigma_\omega = \sigma^2\left(1 + \frac{\hbar^2 t^2}{4m^2\sigma^4}\right), \quad \frac{p_{x0}}{m} = \frac{\hbar k_0}{m} = v
\tag{10.32b}
$$

we obtain the probability density as

$$
\rho = |\psi(x,t)|^2 = \left(\frac{1}{2\pi\sigma_\omega^2}\right)^{1/2} \exp\left[-\frac{1}{2\sigma_\omega^2}(x - vt)^2\right] .
\tag{10.33}
$$

The maximum of the wave packet moves with the group velocity $v_g = v = p_{x0}/m$, like a classical particle. The phase velocity of the individual plane waves in the wave packet are $v_p = \hbar k/(2m)$.

Solved Problem 3:

Compute the expectation values of x, p_x, x^2 and p_x^2 for the $\psi(x,t)$ given by Eq. (10.30).

We obtain

$$
\begin{aligned}
\langle x \rangle &= \int_{-\infty}^{\infty} x |\psi(x,t)|^2 \, dx \\
&= \int_{-\infty}^{\infty} (x - vt)|\psi(x,t)|^2 \, dx + vt \int_{-\infty}^{\infty} |\psi(x,t)|^2 \, dx \\
&= \frac{1}{\sqrt{2\pi}\sigma_\omega} \int_{-\infty}^{\infty} (x - vt) \exp\left[-\frac{1}{2\sigma_\omega^2}(x - vt)^2\right] \, dx + vt \\
&= vt.
\end{aligned}
\tag{10.34a}
$$

Next,

$$
\langle p_x \rangle = mv, \tag{10.34b}
$$

$$
\langle x^2 \rangle = \frac{1}{\sqrt{2\pi}\sigma_\omega} \int_{-\infty}^{\infty} x^2 \exp\left[-\frac{1}{2\sigma_\omega^2}(x - vt)^2\right] \, dx = \sigma_\omega + v^2t^2, \tag{10.34c}
$$

$$
\langle p_x^2 \rangle = m^2v^2 + \frac{\hbar^2}{4\sigma^2}. \tag{10.34d}
$$

The wave packet as a whole moves with the velocity v_0 with the centre of the packet at position vt at time t. The width of the wave packet is given by $(\Delta x)^2 = \langle x^2 \rangle - \langle x \rangle^2 = \sigma_\omega$. Then

$$
\Delta x = \sigma\sqrt{1 + \frac{\hbar^2 t^2}{4m^2\sigma^4}}, \quad \Delta p_x = \frac{\hbar}{2\sigma}. \tag{10.35}
$$

The width of the wave packet increases with time. That is, the wave packet *flattens* or spreads out. This is illustrated in Fig. 10.3. However, dispersion of p_x remains constant. The mean-square width $(\Delta x)^2$ of the packet at $t = 0$ is σ^2. It becomes $2\sigma^2$ (that is doubled) in time $t = 2m\sigma^2/\hbar$. During this time the distance travelled by the wave packet will be $d = vt = 2vm\sigma^2/\hbar = 4\pi\sigma^2/\lambda$, where $\lambda = h/(mv) = 2\pi\hbar/(mv)$ is the de Broglie wavelength. The wave packet essentially evolves by moving more or less classically according to its central momentum p_{x0} but spreading in position.

10.4.4 Gaussian Wave Packet of a Harmonic Oscillator

Let us choose the initial Gaussian wave packet centred at x_0 as the ground state eigenfunction [18,19]

$$
\psi(x,0) = \left(\frac{1}{\sqrt{2\pi}\,\sigma_0}\right)^{1/2} e^{-(x-x_0)^2/(4\sigma_0^2)} e^{ik_0 x}, \tag{10.36}
$$

where σ_0 is the initial width of the wave packet. Write $\psi(x,0)$ as in Eq. (10.20) where C_n's are given by Eq. (10.22) with ϕ_n are the stationary state eigenfunctions of linear harmonic oscillator. Time-dependent solution $\psi(x,t)$ is given by Eq. (10.23). Then (see sec. 4.11)

$$
|\psi(x,t)|^2 = \frac{1}{\sqrt{\pi}\,\sigma_x(t)} e^{-(x-x_0\cos\omega t)^2/\sigma_x^2(t)} \tag{10.37a}
$$

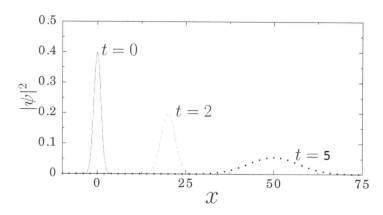

FIGURE 10.3

Time evolution of the Gaussian wave packet $|\psi(x,t)|^2$ (Eq. (10.33)) of a free particle. Here $\sigma = 1$, $\hbar = m = 1$ and $v = 10$.

and

$$|\phi(p,t)|^2 = \frac{1}{\sqrt{\pi}\sigma_p(t)}\,e^{-(p+m\omega x_0 \sin \omega t)^2/\sigma_p^2(t)}\,, \qquad (10.37b)$$

where

$$(\sigma_x(t))^2 = \sigma_0^2 \cos^2 \omega t + \left(\frac{\hbar}{m\omega\sigma_0}\right)^2 \sin^2 \omega t\,, \qquad (10.37c)$$

$$(\sigma_p(t))^2 = \left(\frac{\hbar}{\sigma_0}\right)^2 \cos^2 \omega t + (\sigma_0 m\omega)^2 \sin^2 \omega t\,. \qquad (10.37d)$$

The mean position of the wave packet is $x = x_0 \cos \omega t$. $|\psi(x,t)|^2$ oscillates harmonically at the angular frequency 2ω with time. The wave packet is thus *nonspreading* and this is a special feature of the harmonic oscillator. Figure 10.4 depicts $|\psi|^2$ as a function of time. The expectation value of the wave packet satisfies the classical equation of motion, namely, $\langle x(t)\rangle = x_0 \cos \omega t$ and $\langle p_x(t)\rangle = -m\omega x_0 - \sin \omega t$. The position and momentum uncertainties are given by $\Delta x = \sigma_x(t)/\sqrt{2}$ and $\Delta p_x = \sigma_p(t)/\sqrt{2}$. These uncertainties oscillate with a period twice that of the classical motion regaining the initial values $\Delta x(0)$ and $\Delta p_x(0)$ at two opposing points in phase space except when $\sigma_0^2\hbar/(m\omega) = 1$. When $\sigma^2\hbar = m\omega$ the widths $\sigma_x(t)$ and $\sigma_p(t)$ are independent of time: $\sigma_x(t) = \sigma_0$, $\sigma_p(t) = \hbar/\sigma_0$. Further, $\Delta x \Delta p_x = \hbar/2$, the minimum uncertainty product. Such states are called *coherent states*.

Properties of two-dimensional Gaussian wave packets with fixed mean values of angular momentum are studied with respect to two-dimensional harmonic isotropic oscillator [20]. For a discussion on Gaussian wave packet of a particle subjected to a constant force one may refer to ref. [21].

A function oscillating faster than its fastest Fourier component is termed as a *superoscillatory function* [22–24]. The evolution of superoscillatory wavepackets in harmonic oscillator has been reported [25]. For the Gaussian wave packet dynamics of a quantum bouncing ball one may refer to the refs. [19,26].

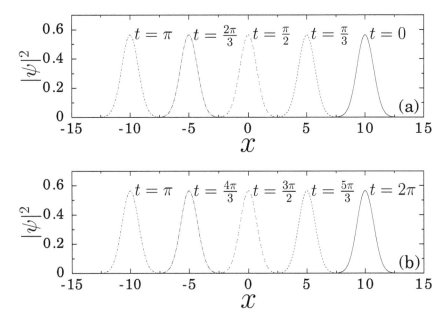

FIGURE 10.4
Time evolution of $|\psi(x,t)|^2$ (Eq. (10.37a)) of a harmonic oscillator. Here $\sigma_x = 1$, $x_0 = 10$ and $\omega = 1$. $|\psi(x,t)|^2$ is periodic with period $2\pi/\omega$.

10.5 Wave Packet Revival

The localized wave packet solution disperses or spreads in time, however, at a later time, it reforms and appears as its initial state. This is called *wave packet revival*. If the probability density $P(x,t) = |\psi(x,t)|^2$ of a nonstationary state satisfies the condition $P(x, t_0) = P(x, t_0 + T)$ for some finite T the wave packet is said to display *exact revivals* at intervals of T.

1. Linear Harmonic Oscillator

Any wave packet $\psi(x,t)$ can be expanded in terms of $\phi_n(x)$, the eigenfunctions of Hamiltonian H, as

$$\psi(x,t) = \sum_n C_n \phi_n(x) \, e^{-iE_n t/\hbar}, \quad C_n = \int_{-\infty}^{\infty} \phi_n^*(x)\psi(x,0)\,\mathrm{d}x. \tag{10.38}$$

For the linear harmonic oscillator, introducing the classical period $T = 2\pi/\omega$ we write the above equation in the form

$$\psi(x,t) = e^{-i\pi t/T} \sum_{n=0}^{\infty} C_n \phi_n(x) \, e^{-i2n\pi t/T}. \tag{10.39}$$

There are two interesting results from Eq. (10.39):

$$\text{(i) } \psi(x, t+T) = -\psi(x,t), \quad \text{(ii) } \psi(x, t = T/2) = -i\psi(-x, 0). \tag{10.40}$$

From (10.40) we have

$$P(x, t+T) = P(x, t), \quad P(x, t = T/2) = P(-x, 0). \tag{10.41}$$

This means that any wave packet $|\psi(x, t)|$ for the linear harmonic oscillator is periodic with period T and that the time T is the shortest interval after which a complete reconstruction of the initial probability density is guaranteed (see Fig. 10.4). Similarly, the time $t = T/2$ is the first instant at which the probability density coincides with the mirror image of the initial probability density (see Fig. 10.4). The wave packet is said to have *exact revival* at $t = T$ and a *mirror revival* at $t = T/2$.

2. Particle in a Box Potential

Referring to sec. 4.6 for a particle in a box potential Eq. (4.83) we can write the eigenfunctions and energy eigenvalues separately for even partity and odd parity. For even parity

$$\phi_{en}(x) = \frac{1}{\sqrt{a}} \cos\left[\frac{(2n-1)\pi x}{2a}\right], \quad E_n = \frac{(2n-1)^2 \pi^2 \hbar^2}{8ma^2}, \quad n = 1, 2, \ldots. \tag{10.42}$$

In the case of odd parity

$$\phi_{on}(x) = \frac{1}{\sqrt{a}} \sin\left[\frac{2n\pi x}{2a}\right], \quad E_n = \frac{(2n)^2 \pi^2 \hbar^2}{8ma^2}, \quad n = 1, 2, \ldots. \tag{10.43}$$

Suppose $\psi(x, 0)$ is a normalized initial nonstationary wave function. $\psi(x, 0)$ can be expanded in terms of the eigenfunctions $\phi_n(x)$ as in Eq. (10.20) and $\psi(x, t)$ is given by Eq. (10.23). For even parity eigenfunctions we write $E_n = (2n-1)^2 2\pi\hbar/T$, where $T = 16ma^2/(\pi\hbar)$. Then

$$\psi(x, t) = \sum C_n \phi_{en}(x) e^{-i(2n-1)^2 2\pi t/T}. \tag{10.44}$$

It is desired to find the lowest $t = t'$ for which $\psi(x, t) \propto \beta\psi(x, 0)$ so that $|\psi(x, t)|^2 = |\psi(x, 0)|^2$ and $\beta^*\beta = 1$. Then exact revivals take place at intervals of t'. As $\psi(x, 0) = \sum C_n \phi_{en}(x)$ we find the lowest value of t for which $e^{-i(2n-1)^2 2\pi t/T} = \beta$ with $\beta^*\beta = 1$. The values of $(2n-1)^2 2\pi t/T$ for $n = 1, 2, 3, \ldots$ are $2\pi t/T$, $18\pi t/T$, $50\pi t/T$, $\ldots$. For $t = T/8$ this sequence becomes $\pi/4$, $2\pi + (\pi/4)$, $6\pi + (\pi/4)$, $\ldots$. Then $\psi(x, t = T/8)$ takes the form

$$\psi(x, t = T/8) = e^{-i\pi/4} \sum C_n \phi_{en}(x) = \frac{1-i}{\sqrt{2}} \psi(x, 0). \tag{10.45}$$

This gives $|\psi(x, T/8)|^2 = |\psi(x, 0)|^2$. The period of the wave packet is $T/8$. For odd parity case the period of the wave packet is $T/4$.

Consider the even parity case. Assume the initial wave packet as the Gaussian form given by Eq. (10.36) with $k_0 = 0$, $\sigma_0 = 0.1$ and $x_0 = 0$. (Can we choose $x_0 \neq 0$?). In Eq. (10.42) choose $a = 1$ units, $\hbar = 1$ units and $m = 1$ units. In Eq. (10.38) the range of the integration is $[-1, 1]$ and it is difficult to obtain an analytical expression for C_n. The integral in the expression for C_n can be numerically computed using, for example, composite trapezoidal rule. Let us calculate C_n and $\psi(x, t)$ numerically. Figure 10.5a shows the variation of numerically computed C_n and $\sum |C_n|^2$ with n. $C_n \approx 0$ for $n > 10$. The quantity $\sum |C_n|^2$ approaches 1 with increase in n. The values of T and $T/8$ are 5.09296 and 0.63662, respectively. Figure 10.5b presents the time evolution of the initial Gaussian wave packet. The shape of $|\psi(x, t)|^2$ varies with time, however, $|\psi(x, t)|^2 = |\psi(x, 0)|^2$ at time values integral multiples of $T/8$ as predicted theoretically. The wave packet revival is

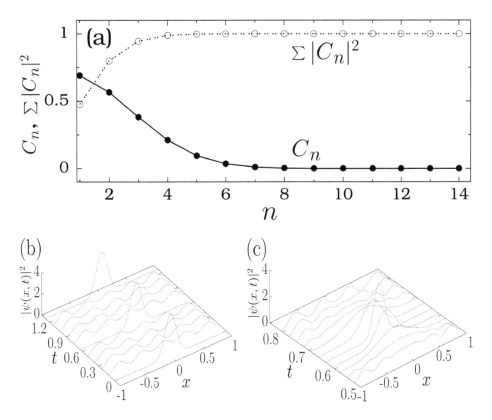

FIGURE 10.5
(a) C_n and $\sum |C_n|^2$ versus n for the even parity case of a particle in a box potential with $\psi(x,0)$ given by Eq. (10.36) with $k_0 = 0$, $\sigma_0 = 0.1$ and $x_0 = 0$. (b) Time evolution of $|\psi(x,0)|^2$ displaying wave packet revivals for a particle in a box potential with even parity case. (c) $|\psi(x,t)|^2$ versus x and t around the time $t = T/8$. For details see the text.

clearly seen in Fig. 10.5b. For clarity in Fig. 10.5c the variation of $|\psi(x,t)|^2$ is shown for a few values of t around $T/8$. The time evolution of initially Gaussian form wave packet is found to be nontrivial.

Compare $|\psi(x,t)|^2$ of a particle in a box with the $|\psi(x,t)|^2$ of linear harmonic osicallator shown in Fig. 10.4. In the case of harmonic oscillator, the shape of $|\psi(x,t)|^2$ remains the same with t but that of a particle in a box potential changes with t, however, retains its initial shape at the integral multiples of $T/8$.

For the case of odd parity case choose $\psi(x,0)$ as an admissible odd function of x (why?). Let us choose

$$\psi(x,0) = N x e^{-\alpha x^2}. \tag{10.46}$$

For $\alpha = 100$ numerically computed $N = 56.4937$. Figure 10.6b shows the normalized $\psi(x,0)$ for $\alpha = 100$. Figure 10.6a shows the variation of C_n and $\sum |C_n|^2$ with n for $a = 1$ units, $\hbar = 1$ units and $m = 1$ units. The values of T and $T/4$ are 5.09296 and 1.27324, respectively. Wave packet revival is clearly evident in Fig. 10.6c.

Revivals in quantum systems were first studied in the Jaynes–Cumming model, describing a two-level atom interacting with a resonant monochromatic field [27,28]. Revivals have

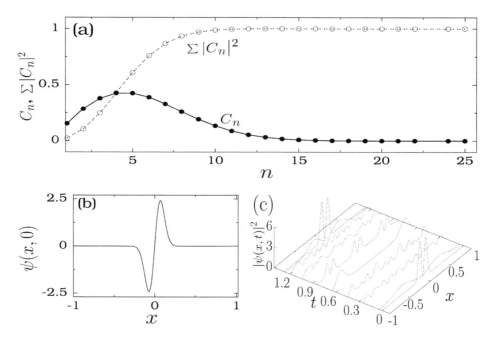

FIGURE 10.6
(a) C_n and $\sum |C_n|^2$ versus n for the odd parity case of a particle in a box potential with $\psi(x,0)$ given by Eq. (10.46). (b) Plot of $\psi(x,0)$ given by Eq. (10.46). (c) $|\psi(x,t)|^2$ versus x and t. For details see the text.

been observed experimentally in a micromaser cavity with rubidium atom [29], in Rydberg electron wave packet [30] and vibrational wave packets in Na$_2$ [31]. The occurrence of wave packet revivals has been studied in systems such as rigid rotator, weakly anharmonic potentials, infinite square-well potential, circular billiards etc. [32–37]. An exact *fractional wave function revival* is said to occur when the wave function of a particle become a superposition of translated initial wave functions. Fractional revivals have been observed in many quantum systems such as Jaynes–Cummings model, Morse-like anharmonic potentials, Na$_2$ etc. [38–45].

10.6   Almost Periodic Wave Packets

As seen earlier the Gaussian wave packet of a free particle, spreads in time while that of the linear harmonic oscillator oscillates with time. Wave packet revival also occurs in certain systems. There are a number of systems for which the time-dependent wave function does not spread in time. The occurrence of nonspreading wave function is found in certain one-dimensional system having discrete energy eigenvalues E_n [45]. The wave function of such systems exhibit almost periodic behaviour in time t. For a periodic function $f(t)$ with period τ we write $f(t+\tau) = f(t)$. By almost periodic function we mean that there is no values of τ strictly satisfying $f(t+\tau) = f(t)$, however, there are infinite number of τ_n, $n = 1, 2, \ldots$, so that $f(t+\tau_n) \approx f(t)$ is satisfied.

Suppose for a particle in a one-dimensional confined potential the stationary state discrete nondegenerate eigenvalues and the eigenfunctions are E_n and ϕ_n, respectively. Then the general time-dependent wave function is written as in Eq. (10.23) with C_n given by Eq. (10.22). It is easy to show that the wave function given by Eq. (10.23) is almost periodic in t [45]. Let the allowed tolerance in f is ϵ, that is $|f(t + \tau_n) - f(t)| < \epsilon$. Almost periodic functions is viewed as a generalization of periodic functions since any almost periodic function is written as $f(t) = \sum_n A_n e^{i\omega_n t}$, where A_n are real or complex while ω_n are real. For a strictly periodic function $\omega_n = n\omega_1$. According to the Riesz–Fischer theorem [46] a trigonometric series is the Fourier series of an almost periodic function as long as $\sum_n |A_n|^2$ converges. Applying this theorem to Eq. (10.23) we say that $\psi(x, t)$ is almost periodic in t provided $\sum |C_n|^2 |\phi_n|^2$ converges. We require

$$\sum_n |C_n|^2 |\phi_n(x)|^2 \le |\phi_m(x)|^2 \sum |C_n|^2 = |\phi_m(x)|^2 , \qquad (10.47)$$

where $|\phi_m(x)|$ is the largest of all of the $\phi_n(x)$ for the particular value of x. Hence, for a discrete energy eigenvalue spectrum $\psi(x, t)$ is an almost periodic function of time t.

For a linear harmonic oscillator $E_n = (n + 1/2)\hbar\omega$. Writing $\Psi(x, t) = e^{i\omega t/2} \psi(x, t)$ Eq. (10.23) becomes

$$\Psi(x, t) = \sum_{n=0}^{\infty} C_n e^{-in\omega t} \phi_n(x) \qquad (10.48)$$

which is a Fourier series in t and is a periodic function of t. The strictly periodic behaviour of Ψ is due to the fact that E_n are equally spaced. For a discussion on fluid-like properties of probability density, interpretation of them for the one-dimensional Gaussian wave packet, stationary states of a particle in a box and hydrogen atom see refs. [47,48].

10.7 Concluding Remarks

Suppose one shines a femtosecond optical pulse on a diatomic molecule that has enough broad bandwidth to cover several vibrational levels in the excited state. Then it is possible to create a nuclear wave packet that is spatially localized and no longer stationary. The wave packet thus created is a linear combination of several eigenstates.

Laser pump-probe techniques in the femtosecond time domain have been applied to temporarily resolve the motion of vibrational wave packet molecules and study the femtochemistry of the transition state of several gas-phase reactions. It has been demonstrated that a wave packet, the coherent superposition of a set of vibrational eigenstates produced by an optical pulse the duration of which is short compared to the vibrational period, can be observed by detecting the molecular species itself as a dissociation product by fluorescence spectroscopy [49]. Since the temporal history of the wave packet bears the signature of the potential well in which it moves, the study on wave packet dynamics has been used to find the potential energy surface of diatomic molecules. All spectral information can be recovered by studying the temporal evolution of a wave packet.

Technological advances led to a significant amount of research being focused at investigating the time-dependent behaviour of electron wave packets. The ultimate theme is to generate wave packets that control electronic process in atoms and molecules. Several control schemes exist which use light to drive atoms or molecules along prescribed pathways. They exploit the coherence of laser light to manipulate the quantum phase relationships between various eigenstates of the system [50].

10.8 Bibliography

[1]   A. Goldberg, H.M. Schey and J.L. Schwartz, *Am. J. Phys.* 35:177, 1967.

[2]   M.H. Bramhall and B.M. Casper, *Am. J. Phys.* 38:1136, 1970.

[3]   B. Diu, *Eur. J. Phys.* 1:231, 1980.

[4]   A. Edgar, *Am. J. Phys.* 63:136, 1995.

[5]   J.S. Boleman and S.B. Haley, *Am. J. Phys.* 43:270, 1975.

[6]   J.V. Greenman, *Am. J. Phys.* 40:1193, 1972.

[7]   C.U. Segre and J.D. Sullivan, *Am. J. Phys.* 44:729, 1976.

[8]   P.A. Deutchman, *Am. J. Phys.* 39:952, 1971.

[9]   E.A. Johnson and H.T. Williams, *Am. J. Phys.* 50:239, 1981.

[10] J.C. Hamilton, J.L. Schwartz and W.A. Bowers, *Am. J. Phys.* 40:1657, 1972.

[11] G. Friedman, W.A. Little, *Am. J. Phys.* 61:835, 1993.

[12] G.W. Ford and R.F. O'Connell, *Am. J. Phys.* 70:319, 2002.

[13] D.F. Styer, *Am. J. Phys.* 58:742, 1990.

[14] L.S. Brown, *Am. J. Phys.* 41:525, 1972.

[15] M. Nauenberg, *Phys. Rev. A* 40:1133, 1989.

[16] M. Nauenberg, C. Stroud and J. Yeazell, *Sci. Am.* 270:44, 1994.

[17] G. Alber and P. Zoller, *Phys. Rep.* 199:231, 1991.

[18] S. Brandt and H.D. Dahmen, *Quantum Mechanics on the Personal Computer.* Springer, New York, 1992.

[19] M.A. Doncheski and R.W. Robinett, *Am. J. Phys.* 69:1084, 2001.

[20] V.V. Dodonov, *J. Phys. A: Math. Theor.* 48:435303, 2015.

[21] R.W. Robinett, *Quantum Mechanics: Classical Results, Modern Systems and Visualized Examples.* Oxford Univ. Press, New York, 1997, pp.102-103.

[22] M.V. Berry, *Waves near zeros.* In Coherence and Quantum Optics. Eds.: N.P. Bigelow, J.H. Eberly and C.R.J. Stroud. Optical Society of America, Washington, DC, 2008. pp.37-41.

[23] M.V. Berry, *J. Opt.* 15:044006, 2013.

[24] M.V. Berry and S. Popescu, *J. Phys. A: Math. Gen.* 39:6965, 2006.

[25] M. Bussell and P. Strange, *Eur. J. Phys.* 36:065028, 2016.

[26] J. Gea-Banacloche, *Am. J. Phys.* 67:776, 1999.

[27] J.H. Eberly, N.B. Narozhny and J.J.S. Mondragon, *Phys. Rev. Lett.* 44:1323, 1980.

[28] N.B. Narozhny, J.J.S. Mondragon and J.H. Eberly, *Phys. Rev. A* 23:236, 1981.

[29] G. Rampe, H. Walther and N. Klein, *Phys. Rev. Lett.* 58:353, 1987.

[30] J.A. Yeazell, M. Mallalieu and C.R. Stroud Jr, *Phys. Rev. Lett.* 64:2007, 1990.

[31] T. Baumert, V. Engel, C. Rottgermann, W.T. Strunz and G. Gerber, *Chem. Phys. Lett.* 191:639, 1992.

[32] R.W. Robinett and S. Heppelmann, *Phys. Rev. A*. 65:062103, 2002.

[33] R. Bluhm, V.A. Kostelecky and J.A. Porter, *Am. J. Phys.* 64:944, 1996.

[34] D.L. Aronstein and C.R. Stroud Jr, *Phys. Rev. A* 55:4526, 1997.

[35] R.W. Robinett, *Am. J. Phys.* 68:410, 2000.

[36] P. Stifter, E.W. Lamb Jr. and W.P. Schleich, *The particle in a box revisited*. In Frontiers of Quantum Optics and Laser Physics: Proceedings of the International Conference on Quantum Optics and Laser Physics. Eds.: S.Y. Zhu, M.S. Zubairy and M.O. Scully. Springer, Singapore, 1997. pp.236-246.

[37] K. Razi, Naqvi and S. Waldenstrom, *Physica Scripta* 62:12, 2000.

[38] B.M. Garraway and K.A. Suominien, *Contemporary Physics* 43:97, 2002.

[39] J. Parker and C.R. Stroud Jr, *Phys. Rev. Lett.* 56:716, 1986.

[40] Z.D. Gaeta and C.R. Stroud Jr, *Phys. Rev. A* 42:6308, 1990.

[41] I.Sh. Averbukh, *Phys. Rev. A* 46:R2205, 1992.

[42] S.I. Vetchinkin and V.V. Eryomin, *Chem. Phys. Lett.* 222:394, 1994.

[43] W.Y. Chen and G.J. Milburn, *Phys. Rev. A* 51:2328, 1995.

[44] D.R. Meacher, P.E. Meyler, I.G. Hughes and P. Ewart, *J. Phys. B* 24:L63, 1991.

[45] M.J.J. Vrakking, D.M. Villeneuve and A. Stolow, *Phys. Rev. A* 54:R37, 1996.

[46] D. Mentrup and M. Luban, *Am. J. Phys.* 71:580, 2003.

[47] A.S. Besicovitch, *Almost Periodic Functions*. Cambridge Univ. Press, Cambridge, 1932.

[48] K. Mitra, *Am. J. Phys.* 71:894, 2003.

[49] M. Dantus, M.H.M. Janssen and A.H. Zewail, *Chem. Phys. Lett.* 181:281, 1991.

[50] J.R.R. Verlet and H.H. Fielding, *Int. Rev. Phys. Chem.* 20:283, 2001.

[51] F. Amaku, F.A.B. Coutinho and F.M. Toyama, *Am. J. Phys.* 85:692, 2017

10.9   Exercises

10.1 Show that the wave packet of electromagnetic waves in free space travels undistorted with speed c.

10.2 What is the relation between the phase velocity v_p of a free particle and the classical free particle velocity v.

10.3 Is there dispersion of wave packet in vacuum? Why?

10.4 For the Gaussian wave packet $\psi(x,0) = \left(\dfrac{1}{\sqrt{2\pi}\,\sigma}\right)^{1/2} e^{-(x-x_0)^2/(4\sigma^2)}$ compute $\Delta x \Delta p_x$.

10.5 Consider an initial wave packet of the form $\psi(x,0) = W(x)e^{i\phi(x)}$, where W and ϕ are real. Find $\langle p_x \rangle$ and $\langle p_x^2 \rangle$ for this wave packet.

10.6 Determine the Gaussian wave packet for a quantum particle undergoing uniform acceleration.

10.7 For the initial wave packet $\psi(x,0) = (1/2\pi)^{1/4}e^{-x^2}e^{i\phi(x)}$ of a free particle of unit mass find $\langle p_x \rangle$, $\langle p_x^2 \rangle$ and $\psi(x,t)$ for $\phi(x) = x^2$.

10.8 Given that the wave packet of a free particle at $t = 0$ as $\psi(x,0) = e^{-a^2/(a^2-x^2)}$ for $|x| < a$ and 0 for $|x| \geq a$ and the time evolution operator as $U(t) = e^{-iHt/\hbar} = \lim_{n\to\infty}\left(1 + \frac{it}{\hbar n}H\right)^{-n}$ show that [51] $\psi(x,t) = \frac{1}{\sqrt{2\pi}}\int_{-\infty}^{\infty}e^{-i\hbar k^2 t/(2m)}e^{ikx}dk$.

10.9 Find the probablity current density associated with the Gaussian wave packet
$$\psi(x,0) = \left(\frac{1}{\sqrt{2\pi}\sigma}\right)^{1/2}e^{-(x-x_0)^2/(4\sigma^2)}e^{ik_0 x} \text{ at } t = 0.$$

10.10 Consider a particle bouncing on a perfectly reflecting surface under the influence of gravity with $V(x) = mgx$ for $x > 0$ and ∞ for $x < 0$. Write the solution of the classical equation of motion of the particle, determine energy eigenfunctions for the quantum case and study the time evolution of a Gaussian wave packet.

10.11 Show that an odd-parity wave packet of a particle in a box potential manifests an exact revival at $t = T/4$, where $T = 16ma^2/(\pi\hbar)$.

10.12 Assume that the initial wave packet of a particle in a box potential $V(x) = 0$ for $|x| \leq a$ and ∞ for $|x| > a$ is $\psi(x,0) = \left(\frac{1}{\sqrt{2\pi}\sigma}\right)^{1/2}e^{-x^2/(4\sigma^2)}$. $\psi(x,t)$ is $\sum_n C_n\phi_n(x)e^{-iE_n t/\hbar}$. For the case of even parity eigenfunctions develop a program to numerically compute C_n and $\psi(x,t)$.

10.13 For a particle in a box potential $V(x) = 0$ for $|x| \leq a$ and ∞ for $|x| > a$ assume $\psi(x,0) = Nxe^{-\alpha x^2}$, $\alpha = 100$ and $\psi(x,t) = \sum C_n\phi_n(x)e^{-iE_n t/\hbar}$, where ϕ_n's are odd parity eigenfunctions. Develop a programme to compute N, C_n and $\psi(x,t)$.

10.14 For a particle in a box potential $V(x) = 0$ for $0 \leq x \leq a$ and ∞ otherwise $\phi_n(x) = \sqrt{2/a}\sin(n\pi x/a)$ and $E_n = n^2\pi^2\hbar^2/(2ma^2)$.

(a) Assume that $\psi(x,0) = Nx(a-x)$ for $0 \leq x \leq a$ and 0 otherwise. Expand $\psi(x,0)$ in terms of ϕ_n and determine the period of $|\psi(x,t)|^2$. Plot C_n versus n and $|\psi(x,t)|^2$ versus x and t.

(b) Assume that $\psi(x,0)$ is such that $C_n \neq 0$ for first some odd and even integer values of n and $\sum |C_n|^2 = 1$. Determine the period of $\psi(x,t)$. Numerically determine N, C_n and $|\psi(x,t)|^2$ for $\psi(x,0) = Ne^{-\alpha(x-x_0)^2}$ with $\alpha = 200$, $x_0 = 0.4$ and $a = 1$ units.

10.15 For a one-dimensional infinite square-well system, with walls at $x = 0$, L a wave packet can be constructed as $\psi(x,t) = \sum_{n=1}^{\infty} a_n u_n(x)e^{-iE_n t/\hbar}$, where $u_n = \sqrt{2/L}\sin(n\pi x/L)$, $E_n = n^2\hbar^2\pi^2/(2mL^2)$. Show that
$$\left|\psi\left(x, t + \frac{T}{2}\right)\right|^2 = |\psi(L-x,t)|^2 \text{ and } \left|\phi\left(p, t + \frac{T}{2}\right)\right|^2 = |\phi(-p,t)|^2$$
where $T = 4mL^2/(\hbar\pi)$ is the wave packet revival time. (This phenomenon is called mirror revival. At half the revival time, any initial wave packet will reform itself at a location mirrored about the centre of the well and its initial momentum profile is also reproduced with the sign of p changing to $-p$.)

11

Theory of Angular Momentum

11.1 Introduction

A particle moving along a straight-line has a linear momentum $m\mathbf{v}$. A revolving particle has angular momentum due to its mass and angular velocity. In classical mechanics, orbital angular momentum is the vector product of position ($\mathbf{r}$) and linear momentum ($\mathbf{p}$):

$$\mathbf{L} = \mathbf{r} \times \mathbf{p}. \tag{11.1}$$

Let us imagine the electron as a small ball, revolving about some axis. It possesses angular momentum. Precisely, the angular momentum vector is a vector pointing along the axis of rotation with magnitude proportional to the speed of rotation. A vector which changes sign under space inversion (parity operation) is called a *polar vector*. $\mathbf{r}$ and $\mathbf{p}$ are polar vectors as they change sign. Since $\mathbf{L} = (-\mathbf{r}) \times (-\mathbf{p})$, $\mathbf{L}$ does not change sign under parity operation. Such vectors which do not change sign under parity operation are called *axial vectors* or *psuedo-vectors*. Classically, the angular momentum of the electron could have any magnitude. In quantum mechanics the orbital angular momentum operator is obtained by replacing $\mathbf{p}$ by $-i\hbar\nabla$. The angular position of a particle is the angle at which it is located in two-dimensional polar coordinates.

Why should one study about angular momentum? Let us point out some significances of angular momentum. Comet tails always point away from the sun because light carries linear momentum. In 1921 Einstein has shown that Planck's black body law and the motion of molecules in a radiation fields could be explained if the linear momentum of a photon is $\hbar k$. In 1909 John Poynting noticed that polarized light has angular momentum – spin angular momentum associated with circular polarization. For a single photon it has a value of $\pm\hbar$. Gibson et al. [1] pioneered the use of orbital angular momentum to encode information in free-space optical communication. Just like an intensity that is switched on or off can represent a value of zero or one (a bit), so different phase-front twists can represent different values. There are infinitely many different twists, and as a single photon can have these twists, photons can carry arbitrary amounts of information. Orbital angular momentum of light beams can be used for super-high density optical data storage [2].

Angular momentum has practical applications in the study of atomic transitions, wireless communication [3,4], quantum experiments [5], electromagnetic vortex image processing [6], nanotechnology [7], radio and congestion [8], holography [9] a few to mention. It has been realized that angular momentum could be carried by light beams with an azimuthal phase dependence [10]. Such light beams are realizable.

For a particle rotating in a two-dimensional plane, we can choose the z-axis as the axis of rotation if the plane of rotation is the xy-plane. In this case $L_x = L_y = 0$ and $L_z = L$. For a central force problem, the force $\mathbf{F}$ acts in the direction of $\mathbf{r}$. Hence, the torque is $\mathbf{N} = d\mathbf{L}/dt = 0$. Therefore, $\mathbf{L}$ is conserved in all central force problems. Since the potential is a spherically symmetric for the central force problem $V(r)$ depends only on the magnitude of $\mathbf{r}$. Because $\mathbf{L}$ is conserved classically for a central force, the motion takes place in a plane

DOI: 10.1201/9781003172178-11

with $\mathbf{L}$ perpendicular to the plane. When the particle rotates in a three-dimensional space, the axis of rotation is perpendicular to the plane of rotation. However, the axis is free to take any direction and $L^2 = L_x^2 + L_y^2 + L_z^2$.

11.2 Scalar Wave Function Under Rotations

A rotation of $\mathbf{r}$ by an angle θ about an axis of unit vector $\mathbf{n}$ changes $\mathbf{r}$ to $\mathbf{r}'$. If the rotation operation in physical space is represented by $R(\mathbf{n}, \theta)$, then $\mathbf{r}' = R(\mathbf{n}, \theta)\mathbf{r}$. For example, if the z-axis is chosen as the rotation axis then we find

$$R(\mathbf{k}, \theta) = \begin{pmatrix} \cos\theta & -\sin\theta & 0 \\ \sin\theta & \cos\theta & 0 \\ 0 & 0 & 1 \end{pmatrix}. \tag{11.2}$$

Since rotation does not change the magnitude of the vector we have $\mathbf{r}' \cdot \mathbf{r}' = \mathbf{r} \cdot \mathbf{r}$. Hence, $RR^T = R^T R = I$. The rotation matrix R is orthogonal. $R(\mathbf{n}, \theta)$ gives the transformation matrix in physical space. We need to find the effect of such rotation in a wave function. As the wave function represents a vector in Hilbert space, the rotational transformation in physical space will lead to a unitary transformation $U_R(\mathbf{n}, \theta)$ in Hilbert space.

11.2.1 The Transformation Operator

Under a rotation through θ about $\mathbf{n}$, $\psi(\mathbf{r}) \to \psi'(\mathbf{r}') = \psi'(x', y', z')$, where ψ' is the rotated function. We know that a rotated scalar function evaluated at the rotated point must be the same as the unrotated scalar function evaluated at the original point. In view of this because ψ is a scalar we have $\psi'(\mathbf{r}') = \psi(\mathbf{r})$. The value of $\psi(\mathbf{r})$ cannot change under rotation of the coordinate system even though our point of viewing is changed. From $\mathbf{r}' = R(\mathbf{n}, \theta)\mathbf{r}$ we write $\mathbf{r} = R(\mathbf{n}, -\theta)\mathbf{r}'$ by which $\psi(\mathbf{r}') = \psi(R(\mathbf{n}, -\theta)\mathbf{r}')$. Dropping the prime on $\mathbf{r}'$, we get $\psi'(\mathbf{r}) = \psi(R(\mathbf{n}, -\theta)\mathbf{r})$. That is, the rotated wave function evaluated at $\mathbf{r}$ is the same as the unrotated function evaluated at the point $R(\mathbf{n}, -\theta)\mathbf{r}$.

If the transformation of $\psi'(\mathbf{r})$ and $\psi(\mathbf{r})$ is given by the operator $U_R(\mathbf{n}, \theta)$ then

$$U_R \psi(\mathbf{r}) = \psi'(\mathbf{r}) = \psi(R(\mathbf{n}, -\theta)\mathbf{r}). \tag{11.3}$$

An expression for $U_R(\mathbf{n}, \theta)$ is obtained by choosing $\mathbf{n}$ along the z-axis and letting θ to be a first-order infinitesimal rotation angle $\Delta\theta$. With the choice $\mathbf{n} = \mathbf{k}$ and $\theta \to \Delta\theta_z$, we obtain from Eq. (11.2)

$$R(\mathbf{k}, -\Delta\theta_z)\mathbf{r} \approx \begin{pmatrix} x + \Delta\theta_z y \\ y - \Delta\theta_z x \\ z \end{pmatrix}. \tag{11.4}$$

Hence, Eq. (11.3) becomes

$$U_R(\mathbf{k}, \Delta\theta_z)\psi(\mathbf{r}) \approx \psi(x + \Delta\theta_z y, y - \Delta\theta_z x, z). \tag{11.5}$$

Using the Taylor series expansion

$$\psi(x + \Delta\theta_z y, y - \Delta\theta_z x, z) \approx \left[1 - \Delta\theta_z \left(x\frac{\partial}{\partial y} - y\frac{\partial}{\partial x}\right)\right]\psi(\mathbf{r}) \tag{11.6}$$

we get

$$U_R(\mathbf{k}, \Delta\theta_z) \approx I - \Delta\theta_z \left(x\frac{\partial}{\partial y} - y\frac{\partial}{\partial x} \right) = I - \frac{\mathrm{i}}{\hbar}\Delta\theta_z \mathbf{k} \cdot \mathbf{L}. \tag{11.7}$$

For a spinless particle, the infinitesimal generators of rotations about the z-axis is simply L_z. Generalizing Eq. (11.7) for an infinitesimal rotation $\Delta\theta$ in the direction $\mathbf{n}$, we get

$$U_R(\mathbf{n}, \Delta\theta) \approx I - \frac{\mathrm{i}}{\hbar}\Delta\theta\, \mathbf{n} \cdot \mathbf{L}. \tag{11.8}$$

Suppose it is desired to find $U_R(\mathbf{n}, \theta)$ for some finite rotation θ. The effect of a small increase in the magnitude of θ from θ to $\theta + \Delta\theta$ is to follow the finite rotation $U_R(\mathbf{n}, \theta)$ by the infinitesimal rotation $I - (\mathrm{i}/\hbar)\Delta\theta\mathbf{n} \cdot \mathbf{L}$ to give

$$U_R(\mathbf{n}, \theta + \Delta\theta) \approx \left(I - \frac{\mathrm{i}}{\hbar}\Delta\theta\, \mathbf{n} \cdot \mathbf{L} \right) U_R(\mathbf{n}, \theta). \tag{11.9}$$

Then $U_R(\mathbf{n}, \theta)$ satisfies the differential equation

$$\frac{\mathrm{d}}{\mathrm{d}\theta}U_R(\mathbf{n}, \theta) = -\frac{\mathrm{i}}{\hbar}\Delta\theta\, \mathbf{n} \cdot \mathbf{L}\, U_R(\mathbf{n}, \theta) \tag{11.10}$$

together with the boundary condition $U_R(0) = I$. Its solution is

$$U_R(\mathbf{n}, \theta) = \mathrm{e}^{-\mathrm{i}\theta\mathbf{n}\cdot\mathbf{L}/\hbar}, \quad \psi'(\mathbf{r}) = \mathrm{e}^{-\mathrm{i}\theta\mathbf{n}\cdot\mathbf{L}/\hbar}\psi(\mathbf{r}). \tag{11.11}$$

11.2.2   Symmetry Groups

In classical mechanics the symmetry of a physical system leads to conservation laws. If the space is isotropic, it possesses rotational symmetry. Consequently, the angular momentum for such a system is conserved. Similarly, the homogeneity of space leads to translational symmetry and it conserves the linear momentum of the system. Invariance of a system under some symmetry operations results in some conservation laws and consequently we have a vast simplification of the mathematical theories of such systems. Associated with every symmetry is a mathematical object known as a *symmetry group* and the set of all the symmetry operations of a system satisfies the postulates of a group[1] [11].

11.2.3   Lie Groups

The number of elements in a group is called its *order*. A group with finite number of elements is called a *finite group* whereas an *infinite group* contains infinite number of elements. An infinite group may further be either discrete or continuous. For example, a group of all integers including zero form a discrete infinite group under algebraic addition. If we consider a group of all real numbers $x \in [-\infty, \infty]$, they form an infinite continuous group under algebraic addition.

The elements of a continuous group is characterized by a set of real, continuous, differentiable parameters $(\alpha_1, \alpha_2, \dots, \alpha_r)$. If r is finite then the group is said to be *finite* and r is called the *order* of the continuous group. For example, the two-dimensional rotation matrix

$$R(\mathbf{k}, \theta) = \begin{pmatrix} \cos\theta & -\sin\theta \\ \sin\theta & \cos\theta \end{pmatrix} \tag{11.12}$$

---

[1]Group theory is the mathematical tool to treat invariants and symmetries.

forms a one-parameter infinite continuous group as θ can vary continuously between 0 to 2π. A continuously connected group in which the parameters of the product of two elements are continuous and differentiable functions of the parameters of the elements is called a *Lie group*.

The dependence of the elements X_1, X_2, ... of a Lie group on its r continuous parameters is written as

$$X_i = X_i\left(\alpha_1, \alpha_2, \ldots, \alpha_r\right), \quad i = 1, 2, \ldots . \tag{11.13}$$

It is convenient to choose the Cartesian parameters of a Lie group such that the image of the identity element is $I = X(0, 0, \ldots, 0)$. With this parametrization an element near the identity is written, due to the analytical properties of the Lie groups as

$$X(0, 0, \ldots, \epsilon_j, \ldots, 0) \approx X(0, 0, \ldots, 0) + \mathrm{i}\epsilon_j T_j(0, 0, \ldots, 0) \tag{11.14}$$

to first-order in ϵ_j. The operator T_j obtained from (11.14) is given by

$$T_j = \lim_{\epsilon_j \to 0} \frac{1}{\mathrm{i}\epsilon_j} \left[X(0, \ldots, \epsilon_j, \ldots, 0) - X(0, \ldots, 0)\right] = \frac{1}{\mathrm{i}}\frac{\partial X}{\partial \epsilon_j}. \tag{11.15}$$

T_j are called the *generators* of the group. If we write $\alpha_j = N\epsilon_j$, where N is a large positive integer so that ϵ_j is a small quantity, then the element for any finite variation of the parameter α_j is obtained from (11.13) as

$$
\begin{aligned}
X\left(0, 0, \ldots, \alpha_j, \ldots, 0\right) &= \lim_{N \to \infty} \left[X(0, 0, \ldots, \epsilon_j, 0, \ldots, 0)\right]^N \\
&= \lim_{N \to \infty} \left[1 + \mathrm{i}\frac{\alpha_j T_j}{N}\right]^N \\
&= \mathrm{e}^{\mathrm{i}\alpha_j T_j}.
\end{aligned} \tag{11.16}
$$

We extend the above result to obtain

$$X\left(\alpha_1, \alpha_2, \ldots, \alpha_r\right) = \exp\left[\sum_{j=1}^{r} \mathrm{i}\alpha_j T_j\right]. \tag{11.17}$$

If T_j are known then all the elements of the Lie group corresponding to various α_j values are found using Eq. (11.17). Hence, T_j are called the *generators* of the Lie group. We can prove that these generators of a group are linearly independent. Hence, they are not unique for a group. Any set of r-linear combination of these generators T_j can also be used as generators of the group. Since $\mathrm{e}^{\mathrm{i}\alpha_k T_k}$ and $\mathrm{e}^{\mathrm{i}\alpha_l T_l}$ are two elements of the group, their product must also belong to the group. Therefore, the multiplication of two group elements will be given by the Baker–Campbell–Hausdorff formula [12]

$$\mathrm{e}^A \mathrm{e}^B = \exp\left\{A + B + \frac{1}{2}[A, B] + \frac{1}{12}\left([A, [A, B]] - [B, [A, B]]\right) + \ldots\right\}. \tag{11.18}$$

Therefore, it is clear that once the commutators of the generators are known, it is possible to evaluate the multiplication of any pair of group elements and the commutators must be a linear combination of the generators T_j. We write $[T_k, T_l] = \sum_j f_{klj} T_j$, where f_{klj} are called the *structure constants* of the group. Since $[T_k, T_l] = -[T_l, T_k]$ we set $f_{klj} = -f_{lkj}$. The group elements follow associative law of combination and the generators satisfy the Jacobi identity

$$[T_k, [T_l, T_m]] + [T_l, [T_m, T_k]] + [T_m, [T_k, T_l]] = 0. \tag{11.19}$$

Moreover, we can prove that if the generators are Hermitian then the structure constants are purely imaginary.

11.3   Orbital Angular Momentum

In three-dimension with $\mathbf{r} = \mathbf{i}x + \mathbf{j}y + \mathbf{k}z$ and $\mathbf{p} = \mathbf{i}p_x + \mathbf{j}p_y + \mathbf{k}p_z$ we write $\mathbf{L} = \mathbf{r} \times \mathbf{p}$ as $\mathbf{L} = \mathbf{i}L_x + \mathbf{j}L_y + \mathbf{k}L_z$, where

$$L_x = yp_z - zp_y, \quad L_y = zp_x - xp_z, \quad L_z = xp_y - yp_x . \qquad (11.20)$$

The above are the classical components of $\mathbf{L}$. The quantum mechanical operators L_x, L_y and L_z are obtained from Eq. (11.20) by replacing the variables by their corresponding operators. We obtain

$$L_x = -\mathrm{i}\hbar \left(y\frac{\partial}{\partial z} - z\frac{\partial}{\partial y} \right) , \qquad (11.21a)$$

$$L_y = -\mathrm{i}\hbar \left(z\frac{\partial}{\partial x} - x\frac{\partial}{\partial z} \right) , \qquad (11.21b)$$

$$L_z = -\mathrm{i}\hbar \left(x\frac{\partial}{\partial y} - y\frac{\partial}{\partial x} \right) . \qquad (11.21c)$$

In general

$$\mathbf{L} = -\mathrm{i}\hbar \mathbf{r} \times \nabla . \qquad (11.22)$$

Is the operator $\mathbf{L}$ *a Hermitian?* The operators x, y, z, $-\mathrm{i}\hbar\partial/\partial x$, $-\mathrm{i}\hbar\partial/\partial y$ and $-\mathrm{i}\hbar\partial/\partial z$ are Hermitian. Product of two operators is Hermitian only if their commutator bracket vanishes. Consider the operator L_x. Since y and p_z, z and p_y are commuting operators their product is also a Hermitian. Hence L_x is Hermitian. Similarly, L_y and L_z are also Hermitian. Consequently, $\mathbf{L}$ is a Hermitian operator as well as $\mathbf{L}^2$. Hence, $\mathbf{L}$ and $\mathbf{L}^2$ are observables.

11.3.1   Components of Angular Momentum in Spherical Polar Coordinates

Let us transform Eq. (11.21) to a system of spherical polar coordinates

$$x = r\sin\theta\cos\phi, \quad y = r\sin\theta\sin\phi, \quad z = r\cos\theta , \qquad (11.23)$$

where

$$r^2 = x^2 + y^2 + z^2, \quad rdr = xdx + ydy + zdz \qquad (11.24)$$

and

$$\frac{\partial r}{\partial x} = \frac{x}{r} = \sin\theta\cos\phi , \quad \frac{\partial r}{\partial y} = \frac{y}{r} = \sin\theta\sin\phi , \qquad (11.25a)$$

$$\frac{\partial r}{\partial z} = \frac{z}{r} = \cos\theta , \quad \tan\theta = \frac{\sqrt{x^2 + y^2}}{z} , \qquad (11.25b)$$

$$\frac{\partial\theta}{\partial x} = \frac{\cos\theta\cos\phi}{r} , \quad \frac{\partial\theta}{\partial y} = \frac{\cos\theta\sin\phi}{r} , \quad \frac{\partial\theta}{\partial z} = -\frac{\sin\theta}{r} , \qquad (11.25c)$$

$$\tan\phi = \frac{y}{x} , \quad \frac{\partial\phi}{\partial x} = -\frac{\sin\phi}{r\sin\theta} , \quad \frac{\partial\phi}{\partial y} = \frac{\cos\phi}{r\sin\theta} , \quad \frac{\partial\phi}{\partial z} = 0 . \qquad (11.25d)$$

Then

$$\frac{\partial}{\partial x} = \frac{\partial r}{\partial x}\frac{\partial}{\partial r} + \frac{\partial \theta}{\partial x}\frac{\partial}{\partial \theta} + \frac{\partial \phi}{\partial x}\frac{\partial}{\partial \phi}$$

$$= \sin\theta\cos\phi\frac{\partial}{\partial r} + \frac{\cos\theta\cos\phi}{r}\frac{\partial}{\partial \theta} - \frac{\sin\phi}{r\sin\theta}\frac{\partial}{\partial \phi}, \tag{11.26a}$$

$$\frac{\partial}{\partial y} = \sin\theta\sin\phi\frac{\partial}{\partial r} + \frac{\cos\theta\sin\phi}{r}\frac{\partial}{\partial \theta} + \frac{\cos\phi}{r\sin\theta}\frac{\partial}{\partial \phi}, \tag{11.26b}$$

$$\frac{\partial}{\partial z} = \cos\theta\frac{\partial}{\partial r} - \frac{\sin\theta}{r}\frac{\partial}{\partial \theta}. \tag{11.26c}$$

Substitution of Eqs. (11.23) and (11.26) in Eq. (11.21) gives

$$L_x = -\mathrm{i}\hbar\left[-\sin\phi\frac{\partial}{\partial \theta} - \cot\theta\cos\phi\frac{\partial}{\partial \phi}\right], \tag{11.27a}$$

$$L_y = -\mathrm{i}\hbar\left[\cos\phi\frac{\partial}{\partial \theta} - \cot\theta\sin\phi\frac{\partial}{\partial \phi}\right], \tag{11.27b}$$

$$L_z = -\mathrm{i}\hbar\frac{\partial}{\partial \phi}. \tag{11.27c}$$

Using Eqs. (11.27) $\mathbf{L}^2$ is worked out as

$$\mathbf{L}^2 = -\hbar^2\left[\frac{1}{\sin\theta}\frac{\partial}{\partial \theta}\left(\sin\theta\frac{\partial}{\partial \theta}\right) + \frac{1}{\sin^2\theta}\frac{\partial^2}{\partial \phi^2}\right]. \tag{11.28}$$

Using the Eqs. (11.21) and (11.26) we get two new operators

$$L_\pm = L_x \pm \mathrm{i}L_y = \hbar\,\mathrm{e}^{\pm\mathrm{i}\phi}\left(\pm\frac{\partial}{\partial \theta} + \mathrm{i}\cot\theta\frac{\partial}{\partial \phi}\right). \tag{11.29}$$

The components of $\mathbf{L}$ and $\mathbf{L}^2$ commute with r and $\partial/\partial r$. L_z commutes with any operator and with ∇^2 and with any spherically symmetric Hamiltonian, H, that is $[\mathbf{L}, H] = 0$ and $[\mathbf{L}^2, H] = 0$. This implies that (i) an eigenfunction of a spherically symmetric H is also an eigenfunction of L_x or L_y or L_z or of any linear combination of these operators and (ii) the law of conservation of angular momentum holds in both quantum and classical mechanics.

Usually, only one of the three Cartesian components of angular momentum is specified in a quantum mechanical state. The remaining two are determined from the known third. Generally, the z-component of $\mathbf{L}$ is specified due to the simplicity of its operator form.

11.3.2   Eigenpairs of $\mathbf{L}^2$ and L_z

The eigenvalue equation for $\mathbf{L}^2$ in spherical polar coordinates is

$$\mathbf{L}^2\psi = -\hbar^2\left[\frac{1}{\sin\theta}\frac{\partial}{\partial \theta}\left(\sin\theta\frac{\partial}{\partial \theta}\right) + \frac{1}{\sin^2\theta}\frac{\partial^2}{\partial^2\phi}\right]\psi = \lambda\hbar^2\psi, \tag{11.30}$$

where $\lambda\hbar^2$ is the eigenvalue of $\mathbf{L}^2$. Equation (11.30) is solved by the method of separation of variables. Equation (11.30) is same as the Eq. (4.226) which we considered earlier in the study of rigid rotator. The only difference is that instead of $\hbar^2$ we have $\hbar^2/(2I)$ in the left-side of Eq. (4.226) and instead of $\lambda\hbar^2$ we have E in the right-side of Eq. (4.226). Referring

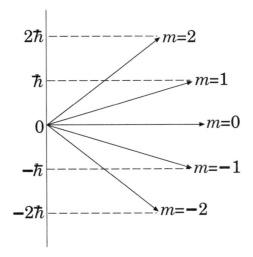

FIGURE 11.1
Orientations of **L** for $l = 2$.

to our discussion on rigid rotator the eigenfunctions ψ_{lm} of $\mathbf{L}^2$ are written as

$$
\begin{aligned}
\psi_{lm} &= \Theta_{lm}(\theta)\Phi_m(\phi) \\
&= \left[\frac{(2l+1)}{4\pi}\frac{(l-|m|)!}{(l+|m|)!}\right]^{1/2} P_l^{|m|}(\cos\theta)e^{im\phi} \\
&= Y_{lm}(\theta,\phi).
\end{aligned}
\tag{11.31}
$$

The eigenvalues of $\mathbf{L}^2$ are $\lambda\hbar^2 = l(l+1)\hbar^2$. As $\mathbf{L}^2$ and L_z are commuting operators they have same set of eigenfunctions. Y_{lm} are the eigenfunctions of $\mathbf{L}^2$. Therefore, Y_{lm} are the eigenfunctions of L_z also. Now, from the eigenvalue equation of L_z, $L_z Y_{lm} = -i\hbar\partial Y_{lm}/\partial\phi$ we obtain $L_z Y_{lm} = m\hbar Y_{lm}$.

To summarize,

$$
\begin{aligned}
\mathbf{L}^2 Y_{lm}(\theta,\phi) &= l(l+1)\hbar^2 Y_{lm}(\theta,\phi), \quad l = 0, 1, 2, \ldots && \text{(11.32a)} \\
L_z Y_{lm}(\theta,\phi) &= m\hbar Y_{lm}(\theta,\phi), \quad m = l, l-1, \ldots, -l. && \text{(11.32b)}
\end{aligned}
$$

The observation is that angular momentum is quantized. The eigenvalues of $L = |\mathbf{L}| = \sqrt{\mathbf{L}^2}$ are discrete, quantized. L can assume only one of the discrete set of values, $\sqrt{l(l+1)}\hbar$, $l = 0, 1, 2, \ldots$. (Here l measures the magnitude of $\mathbf{L}$.) But as noted in sec. 1.4 the Bohr's prediction is $L = n\hbar$. The point is that we have to discard the planetary model of electrons moving about a nucleus. The angle ϕ is a measure of the rotation about the z-axis. The z-component of angular momentum L_z has eigenvalues $m\hbar$. Since m takes only integer values we say that only certain orientations of **L** alone are possible. This is called *space quantization* in vector atom model[2]. Figure 11.1 shows orientation of **L** for $l = 2$.

11.3.3 Properties of Components of L and $\mathbf{L}^2$

In the following we present some of the useful properties of the components of **L** and $\mathbf{L}^2$.

---

[2]When an atom is placed in a magnetic field applied along z-axis, the energy of the atom is changed by an amount proportional to the z-component of its magnetic moment. This change is related to L_z. For this reason m is called the *magnetic quantum number*.

11.3.3.1   Commutation Relations

It is easy to show that

$$[L_x, x] = 0, \quad [L_x, y] = i\hbar z, \quad [L_x, z] = -i\hbar y, \tag{11.33a}$$

$$[L_x, p_x] = 0, \quad [L_x, p_y] = i\hbar p_z, \quad [L_x, p_z] = -i\hbar p_y, \tag{11.33b}$$

$$[L_x, \mathbf{p}^2] = 0, \ [L_y, \mathbf{p}^2] = 0, \ [L_z, \mathbf{p}^2] = 0, \tag{11.33c}$$

$$[L_x, L_y] = i\hbar L_z, \quad [L_y, L_z] = i\hbar L_x, \quad [L_z, L_x] = i\hbar L_y, \tag{11.33d}$$

$$[L_x, \mathbf{L}^2] = [\mathbf{L}^2, L_x] = 0, \quad [L_y, \mathbf{L}^2] = [\mathbf{L}^2, L_y] = 0, \tag{11.33e}$$

$$[L_z, \mathbf{L}^2] = [\mathbf{L}^2, L_z] = 0, \ \mathbf{L} \times \mathbf{L} = i\hbar \mathbf{L}. \tag{11.33f}$$

We prove a few of the above results. Let us calculate $[L_x, L_y]$. We obtain

$$
\begin{aligned}
[L_x, L_y] &= [yp_z - zp_y, zp_x - xp_z] \\
&= [yp_z, zp_x] - [yp_z, xp_z] - [zp_y, zp_x] + [zp_y, xp_z] \\
&= yp_z(zp_x) - zp_x(yp_z) - yp_z(xp_z) + xp_z(yp_z) \\
&\quad -zp_y(zp_x) + zp_x(zp_y) + zp_y(xp_z) - xp_z(zp_y) \\
&= -i\hbar yp_x + yzp_zp_x - yzp_xp_z - xyp_zp_z + xyp_zp_z \\
&\quad -z^2p_yp_x + z^2p_xp_y + zxp_yp_z + i\hbar xp_y - xzp_zp_y \\
&= i\hbar (xp_y - yp_x) \\
&= i\hbar L_z. \tag{11.34}
\end{aligned}
$$

Next, compute $[\mathbf{L}^2, L_z]$. We get

$$[\mathbf{L}^2, L_z] = [L_x^2 + L_y^2 + L_z^2, L_z] = [L_x^2, L_z] + [L_y^2, L_z]. \tag{11.35}$$

Using the relations $[AB, C] = A[B, C] + [A, C]B$ we obtain

$$
\begin{aligned}
[\mathbf{L}^2, L_z] &= L_x [L_x, L_z] + [L_x, L_z] L_x + L_y [L_y, L_z] + [L_y, L_z] L_y \\
&= -i\hbar L_x L_y - i\hbar L_y L_x + i\hbar L_y L_x + i\hbar L_x L_y \\
&= 0. \tag{11.36}
\end{aligned}
$$

The nonzero values of $[L_x, L_y]$, $[L_y, L_z]$ and $[L_z, L_x]$ imply that L_x, L_y and L_z cannot be measured simultaneously with desired accuracy, that is, we cannot simultaneously have definite values of L_x, L_y and L_z. The operator $\mathbf{L}^2$ commutes with every component of $\mathbf{L}$. However, the components of $\mathbf{L}$ are not commuting operators. Hence, it is impossible to form a simultaneous eigenfunction of two different components of $\mathbf{L}$.

11.3.3.2   What is the Significance of $[\mathbf{L}^2, L_z] = 0$?

In order to understand the significance of the relation $[\mathbf{L}^2, L_z] = 0$ consider $\mathbf{L}^2$ and L_z. *What is the condition for ψ to be a simultaneous eigenfunction of $\mathbf{L}^2$ and L_z?* If ψ is an eigenfunction of $\mathbf{L}^2$ and L_z with eigenvalues λ and λ', respectively, then

$$\mathbf{L}^2\psi = \lambda\psi, \tag{11.37a}$$

$$L_z\psi = \lambda'\psi. \tag{11.37b}$$

Operating $-L_z$ from left on Eq. (11.37a) we have

$$-L_z\mathbf{L}^2\psi = -\lambda\lambda'\psi. \tag{11.38a}$$

Operating $\mathbf{L}^2$ from left on Eq. (11.37b) we get

$$\mathbf{L}^2 L_z \psi = \lambda \lambda' \psi. \tag{11.38b}$$

Adding Eqs. (11.38a) and (11.38b) we obtain

$$\mathbf{L}^2 L_z \psi - L_z \mathbf{L}^2 \psi = [\mathbf{L}^2, L_z] \psi = 0. \tag{11.39}$$

That is, $[\mathbf{L}^2, L_z] = 0$. Thus, the above relation implies that if ψ is an eigenfunction of $\mathbf{L}^2$ then it is an eigenfunction of L_z also and vice-verse. Further, $\mathbf{L}^2$ and L_z can be measured simultaneously.

11.3.3.3 Ladder Operators $L_\pm$

Earlier we defined the ladder operators $L_\pm$ as $(L_x \pm iL_y)$. Then

$$
\begin{aligned}
[L_z, L_\pm] &= [L_z, L_x] \pm i[L_z, L_y] \\
&= i\hbar L_y \pm \hbar L_x \\
&= \pm \hbar (L_x \pm iL_y) \\
&= \pm \hbar L_\pm.
\end{aligned}
\tag{11.40}
$$

Solved Problem 1:

Find the effect of $L_\pm$ on ψ.

Consider $L_z(L_\pm \psi)$. We obtain

$$
\begin{aligned}
L_z(L_\pm \psi) &= (L_z L_\pm) \psi \\
&= (L_z L_\pm - L_\pm L_z + L_\pm L_z) \psi \\
&= ([L_z, L_\pm] + L_\pm L_z) \psi \\
&= (\pm \hbar L_\pm + L_\pm L_z) \psi \\
&= (\pm \hbar L_\pm + m\hbar L_\pm) \psi \\
&= (m \pm 1) \hbar L_\pm \psi.
\end{aligned}
\tag{11.41}
$$

Notice that $L_\pm \psi$ is an eigenfunction of L_z. We know that the effect of L_z on ψ is to yield the eigenvalues $m\hbar$. When $L_\pm$ is also operated then the eigenvalue is changed to $(m \pm 1)\hbar$ as seen from Eq. (11.41). Thus, the effect of $L_\pm$ is to change the eigenvalue by $\pm \hbar$. L_+ increases the eigenvalue by $+\hbar$ while L_- decreases the eigenvalue by $-\hbar$. That is, L_+ (L_-) increases (decreases) z-component of the angular momentum by one unit of $\hbar$.

11.3.3.4 Expectation Values

Because of the symmetry properties of ψ_{nlm} in angular coordinates, we obtain

$$\langle L_{x,y} \rangle_{lm} = (\psi_{nlm}, L_{x,y} \psi_{nlm}) = (Y_{lm}, L_{x,y} Y_{lm}) = 0. \tag{11.42}$$

Further,

$$L_\pm \psi_{jm} = (\lambda_j \pm \hbar) \psi_{jm} = \psi_{jm \pm 1} \tag{11.43}$$

and

$$
\begin{aligned}
\langle L_x^2 \rangle_{lm} &= \langle L_y^2 \rangle_{lm} = (Y_{lm}, L_x^2 Y_{lm}) = \frac{1}{2} (Y_{lm}, (\mathbf{L}^2 - L_z^2) Y_{lm}) \\
&= \frac{1}{2} \hbar^2 [l(l+1) - m^2].
\end{aligned}
\tag{11.44}
$$

Then

$$\Delta L_x = \Delta L_y = \left[\langle L_x^2 \rangle_{lm} - \langle L_x \rangle_{lm}^2\right]^{1/2} = \hbar \left\{\frac{1}{2}\left[l(l+1) - m^2\right]\right\}^{1/2}. \qquad (11.45)$$

For Y_{ll} states $\Delta L_x = \Delta L_y = \hbar\sqrt{l/2}$. We note that $\langle L_x^2 \rangle \neq 0$. If $\langle L_x^2 \rangle = 0$ then all the L_x's are zero because L_x^2 is positive. In this case the system is in a definite state for which $L_x = 0$ which means the uncertainty in L_x is zero. Similarly, for L_y also the uncertainty is zero. But they are noncommuting operators. Therefore, the uncertainty relation is $(\Delta L_x)(\Delta L_y) \geq \hbar/2$. Hence, $\langle L_x^2 \rangle = 0$ is not possible.

We can find [13]

$$\langle \mathbf{L} \rangle = \int_V \psi^* L \psi \, d^3 r = m \int \mathbf{r} \times \mathbf{J} \, d^3 r, \qquad (11.46)$$

where $\mathbf{J}$ is the probability current density. In terms of matrix representation $\langle \mathbf{L} \rangle = \langle \Psi_l | \mathbf{L} | \Psi_l \rangle$. Expanding Ψ_l as

$$\Psi_l = \sum_{m=-l}^{l} a_m \psi_{lm} \qquad (11.47)$$

we write

$$\langle \mathbf{L} \rangle = \sum_{m,m'} a_m^* L_{mm'} a_{m'}, \quad L_{mm'} = \langle \psi_{lm} | \mathbf{L} | \psi_{lm'} \rangle. \qquad (11.48)$$

ΔL_x and ΔL_y are minimum when $m = l$. In this case $\Delta L_x = \Delta L_y = \hbar\sqrt{l/2}$. This is the highest precision of a measurement on L_x and L_y in a state Y_{lm}.

11.3.3.5 L_z^n and e^{L_z} are Hermitians

In spherical polar coordinates

$$\int_0^{2\pi} \psi^* L_z \Phi \, d\phi = \int_0^{2\pi} \psi^* \left[-i\hbar \frac{\partial \Phi}{\partial \phi}\right] d\phi$$

$$= \left[-i\hbar \psi^* \Phi\right]_{\phi=0}^{2\pi} + i\hbar \int_0^{2\pi} \Phi \frac{\partial \psi^*}{\partial \phi} \, d\phi. \qquad (11.49)$$

The first term in the above equation vanishes. Then

$$\int_0^{2\pi} \psi^* L_z \Phi \, d\phi = \int_0^{2\pi} \left[-i\hbar \frac{\partial \psi}{\partial \phi}\right]^* \Phi \, d\phi = \int_0^{2\pi} (L_z \psi)^* \Phi d\phi. \qquad (11.50)$$

Therefore, L_z is Hermitian. Since nth power of a Hermitian operator is also a Hermitian, L_z^n are also Hermitians. Further, we write

$$e^{L_z} = \sum_{k=0}^{\infty} \frac{1}{k!} L_z^k. \qquad (11.51)$$

Since L_z^k, $k = 0, 1, 2, \ldots$ are Hermitians, e^{L_z} is also a Hermitian.

11.3.4 Eigenspectra Through Commutation Relations

In sec. 11.3.2 we obtained the eigenvalues of $\mathbf{L}^2$ and L_z by solving their eigenvalue equations which are differential equations. We now determine the eigenvalues of $\mathbf{L}^2$ and L_z using commutation relations together with the ladder operators $L_\pm$. The procedure is similar to the method which we used in sec. 4.4 for linear harmonic oscillator.

11.3.4.1 Determination of Eigenvalues

Let ψ_{lm} denote the state vector belonging to the eigenvalues $\lambda_l \hbar^2$ of $\mathbf{L}^2$ and $m\hbar$ of L_z. (This state can be represented as $|l, m\rangle$ or $|lm\rangle$.) Then

$$L_z \psi_{lm} = m\hbar \psi_{lm}, \quad \mathbf{L}^2 \psi_{lm} = \lambda_l \hbar^2 \psi_{lm}. \tag{11.52}$$

Introduce a function $\phi_{lm}^{\pm}$ through the equation $L_{\pm} \psi_{lm} = \phi_{lm}^{\pm}$. Since the effect of $L_{\pm}$ on ψ_{lm} is to change the eigenvalue by $\pm\hbar$ the function $\phi_{lm}^{\pm}$ is proportional to $\psi_{lm\pm1}$. That is, $\phi_{lm}^{\pm} = N_{\pm} \hbar \psi_{lm\pm1}$, where $N_{\pm}$ is a constant to be determined.

We denote the minimum of the eigenvalue as $m_{\min}$ and the maximum as $m_{\max}$. We determine their values. If L_+ is operated on $\psi_{lm_{\max}}$ then the eigenvalue must be $m_{\max}+1$ which is not possible. Thus, $L_+ \psi_{lm_{\max}} = 0$. Similarly, $L_- \psi_{lm_{\min}} = 0$. Now, consider

$$L_- (L_+ \psi_{lm_{\max}}) = 0. \tag{11.53}$$

Expanding the left-side of Eq. (11.53) we get

$$\begin{aligned}
0 &= (L_x - iL_y)(L_x + iL_y) \psi_{lm_{\max}} \\
&= (\mathbf{L}^2 - L_z^2 - \hbar L_z) \psi_{lm_{\max}} \\
&= \hbar^2 (\lambda_l - m_{\max}^2 - m_{\max}) \psi_{lm_{\max}}. \tag{11.54}
\end{aligned}$$

The above equation gives

$$\lambda_l - m_{\max}(m_{\max}+1) = 0. \tag{11.55}$$

Similarly, from $L_+(L_- \psi_{lm_{\min}}) = 0$ we have

$$\lambda_l - m_{\min}(m_{\min}-1) = 0. \tag{11.56}$$

Comparison of Eqs. (11.55) and (11.56) gives $m_{\min}^2 - m_{\min} = m_{\max}^2 + m_{\max}$ or

$$(m_{\max} + m_{\min})(m_{\max} - m_{\min} + 1) = 0. \tag{11.57}$$

From Eq. (11.57), $m_{\max}$ is $-m_{\min}$ or $m_{\max} = m_{\min} - 1$. Because $m_{\max}$ is the maximum value it cannot be $m_{\min} - 1$. Therefore, we choose $m_{\max} = -m_{\min}$. Since the values of m are integers the number $m_{\max} - m_{\min}$ must be an integer and we denote it as $2l$. The possible values of m are $m = l, l-1, \ldots, -l+1, -l$. Then the possible values of l are $0, 1/2, 1, \ldots$. Substitution of $m_{\max} = -m_{\min}$ in the expression $m_{\max} - m_{\min} = 2l$ gives $m_{\max} = l$. Further, from Eq. (11.55) λ_l is obtained as

$$\lambda_l = m_{\max}(m_{\max}+1) = l(l+1). \tag{11.58}$$

Next, we determine $N_{\pm}$. Using Eq. (11.50), we write

$$\int (\phi_{lm}^{\pm})^* (\phi_{lm}^{\pm}) \, d\tau = |N_{\pm}|^2 \hbar^2 \int \psi_{lm\pm1}^* \psi_{lm\pm1} \, d\tau. \tag{11.59}$$

Since ψ_{lm} form a complete set, the integral in the right-side of the above equation is 1. Replacing $\phi_{lm}^{\pm}$ using $L_{\pm} \psi_{lm} = \phi_{lm}^{\pm}$, we have

$$|N_{\pm}|^2 \hbar^2 = \int (L_{\pm} \psi_{lm})^* (L_{\pm} \psi_{lm}) \, d\tau = \hbar^2 [(l \mp m)(l \pm m + 1)]. \tag{11.60}$$

Now, the eigenvalue equations for L_z, $\mathbf{L}^2$ and $L_{\pm}$ are

$$\begin{aligned}
L_z \psi_{lm} &= m\hbar \psi_{lm}, \tag{11.61a} \\
\mathbf{L}^2 \psi_{lm} &= \hbar^2 l(l+1) \psi_{lm}, \tag{11.61b} \\
L_{\pm} \psi_{lm} &= \hbar [(l \mp m)(l \pm m + 1)]^{1/2} \psi_{lm\pm1}. \tag{11.61c}
\end{aligned}$$

11.3.4.2   Properties of Eigenvalues

From (11.61b) we state the following features of eigenvalue spectra of $\mathbf{L}^2$:

1. The eigenvalues of $\mathbf{L}^2$ are $\neq (l\hbar)^2$ but $= l(l+1)\hbar^2$ and are greater than $(l\hbar)^2$. The reason for this is as follows [14]. For a state ψ_{lm} we have $\langle \mathbf{L}^2 \rangle = \lambda \hbar^2$ and $\langle L_z^2 \rangle = m^2 \hbar^2$. Then

$$\langle \mathbf{L}^2 \rangle = \langle L_x^2 \rangle + \langle L_y^2 \rangle + \langle L_z^2 \rangle \tag{11.62}$$

which gives

$$\lambda \hbar^2 = \langle L_x^2 \rangle + \langle L_y^2 \rangle + (m\hbar)^2 . \tag{11.63}$$

The expectation values of L_x^2 and L_y^2 are nonnegative. This means $\lambda \geq m^2$. If we want to have a state with some positive value of l for m then $\lambda \geq l^2$. The eigenvalues of $\mathbf{L}^2$ are $\geq (l\hbar)^2$. To have equality sign we must have $\langle L_x^2 \rangle = \langle L_y^2 \rangle = 0$. If the projection of the angular momentum on the z-axis were the same length as the angular momentum vector then the projections on the x and y axes would be zero. This violates the uncertainty relation. This is because the components of angular momentum do not commute with each other as seen earlier. Thus, the eigenvalue of $\mathbf{L}^2 > (l\hbar)^2$.

2. For a given l the m can take $(2l+1)$ distinct values. Therefore, eigenvalues of $\mathbf{L}^2$ are $(2l+1)$-fold degenerate.

3. *What is the fundamental difference between the states with integer and half-integer values of l?* Consider the equation $\mathrm{e}^{-\mathrm{i}\theta L_z}\psi_{lm} = \mathrm{e}^{-\mathrm{i}m\theta}\psi_{lm}$. For integer values of l the states ψ_{lm} will not change under a rotation through $\theta = 2\pi$ and are single-valued. But the states will change sign for half-integer angular momentum and are hence double-valued under rotation.

Next, consider the ψ_{lm} given by

$$\begin{aligned} \psi_{lm} &= (-1)^m Y_{lm}(\theta, \phi) \\ &= \left[\frac{(2l+1)(l-|m|)!}{4\pi(l+|m|)!} \right]^{1/2} P_l^{|m|}(\cos\theta)\mathrm{e}^{\mathrm{i}m\phi}. \end{aligned} \tag{11.64}$$

The above solution is valid only for integer values of l and m. For example, consider $l = 1/2$, $m = \pm 1/2$. We have

$$\psi_{\frac{1}{2}\pm\frac{1}{2}} = \frac{1}{\sqrt{4\pi}}\mathrm{e}^{\pm\mathrm{i}\phi/2}\sqrt{\sin\theta}. \tag{11.65}$$

Consider $L_\pm \psi_{\frac{1}{2}\frac{1}{2}}$. We expect $L_+\psi_{\frac{1}{2}\frac{1}{2}} = 0$. But

$$\begin{aligned} L_+\psi_{\frac{1}{2}\frac{1}{2}} &= \mathrm{e}^{\mathrm{i}\phi}\left(\frac{\partial}{\partial\theta} + \mathrm{i}\frac{\cos\theta}{\sin\theta}\frac{\partial}{\partial\phi} \right)\mathrm{e}^{\mathrm{i}\phi/2}\sqrt{\sin\theta} \\ &= \mathrm{e}^{\mathrm{i}3\phi/2}\left(\frac{\partial}{\partial\theta} - \frac{1}{2}\cot\theta \right)\sqrt{\sin\theta} \\ &\neq 0 . \end{aligned} \tag{11.66a}$$

Similarly,

$$L_-\psi_{\frac{1}{2}-\frac{1}{2}} = \mathrm{e}^{-\mathrm{i}\phi}\left(-\frac{\partial}{\partial\theta} + \mathrm{i}\cot\theta\frac{\partial}{\partial\phi} \right)\mathrm{e}^{-\mathrm{i}\phi/2}\sqrt{\sin\theta} \neq 0 . \tag{11.66b}$$

Therefore, the functions (11.64) corresponding to half-integer values of l are not admissible eigenfunctions of $\mathbf{L}$. Half-integer values of l have to be discarded.

4. In sec. 11.3.2 the eigenfunctions and eigenvalues of $\mathbf{L}^2$ are obtained by solving the eigenvalue Eq. (11.30). In this case the eigenvalues are $l(l+1)\hbar^2$, where l is a nonnegative integer, $l = 0, 1, 2, \ldots$. In contrast to this, when the commutation relations are alone used as above, the eigenvalues are $l(l+1)\hbar^2$, where $l = 0, 1/2, 1, \ldots$. In addition to integer l, half-integer values also exist, however, they do not correspond to orbital angular momentum. In working with the commutation relations instead of the quantum number l suppose we use a different quantum number j. Then the eigenvalues are $j(j+1)\hbar$, where $j = 0, 1/2, 1, \ldots$. These j quantum numbers will be representing the total angular momentum $\mathbf{J} = \mathbf{L} + \mathbf{S}$, where $\mathbf{S}$ is spin angular momentum.

Solved Problem 2:

Show that $\psi_\pm = N(x \pm iy)f(r)$ is an eigenfunction both of $\mathbf{L}^2$ and L_z and state the corresponding eigenvalues.

We have $x + iy = r\sin\theta(\cos\phi + i\sin\phi) = r\sin\theta\, e^{i\phi} = -r\sqrt{8\pi/3}\, Y_{11}$. The given ψ_+ can be rewritten as $\psi_+ = -N\sqrt{8\pi/3}\, rf(r)Y_{11}$. Since $\mathbf{L}^2 Y_{lm} = l(l+1)\hbar^2 Y_{lm}$ and $L_z Y_{lm} = m\hbar$ we get

$$\mathbf{L}^2\psi_+ = 2\hbar^2\psi_+\,, \quad L_z\psi_+ = \hbar\psi_+\,. \tag{11.67}$$

The eigenvalues of $\mathbf{L}^2$ and L_z corresponding to ψ_+ are $2\hbar^2$ and $\hbar$, respectively.

Next, we have $x - iy = r\sin\theta(\cos\phi - i\sin\phi) = r\sin\theta\, e^{-i\phi} = r\sqrt{8\pi/3}\, Y_{1-1}$. Then $\psi_- = N\sqrt{8\pi/3}\, rf(r)Y_{1-1}$. We obtain

$$\mathbf{L}^2\psi_- = 2\hbar^2\psi_-\,, \quad L_z\psi_- = -\hbar\psi_-\,. \tag{11.68}$$

The eigenvalues of $\mathbf{L}^2$ and L_z corresponding to ψ_- are $2\hbar^2$ and $-\hbar$, respectively.

11.3.5 Matrix Representation of $\mathbf{L}^2$, L_z and $L_\pm$

Suppose the wave function ψ is a vector. For example, ψ of an atom is complex and we can treat it as a two-component vector function. If ψ is regarded as a vector then the operator acting on it must be represented in matrix form. For illustrative purpose consider $f(x)$ and $g(x)$ such that $Af(x) = g(x)$. Since the wave function has the completeness property, we write

$$f(x) = \sum_m a_m\psi_m\,, \quad g(x) = \sum_n b_n\psi_n\,. \tag{11.69}$$

Then $Af(x) = g(x)$ is rewritten as $\sum_m a_m A\psi_m = \sum_n b_n\psi_n$. a_m can be determined using normalization condition. To find b_n, multiply the above expression by ψ_l^* and integrate over entire space which gives

$$\sum a_m A_{lm} = \sum a_m \int \psi_l^* A\psi_m\, d\tau = \sum_n b_n\delta_{ln} = b_l\,, \tag{11.70}$$

where we have used

$$A_{lm} = \int \psi_l^* A\psi_m\, d\tau\,. \tag{11.71}$$

A_{lm} form a matrix:

$$A_{lm} = \begin{pmatrix} A_{11} & A_{12} & \ldots & A_{1n} \\ A_{21} & A_{22} & \ldots & A_{2n} \\ & & \ldots & \\ A_{n1} & A_{n2} & \ldots & A_{nn} \end{pmatrix}. \tag{11.72}$$

This is called *matrix representation*.

The matrix elements of an operator A in a given l state are given by

$$A_{mm'} = \langle lm|A|lm'\rangle = \int \psi_{lm}^* A \psi_{lm'}\, \mathrm{d}\tau, \quad m, m' = l, l-1, \ldots, -l \qquad (11.73)$$

where m and m' are the allowed range of eigenvalues of L_z. The matrices for any values of l are represented by rows labeled by m and columns by m'. The matrix form of A is written as

$$A = \begin{pmatrix} \langle ll|A|ll\rangle & \langle ll|A|ll-1\rangle & \cdots & \langle ll|A|l-l\rangle \\ \langle ll-1|A|ll\rangle & \langle ll-1|A|ll-1\rangle & \cdots & \langle ll-1|A|l-l\rangle \\ & & \cdots & \\ \langle l-l|A|ll\rangle & \langle l-l|A|ll-1\rangle & \cdots & \langle l-l|A|l-l\rangle \end{pmatrix}. \qquad (11.74)$$

The elements of $\mathbf{L}^2$ are

$$\langle lm|\mathbf{L}^2|lm'\rangle = \int \psi_{lm}^* \mathbf{L}^2 \psi_{lm'}\, \mathrm{d}\tau = \hbar^2 l(l+1)\delta_{mm'}. \qquad (11.75)$$

Similarly,

$$\langle lm|L_z|lm'\rangle = m\hbar\delta_{mm'}, \qquad (11.76)$$

$$\langle lm|L_\pm|lm'\rangle = \hbar\left[(l \mp m')(l \pm m' + 1)\right]^{1/2}\delta_{mm'\pm 1}. \qquad (11.77)$$

For $l = 0$ the matrix elements of $\mathbf{L}^2$, L_z, $L_\pm$, L_x and L_y all are zero.

11.4   Spin Angular Momentum

So far, we have not introduced the term spin in quantum mechanics. In this section, first the notion of spin [15] is introduced and a few examples of necessity of spin angular momentum are mentioned. Then spin states of an electron are obtained. The spin-orbit interaction is brought out.

11.4.1   Physical Picture of Spin [15]

Soon after Bohr came up with his atom model, it was assumed that the electron revolving around the nucleus was also spinning about an axis. Later, it was realized that this assumption had no validity. Spin of particle is not a measure of how it is spinning. In fact, it is impossible to tell whether an electron is spinning. The word spin is a convenient way of representing the intrinsic angular momentum of a particle.

The spin is said to be a *nonorbital, intrinsic* or *inherent angular momentum*. Also, spin is thought of as due to an internal structure of the electron. Further, the spin arises in a natural way from Dirac equation. We can regard spin as due to a circulating flow of energy or a momentum density, in the electron wave field. This picture of spin is valid not only for electrons, but also for photons, vector mesons and gravitons. In all these cases the spin angular momentum is due to a circulating energy flow in the fields.

The spin of the electron has a close classical analog. It is an angular momentum of the same kind as possessed by the fields of a circularly polarized electromagnetic wave. Gordon showed that the magnetic moment of the electron is due to the circulating flow of charge in the electron wave field. This means that spin and the magnetic moment are not due to the internal properties of the electron but with the structure of its wave field.

The momentum density in the Dirac field is [15]

$$G = \frac{\hbar}{2i}[\psi^*\nabla\psi - (\nabla\psi^*)\psi] + \frac{\hbar}{4}\nabla \times (\psi^*\sigma\psi), \tag{11.78}$$

where $\sigma_1 = -i\alpha_2\alpha_3$, $\sigma_2 = -i\alpha_3\alpha_1$ and $\sigma_3 = -i\alpha_1\alpha_2$. Here σ_i's are Pauli matrices and α_i's are Dirac matrices (see sec. 19.3.1). The first term in G is associated with the translational motion of the electron. The second term is associated with circulating flow of energy. For a Gaussian wave packet

$$\psi = \frac{1}{(\pi d^2)^{3/4}}e^{-r^2/2d^2}, \quad d \gg \frac{\hbar}{mc} \tag{11.79}$$

the first term in Eq. (11.78) is zero and

$$G = \frac{\hbar}{4d^2}\left(\frac{1}{\pi d^2}\right)^{3/2}e^{-r^2/d^2}(-2y\hat{x} + 2x\hat{y}). \tag{11.80}$$

The flow lines for the energy are circular. Such a circulating flow of energy will give an angular momentum and is the spin. For historical continuity this property was called *spin* and not means that the electron is spinning. For an arbitrary wave packet, the net angular momentum is

$$\begin{aligned} \mathbf{J} &= \int \hat{\mathbf{X}} \times \mathbf{G}\,d^3\mathbf{X} \\ &= \int \frac{\hbar}{2i}\hat{\mathbf{X}} \times [\psi^*\nabla\psi - (\nabla\psi^*)\psi]\,d^3\mathbf{X} + \int \frac{\hbar}{4}\hat{\mathbf{X}} \times [\nabla \times (\psi^*\sigma)\psi]\,d^3\mathbf{X}. \end{aligned} \tag{11.81}$$

Expanding the cross product in the second integral into two dot products and then integrating both of these by parts we obtain

$$\begin{aligned} \mathbf{J} &= \frac{\hbar}{2i}\int \hat{\mathbf{X}} \times [\psi^*\nabla\psi - (\nabla\psi^*)\psi]\,d^3\mathbf{X} + \frac{\hbar}{2}\int \psi^*\sigma\psi\,d^3\mathbf{X} \\ &= \mathbf{L} + \frac{\hbar}{2}\langle\sigma\rangle \\ &= \mathbf{L} + \mathbf{S}. \end{aligned} \tag{11.82}$$

The first term is the orbital angular momentum, and the second term is the spin $\mathbf{S}$. The first term is independent of the spin-orbital state. The second term is dependent on the spin-orbital state. Because the spin $\mathbf{S}$ given by Eq. (11.82) is the expectation value of σ, the operator representing the spin must be

$$\mathbf{S}_{op} = \frac{\hbar}{2}\sigma. \tag{11.83}$$

Thus, the spin operator obeys all the usual commutation relations. In particular, the eigenvalues of any component of the spin, say z-component, are $\pm\hbar/2$ (or $\pm1/2$ in units of $\hbar$).

Whatever we have said in the earlier sections of the present chapter for orbital angular momentum $\mathbf{L}$ is also true for spin angular momentum and also true for the total angular momentum $\mathbf{J} = \mathbf{L} + \mathbf{S}$.

11.4.2   Components of Spin Operator

In matrix form the components of the spin operator are (the eigenvalues of components of spin are in units of $\hbar$)

$$S_z = \frac{\hbar}{2}\begin{pmatrix} 1 & 0 \\ 0 & -1 \end{pmatrix}, \quad S_+ = \hbar\begin{pmatrix} 0 & 1 \\ 0 & 0 \end{pmatrix}, \tag{11.84a}$$

$$S_- = \hbar\begin{pmatrix} 0 & 0 \\ 1 & 0 \end{pmatrix}, \quad S_x = \frac{1}{2}(S_+ + S_-) = \frac{\hbar}{2}\begin{pmatrix} 0 & 1 \\ 1 & 0 \end{pmatrix}, \tag{11.84b}$$

$$S_y = \frac{1}{2i}(S_- - S_+) = \frac{\hbar}{2}\begin{pmatrix} 0 & -i \\ i & 0 \end{pmatrix}, \tag{11.84c}$$

Then

$$\mathbf{S}^2 = S_x^2 + S_y^2 + S_z^2 = \frac{3}{4}\hbar^2\begin{pmatrix} 1 & 0 \\ 0 & 1 \end{pmatrix}. \tag{11.85}$$

The eigenvalues of S_x, S_y, S_z and $\mathbf{S}^2$ are $\pm\hbar/2$, $\pm\hbar/2$, $\pm\hbar/2$ and $3\hbar^2/4$, respectively.

The components of $\mathbf{S}$ satisfy orbital angular momentum type commutation relations:

$$[S_x, S_y] = i\hbar S_z, \quad [S_y, S_z] = i\hbar S_x, \quad [S_z, S_x] = i\hbar S_y. \tag{11.86}$$

Verify that

$$S_+|\alpha\rangle = 0, \quad S_-|\alpha\rangle = \hbar|\beta\rangle, \quad S_+|\beta\rangle = \hbar|\alpha\rangle, \quad S_-|\beta\rangle = 0, \tag{11.87}$$

where $|\alpha\rangle$ and $|\beta\rangle$ are the eigenkets of S_z with eigenvalues $\hbar/2$ and $-\hbar/2$, respectively. Using the above relations it is easy to prove that

$$\sigma_x|\alpha\rangle = |\beta\rangle, \quad \sigma_x|\beta\rangle = |\alpha\rangle, \quad \sigma_y|\alpha\rangle = i|\beta\rangle, \tag{11.88a}$$
$$\sigma_y|\beta\rangle = -i|\alpha\rangle, \quad \sigma_z|\alpha\rangle = |\alpha\rangle, \quad \sigma_z|\beta\rangle = -|\beta\rangle. \tag{11.88b}$$

Because $S_x = (S_+ + S_-)/2$ we obtain

$$S_x|\alpha\rangle = \frac{1}{2}(S_+ + S_-)|\alpha\rangle = \frac{1}{2}(S_+|\alpha\rangle + S_-|\alpha\rangle) = \frac{\hbar}{2}|\beta\rangle \tag{11.89a}$$

and

$$S_x|\beta\rangle = \frac{1}{2}(S_+ + S_-)|\beta\rangle = \frac{1}{2}(S_+|\beta\rangle + S_-|\beta\rangle) = \frac{\hbar}{2}|\alpha\rangle. \tag{11.89b}$$

Similarly, we obtain

$$S_y|\alpha\rangle = \frac{\hbar}{2}|\beta\rangle, \quad S_y|\beta\rangle = -\frac{i\hbar}{2}|\alpha\rangle. \tag{11.89c}$$

Using the relation $\sigma_x = (2/\hbar)S_x$ we get

$$\sigma_x|\alpha\rangle = \frac{2}{\hbar}S_x|\alpha\rangle = |\beta\rangle, \quad \sigma_x|\beta\rangle = |\alpha\rangle. \tag{11.90a}$$

In a similar manner we get

$$\sigma_y|\alpha\rangle = i|\beta\rangle, \quad \sigma_y|\beta\rangle = -i|\alpha\rangle, \tag{11.90b}$$
$$\sigma_z|\alpha\rangle = |\alpha\rangle, \quad \sigma_z|\beta\rangle = -|\beta\rangle. \tag{11.90c}$$

11.4.3 Normalized Eigenfunctions of σ_x, σ_y and σ_z

We obtain the normalized eigenfunctions of σ_x, σ_y and σ_z. Since $\sigma_x|\alpha\rangle = |\beta\rangle$, $\sigma_x|\beta\rangle = |\alpha\rangle$ we have

$$\sigma_x \left(|\alpha\rangle \pm |\beta\rangle\right) = \pm|\alpha\rangle + |\beta\rangle. \tag{11.91}$$

Similarly, from Eqs. (11.90b) we obtain

$$\sigma_y \left(|\alpha\rangle \pm \mathrm{i}|\beta\rangle\right) = \pm|\alpha\rangle + \mathrm{i}|\beta\rangle. \tag{11.92}$$

The normalization gives the factor $1/\sqrt{2}$. *What is the major difference between the forms of operators and eigenfunctions of orbital angular momentum and spin angular momentum?* For a particle with spin $\hbar/2$ we have [13]

$$\langle \mathbf{S} \rangle = \int \psi^* \mathbf{S} \psi \, \mathrm{d}\tau = m \int_V \mathbf{r} \times \mathbf{J}_\mathrm{s} \, \mathrm{d}^3 r, \quad \mathbf{J}_\mathrm{s} = \nabla \times \frac{\hbar}{4m} \psi^* \boldsymbol{\sigma} \psi. \tag{11.93}$$

11.4.4 Examples of Necessity of Spin Angular Momentum

Without spin the quantum theory cannot explain many observed phenomena. Some examples are given below:

1. The energy levels of hydrogen atom are given by Eq. (12.35). Since E_n depends only on n and independent of l and m we say that the energy levels are degenerate with respect to l and m. The number of degeneracy is $2n^2$ (see Eq. (12.65)). To distinguish these states we use spin angular momentum.

2. The magnetic dipole moments and susceptibility of ferromagnets are explained with spin angular momentum.

3. The D-lines of sodium obtained at $5890\,\text{Å}$ and $5896\,\text{Å}$ show the existence of fine structure and is explained using spin.

4. A beam of atoms passing through an inhomogeneous magnetic field should split into $(2l+1)$ lines. In an experiment with $l = 0$ (ground state), the beam splits into two separate beams. Therefore, some additional angular momentum is required to explain the above observation.

Solved Problem 3:

Show that $\mathbf{S}_1 \cdot \mathbf{S}_2 = S_{1z}S_{2z} + (S_{1+}S_{2-} + S_{1-}S_{2+})/2$.

We have

$$
\begin{aligned}
\mathbf{S}_1 \cdot \mathbf{S}_2 &= S_{1x}S_{2x} + S_{1y}S_{2y} + S_{1z}S_{2z}, \\
S_{1x} &= (S_{1+} + S_{1-})/2, \quad S_{2x} = (S_{2+} + S_{2-})/2, \\
S_{1x}S_{2x} &= \left(\frac{S_{1+} + S_{1-}}{2}\right)\left(\frac{S_{2+} + S_{2-}}{2}\right) \\
&= \left(S_{1+}S_{2+} + S_{1+}S_{2-} + S_{1-}S_{2+} + S_{1-}S_{2-}\right)/4, \\
S_{1y}S_{2y} &= \left(\frac{S_{1+} - S_{1-}}{2\mathrm{i}}\right)\left(\frac{S_{2+} - S_{2-}}{2\mathrm{i}}\right) \\
&= -\left(S_{1+}S_{2+} + S_{1+}S_{2-} + S_{1-}S_{2+} - S_{1-}S_{2-}\right)/4. \tag{11.94}
\end{aligned}
$$

Substituting these in the expression for $\mathbf{S}_1 \cdot \mathbf{S}_2$ we get $\mathbf{S}_1 \cdot \mathbf{S}_2 = S_{1z}S_{2z} + (S_{1+}S_{2-} + S_{1-}S_{2+})/2$.

11.4.5 Spin States of an Electron

Since the spin quantum number for an electron is found to be $1/2$ the components of $\mathbf{S}$ after setting $\hbar = 1$ are

$$S_x = \frac{1}{2}\sigma_x, \quad S_y = \frac{1}{2}\sigma_y, \quad S_z = \frac{1}{2}\sigma_z, \tag{11.95}$$

where the Pauli's spin matrices σ's are given by

$$\sigma_x = \begin{pmatrix} 0 & 1 \\ 1 & 0 \end{pmatrix}, \quad \sigma_y = \begin{pmatrix} 0 & -i \\ i & 0 \end{pmatrix}, \quad \sigma_z = \begin{pmatrix} 1 & 0 \\ 0 & -1 \end{pmatrix}. \tag{11.96}$$

We have

$$\mathbf{S}^2 = S_x^2 + S_y^2 + S_z^2 = \frac{3}{4}\begin{pmatrix} 1 & 0 \\ 0 & 1 \end{pmatrix} = s(s+1)I, \tag{11.97}$$

where $s = 1/2$. The operators $\mathbf{S}^2$ and S_z are diagonal implying that we can represent the spin-up and spin-down states of them by a two-component column vector called *spinor*. We write

$$S_z \begin{pmatrix} u \\ v \end{pmatrix} = \pm \frac{1}{2}\begin{pmatrix} u \\ v \end{pmatrix}. \tag{11.98}$$

That is,

$$\begin{pmatrix} 1 & 0 \\ 0 & -1 \end{pmatrix}\begin{pmatrix} u \\ v \end{pmatrix} = \pm \begin{pmatrix} u \\ v \end{pmatrix} \tag{11.99}$$

and get

$$\begin{pmatrix} u \\ -v \end{pmatrix} = \pm \begin{pmatrix} u \\ v \end{pmatrix}. \tag{11.100}$$

For plus sign $v = 0$ and for minus sign $u = 0$. *What are the eigenvectors of S_x and S_y?*

11.4.6 Eigenvectors of S_z

We have $\sigma_z \chi_\pm = \pm \chi_\pm$. The following relations for Pauli spin matrices are easy to verify:

$$\sigma_x^2 = \sigma_y^2 = \sigma_z^2 = 1, \quad \sigma_x\sigma_y = i\sigma_z, \quad \sigma_y\sigma_z = i\sigma_x, \quad \sigma_z\sigma_x = i\sigma_y, \tag{11.101a}$$
$$\sigma_x\sigma_y + \sigma_y\sigma_x = 0, \quad \sigma_y\sigma_z + \sigma_z\sigma_y = 0, \quad \sigma_z\sigma_x + \sigma_x\sigma_z = 0. \tag{11.101b}$$

and $\sigma_x\sigma_y\sigma_z = i$. We obtain

$$\chi_+ = \begin{pmatrix} 1 \\ 0 \end{pmatrix}, \quad \chi_- = \begin{pmatrix} 0 \\ 1 \end{pmatrix}. \tag{11.102}$$

These two eigenstates are the *spin-up* and *spin-down states*, respectively. They are the states where z-component of the spin is parallel ($m_s = 1/2$) and anti-parallel ($m_s = -1/2$) to the z-axis.

In terms of χ_+ and χ_- we write the arbitrary spinor as

$$\begin{pmatrix} \alpha_+ \\ \alpha_- \end{pmatrix} = \alpha_+ \begin{pmatrix} 1 \\ 0 \end{pmatrix} + \alpha_- \begin{pmatrix} 0 \\ 1 \end{pmatrix}, \quad |\alpha_+|^2 + |\alpha_-|^2 = 1. \tag{11.103}$$

11.4.7 Expectation Value of S

Let us calculate the expectation value of $\mathbf{S}$. We write

$$\langle\alpha|\mathbf{S}|\alpha\rangle = \sum_i \sum_j \langle\alpha|i\rangle\langle i|\mathbf{S}|j\rangle\langle j|\alpha\rangle. \tag{11.104}$$

We write the right-side of the above equation as $(\alpha_+^*, \alpha_-^*)\, \mathbf{S}\binom{\alpha_+}{\alpha_-}$. Then

$$
\begin{aligned}
\langle S_x\rangle &= \frac{1}{2}\left(\alpha_+^*, \alpha_-^*\right)\begin{pmatrix} 0 & 1 \\ 1 & 0 \end{pmatrix}\begin{pmatrix} \alpha_+ \\ \alpha_- \end{pmatrix} \\
&= \frac{1}{2}\left(\alpha_+^*\alpha_- + \alpha_-^*\alpha_+\right), & (11.105\text{a}) \\
\langle S_y\rangle &= \frac{1}{2}\left(\alpha_+^*, \alpha_-^*\right)\begin{pmatrix} 0 & -i \\ i & 0 \end{pmatrix}\begin{pmatrix} \alpha_+ \\ \alpha_- \end{pmatrix} \\
&= -\frac{i}{2}\left(\alpha_+^*\alpha_- - \alpha_-^*\alpha_+\right), & (11.105\text{b}) \\
\langle S_z\rangle &= \frac{1}{2}\left(\alpha_+^*, \alpha_-^*\right)\begin{pmatrix} \alpha_+ \\ -\alpha_- \end{pmatrix} = \frac{1}{2}\left(|\alpha_+|^2 - |\alpha_-|^2\right). & (11.105\text{c})
\end{aligned}
$$

Solved Problem 4:

Determine the eigenvalues and the eigenvectors of the operator $\sigma \cdot \hat{n}$, where $\hat{n}$ is a unit vector in an arbitrary direction.

We have

$$(\sigma \cdot \hat{n})^2 = (\sigma_x n_x + \sigma_y n_y + \sigma_z n_z)^2 = 1. \tag{11.106}$$

The eigenvalues of $\sigma \cdot \hat{n}$ are thus ± 1. Denote the eigenvectors as χ_{n+} and χ_{n-}. To determine $\chi_{n\pm}$, construct the vectors

$$\frac{1}{2}(1 + \sigma \cdot \hat{n})\chi, \quad \frac{1}{2}(1 - \sigma \cdot \hat{n})\chi, \tag{11.107}$$

where $\chi = a_+\chi_+ + a_-\chi_-$ is an arbitrary spin vector. Operation of $\sigma \cdot \hat{n}$ on each of these vectors gives

$$
\begin{aligned}
\sigma \cdot \hat{n}\,\frac{1}{2}(1 + \sigma \cdot \hat{n})\chi &= \frac{1}{2}(1 + \sigma \cdot \hat{n})\chi, & (11.108\text{a}) \\
\sigma \cdot \hat{n}\,\frac{1}{2}(1 - \sigma \cdot \hat{n})\chi &= -\frac{1}{2}(1 - \sigma \cdot \hat{n})\chi. & (11.108\text{b})
\end{aligned}
$$

For $\chi = \chi_+$ these two equations become

$$
\begin{aligned}
\frac{1}{2}(1 + \sigma \cdot \hat{n})\chi_+ &= \frac{1}{2}(1 + \sigma_x n_x + \sigma_y n_y + \sigma_z n_z)\chi_+ \\
&= \frac{1}{2}\left(\chi_+ + n_x\sigma_x\chi_+ + n_y\sigma_y\chi_+ + n_z\sigma_z\chi_+\right) \\
&= \frac{1}{2}\left[(1 + n_z)\chi_+ + (n_x + in_y)\chi_-\right] & (11.109\text{a}) \\
\frac{1}{2}(1 - \sigma \cdot \hat{n})\chi_+ &= \frac{1}{2}\left[(1 - n_z)\chi_+ - (n_x + in_y)\chi_-\right]. & (11.109\text{b})
\end{aligned}
$$

The square of norms of these two vectors are obtained as $(1 + n_z)/2$ and $(1 - n_z)/2$. Then the normalized eigenfunctions are

$$\chi_{n+} = \frac{1}{\sqrt{2(1 + n_z)}} [(1 + n_z)\chi_+ + (n_x + in_y)\chi_-] , \qquad (11.110a)$$

$$\chi_{n-} = \frac{1}{\sqrt{2(1 - n_z)}} [(1 - n_z)\chi_+ - (n_x + in_y)\chi_-] . \qquad (11.110b)$$

In the state χ_{n+}, a measurement of the z-component of the spin will give the values $\pm 1/2$ with the relative probabilities

$$(1 + n_z)^2 : (n_x^2 + n_y^2) = \cos^2(\theta/2) : \sin^2(\theta/2) , \qquad (11.111)$$

where θ is the angle between $\hat{n}$ and z.

11.4.8   Density Matrix of a Spin-1/2 System

Let us consider a spin-1/2 system whose spin states are labeled as $| \uparrow \rangle$ and $| \downarrow \rangle$. Suppose we are dealing with electron beams. The density matrix of an electron in the spin state $| \uparrow \rangle$ is $\rho_\uparrow = | \uparrow \rangle \langle \uparrow |$ while the density matrix for an electron in the spin state $| \downarrow \rangle$ is $\rho_\downarrow = | \downarrow \rangle \langle \downarrow |$. Suppose the two beams are mixed. For a pure state

$$|\psi\rangle = c_1| \uparrow \rangle + c_2| \downarrow \rangle . \qquad (11.112)$$

The condition $\langle \psi | \psi \rangle = 1$ gives $|c_1|^2 + |c_2|^2 = 1$. The density matrix is

$$
\begin{aligned}
\overline{\rho}_{\text{P}} &= |\psi\rangle\langle\psi| \\
&= (c_1| \uparrow \rangle + c_2| \downarrow \rangle)(c_1^*\langle \uparrow | + c_2^*\langle \downarrow |) \\
&= |c_1|^2| \uparrow \rangle\langle \uparrow | + c_1 c_2^*| \uparrow \rangle\langle \downarrow | + c_2 c_1^*| \downarrow \rangle\langle \uparrow | + |c_2|^2| \downarrow \rangle\langle \downarrow | . \qquad (11.113)
\end{aligned}
$$

In matrix representation

$$\overline{\rho}_{mn} = \langle n|\overline{\rho}|m\rangle, \quad n, m = \uparrow, \downarrow, \quad \overline{\rho}_{\text{P}} = \begin{pmatrix} |c_1|^2 & c_1 c_2^* \\ c_2 c_1^* & |c_2|^2 \end{pmatrix} . \qquad (11.114)$$

We can easily verify that $\text{Tr}\overline{\rho}_{\text{P}} = 1$, $\overline{\rho}_{\text{P}}^2 = \overline{\rho}_{\text{P}}$ and $\langle \psi |\overline{A}|\psi\rangle = \text{Tr}(\overline{\rho}_{\text{P}} \overline{A}) = \text{Tr}(\overline{A} \overline{\rho}_{\text{P}})$ for any operator.

For a mixed state

$$\overline{\rho}_{\text{M}} = p_1| \uparrow \rangle\langle \uparrow | + p_2| \downarrow \rangle\langle \downarrow | , \quad p_1 + p_2 = 1 . \qquad (11.115)$$

Then

$$\overline{\rho}_{\text{M}} = \begin{pmatrix} p_1 & 0 \\ 0 & p_2 \end{pmatrix} \qquad (11.116)$$

and $\overline{\rho}_{\text{M}}^2 = p_1^2| \uparrow \rangle\langle \uparrow | + p_2^2| \downarrow \rangle\langle \downarrow | \neq \overline{\rho}_{\text{M}}$.

Since any 2×2 matrix can be written as a linear combination of the unit matrix and the Pauli matrices σ_x, σ_y, σ_z and because of $\text{Tr}I = 2$ and $\text{Tr}\sigma_{x,y,z} = 0$ we can write the general spin density matrix as $\rho = (P_0 I + \mathbf{P} \cdot \sigma)/2$. We have

$$
\begin{aligned}
\rho &= \frac{1}{2}[P_0 I + P_x\sigma_x + P_y\sigma_y + P_z\sigma_z] \\
&= \frac{1}{2}\begin{pmatrix} P_0 + P_z & P_x - iP_y \\ P_x + iP_y & P_0 - P_z \end{pmatrix} . \qquad (11.117)
\end{aligned}
$$

P_0 is determined from the condition $\text{Tr}\rho = 1$. We obtain $P_0 = 1$. Then

$$\rho = \frac{1}{2}(I + \mathbf{P}\cdot\sigma) = \frac{1}{2}\left(\begin{array}{cc} 1+P_z & P_x - iP_y \\ P_x + iP_y & 1-P_z \end{array}\right). \tag{11.118}$$

The eigenvalues of ρ must be real and nonnegative. The eigenvalue equation for the matrix ρ is $\det(\rho - \lambda I) = 0$ which gives

$$\lambda^2 - \lambda + \frac{1}{4}\left(1 - \mathbf{P}^2\right) = 0. \tag{11.119}$$

Then $\lambda_{1,2} = \left(1 \pm \mathbf{P}^2\right)/2$. Since λ_1, $\lambda_2 > 0$, we have $\mathbf{P}^2 \geq 1$.

We can easily obtain the expectation value of a spin operator $\langle\mathbf{S}\rangle$, where $\mathbf{S} = (\hbar/2)\sigma$. We write

$$\langle\sigma_x\rangle = \text{Tr}(\sigma_x P) = \text{Tr}\frac{1}{2}(\sigma_x + P_x + i\sigma_z P_y - i\sigma_y P_z) = P_x. \tag{11.120}$$

Similarly, we obtain $\langle\sigma_y\rangle = P_y$ and $\langle\sigma_z\rangle = P_z$. Thus, we can write $\langle\sigma\rangle = \mathbf{P}$ and $\langle\mathbf{S}\rangle = (\hbar/2)\langle\sigma\rangle = (\hbar/2)\mathbf{P}$. We note that $\mathbf{P}$ gives the average spin vector of the ensemble and is called the *polarization vector*. The spin is completely polarized (a pure state) if $\mathbf{P}^2 = 1$. The spin is said to be in a pure state corresponding to the eigenstate of $\mathbf{P}\cdot\mathbf{S}$ with the eigenvalue $\hbar/2$. $\mathbf{P} = 0$ corresponds to an unpolarized spin. Further, we obtain

$$\text{Tr}\rho^2 = \text{Tr}\frac{1}{4}\left(1 + \mathbf{P}^2 + 2\sigma\cdot\mathbf{P}\right) = \frac{1}{2}\left(1 + \mathbf{P}^2\right). \tag{11.121}$$

Therefore, $\text{Tr}\rho^2$ is maximum when $\mathbf{P}^2 = 1$. In other words, $\text{Tr}\rho^2 = 1$ represents a full polarization and $\text{Tr}\rho^2 = 1/2$ corresponds to null polarization.

11.5 Spin-Orbit Coupling

There is a magnetic moment (μ) associated with the intrinsic rotation of the electron. It is proportional to the spin $\mathbf{S}$. The magnetic moment interacts with the magnetic field created by the orbital motion of the electron. This is called *spin-orbital interaction*. The energy of this interaction is proportional to $\mathbf{L}\cdot\mathbf{S}$. For an electron in a spherically symmetric potential the proportionality factor is shown to be $(e/2m^2c^2r)\mathrm{d}V/\mathrm{d}r$. That is,

$$H_{\text{SOI}} = \frac{e}{2m^2c^2r}\frac{\mathrm{d}V}{\mathrm{d}r}\mathbf{L}\cdot\mathbf{S}. \tag{11.122}$$

In the presence of any external electromagnetic field, the perturbed Hamiltonian is

$$H_{\text{orb}} = \frac{1}{2m}\left(\mathbf{p} - \frac{e\mathbf{A}}{c}\right)^2 + V(r) + e\phi. \tag{11.123}$$

If $\phi = \nabla\cdot\mathbf{A} = 0$ then

$$H_{\text{orb}} = \frac{1}{2m}\mathbf{p}^2 - \frac{e}{mc}\mathbf{A}\cdot\mathbf{p} + V(r) + \frac{e^2}{2mc^2}\mathbf{A}^2. \tag{11.124}$$

For a homogeneous magnetic field $\mathbf{H}_{\text{B}}$ we choose $\mathbf{A} = (\mathbf{H}_{\text{B}}\times\mathbf{r})/2$. Then

$$-\frac{e}{mc}\mathbf{A}\cdot\mathbf{p} = -\frac{e}{2mc}(\mathbf{H}_{\text{B}}\times\mathbf{r})\cdot\mathbf{p} = -\mathbf{H}_{\text{B}}\cdot\frac{e}{2mc}\mathbf{L} = -\mathbf{H}_{\text{B}}\cdot\mu_L, \tag{11.125}$$

where $\mu_L = e\mathbf{L}/(2mc)$. If we include spin then we have the spin-magnetic moment interaction energy H_{SM} given by

$$H_{\text{SM}} = -\mu_{\text{S}} \cdot \mathbf{H}_{\text{B}} , \quad \mu_{\text{S}} = \frac{e}{mc}\mathbf{S} = \frac{e\hbar}{2mc}\sigma . \tag{11.126}$$

Now, the total Hamiltonian is

$$H = H_0 + H_{\text{mag}} + H_{\text{SOI}} , \quad H_0 = \frac{1}{2m}\mathbf{p}^2 + V , \tag{11.127a}$$

and

$$H_{\text{mag}} = -\frac{e}{2mc}\mathbf{H}_{\text{B}} \cdot (\mathbf{L} + 2\mathbf{S}) . \tag{11.127b}$$

The spin-orbit interaction or coupling gives a shift in energy due to the electron possessing both orbital and spin angular momenta.

Spin-orbit coupling is found to be important in the field of spintronics [16]. Spin-splitting in energy spectrum due to spin-orbit coupling in nanotubes and a model of spin-orbit interaction in graphene have been reported [17]. Spin-orbit coupling induced spin-Hall effect [18], BE condensation [19,20] and new exotic phases of matter [21] were realized. The effect of spin-orbit interaction on band structure in quantum wires and graphene was investigated [22,23]. The spin and orbital entanglement is found to give rise new and interesting phenomena such as relativistic Mott-insulating behaviour in transition metal oxides, topological nontrivial states and superexchange borals and spin liquid states. Spin-orbit interaction has received a great interest in doped spin-orbital systems, Kondo systems, Majorana and Weyl fermions, multiferroics, spin Hall effect and skyrmion lattices.

What is the form of the probability current density for a particle with spin and experiencing the spin-orbit interaction? We obtain the continuity equation for spin-1/2 electrons. Choose the set $\{\mathbf{X}, S_z\}$ as a complete set of commuting operators and generalize ψ as $\psi(\mathbf{X}, \sigma, t)$. That is,

$$\psi(\mathbf{X}, \sigma, t) = \begin{bmatrix} \psi(\mathbf{X}, \uparrow, t) \\ \psi(\mathbf{X}, \downarrow, t) \end{bmatrix} , \tag{11.128}$$

where $\uparrow$ and $\downarrow$ take the value $1/2$ and $-1/2$, respectively. Consequently, the probability density function $\rho(\mathbf{X}, t)$ is given by

$$\rho(\mathbf{X}, t) = \sum_{\sigma=\uparrow,\downarrow} |\psi(\mathbf{X}, \sigma, t)|^2 . \tag{11.129}$$

The Hamiltonian including the spin-orbit interaction is

$$H = \frac{1}{2m}\mathbf{p}^2 + V + \frac{1}{2m^2c^2}\mathbf{S} \cdot \nabla V \times \mathbf{p}. \tag{11.130}$$

Denoting $\psi_\sigma = \psi(\mathbf{X}, \sigma, t)$ the time-dependent Schrödinger equation takes the form

$$i\hbar\frac{\partial\psi_\sigma}{\partial t} = -\frac{\hbar^2}{2m}\nabla^2\psi_\sigma + V\psi_\sigma + \frac{1}{2m^2c^2}\sum_{\sigma'=\uparrow,\downarrow}\mathbf{S}_{\sigma,\sigma'} \cdot \nabla V \times \frac{\hbar}{i}\nabla\psi_{\sigma'}. \tag{11.131}$$

Then $\mathbf{J}$ is obtained as [24]

$$\mathbf{J} = \mathbf{J}_{\text{P}} + \mathbf{J}_{\text{SOI}}, \tag{11.132a}$$

where

$$\mathbf{J}_{\mathrm{P}} \;=\; \frac{\hbar}{2mi} \sum_{\sigma=\uparrow,\downarrow} (\psi_\sigma^* \nabla \psi_\sigma - \psi_\sigma \nabla \psi_\sigma^*)\,, \tag{11.132b}$$

$$\mathbf{J}_{\mathrm{SOI}} \;=\; \frac{1}{2m^2 c^2} \sum_{\sigma,\sigma'=\uparrow,\downarrow} \psi_\sigma^* \mathbf{S}_{\sigma,\sigma'} \psi_{\sigma'} \times \nabla V\,. \tag{11.132c}$$

In addition to $\mathbf{J}_{\mathrm{P}}$ and $\mathbf{J}_{\mathrm{SOI}}$ there is a magnetization probability current

$$\mathbf{J}_{\mathrm{M}} \;=\; -\frac{1}{e}\mathbf{J} = \frac{g}{2m}\nabla \times (\psi^\dagger \mathbf{S}\psi)\,, \tag{11.133}$$

where g is the electron-g factor ≈ 2.

11.6   Addition of Angular Momenta

Let us introduce J_x, J_y and J_z as the operators corresponding to the components of an angular momentum which may be an orbital angular momentum or a spin angular momentum or even an addition of two angular momenta. Coupling angular momenta may arise due to an interaction. We may think of coupling orbital and spin angular momenta of a particle or coupling of the angular momenta of different particles.

Suppose we consider a system of particles. The properties obeyed by them for a single particle are also satisfied by the respective operators for a system of particles. Let us consider the orbital angular momentum. The total angular momentum operator for a system of particles is defined as

$$\mathbf{J} \;=\; \sum_i \mathbf{J}(i)\,, \quad \mathbf{J}^2 = J_x^2 + J_y^2 + J_z^2\,, \tag{11.134a}$$

$$J_x \;=\; \sum_i J_x(i)\,, \quad J_y = \sum_i J_y(i)\,, \quad J_z = \sum_i J_z(i)\,. \tag{11.134b}$$

We obtain

$$[J_x, J_y] = \left[\sum_i J_x(i), \sum_k J_y(k)\right] = \sum_i \sum_k [J_x(i), J_y(k)] = i\hbar J_z\,. \tag{11.135a}$$

Similarly,

$$[J_y, J_z] \;=\; i\hbar J_x\,, \quad [J_z, J_x] = i\hbar J_y\,, \tag{11.135b}$$

$$[\mathbf{J}^2, J_x] \;=\; 0\,, \quad [\mathbf{J}^2, J_y] = 0\,, \quad [\mathbf{J}^2, J_z] = 0\,. \tag{11.135c}$$

The eigenvalue equations for $\mathbf{J}^2, J_z$ and $J_\pm$ are

$$\mathbf{J}^2 |jm\rangle \;=\; j(j+1)\hbar^2|jm\rangle\,, \tag{11.136a}$$

$$J_z|jm\rangle \;=\; m\hbar|jm\rangle\,, \tag{11.136b}$$

$$J_\pm|jm\rangle \;=\; [(j \mp m)(j \pm m + 1)]^{1/2}|jm \pm 1\rangle\,. \tag{11.136c}$$

The eigenvalues of $\mathbf{J}^2$ can also be obtained based on the uncertainty relation for the components of J [14].

11.6.1   Clebsch–Gordan Coefficients

Now, we describe the combining of the angular momenta associated with various subsystems of a given system. For simplicity, consider the addition of two angular momenta $\mathbf{J}_1$ and $\mathbf{J}_2$ of two noninteracting systems 1 and 2. Here $\mathbf{J}_{1,2}$ is the sum of the orbital angular momenta $\mathbf{L}_{1,2}$ and spin angular momenta $\mathbf{S}_{1,2}$. $\mathbf{J}_1$ and $\mathbf{J}_2$ are combined to form a total angular momentum $\mathbf{J}$ as $\mathbf{J} = \mathbf{J}_1 + \mathbf{J}_2$. The individual eigenstates of 1 and 2 satisfy

$$\mathbf{J}_1^2|j_1m_1\rangle = j_1(j_1+1)\hbar^2|j_1m_1\rangle, \tag{11.137a}$$
$$\mathbf{J}_{1z}|j_1m_1\rangle = m_1\hbar|j_1m_1\rangle, \tag{11.137b}$$
$$\mathbf{J}_2^2|j_2m_2\rangle = j_2(j_2+1)\hbar^2|j_2m_2\rangle, \tag{11.137c}$$
$$J_{2z}|j_2m_2\rangle = m_2\hbar|j_2m_2\rangle. \tag{11.137d}$$

For noninteracting systems $\mathbf{J}_1$, $\mathbf{J}_2$ and $\mathbf{J}$ are fixed. The eigenstates $|j_1m_1; j_2m_2\rangle$ separate into $|j_1m_1\rangle$ and $|j_2m_2\rangle$. We write $|j_1m_1; j_2m_2\rangle$ as a tensor product of $|j_1m_1\rangle$ and $|j_2m_2\rangle$. We define the tensor product or basis vectors as

$$|j_1m_1; j_2m_2\rangle = |j_1m_1\rangle|j_2m_2\rangle, \tag{11.138}$$

where $m_1 = -j_1, \ldots, j_1$, $m_2 = -j_2, \ldots, j_2$.

Let us denote the new basis set as $|jm\rangle$ with

$$\mathbf{J}^2|jm\rangle = j(j+1)\hbar^2|jm\rangle, \tag{11.139a}$$
$$J_z|jm\rangle = m\hbar|jm\rangle. \tag{11.139b}$$

Any $|jm\rangle$ is written as a linear combination of the states $|j_1m_1; j_2m_2\rangle$:

$$|jm\rangle = \sum_{m_1}\sum_{m_2} |j_1m_1; j_2m_2\rangle\langle j_1m_1; j_2m_2|jm\rangle. \tag{11.140}$$

The coefficients $\langle j_1m_1; j_2m_2|jm\rangle$ are called *Clebsch–Gordan*[3] *(CG) coefficients* or vector coupling coefficients. They are determined by a systematic procedure and are required to construct the states $|jm\rangle$. For simplicity, we use the notation $\langle m_1; m_2|$ for $\langle j_1m_1; j_2m_2|$.

Because $J_z = J_{1z} + J_{2z}$ from Eqs. (11.137b), (11.137d) and (11.139b) we get $m = m_1 + m_2$. That is,

$$\langle m_1; m_2|jm\rangle = 0, \quad \text{unless } m_2 = m - m_1. \tag{11.141}$$

This restricts the summation in Eq. (11.140):

$$|jm\rangle = \sum_{m_1} |m_1; m-m_1\rangle\langle m_1; m-m_1|jm\rangle. \tag{11.142}$$

The maximum value of m is $j_1 + j_2$ since the maximum values of m_1 and m_2 are j_1 and j_2, respectively. $j_1 + j_2$ is also the maximum value of j. The minimum value of j is $|j_1 - j_2|$. The possible values of j are

$$j_1 + j_2, \; j_1 + j_2 - 1, \; \ldots, \; |j_1 - j_2|. \tag{11.143}$$

---

[3]Paul Albert Gordan was born on 27 April 1837 in Breslau, Germany (now Wroclaw, Poland) and died on 21 December 1912 in Erlangen, Germany. Rudolf Friedrich Alfred Clebsch was a German mathematician born on 19 January 1833 and died on 7 November 1872.

We prove that if Eq. (11.143) holds then the dimensionality of the space spanned by $|j_1 m_1; j_2 m_2\rangle$ is the same as that of the space spanned by $|jm\rangle$. Since m_1 takes $2j_1 + 1$ values and m_2 takes $2j_2 + 1$ values, the dimensionality of the space spanned by $|j_1 m_1; j_2 m_2\rangle$ is

$$N = (2j_1 + 1)(2j_2 + 1) . \tag{11.144a}$$

To find the dimensionality of the space spanned by $|jm\rangle$ kets, we write (assuming $j_1 > j_2$)

$$
\begin{aligned}
N &= \sum_{j=j_1-j_2}^{j_1+j_2} (2j+1) \\
&= \frac{1}{2} \left[2(j_1 - j_2) + 1 + 2(j_1 + j_2) + 1 \right] (2j_2 + 1) \\
&= (2j_1 + 1)(2j_2 + 1) .
\end{aligned} \tag{11.144b}
$$

Since both the space have the same dimensionality Eq. (11.140) holds.

11.6.2 Recursion Relations for the Clebsch–Gordan Coefficients

The CG coefficients form a unitary matrix. Further more, the matrix elements are taken to be real by convention. The unitary matrix becomes an orthogonal matrix. We can easily verify that the CG coefficients are real and orthogonal:

$$\langle m_1; m_2 | jm \rangle = \langle jm | m_1; m_2 \rangle , \tag{11.145a}$$

$$\sum_{m_1} \sum_{m_2} \langle jm | m_1; m_2 \rangle \langle m_1; m_2 | j'm' \rangle = \delta_{jj'} \delta_{mm'} , \tag{11.145b}$$

$$\sum_{j} \sum_{m} \langle m_1; m_2 | jm \rangle \langle jm | m_1'; m_2' \rangle = \delta_{m_1 m_1'} \delta_{m_2 m_2'} . \tag{11.145c}$$

Consider the special case: $j_1 = j$, $m' = m = m_1 + m_2$. Then we obtain from (11.145b)

$$\sum_{m_1} \sum_{m_2} |\langle m_1; m_2 | jm \rangle|^2 = 1 \tag{11.145d}$$

which is the normalization condition for $|jm\rangle$.

With j_1, j_2 and j fixed, the coefficients with different m_1 and m_2 are related to each other by recursion relations. Using Eq. (11.139) we get

$$J_{\pm} |jm\rangle = (J_{1\pm} + J_{2\pm}) \sum_{m_1} \sum_{m_2} |m_1 m_2\rangle \langle m_1 m_2 | jm \rangle . \tag{11.145e}$$

Using Eq. (11.136c), we obtain (with $m_1 \to m_1'; m_2 \to m_2'$)

$$
\begin{aligned}
[j(j+1) &- m(m \pm 1)]^{1/2} |jm \pm 1\rangle \\
&= \sum_{m_1'} \sum_{m_2'} [j_1(j_1+1) - m_1(m_1 \pm 1)]^{1/2} |m_1' \pm 1; m_2'\rangle \\
&\quad + [j_2(j_2+1) - m_2(m_2 \pm 1)]^{1/2} |m_1'; m_2' \pm 1\rangle .
\end{aligned} \tag{11.145f}
$$

Premultiplying with $\langle m_1; m_2 |$ and using orthogonality condition which means nonvanishing contributions from the right-side are possible only with $m_1 = m_1' \pm 1$, $m_2 = m_2'$ for the first term and $m_1 = m_1'$, $m_2 = m_2' \pm 1$ for the second term we obtain the desired recursion relations

$$[j(j+1) - m(m \pm 1)]^{1/2} \langle m_1; m_2 | jm \pm 1 \rangle$$
$$= [j_1(j_1+1) - m_1(m_1 \mp 1)]^{1/2} \langle m_1 \pm 1; m_2 | jm \rangle$$
$$+ [j_2(j_2+1) - m_2(m_2 \mp 1)]^{1/2} \langle m_1; m_2 \mp 1 | jm \rangle. \qquad (11.146)$$

These recursion relations and the normalization condition determine all the CG coefficients.

11.6.3 General Formula for the Clebsch–Gordan Coefficients

The general formula for the CG coefficients is

$$\langle m_1; m_2 | jm \rangle = \delta_{m_1+m_2;m} \, \alpha \beta \,, \qquad (11.147a)$$

where

$$\alpha = (2j+1) \frac{(s-2j)! \, (s-2j_2)! \, (s-2j_1)!}{(s+1)!} (j_1+m_1)! \, (j_1-m_1)!$$
$$\times (j_2+m_2)! \, (j_2-m_2)! \, (j+m)! \, (j-m)! \,, \qquad (11.147b)$$

$$\beta = \sum_\nu \left\{ \frac{(-1)^\nu}{\nu! (j_1+j_2-j-\nu)! \, (j_1-m_1-\nu)! \, (j_2+m_2-\nu)!} \right.$$
$$\left. \times \frac{1}{(j-j_2+m_1+\nu)! \, (j-j_2-m_2+\nu)!} \right\}. \qquad (11.147c)$$

In Eqs. (11.147) $s = j_1 + j_2 + j$ and the ν summation runs over all positive integer values for which all of the factorial arguments are ≥ 0. The above formula is difficult to work with, so we present three special cases.

Case (i): $m_1 = j_1$, $m_2 = j_2$

$$\langle m_1; m_2 | jm \rangle = \langle j_1 j_1; j_2 j_2 | jj \rangle = 1 \,. \qquad (11.148)$$

Case (ii): $m_1 = \pm j_1$ or $m_2 = \pm j_2$ and $m = \pm j$

$$\langle j_1 - m_1; j_2 - m_2 | j - j \rangle$$
$$= (-1)^{j_1-m_1} \left[\frac{(2j+1)! \, (j_1+j_2-j)!}{(j_1+j_2+j+1)! \, (j+j_1-j_2)! \, (j-j_1+j_2)!} \right]^{1/2}$$
$$\times \left[\frac{(j_1+m_1)! \, (j_2+m_2)!}{(j_1-m_1)! \, (j_2-m_2)!} \right]^{1/2} . \qquad (11.149)$$

Case (iii): $j_1 + j_2 = j$

$$\langle j_1 m_1; j_2 m_2 | jm \rangle = \left[\frac{(j+m)! \, (j-m)! \, (2j_1)! \, (2j_2)!}{(2j)! \, (j_1+m_1)! \, (j_1-m_1)! \, (j_2+m_2)! \, (j_2-m_2)!} \right]^{1/2} . \qquad (11.150)$$

11.6.4 Example

Consider the addition of spin-1/2 angular momenta. We have $j_1 = j_2 = 1/2$. The possible values of j are 1 and 0. The values of m_1 and m_2 are $-1/2$ and $1/2$. The values of m are 1, 0 and -1. The states $|jm\rangle$ are

$$|jm\rangle = |11\rangle, \ |10\rangle, \ |1-1\rangle, \ |00\rangle \,. \qquad (11.151)$$

Consider the state $|11\rangle$. It is expanded as

$$|11\rangle = \sum_{m_1=-1/2}^{1/2} \sum_{m_2=-1/2}^{1/2} \left|\frac{m_1}{2}; \frac{m_2}{2}\right\rangle \left\langle \frac{m_1}{2}; \frac{m_2}{2}\middle|1\right\rangle. \tag{11.152}$$

Since $m_1 = m_2 = 1/2$ alone gives $m_1 + m_2 = 1$ we have

$$|11\rangle = \left|\frac{1}{2}\frac{1}{2}; \frac{1}{2}\frac{1}{2}\right\rangle \left\langle \frac{1}{2}\frac{1}{2}; \frac{1}{2}\frac{1}{2}\middle|1\right\rangle. \tag{11.153}$$

Here $m_1 = j_1$, $m_2 = j_2$ and this belongs to the case (i). The CG coefficient $\left\langle \frac{1}{2}\frac{1}{2}; \frac{1}{2}\frac{1}{2}\middle|1\right\rangle = 1$. Then

$$|11\rangle = \left|\frac{1}{2}\frac{1}{2}; \frac{1}{2}\frac{1}{2}\right\rangle. \tag{11.154}$$

The state $|10\rangle$ is expanded as

$$|10\rangle = \sum_{m_1} \sum_{m_2} \left|\frac{m_1}{2}; \frac{m_2}{2}\right\rangle \left\langle \frac{m_1}{2}; \frac{m_2}{2}\middle|0\right\rangle. \tag{11.155}$$

$m = m_1 + m_2 = 0$ is possible for (i) $m_1 = 1/2$, $m_2 = -1/2$ and (ii) $m_1 = -1/2$, $m_2 = 1/2$. Therefore,

$$|10\rangle = \left|\frac{1}{2}\frac{1}{2}; \frac{1}{2}\frac{-1}{2}\right\rangle \left\langle \frac{1}{2}\frac{1}{2}; \frac{1}{2}\frac{-1}{2}\middle|0\right\rangle + \left|\frac{1}{2}\frac{-1}{2}; \frac{1}{2}\frac{1}{2}\right\rangle \left\langle \frac{1}{2}\frac{-1}{2}; \frac{1}{2}\frac{1}{2}\middle|0\right\rangle. \tag{11.156}$$

Since $j = j_1 + j_2 = 1$ we use the result of the case (iii). We obtain

$$\left\langle \frac{1}{2}; \frac{1}{2}\frac{-1}{2}\middle|0\right\rangle = \sqrt{\frac{1!\,1!}{2!}} \sqrt{\frac{1!\,1!}{1!\,0!\,0!\,1!}} = \frac{1}{\sqrt{2}}, \tag{11.157a}$$

$$\left\langle \frac{1}{2}\frac{-1}{2}; \frac{1}{2}\frac{1}{2}\middle|0\right\rangle = \sqrt{\frac{1!\,1!}{2!}} \sqrt{\frac{1!\,1!}{0!\,1!\,1!\,0!}} = \frac{1}{\sqrt{2}} \tag{11.157b}$$

and

$$|10\rangle = \frac{1}{\sqrt{2}} \left(\left|\frac{1}{2}\frac{1}{2}; \frac{1}{2}\frac{-1}{2}\right\rangle + \left|\frac{1}{2}\frac{-1}{2}; \frac{1}{2}\frac{1}{2}\right\rangle \right). \tag{11.157c}$$

The state $|00\rangle$ is written as

$$\begin{aligned}
|00\rangle &= \sum_{m_1} \sum_{m_2} \left|\frac{m_1}{2}; \frac{m_2}{2}\right\rangle \left\langle \frac{m_1}{2}; \frac{m_2}{2}\middle|0\right\rangle \\
&= \left\langle \frac{1}{2}\frac{1}{2}; \frac{1}{2}\frac{-1}{2}\middle|0\right\rangle + \left\langle \frac{1}{2}\frac{-1}{2}; \frac{1}{2}\frac{1}{2}\middle|0\right\rangle.
\end{aligned} \tag{11.158}$$

Since $m = j$ from the case (iii) we get

$$\left\langle \frac{1}{2}\frac{1}{2}; \frac{1}{2}\frac{-1}{2}\middle|0\right\rangle = (-1)^{\frac{1}{2}-\frac{1}{2}} \sqrt{\frac{1!\,1!}{2!\,0!\,0!}} \sqrt{\frac{1!\,0!}{0!\,1!}} = \frac{1}{\sqrt{2}}, \tag{11.159a}$$

$$\left\langle \frac{1}{2}\frac{-1}{2}; \frac{1}{2}\frac{1}{2}\middle|0\right\rangle = -\frac{1}{\sqrt{2}}. \tag{11.159b}$$

TABLE 11.1
The CG coefficients for $j_2 = 1/2$ or 1.

	m_2	
j	$1/2$	$-1/2$
$j_1 + \frac{1}{2}$	$\sqrt{\dfrac{j_1 + m + \frac{1}{2}}{2j_1 + 1}}$	$\sqrt{\dfrac{j_1 - m + \frac{1}{2}}{2j_1 + 1}}$
$j_1 - \frac{1}{2}$	$-\sqrt{\dfrac{j_1 - m + \frac{1}{2}}{2j_1 + 1}}$	$\sqrt{\dfrac{j_1 + m + \frac{1}{2}}{2j_1 + 1}}$

TABLE 11.2
The CG coefficients $\langle j_1\, m - m_2;\, 1\, m_2 | jm \rangle$. Here $\alpha_\pm = j_1 \pm m + 1$.

j	$m_2 = 1$	$m_2 = 0$	$m_2 = -1$
$j_1 + 1$	$\left[\dfrac{(j_1+m)\alpha_+}{(2j_1+1)(2j_1+2)}\right]^{1/2}$	$\left[\dfrac{\alpha_-\alpha_+}{(2j_1+1)(j_1+1)}\right]^{1/2}$	$\left[\dfrac{(j_1-m)\alpha_-}{(2j_1+1)(2j_1+2)}\right]^{1/2}$
j_1	$-\left[\dfrac{(j_1+m)\alpha_-}{2j_1(j_1+1)}\right]^{1/2}$	$\left[\dfrac{m^2}{j_1(j_1+1)}\right]^{1/2}$	$\left[\dfrac{(j_1-m)\alpha_+}{2j_1(j_1+1)}\right]^{1/2}$
$j_1 - 1$	$\left[\dfrac{(j_1-m)\alpha_-}{2j_1(2j_1+1)}\right]^{1/2}$	$-\left[\dfrac{(j_1-m)(j_1+m)}{j_1(2j_1+1)}\right]^{1/2}$	$\left[\dfrac{\alpha_+(j_1+m)}{2j_1(2j_1+1)}\right]^{1/2}$

We obtain

$$|1 - 1\rangle = \left|\frac{1}{2}\frac{-1}{2}; \frac{1}{2}\frac{-1}{2}\right\rangle. \tag{11.160}$$

Tables 11.1 and 11.2 present the CG coefficients where $j_2 = 1/2$ or 1.

The eigenvectors of $\mathbf{L}^2$ and L_z in coordinate representation are $\langle \theta\phi|lm \rangle = Y_{lm}(\theta, \phi)$. The coordinate representation for the composite state $|l_1 m_1; l_2 m_2\rangle$ is given by the projection

$$\langle \theta_1\phi_1; \theta_2\phi_2 | l_1 m_1; l_2 m_2 \rangle = Y_{l_1 m_1}(\theta_1, \phi_1) Y_{l_2 m_2}(\theta_2, \phi_2). \tag{11.161}$$

Therefore, $\langle \theta\phi|lm \rangle = Y_{lm}(\theta, \phi)$ gives

$$Y_{lm}(\theta, \phi) = \sum_{m_1} \sum_{m_2} \langle l_1 m_1; l_2 m_2 | lm \rangle Y_{l_1 m_1}(\theta_1, \phi_1) Y_{l_2 m_2}(\theta_2, \phi_2). \tag{11.162}$$

Thus, we notice that the CG expansion affords a way of obtaining the coordinate representation of the composite state $Y_{lm}(\theta, \phi)$.

Solved Problem 5:

Obtain the matrix representation of the components of angular momentum for $j = 1/2$ and $l = 1$.

For $j = 1/2$, the values of m and m' are $-j$, j, that is $-1/2$, $1/2$. The eigenvalues of $\mathbf{J}^2$ are $j(j+1)\hbar^2 = 3\hbar^2/4$, whereas the eigenvalues of J_z are $m\hbar = -\hbar/2$, $\hbar/2$. Now,

$$
\mathbf{J}^2 = \begin{pmatrix} \langle jj|\mathbf{J}^2|jj\rangle & \langle jj|\mathbf{J}^2|j-j\rangle \\ \langle j-j|\mathbf{J}^2|jj\rangle & \langle j-j|\mathbf{J}^2|j-j\rangle \end{pmatrix}
$$

$$
= \begin{pmatrix} \langle \tfrac{1}{2}\tfrac{1}{2}|\mathbf{J}^2|\tfrac{1}{2}\tfrac{1}{2}\rangle & \langle \tfrac{1}{2}\tfrac{1}{2}|\mathbf{J}^2|\tfrac{1}{2}\tfrac{-1}{2}\rangle \\ \langle \tfrac{1}{2}\tfrac{-1}{2}|\mathbf{J}^2|\tfrac{1}{2}\tfrac{1}{2}\rangle & \langle \tfrac{1}{2}\tfrac{-1}{2}|\mathbf{J}^2|\tfrac{1}{2}\tfrac{-1}{2}\rangle \end{pmatrix}. \tag{11.163}
$$

The matrix elements are determined from Eq. (11.75). They are

$$
\left\langle \tfrac{1}{2}\tfrac{1}{2}\Big|\mathbf{J}^2\Big|\tfrac{1}{2}\tfrac{1}{2}\right\rangle = \frac{3}{4}\hbar^2, \quad \left\langle \tfrac{1}{2}\tfrac{1}{2}\Big|\mathbf{J}^2\Big|\tfrac{1}{2}\tfrac{-1}{2}\right\rangle = 0, \tag{11.164a}
$$

$$
\left\langle \tfrac{1}{2}\tfrac{-1}{2}\Big|\mathbf{J}^2\Big|\tfrac{1}{2}\tfrac{1}{2}\right\rangle = 0, \quad \left\langle \tfrac{1}{2}\tfrac{-1}{2}\Big|\mathbf{J}^2\Big|\tfrac{1}{2}\tfrac{-1}{2}\right\rangle = \frac{3}{4}\hbar^2. \tag{11.164b}
$$

Then

$$
\mathbf{J}^2 = \frac{3}{4}\hbar^2 \begin{pmatrix} 1 & 0 \\ 0 & 1 \end{pmatrix} = \frac{3}{4}\hbar^2 I. \tag{11.165}
$$

Similarly, we obtain

$$
J_z = \frac{\hbar}{2}\begin{pmatrix} 1 & 0 \\ 0 & -1 \end{pmatrix} = \frac{\hbar}{2}\sigma_z, \tag{11.166a}
$$

$$
J_+ = \hbar\begin{pmatrix} 0 & 1 \\ 0 & 0 \end{pmatrix}, \quad J_- = \hbar\begin{pmatrix} 0 & 0 \\ 1 & 0 \end{pmatrix} \tag{11.166b}
$$

and

$$
J_x = \frac{1}{2}(J_+ + J_-) = \frac{\hbar}{2}\begin{pmatrix} 0 & 1 \\ 1 & 0 \end{pmatrix} = \frac{\hbar}{2}\sigma_x, \tag{11.166c}
$$

$$
J_y = \frac{1}{2i}(J_+ - J_-) = \frac{\hbar}{2}\begin{pmatrix} 0 & -i \\ i & 0 \end{pmatrix} = \frac{\hbar}{2}\sigma_y. \tag{11.166d}
$$

As already discussed $l = 1/2$ cannot represent orbital angular momentum, the above represents spin angular momentum with quantum number $s = 1/2$.

For $l = 1$ we obtain

$$
\mathbf{L}^2 = 2\hbar^2\begin{pmatrix} 1 & 0 & 0 \\ 0 & 1 & 0 \\ 0 & 0 & 1 \end{pmatrix}, \quad L_z = \hbar\begin{pmatrix} 1 & 0 & 0 \\ 0 & 0 & 0 \\ 0 & 0 & -1 \end{pmatrix}, \tag{11.167a}
$$

$$
L_x = \frac{\hbar}{\sqrt{2}}\begin{pmatrix} 0 & 1 & 0 \\ 1 & 0 & 1 \\ 0 & 1 & 0 \end{pmatrix}, \quad L_y = \frac{i\hbar}{\sqrt{2}}\begin{pmatrix} 0 & -1 & 0 \\ 1 & 0 & -1 \\ 0 & 1 & 0 \end{pmatrix}. \tag{11.167b}
$$

11.7 Rotational Transformation of a Vector Wave Function and Spin

For a scalar field ψ the operators L_x, L_y and L_z are the generators of the infinitesimal rotation about the three coordinate axes through the angles $\Delta\theta_x$, $\Delta\theta_y$ and $\Delta\theta_z$, respectively.

A particle possessing internal degrees of freedom like *spin* is classified by considering the transformation properties of the wave function under rotation. For an infinitesimal rotation

$$\Delta\theta\mathbf{n} = \mathbf{i}\Delta\theta_x + \mathbf{j}\Delta\theta_y + \mathbf{k}\Delta\theta_z \tag{11.168}$$

the relation $\mathbf{r}' = R(\mathbf{n}, \theta)\mathbf{r}$ up to the first-order in $\Delta\theta$ is

$$\mathbf{r}' = \mathbf{r} + \Delta\theta\mathbf{n} \times \mathbf{r}, \quad R(\mathbf{n}, \Delta\theta) \approx \begin{pmatrix} 1 & -\Delta\theta_z & \Delta\theta_y \\ \Delta\theta_z & 1 & -\Delta\theta_x \\ -\Delta\theta_y & \Delta\theta_x & 1 \end{pmatrix}. \tag{11.169}$$

If $\psi(\mathbf{r})$ transforms like a scalar then we have the Eqs. (11.3) and (11.8).

For a particle described by a vector wave function $\Psi(\mathbf{r})$ instead of a scalar wave function $\psi(\mathbf{r})$, both $\mathbf{r}$ and $\Psi(\mathbf{r})$ will transform like a vector due to a rotation. Hence, the transformation must be given as

$$\begin{aligned} \Psi'(\mathbf{r}) &= R\Psi\left(R^{-1}\mathbf{r}\right) \\ &= \Psi\left(R^{-1}\mathbf{r}\right) + \Delta\theta\mathbf{n} \times \Psi\left(R^{-1}\mathbf{r}\right) \\ &= \Psi(R(\mathbf{n}, -\Delta\theta)\mathbf{r}) + \Delta\theta\mathbf{n} \times \Psi(\mathbf{r}) \\ &= \Psi(\mathbf{r}) - \frac{i}{\hbar}(\Delta\theta\mathbf{n} \cdot \mathbf{L})\Psi(\mathbf{r}) + \Delta\theta\mathbf{n} \times \Psi(\mathbf{r}). \end{aligned} \tag{11.170}$$

Using

$$\Delta\theta\,\mathbf{n} \times \Psi(\mathbf{r}) = -\frac{i}{\hbar}(\Delta\theta\,\mathbf{n} \cdot \mathbf{S})\Psi_\alpha(\mathbf{r}) \tag{11.171}$$

we get the components of $\mathbf{S}$ as the following 3×3 matrices

$$S_x = i\hbar \begin{pmatrix} 0 & 0 & 0 \\ 0 & 0 & -1 \\ 0 & 1 & 0 \end{pmatrix}, \quad S_y = i\hbar \begin{pmatrix} 0 & 0 & 1 \\ 0 & 0 & 0 \\ -1 & 0 & 0 \end{pmatrix},$$

$$S_z = i\hbar \begin{pmatrix} 0 & -1 & 0 \\ 1 & 0 & 0 \\ 0 & 0 & 0 \end{pmatrix}. \tag{11.172}$$

The transformation is given as

$$\Psi'(\mathbf{r}) = U_R(\mathbf{n}, \Delta\theta)\psi(\mathbf{r}). \tag{11.173}$$

From Eqs. (11.171)–(11.173) we get the transformation equation for the infinitesimal rotation of a vector wave function as

$$U_R(\mathbf{n}, \Delta\theta) = I - \frac{i}{\hbar}\Delta\theta\,\mathbf{n} \cdot (\mathbf{L} + \mathbf{S}). \tag{11.174}$$

If the wave function transforms like a vector then the three generators of infinitesimal rotations are

$$J_x = L_x + S_x, \quad J_y = L_y + S_y \text{ and } J_z = L_z + S_z, \tag{11.175}$$

where $\mathbf{J} = \mathbf{L} + \mathbf{S}$ is identified as the total angular momentum of the particle. Since S_x, S_y and S_z are independent of x, y and z, all the components of $\mathbf{L}$ commute with all the components of $\mathbf{S}$. $\mathbf{L}$ acts only on the $\mathbf{r}$ dependence of $\Psi(\mathbf{r})$ and the operator $\mathbf{S}$ rearranges the components of the vector wave function $\Psi(\mathbf{r})$ without affecting its $\mathbf{r}$ dependence. In case where H contains terms coupling $\mathbf{S}$ and $\mathbf{L}$ to each other, $\mathbf{L}$ and $\mathbf{S}$ do not separately commutes with H but $\mathbf{J}$ commutes with it.

We find from Eq. (11.172) that the total spin operator $\mathbf{S}^2 = S_x^2 + S_y^2 + S_z^2$ has the eigenvalue $2\hbar^2$. Since the eigenvalue of the angular momentum operator is $s(s+1)\hbar^2$, we find $s = 1$. If the spin quantum number of a particle is 1 then its wave function transforms like a vector. For particles like electrons having a spin quantum number $1/2$, the wave function transforms like a two-component spinor. We conclude that different transformation properties of the wave function lead to different values of the spin.

11.8 Rotational Properties of Vector Operators

A vector in classical physics transform like $V_i \rightarrow \sum_j R_{ij} V_j$, $i, j = 1, 2, 3$, where $R_{ij} = R(\mathbf{n}, \theta)$ is the 3×3 rotational matrix. It is reasonably assumed that $\langle \mathbf{V} \rangle$ in quantum mechanics transform like a classical vector under rotation. Specifically, as the state ket $|\alpha\rangle$ is changed under a finite rotation according to $|\alpha\rangle \rightarrow U_{\mathrm{R}}(\mathbf{n}, \theta)|\alpha\rangle$ the $\langle \mathbf{V} \rangle$ is assumed to change as

$$\langle \alpha | V_i | \alpha \rangle \rightarrow \langle \alpha | U_{\mathrm{R}}^\dagger(\mathbf{n}, \theta) V_i U_{\mathrm{R}}(\mathbf{n}, \theta) | \alpha \rangle = \sum_j R_{ij} \langle \alpha | V_j | \alpha \rangle . \qquad (11.176)$$

This is true for any arbitrary $|\alpha\rangle$. Therefore,

$$U_{\mathrm{R}}^\dagger(\mathbf{n}, \theta) V_i U_{\mathrm{R}}(\mathbf{n}, \theta) = \sum_j R_{ij} V_j \qquad (11.177)$$

must hold as an operator equation. Let us consider a specific case, an infinitesimal rotation $\Delta\theta$ in the direction $\mathbf{n}$. When the rotation is infinitesimal, we have

$$U_{\mathrm{R}}(\mathbf{n}, \Delta\theta) = I - \mathrm{i}\Delta\theta \, \mathbf{n} \cdot \mathbf{J}/\hbar . \qquad (11.178)$$

We now write (11.177) as

$$V_i + \frac{\Delta\theta}{\mathrm{i}\hbar} [V_i, \mathbf{n} \cdot \mathbf{J}] = \sum_j R_{ij}(\mathbf{n}, \Delta\theta) V_j . \qquad (11.179)$$

In particular, for $\mathbf{n}$ along the z-axis, we have from Eq. (11.2)

$$R(\mathbf{n}, \Delta\theta_z) = \begin{pmatrix} 1 & -\Delta\theta_z & 0 \\ \Delta\theta_z & 1 & 0 \\ 0 & 0 & 1 \end{pmatrix} . \qquad (11.180)$$

Therefore,

$$i = 1 : \ V_x + \frac{\Delta\theta_z}{\mathrm{i}\hbar} [V_x, J_z] = V_x - \Delta\theta_z V_y . \qquad (11.181a)$$

$$i = 2 : \ V_y + \frac{\Delta\theta_z}{\mathrm{i}\hbar} [V_y, J_z] = V_y + \Delta\theta_z V_x . \qquad (11.181b)$$

$$i = 3 : \ V_z + \frac{\Delta\theta_z}{\mathrm{i}\hbar} [V_z, J_z] = V_z . \qquad (11.181c)$$

This means that the vector operator $\mathbf{V}$ must satisfy the commutation relations $[V_i, J_j] = \mathrm{i}\epsilon_{ijk}\hbar V_k$, where ϵ_{ijk}, which is known as the totally antisymmetric unit tensor, is defined by $\epsilon_{123} = \epsilon_{231} = \epsilon_{312} = 1$ and $\epsilon_{321} = \epsilon_{213} = \epsilon_{132} = -1$ with all other components equal to zero.

Clearly, the behaviour of any vector operator $\mathbf{V}$ under a finite rotation is completely determined by the commutation relations (11.181). We use (11.181) as the defining property of a vector operator. Notice that the angular momentum commutation relations (11.135a) and (11.135b) are a special case of (11.181) in which we let $V_i \to J_i$, $V_k \to J_k$. Other special cases $V_i \to \mathbf{r}$, $J_k \to L_k$ give $[y, L_z] = \mathrm{i}\hbar x$, $[x, L_z] = -\mathrm{i}\hbar y$, etc. Similarly, if $V_i \to p_i$, we get $[p_x, L_z] = -\mathrm{i}\hbar p_y$, $[p_y, L_z] = \mathrm{i}\hbar p_z$, etc.

11.9   Tensor Operators and the Wigner–Eckart Theorem

In the previous sections, we examined the rotational properties of scalar and vector operators. Operators with rotational properties of higher-order than scalars and vectors can also be constructed.

11.9.1   Tensor Operators

A Cartesian tensor transforms like

$$T_{ijk...} = \sum_{i'j'k'} \ldots R_{ii'} R_{jj'} R_{kk'} \ldots T'_{i'j'k'} \tag{11.182}$$

under a rotation where R is the 3×3 rotation matrix. The number of indices is called the rank of a tensor. For example, a Cartesian tensor of rank two (called *dyadic*) is obtained as $T_{ij} = U_i V_j$, where U_i and V_j are the components of a Cartesian vector. T_{ij} has nine components. But Cartesian tensors are not in a convenient form as they can be decomposed into objects that transform differently under rotation. In other words, Cartesian tensors are reducible. For example, the dyadic in $T_{ij} = U_i V_j$ can be decomposed as

$$U_i V_j = \frac{1}{3}(\mathbf{U} \cdot \mathbf{V})\delta_{ij} + \frac{1}{2}(U_i V_j - U_j V_i) + \left[\frac{1}{2}(U_i V_j + U_j V_i) - \frac{1}{3}\mathbf{U} \cdot \mathbf{V}\delta_{ij}\right]. \tag{11.183}$$

The first term on the right-hand side, $\mathbf{U} \cdot \mathbf{V}$, is a scalar product invariant under rotations. The second term is an antisymmetric tensor and is written as $\epsilon_{ijk}(\mathbf{U} \times \mathbf{V})_k$. There are 3 independent components. The last term is a 3×3 symmetric traceless tensor with 5 ($6 - 1$, where 1 comes from the traceless condition) independent components, totally we get the required 9 independent components of T_{ij}. We note that the spherical harmonics $Y_{lm}(\theta, \phi)$ has 1 independent component for $l = 0$, 3 independent components for $l = 1$ and 5 independent components for $l = 2$. This suggests that the dyadic T_{ij} can be decomposed into tensor that transforms like spherical harmonics with $l = 0$, 1 and 2. In fact Eq. (11.183) illustrates an example of reducing a Cartesian tensor into irreducible spherical tensors as $Y_{lm}(\theta, \phi)$ is an example of spherical tensor.

We have seen that $|\alpha\rangle$ is any state, and $\hat{O}$, an operator representing some dynamical variable of a quantum system, then rotation through an infinitesimal angle $\Delta\theta$ about the direction $\mathbf{n}$ takes

$$|\alpha\rangle \to (I - \mathrm{i}\Delta\theta \mathbf{n} \cdot \mathbf{J})|\alpha\rangle \tag{11.184}$$

and

$$\hat{O} \quad \to \quad (I - \mathrm{i}\Delta\theta \mathbf{n} \cdot \mathbf{J})\hat{O}(I + \mathrm{i}\Delta\theta \mathbf{n} \cdot \mathbf{J}) \approx \hat{O} - \mathrm{i}\Delta\theta \mathbf{n} \cdot [\mathbf{J}, \hat{O}]. \tag{11.185}$$

Thus, the changes in $|\alpha\rangle$ and $\hat{O}$, brought about by infinitesimal rotations, are directly expressed in terms of $\mathbf{J}|\alpha\rangle$ and $[\mathbf{J}, \hat{O}]$, respectively. The commutator with $\mathbf{J}$ determines the rotation properties of an operator. Since $Y_{lm}(\theta, \phi)$ are tensor operators, let us study their commutation relation with $\mathbf{J}$. Since $J_z = -i\hbar \partial/\partial\phi$ we have

$$J_z Y_{lm} f(\theta, \phi) = f(\theta, \phi) J_z Y_{lm} + Y_{lm} J_z f(\theta, \phi) \tag{11.186}$$

and

$$[J_z Y_{lm} - Y_{lm} J_z] f = m\hbar Y_{lm} f. \tag{11.187}$$

Since $f(\theta, \phi)$ is arbitrary, we obtain the commutation relation

$$[J_z, Y_{lm}] = m\hbar Y_{lm}. \tag{11.188a}$$

In a similar fashion

$$[J_{\pm}, Y_{lm}] = [l(l+1) - m(m \pm 1)]^{1/2} \hbar Y_{l, m\pm 1}. \tag{11.188b}$$

Now, define an irreducible tensor operator $T(k)$, where $k = 0, 1/2, 1, 3/2, \ldots$, to be a set of $2k+1$ operators $T(k, q)$, where $q = k, k-1, \ldots, -k$ that have commutations with $\mathbf{J}$ similar to those in Eqs. (11.188):

$$[J_z, T(k, q)] = q\hbar T(k, q), \tag{11.189a}$$

$$[J_{\pm}, T(k, q)] = [k(k+1) - q(q \pm 1)]^{1/2} \hbar T(k, q \pm 1). \tag{11.189b}$$

These relations can be considered as a definition of spherical tensors. Since the tensor operator $T(k, q)$ and the spherical harmonics $|jm\rangle$ satisfy the same commutation relations (11.188) and (11.189), tensor operators can be combined in accordance with the same rule as angular momentum eigenstates.

We have the basis $|j_1 m_1, j_2 m_2\rangle$ which is the tensor product of the basis $|j_1 m_1\rangle$ and $|j_2 m_2\rangle$. We have seen in Eq. (11.140) that a new basis $|jm\rangle$ can be given as a linear coefficient of the tensor product basis $|j_1 m_1, j_2 m_2\rangle$ using the CG coefficients $\langle j_1 m_1, j_2 m_2 | jm\rangle$. Hence, an analog to Eq. (11.140) using the correspondence $T_1(k_1, q_1) \to |j_1 m_1\rangle$, $T_2(k_2, q_2) \to |j_2 m_2\rangle$ and $T(k, q) \to |jm\rangle$ gives the product of tensors relation

$$T(k, q) = \sum_{q_1 = -k_1}^{k_1} \sum_{q_2 = -k_2}^{k_2} T_1(k_1, q_1) T_2(k_2, q_2) \langle q_1 q_2 | kq \rangle, \tag{11.190}$$

where the CG coefficients $\langle q_1 q_2 | kq \rangle$ corresponds to $j_1 = k_1$ and $j_2 = k_2$. We can prove that if $T_1(k_1, q_1)$ and $T_2(k_2, q_2)$ are irreducible spherical tensors of rank k_1 and k_2, respectively, then $T(k, q)$ is also an irreducible tensor of rank k by verifying that $T(k, q)$ satisfies Eq. (11.189).

Using $\mathbf{J} = \mathbf{J}_1 + \mathbf{J}_2$ and $[\mathbf{J}, T_1 T_2] = [\mathbf{J}, T_1] T_2 + T_1 [\mathbf{J}, T_2]$ we obtain

$$[J_z, T(k, q)] = \sum_{q_1} \sum_{q_2} \{[J_z, T_1] T_2 + T_1 [J_z, T_2]\} \langle q_1 q_2 | kq \rangle. \tag{11.191}$$

Since T_1 and T_2 satisfy Eq. (11.189), we get

$$[J_z, T(k, q)] = \sum_{q_1} \sum_{q_2} (q_1 + q_2) \hbar T_1(k_1, q_1) T_2(k_2, q_2) \langle q_1 q_2 | kq \rangle. \tag{11.192}$$

Because $\langle q_1 q_2 | kq \rangle$ is zero unless $q_1 + q_2 = q$, we get from (11.191)

$$[J_z, T(k, q)] = q\hbar T(k, q). \tag{11.193}$$

It is easy to show that

$$[J_\pm, T(k,q)] = [k(k+1) - q(q\pm1)]^{1/2}\,\hbar T(k, q\pm1)\,. \tag{11.194}$$

From Eqs. (11.193) and (11.194) we find that the product $T(k,q)$ is also an irreducible tensor operator. The construction of the operator $T(k,q)$ is in exact analogy with that of the states $|jm\rangle$ from states $|j_1 m_1; j_2 m_2\rangle$.

Solved Problem 6:

Show that $\langle jm, \alpha'|T(k_1, q_1)|\alpha, j_2 m_2\rangle = 0$ if $m \neq q_1 + m_2$.

The tensor operator $T(k_1, q_1)$ satisfies the relation

$$[T_z, T(k_1, q_1)] = \hbar q_1 T(k_1, q_1)\,. \tag{11.195}$$

Then

$$
\begin{aligned}
\langle jm, \alpha'|\,[J_z, T(k_1,q_1)]\,\alpha, j_2 m_2\rangle &= \langle jm, \alpha'|J_z T(k_1,q_1)|\alpha, j_2 m_2\rangle \\
&\quad -\langle jm, \alpha|TJ_z|\alpha, j_2 m_2\rangle \\
&= m\hbar\langle jm, \alpha'|T|\alpha, j_2 m_2\rangle \\
&\quad -m_2\hbar\langle jm, \alpha'|T|\alpha, j_2 m_2\rangle \\
&= (m - m_2)\hbar\langle jm, \alpha'|T|\alpha, j_2 m_2\rangle\,.
\end{aligned}
\tag{11.196}
$$

From Eqs. (11.195) and (11.196) we get

$$(m - m_2)\hbar\langle jm, \alpha'|T|\alpha, j_2 m_2\rangle = \hbar q_1\langle jm, \alpha'|T|\alpha, j_2 m_2\rangle\,. \tag{11.197}$$

We find $m - m_2 = q_1$, that is, $m = q_1 + m_2$. Therefore, $\langle jm, \alpha'|T(k_1, q_1)|\alpha, j_2 m_2\rangle = 0$ if $m \neq q_1 + m_2$.

11.9.2 The Wigner–Eckart Theorem

In considering the interactions of an electromagnetic field with atoms and nuclei, it is often necessary to evaluate matrix elements of tensor operators with respect to angular momentum eigenstates $|jm\rangle$. Since $|jm\rangle$ are also tensor operators, we write from Eq. (11.190)

$$|jm\rangle = \sum T_1(k_1, q_1)|j_2 m_2\rangle\,\langle q_1 m_2|jm\rangle\,. \tag{11.198}$$

The series expansion of $|jm\rangle$ in terms of $T_1(k_1, q_1)|j_2 m_2\rangle$, given in Eq. (11.198) can be inverted by making use of the orthonormality of the CG coefficients:

$$T_1(k_1, q_1)|j_2 m_2\rangle = \sum_{m=-j}^{j} \sum_{j=|k_1-j_2|}^{k_1+j_2} |jm\rangle\langle q_1 m_2|jm\rangle\,, \tag{11.199}$$

where we have used $\langle jm|q_1 m_2\rangle = \langle q_1 m_2|jm\rangle$ as CG coefficients and are taken as real. If we include explicitly quantum numbers other than j and m for the state by α, then Eq. (11.199) is given as

$$T_1(k_1, q_1)|\alpha, j_2 m_2\rangle = \sum_{m=-j}^{j} \sum_{j=|k_1-j_2|}^{k_1+j_2} |\alpha, jm\rangle\langle q_1 m_2|\alpha, jm\rangle\,, \tag{11.200}$$

We may now multiply the above equation from left with $\langle j'm', \alpha'|$ to get the matrix element

$$\langle j'm', \alpha'|T_1(k_1, q_1)|\alpha, j_2 m_2\rangle = \sum_{m=-j}^{j} \sum_{j=|k_1-j_2|}^{k_1+j_2} \langle j'm', \alpha'|\alpha, jm\rangle \langle q_1 m_2|jm\rangle. \quad (11.201)$$

On making use of the orthogonality relation

$$\langle j'm', \alpha'|\alpha, jm\rangle = 0 \text{ unless } j = j', \ m = m' \quad (11.202)$$

we obtain

$$\langle j'm', \alpha'|T_1(k_1, q_1)|\alpha, j_2 m_2\rangle = N\langle q_1 m_2|j'm'\rangle, \quad (11.203)$$

where $N = \langle j'm', \alpha|\alpha', j'm'\rangle$ is the normalization factor. Dropping the primes and writing in the standard notation for N, the matrix element of tensor operators with respect to angular momentum eigenstates are given as

$$\langle jm, \alpha'|T_1(k_1, q_1)|\alpha, j_2 m_2\rangle$$
$$= (-1)^{k_1-j_2+j}(2j+1)^{-1/2}\langle j, \alpha'||T(k)||\alpha, j_2\rangle\langle q_1 m_2|jm\rangle. \quad (11.204)$$

The above equation is known as *Wigner–Eckart theorem*. This theorem states that the matrix element is the product of two terms. The physical properties of the matrix element are contained in $\langle j, \alpha'||T(k)||\alpha, j_2\rangle$. The second term is a CG coefficient for adding j_2 and k to get j. It depends only on the way the system is oriented with respect to the z-axis. The first factor depends on the dynamics, for instance, α may stand for the radial quantum number. On the other hand, it is independent of magnetic quantum numbers m_1, m_2 and q_1 which specify the orientation of the physical system. To evaluate the matrix element of the tensor operator with various combination of m_1, m_2 and q_1, it is sufficient to know one of them; all others can be related geometrically because they are proportional to CG coefficients.

The selection rule for tensor operator matrix element is read-off from the selection rules for adding angular momentum. Indeed, from the requirement that the CG coefficients can be nonvanishing we obtain the m-selection rule as $m = q_1 + m_2$ and the triangular relation $|j_2 - k_1| \leq j \leq j_2 + k_1$.

11.10   Concluding Remarks

We note that angular position is restricted to values between $-\pi$ to π whereas linear position can take any value. A consequence of this is that the uncertainty in the angular position is always finite. Another implication is that the angular momentum is quantized and can only have its eigenvalues whereas linear position can take any value. For a big fly-wheel or cycle-wheel one could think that by giving it various angular velocities one can make the fly-wheel to have any angular momentum. But this is not true. Angular momentum can only be increased in discrete steps of $\hbar$. Because $\hbar = 1.0546 \times 10^{-27}$ergs-sec is very small that in the example of fly-wheel the step increased is hardly visible. Another point is that Y_{00} is an eigenfunction of the three operators L_x, L_y and L_z. It is a simultaneous eigenfunction of these three operators which are noncommuting. In this sense Y_{00} is a special eigenfunction. If a system is in Y_{00} state then we know the three components of angular momentum with certainty—they are all zero.

We note that the effect of $L_\pm$ on a spherical eigenstate for any central potential is to change the value of m. The other quantum numbers l and n are unchanged. This is due to the symmetry of the central potential. Thus, $L_\pm$ are ladder operators for any central potential.

According to Eq. (11.82), the mechanism underlying spin is not essentially different from that underlying orbital angular momentum. Both forms of angular momentum arise from the momentum density in the fields. The distinction between spin and orbital contributions to the total angular momentum merely results from the independence of these contributions. The position of the momentum density that gives rise to the spin can be reversed independently of the position that gives rise to the orbital angular momentum.

11.11 Bibliography

[1] G. Gibson, J. Courtial, M.J. Padgett, M. Vasnetsov, V. Pas'ko, S.M. Barnett and S. Franke-Arnold, *Opt. Express* 12:5448, 2004.

[2] R.J. Voogd, M. Singh, S.F. Pereira, A.S. van de Nes and J.J.M. Braat, *The use of orbital angular momentum of light beams for super-high density optical data storage*. In Frontiers in Optics. Eds.: B.V.K. Vijaya Kumar and H. Kobori. OSA Technical Digest Series, Optical Society of America, 2004.

[3] S.M. Mohammadi, L.K.S. Daldorff, J.E.S. Bergman and R.L. Karlsson, *IEEE Trans. Antenn. Propag.* 58:565, 2010.

[4] F. Tamburini, E. Mari, A. Sponselli, B. Thide, A. Bianchini and F. Romanato, *New. J. Phys.* 14:033001, 2012.

[5] A. Vaziri, G. Weihs and A. Zeilinger, *J. Opt. B: Quantum Semiclass. Opt.* 4:S47, 2002.

[6] K. Liu, Y. Cheng, Z. Yang, H. Wang, Y. Qin and X. Li, *IEEE Antenn. Wireless Propag. Letts.* 14:711, 2015.

[7] D.G. Grier, *Nature* 424:810, 2003.

[8] B. Thide, H. Then, J. Sjoholm, K. Palmer, J. Bergman, T.D. Carozzi, Ya.N. Istomin, N.H. Ibragimov and R. Khamitova, *Phys. Rev. Lett.* 99:087701, 2007.

[9] H. Ren, G. Briere, X. Fang, P. Ni, R. Sawant, S. Heron, S. Chenot, S. Vezian, B. Damilano, V. Brandli, S.A. Maier and P. Genevet, *Nature Commun.* 10:2986, 2019.

[10] L. Allen, M.W. Beijersbergen, R.J.C. Spreeuw and J.P. Woerdman, *Phys. Rev. A* 45:8185, 1992.

[11] G. Burns and A.M. Glazer, *Space Groups for Scientists and Engineers*. Academic Press, Boston, 1990.

[12] Yu.A. Bakhturin, *Campbell–Hausdorff formula*. In Encyclopaedia of Mathematics. Kluwer Academic Publishers, Dordrecht, 2000.

[13] K.M. Mita, *Am. J. Phys.* 68:259, 2000.

[14] D.H. Kobe, *Eur. J. Phys.* 19:119, 1998.

[15] H.C. Ohanian, *Am. J. Phys.* 54:500, 1986.

[16] J.D. Koralek, C.P. Weber, J. Orenstein, B.A. Bernevig, S.C. Zhang, S. Mack and D.D. Awschalom, *Nature* 458:610, 2009.

[17] D.H. Hernando, F. Gulnea and A. Brataas, *Phys. Rev. B* 74:155426, 2006.

[18] J. Wunderllch, B. Kaestner, J. Sinova and T. Jungwirth, *Phys. Rev. Lett.* 94:047204, 2005.

[19] W.C. Jun, I.M. Shen and Z.X. Fa, *Chin. Phys. Lett.* 28:097102, 2011.

[20] Y.J. Lin, K.J. Garcia and I.B. Spielman, *Nature* 471:83, 2011.

[21] L.W. Cheuk, A.T. Sommer, Z. Hadzibabic, T. Yefsah, W.S. Bakr and M.W. Zwieriein, *Phys. Rev. Lett.* 109:095302, 2012.

[22] M. Governale and U. Zullcke, *Phys. Rev. B* 66:073311, 2002.

[23] M. Gmitra, S. Konschuh, C. Ertier, C.A. Drazi and J. Fabean, *Phys. Rev. B* 80:235431, 2009.

[24] W.B. Hodge, S.V. Migirditch and W.C. Kerr, *Am. J. Phys.* 82:681, 2014.

11.12 Exercises

11.1 Show that $R(\mathbf{k},\theta) = \begin{pmatrix} \cos\theta & -\sin\theta & 0 \\ \sin\theta & \cos\theta & 0 \\ 0 & 0 & 1 \end{pmatrix}$.

11.2 Show that the rotation matrices $R(\mathbf{k},\theta) = \begin{pmatrix} \cos\theta & -\sin\theta & 0 \\ \sin\theta & \cos\theta & 0 \\ 0 & 0 & 1 \end{pmatrix}$ form a continuous group.

11.3 Show that $\mathbf{p}$ and $-H$ are the generators of infinitesimal displacements in space and time, respectively.

11.4 Show that a state of a system corresponding to finite angular momentum cannot be a simultaneous eigenstate of any two of the Cartesian components of $\mathbf{L}$.

11.5 Show that if an operator $\mathbf{O}$ commutes with L_x, L_y and L_z then $\mathbf{O}^2$ commutes with $\mathbf{L}^2$.

11.6 Find the values of $[L_x,x]$ and $[L_x,y]$.

11.7 Show that if a particle has the normalized eigenfunction $\psi = Ne^{ikz}$ then the z component of its angular momentum is zero.

11.8 What are the expectation values of L_x and L_y for a system which is in an eigenstate of L_z?

11.9 Show that $[\mathbf{L}^2,\mathbf{r}] = -2i\hbar(\mathbf{L}\times\mathbf{r} - i\hbar\mathbf{r})$.

11.10 Simplify $\mathbf{L}\times\mathbf{r} + \mathbf{r}\times\mathbf{L}$.

11.11 Show that $\mathbf{L}\cdot\mathbf{p} = \mathbf{L}\cdot\mathbf{r} = 0$.

11.12 Determine the values of $[L_x,r^2]$, $[L_y,r^2]$ and $[L_z,r^2]$.

11.13 Suppose an atom or a particle has the angular momentum $\mathbf{L}$. If a magnetic field $\mathbf{B}$ is applied, then the magnetic moment is $\mu = \gamma\mathbf{L}$, where γ is the gyromagnetic ratio. Then its Hamiltonian operator is $H_0 = -\mu\cdot\mathbf{B}$. Prove that the equation of motion for the $\mathbf{L}$ operator is $\dot{\mathbf{L}} = -\gamma\mathbf{B}\times\mathbf{L}$.

11.14 Let $\psi(\theta, \phi)$ be the eigenfunctions of L_x operator corresponding to $l = 1$. Writing $\psi = a_1 Y_{11} + a_2 Y_{10} + a_3 Y_{1-1}$, find the eigenvalues and eigenfunctions of L_x.

11.15 Show that $|ll\rangle = N_l e^{il\phi} \sin^l \theta$ is a solution of the differential equation $L_+|ll\rangle = 0$.

11.16 If $\mathbf{J} = \mathbf{J}_1 + \mathbf{J}_2 + \mathbf{J}_3$ where all the $\mathbf{J}_i$ are for $j = 1$ and let $J(J+1)$ be the eigenvalues of $\mathbf{J}^2$. (a) What are the possible values of J? (b) How many linearly independent states are there for each of these values?

11.17 A rigid rotator is in the state $\psi(\theta, \phi) = N \sin^2 \theta \cos 2\phi$. What are the values of $\mathbf{L}^2$ and L_z found in measurement? Find the measurement probabilities for each of these values.

11.18 Find (a) the total number of eigenstates of $\mathbf{J} = \mathbf{J}_1 + \mathbf{J}_2 + \ldots + \mathbf{J}_N$ and then (b) the number of eigenstates for the coupling of one p, one d and one f electrons (excluding spin).

11.19 Find the transformation matrix between the Cartesian components of the position (x, y, z) and the spherical basis $(Y_{11}, Y_{10}, Y_{1-1})$.

11.20 The rotational Hamiltonian for a symmetric top molecule is given by $H = \left(L_x^2 + L_y^2\right)/(2I_1) + L_z^2/(2I_2)$. Find its (a) eigenvalues and (b) eigenfunctions.

11.21 Show that the rotation transformation operator for the spin angular momentum $e^{-i\mathbf{S}\cdot\mathbf{n}\phi/\hbar} = I\cos(\phi/2) - i\sigma \cdot \mathbf{n}\sin(\phi/2)$, where $\mathbf{S} = \hbar\sigma/2$ and σ are the Pauli's spin matrices. Note that $(\sigma \cdot \mathbf{n})^k = 1$ for k even and $\sigma \cdot \mathbf{n}$ for k odd.

11.22 An electron is in the spin state $|\phi\rangle = \dfrac{1}{\sqrt{3}}\begin{bmatrix} \sqrt{2} \\ 1 \end{bmatrix}$. (a) What is the probability that a measurement of S_z will give $\hbar/2$? (b) What is the probability that a measurement of S_x will give $-\hbar/2$?

11.23 Using the recursion relation

$$[j(j+1) - m(m \pm 1)]^{1/2} \langle m_1; m_2 | jm \pm 1\rangle$$
$$= [j_1(j_1+1) - m_1(m_1 \mp 1)]^{1/2} \langle m_1 \pm 1; m_2 | jm\rangle$$
$$+ [j_2(j_2+1) - m_2(m_2 \mp 1)]^{1/2} \langle m_1; m_2 \mp 1 | jm\rangle$$

show that

$$\langle j_1, j_2 - 1 | j_1 + j_2, j_1 + j_2 - 1\rangle = (j_2/(j_1+j_2))^{1/2} \text{ and}$$
$$\langle j_1 - 1, j_2 | j_1 + j_2, j_1 + j_2 - 1\rangle = (j_1/(j_1+j_2))^{1/2}.$$

11.24 Find the CG matrix for the case $j_1 = 1/2$ and $j_2 = 1/2$.

11.25 Show that the energy eigenvalue of a rigid rotator $l(l+1)\hbar^2/(2I)$ follows from the Bohr–Sommerfeld quantization rule $\oint L\, d\phi = (l + 1/2)h$ in the limit of large quantum number.

11.26 Obtain the value of $\langle\{L_+, L_-\}\rangle = \langle L_+L_- + L_-L_+\rangle$, where $\{\}$ is used for anti-commutator.

11.27 The Hamiltonian for a particle with orbital angular momentum unity ($l = 1$) is given by $H = aL_z^2 + b(L_x^2 - L_y^2)$, $a \neq b$. Determine the eigenvalues and eigenvectors of H.

11.28 A measurement $L_{xy} = L_x + L_y$ is done on a system with $l = 1$. What are the possible results of such a measurement?

11.29 Construct the state with $j = 3/2$ and $m_j = -1/2$ for a 2p-electron.

11.30 Given the normalized eigenvectors of L_x for $l = 1$ as

$$\psi_{11} = (Y_{11} + \sqrt{2}Y_{10} + Y_{1-1})/2, \quad \psi_{10} = (Y_{11} - Y_{1-1})/\sqrt{2},$$
$$\psi_{1-1} = (Y_{11} - \sqrt{2}Y_{10} + Y_{1-1})/2$$

find the matrix of L_y and L_z in this representation.

11.31 For a spin-1/2 system the eigenstates of S_z (z-component of spin angular momentum) are $|a\rangle$ and $|b\rangle$ with the eigenvalues $\pm\hbar/2$. Write the density matrix and find its trace corresponding to the pure state $|\psi\rangle = c_1|a\rangle + c_2|b\rangle$, where $|c_1|^2 + |c_2|^2 = 1$.

11.32 Assume that the eigenstates of S_z of a spin-1/2 system are $|a\rangle$ and $|b\rangle$ with the eigenvalues $\pm\hbar/2$. Set up the density matrix and obtain its trace for the mixed state specified as: $|a\rangle$ with probability 2/3 and $|b\rangle$ with probability 1/3.

11.33 Consider the exercises 11.31 and 11.32. Calculate $\langle S_z \rangle$ in the pure and mixed states.

11.34 For a spin-1 system the normalized eigenstates of J_z are $|a\rangle$, $|b\rangle$ and $|c\rangle$ with the eigenvalues $\hbar$, 0 and $-\hbar$, respectively. Determine the density matrix for the system in the mixed state $|a\rangle$, $(|a\rangle + |b\rangle)/\sqrt{2}$ and $(|b\rangle + |c\rangle)/\sqrt{2}$, each with probability 1/3. Also, find $\langle J_z \rangle$, $\langle J_z^2 \rangle$ and ΔJ_z.

11.35 The states $|11\rangle$, $|10\rangle$ and $|1-1\rangle$ are called triplet states and $|00\rangle$ is called a singlet state. Show that $P_1 = (3/4) + (1/\hbar^2)\mathbf{S}_1 \cdot \mathbf{S}_2$ is the projection operator for triplet state and $P_0 = (1/4) - (1/\hbar^2)\mathbf{S}_1 \cdot \mathbf{S}_2$ is the projection operator for singlet state.

12

Hydrogen Atom

12.1 Introduction

The hydrogen atom is the simplest of all atoms consisting of proton of charge e and the electron of charge $-e$. It is a two particle system. A two-body problem can always be reduced to a one-body problem by choosing the centre of mass coordinate and the relative coordinates. In the hydrogen atom, the attractive force exerted by the proton on electron prevents it from escape. The potential energy V of the system is $-e^2/r$ (or $-e^2/(4\pi\epsilon_0 r)$), where r is the distance between the electron and the proton. Note that the potential energy approaches $-\infty$ when the electron becomes near the location of proton. Suppose the nucleus is held fixed at the origin of coordinates. $V(r)$ is independent of θ and ϕ, that is, insensitive to the direction of $\mathbf{r}$ and depends only on the scalar magnitude r. The potential energy remains invariant under rotation and is said to be *spherically symmetric*. Figure 12.1 shows $V(r)$ as a function of r. In this chapter, the hydrogen atom problem in three-dimension is solved and momentum eigenfunction is obtained. The problem in D-dimension is analyzed. The magnetic field produced by a hydrogen atom is worked out. Finally, the system in parabolic coordinates is discussed.

12.2 Hydrogen Atom in Three-Dimension

In the hydrogen atom, an electron of mass m_e is coupled to a proton of mass m_p. The force is the Coulomb force $-e^2/|\mathbf{r}_1 - \mathbf{r}_2|$. The Schrödinger equation of hydrogen atom is

$$i\hbar\frac{\partial\psi}{\partial t} = \left(-\frac{\hbar^2}{2m_e}\nabla_1^2 - \frac{\hbar^2}{2m_p}\nabla_2^2 - \frac{e^2}{|\mathbf{r}_1 - \mathbf{r}_2|}\right)\psi, \tag{12.1}$$

where ∇_i, $i = 1, 2$ is the Laplacian operator and $\psi = \psi(\mathbf{r}_1, \mathbf{r}_2, t)$ is the joint probability amplitude function of finding the electron at $\mathbf{r}_1$ and the proton at $\mathbf{r}_2$ at time t.

For atoms which are isoelectronic with hydrogen atom such as He$^+$, Li^{++}, Be^{+++}, etc, the potential is $-Ze^2/r$, where Z is 2, 3, 4, etc., respectively. For these atoms we have Eq. (12.1) with $e^2/|\mathbf{r}_1-\mathbf{r}_2|$ replaced by $-Ze^2/|\mathbf{r}_1-\mathbf{r}_2|$. Introducing the relative displacement $\mathbf{r} = \mathbf{r}_1 - \mathbf{r}_2$, and the centre of mass displacement $\mathbf{R} = (m_e\mathbf{r}_1 + m_p\mathbf{r}_2)/(m_e + m_p)$, the Schrödinger Eq. (12.1) becomes

$$i\hbar\frac{\partial\psi}{\partial t} = \left(-\frac{\hbar^2}{2M}\nabla_R^2 - \frac{\hbar^2}{2\mu}\nabla_r^2 - \frac{e^2}{r}\right)\psi, \tag{12.2}$$

where $M = m_e + m_p$ is the total mass, $\mu = m_e m_p/(m_e + m_p)$ is the reduced mass and $\psi = \psi(\mathbf{r}, \mathbf{R}, t)$. The advantage of transforming Eq. (12.1) to Eq. (12.2) is evident as the eigenfunctions of Eq. (12.1), $\psi(r_1, r_2, t)$ cannot be written as a product $\psi_1(r_1)\psi_2(r_2)$,

DOI: 10.1201/9781003172178-12

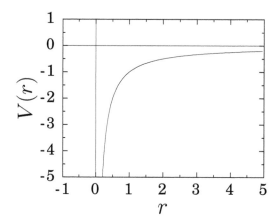

FIGURE 12.1

Potential of hydrogen atom. The y-axis is $V(r) = -1/r$.

whereas the eigenfunctions of Eq. (12.2) can be written as the product of two eigenfunctions as Eq. (12.2) can be separated into two equations one depending on r and the other on R.

We write the total energy of the system as $E_{\text{tot}} = E + E_{\text{cm}}$, where E_{cm} is the centre of mass energy and E is the energy of relative motion. Now, writing

$$\psi(\mathbf{r}, \mathbf{R}, t) = \psi_E(\mathbf{r})\psi_{\text{cm}}(\mathbf{R})\, e^{-i(E+E_{\text{cm}})t/\hbar}\,, \tag{12.3}$$

where $\psi_{\text{cm}}(\mathbf{R})$ represents a free particle wave function for the centre of mass and $\psi_E(\mathbf{r})$ is the probability amplitude function to find the electron and proton with a relative displacement $\mathbf{r}$ and relative energy E. The Schrödinger equation for the relative motion of the electron and proton takes the form

$$\left(-\frac{\hbar^2}{2\mu}\nabla_r^2 - \frac{e^2}{r}\right)\psi_E(\mathbf{r}) = E\psi_E(\mathbf{r})\,. \tag{12.4}$$

Writing ∇_r and ψ_E as simply ∇ and ψ, respectively, and defining $V = -e^2/r$, Eq. (12.4) is rewritten as

$$\left(-\frac{\hbar^2}{2\mu}\nabla^2 + V\right)\psi = E\psi\,. \tag{12.5}$$

It is convenient to solve the Eq. (12.5) in spherical coordinates, because it is generally difficult to obtain the solution of it in Cartesian coordinates.

Solved Problem 1:

Show that the time-independent Schrödinger equation

$$\left(-\frac{\hbar^2}{2m}\nabla_R^2 - \frac{\hbar^2}{2\mu}\nabla_r^2 - \frac{e^2}{r}\right)\psi(\mathbf{r}, \mathbf{R}) = E_T\psi(\mathbf{r}, \mathbf{R})$$

can be separated into two independent equations for the variables $\mathbf{r}$ and $\mathbf{R}$ and find the solution corresponding to the $\mathbf{R}$-part equation.

Assume the solution of the given equation as $\psi = \psi_E(\mathbf{r})\psi_{\text{CM}}(\mathbf{R})$. Substituting this in the given equation we get

$$-\frac{\hbar^2}{2m}\psi_E\nabla_R^2\psi_{\text{CM}} - \frac{\hbar^2}{2\mu}\psi_{\text{CM}}\nabla_r^2\psi_E - \frac{e^2}{r}\psi_E\psi_{\text{CM}} = E_T\psi_E\psi_{\text{CM}}\,. \tag{12.6}$$

Dividing throughout by $\psi_E\psi_{CM}$ we get

$$-\frac{\hbar^2}{2m}\frac{1}{\psi_{CM}}\nabla_R^2\psi_{CM} = E_T + \frac{\hbar^2}{2\mu}\frac{1}{\psi_E}\nabla_r^2\psi_E + \frac{e^2}{r}. \tag{12.7}$$

Since $\mathbf{r}$ and $\mathbf{R}$ are independent variables the above equation must be a constant and say that it is E_{CM}. We write

$$-\frac{\hbar^2}{2m}\nabla_R^2\psi_{CM} = E_{CM}\psi_{CM} \tag{12.8}$$

and

$$-\frac{\hbar^2}{2\mu}\nabla_r^2\psi_E - \frac{e^2}{r}\psi_E = (E_T - E_{CM})\psi_E = E\psi_E. \tag{12.9}$$

Equation (12.8) gives the Schrödinger equation for the centre of mass coordinate $\mathbf{R}$ and Eq. (12.9) gives the Schrödinger equation for the relative coordinate $\mathbf{r}$. The solution of Eq. (12.8) is

$$\psi_{CM} = A\,e^{\pm i\mathbf{K}\cdot\mathbf{R}}, \quad K^2 = \frac{2mE_{CM}}{\hbar^2} \tag{12.10}$$

which is a free particle eigenfunction.

12.2.1 Application of Variable Separable Method

Let us assume the solution of the form

$$\psi(r,\theta,\phi) = R(r)\Theta(\theta)\Phi(\phi) = R(r)Y(\theta,\phi) \tag{12.11}$$

and obtain the differential equations for R, Θ and Φ. In spherical coordinates

$$x = r\sin\theta\cos\phi, \quad y = r\sin\theta\sin\phi, \quad z = r\cos\theta, \tag{12.12a}$$

$$\nabla = \hat{r}\frac{\partial}{\partial r} + \hat{\theta}\frac{1}{r}\frac{\partial}{\partial\theta} + \hat{\phi}\frac{1}{r\sin\theta}\frac{\partial}{\partial\phi}, \tag{12.12b}$$

$$\nabla^2 = \frac{1}{r^2}\frac{\partial}{\partial r}\left(r^2\frac{\partial}{\partial r}\right) + \frac{1}{r^2\sin\theta}\frac{\partial}{\partial\theta}\left(\sin\theta\frac{\partial}{\partial\theta}\right) + \frac{1}{r^2\sin^2\theta}\frac{\partial^2}{\partial\phi^2}, \tag{12.12c}$$

where $\hat{r}$, $\hat{\theta}$ and $\hat{\phi}$ are the usual unit vectors. Using Eqs. (12.11) and (12.12) in Eq. (12.5) the Schrödinger equation in spherical polar coordinates is

$$\frac{1}{R}\frac{d}{dr}\left(r^2\frac{dR}{dr}\right) + \frac{2\mu r^2}{\hbar^2}(E-V)$$
$$= -\frac{1}{Y}\left[\frac{1}{\sin\theta}\frac{\partial}{\partial\theta}\left(\sin\theta\frac{\partial Y}{\partial\theta}\right) + \frac{1}{\sin^2\theta}\frac{\partial^2 Y}{\partial\phi^2}\right]. \tag{12.13}$$

The left-side of Eq. (12.13) is a function of r only while the right-side is a function of θ and ϕ only. Therefore, they must be equal to a same constant, say, λ. Hence, from Eq. (12.13) we write

$$\frac{1}{r^2}\frac{d}{dr}\left(r^2\frac{dR}{dr}\right) + \left[\frac{2\mu}{\hbar^2}(E-V) - \frac{\lambda}{r^2}\right]R = 0, \tag{12.14}$$

$$\frac{1}{\sin\theta}\frac{\partial}{\partial\theta}\left(\sin\theta\frac{\partial Y}{\partial\theta}\right) + \frac{1}{\sin^2\theta}\frac{\partial^2 Y}{\partial\phi^2} + \lambda Y = 0. \tag{12.15}$$

For all central force problems with spherically symmetric potential the energy term occurs only in the radial equation. Equation (12.15) is solved without knowing the form of V as follows.

Equation (12.15) is separated by substituting $Y = \Theta(\theta)\Phi(\phi)$ which gives

$$-\frac{1}{\Phi}\frac{d^2\Phi}{d\phi^2} = \frac{\sin\theta}{\Theta}\frac{d}{d\theta}\left(\sin\theta\frac{d\Theta}{d\theta}\right) + \lambda\sin^2\theta. \tag{12.16}$$

Left-side of Eq. (12.16) is a function of ϕ only and the right-side is a function of θ only and hence they must be equal to a same constant which we set as m^2. Now, from Eq. (12.16) we write

$$\frac{d^2\Phi}{d\phi^2} + m^2\Phi = 0, \tag{12.17a}$$

$$\frac{1}{\sin\theta}\frac{d}{d\theta}\left(\sin\theta\frac{d\Theta}{d\theta}\right) + \left(\lambda - \frac{m^2}{\sin^2\theta}\right)\Theta = 0. \tag{12.17b}$$

12.2.1.1 Solution of Φ Equation

The solution of Eq. (12.17a) is $\Phi = Ae^{im\phi} + Be^{-im\phi}$. Applying the condition $\Phi(0) = \Phi(2\pi)$, we observe that $m = 0, \pm1, \pm2,\ldots$ and thus

$$\Phi = Ae^{im\phi}. \tag{12.18}$$

The Φ given above is also an eigenfunction of L_z with the eigenvalue $m\hbar$. Since the Hamiltonian for a magnetic field applied along the z-direction is $H_B = -eBL_z/(2m)$, the first-order perturbation due to a magnetic field shifts energy levels proportional to m and hence m is some times called a *magnetic quantum number*. Normalization of Φ gives $A = 1/\sqrt{2\pi}$ (verify).

12.2.1.2 Solution of Θ-Equation

Introducing the change of variable as $x = \cos\theta$, rewrite the Eq. (12.17b) as

$$\frac{d}{dx}\left[(1-x^2)\frac{d\Theta}{dx}\right] + \left[\lambda - \frac{m^2}{1-x^2}\right]\Theta = 0. \tag{12.19}$$

When $\lambda = l(l+1)$ with $l = 0,1,2,\ldots$ the above equation is the associated Legendre differential equation. To ensure a finite probability density for all $x = \cos\theta$, only polynomial solutions to Eq. (12.19) have to be considered. Then we get the equation

$$\frac{d}{dx}\left[(1-x^2)\frac{d}{dx}P_l^m(x)\right] + \left[l(l+1) - \frac{m^2}{1-x^2}\right]P_l^m(x) = 0, \tag{12.20a}$$

where $P_l^m(x)$ is called *associated Legendre polynomial* which is given by the Rodrigues formula

$$P_l^m(x) = \frac{1}{2^l l!}(1-x^2)^{|m|/2}\frac{d^{l+|m|}}{dx^{l+|m|}}(x^2-1)^l. \tag{12.20b}$$

The solution of Eq. (12.19) is written as $\Theta = BP_l^m(x) = BP_l^m(\cos\theta)$ provided $\lambda = l(l+1)$ and $|m| \leq l$. The normalization condition is $\int_0^\pi \Theta^*\Theta\sin\theta d\theta = 1$, that is, $\int_{-1}^1 B^2 P_l^m(x)P_l^m(x)dx = 1$. This gives

$$B = \epsilon\left[\frac{(2l+1)(l-|m|)!}{2(l+|m|)!}\right]^{1/2}, \tag{12.21}$$

where $\epsilon = (-1)^m$ for $m > 0$ and $\epsilon = 1$ for $m \le 0$. Thus,

$$Y_{lm}(\theta, \phi) = \Theta\Phi = (-1)^m \left[\frac{(2l+1)(l-m)!}{4\pi(l+m)!}\right]^{1/2} e^{im\phi} P_l^m(\cos\theta), \qquad (12.22)$$

where $l = 0, 1, 2, \ldots$, $m = l, l-1, \ldots, -l+1, -l$. l is called the *orbital quantum number* since $Y_{lm}(\theta, \phi)$ are the eigenfunctions of L^2 operator with eigenvalues $l(l+1)\hbar^2$. Expression for some of the Y_{lm} calculated using (12.22) and (12.20b) are given below (verify):

$$Y_{00} = \left(\frac{1}{4\pi}\right)^{1/2}, \quad Y_{10} = \left(\frac{3}{4\pi}\right)^{1/2}\cos\theta = \left(\frac{3}{4\pi}\right)^{1/2}\frac{z}{r}, \qquad (12.23a)$$

$$Y_{11} = -\left(\frac{3}{8\pi}\right)^{1/2}\sin\theta\, e^{i\phi} = -\left(\frac{3}{8\pi}\right)^{1/2}\frac{(x+iy)}{r}, \qquad (12.23b)$$

$$Y_{1-1} = \left(\frac{3}{8\pi}\right)^{1/2}\sin\theta\, e^{-i\phi} = \left(\frac{3}{8\pi}\right)^{1/2}\frac{(x-iy)}{r}, \qquad (12.23c)$$

$$Y_{20} = \left(\frac{5}{16\pi}\right)^{1/2}(3\cos^2\theta - 1) = \left(\frac{5}{16\pi}\right)^{1/2}\left(\frac{3z^2}{r^2} - 1\right). \qquad (12.23d)$$

12.2.1.3 Solution of Radial Equation

Having obtained the forms of Φ and Θ, next we solve the radial equation with $V(r) = -e^2/r$. The radial equation is

$$\frac{d^2R}{dr^2} + \frac{2}{r}\frac{dR}{dr} + \frac{2\mu}{\hbar^2}\left[E + \frac{e^2}{r} - \frac{l(l+1)\hbar^2}{2\mu r^2}\right]R = 0. \qquad (12.24)$$

With the change of variable $R = f(r)/r$ Eq. (12.24) becomes

$$\frac{d^2f}{dr^2} + \frac{2\mu}{\hbar^2}\left[E + \frac{e^2}{r} - \frac{l(l+1)\hbar^2}{2\mu r^2}\right]f = 0. \qquad (12.25)$$

Compare this equation with the one-dimensional Schrödinger equation. The effective potential energy is

$$V_{\text{eff}}(r) = -\frac{e^2}{r} + \frac{l(l+1)\hbar^2}{2\mu r^2}. \qquad (12.26)$$

Then Eq. (12.25) is the Schrödinger equation for a particle of mass μ moving in the effective potential V_{eff}. From $E = H = p^2/(2\mu) + V_{\text{eff}}$ we have $E - V_{\text{eff}} = p^2/(2\mu) \ge 0$. The region with $E - V_{\text{eff}} \ge 0$ is the classically allowed region. $E - V_{\text{eff}} = 0$ gives the turning points at which the velocity of the particle is zero. Figure 12.2 shows r versus V_{eff} for a few values of l. We see from Fig. 12.2 that for $l \ne 0$, $V_{\text{eff}}(r \to 0) \to +\infty$. An infinite potential at some point means that the wave function must vanish there, that is, $R(r) \to 0$ for $l \ne 0$.

The turning points are the roots of the equation

$$E + \frac{e^2}{r} - \frac{l(l+1)\hbar^2}{2\mu r^2} = 0. \qquad (12.27)$$

The roots of the above equation are

$$r_\pm = -\frac{e^2}{2E} \pm \sqrt{\frac{e^4}{4E^2} + \frac{l(l+1)\hbar^2}{2\mu E}}. \qquad (12.28)$$

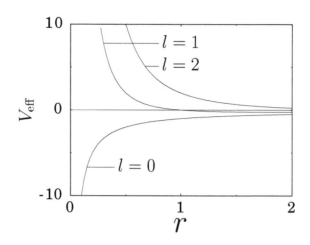

FIGURE 12.2
The effective potential of hydrogen atom for a few values of l where we have set $e^2 = 1$, $\hbar = 1$ and $\mu = 1$.

If $E > 0$ then $r_+ \geq 0$ and $r_- < 0$. Because $r \in [0, \infty]$ r_+ is the only turning point for $E > 0$ implying that the classical particle does not exhibit bounded motion. The two turning points are real and positive when $-e^4\mu/(l(l+1)\hbar^2) < E < 0$ and the motion is bounded. Substituting the Bohr energy relation $E_n = -e^4\mu/(2n^2\hbar^2)$ for E in Eq. (12.28), we obtain

$$
\begin{aligned}
r_{\pm} &= \frac{n^2\hbar^2}{e^2\mu} \pm \sqrt{\frac{n^4\hbar^4}{e^4\hbar^2} - \frac{n^2 l(l+1)\hbar^4}{e^4\mu^2}} \\
&= na_0 \left[n \pm \sqrt{n^2 - l(l+1)} \right], \quad a_0 = \frac{\hbar^2}{e^2\mu}
\end{aligned}
\tag{12.29}
$$

where a_0 is the Bohr radius. The classically allowed region is $r_- < r < r_+$. With the change of variables

$$
\rho^2 = -\frac{8r^2\mu E}{\hbar^2}, \quad R = \frac{f(\rho)}{r} = \sqrt{-8E\mu/\hbar^2} \, f(\rho)
\tag{12.30}
$$

Eq. (12.25) becomes

$$
f'' + \left[\frac{e^2\sqrt{\mu}}{\hbar\rho\sqrt{-2E}} - \frac{(2l+1)^2 - 1}{4\rho^2} - \frac{1}{4} \right] f = 0.
\tag{12.31}
$$

The physically acceptable solution of Eq. (12.31) must vanish at $\rho = 0$ (that is at $r = 0$) since $R = f(\rho)/r$. When the electron is far from the nucleus we assume ρ be very large and neglect the terms containing $1/\rho$ and $1/\rho^2$ in Eq. (12.31). In this case $f'' - f/4 = 0$ whose solution is $f(\rho) = \mathrm{e}^{-\rho/2}$. Note that $f(\rho) \to 0$ as $\rho \to \infty$ and is therefore a good approximation. For more general situation choose $f(\rho) = F(\rho)\mathrm{e}^{-\rho/2}$, where $F(\rho)$ take care of small values of ρ. The solution of the full Eq. (12.31) is obtained by comparing it with the equation [1]

$$
y'' + \left[\frac{(2q+p+1)}{2\rho} + \frac{(1-p^2)}{4\rho^2} - \frac{1}{4} \right] y = 0.
\tag{12.32}
$$

The solution of the above equation is

$$
y = \mathrm{e}^{-\rho/2} \rho^{(p+1)/2} L_q^{(p)}(\rho),
\tag{12.33}
$$

where $L_q^{(p)}(\rho)$ is a generalized Laguerre polynomial. For a finite wave function q must be 0, 1, 2, ... and we get a polynomial solution. We define $q = n - l - 1$. Comparing Eq. (12.31) with Eq. (12.32) we obtain $p = \pm(2l + 1)$ and

$$\frac{e^2\sqrt{\mu}}{h\rho\sqrt{-2E}} = \frac{2q + p + 1}{2\rho} = \frac{2(n - l - 1) + (2l + 1) + 1}{2\rho} = n \qquad (12.34)$$

The result is

$$E_n = -\frac{e^4\mu}{2\hbar^2 n^2}, \quad n = 1, 2, \dots \qquad (12.35)$$

where n is called *principal quantum number*. E is independent of m and l. Usually, the non-dependency of E on the quantum number m is attributed to the spherical symmetry of hydrogen atom. The non-dependency of E on l is attributed to the non-obvious symmetry known as *hidden* or *dynamical symmetry*.

E_n given by the Eq. (12.35) is in CGS system. In SI system

$$E_n = -\frac{\mu}{2\hbar^2 n^2}\left(\frac{e^2}{4\pi\epsilon_0}\right)^2, \quad n = 1, 2, \dots. \qquad (12.36)$$

In atomic units ($\hbar = m = e = 1$) $E_n = -1/(2n^2)$. For hydrogen-like atoms the results are obtained by replacing e^2 by Ze^2. Then $E_n = -Z^2 e^4 \mu/(2\hbar^2 n^2)$. For hydrogen atom $p = (2l + 1)$ is physically acceptable and $p = -(2l + 1)$ is unacceptable since $R \to \infty$ as $r \to 0$. Since $q \geq 0$ we get from $q = n - l - 1$ the result $l \leq n - 1$. Hence, the value of the orbital quantum number l cannot exceed $n - 1$ for an acceptable solution.

12.2.2 Orthogonality Relation and Radial Wave Function

From Eq. (12.32) we write

$$\rho\frac{d^2 y_q}{d\rho^2} + \left(q + \frac{p+1}{2} + \frac{1-p^2}{4\rho} - \frac{\rho}{4}\right)y_q = 0 \qquad (12.37)$$

and

$$\rho\frac{d^2 y_{q'}}{d\rho^2} + \left(q' + \frac{p+1}{2} + \frac{1-p^2}{4\rho} - \frac{\rho}{4}\right)y_{q'} = 0. \qquad (12.38)$$

For $q \neq q'$ multiplication of Eq. (12.37) by $y_{q'}$ and Eq. (12.38) by y_q and subtracting one from another lead to

$$\frac{d}{d\rho}\left(y_{q'}\frac{dy_q}{d\rho} - y_q\frac{dy_{q'}}{d\rho}\right) = \frac{q' - q}{\rho}y_q y_{q'}. \qquad (12.39)$$

Integration of this equation gives

$$\left[y_{q'}\frac{dy_q}{d\rho} - y_q\frac{dy_{q'}}{d\rho}\right]_0^\infty = (q' - q)\int_0^\infty \frac{y_q y_{q'}}{\rho}\,d\rho. \qquad (12.40)$$

For $\rho > -1$, y_q and $y_{q'}$ vanish at $\rho = 0$ and ∞. Therefore,

$$(q' - q)\int_0^\infty \frac{y_q y_{q'}}{\rho}\,d\rho = 0. \qquad (12.41)$$

Substituting for y_q and $y_{q'}$ we find

$$\int_0^\infty e^{-\rho}\rho^p L_q^p(\rho)L_{q'}^p(\rho)\,d\rho = 0, \quad (q \neq q'). \qquad (12.42)$$

The polynomials $L_q^p(\rho)$ are orthogonal in $(0, \infty)$ with respect to $e^{-\rho}\rho^p$.

The integral (12.42) for $q = q'$ is evaluated with the help of the generating function

$$G(\rho, z) = \frac{e^{-\rho z/(1-z)}}{(1-z)^{p+1}} = \sum_{q=0}^{\infty} L_q^p(\rho) z^q. \tag{12.43}$$

Then the integral in Eq. (12.42) becomes

$$\int_0^\infty e^{-\rho} \rho^p G^2 \, d\rho = \sum_{q=0}^{\infty} \sum_{q'=0}^{\infty} \int_0^\infty e^{-\rho} e^\rho L_q^p L_{q'}^p z^q z^{q'} \, d\rho. \tag{12.44}$$

Using Eq. (12.42) the above equation becomes

$$\int_0^\infty e^{-t} \rho^p G^2 \, d\rho = \sum_{q=0}^{\infty} \int_0^\infty e^{-\rho} e^\rho \left(L_q^p \right)^2 z^{2q} \, d\rho. \tag{12.45}$$

Substituting the generating function we obtain

$$\int_0^\infty e^{-t} \rho^p G^2 \, d\rho = \frac{1}{(1-z)^{2p+2}} \int_0^\infty t^p e^{-\rho(1+z)/(1-z)} \, d\rho$$

$$= \frac{\Gamma(p+1)}{(1-z^2)^{p+1}}. \tag{12.46}$$

We have

$$\frac{1}{(1-z^2)^{p+1}} = \sum_{q=0}^{\infty} \frac{\Gamma(q+p+1)}{\Gamma(q+1)\Gamma(p+1)} z^{2q}. \tag{12.47}$$

Using (12.46) and (12.47) in (12.45) the result is

$$\sum_{q=0}^{\infty} \frac{\Gamma(q+p+1)}{\Gamma(q+1)} z^{2q} = \sum_{q=0}^{\infty} \int_0^\infty e^{-\rho} \rho^p \left(L_q^p \right)^2 z^{2q} \, d\rho. \tag{12.48}$$

That is,

$$\int_0^\infty e^{-\rho} \rho^p \left(L_q^p \right)^2 \, d\rho = \frac{\Gamma(q+p+1)}{\Gamma(q+1)}. \tag{12.49}$$

Using the following recurrence relation of Laguerre polynomial

$$\rho L_q^p = (2q+p+1) L_q^p - (q+1) L_{q+1}^p - (q+p) L_{q-1}^p \tag{12.50}$$

we evaluate the integral is

$$\int_0^\infty e^{-\rho} \rho^{p+1} \left(L_q^p \right)^2 \, d\rho = (2q+p+1) \int_0^\infty e^{-\rho} \rho^p \left(L_q^p \right)^2 \, d\rho$$

$$- (q+1) \int_0^\infty e^{-\rho} \rho^p L_q^p L_{q+1}^p \, d\rho$$

$$- (q+p) \int_0^\infty e^{-\rho} \rho^p L_q^p L_{q-1}^p \, d\rho. \tag{12.51}$$

The use of the orthogonality relation (12.42) and the Eq. (12.49) gives

$$\int_0^\infty e^{-\rho} \rho^{p+1} \left(L_q^p \right)^2 \, d\rho = (2q+p+1) \frac{\Gamma(p+q+1)}{\Gamma(q+1)}. \tag{12.52}$$

We have

$$y = e^{-\rho/2} \rho^{(p+1)/2} L_q^p(\rho) . \tag{12.53}$$

Comparing Eq. (12.31) for the radial part of the hydrogen atom with that for y, we find that with $q = n - l - 1$ and $p = 2l + 1$

$$f \sim e^{-\rho/2} \rho^{l+1} L_{n-l-1}^{2l+1}(\rho) \tag{12.54}$$

and hence

$$R_{nl}(\rho) \sim \frac{f(\rho)}{r} = N_{nl} e^{-\rho/2} \rho^l L_{n-l-1}^{2l+1}(\rho) \tag{12.55}$$

with N_{nl} is the normalization constant and

$$\rho = \frac{\sqrt{8\mu|E_n|}}{\hbar} r = \frac{2r}{na_0} , \tag{12.56}$$

where a_0 is the Bohr radius. Using the relation $\int_0^\infty R_{nl}^2(r) r^2 \, dr = 1$ the normalization condition takes the form

$$1 = \left(\frac{na_0}{2} \right)^3 N_{nl}^2 \int_0^\infty e^{-\rho} \rho^{2l+2} \left(L_{n-l-1}^{2l+1} \right)^2 d\rho \tag{12.57}$$

Defining $p = 2l + 1$, $q = n - l - 1$ we get

$$\int_0^\infty e^{-\rho} \rho^{2l+2} \left(L_{n-l-1}^{2l+1} \right)^2 d\rho = \frac{2n \cdot (n+l)!}{(n-l-1)!} . \tag{12.58}$$

Then

$$N_{nl} = \left[\left(\frac{2}{na_0} \right)^3 \frac{(n-l-1)!}{2n \cdot (n+l)!} \right]^{1/2} . \tag{12.59}$$

Hence,

$$R_{nl}(\rho) = \left[\left(\frac{2}{na_0} \right)^3 \frac{(n-l-1)!}{2n \cdot (n+l)!} \right]^{1/2} e^{-\rho/2} \rho^l L_{n-l-1}^{2l+1} . \tag{12.60}$$

12.2.3 Normalized Eigenfunctions

The normalized energy eigenfunctions for the hydrogen atom are

$$\psi_{nlm}(r, \theta, \phi) = R_{nl} \left(\frac{2r}{na_0} \right) Y_{lm}(\theta, \phi) . \tag{12.61}$$

The radial eigenfunction $R_{nl}(\rho)$ is evaluated using the Rodrigues formula

$$L_q^p(x) = e^x x^{-p} \frac{d^q}{dx^q} \left(e^{-x} x^{p+q} \right) . \tag{12.62}$$

For $n = 1$ and $l = 0$ with $L_0^1(\rho) = 1$

$$R_{10} = 2 \left(\frac{1}{a_0} \right)^{3/2} e^{-r/a_0} . \tag{12.63a}$$

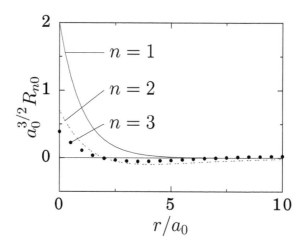

FIGURE 12.3
Radial wave function R_{nl} with $l = 0$ and $n = 1$, 2 and 3.

Similarly, we obtain

$$R_{20} = \frac{1}{\sqrt{8}} \left(\frac{1}{a_0} \right)^{3/2} \left(2 - \frac{r}{a_0} \right) e^{-r/(2a_0)}, \tag{12.63b}$$

$$R_{21} = \frac{1}{\sqrt{24}} \left(\frac{1}{a_0} \right)^{3/2} \frac{r}{a_0} e^{-r/(2a_0)}, \tag{12.63c}$$

$$R_{30} = \frac{2}{81\sqrt{3}} \left(\frac{1}{a_0} \right)^{3/2} \left[27 - 18\frac{r}{a_0} + 2\left(\frac{r}{a_0} \right)^2 \right] e^{-r/(3a_0)}, \tag{12.63d}$$

$$R_{31} = \frac{4}{81\sqrt{6}} \left(\frac{1}{a_0} \right)^{3/2} \frac{r}{a_0} \left(6 - \frac{r}{a_0} \right) e^{-r/(3a_0)}, \tag{12.63e}$$

$$R_{32} = \frac{4}{81\sqrt{30}} \left(\frac{1}{a_0} \right)^{3/2} \left(\frac{r}{a_0} \right)^2 e^{-r/(3a_0)}. \tag{12.63f}$$

Figure 12.3 shows $R_{n0}(r)$ for a few values of n where a_0 is set to unity. For hydrogen-like atoms the wave functions is obtained from Eq. (12.61) by replacing a_0 by a_0/Z. Some Y_{lm} are given in Eqs. (12.23).
    The first few ψ_{nlm} are given by

$$\psi_{100} = R_{10}Y_{00} = \frac{1}{\sqrt{\pi}} \left(\frac{1}{a_0} \right)^{3/2} e^{-r/a_0}, \tag{12.64a}$$

$$\psi_{200} = R_{20}Y_{00} = \frac{1}{\sqrt{32\pi}} \left(\frac{1}{a_0} \right)^{3/2} \left(2 - \frac{r}{a_0} \right) e^{-r/(2a_0)}, \tag{12.64b}$$

$$\psi_{210} = R_{21}Y_{10} = \frac{1}{\sqrt{32\pi}} \left(\frac{1}{a_0} \right)^{3/2} \frac{r}{a_0} \cos\theta\, e^{-r/(2a_0)}, \tag{12.64c}$$

$$\psi_{211} = R_{21}Y_{11} = -\frac{1}{8\sqrt{\pi}} \left(\frac{1}{a_0} \right)^{3/2} \frac{r}{a_0} \sin\theta\, e^{-r/(2a_0)} e^{i\phi}, \tag{12.64d}$$

$$\psi_{21-1} = R_{21}Y_{1-1} = \frac{1}{8\sqrt{\pi}} \left(\frac{1}{a_0} \right)^{3/2} \frac{r}{a_0} \sin\theta\, e^{-r/(2a_0)} e^{-i\phi}. \tag{12.64e}$$

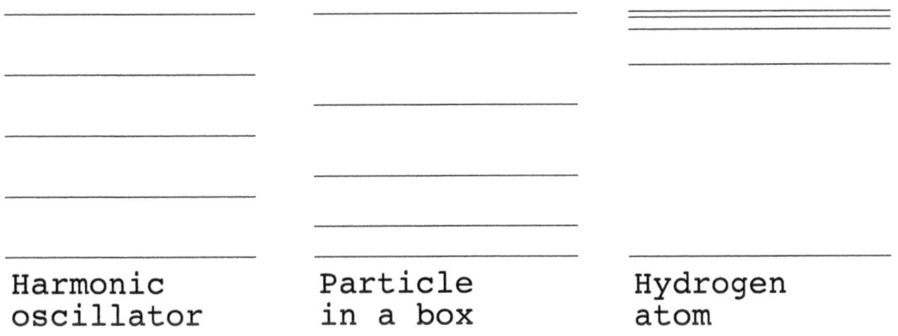

———————		══════
	———————	———————
———————		
	———————	
———————		
	———————	
———————		
	———————	
———————	———————	———————

| Harmonic | Particle | Hydrogen |
| oscillator | in a box | atom |

FIGURE 12.4
First few energy levels of linear harmonic oscillator, particle in a box potential and the hydrogen atom.

12.2.4 Features of Energy Eigenstates

Let us enumerate the features of eigenstates of hydrogen atom in the following.

1. There are infinite number of energy levels as n takes values 1, 2, ...

2. As n increases $|E_n|$ decreases and $E_n \to 0$ as $n \to \infty$. The lowest energy corresponds to $n = 1$. An energy of 0 corresponds to ionization of the hydrogen atom, that is complete separation of the proton and the electron with both the particles at rest. Figure 12.4 shows first few energy levels of hydrogen atom along with the few energy levels of linear harmonic oscillator and particle in a box for comparison.

3. We write $E_n = -E_1/n^2$, where $E_1 = \mu e^4/(2\hbar^2) \approx -13.6\,\text{eV}$.

4. The energy difference between nth and $(n+1)$th levels is given by $\Delta E_n = E_1(2n+1)/(n^2(n+1)^2)$. As n increases ΔE_n decreases.

5. E_n depends on n and independent of l and m. Therefore, each of the functions RY_{lm} $(m = -l,\ldots,l)$ is a solution of the Schrödinger equation of the hydrogen atom for the same value of E. The energy levels are thus degenerate. *How many energy levels do exist with the same E_n?* We know that l takes the values 0, 1, ..., $n-1$ and for each l, $m = l, l-1, \ldots, -l+1, -l$. There are $(2l+1)$ values of m labelled by n different values of l. Thus, the total number of degeneracy is given by

$$\sum_{l=0}^{n-1}(2l+1) = \sum 2l + \sum l = \frac{2n(n-1)}{2} + n = n^2. \tag{12.65}$$

6. Since electrons have two spin states, two electrons can be accommodated in a single state. The ground state $(n = 1)$ is nondegenerate. All other states are degenerate states. The existence of degenerate energy eigenvalues implies that linear combination of the corresponding eigenfunctions are solutions of the wave equation with the same energy. Hence, the number of electrons that can be placed in the shell corresponding to a particular n is $\sum_{l=0}^{n-1} 2(2l+1) = 2n^2$.

7. In the s-state $(l = 0)$ R is nonzero at the origin $r = 0$. On the other hand, for $l \neq 0$, $R_{nl}(0) = 0$. Since $Y_{00}(\theta,\phi) = 1/\sqrt{4\pi}$ we find that

$$|\psi_{n00}(0)|^2 = \frac{|R_{n0}(0)|^2}{4\pi} = \frac{Z^3}{\pi a_0^3 n^3}. \tag{12.66}$$

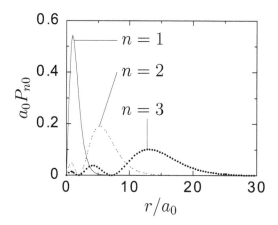

FIGURE 12.5
Radial probability distribution of hydrogen atom for three values of n with $l = 0$.

8. The probability distribution of the position of the electron is $P_{nl}(r) = |R_{nl}|^2 r^2$.
    For $n = 1$, $l = 0$ (1s (ground)state) $P = 4(r/a_0)^2 e^{-2r/a_0}/a_0$. For $n = 2$, $l = 0$
    (2s-state) $P = (r/a_0)^2 (1 - (r/2a_0))^2 e^{-r/a_0}/(2a_0)$. For $n = 2$, $l = 1$ (2p-state)
    $P = (r/a_0)^4 e^{-r/a_0}/(24a_0)$. Figure 12.5 depicts the variation of P_{n0} for three
    values of n.

9. The probability density

    $$\begin{aligned}
    P_{nlm} &= |\psi_{nlm}|^2 \\
    &= |R_{nl}(r)|^2 \, |Y_{lm}(\theta, \phi)|^2 \\
    &= |R_{nl}(r)|^2 \, \frac{(2l+1)(l-m)!}{4\pi(l+m)!} \left(P_l^{|m|}(\cos\theta) \right)^2 \qquad (12.67)
    \end{aligned}$$

    is independent of ϕ and function of r and θ only. $|R_{nl}(r)|^2$ gives the electron
    density as a function of r along a given direction θ and the remaining term in
    P_{nlm} gives the angular distribution for a given r.

10. The action of parity operator on the hydrogen-like wave function is

    $$P\psi_{nlm} = P[R_{nl}Y_{lm}] = R_{nl}(-1)^l Y_{lm} = (-1)^l \psi_{nlm}. \qquad (12.68)$$

    The states ψ_{nlm} have the parity of l.

    An algebraic method for solving hydrogen-like atoms is described in ref.[2]. For the
construction of necessary and sufficient conditions for the existence of bound states in
central potentials one may refer to refs. [3, 4].

Solved Problem 2:

What is the probability that an electron in the ground state of hydrogen atom be found in
the classically forbidden region (that is, in the region where $V(r) > E$)?

We have $E_1 = -\mu e^4/(2\hbar^2)$. The classical turning point can be found from $-e^2/r_e = -\mu e^4/(2\hbar^2)$. We find $r_e = 2\hbar^2/(\mu e^2) = 2a_0$. The ground state radial eigenfunction is

$R_{10} = 2(1/a_0)^{3/2} e^{-r/a_0}$. Hence, the probability for an electron to be found in the classically forbidden region is

$$
\begin{aligned}
\int_{2a_0}^{\infty} R_{10}^2 r^2 \, dr &= \frac{4}{a_0^3} \int_{2a_0}^{\infty} e^{-2r/a_0} r^2 \, dr \\
&= \frac{4}{a_0^3} \left[\frac{e^{-2r/a_0} r^2}{-2/a_0} \right]_{2a_0}^{\infty} + \frac{4}{a_0^2} \int_{2a_0}^{\infty} e^{-2r/a_0} r \, dr \\
&= 8e^{-4} + \frac{4}{a_0^2} \left[\frac{e^{-2r/a_0} r}{-2/a_0} \right]_{2a_0}^{\infty} + \frac{2}{a_0} \int_{2a_0}^{\infty} e^{-2r/a_0} \, dr \\
&= 8e^{-4} + 4e^{-4} + e^{-4} \\
&= 13e^{-4}.
\end{aligned}
\tag{12.69}
$$

Solved Problem 3:

Calculate the momentum eigenfunction for the ground state of hydrogen atom.

From the Eqs. (12.64a) and (9.12b) we obtain

$$
C(\mathbf{K}) = \frac{1}{(2\pi)^{3/2}} \frac{1}{(\pi a_0^3)^{1/2}} \int e^{-r/a_0} e^{-i\mathbf{K}\cdot\mathbf{r}} \, d\mathbf{r}.
\tag{12.70}
$$

Choosing the direction of $\mathbf{K}$ along z-axis we have $\mathbf{K}\cdot\mathbf{r} = kr\cos\theta$ and $d\mathbf{r} = r^2 \sin\theta \, dr \, d\theta \, d\phi$. Then

$$
\begin{aligned}
C(\mathbf{K}) &= \frac{1}{(2\pi)^{3/2}} \frac{1}{(\pi a_0^3)^{1/2}} \int_0^{\infty} \int_0^{\pi} \int_0^{2\pi} e^{-r/a_0} e^{-ikr\cos\theta} r^2 \sin\theta \, dr \, d\theta \, d\phi \\
&= \frac{(2a_0)^{3/2}}{\pi(1 + k^2 a_0^2)^2}.
\end{aligned}
\tag{12.71}
$$

Next, consider the eigenfunction ψ_{200} given by Eq. (12.64b). We get (with $A = 1/(\sqrt{32}\pi a_0^{3/2})$)

$$
\begin{aligned}
C(\mathbf{K}) &= A \int_0^{\infty} \int_0^{\pi} \int_0^{2\pi} \left(2 - \frac{r}{a_0} \right) e^{-r/a_0} e^{-ikr\cos\theta} r^2 \sin\theta \, dr \, d\theta \, d\phi \\
&= \frac{2\pi A}{ik} \int_0^{\infty} \left(2r - \frac{r^2}{a_0} \right) e^{-r/a_0} \left[e^{ikr} - 1 \right] \, dr.
\end{aligned}
\tag{12.72}
$$

Using the integral formula $\int_0^{\infty} x^n e^{-\mu x} \, dx = n!/\mu^{n+1}$ we obtain

$$
C(\mathbf{K}) = i4\pi A \left[\frac{a_0^2 - 1}{a_0^4} + \frac{ika_0^3}{(1 - ika_0)^3} \right].
\tag{12.73}
$$

12.3 Hydrogen Atom in D-Dimension

In this section, we are concerned with the bound states ($E < 0$) of the hydrogen atom in D-dimension, where D may be noninteger [5]. We restrict ourselves to radial part of the problem.

The Coulomb potential is taken as $V = -e^2/r$ for all values of D. For $D = 1$, r is to be interpreted as $|r|$. For the three-dimensional case the radial Eq. (12.24) is rewritten as

$$\nabla^2 R + \left[\frac{2\mu E}{\hbar^2} + \frac{2e^2\mu}{\hbar^2 r}\right] R = 0, \tag{12.74a}$$

where

$$\nabla^2 = \frac{d^2}{dr^2} + \frac{2}{r}\frac{d}{dr} - \frac{l(l+1)}{r^2}. \tag{12.74b}$$

For D-dimension the volume element is $d\tau = r^{D-1}dr$ and the Laplacian is

$$\nabla^2 = \frac{d^2}{dr^2} + \frac{D-1}{r}\frac{d}{dr} - \frac{l(l+D-2)}{r^2}. \tag{12.75}$$

Consequently, the Schrödinger equation in D-dimension is

$$\frac{d^2 R}{dr^2} + \frac{(D-1)}{r}\frac{dR}{dr} + \left[\frac{2e^2\mu}{\hbar^2 r} + \frac{2\mu E}{\hbar^2} - \frac{l(l+D-2)}{r^2}\right] R = 0. \tag{12.76}$$

With the change of variables

$$\rho^2 = -\frac{8\mu E r^2}{\hbar^2}, \quad R = \frac{f(\rho)}{r^{-(D-1)/2}} \tag{12.77}$$

Eq. (12.76) becomes

$$f'' + \left\{\frac{e^2\sqrt{\mu}}{\hbar\rho\sqrt{-2E}} + \left[\frac{1 + (2l+D-2)^2}{4\rho^2}\right] - \frac{1}{4}\right\} f = 0. \tag{12.78}$$

Defining $q = n + l$ and comparing the above equation with Eq. (12.32) give

$$p = \pm(2l+D-2), \quad E = -\frac{2\mu e^4}{\hbar^2(2n+2l+p+1)^2}. \tag{12.79}$$

The eigenfunction is

$$R = e^{-\rho/2}\rho^{(p-D+2)/2}L_q^p(\rho). \tag{12.80}$$

For $D = 1$, the only value that l can take is 0. Then $p = \pm 1$. For $p = 1$

$$E = -\frac{\mu e^4}{2\hbar^2(n+1)^2}, \quad R = e^{-\rho/2}\rho L_n^1(\rho), \quad \rho = \frac{2\mu e^2 r}{\hbar^2(n+1)}. \tag{12.81}$$

For $p = -1$

$$E = -\frac{\mu e^4}{2\hbar^2 n^2}, \quad R = e^{-\rho/2}\rho L_n^{-1}(\rho), \quad \rho = \frac{2\mu e^2 r}{\hbar^2 n}. \tag{12.82}$$

The results obtained earlier for three-dimensional case is recovered from Eqs. (12.79) and (12.80) by the substitution $D = 3$.

12.4   Field Produced by a Hydrogen Atom

In this section, we calculate the magnetic field in a hydrogen atom [6]. The contribution from both the orbital and spin motion of the electron are considered. The calculations are nonrelativistic and the energy eigenstates $|nlm_lm_s\rangle$ and $|nljm_j\rangle$ are alone considered. The calculation is tedious for a state of superposition of these eigenfunctions. This is because it is incorrect simply to add the fields due to each eigenstate.

12.4.1 General Considerations

For $m_l \neq 0$ states, we treat the electron cloud as circulating round the z-axis. This orbital motion produces the magnetic field. The different parts of the cloud may be circulating in opposite directions. Now, the probability current density is given by

$$\mathbf{J} = \frac{\hbar}{2mi} \left[\psi^* \nabla \psi - (\nabla \psi^*) \psi \right] , \tag{12.83}$$

where m is mass of the electron. The electric current density $\mathbf{j}$ is

$$\mathbf{j} = -e\mathbf{J} = i\mu_B \left[\psi^* \nabla \psi - (\nabla \psi^*) \psi \right] , \tag{12.84}$$

where $\mu_B = e\hbar/(2m)$ is the Bohr magnetron .

Assuming $\psi = F(r,\theta) e^{im_l \phi}$ for an energy eigenstate $|nlm_l\rangle$, Eq. (12.84) in spherical polar coordinates gives (the proof is left as an exercise to the reader)

$$\mathbf{j} = (0)\mathbf{r}_1 + (0)\hat{\theta}_1 - \frac{2\mu_B |F|^2 m_l}{r\sin\theta} \hat{\phi}_1 . \tag{12.85}$$

Thus,

$$j_r = 0, \quad j_\theta = 0, \quad j_\phi = -\frac{2\mu_B |F|^2 m_l}{r\sin\theta} = -\frac{2\mu_B m_l}{r\sin\theta} \psi^* \psi . \tag{12.86}$$

Using Eq. (12.86) the vector potential $\mathbf{A}$ and the field is calculated. Since $\mathbf{A}$ will be directed in the direction of the current density we have $A_r = A_\theta = 0$ and $\partial A_\phi/\partial\phi = 0$. Hence, r and θ components of $\nabla^2 \mathbf{A}$ vanish. Then

$$\nabla^2 \mathbf{A} = \left(\nabla^2 A_\phi - \frac{A_\phi}{r^2 \sin\theta} \right) \hat{\phi} . \tag{12.87}$$

Since $\partial A_\phi/\partial\phi = 0$ and

$$\nabla^2 A_\phi = \frac{1}{r^2} \frac{\partial}{\partial r} \left(r^2 \frac{\partial A_\phi}{\partial r} \right) + \frac{1}{r^2 \sin\theta} \frac{\partial}{\partial\theta} \left(\sin\theta \frac{\partial A_\phi}{\partial\theta} \right) \tag{12.88}$$

$\nabla^2 \mathbf{A}$ takes the form

$$\nabla^2 \mathbf{A} = \frac{1}{r^2} \left[\frac{\partial}{\partial r} \left(r^2 \frac{\partial A_\phi}{\partial r} \right) + \frac{1}{\sin\theta} \frac{\partial}{\partial\theta} \left(\sin\theta \frac{\partial A_\phi}{\partial\theta} \right) - \frac{A_\phi}{\sin\theta} \right] \hat{\phi}. \tag{12.89}$$

Then $\nabla^2 \mathbf{A} = -\mu_0 \mathbf{j}$ becomes

$$\frac{\partial}{\partial r} \left(r^2 \frac{\partial A_\phi}{\partial r} \right) + \frac{1}{\sin\theta} \frac{\partial}{\partial\theta} \left(\sin\theta \frac{\partial A_\phi}{\partial\theta} \right) - \frac{A_\phi}{\sin\theta} = -\mu_0 r^2 j_\phi . \tag{12.90}$$

The field will involve the alternate multipoles, that is, magnetic dipole, octupole, ... The corresponding θ dependence of j_ϕ are, respectively, $\sin\theta$, $4\cos^2\theta\sin\theta - \sin^3\theta$, ... Therefore, we write

$$j_\phi = R_{\text{dip}}(r)\sin\theta + R_{\text{oct}}(r)(4\cos^2\theta\sin\theta - \sin^3\theta) + \dots . \tag{12.91}$$

12.4.2 Magnetic Dipole Contribution

Substitution of Eq. (12.91) with the dipole term alone in Eq. (12.90) gives

$$\frac{\partial}{\partial r} \left(r^2 \frac{\partial A_\phi}{\partial r} \right) + \frac{1}{\sin\theta} \frac{\partial}{\partial\theta} \left(\sin\theta \frac{\partial A_\phi}{\partial\theta} \right) - \frac{A_\phi}{\sin^2\theta} = -\mu_0 r^2 R_{\text{dip}} \sin\theta . \tag{12.92}$$

Assume a solution of this equation as $A_\phi = S(r) \sin \theta$. The differential equation for S is obtained as

$$r^2 S'' + 2rS' - 2S = -\mu_0 R_{\text{dip}} r^2 . \tag{12.93}$$

Solutions of the homogeneous part of Eq. (12.93) are r and $1/r^2$. Therefore, $A_\phi \to 0$ as $r \to \infty$, we choose $S = T/r^2$. Then Eq. (12.93) becomes

$$\frac{d}{dr}\left(\frac{T'}{r^2}\right) = -\mu_0 R_{\text{dip}} . \tag{12.94}$$

Integration of Eq. (12.94) once gives

$$\frac{T'}{r^2} = -\mu_0 \int R_{\text{dip}}\, dr + \text{constant} . \tag{12.95}$$

12.4.3 Magnetic Octupole Contribution

Proceeding as before, we find

$$A_\phi = \frac{V(r)}{r^4}\left(4\cos^2\theta \sin\theta - \sin^3\theta\right) , \tag{12.96a}$$

where V is obtained from

$$\frac{V'}{r^6} = -\mu_0 \int \frac{R_{\text{oct}}}{r^2}\, dr + \text{constant} . \tag{12.96b}$$

12.4.4 Application to Particular State

When $m_l = 0$, Eq. (12.86) gives $j_\phi = 0$ and therefore the field vanishes. On the other hand, from Eq. (12.86) we note that if the sign of m_l is changed, then $\mathbf{j}$ changes sign and hence so do $\mathbf{A}$ and $\mathbf{B}$. Therefore, we consider only the states with $m_l > 0$.

For the state $|2\,1\,1\rangle$ with the eigenfunction ψ_{211}

$$j_\phi = -\frac{\mu_B}{32\pi a_0^5}\, re^{-r/a_0} \sin\theta . \tag{12.97}$$

Comparison of Eq. (12.97) with Eq. (12.91) gives

$$R_{\text{dip}} = -\frac{\mu_B}{32\pi a_0^5}\, re^{-r/a_0} . \tag{12.98}$$

Substituting the above expression for R_{dip} in Eq. (12.95) we obtain T. Then A_ϕ is worked out to be

$$A_\phi = \frac{\mu_0 \mu_B \sin\theta}{32\pi a_0^2}\left[e^{-r/a_0}\left(\frac{r}{a_0} + 4 + \frac{8a_0}{r} + \frac{8a_0^2}{r^2}\right) + k_1 r + \frac{k_2}{r^2}\right] , \tag{12.99}$$

where k_1 and k_2 are arbitrary constants. To determine k_1 and k_2 we apply the conditions $\mathbf{A} \to 0$ as $r \to \infty$ and $A \to$ finite as $r \to 0$. These gives $k_1 = 0$ and $k_2 = -8a_0^2$. Then

$$\mathbf{A} = \frac{\mu_0 \mu_B \sin\theta}{4\pi a_0^2}\left[e^{-r/a_0}\left(\frac{r}{8a_0} + \frac{1}{2} + \frac{a_0}{r} + \frac{a_0^2}{r^2}\right) - \frac{a_0^2}{r^2}\right]\hat{\phi} . \tag{12.100a}$$

We find $A_r = A_\theta = 0$. Then the components of $\mathbf{B}$ are

$$B_r = \frac{\mu_0 \mu_B \cos\theta}{2\pi a_0^3}\left[e^{-r/a_0}\left(\frac{1}{8} + \frac{a_0}{2r} + \frac{a_0^2}{r^2} + \frac{a_0^3}{r^3}\right) - \frac{a_0^3}{r^3}\right] , \tag{12.100b}$$

$$B_\theta = \frac{\mu_0 \mu_B \sin\theta}{4\pi a_0^3}\left[e^{-r/a_0}\left(\frac{r}{8a_0} + \frac{1}{4} + \frac{a_0}{2r} + \frac{a_0^2}{r^2} + \frac{a_0^3}{r^3}\right) - \frac{a_0^3}{r^3}\right] . \tag{12.100c}$$

12.4.5 Contribution of Electron Spin

The inclusion of spin involves an additional term in the probability current density. This term is $\mathbf{J}^{(s)} = (\hbar/2m)\nabla \times (\psi^* \boldsymbol{\sigma} \psi)$, where $\boldsymbol{\sigma} = (\sigma_x, \sigma_y, \sigma_z)$ are Pauli spin matrices. If $m_s = 1/2$ then $\psi = \psi \begin{pmatrix} 1 \\ 0 \end{pmatrix}$ and $\psi^* \boldsymbol{\sigma} \psi = \psi^* \psi \mathbf{K}$. Similarly, if $m_l = -1/2$ then $\psi^* \boldsymbol{\sigma} \psi = -\psi^* \psi \mathbf{K}$.

The spin current density is

$$
\begin{aligned}
\mathbf{j}^{(s)} &= -e\mathbf{J}^{(s)} = -2\mu_B m_s [\nabla(\psi^* \psi)] \times \mathbf{K} \\
&= 2m_s \mu_B \left(\sin\theta \frac{\partial}{\partial r} + \frac{\cos\theta}{r} \frac{\partial}{\partial\theta} \right) (\psi^* \psi)\hat{\phi}.
\end{aligned}
\tag{12.101}
$$

The case $m_s = -1/2$ gives the same fields of $m_s = 1/2$ but with reversed sign. For the ground state the orbit contribution is zero for the magnetic field and

$$
j_\phi^{(s)} = -\frac{2\mu_B}{\pi a_0^4} e^{-2r/a_0} \sin\theta.
\tag{12.102}
$$

Since, $j_\phi^{(s)}$ has only dipole term we use Eq. (12.95) and obtain

$$
\mathbf{A}^{(s)} = \frac{\mu_0 \mu_B \sin\theta}{4\pi a_0^2} \left[e^{-2r/a_0} \left(2 + 2\frac{a_0}{r} + \frac{a_0^2}{r^2} \right) - \frac{a_0^2}{r^2} \right] \hat{\phi},
\tag{12.103a}
$$

$$
B_r^{(s)} = \frac{\mu_0 \mu_B \cos\theta}{2\pi a_0^3} \left[e^{-2r/a_0} \left(\frac{2a_0}{r} + \frac{2a_0^2}{r^2} + \frac{a_0^3}{r^3} \right) - \frac{a_0^3}{r^3} \right],
\tag{12.103b}
$$

$$
B_\theta^{(s)} = \frac{\mu_0 \mu_B \sin\theta}{4\pi a_0^3} \left[e^{-2r/a_0} \left(4 + \frac{2a_0}{r} + \frac{2a_0^2}{r^2} + \frac{a_0^3}{r^3} \right) - \frac{a_0^3}{r^3} \right].
\tag{12.103c}
$$

At large r, $\mathbf{A} = -\mu_0 \mu_B \sin\theta/(4\pi r^2)\hat{\phi}$ which implies that the dipole moment due to the spin is $M_s = -\mu_B$ as expected. At $r = 0$, $\mathbf{B}$ is in the z-direction with $B_z^{(s)}(0) = -2\mu_0 \mu_B/(3\pi a_0^5)$.

12.5 System in Parabolic Coordinates

Consider the hydrogen atom in the presence of an external field. Let the applied constant electric field $\mathbf{E}_0$ is along the z-axis so that $\mathbf{E}_0 = E_0 \mathbf{z}$. Then the electron and proton are stretched apart in the z-direction. Now, the atom becomes elongated along the z-axis and the parabolic coordinates system is better suited than spherical polar coordinates to describe the system. The Schrödinger equation of the system is

$$
\left(-\frac{\hbar^2}{2\mu} \nabla^2 - \frac{e^2}{r} - e\mathbf{r} \cdot \mathbf{E}_0 \right) \psi = E\psi.
\tag{12.104}
$$

Dimensionless orthogonal parabolic coordinates (ξ, η, ϕ) are given by:

$$
x = a_0 \sqrt{\xi\eta} \cos\phi, \quad y = a_0 \sqrt{\xi\eta} \sin\phi, \quad z = \frac{a_0}{2}(\xi - \eta),
\tag{12.105a}
$$

$$
r = \frac{a_0}{2}(\xi + \eta), \quad r^2 = x^2 + y^2 + z^2,
\tag{12.105b}
$$

where x, y and z are the Cartesian coordinates, $0 \leq \xi \leq \infty$, $0 \leq \eta \leq \infty$, $0 \leq \phi \leq 2\pi$ and a_0 is the radius of first Bohr orbit. The differential elements of length ds and volume element $d\tau$ are written as

$$(ds)^2 = a_0^2 \left[\frac{\xi + \eta}{4\xi} (d\xi)^2 + \frac{\xi + \eta}{4\eta} (d\eta)^2 + \xi\eta(d\phi)^2 \right], \qquad (12.106a)$$

$$d\tau = \frac{a_0^3}{4} (\xi + \eta) \, d\xi \, d\eta \, d\phi. \qquad (12.106b)$$

The operator ∇^2 is given by

$$\nabla^2 = \frac{1}{a_0^2} \left\{ \frac{4}{(\xi + \eta)} \left[\frac{\partial}{\partial \xi} \left(\xi \frac{\partial}{\partial \xi} \right) + \frac{\partial}{\partial \eta} \left(\eta \frac{\partial}{\partial \eta} \right) \right] + \frac{1}{\xi\eta} \frac{\partial^2}{\partial \phi^2} \right\}. \qquad (12.107)$$

Then the Schrödinger equation in parabolic coordinates is

$$\left\{ -\frac{4}{\xi + \eta} \left[\frac{\partial}{\partial \xi} \left(\xi \frac{\partial}{\partial \xi} \right) + \frac{\partial}{\partial \eta} \left(\eta \frac{\partial}{\partial \eta} \right) \right] - \frac{1}{\xi\eta} \frac{\partial^2}{\partial \phi^2} \right.$$
$$\left. - \frac{4}{\xi + \eta} + E_0(\xi - \eta) \right\} \psi = 2E\psi. \qquad (12.108)$$

Assume the solution of Eq. (12.108) as

$$\psi = f_1(\xi) f_2(\eta) e^{im\phi}. \qquad (12.109)$$

Multiplication of Eq. (12.108) by $(\xi + \eta)/4$ and substitution of Eq. (12.109) in the resulting equation decouples it into the equations

$$\frac{d}{d\xi} \left(\xi \frac{df_1}{d\xi} \right) + \left[\frac{1}{2} E_0 \xi - \frac{m^2}{4\xi} - \frac{1}{4} E_0 \xi^2 + \beta_1 \right] f_1 = 0, \qquad (12.110a)$$

$$\frac{d}{d\eta} \left(\eta \frac{df_2}{d\eta} \right) + \left[\frac{1}{2} E_0 \eta - \frac{m^2}{4\eta} - \frac{1}{4} E_0 \eta^2 + \beta_2 \right] f_2 = 0, \qquad (12.110b)$$

where $\beta_1 + \beta_2 = 1$.

12.5.1   Solution for $E_0 = 0$

For $E_0 = 0$ the problem reduces to that of the hydrogen atom. As shown in sec. 12.2 the energy of a bound state is $E_n = -1/(2n^2)$. Now, introduce two new coordinates $\rho_1 = \xi/n$, $\rho_2 = \eta/n$ and write $n\beta_i = \alpha_i + (|m| + 1)/2$, $i = 1, 2$. Then Eqs. (12.110) take the form

$$\frac{d^2 f_i}{d\rho_i^2} + \frac{1}{\rho_i} \frac{df_i}{d\rho_i} + \left[-\frac{1}{4} + \frac{1}{\rho_i} \left(\alpha_i + \frac{1}{2}(|m| + 1) \right) \right.$$
$$\left. - \frac{m^2}{4\rho_i^2} - \frac{n^3 E_0 \rho_i}{4} \right] f_i = 0, \quad i = 1, 2. \qquad (12.111)$$

Solution of Eq. (12.111) is

$$f_i(\rho_i) = e^{-\rho_i/2} \rho_i^{|m|/2} W_i(\rho_i), \quad i = 1, 2 \qquad (12.112a)$$

$$W_i(\rho_i) = \frac{|m|! \alpha_i!}{(|m| + \alpha_i)!} L_{\alpha_i}^{|m|}(\rho_i) \qquad (12.112b)$$

with $L_{\alpha_i}^{|m|}$ being an associated Laguerre polynomial.

In Eq. (12.112b), α_1 and α_2 must take the integer values $0, 1, 2, \ldots$ in order to have normalizable solutions. Because

$$\beta_1 + \beta_2 = \frac{1}{n}\left[\alpha_1 + \frac{1}{2}(|m| + 1)\right] + \frac{1}{n}\left[\alpha_2 + \frac{1}{2}(|m| + 1)\right] = 1 \qquad (12.113)$$

n is obtained as $n = \alpha_1 + \alpha_2 + |m| + 1$, where $0 \leq |m| \leq n - 1$, $0 \leq \alpha_i \leq (n - |m| - 1)$, $m = \pm|m|$. Now,

$$\psi_{\alpha_1, \alpha_2, m} = C(\xi\eta)^{|m|/2} e^{-(\xi+\eta)/2n} L_{\alpha_1}^{|m|}(\xi/n) L_{\alpha_2}^{|m|}(\eta/n)\, e^{im\phi}, \qquad (12.114)$$

where C is a normalization constant and is determined from the condition

$$\frac{1}{4}\int_0^\infty d\xi \int_0^\infty d\eta \int_0^{2\pi} d\phi (\xi + \eta) |\psi_{\alpha_1, \alpha_2, m}|^2 = 1. \qquad (12.115)$$

C is worked out as

$$C^2 = \frac{\alpha_1!\, \alpha_2!}{\pi n^4 n^{2|m|}(|m| + \alpha_1)!\,(|m| + \alpha_2)!}. \qquad (12.116)$$

The bound state energy is $E_n = -1/(2n^2) = -1/[2(\alpha_1 + \alpha_2 + |m| + 1)^2]$.

12.5.2 Solution for Large E_0

For large enough external field, E_0, the motion is dominated by the states, $\alpha_1 = 0$, $\alpha_2 > 0$ and $m = 0$. In this case $n = \alpha_2 + 1$. The eigenfunction is

$$\psi_{0, \alpha_2, 0} = \frac{1}{n^2}\sqrt{\frac{1}{\pi}}\, e^{-(\xi+\eta)/2n} L_{\alpha_2}^0(\eta/n). \qquad (12.117)$$

The most probable value of ξ is 0. Thus, $x \approx y \approx 0$ and $z \approx -\eta/2$. Some of the dipole matrix elements are obtained as [7]

$$\langle n; 0, \alpha_2, 0|1/a_0|n; 0, \alpha_2, 0| = -\frac{3}{2}n(n - 1), \qquad (12.118a)$$

$$\langle n; 0, n - 1, 0|1/a_0|n \pm 1; 0, n - 1 \pm 1, 0\rangle \approx 0.32n^2, \qquad (12.118b)$$

$$\langle n; 0, n - 1, 0|1/a_0|n \pm 2; 0, n - 1 \pm 2, 0\rangle \approx 0.11n^2. \qquad (12.118c)$$

12.6 Concluding Remarks

Since the potential for a hydrogen-like atom is given by $V(r) = -Ze^2/r$, and the nuclear mass is different, the energy eigenvalues and the position of maximum radial probability distribution will all be different. The hydrogenic ions $He^+(Z = 2)$, $Li^{2+}(Z = 3)$, $Be^{3+}(Z = 4)$ etc. are examples of such systems. The (neutral) isotopes of atomic hydrogen, deuterium and tritium also provide examples of hydrogenic systems. Since the mass of deuterium is $m_d \approx 2m_p$ and that of tritium is $m_t \approx 3m_p$, the reduced masses of them will be slightly different from that of hydrogen. These small differences in the values of μ give rise to isotopic shifts of the spectral lines.

In addition to deuterium and tritium, there exists other hydrogenic systems in which the role of proton is played by another positive particle. For example, positronium is a bound hydrogenic system made of a positron e^+ and an electron. Muonium is another example of such a system in which the proton is replaced by a positive muon μ^+, a particle which is very smaller to the positron, except that it has a mass $m_\mu \approx 207 m_e$ and is unstable with a life-time of about 2.2×10^{-6} sec.

In all the above mentioned hydrogenic atoms the negative particle is an electron. Other negative particles could form a bound system with a positive nucleon. For example, a negative muon μ^- can be captured by the Coulombic attraction of a nucleon of charge Ze, thus forming a muonic atom. The simplest example of muonic atom is the bound system $(p^+ \mu^-)$ consisting of a proton p^+ and a negative μ^-. The radius a of muonic hydrogen $(p^+ \mu^-)$ is nearly 186 times smaller than that of ordinary hydrogen atom. The reduced mass of muonic hydrogen is $\approx 186 m_e$.

12.7 Bibliography

[1] M. Abramowitz and I.A. Stegun, *Handbook of Mathematical Functions*. Dover, New York, 1965, pp. 773-778.

[2] T.H. Cooke and J.L. Wood, *Am. J. Phys.* 70:945, 2002.

[3] V. Glaser, H. Grosse, A. Martin and W. Thirring, *Essays in Honour of Valentine Bargmann*. Princeton University Press, Princeton, New Jersey, 1976.

[4] F. Brau, *J. Phys. A* 37:6687, 2004.

[5] R.E. Moss, *Am. J. Phys.* 55:397, 1987.

[6] W. Gough, *Eur. J. Phys.* 17:208, 1996.

[7] J.N. Bardsley and B. Sundaram, *Phys. Rev. A* 32:689, 1985.

12.8 Exercises

12.1 Show that the equation

$$i\hbar \frac{\partial \psi}{\partial t} = -\frac{\hbar^2}{2m_e} \nabla_1^2 - \frac{\hbar^2}{2m_p} \nabla_2^2 - \frac{e^2}{|\mathbf{r}_1 - \mathbf{r}_2|} \psi$$

reduces to

$$i\hbar \frac{\partial \psi}{\partial t} = \left(-\frac{\hbar^2}{2M} \nabla_R^2 - \frac{\hbar^2}{2\mu} \nabla_r^2 - \frac{e^2}{r} \right) \psi,$$

where $M = m_e + m_p$, $\mu = m_e m_p / (m_e + m_p)$ under the centre of mass and relative coordinate transformation $\mathbf{r} = \mathbf{r}_1 - \mathbf{r}_2$ and $(m_e + m_p)\mathbf{R} = m_e \mathbf{r}_1 + m_p \mathbf{r}_2$.

12.2 The normalized eigenfunction of 1s electron is $\psi_{100} = (1/(\pi a_0^3))^{1/2}\, e^{-r/a_0}$. Calculate the probability of finding the electron in a sphere of radius $r = a_0$.

12.3 Show that the total angular momentum is unchanged in the transformation $(\mathbf{r}_1, \mathbf{r}_2) \to (\mathbf{R}, \mathbf{r})$.

12.4 For the equation

$$\frac{d^2R}{dr^2} + \frac{2}{r}\frac{dR}{dr} + \left[\frac{2\mu E}{\hbar^2} + \frac{2\mu e^2}{\hbar^2 r} - \frac{l(l+1)}{r^2}\right]R = 0$$

obtain the asymptotic solution of bound states for $n \to \infty$.

12.5 Using the equation

$$f'' + \frac{2\mu}{\hbar^2}\left[E + \frac{e^2}{r} - \frac{l(l+1)\hbar^2}{2\mu r^2}\right]f = 0$$

show that near the origin $r \to 0$, the physically acceptable solution behaves like $f(r) \sim r^{l+1}$ for $l \neq 0$.

12.6 What is the probability distribution of the radial position r?

12.7 At time $t = 0$, a hydrogen atom is described by the superposition state

$$\psi_T(\mathbf{r},0) = \frac{1}{\sqrt{2}}\psi_{100}(\mathbf{r}) - \frac{1}{\sqrt{6}}\psi_{210}(\mathbf{r}) + \frac{1}{\sqrt{3}}\psi_{211}(\mathbf{r}).$$

What is $\psi_T(\mathbf{r},t)$ for $t > 0$?

12.8 Consider the wave function given in the exercise 12.7. Determine the probability that at $t > 0$, a measurement would find (a) $E = E_2$, (b) $L^2 = 2\hbar^2$ and (c) $L_z = 0$.

12.9 Determine whether the probability density associated with the wave function given in the exercise 12.7 is periodic in time and if it is so find the period.

12.10 For a hydrogen atom in the ground state calculate the average distance of the electron from the nucleus.

12.11 If the interaction potential V is spherically symmetric and proportional to r^k then the virial theorem states $2\langle T\rangle = k\langle V\rangle$. Prove the virial theorem for hydrogen atom.

12.12 According to Bohr's correspondence principle the quantum mechanical results reduce to classical results at large quantum numbers. For large values of n find the location of the electron.

12.13 Show that the probability distribution of the radial function of the electron for the hydrogen atom $P_{nl}(r) = |R_{nl}(r)|^2 r^2$ has $n - l$ maxima and for the lowest value of $l = n - 1$, there is just one maximum at the value given by the Bohr model.

12.14 Find the lowest energy and the classical turning radii of hydrogen atom in the $l = 4$ state.

12.15 Assuming the radius of the Li^{++} nucleus as 3fm calculate the probability of finding the 1s electron inside the nucleus.

12.16 Find the expectation value of the position vector of hydrogen atom.

12.17 Show that the transition between $n = 2$ to $n = 1$ levels of muonic hydrogen atom lies in the X-ray region.

12.18 A hydrogen atom is subjected to a potential a/r^2 in addition to the potential $-e^2/r$. Assume the ground state eigenfunction as $\psi(r) = r^\alpha e^{-\beta r}$. Find the ground state energy using the radial equation and the condition on the potential constant a for the admissible real energy value.

12.19 According to Bohr's correspondence principle the quantum mechanical results reduce to classical results at large quantum numbers. Show that for large values of n, the electron has energy, which is the same as that of a classical electron in a circular orbit.

12.20 Assume that the admissible wave function in the neighbourhood of the region for a system with a potential $V(r)$, which is no more singular than r^{-2} as $\psi(\pi, \theta, 0) = r^\alpha f(\theta, \phi)$. Find the condition on α for the existence of $\langle H \rangle$.

13

Approximation Methods I: Time-Independent Perturbation Theory

13.1 Introduction

The starting point in a quantum mechanical problem is the construction of a Schrödinger equation. We try to solve it exactly to obtain wave function. A Schrödinger equation is exactly solvable only for a certain number of simple potentials. In most of the cases the Schrödinger equation cannot be solved exactly. The solution of the Schrödinger equation for the real systems is usually complicated and often exact solutions are very difficult to find or solutions do not exist. Examples include as anharmonic oscillator, hydrogen atom in an electric field, a plane rotator in the presence of an applied electric field and so on. Therefore, we need to rely on approximation methods. The simple and most interesting and powerful approximate method is the perturbation theory developed by Schrödinger in 1926 [1]. Earlier Lord Rayleigh [2] analyzed harmonic vibrations of a string perturbed by small inhomogeneities. The work of Rayleigh was pointed out by Schrödinger. Consequently, the method is also called *Rayleigh–Schrödinger perturbation theory*. This method is powerful for determining the changes in the discrete energy values and the associated eigenfunctions of a system due to a weak perturbation, provided the eigenfunctions of the undisturbed system are available. The approximations obtained from the perturbation theory are used to calculate energy level diagrams and to understand the splitting and assignments of atomic and molecular spectra [3–5]. The perturbation theory has been proven to be useful in several fields of theoretical physics and also in chemistry. It provides considerable insight into various phenomena. The main disadvantage of this method is that in certain problems the perturbation series converges slowly or diverges.

In perturbation problems the Hamiltonian is split into two parts. One part called *unperturbed part* is large and the corresponding Schrödinger equation is assumed to be exactly solvable. The other part is treated as perturbation. The changes in the energies and the eigenfunctions are represented as a power series in the perturbation parameter. The approximation consists of neglecting terms in the series after the first few terms. Approximating the series to the first n terms gives the n-th order approximation.

In this chapter, we develop the perturbation theory for the time-independent perturbation. We consider the theory for both degenerate and nondegenerate energy levels. We apply the theory to a few physically interesting problems. The problem of time-dependent perturbation will be discussed in the next chapter.

DOI: 10.1201/9781003172178-13

13.2   Theory for Nondegenerate Case

Let $H^{(0)}$ be the unperturbed Hamiltonian operator and its energy eigenvalues are discrete and nondegenerate. As noted earlier in chapter 4, by a *nondegenerate* energy level we mean that there is only one linearly independent eigenfunction corresponding to that energy level. In case there are k-linearly independent eigenfunctions corresponding to an energy level then the energy level is said to be *k-fold degenerate*. An example of a system with two-fold degeneracy is a free particle (see sec. 4.3).

Let the total Hamiltonian of the system be

$$H = H^{(0)} + \lambda H^{(1)} , \tag{13.1}$$

where $H^{(0)}$ is the Hamiltonian of the unperturbed system and $H^{(1)}$ is the perturbation. It is desired to solve $H\phi_n = E_n\phi_n$, that is,

$$\left(H^{(0)} + \lambda H^{(1)} \right) \phi_n = E_n\phi_n . \tag{13.2}$$

When $\lambda = 0$, Eq. (13.2) reduces to

$$H^{(0)}\phi_n^{(0)} = E_n^{(0)}\phi_n^{(0)} . \tag{13.3}$$

The system (13.3) is assumed to be exactly solvable so that $\phi_n^{(0)}$ and $E_n^{(0)}$ are explicitly known. The goal is to express E_n and ϕ_n of the perturbed system as a power series in terms of λ as

$$
\begin{aligned}
E_n &= E_n^{(0)} + \lambda E_n^{(1)} + \lambda^2 E_n^{(2)} + \dots , & (13.4) \\
\phi_n &= \phi_n^{(0)} + \lambda \phi_n^{(1)} + \lambda^2 \phi_n^{(2)} + \dots . & (13.5)
\end{aligned}
$$

Here $E_n^{(k)}$ and $\phi_n^{(k)}$ are the k-th order corrections. We restrict ourselves with the calculation of $E_n^{(1)}$, $\phi_n^{(1)}$, $E_n^{(2)}$ and $\phi_n^{(2)}$. Substitution of Eqs. (13.4) and (13.5) in Eq. (13.2) gives

$$
\left(H^{(0)} + \lambda H^{(1)} \right) \left(\phi_n^{(0)} + \lambda \phi_n^{(1)} + \lambda^2 \phi_n^{(2)} + \dots \right)
$$
$$
= \left(E_n^{(0)} + \lambda E_n^{(1)} + \lambda^2 E_n^{(2)} + \dots \right) \left(\phi_n^{(0)} + \lambda \phi_n^{(1)} + \lambda^2 \phi_n^{(2)} + \dots \right) . \tag{13.6}
$$

For this equation to be satisfied for all values of λ the coefficients of various powers of λ must be equal to zero. The first few equations obtained in this way are

$$
\text{Coeff. of } \lambda^0 : \ H^{(0)}\phi_n^{(0)} = E_n^{(0)}\phi_n^{(0)} . \tag{13.7}
$$
$$
\text{Coeff. of } \lambda^1 : \ H^{(0)}\phi_n^{(1)} + H^{(1)}\phi_n^{(0)} = E_n^{(0)}\phi_n^{(1)} + E_n^{(1)}\phi_n^{(0)} . \tag{13.8}
$$
$$
\text{Coeff. of } \lambda^2 : \ H^{(0)}\phi_n^{(2)} + H^{(1)}\phi_n^{(1)} = E_n^{(0)}\phi_n^{(2)} + E_n^{(1)}\phi_n^{(1)} + E_n^{(2)}\phi_n^{(0)}. \tag{13.9}
$$

Equation (13.7) is assumed as already solved. We proceed to solve the Eqs. (13.8) and (13.9). First, we solve the Eq. (13.8) to find $E_n^{(1)}$ and $\phi_n^{(1)}$. Then we find $E_n^{(2)}$ and $\phi_n^{(2)}$.

13.2.1   First-Order Corrections $E_n^{(1)}$ and $\phi_n^{(1)}$

To solve the Eq. (13.8) we write $\phi_n^{(1)}$ as

$$\phi_n^{(1)} = \sum_m c_m^{(1)} \phi_m^{(0)} . \tag{13.10}$$

The reason for the above form is that the various possible states $\phi_m^{(0)}$ are known and they form an orthonormal set and any arbitrary wave function can be written as a linear combination of these functions. Now, Eq. (13.8) becomes

$$\sum c_m^{(1)} H^{(0)} \phi_m^{(0)} + H^{(1)} \phi_n^{(0)} = E_n^{(0)} \sum c_m^{(1)} \phi_m^{(0)} + E_n^{(1)} \phi_n^{(0)} . \tag{13.11}$$

Using Eq. (13.3) the above equation is rewritten as

$$\sum c_m^{(1)} \left(E_m^{(0)} - E_n^{(0)} \right) \phi_m^{(0)} = \left(E_n^{(1)} - H^{(1)} \right) \phi_n^{(0)} . \tag{13.12}$$

Multiplication of Eq. (13.12) by $\phi_r^{(0)*}$ and integrating over all space give

$$c_r^{(1)} \left(E_r^{(0)} - E_n^{(0)} \right) = E_n^{(1)} \delta_{rn} - H_{rn}^{(1)} , \tag{13.13}$$

where

$$H_{rn}^{(1)} = \int_{-\infty}^{\infty} \phi_r^{(0)} H^{(1)} \phi_n^{(0)} \, \mathrm{d}\tau \tag{13.14}$$

and we used the result $\int_{-\infty}^{\infty} \phi_r^{(0)*} \phi_m^{(0)} \, \mathrm{d}\tau = \delta_{rm}$.

13.2.2 Determination of $E_n^{(1)}$

When $r = n$ Eq. (13.13) gives

$$E_n^{(1)} = H_{nn}^{(1)} = \langle H^{(1)} \rangle . \tag{13.15}$$

The first-order correction to the nth energy eigenvalue is the expectation value of perturbation $H^{(1)}$ over the nth unperturbed state.

13.2.3 Determination of $c_m^{(1)}$

For $r \neq n$, Eq. (13.13) gives

$$c_r^{(1)} = \frac{H_{rn}^{(1)}}{\left(E_n^{(0)} - E_r^{(0)} \right)} , \quad r \neq n . \tag{13.16}$$

Further, note that any of the function $\phi_n^{(s)}$ can have an arbitrary multiple of $\phi_n^{(0)}$ added to it without affecting the Eqs. (13.8) and (13.9) and hence without affecting the determination of $\phi_n^{(s)}$ in terms of lower-order function. Hence, the multiple can be chosen in such a way that $c_n^{(s)} = \langle \phi_n^{(0)} | \phi_n^{(s)} \rangle = 0$ for all $s > 0$. Hence, expand $\phi_n^{(s)}$ as (with $c_n^{(s)} = 0$)

$$\phi_n^{(s)} = \sum_{m \neq n} c_m^{(s)} \phi_m^{(0)} . \tag{13.17}$$

Now,

$$\phi_n = \phi_n^{(0)} + \lambda \sum_{m \neq n} \frac{H_{mn}^{(1)}}{\left(E_n^{(0)} - E_m^{(0)} \right)} \phi_m^{(0)} , \tag{13.18a}$$

$$E_n = E_n^{(0)} + \lambda H_{nn}^{(1)} . \tag{13.18b}$$

In order to use it we require

$$\left|\lambda H_{nn}^{(1)}\right| \ll \left|E_n^{(0)} - E_m^{(0)}\right|, \quad \text{for all } n \neq m. \tag{13.19}$$

Another criterion is that the change in energy $\lambda H_{nn}^{(1)}$ should be small compared to the spacing between $E_n^{(0)}$ and the energies nearest to it. From Eq. (13.18a) we note that the first-order perturbation theory breaks down for degenerate energy levels since $E_n^{(0)} - E_m^{(0)}$ becomes zero.

Solved Problem 1:

Determine the second-order corrections $E_n^{(2)}$ and $\phi_n^{(2)}$.

We write

$$\phi_n^{(2)} = \sum_m c_m^{(2)} \phi_m^{(0)}. \tag{13.20}$$

Substituting $\phi_n^{(1)}$ and $\phi_n^{(2)}$ in Eq. (13.9) and multiplying the resulting equation by $\phi_r^{(0)*}$ and integrating over all space we get

$$c_r^{(2)}\left(E_r^{(0)} - E_n^{(0)}\right) = E_n^{(1)} \sum_{m \neq n} c_m^{(1)} \delta_{rm} + E_n^{(2)} \delta_{rn} - \sum_{m \neq n} c_m^{(1)} H_{rm}^{(1)}. \tag{13.21}$$

When $r = n$ from the above equation we obtain

$$E_n^{(2)} = \sum_{m \neq n} c_m^{(1)} H_{nm}^{(1)} = \sum_{m \neq n} \frac{\left|H_{nm}^{(1)}\right|^2}{\left(E_n^{(0)} - E_m^{(0)}\right)}. \tag{13.22}$$

When $r \neq n$ from Eq. (13.21) we get

$$c_r^{(2)} = -\frac{E_n^{(1)} H_{rn}^{(1)}}{\left(E_r^{(0)} - E_n^{(0)}\right)^2} - \sum_{m \neq n} \frac{H_{rm}^{(1)} H_{mn}^{(1)}}{\left(E_r^{(0)} - E_n^{(0)}\right)\left(E_n^{(0)} - E_m^{(0)}\right)}. \tag{13.23}$$

The normalization condition gives

$$c_n^{(2)} = -\frac{1}{2} \sum_{r \neq n} \left|c_r^{(1)}\right|^2. \tag{13.24}$$

Now,

$$\phi_n^{(2)} = \sum_{r \neq n}\left\{\left[\sum_{m \neq n} \frac{H_{rm}^{(1)} H_{mn}^{(1)}}{\left(E_n^{(0)} - E_r^{(0)}\right)\left(E_n^{(0)} - E_m^{(0)}\right)}\right.\right.$$
$$\left.\left. -\frac{E_n^{(1)} H_{rn}^{(1)}}{\left(E_r^{(0)} - E_n^{(0)}\right)^2}\right]\phi_r^{(0)} - \frac{\frac{1}{2}\left|H_{rn}^{(1)}\right|^2}{\left(E_r^{(0)} - E_n^{(0)}\right)^2}\phi_n^{(0)}\right\}. \tag{13.25}$$

We have found first- and second-order corrections to the unperturbed energy eigenvalues and eigenfunctions. If the Hamiltonian has both discrete and continuous states then we must include both the sum over all discrete states and the integral over all the continuous states.

We notice several important points from Eqs. (13.22) and (13.25):

1. The ground state energy is always lowered in the second-order perturbation because in Eq. (13.22) the quantity $E_n^{(0)} - E_m^{(0)}$ is negative.

2. Energies of two states are pushed apart in second-order perturbation.

3. Due to the term $1 \left/ \left(E_n^{(0)} - E_m^{(0)} \right) \right.$ if the energy levels of unperturbed systems are close, the levels are more affected.

4. The theory breaks down if the energy levels are degenerate.

We can proceed further to obtain higher-order correction. In practice we may stop the calculation after second-order correction.

13.2.4 Dalgarno and Lewis Method of Calculation of $E_n^{(2)}$

The $E_n^{(2)}$ given by Eq. (13.22) involves evaluation of infinite terms in the summation. In some cases computation of them is a difficult task. However, this is simplified by the method of Dalgarno and Lewis [6,7].

13.2.4.1 Formula for $E_n^{(2)}$

Suppose F_n is a state-dependent operator with the commutator relation

$$\left[F_n, H^{(0)} \right] \phi_n^{(0)} = \left(F_n H^{(0)} - H^{(0)} F_n \right) \phi_n^{(0)} = \left(H^{(1)} - E_n^{(1)} \right) \phi_n^{(0)}. \qquad (13.26)$$

Multiplying Eq. (13.26) by $\phi_m^{(0)*}$ and integrating over all space we get

$$\begin{aligned} \left\langle m \left| \left[F_n, H^{(0)} \right] \right| n \right\rangle &= \left\langle m \left| F_n H^{(0)} \right| n \right\rangle - \left\langle m \left| H^{(0)} F_n \right| n \right\rangle \\ &= \left\langle m \left| H^{(1)} - E_n^{(1)} \right| n \right\rangle. \end{aligned} \qquad (13.27)$$

When $m = n$

$$\begin{aligned} \left\langle n \left| \left[F_n, H^{(0)} \right] \right| n \right\rangle &= \left\langle n \left| F_n H^{(0)} - H^{(0)} F_n \right| n \right\rangle \\ &= \left\langle n \left| F_n E_n^{(0)} - E_n^{(0)} F_n \right| n \right\rangle \\ &= 0. \end{aligned} \qquad (13.28)$$

That is,

$$\left\langle n \left| \left[F_n, H^{(0)} \right] \right| n \right\rangle = \left\langle n \left| H^{(1)} - E_n^{(1)} \right| n \right\rangle = 0. \qquad (13.29)$$

When $m \neq n$ we obtain from Eq. (13.27)

$$\left\langle m \left| F_n H^{(0)} \right| n \right\rangle - \left\langle m \left| H^{(0)} F_n \right| n \right\rangle = \left\langle m \left| H^{(1)} \right| n \right\rangle - \left\langle m \left| E_n^{(1)} \right| n \right\rangle. \qquad (13.30)$$

That is,

$$\left\langle m \left| F_n \right| n \right\rangle \left(E_n^{(0)} - E_m^{(0)} \right) = \left\langle m \left| H^{(1)} \right| n \right\rangle. \qquad (13.31)$$

Rewrite (13.31) as

$$\left\langle m \left| F_n \right| n \right\rangle = \frac{H_{mn}^{(1)}}{\left(E_n^{(0)} - E_m^{(0)} \right)}, \quad m \neq n. \qquad (13.32)$$

Substituting Eq. (13.32) in Eq. (13.22), we get

$$E_n^{(2)} = \sum_{m \neq n} \frac{H_{nm}^{(1)} H_{mn}^{(1)}}{\left(E_n^{(0)} - E_m^{(0)}\right)} = \sum_{m \neq n} \left\langle n \left| H^{(1)} \right| m \right\rangle \langle m \left| F_n \right| n \rangle . \tag{13.33}$$

In Eq. (13.33) denominator disappeared. Using the closure property $\sum_m |m\rangle\langle m| + \int_{-\infty}^{\infty} |k\rangle\langle k| \, \mathrm{d}k = 1$ we get

$$E_n^{(2)} = \left\langle n \left| H^{(1)} F_n \right| n \right\rangle - E_n^{(1)} \langle n \left| F_n \right| n \rangle . \tag{13.34}$$

The point is that instead of the infinite number of summations in Eq. (13.22), now we need to calculate only at most two integrals $\langle n \left| H^{(1)} F_n \right| n \rangle$, $\langle n \left| F_n \right| n \rangle$ or only one if $E_n^{(1)}$ is zero. Similarly, we obtain

$$E_n^{(3)} = \left\langle n \left| F_n H^{(1)} F_n \right| n \right\rangle - 2E_n^{(2)} \langle n \left| F_n \right| n \rangle - E_n^{(1)} \langle n \left| F_n \right| n \rangle . \tag{13.35}$$

13.2.4.2 Expression for F_n

If $H^{(1)}$ has no differential operators then from Eq. (13.26) we obtain

$$\left\{ F_n \left(\frac{1}{2m} p_x^2 + V \right) - \left(\frac{1}{2m} p_x^2 + V \right) F_n \right\} \phi_n^{(0)} = \left(H^{(1)} - E_n^{(1)} \right) \phi_n^{(0)} . \tag{13.36}$$

Substituting $p_x = -i\hbar \mathrm{d}/\mathrm{d}x$, Eq. (13.36) becomes

$$\phi_n^{(0)} F_n'' + 2F_n' \phi_n^{(0)\prime} = \frac{2m}{\hbar^2} \left(H^{(1)} - E_n^{(1)} \right) \phi_n^{(0)} . \tag{13.37}$$

After simple mathematics, from Eq. (13.37), we get

$$\left[\left(\phi_n^{(0)}(x) \right)^2 F_n'(x) \right]_a^x = \frac{2m}{\hbar^2} \int_a^x \left(H^{(1)} - E_n^{(1)} \right) \left(\phi_n^{(0)} \right)^2 \mathrm{d}x . \tag{13.38}$$

Suppose a is such that $\phi_n^{(0)}(a) = 0$. Then

$$F_n'(x) = \frac{2m}{\hbar^2 \left(\phi_n^{(0)} \right)^2} \int_a^x \left(H^{(1)} - E_n^{(1)} \right) \left(\phi_n^{(0)} \right)^2 \mathrm{d}x . \tag{13.39}$$

Integration of the above equation gives

$$F_n(x) = \int^x \frac{1}{\left(\phi_n^{(0)} \right)^2} \left\{ \frac{2m}{\hbar^2} \int_a^x \left(H^{(1)} - E_n^{(1)} \right) \left(\phi_n^{(0)}(\xi) \right)^2 \mathrm{d}\xi \right\} \mathrm{d}x \tag{13.40}$$

Solved Problem 2:

Consider a particle in a box with the Hamiltonian $H = H^{(0)} + H^{(1)}$, $H^{(0)} = p_x^2/(2m) + V(x)$ where

$$V(x) = \begin{cases} 0, & \text{for } |x| \leq \pi/2 \\ \infty, & \text{for } |x| > \pi/2 , \end{cases} \qquad H^{(1)} = \begin{cases} \alpha x, & \text{for } |x| \leq \pi/2 \\ 0, & \text{for } |x| > \pi/2 . \end{cases} \tag{13.41}$$

Determine $E_0^{(2)}$ of the particle.

The first few unperturbed eigenfunctions are

$$\phi_0^{(0)} = \sqrt{2/\pi}\,\cos x\,, \quad \phi_1^{(0)} = \sqrt{2/\pi}\,\sin 2x\,, \tag{13.42a}$$

$$\phi_3^{(0)} = \sqrt{2/\pi}\,\sin 4x\,, \quad E_n^{(0)} = (n+1)^2 \frac{\hbar^2}{2m}\,. \tag{13.42b}$$

Let us calculate $E_0^{(2)}$ using Dalgarno–Lewis method. For the ground state we can easily find that $E_n^{(1)} = \langle H_{nn}^{(1)}\rangle = 0$. Then F_0 is calculated from Eq. (13.40) as

$$
\begin{aligned}
F_0(x) &= \frac{2m\alpha}{\hbar^2}\int^x \frac{1}{\cos^2 x}\left(\int_a^x \xi\cos^2\xi\,\mathrm{d}\xi\right)\mathrm{d}x \\
&= \frac{\alpha m}{\hbar^2}\int^x \sec^2 x\left[\frac{\xi^2}{2} + \frac{1}{4}(2\xi\sin 2\xi + \cos 2\xi)\right]_a^x \mathrm{d}x \\
&= \frac{\alpha m}{2\hbar^2}\int^x \left[x^2\sec^2 x + 2x\tan x + 1 - \frac{1}{2}\sec^2 x\right. \\
&\qquad\qquad \left. - \sec^2 x\left(a^2 + a\sin 2a + \frac{1}{2}\cos 2a\right)\right]\mathrm{d}x\,.
\end{aligned}
\tag{13.43}
$$

The condition $\phi_0^{(0)}(a) = 0$ gives $a = \pm\pi/2$. We identify the correct choice of a using Eq. (13.32). Both the values of a gives

$$F_0(x) = \frac{\alpha m}{2\hbar^2}\left[\left(x^2 - \frac{\pi^2}{4}\right)\tan x + x\right]\,. \tag{13.44}$$

Consider $\langle 0|F_0|1\rangle$. We get

$$\langle 0|F_0|1\rangle = \frac{\langle 0|H^{(1)}|1\rangle}{E_0^{(0)} - E_1^{(0)}}\,. \tag{13.45}$$

The right-side of the Eq. (13.45) gives

$$\frac{\langle 0|H^{(1)}|1\rangle}{E_0^{(0)} - E_1^{(0)}} = -\frac{32\alpha m}{27\pi\hbar^2}\,. \tag{13.46}$$

The correct choice of a is determined by computing $\langle 0|F_0|1\rangle$ for $a = \pi/2$ and $a = -\pi/2$ and comparing them with Eq. (13.46). We find that both the choices lead to $\langle 0|F_0|1\rangle = -32\alpha m/(27\pi\hbar^2)$. Then

$$E_0^{(2)} = \frac{\langle 0|H^{(1)}|1\rangle\langle 1|H^{(1)}|0\rangle}{E_0^{(0)} - E_1^{(0)}} + \cdots = -\frac{2^9\alpha^2 m}{3^5\pi^2\hbar^2} + \cdots\,. \tag{13.47}$$

Next, we have $\langle 0|F_0|2\rangle = 0$ and $\langle 0|F_0|3\rangle = 2^6\alpha m\pi/(3^3 5^3\hbar^2)$. Thus,

$$E_0^{(2)} = -\frac{2^9\alpha^2 m}{3^5\pi^2\hbar^2} - \frac{2^{11}\alpha^2 m}{3^5 5^5\pi^2\hbar^2} - \cdots \approx -\frac{2.1097\alpha^2 m}{\hbar^2\pi^2}\,. \tag{13.48}$$

We compare this result with the exact second-order correction given by Eq. (13.34), we get

$$
\begin{aligned}
E_0^{(2)} &= \frac{\alpha^2 m}{\hbar^2\pi^2}\int_{-\pi/2}^{\pi/2} x\left[\left(x^2 - \frac{\pi^2}{4}\right)\tan x + x\right]\cos^2 x\,\mathrm{d}x \\
&= -\left(15 - \pi^2\right)\frac{\pi^2\alpha^2 m}{24\hbar^2\pi^2}\,.
\end{aligned}
\tag{13.49}
$$

13.3 Applications to Nondegenerate Levels

To demonstrate the use of perturbation theory let us apply it to some interesting systems. First, apply it to the linearly perturbed harmonic oscillator whose unperturbed and as well as the perturbed parts are exactly solvable.

13.3.1 Perturbed Linear Harmonic Oscillator

Consider the one-dimensional harmonic oscillator in an electric field. The potential and the Hamiltonian of the system are $V(x) = m\omega^2 x^2/2 - eEx$ and

$$H = H^{(0)} + \lambda H^{(1)}, \quad H^{(0)} = \frac{1}{2m}p_x^2 + \frac{1}{2}m\omega^2 x^2, \quad H^{(1)} = -eEx. \tag{13.50}$$

The system with $\lambda = 0$ is exactly solved and its energy levels and eigenfunctions are given by Eqs. (4.33) and (4.29), respectively. The eigenpairs of the system with $\lambda \neq 0$ are also explicitly known.

Let us obtain E_n(exact) of the perturbed system. Introducing the change of variable $x = y + \alpha$ the Hamiltonian H becomes

$$H = \frac{1}{2m}p_y^2 + \frac{1}{2}m\omega^2 \left(y^2 + 2\alpha y + \alpha^2\right) - \lambda e E \left(y + \alpha\right). \tag{13.51}$$

We set the coefficient of y to zero. This gives $\alpha = \lambda eE/(m\omega^2)$. Then

$$H = \frac{1}{2m}p_y^2 + \frac{1}{2}m\omega^2 y^2 - \frac{\lambda^2 e^2 E^2}{2m\omega^2} \tag{13.52}$$

which we rewrite as

$$H + \frac{\lambda^2 e^2 E^2}{2m\omega^2} = H' = \frac{1}{2m}p_y^2 + \frac{1}{2}m\omega^2 y^2. \tag{13.53}$$

H' is the Hamiltonian of the harmonic oscillator with the position operator y. Thus, for the perturbed oscillator

$$E_n(\text{exact}) = \left(n + \frac{1}{2}\right)\hbar\omega - \frac{\lambda^2 e^2 E^2}{2m\omega^2}. \tag{13.54}$$

The above equation shows that the change in energy is second-order in λ. In the following we apply the perturbation theory to find the correction to the energy eigenvalues. We expect the first-order correction to energy as zero.

The first-order correction to E_n is given by

$$E_n^{(1)} = H_{nn}^{(1)} = \int_{-\infty}^{\infty} \phi_n^{(0)*} H^{(1)} \phi_n^{(0)} \, dx. \tag{13.55}$$

The value of the above integral can be easily worked out if we replace the position operator x in terms of the raising and lowering operators $a^\dagger$ and a introduced in sec. 4.4 and expressing the eigenstates and the integral in Eq. (13.55) in Dirac notation.

In terms of a and $a^\dagger$ the operator x is expressed as $x = \sqrt{\hbar/(2m\omega)}\left(a + a^\dagger\right)$. We denote the eigenstates as $|n\rangle$. We have the relation

$$a|n\rangle = \sqrt{n}|n - 1\rangle, \quad a^\dagger|n\rangle = \sqrt{n+1}|n+1\rangle. \tag{13.56}$$

The orthogonality of two eigenstates $|n'\rangle$ and $|n\rangle$ is written as $\langle n'|n\rangle = \delta_{n'n}$.

Now, $E_n^{(1)}$ is obtained as

$$
\begin{aligned}
E_n^{(1)} &= \langle n|H^{(1)}|n\rangle \\
&= -eE\langle n|x|n\rangle \\
&= -eE\left(\frac{\hbar}{2m\omega}\right)^{1/2}\langle n|a+a^\dagger|n\rangle \\
&= -eE\left(\frac{\hbar}{2m\omega}\right)^{1/2}\left[\sqrt{n}\,\langle n|n-1\rangle + \sqrt{n+1}\,\langle n|n+1\rangle\right] \\
&= 0.
\end{aligned}
\tag{13.57}
$$

Since $E_n^{(1)} = 0$ we find the second-order correction. $E_n^{(2)}$ is calculated as

$$
\begin{aligned}
E_n^{(2)} &= \sum_{m\neq n}\frac{\left|\langle n\left|H^{(1)}\right|m\rangle\right|^2}{E_n^{(0)} - E_m^{(0)}} \\
&= \frac{1}{\hbar\omega}\sum_{m\neq n}\frac{\left|\langle n\left|-eEx\right|m\rangle\right|^2}{n-m} \\
&= \frac{e^2E^2}{\hbar\omega}\left(\frac{\hbar}{2m\omega}\right)\sum_{m\neq n}\frac{\left|\langle n\left|a+a^\dagger\right|m\rangle\right|^2}{n-m} \\
&= \frac{e^2E^2}{2m\omega^2}\sum_{m\neq n}\frac{\left|\sqrt{m}\delta_{n,m-1}+\sqrt{m+1}\delta_{n,m+1}\right|^2}{n-m} \\
&= \frac{e^2E^2}{2m\omega^2}\left[-\frac{n+1}{1}+\frac{n}{1}\right] \\
&= -\frac{e^2E^2}{2m\omega^2}
\end{aligned}
\tag{13.58}
$$

which is exactly the correction given in Eq. (13.54).

13.3.2   Hyperfine Splitting of the Ground State of Hydrogen Atom [8]

The splitting of an energy level due to the interaction of the nuclear magnetic dipole and the magnetic field $\mathbf{B}_e$ produced by the electron orbit current is called the *fine structure*. The term *hyperfine* structure is used to describe extremely small splitting of spectral lines. In the following we derive the hyperfine structure of hydrogen by employing first-order perturbation theory. We consider the hyperfine splitting of the ground state [8]. This splitting is associated with the interaction of the magnetic moments of the proton and the electron.

The electron and proton in atomic hydrogen constitute tiny magnetic dipoles. The magnetic field of a magnetic dipole kept at the centre of a sphere of radius R is given by

$$
\mathbf{B}(\mathbf{r}) = \frac{\mu_0}{4\pi r^3}\left[3\left(\mathbf{m}\cdot\mathbf{r}\right)\hat{\mathbf{r}} - \mathbf{m}\right] + \frac{2}{3}\mu_0\mathbf{m}\delta^3(\mathbf{r}),
\tag{13.59}
$$

where the first term is only to the region outside an infinitesimal sphere about $r = 0$. Then the energy of a magnetic dipole $\mathbf{m}$ in the presence of a magnetic field $\mathbf{B}$ is $H = -\mathbf{m}\cdot\mathbf{B}$. Now, the energy of one magnetic dipole ($\mathbf{m}_1$) in the field of another magnetic dipole ($\mathbf{m}_2$) is

$$
H = -\frac{\mu_0}{4\pi r^3}\left[3\left(\mathbf{m}_1\cdot\mathbf{r}_1\right)\left(\mathbf{m}_2\cdot\mathbf{r}_2\right) - \mathbf{m}_1\cdot\mathbf{m}_2\right] - \frac{2}{3}\mu_0\mathbf{m}_1\cdot\mathbf{m}_2\delta^3(\mathbf{r}),
\tag{13.60}
$$

where $\mathbf{r}$ is their separation.

According to the first-order perturbation theory, the change in the ground state energy is

$$E' = \int_{-\infty}^{\infty} \phi_0^* H \phi_0 \, d\tau \,, \quad \phi_0 = \left(\frac{1}{\pi a_0^3}\right)^{1/2} e^{-r/a_0} s \,, \tag{13.61}$$

with $a_0 = 0.52917706\,\text{Å}$ is the Bohr radius and s denotes the spin of the electron. Because ϕ_0 is spherically symmetrical, the θ integral is 0 and

$$E' = -\frac{2}{3}\frac{\mu_0}{\pi a_0^3}\langle \mathbf{m}_1 \cdot \mathbf{m}_2 \rangle \,. \tag{13.62}$$

Here $\mathbf{m}_1$ and $\mathbf{m}_2$ are the magnetic dipole moments of the proton and electron, respectively. They are given by $\mathbf{m}_1 = \gamma_p \mathbf{S}_p$ and $\mathbf{m}_2 = -\gamma_e \mathbf{S}_e$, where γ's are the two gyromagnetic ratios. Thus,

$$E' = \frac{2}{3}\frac{\mu_0}{\pi a_0^3}\gamma_p \gamma_e \langle \mathbf{S}_e \cdot \mathbf{S}_p \rangle \,. \tag{13.63}$$

The good quantum numbers of the system are the eigenvalues of the total angular momentum $\mathbf{J} = \mathbf{S}_e + \mathbf{S}_p$. Now $\mathbf{J}^2 = (\mathbf{S}_e + \mathbf{S}_p)^2 = \mathbf{S}_e^2 + \mathbf{S}_p^2 + 2\mathbf{S}_e \cdot \mathbf{S}_p$ so that $\mathbf{S}_e \cdot \mathbf{S}_p = \left(\mathbf{J}^2 - \mathbf{S}_e^2 - \mathbf{S}_p^2\right)/2$. Since the electrons and protons carry spin-1/2 the eigenvalues of $\mathbf{S}_e^2$ and $\mathbf{S}_p^2$ are $3\hbar^2/4$. The two spins form a spin-1 triplet ($\mathbf{J}^2 = 2\hbar^2$) and a spin-0 singlet ($\mathbf{J}^2 = 0$). Thus,

$$\langle \mathbf{S}_e \cdot \mathbf{S}_p \rangle = \begin{cases} \hbar^2/4 \text{ (triplet)} \\ -3\hbar^2/4 \text{ (singlet)} \,. \end{cases} \tag{13.64}$$

The singlet level is lower because in this state the proton and electron spins are antiparallel (their magnetic moments are parallel). The energy gap between the triplet and singlet states is

$$\Delta E_{\text{hyd}} = \frac{2\mu_0 \hbar^2}{3\pi a^3}\gamma_p \gamma_e = 5.884 \times 10^{-6}\,\text{eV} \,, \tag{13.65}$$

where we have substituted $\gamma_e = eg_e/(2m)$, $\gamma_p = eg_p/(2m)$, $g_e = 2.0023$ and $g_p = 5.5857$. The frequency of the photon emitted in a transition from the triplet state to the singlet state is $\gamma = \Delta E/h = 1422.8\,\text{MHz}$ while the experimental value is $1420.4\,\text{MHz}$.

13.4 Theory for Degenerate Levels

The first-order perturbation theory developed for nondegenerate levels is not applicable when the energy level of the unperturbed system is degenerate. This is because, if the energy values of $\phi_m^{(0)}$ and $\phi_n^{(0)}$ are the same, say, $E^{(0)}$ then $E_m^{(0)} - E_n^{(0)} = 0$ making the correction term in Eq. (13.18a) infinity. Therefore, we require a separate perturbation theory for degenerate levels. Note that degeneracy in any quantum mechanical system stems from symmetries inherent to the system. If any perturbation distorts the symmetry of the system, then it will tend to remove the related degeneracy.

Suppose the eigenvalues of unperturbed Hamiltonian $H^{(0)}$ are m-fold degenerate. Then there are m linearly independent orthogonal eigenfunctions $\phi_1^{(0)}$, $\phi_2^{(0)}$,...,$\phi_m^{(0)}$ corresponding to $E_1^{(0)} = E_2^{(0)} = \ldots = E_m^{(0)} = E^{(0)}$. Given $E^{(0)}$ the correct choice of the eigenfunction out of the above m eigenfunctions is not known. Therefore, choose the unperturbed eigenfunction of a degenerate level as a linear combination of m eigenfunctions and allow the perturbation to choose the appropriate one. We write

$$\phi^{(0)} = \sum_{i=1}^{m} c_i \phi_i^{(0)},\qquad(13.66)$$

where c_i's are constants. The perturbed eigenfunction and the corresponding energy are written as

$$\phi = \phi^{(0)} + \lambda \phi^{(1)} + \lambda^2 \phi^{(2)} + \dots,\qquad(13.67a)$$
$$E = E^{(0)} + \lambda E^{(1)} + \lambda^2 E^{(2)} + \dots.\qquad(13.67b)$$

Substitution of Eqs. (13.67) in the equation $\left(H^{(0)} + \lambda H^{(1)}\right)\phi = E\phi$ gives

$$\left(H^{(0)} + \lambda H^{(1)}\right)\left(\phi^{(0)} + \lambda \phi^{(1)} + \dots\right)$$
$$= \left(E^{(0)} + \lambda E^{(1)} + \dots\right)\left(\phi^{(0)} + \lambda \phi^{(1)} + \dots\right).\qquad(13.68)$$

Equating the coefficients of various powers of λ to zero we get

$$\lambda^0 \;:\; H^{(0)}\phi^{(0)} = E^{(0)}\phi^{(0)},\qquad(13.69a)$$
$$\lambda^1 \;:\; H^{(0)}\phi^{(1)} + H^{(1)}\phi^{(0)} = E^{(0)}\phi^{(1)} + E^{(1)}\phi^{(0)},\qquad(13.69b)$$

and so on. Next, write

$$\phi^{(1)} = \sum_i a_i \phi_i^{(0)}.\qquad(13.70)$$

Then Eq. (13.69b) becomes

$$\sum_i \left(E_i^{(0)} - E^{(0)}\right) a_i \phi_i^{(0)} = \left(E^{(1)} - H^{(1)}\right)\sum_{i=1}^{m} c_i \phi_i^{(0)}.\qquad(13.71)$$

Multiplication of Eq. (13.71) by $\phi_r^{(0)*}$ and integration over all space lead to

$$a_r \left(E_r^{(0)} - E^{(0)}\right) = E^{(1)}c_r - \sum_{i=1}^{m} H_{ri}^{(1)} c_i,\quad r = 0, 1, \dots.\qquad(13.72)$$

For $r \le m$, $E_r^{(0)} = E^{(0)}$, therefore Eq. (13.72) gives

$$\sum_{i=1}^{m} \left(H_{ri}^{(1)} - E^{(1)}\delta_{ri}\right) c_i = 0,\quad r \le m.\qquad(13.73)$$

Equation (13.73) generates m simultaneous equations for the c_i's. For nontrivial solution the determinant of coefficient matrix of c must be zero:

$$\begin{vmatrix} H_{11}^{(1)} - E^{(1)} & H_{12}^{(1)} & \dots & H_{1m}^{(1)} \\ H_{21}^{(1)} & H_{22}^{(1)} - E^{(1)} & \dots & H_{2m}^{(1)} \\ \vdots & \vdots & \dots & \vdots \\ H_{m1}^{(1)} & H_{m2}^{(1)} & \dots & H_{mm}^{(1)} - E^{(1)} \end{vmatrix} = 0.\qquad(13.74)$$

This is known as *secular equation*. The m real roots $E_1^{(1)}$, $E_2^{(1)}$, ..., $E_m^{(1)}$ of the Eq. (13.74) are the required corrections to the eigenvalues in the first approximation. When all these

$E_i^{(1)}$ are distinct then the perturbed eigenvalues are distinct. The perturbation removed the degeneracy completely. On the other hand, if some of them are identical then the removal is partial.

The first-order eigenfunction is determined by substituting $c_r = 0$ for $r > m$ in Eq. (13.72). This gives

$$a_r = \frac{\sum_{i=1}^{m} H_{ri}^{(1)} c_i}{\left(E^{(0)} - E_r^{(0)}\right)}, \quad r > m. \tag{13.75}$$

In Eqs. (13.72), equations with $r \leq m$ are used for the calculation of $E^{(1)}$. We are thus left with equations with $r > m$. Therefore, we set $a_r = 0$ for $r \leq m$. Then

$$E_k = E_k^{(0)} + \lambda E_k^{(1)}, \tag{13.76a}$$

$$\phi_k = \phi_k^{(0)} + \lambda \sum_{r>m} \frac{\sum_{i=1}^{m} H_{ri}^{(1)} c_i}{\left(E^{(0)} - E_r^{(0)}\right)} \phi_r^{(0)}. \tag{13.76b}$$

In the next section, we apply the degenerate perturbation theory to hydrogen atom in an applied electric field.

Solved Problem 3:

Consider a two-dimensional harmonic oscillator with the Hamiltonian $H^{(0)} = (1/2m)p_x^2 + (1/2m)p_y^2 + (k/2)\left(x^2 + y^2\right)$. Its eigenstates are denoted as $|n_x n_y\rangle$. Analyze the effect of the perturbation $H^{(1)} = \lambda xy$ on the two-fold degenerate states $|10\rangle$ and $|01\rangle$.

In terms of ladder operators $H^{(0)}$ is expressed as

$$H^{(0)} = \hbar\omega\left(a^\dagger a + b^\dagger b + 1\right), \tag{13.77}$$

where

$$a = \frac{1}{\sqrt{2}}\left(\alpha x + \frac{1}{\alpha}\frac{d}{dx}\right), \quad a^\dagger = \frac{1}{\sqrt{2}}\left(\alpha x - \frac{1}{\alpha}\frac{d}{dx}\right), \tag{13.78a}$$

$$b = \frac{1}{\sqrt{2}}\left(\alpha y + \frac{1}{\alpha}\frac{d}{dy}\right), \quad b^\dagger = \frac{1}{\sqrt{2}}\left(\alpha y - \frac{1}{\alpha}\frac{d}{dy}\right), \tag{13.78b}$$

$$x = \frac{1}{\sqrt{2}\,\alpha}\left(a + a^\dagger\right), \quad y = \frac{1}{\sqrt{2}\,\alpha}\left(b + b^\dagger\right), \quad \alpha = \frac{m\omega}{\hbar}, \tag{13.78c}$$

$$p_x = -\frac{i\hbar\alpha}{\sqrt{2}}(a - a^\dagger), \quad p_y = \frac{i\hbar\alpha}{\sqrt{2}}(b - b^\dagger). \tag{13.78d}$$

The eigenstates of $H^{(0)}$ are given by the product of the kets $|n_x\rangle$ and $|n_y\rangle$: $|n_x\rangle|n_y\rangle = |n_x n_y\rangle$. The corresponding energy is

$$E_{n_x, n_y} = (n_x + n_y + 1)\hbar\omega \tag{13.79}$$

which is $(n_x + n_y + 1)$-fold degenerate. The state $|10\rangle$ and $|01\rangle$ are degenerate both having the same energy eigenvalue $2\hbar\omega$.

For the case of $H^{(1)} = \lambda xy$, $|10\rangle$ and $|01\rangle$ the secular matrix is

$$\begin{vmatrix} \langle 10|\lambda xy|10\rangle - E^{(1)} & \langle 10|\lambda xy|01\rangle \\ \\ \langle 01|\lambda xy|10\rangle & \langle 01|\lambda xy|01\rangle - E^{(1)} \end{vmatrix} = 0. \tag{13.80}$$

We have $\langle 10|\lambda xy|10\rangle = \lambda\langle 1|x|1\rangle\langle 0|y|0\rangle = 0$ since $\langle 1|x|1\rangle = \langle 0|y|0\rangle = 0$. Next,

$$\langle 10|\lambda xy|01\rangle = \lambda\langle 1|x|0\rangle\langle 0|y|1\rangle = \frac{\lambda}{2\alpha^2}\,. \tag{13.81}$$

Then Eq. (13.80) gives $E^{(1)} = \pm\lambda/(2\alpha^2)$. The degenerate states $|10\rangle$ and $|01\rangle$ with energy $2\hbar\omega$ splits into two levels with energies $2\hbar\omega+\lambda/(2\alpha^2)$ and $2\hbar\omega-\lambda/(2\alpha^2)$. The corresponding new eigenfunctions are obtained by substituting $E^{(1)} = \pm\lambda/(2\alpha^2)$ in the equation

$$\begin{pmatrix} -E^{(1)} & \lambda/2\alpha^2 \\ \lambda/2\alpha^2 & -E^{(1)} \end{pmatrix}\begin{pmatrix} u_1 \\ u_2 \end{pmatrix} = 0\,. \tag{13.82}$$

For $E^{(1)} = \lambda/2\alpha^2$ we get $u_1 = u_2$ so the normalized eigenfunction is $\dfrac{1}{\sqrt{2}}\begin{pmatrix} 1 \\ 1 \end{pmatrix}$ which gives $\phi_1 = \dfrac{1}{\sqrt{2}}(|10\rangle + |01\rangle)$. For $E^{(1)} = -\lambda/2\alpha^2$ we get $u_1 = -u_2$ and so $\phi_2 = \dfrac{1}{\sqrt{2}}(|10\rangle - |01\rangle)$.

13.5 First-Order Stark Effect in Hydrogen

The splitting of energy levels of an atom by a uniform external field E is known as *Stark effect*. The effect was discovered by Stark in 1913 during the observation of Balmer series of hydrogen.

Suppose the hydrogen atom is subjected to an electric field along z-direction. The first excited energy level is found to split into three energy levels. This is explained using perturbation theory. The perturbation is $\lambda H^{(1)} = -e\mathcal{E}r\cos\theta$, where $\mathcal{E}$ is the strength of the field. Choosing $\lambda = e\mathcal{E}$ we write $H^{(1)} = -r\cos\theta$. In sec. 12.2 we solved the unperturbed hydrogen atom. The normalized eigenfunctions are designated as ψ_{nlm} and are given by Eq. (12.61). Applying the perturbation theory we calculate the change in the energy of the ground state and first excited state due to the applied field.

13.5.1 Ground State: $n = 1$

The ground state eigenfunction is $\psi_{100} = \left(1/\sqrt{\pi a_0^3}\right)e^{-r/a_0}$. According to the first-order perturbation theory for nondegenerate case the change in (ground state)energy E_1 is given by

$$\begin{aligned} E_1^{(1)} &= \lambda\langle H_{100}^{(1)}\rangle \\ &= \lambda\int \psi_{100}^* H_{100}^{(1)}\psi_{100}\,d\tau \\ &= -\frac{e\mathcal{E}}{\pi a_0^3}\int_0^\infty\int_0^\pi\int_0^{2\pi} r^3 e^{-2r/a_0}\cos\theta\sin\theta\,dr\,d\theta\,d\phi\,. \end{aligned} \tag{13.83}$$

In the above integral, integration with respect to θ becomes zero. Therefore, $E_1^{(1)} = 0$. Thus, there is no change in ground state energy according to first-order perturbation theory.

Since the first-order correction to E_1 is zero we proceed to calculate the second-order correction [9,10] by the Dalgarno and Lewis method described earlier. According to this method

$$E_1^{(2)} = -\left(H^{(1)}F\right)_{11} = \langle 1|H^{(1)}F|1\rangle\,, \tag{13.84a}$$

where F is to be determined from

$$H^{(1)}\psi_{100} = \left[F, H^{(0)}\right]\psi_{100} . \tag{13.84b}$$

Substituting for $H^{(0)}$ and $H^{(1)}$ we get

$$\begin{aligned}
-r\cos\theta\psi_{100} &= -\left(H^{(0)}F - FH^{(0)}\right)\psi_{100} \\
&= \left(\frac{\hbar^2}{2m}\nabla^2 + \frac{e^2}{r}\right)F\psi_{100} + FH^{(0)}\psi_{100} \\
&= \frac{\hbar^2}{2m}\left(2\nabla\psi_{100}\cdot\nabla F + \psi_{100}\nabla^2 F\right) .
\end{aligned} \tag{13.85}$$

Substituting for ψ_{100} in the above equation we get

$$\nabla^2 F - \frac{2}{a_0}\left(\frac{\partial F}{\partial r}\right) + \frac{2m}{\hbar^2}r\cos\theta = 0 . \tag{13.86}$$

Seeking a solution of the form $F = f(r)\cos\theta$ we obtain

$$\frac{1}{r^2}\frac{\mathrm{d}}{\mathrm{d}r}\left(r^2\frac{\mathrm{d}f}{\mathrm{d}r}\right) - \frac{2f}{r^2} - \frac{2}{a_0}\frac{\mathrm{d}f}{\mathrm{d}r} + \frac{2mr}{\hbar^2} = 0 . \tag{13.87}$$

Its solution is

$$f = \frac{ma_0}{\hbar^2}\left(a_0 + \frac{r}{2}\right)r\cos\theta . \tag{13.88}$$

Then

$$\begin{aligned}
E_1^{(2)} &= -\frac{ma_0}{\hbar^2\pi a_0^3}\int_0^\infty\int_0^\pi\int_0^{2\pi} e^{-2r/a_0}\left(a_0 + \frac{r}{2}\right)r^4\cos^2\theta\sin\theta\,\mathrm{d}r\,\mathrm{d}\theta\,\mathrm{d}\phi \\
&= -\frac{9ma_0^4}{4\hbar^2} ,
\end{aligned} \tag{13.89}$$

where we used $\int_0^\infty x^n e^{-x}\,\mathrm{d}x = n!$. Including $\lambda = e\mathcal{E}$ in $E_1^{(2)}$ we write

$$E_1^{(2)} = -\frac{9ma_0^4 e^2\mathcal{E}^2}{4\hbar^2} = -\frac{9}{4}a_0^3\mathcal{E}^2 . \tag{13.90}$$

This shift in energy is interpreted in terms of the dipole moment $p = \alpha\mathcal{E}$ acquired by the hydrogen atom, where α is called *electric polarizability* of hydrogen atom. Here $p = -\partial E_1^{(2)}/\partial\mathcal{E}$ and α is obtained as

$$\alpha = -\frac{1}{\mathcal{E}}\frac{\partial E_1^{(2)}}{\partial\mathcal{E}} = \frac{9}{2}a_0^3 . \tag{13.91}$$

The shift in the energy level, $-9a_0^3\mathcal{E}^2/4$, is due to the electric polarization of the atom. Before perturbation, that is, before applying the electric field, the system has one ground state eigenfunction with an eigenvalue $E_1^{(0)}$. Now, the energy is changed to $E_1^{(0)} + E_1^{(1)}$ by the applied field. There is no Stark effect in the ground state. Classically, an energy shift of $-\mathbf{p}\cdot\mathcal{E}$ will be realized if a system possesses an electric dipole moment $\mathbf{p}$. Therefore, we say that the atom in its ground state has no permanent dipole moment. Generally, systems in nondegenerate states have no permanent dipole moments.

13.5.2 First Excited State: $n = 2$

Next, work out the effect of the applied field on the first excited state corresponding to $n = 2$. The possible values of (l, m) are $(0,0)$, $(1,0)$, $(1,1)$, $(1,-1)$ (since $l = 0$, 1 and $m = -l$ to l). There are four states given by $\psi_{nlm} \to \psi_{200}$, ψ_{210}, ψ_{211} and ψ_{21-1}. The first excited state is thus four-fold degenerate. In this case the zeroth-order wave function is

$$\psi^{(0)} = c_{00}\psi_{200} + c_{10}\psi_{210} + c_{11}\psi_{211} + c_{1-1}\psi_{21-1}. \tag{13.92}$$

The secular equation is written as

$$\begin{vmatrix} H^{(1)}_{200,200} - E^{(1)}_2 & H^{(1)}_{200,210} & H^{(1)}_{200,211} & H^{(1)}_{200,21-1} \\ H^{(1)}_{210,200} & H^{(1)}_{210,210} - E^{(1)}_2 & H^{(1)}_{210,211} & H^{(1)}_{210,21-1} \\ H^{(1)}_{211,200} & H^{(1)}_{211,210} & H^{(1)}_{211,211} - E^{(1)}_2 & H^{(1)}_{211,21-1} \\ H^{(1)}_{21-1,200} & H^{(1)}_{21-1,210} & H^{(1)}_{21-1,211} & H^{(1)}_{21-1,21-1} - E^{(1)}_2 \end{vmatrix}$$

$$= 0, \tag{13.93}$$

where

$$H^{(1)}_{nlm,nl'm'} = \int \psi^*_{nlm} H^{(1)} \psi_{nl'm'} \, d\tau. \tag{13.94}$$

The eigenfunctions ψ_{200}, ψ_{210}, ψ_{211}, ψ_{21-1} are given by

$$\psi_{200} = N\left(2 - \frac{r}{a_0}\right) e^{-r/2a_0} e^{im\phi}, \quad m = 0 \tag{13.95a}$$

$$\psi_{210} = N\frac{r}{a_0} e^{-r/2a_0} \cos\theta \, e^{im\phi}, \quad m = 1 \tag{13.95b}$$

$$\psi_{211} = N\frac{r}{\sqrt{2}\,a_0} e^{-r/2a_0} \sin\theta \, e^{im\phi}, \quad m = 1 \tag{13.95c}$$

$$\psi_{21-1} = N\frac{r}{\sqrt{2}\,a_0} e^{-r/2a_0} \sin\theta \, e^{im\phi}, \quad m = -1 \tag{13.95d}$$

where $N = 1/\sqrt{32\pi a_0^3}$. The quantities $H^{(1)}_{2lm,2l'm'}$ are given by

$$H^{(1)}_{2lm,2l'm'} = \int_0^\infty \int_0^\pi \int_0^{2\pi} \psi^*_{2lm} H^{(1)} \psi_{2l'm'} r^2 \sin\theta \, dr \, d\theta \, d\phi$$

$$= -\int_0^\infty \int_0^\pi \int_0^{2\pi} \psi^*_{2lm} \psi_{2l'm'} r^3 \cos\theta \sin\theta \, dr \, d\theta \, d\phi. \tag{13.96}$$

First, perform the integration with respect to ϕ. The result is

$$H^{(1)}_{2lm,2l'm'}(\phi) = \int_0^{2\pi} e^{i(m'-m)\phi} \, d\phi = \begin{cases} 2\pi, & \text{if } m' = m \\ 0, & \text{if } m' \neq m. \end{cases} \tag{13.97}$$

Hence, the quantities $H^{(1)}_{2lm,2l'm'}$ with $m' \neq m$ are zero. Now, we need to evaluate the quantities $H^{(1)}_{200,200}$, $H^{(1)}_{200,210}$, $H^{(1)}_{210,200}$, $H^{(1)}_{210,210}$, $H^{(1)}_{211,211}$, $H^{(1)}_{21-1,21-1}$. Integration with respect to θ makes some of these integrals vanish:

$$H^{(1)}_{200,200}(\theta) = \int_0^\pi \sin\theta \cos\theta \, d\theta = \left.\frac{\sin^2\theta}{2}\right|_0^\pi = 0, \qquad (13.98a)$$

$$H^{(1)}_{210,210}(\theta) = \int_0^\pi \cos^3\theta \sin\theta \, d\theta = \left.-\frac{\cos^4\theta}{4}\right|_0^\pi = 0, \qquad (13.98b)$$

$$H^{(1)}_{211,211}(\theta) = \int_0^\pi \sin^3\theta \cos\theta \, d\theta = \left.\frac{\sin^4\theta}{4}\right|_0^\pi = 0, \qquad (13.98c)$$

$$H^{(1)}_{21-1,21-1}(\theta) = \int_0^\pi \sin^3\theta \cos\theta \, d\theta = \left.\frac{\sin^4\theta}{4}\right|_0^\pi = 0. \qquad (13.98d)$$

Thus, only the integrals $H^{(1)}_{200,210}$ and $H^{(1)}_{210,200}$ have to be evaluated. We obtain

$$\begin{aligned}
H^{(1)}_{200,210} &= \left(H^{(1)}_{210,200}\right)^* \\
&= \int_0^\infty \int_0^\pi \int_0^{2\pi} \psi^*_{210} H^{(1)} \psi_{200} r^2 \sin\theta \, dr \, d\theta \, d\phi \\
&= -\frac{1}{24a_0^4} \int_0^\infty r^4 \left(2 - \frac{r}{a_0}\right) e^{-r/a_0} \, dr \\
&= 3a_0 .
\end{aligned} \qquad (13.99)$$

Now, the secular equation is written as

$$\begin{vmatrix}
-E_2^{(1)} & 3a_0 & 0 & 0 \\
3a_0 & -E_2^{(1)} & 0 & 0 \\
0 & 0 & -E_2^{(1)} & 0 \\
0 & 0 & 0 & -E_2^{(1)}
\end{vmatrix} = 0. \qquad (13.100)$$

Expanding the determinant we obtain

$$\left(E_2^{(1)}\right)^2 \left[\left(E_2^{(1)}\right)^2 - 9a_0^2\right] = 0. \qquad (13.101)$$

The roots of the above equation are

$$E_2^{(1)} = 0, \, 0, \, 3a_0, \, -3a_0. \qquad (13.102)$$

The energy E_2 to the first-order correction is

$$E_2 = E_2^{(0)}, \, E_2^{(0)}, \, E_2^{(0)} + 3\lambda a_0, \, E_2^{(0)} - 3\lambda a_0. \qquad (13.103)$$

The energy of the first excited state level in the presence of the applied field becomes $E_2^{(0)} + 0$, $E_2^{(0)} + 0$, $E_2^{(0)} + 3\lambda a_0$, $E_2^{(0)} - 3\lambda a_0$. That is, the level $n = 2$ is splitted into three levels. The middle level $E_2^{(0)} + 0$ is still doubly degenerate. Since the energy of an electric dipole with moment $\mathbf{p}$, oriented relative to the electric field is $-\mathbf{p} \cdot \lambda$, we note that the energy perturbations $+3\lambda a_0$, $-3\lambda a_0$, 0 are those of a dipole with moment $3\lambda a_0$ oriented antiparallel, parallel and perpendicular to the electric field, respectively.

Next, determine the zeroth-order eigenfunction given by Eq. (13.92). We have to determine c_{00}, c_{10}, c_{11} and c_{1-1}. They are obtained by solving the set of equations (refer Eq. (13.73))

$$\sum_{i=1}^4 H^{(1)}_{ri} c_i = E_2^{(1)} c_r, \quad r = 1, 2, 3, 4. \qquad (13.104)$$

In matrix form the above equation is written as, after using the values of $H_{ri}^{(1)}$,

$$
\begin{bmatrix}
0 & 3a_0 & 0 & 0 \\
3a_0 & 0 & 0 & 0 \\
0 & 0 & 0 & 0 \\
0 & 0 & 0 & 0
\end{bmatrix}
\begin{bmatrix}
c_{00} \\
c_{10} \\
c_{11} \\
c_{1-1}
\end{bmatrix}
= E_2^{(1)}
\begin{bmatrix}
c_{00} \\
c_{10} \\
c_{11} \\
c_{1-1}
\end{bmatrix} .
\tag{13.105}
$$

For $E_2^{(1)} = 0$, from Eq. (13.105) we obtain $3a_0c_{10} = 0$ and $3a_0c_{00} = 0$ which imply $c_{10} = 0$, $c_{00} = 0$, $c_{11} =$ arbitrary and $c_{1-1} =$ arbitrary. The corresponding eigenfunctions are ψ_{211} and ψ_{21-1} where we have chosen $c_{11} = c_{1-1} = 1$ so that the eigenfunctions are normalized. These two eigenfunctions have $E_2 = E_2^{(0)} + 0$. For $E_2^{(1)} = \pm 3a_0$ Eq. (13.105) gives

$$
3a_0c_{10} = \pm 3a_0c_{00}, \quad 3a_0c_{00} = \pm 3a_0c_{10}, \quad 3a_0c_{11} = 0, \quad 3a_0c_{1-1} = 0 .
\tag{13.106}
$$

From Eq. (13.106) we have $c_{10} = \pm c_{00}$, $c_{11} = c_{1-1} = 0$. We choose $c_{00} = 1/\sqrt{2}$ and $c_{10} = \pm 1/\sqrt{2}$ so that the eigenfunction $\psi^{(0)} = c_{00}\psi_{200} + c_{10}\psi_{210}$ is normalized.

The eigenfunctions for $E_2^{(1)} = 3a_0$ and $-3a_0$ are given by

$$
\psi_+^{(0)} = \frac{1}{\sqrt{2}}(\psi_{200} + \psi_{210}) , \quad \psi_-^{(0)} = \frac{1}{\sqrt{2}}(\psi_{200} - \psi_{210}) ,
\tag{13.107}
$$

respectively. The $\psi_+^{(0)}$ state with the lowest energy $E_2^{(0)} - 3a_0$ is the most stable state.

Now, one may ask: *What does happen in $n = 3$ and higher excited states?* Particularly, we want to know whether the dipolar nature found in $n = 2$ states is preserved in the higher excited states. The dipolar nature is found to be preserved in the higher excited states also [10].

13.6 Alternate Perturbation Theories

Perturbation theory is one of the most widely used approximate methods in quantum mechanics due to its transparency which greatly facilitates physical insight. The calculation of sufficient perturbation corrections is not trivial in many cases of interest and for this reason several alternate implementations have been suggested. We have already seen in sec. 13.2.4 the Dalgarno and Lewis method. It consists of writing the nth perturbation correction to the eigenfunction ϕ_n as $F_n\phi_0$ and then solving the resulting equation for F_n. In this section, we discuss another perturbation theory which has been successfully applied to a variety of problems.

13.6.1 Hypervirial Perturbation Theory

Direct application of Rayleigh–Schrödinger perturbation theory is tedious for the perturbation containing a polynomial function like the anharmonic potentials of a diatomic molecule. For such systems, analytic perturbation calculation of higher-order is straight-forward when we take into account the hypervirial and Hellmann–Feynman theorems. The method, commonly termed perturbation theory without eigenfunctions or the hypervirial perturbation method, has been applied to many quantum mechanical models [11].

13.6.2   Quantum Mechanical Hypervirial Theorem (QMHT)

Let H be the time-independent and self-adjoint operator with eigenfunction ψ. Then for any time-independent linear operator A on the space H, the QMHT states that the following expectation value vanishes:

$$\langle \psi, [A, H]\psi \rangle = 0. \tag{13.108}$$

We illustrate the usefulness of this theorem with the Hamiltonian

$$H = -\frac{\hbar^2}{2m} D_x^2 + V(x), \quad D_x = \frac{\mathrm{d}}{\mathrm{d}x}. \tag{13.109}$$

With $A = x^k D_x$, $k = 1, 2, \ldots$ if we evaluate the commutator explicitly, then a set of recursion relations involving the expectation values,

$$\langle x^m \rangle = \langle \psi, x^m \psi \rangle, \quad m = 0, 1, 2, \ldots, \quad \langle x^0 \rangle = 1, \tag{13.110}$$

are obtained. These relations are known as the *hypervirial relations*. In particular, the case $k = 1$ corresponds to the usual virial theorem. To evaluate the commutators, the following relations are useful:

$$[D_x, H] = \frac{\mathrm{d}V}{\mathrm{d}x}, \tag{13.111a}$$

$$[x^k, H] = \frac{\hbar^2}{m}\left(k x^{k-1} D_x + \frac{1}{2}k(k-1)x^{k-2} \right), \tag{13.111b}$$

$$[x^k D_x, H] = x^k [D_x, H] + [x^k, H] D_x. \tag{13.111c}$$

The identity Eq. (13.111c) is used to rewrite the commutator, while the appearance of D_x and D_x^2 are eliminated using Eqs. (13.111a)–(13.111b) and the time-independent Schrödinger equation. Taking expectation values with respect to an eigenstate ψ with energy E yields the hypervirial relations

$$2kE\langle x^{k-1} \rangle = 2k\langle x^{k-1}V \rangle + \langle x^k D_x V \rangle - \frac{\hbar^2}{4m}k(k-1)(k-2)\langle x^{k-3} \rangle. \tag{13.112}$$

In the case of solvable Hamiltonians, like harmonic oscillator, (radial) hydrogen atom, these relations necessarily coincide with those obtained from the knowledge of the recursion relation between the orthogonal polynomials (Hermite and Laguerre, respectively) which comprise the eigenfunctions. For harmonic oscillator, Eq. (13.112) produces the following recursion relations:

$$\langle x^{k+2} \rangle = \frac{2(k+1)\hbar}{(k+2)m\omega}\left(n + \frac{1}{2} \right)\langle x^k \rangle + \frac{k(k-1)(k+1)\hbar}{4(k+2)m\omega}\langle x^{k-2} \rangle, \tag{13.113}$$

where $\langle x^0 \rangle = 1$. Since $\langle x \rangle = 0$, all the expectation values of the odd powers of x vanish.

13.6.3   Quantum Mechanical Hellmann–Feynman Theorem

Let H_λ be a time-independent, self-adjoint operator that depends explicitly on a scalar parameter λ and

$$H_\lambda \psi = E(\lambda)\psi, \quad \langle \psi, \psi \rangle = 1 \tag{13.114a}$$

then the *Hellmann–Feynman theorem* states that

$$\frac{\delta E}{\delta \lambda} = \left\langle \psi, \frac{\delta H_\lambda}{\delta \lambda}\psi \right\rangle. \tag{13.114b}$$

Solved Problem 4:

If $H_\lambda = H^{(0)} + \lambda H^{(1)}$, prove that $\dfrac{\partial \langle E \rangle}{\partial \lambda} = \left\langle \psi, \dfrac{\partial H_\lambda}{\partial \lambda} \psi \right\rangle = \left\langle H^{(1)} \right\rangle$.

We have $H_\lambda \psi = E(\lambda) \psi$, $\langle \psi, \psi \rangle = 1$ and $\langle E \rangle = \langle \psi, H_\lambda \psi \rangle$. Then

$$
\begin{aligned}
\frac{\partial \langle E \rangle}{\partial \lambda} &= \left\langle \frac{\partial \psi}{\partial \lambda}, H_\lambda \psi \right\rangle + \left\langle \psi, \frac{\partial H_\lambda}{\partial \lambda} \psi \right\rangle + \left\langle \psi, H_\lambda \frac{\partial \psi}{\partial \lambda} \right\rangle \\
&= E \left\langle \frac{\partial \psi}{\partial \lambda}, \psi \right\rangle + \left\langle \psi, \frac{\partial H_\lambda}{\partial \lambda} \psi \right\rangle + E \left\langle \psi, \frac{\partial \psi}{\partial \lambda} \right\rangle \\
&= E \left[\left\langle \frac{\partial \psi}{\partial \lambda}, \psi \right\rangle + \left\langle \psi, \frac{\partial \psi}{\partial \lambda} \right\rangle \right] + \left\langle \psi, \frac{\partial H_\lambda}{\partial \lambda} \psi \right\rangle \\
&= E \frac{\partial}{\partial \lambda} \langle \psi, \psi \rangle + \left\langle \psi, \frac{\partial H_\lambda}{\partial \lambda} \psi \right\rangle \\
&= \left\langle \psi, \frac{\partial H_\lambda}{\partial \lambda} \psi \right\rangle \\
&= \left\langle \psi, H^{(1)} \psi \right\rangle \\
&= \left\langle H^{(1)} \right\rangle.
\end{aligned}
$$

13.6.4 Perturbation Expansion

The quantum mechanical hypervirial and Hellmann-Feymann theorems are used to provide Rayleigh–Schrödinger perturbation expansions. The first step in the treatment of perturbed Hamiltonians is to obtain the hypervirial relations for $\langle x^k \rangle$ and E for a particular perturbed eigenstate. Next, the following series expansions are assumed:

$$
E(\lambda) = \sum_{n=0}^{\infty} E^{(n)} \lambda^n, \quad \langle x^k \rangle (\lambda) = \sum_{n=0}^{\infty} C_k^{(n)} \lambda^n. \tag{13.115}
$$

The unperturbed energy $E^{(0)}$ is known. As well, the normalized condition $\langle x^0 \rangle = 1$ is imposed, implying that $C_0^{(n)} = \delta_{0n}$. Substitution of these into Eq. (13.113) and collecting like powers of λ^n yield a system of difference equations involving $C_k^{(i)}$ and $E^{(j)}$. However, this system is not closed, since the relation between these two sets of coefficients has not yet been defined. The closure is provided by the Hellmann–Feynman relation $\partial E / \partial \lambda = \langle H^{(1)} \rangle$. The series for $E(\lambda)$, Eq. (13.115), is differentiated and matched term-wise in powers of λ with the series for $\langle H^{(1)} \rangle$. The $C_k^{(i)}$ array is generally calculated column-wise. A knowledge of the nth column in $C_k^{(i)}$ array permits calculation of $E^{(n+1)}$ in analogy with Rayleigh–Schrödinger perturbation theory using the wave function.

13.7 Concluding Remarks

In this chapter, we demonstrated the power of perturbation theory in calculating the small shifts in the spectral lines of some important systems due to the applied perturbation. Though the corrections to the energy eigenvalues given by the perturbation theory are not exact, the results are accurate as long as the perturbation parameter λ is very small. However, in certain cases the perturbation theory is undesirable, particularly, when the systems under considerations cannot be splitted into unperturbed and perturbed parts and also when the strength of the perturbation is very large. There are exactly solvable systems which when subjected to certain interactions posses new set of eigenstates and they cannot be treated as the unperturbed states + corrections. In such cases other approximation methods such as variational method and WKB approximations are useful. An alternative perturbation method on the application of a factorization method to a Riccati equation derived from the Schrödinger equation has been developed and studied [12,13].

13.8   Bibliography

[1] E. Schrödinger, *Annalen der Physik* 80:437, 1926.

[2] J.W.S. Rayleigh, *Theory of Sound*. Macmillan, London, 1894.

[3] E. O'Reilly, *Quantum Theory of Solids*. CRC Press, Florence, 2002.

[4] V. Magnasco, *Elementary Methods of Molecular Quantum Mechanics*. Elsevier, Amsterdam, 2006.

[5] M.A. Parker, *Solid State and Quantum Theory for Optoelectronics*. CRC Press, Boco Raton, 2009.

[6] A. Dalgarno and J.T. Lewis, *Proc. Royal Soc. London A* 233:70, 1955.

[7] H.A. Mavromatis, *Am. J. Phys.* 59:738, 1991.

[8] D.J. Griffiths, *Am. J. Phys.* 50:698, 1982.

[9] R.K. Prasad, *Quantum Chemistry*. New Age International Publications, New Delhi, 1997.

[10] C. Barratt, *Am. J. Phys.* 51:610, 1983.

[11] F.M. Fernandez and E.A. Castro, *Hypervirial Theorems*. Springer, Berlin, 1987.

[12] C.N. Bessis and G. Bessis, *J. Math. Phys.* 38:5483, 1997.

[13] F.M. Fernandez, *J. Phys. A* 39:1683, 2006.

13.9   Exercises

13.1 Evaluate the third-order energy term $E_n^{(3)}$ for a nondegenerate energy level.

13.2 In nondegenerate time-independent perturbation theory, what is the probability of finding a system in a perturbed eigenstate ϕ_n if the corresponding unperturbed eigenstate is $\phi_n^{(0)}$? Solve this up to terms of order λ^2.

13.3 A harmonic oscillator potential is perturbed by λbx^2. Calculate the first and second-orders corrections to the energy eigenvalues.

13.4 Calculate the first and second-orders corrections to the energy eigenvalues of a linear harmonic oscillator with the cubic term $-\lambda\mu x^3$ added to the potential. Discuss the condition for the validity of the approximation.

13.5 A harmonic oscillator Hamiltonian is perturbed by a quartic term in x as $\lambda b x^4$, where b has units of energy/length^4. Calculate the first-order correction to the energy eigenvalues of the perturbed system treating the harmonic oscillator as the unperturbed system.

13.6 Consider the perturbed harmonic oscillator Hamiltonian $H = p_x^2/(2m) + x^2/(2m) + \lambda\hbar\sqrt{m/\hbar}\,x$, $-\infty \le x \le \infty$. Applying Dalgarno–Lewis method compute $E_0^{(2)}$.

13.7 Show that a constant perturbation $H^{(1)} = \alpha$ to a particle in a box of one-dimension with walls at $x = 0$ and $x = L$ shifts the energy levels without changing the eigenfunctions.

13.8 Calculate the first-order correction to $E_3^{(0)}$ for a particle in a one-dimensional box with walls at $x = 0$ and $x = L$ due to the perturbation $H^{(1)} = ax^2$.

13.9 Consider a particle of mass m in a one-dimensional infinite square-well of length $2L$ with its centre at the origin. The perturbation is given by $H^{(1)} = V(x) = E_0 L \delta(x)$, where E_0 has the units of energy. Obtain the first-order corrections to energy eigenvalues and eigenfunctions [I.R. Lapidus, *Am. J. Phys.* 55:172, 1987].

13.10 If the Hamiltonian of a particle in a box of length L is subjected to a uniform electric field given by $H^{(1)} = -eEx$ calculate the first-order correction to the energy eigenvalues.

13.11 A particle in a box potential of width L is perturbed by the term V_0 for $0 < x < L/2$ and zero otherwise. Compute the first-order correction to the energy eigenvalues.

13.12 An electron in the potential $V(x) = -a/x$ for $x > 0$ and ∞ for $x \le 0$ is in its ground state characterized by the eigenfunction $\phi_0 = Nxe^{-\beta x}$. The particle is subjected to an applied electric field E_e in the x-direction. Calculate the first-order correction to ground state energy eigenvalue.

13.13 Assume that the perturbation added to the infinite height potential is
$$V_p(x) = \begin{cases} 0, & 0 < x < L/2 \\ V_0(x - L/2), & L/2 < x < L. \end{cases}$$
Calculate the first-order correction to the energy eigenvalues.

13.14 A particle in a box potential of width L is given by $V(x) = 0$ for $0 < x < L$ and 0 elsewhere. It is perturbed by the additional term $V_0 \sin(\pi x/L)$ in the interval 0 to L. Applying the perturbation theory, compute the first-order correction to the energy levels.

13.15 Calculate the first-order correction to the energy eigenvalues if the infinite height potential of width L is subjected to the perturbation
$$V_p(x) = \begin{cases} V_0, & L/4 < x < 3L/4 \\ 0, & \text{otherwise.} \end{cases}$$

13.16 A particle of mass m is in a two-dimensional quantum box whose infinite height sides are at $x = y = 0$, $x = y = L$. The particle is subjected to the perturbation λxy. Determine the first-order perturbed energy eigenvalues in the $|n_x, n_y\rangle$ state.

13.17 The exact eigenfunctions and energy eigenvalues of a plane rotator are given by $\phi_n = (1/\sqrt{2\pi})e^{in\phi}$, $E_n = n^2\hbar^2/(8\pi^2 I)$, where I is the moment of inertia. The system is subjected to an electric field and $-F\mu\cos\phi$ represents the perturbed term, where F is the strength of the field and μ is the electric moment. Determine the first- and second-order corrections to the energy levels.

13.18   A hydrogen atom Hamiltonian is perturbed by the potential

$$V(x) = \begin{cases} 0, & r \geq \rho_0 \\ \dfrac{e^2}{2\rho_0}\left(\dfrac{r}{\rho_0} - 3\right) + \dfrac{e^2}{r}, & r < \rho_0, \end{cases}$$

where $\rho_0 \ll a_0$ and a_0 is the Bohr radius. The perturbed term can be thought of as the change in the potential due to the finite value of a radius ρ_0 occupied by the nucleus. Calculate the first-order correction to the ground state energy. Assume that $\int_0^b r^n e^{-\alpha r}\, dr = (-\partial/\partial\alpha)^n \int_0^b e^{-\alpha r}\, dr$ and $e^{-\beta\rho_0} \approx 1$ since $\rho_0 \ll a_0$.

13.19   Applying the time-independent perturbation theory calculate the first-order change in the ground state and first excited state energy levels of hydrogen atom due to the perturbation $-eF$. Compare the results with the exact values of energy eigenvalues. Can you observe Stark effect due to the application of the above field?

13.20   Calculate the first-order change in the ground state and first excited state energy levels of hydrogen atom due to the perturbation $-\alpha/r$.

13.21   Consider the hydrogen atom with the perturbation $V_p(r) = -\alpha/r^2$. Calculate the correction to the ground state energy. Show that the perturbation partially removes the degeneracy of first excited state.

13.22   The Hamiltonian matrix of a two-state system is $H = \begin{pmatrix} E_1^{(0)} & \lambda\Delta \\ \lambda\Delta & E_2^{(0)} \end{pmatrix}$. The energy eigenfunctions for the unperturbed Hamiltonian ($\lambda = 0$) are given by $|1^{(0)}\rangle = \begin{pmatrix} 1 \\ 0 \end{pmatrix}$, $|2^{(0)}\rangle = \begin{pmatrix} 0 \\ 1 \end{pmatrix}$ with eigenvalues $E_1^{(0)}$ and $E_2^{(0)}$, respectively.

   (a) Find the exact energy eigenvalues of H.

   (b) Assuming $\lambda|\Delta| \ll \left|E_1^{(0)} - E_2^{(0)}\right|$ and using the time-independent perturbation theory, evaluate the first-order and second-order energy corrections.

13.23   Assuming the degenerate eigenfunctions for a p-electron characterized by $|n, l = 1, m = \pm 1, 0\rangle$ to be $\phi_x^{(0)} = xf(r)$, $\phi_y^{(0)} = yf(r)$ and $\phi_z^{(0)} = zf(r)$, show that the three-fold degeneracy is completely removed, where it is subjected to a potential $V = \lambda(x^2 - y^2)$. You need not evaluate the energy shifts in detail.

13.24   Suppose the Hamiltonian of a rigid rotator in a magnetic field perpendicular to the axis is of the form $AL^2 + BL_z + CL_y$ if terms quadratic in the field are neglected. Assuming $B \gg C$, use perturbation theory to lowest nonvanishing order to get approximate energy eigenvalues for $l = 1$ states. Use the following relations.

$$L_y = (L_+ - L_-)/2i, \quad L_+|lm\rangle = \sqrt{(l-m)(l+m+1)}\,\hbar|lm+1\rangle \text{ and}$$
$$L_-|lm\rangle = \sqrt{(l+m)(l-m+1)}\,\hbar|lm-1\rangle.$$

13.25   For a perturbed system $H = \begin{pmatrix} 1 & 2\epsilon & 0 \\ 2\epsilon & 2+\epsilon & 3\epsilon \\ 0 & 3\epsilon & 3+2\epsilon \end{pmatrix}$, where $\epsilon \ll 1$. Work out the first-order eigenvalues and eigenvectors using the perturbation theory.

14

Approximation Methods II: Time-Dependent Perturbation Theory

14.1 Introduction

Approximation methods are developed for systems where the time dependence of the Hamiltonian is small, slowly varying or rapidly varying. These correspond to perturbation, adiabatic and sudden approximations, respectively. In this chapter, we consider the time-dependent perturbation theory applicable for small time-dependent perturbation. This is the case when we wish to investigate the response of systems to incident radiation. The incident electromagnetic wave makes the electric and magnetic potentials to oscillate and the perturbation is usually very small compared with the static Hamiltonian. The theory of time-dependent perturbation was developed by Dirac in 1927 [1].

In the case of time-independent perturbation, the perturbation may vary in space but at each space point the perturbation remains the same. Hence, we assumed that the state of the system remains the same and the effect of a weak perturbation is to slightly shift the energy eigenvalues and the change in the eigenfunction is small. In contrast to this, in the time-dependent perturbation as the perturbation varies with time the system may undergo transiton from one state to another state and this can take place as long as the perturbation is present. Thus, in the case of time-dependent perturbation instead of finding the correction to the energy eigenvalues and the eigenfunctions we consider the probability of finding the system in a state at time t. Determination of this probability is also generally difficult if the perturbation is present forever. Therefore, we assume that the perturbation is switched-on for a finite interval of time.

In this chapter, we discuss the theory for constant and harmonic perturbations applied during a finite time interval, adiabatic and sudden perturbations. Also, we present the semi-classical theory of radiation. The sudden and the adiabatic approximations are at opposite ends of the time specturm, the former describing what in essence is an instantaneous change, while the latter is applicable to those events for which $\mathrm{d}H/\mathrm{d}t \approx 0$.

14.2 Transition Probability

The problem is to solve the time-dependent Schrödinger equation

$$i\hbar \frac{\partial \psi}{\partial t} = \left(H^{(0)} + \lambda H^{(1)} \right) \psi = H\psi \,. \tag{14.1}$$

Let (E_n, ϕ_n) are the stationary state eigenpairs of the unperturbed system. The time evolution of $\phi_n(x)$ is given by

$$\phi_n(x,t) = e^{-iE_n t/\hbar} \phi_n(x) \,. \tag{14.2}$$

DOI: 10.1201/9781003172178-14

Assume that the perturbation is switched-on at $t = 0$ and H is very close to $H^{(0)}$ so that the set $\{\phi_n(x,t)\}$ is also a complete set of eigenstates of the total Hamiltonian. We write the solution of Eq. (14.1) as

$$\psi(x,t) = \sum_n a_n(t) \mathrm{e}^{-\mathrm{i}E_n t/\hbar} \phi_n(x), \qquad (14.3)$$

where $a_n(t)$ to be determined. Here $a_n(t)$ is the probability amplitude of finding the system in the state $\phi_n(x,t)$ at time t. If the perturbation is absent then $a_n(t) = a_n(0)$.

To determine $a_n(t)$ we substitute Eq. (14.3) in Eq. (14.1) and obtain

$$\mathrm{i}\hbar \sum \dot{a}_n \mathrm{e}^{-\mathrm{i}E_n t/\hbar} \phi_n = \lambda \sum a_n H^{(1)} \mathrm{e}^{-\mathrm{i}E_n t/\hbar} \phi_n . \qquad (14.4)$$

Multiplication of Eq. (14.4) by ϕ_f^* and integrating overall space lead to

$$\mathrm{i}\hbar \dot{a}_f = \lambda \sum_n a_n \mathrm{e}^{\mathrm{i}\omega_{fn} t} H_{fn} , \qquad (14.5a)$$

where

$$\omega_{fn} = (E_f - E_n)/\hbar , \quad H_{fn} = \int_{-\infty}^{\infty} \phi_f^* H^{(1)}(t) \phi_n \, \mathrm{d}x . \qquad (14.5b)$$

In the time-independent perturbation theory, we expanded energy eigenvalues and the eigenfucntions in power series and calculated the first-order and second-order corrections. In the case of time-dependent perturbation, we expand a_f in power series and determine the first-order correction in a_f. We expand $a_f(t)$ as

$$a_f(t) = a_f^{(0)} + \lambda a_f^{(1)} + \lambda^2 a_f^{(2)} + \ldots . \qquad (14.6)$$

The obvious criterion for the applicability of the above expansion of $a_f(t)$ is $a_f^{(1)} \ll a_f^{(0)}$. Substituting the above series in Eq. (14.5) and collecting equal powers of λ and equating them to zero individually we obtain

$$\text{Coeff. of } \lambda^0 : \; \mathrm{i}\hbar \dot{a}_f^{(0)} = 0 . \qquad (14.7a)$$

$$\text{Coeff. of } \lambda^1 : \; \mathrm{i}\hbar \dot{a}_f^{(1)} = \sum_n a_n^{(0)} \mathrm{e}^{\mathrm{i}\omega_{fn} t} H_{fn} . \qquad (14.7b)$$

$$\text{Coeff. of } \lambda^2 : \; \mathrm{i}\hbar \dot{a}_f^{(2)} = \sum_n a_n^{(1)} \mathrm{e}^{\mathrm{i}\omega_{fn} t} H_{fn} . \qquad (14.7c)$$

$$\text{Coeff. of } \lambda^{s+1} : \; \mathrm{i}\hbar \dot{a}_f^{(s+1)} = \sum_n a_n^{(s)} \mathrm{e}^{\mathrm{i}\omega_{fn} t} H_{fn} . \qquad (14.7d)$$

We note that to zeroth-order (no perturbation), the probability amplitudes $a_f^{(0)}$ are constants in time: $a_f^{(0)}(t) = a_f^{(0)}(0)$ and specify the state of the system before the perturbation is applied. Now, we analyze the first-order effect on the system.

Equation (14.7) is a set of coupled integro-differential equations. If the system has m eigenstates then Eq. (14.7) generates m coupled equations and are difficult to solve. Therefore, we slightly modify the problem. Suppose the perturbation is switched-on at $t = 0$ and switched-off at $t = T$. The perturbation is applied during a finite time interval T. Assume that the system is initially in a state i with the eigenfunction ϕ_i. Then at $t = 0$ the probability of finding the system in ϕ_i is unity and the probability of finding the system in other states $\phi_n(n \neq i)$ is zero: $a_n^{(0)} = \delta_{ni}$. Due to the applied perturbation transition from the state i to an another state after the time T can happen. Once the perturbation is switched-off the system settles down to a stationary state and denote this final state as f. *What is the probability for the transition from ϕ_i to ϕ_f state?* It is simply the probability of finding the system in ϕ_f: $P_{fi} = a_f^* a_f$. Now, Eq. (14.7b) is

$$\dot{a}_f^{(1)} = \frac{1}{i\hbar} \sum_n \delta_{ni} e^{i\omega_{fn}t} H_{fn} = \frac{1}{i\hbar} e^{i\omega_{fi}t} H_{fi} . \tag{14.8}$$

Integrating the above equation from 0 to T we obtain

$$a_f^{(1)} = \frac{1}{i\hbar} \int_0^T e^{i\omega_{fi}t} H_{fi}(t) \, dt . \tag{14.9}$$

The transition probability for ith state to fth state when $f \neq i$ is given by $P_{fi} = |a_f^{(1)}|^2$.
We consider the cases of

$$
\begin{align}
H^{(1)}(t) &= W(x), \tag{14.10} \\
H^{(1)}(t) &= W(x) e^{-i\omega t}, \tag{14.11} \\
H^{(1)}(t) &= W(x) \sin \omega t, \tag{14.12} \\
H^{(1)}(t) &= W(x) \cos \omega t \tag{14.13}
\end{align}
$$

separately in the following sections.

14.3   Constant Perturbation

For the constant perturbation, $H^{(1)}(t) = W(x)$, Eq. (14.8) becomes

$$\dot{a}_f^{(1)} = \frac{1}{i\hbar} e^{i\omega_{fi}t} H_{fi} , \qquad H_{fi} = \int_{-\infty}^{\infty} \phi_f^* W(x) \phi_i \, dx . \tag{14.14}$$

Its solution is obtained as

$$a_f^{(1)} = \frac{1}{i\hbar} H_{fi} \int_0^T e^{i\omega_{fi}t} \, dt = \frac{H_{fi}}{\hbar \omega_{fi}} \left(1 - e^{i\omega_{fi}T}\right) . \tag{14.15}$$

Then the first-order transition probability is

$$P_{fi} = a_f^{(1)*} a_f^{(1)} = \frac{4|H_{fi}|^2}{\hbar^2 \omega_{fi}^2} \sin^2(\omega_{fi}T/2) . \tag{14.16}$$

14.3.1   Features of First-Order Transition Probability

Some of the features of Eq. (14.16) are summarized below.

1. The P_{fi} given by Eq. (14.16) is the probability for the system to undergo a transition from a state i to f. This transition is due to the applied perturbation. P_{fi} depends on H_{fi}, T and ω_{fi} (that is, E_f and E_i).

2. **Selection rule**: The transition from ith state to fth state is forbidden if $H_{fi} = 0$. The selection rule is thus $H_{fi} \neq 0$.

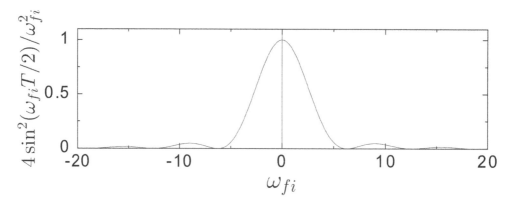

FIGURE 14.1
Plot of $[4\sin^2(\omega_{fi}T/2)]/\omega_{fi}^2$ as a function of ω_{fi} for $T = 1$.

3. Figure 14.1 shows the plot of $s = [4\sin^2(\omega_{fi}T/2)]/\omega_{fi}^2$ versus ω_{fi} for fixed T. The function s has a sharp peak centred at $\omega_{fi} = 0$. For small ω_{fi}, $s = T^2$. The height of the main peak is T^2. In this case $P_{fi} \propto T^2$. Hence, the probability per unit time or transition rate Γ_{fi} is $\propto T$.

4. For large ω_{fi}, $P_{fi} \to 0$.

5. P_{fi} is zero when $\sin^2(\omega_{fi}T/2) = 0$, that is, $\omega_{fi} \approx 2n\pi/T$.

6. A striking feature of P_{fi} from Fig. 14.1 is that it does not change monotonically with ω_{fi}. It neither increases nor decreases continuously with ω_{fi} but oscillates with rapidly decreasing amplitude.

7. P_{fi} **for closely spaced levels**: Equation (14.16) represents the probability for transition to one particular state ϕ_f. In real systems we wish to know P_{fi} for transitions to a group of closely spaced states. In this case $P_{fi} = \int_{-\infty}^{\infty} |a_f|^2 \rho_f(E_f)\,\mathrm{d}E_f$, where $\rho_f \mathrm{d}E$ represents number of states in the range $\mathrm{d}E$. Assume that ρ is almost constant in the range $\mathrm{d}E$. Then

$$P_{fi} = \rho_f \int_{-\infty}^{\infty} |a_f|^2 \,\mathrm{d}E_f = \frac{2\pi}{\hbar}\rho_f|H_{fi}|^2 T, \qquad (14.17)$$

where we have used the result $\int_{-\infty}^{\infty} (\sin^2 \xi)/\xi^2 \,\mathrm{d}\xi = \pi$. Thus, the total transition probability is proportional to T.

8. The transition rate per unit time is

$$\frac{\mathrm{d}P_{fi}}{\mathrm{d}T} = \Gamma_{fi} = \frac{2\pi}{\hbar}\rho_f|H_{fi}|^2 \qquad (14.18)$$

which is independent of T. The general form of the above formula is applicable to atomic transitions, nuclear decay and many physical transitions and hence Fermi [2] called it the *(Fermi) golden rule*. This rule is often used in spectroscopy to describe the transition rate between states. Calculations of intensities of spectral lines and many of cross-sections for various processes are based on the use of this rule.

9. For large T the area under the central peak is large compared to the other peaks. Therefore, the transitions to the states within this peak occur mostly. In this

case $|\omega_{fi}| \lesssim 2\pi/T$. Suppose, we wish to measure ω_{fi} by measuring P_{fi}. Then the uncertainty $\Delta\omega \sim 2\pi/T$ or $\Delta E \sim 2\pi\hbar/T$ or $\Delta E \sim h/T$. The energy change ΔE lasts for a time $\Delta t = T$. Therefore, $\Delta E \Delta T \approx h$. This is known as *time-energy uncertainty*. If T is small, we have a broader peak in Fig. 14.1 and as a result the energy is not conserved. On the other hand, if the perturbation is applied for a very long time, we have a very narrow peak, and approximate energy conservation is required for a transition with appreciable probability.

10. From Eq. (14.16) we note that the transition probability varies for various states. *In experiments, how do we account for this?* If we look at a spectrum we will notice that the intensity of spectral lines varies — some are very dark while some are very faint and so on. If the intensity of a line is large then that line corresponds to a transition which takes place very often, that is, this corresponds to large P_{ji}.

Solved Problem 1:

A linear harmonic oscillator is subjected to an applied electric field during the time $t = 0$ to T. Assume that the initial state of the system is ϕ_i and the perturbation to the Hamiltonian is $H^{(1)} = -eEx$. Find the probability for the transition to the state ϕ_f in time T.

The transition probability is given by Eq. (14.16). We obtain

$$
\begin{aligned}
H_{fi} &= -eE\langle f|x|i\rangle \\
&= -eE\left(\frac{\hbar}{2m\omega}\right)^{1/2}\langle f|a + a^\dagger|i\rangle \\
&= -eE\left(\frac{\hbar}{2m\omega}\right)^{1/2}\left[\sqrt{i+1}\,\delta_{f,i+1} + \sqrt{i}\,\delta_{f,i-1}\right].
\end{aligned}
\tag{14.19}
$$

According to first-order perturbation theory, the possible transitions are (i) $i \to i$, (ii) $i \to i+1$ and (iii) $i \to i-1$. In the first case the probability is 1 because $a_i^{(0)} = 1$ and $a_i^{(1)} = 0$. However, we are interested in the transition from ith state to f with $f \neq i$. Therefore, the selection rule is $f = i \pm 1$. For the transition i to $i+1$ (absorption process)

$$
\left|H_{i+1,i}^{(1)}\right|^2 = \frac{\hbar e^2(i+1)E^2}{2m\omega}.
\tag{14.20}
$$

For i to $i-1$ (emission process)

$$
\left|H_{i-1,i}^{(1)}\right|^2 = \frac{\hbar e^2 i E^2}{2m\omega}.
\tag{14.21}
$$

Solved Problem 2:

The motion of a particle of mass m and electric charge e confined to an interval of length L is subjected to a uniform electric field $\mathcal{E}$ from $t = 0$ to T. Assume that the particle is initially in the state i. Calculate the probability that after a time T the system is in the state f where $f > i$.

The Hamiltonian of the unperturbed system is $H^{(0)} = p_x^2/(2m) + V$, $V = 0$ for $0 < x < L$ and $V = \infty$ elsewhere. The unperturbed eigenstates are

$$
\phi_n = \sqrt{\frac{2}{L}}\sin\left(\frac{n\pi x}{L}\right), \quad E_n = \frac{n^2\pi^2\hbar^2}{2mL^2}.
\tag{14.22}
$$

The perturbation due to the applied field is $H^{(1)} = -e\mathcal{E}x$. The transition probability P_{fi} is given by

$$P_{fi} = \frac{4\left|H_{fi}^{(1)}\right|^2}{\hbar^2\omega_{fi}^2}\sin^2(\omega_{fi}T/2)\,, \quad \omega_{fi} = \frac{\pi^2\hbar^2(f^2 - i^2)}{2mL^2}\,. \tag{14.23}$$

We obtain $H_{fi}^{(1)} = \langle\phi_f|H^{(1)}|\phi_i\rangle$ as

$$\begin{aligned}H_{fi}^{(1)} = {}& -\frac{e\mathcal{E}L}{\pi^2(f^2 - i^2)^2}\left[(f + i)^2(\cos(f - i)\pi - 1)\right.\\&\left. -(f - i)^2(\cos(f + i)\pi - 1)\right].\end{aligned} \tag{14.24}$$

We have the results

$$\cos(f + i)\pi = \cos(f - i)\pi = 1, \quad \text{for } f + i = \text{even} \tag{14.25a}$$
$$\cos(f + i)\pi = \cos(f - i)\pi = -1, \quad \text{for } f + i = \text{odd}. \tag{14.25b}$$

Therefore, for $f + i = \text{even}$ $H_{fi}^{(1)} = 0$. For $f + i = \text{odd}$

$$H_{fi}^{(1)} = \frac{8e\mathcal{E}Lfi}{\pi^2(f^2 - i^2)^2}\,. \tag{14.26}$$

Thus, for $f + i = \text{even}$ we have $P_{fi} = 0$ and for $f + i = \text{odd}$ it is given by Eq. (14.23) with $H_{fi}^{(1)}$ given by Eq. (14.26).

14.4 Harmonic Perturbation

Suppose the perturbation $H^{(1)}$ is a slowly varying harmonic function of time: $H^{(1)}(t) = W(x)\,\mathrm{e}^{-\mathrm{i}\omega t}$, where ω is the frequency of the applied force. Assume that the perturbation is switched on at $t = 0$ and switched-off at $t = T$. Following the treatment of previous section, we have

$$\dot{a}_f = \frac{1}{\mathrm{i}\hbar}\mathrm{e}^{\mathrm{i}(\omega_{fi} - \omega)t}H_{fi}\,, \quad H_{fi} = \int_{-\infty}^{\infty}\phi_f^* W(x)\phi_i\,\mathrm{d}\tau\,. \tag{14.27}$$

Integration of the above equation from $t = 0$ to T gives

$$a_f(T) = \frac{H_{fi}}{\hbar(\omega_{fi} - \omega)}\left(1 - \mathrm{e}^{\mathrm{i}(\omega_{fi} - \omega)T}\right). \tag{14.28}$$

Then

$$P_{fi} = a_f^* a_f = \frac{4\left|H_{fi}\right|^2}{\hbar^2(\omega_{fi} - \omega)^2}\sin^2\left[(\omega_{fi} - \omega)T/2\right]. \tag{14.29}$$

This expression of P_{fi} differs from that of the constant perturbation by the term $(\omega_{fi} - \omega)$. With this replacement the entire discussion on P_{fi} of constant perturbation is valid for the harmonic perturbation.

We now list some of the differences between constant and harmonic perturbations:

1. For constant perturbation $P_{fi} \propto \omega_{fi}$ whereas for harmonic perturbation it is $\propto \omega_{fi} - \omega$.

2. For constant perturbation P_{fi} is large when $\omega_{fi} \approx 0$, that is, $E_f \approx E_i$. For harmonic perturbation P_{fi} is large when $\omega_{fi} - \omega \approx 0$, that is, $E_f - E_i \approx \hbar\omega$. Therefore, most of the transitions take place between the states for which $E_f - E_i \approx \hbar\omega$.

3. For constant perturbation P_{fi} is proportional to $\sin^2(\omega_{fi}T/2)$. For harmonic perturbation we have all frequencies ω. Integration over all frequencies gives P_{fi} proportional to T.

14.4.1 Perturbation with Whole Spectrum of Frequencies

In the above we considered the perturbation with a single frequency ω. Suppose the perturbation contains a whole spectrum of frequencies ω and the phases of the frequency components are unrelated. Then the total transition probability is

$$P_{fi} = \int |a_f^2| \rho(\omega)\, d\omega$$

$$= \frac{4}{\hbar^2} \int \frac{\left|H_{fi}^{(\omega)}\right|^2 \sin^2\left[(\omega_{fi}-\omega)T/2\right]}{(\omega_{fi}-\omega)^2} \rho(\omega)\, d\omega. \tag{14.30}$$

Referring to Eq. (14.17) we obtain

$$\Gamma_{fi} = \frac{2\pi}{\hbar^2}\left|H_{fi}(\omega_{fi})\right|^2 \rho(\omega_{fi}). \tag{14.31}$$

The upward transition rate is $(i = n,\ f = m,\ E_m > E_n)$

$$\Gamma_{mn} = \frac{2\pi}{\hbar^2}\left|H_{mn}(\omega_{mn})\right|^2 \rho(\omega_{mn}). \tag{14.32}$$

The transition rate for downward transition is $(i = m,\ f = n,\ E_m > E_n)$

$$\Gamma_{nm} = \frac{2\pi}{\hbar^2}\left|H_{nm}(\omega_{nm})\right|^2 \rho(\omega_{nm})$$

$$= \frac{2\pi}{\hbar^2}\left|H_{nm}(-\omega_{mn})\right|^2 \rho(-\omega_{mn}). \tag{14.33}$$

Because of $\rho(-\omega) = \rho(\omega)$ and $H_{fi}(-\omega) = (H_{if}(\omega))^\dagger$ we have $\Gamma_{nm} = \Gamma_{mn}$. That is, the transition rate for upward and downward transitions are the same.

Suppose $H^{(1)} = W(x)\sin\omega t$. Then

$$\dot{a}_f = \frac{1}{i\hbar}H_{fi}^{(1)}e^{i\omega_{fi}t} = \frac{1}{2\hbar}H_{fi}\left[e^{i(\omega_{fi}-\omega)t} - e^{i(\omega_{fi}+\omega)t}\right]. \tag{14.34}$$

Integrating the above equation from $t = 0$ to T we obtain

$$a_f = -\frac{iH_{fi}}{2\hbar}\left[\frac{1 - e^{i(\omega_{fi}+\omega)T}}{\omega_{fi}+\omega} - \frac{1 - e^{i(\omega_{fi}-\omega)T}}{\omega_{fi}-\omega}\right] \tag{14.35}$$

and

$$P_{fi}(T) = \frac{\left|H_{fi}\right|^2}{4\hbar^2}\left|\frac{1 - e^{i(\omega_{fi}+\omega)T}}{\omega_{fi}+\omega} - \frac{1 - e^{i(\omega_{fi}-\omega)T}}{\omega_{fi}-\omega}\right|^2. \tag{14.36}$$

Similarly, for $H^{(1)} = W(x) \cos \omega t$ we have

$$a_f = \frac{H_{fi}}{2\hbar} \left[\frac{1 - e^{i(\omega_{fi}+\omega)T}}{\omega_{fi} + \omega} + \frac{1 - e^{i(\omega_{fi}-\omega)T}}{\omega_{fi} - \omega} \right], \tag{14.37a}$$

$$P_{fi}(T) = \frac{|H_{fi}|^2}{4\hbar^2} \left| \frac{1 - e^{i(\omega_{fi}+\omega)T}}{\omega_{fi} + \omega} + \frac{1 - e^{i(\omega_{fi}-\omega)T}}{\omega_{fi} - \omega} \right|^2. \tag{14.37b}$$

For the harmonic perturbations $\sin \omega t$ and $\cos \omega t$, the term $| \cdot |^2$ in Eqs. (14.36) and (14.37b) has an appreciable amplitude only if the denominator of one of the two terms in it is ≈ 0: $\omega_{fi} = \pm \omega$ or $E_f = E_i \pm \hbar \omega$. Thus, according to first-order perturbation theory for the harmonic perturbation the system must either absorb or emit a quanta of energy $\hbar \omega_{fi}$ from the perturbation.

Suppose the system is in the eigenstate ϕ_i at $t = a$ (for example at $t = -\infty$ or 0). The perturbation is switched-on at $t = a$ and left forever. It is desired to know the state at $t = b$ (for example $t = \infty$ or T). In this case $a_f = (1/i\hbar) \int_a^b H_{fi} e^{i\omega_{fi}t} \, dt$ and the probability of finding the particle in the state f at $t = b$ is

$$|a_f|^2 = \frac{1}{\hbar^2} \left| \int_a^b H_{fi} e^{i\omega_{fi}t} \, dt \right|^2. \tag{14.38}$$

Solved Problem 3:

A linear harmonic oscillator is subjected to the time varying field described by $H^{(1)} = -e\mathcal{E}xe^{-i\Omega t}$ during the time $t = 0$ to T. If the system is initially in its ground state calculate the probability of finding the system in the first-excited state for $t > T$. Find the probability if (i) $H^{(1)} = -e\mathcal{E}x \sin \Omega t$ and (ii) $H^{(1)} = -e\mathcal{E}x \cos \Omega t$.

The transition probability is given by

$$P_{10} = \frac{4|H_{10}|^2}{\hbar^2(\omega_{10} - \Omega)^2} \sin^2 \left[(\omega_{10} - \Omega)T/2 \right]. \tag{14.39}$$

We obtain $\omega_{10} = \left(\dfrac{3}{2}\hbar\omega - \dfrac{1}{2}\hbar\omega \right) / \hbar = \omega$ and

$$\begin{aligned} H_{10} &= -e\mathcal{E}\langle 1|x|0 \rangle \\ &= -e\mathcal{E} \left(\frac{\hbar}{2m\omega} \right)^{1/2} \langle 1|a + a^\dagger|0 \rangle \\ &= -e\mathcal{E} \left(\frac{\hbar}{2m\omega} \right)^{1/2}. \end{aligned} \tag{14.40}$$

Then

$$P_{10} = \frac{2e^2\mathcal{E}^2}{\hbar m\omega(\omega - \Omega)^2} \sin^2 \left[(\omega - \Omega)T/2 \right]. \tag{14.41}$$

In the limit $\Omega \to \omega$, $P_{10} = e^2\mathcal{E}^2T^2/(2m\hbar\omega)$. For $H^{(1)} = -e\mathcal{E}x \sin \Omega t$

$$P_{10} = \frac{e^2\mathcal{E}^2}{8\hbar m\omega} \left| \frac{1 - e^{i(\omega+\Omega)T}}{(\omega + \Omega)} - \frac{1 - e^{i(\omega-\Omega)T}}{(\omega - \Omega)} \right|^2 \tag{14.42a}$$

and for $H^{(1)} = -e\mathcal{E}x\cos\omega t$

$$P_{10} = \frac{e^2\mathcal{E}^2}{8\hbar m\omega} \left| \frac{1 - e^{i(\omega+\Omega)T}}{(\omega+\Omega)} + \frac{1 - e^{i(\omega-\Omega)T}}{(\omega-\Omega)} \right|^2 . \tag{14.42b}$$

Solved Problem 4:

Assume that a hydrogen atom with an electron and an infinitely heavy proton is subjected to a spatially uniform and a harmonically time-varying electric field $\mathcal{E}(t) = 2\mathbf{k}\mathcal{E}_0\cos\omega t$, $t > t_0$. If the perturbation $H^{(1)}(t)$ is taken as

$$H^{(1)}(t) = W(x)\left(e^{i\omega t} + e^{-i\omega t}\right) \tag{14.43}$$

calculate the differential ionization probability per unit solid angle.

The time-independent part of the hydrogen atom is given by $W(x) = e\mathcal{E}_0 z = e\mathcal{E}_0 r\cos\theta$. The initial ground state of the hydrogen atom is given by $\psi_{100} = (1/\sqrt{\pi a_0^3})e^{-r/a_0}$. The final states should correspond to a positive energy electron in Coulomb field of the proton. Since the eigenfunction for such a state is quite complicated, a simple eigenfunction for a free particle confined in a box of side L can be assumed. Then the final state k is given as $\psi_k = L^{-3/2}e^{i\mathbf{k}\cdot\mathbf{r}}$. The quantities $|H_{fi}^{(1)}|^2$ in (14.37b) will depend on the direction of $\mathbf{k}$. Hence, the range of $\mathbf{k}$ is assumed to be in a solid angle. Since $k_x = 2\pi n_x/L$, $k_y = 2\pi n_y/L$ and $k_z = 2\pi n_z/L$, the number of states in the range $dk_x dk_y dk_z$ becomes $(L/2\pi)^3 dk_x dk_y dk_z$ which expressed in spherical polar coordinates gives

$$\left(\frac{L}{2\pi}\right)^3 k^2 \sin\theta\, dk d\theta d\phi, \tag{14.44}$$

where θ and ϕ are the polar angles of $\mathbf{k}$ with respect to the electric field $\mathcal{E}$. We have

$$\rho(k)\, dE_k = \left(\frac{L}{2\pi}\right)^3 k^2\, dk \sin\theta d\theta d\phi. \tag{14.45}$$

Since $E_k = \hbar^2 k^2/(2m_e)$ we get $dE_k = \hbar^2 k/m_e$. Then

$$\rho(k) = \frac{m_e L^3 k \sin\theta\, d\theta\, d\phi}{8\pi^3\hbar^2}. \tag{14.46}$$

The initial state is the state of hydrogen atom $\rho(\omega_{fi}) = \rho(k)$ and

$$H_{fi}^{(1)} = -\frac{i32\pi e\mathcal{E}_0 k a_0^5 \cos\theta}{\left(\pi a_0^3 L^3\right)^{1/3}\left(1 + k^2 a_0^2\right)^3}. \tag{14.47}$$

Then the probability per unit time that the electron of the hydrogen atom is ejected into the solid angle $\sin\theta\, d\theta\, d\phi$ is

$$\Gamma_{fi} = \frac{256 m_e \mathcal{E}_0^2 k^3 e^2 a_0^7}{\pi\hbar^3\left(1 + k^2 a_0^2\right)^6}\cos^2\theta\sin\theta\, d\theta\, d\phi. \tag{14.48}$$

The differential ionization probability per unit solid angle $\Gamma_{fi}/(\sin\theta\, d\theta\, d\phi)$ is $\propto \cos^2\theta$ as the driving force in the direction of ejection is equal to $e\mathcal{E}_0\cos\theta$.

14.5 Adiabatic Perturbation

What will happen if a potential changes very slowly? For each differential change in the potential, a differential change say ϵ is caused in the wave function. But, if the changes take place at different times, the differential amplitudes contribute with uncorrelated phases. Then the total change in the probability is $\sum |\epsilon|^2 = 0$. Hence, the probability associated with each state remains the same. This is known as *adiabatic approximation*[1]. For example, consider a pendulum exhibiting oscillation in a vertical plane. Suppose we move the support. The mode of oscillation of the system change. *What will happen if the support is moved very slowly?* The motion of the pendulum relative to the support will be unchanged. That is, when the perturbation is sufficiently slow then the state of the system remains the same. This is known as adiabatic process.

14.5.1 Transition Probability

Suppose the perturbation applied to a quantum mechanical system varies slowly with time and is written as $H^{(1)}(t) = W(x)V(t)$. An example is $V(t) = e^{\eta t}$, where $0 < \eta \ll 1$. The perturbation is switched-on at $t = -\infty$ and the initial state of the system is ϕ_i. The perturbation $H^{(1)}(t)$ is assumed small in comparison with E/τ, where E and τ are energy and time typically characteristic of the system. We wish to find the probability of finding the system in the state ϕ_f at time t. From Eq. (14.8) the evolution equation for transition amplitude is written as

$$\dot{a}_f(t) = \frac{1}{i\hbar} H^{(1)}_{fi} V(t) \, e^{i\omega_{fi}t} \,. \tag{14.49}$$

Integrating the above equation with respect to time from $t = -\infty$ we get

$$a_f(t) = \frac{H^{(1)}_{fi}}{i\hbar} \int_{-\infty}^{t} V(t) \, e^{i\omega_{fi}t} \, dt \,. \tag{14.50}$$

Since $V(t)$ is a slowly varying function of t we assume that it is almost constant during a time interval dt and obtain

$$a_f(t) = -\frac{H^{(1)}_{fi} V(t) e^{i\omega_{fi}t}}{\hbar \omega_{fi}} \,. \tag{14.51}$$

We require $|a_f| \ll 1$. That is, $|H^{(1)}_{fi}| \ll \hbar |\omega_{fi}|$. Next,

$$P_{fi} = \frac{\left|H^{(1)}_{fi}\right|^2 V^2}{\hbar^2 \omega_{fi}^2} \,. \tag{14.52}$$

When $V(t) = e^{\eta t}$ from Eq. (14.52)

$$P_{fi} = \frac{\left|H^{(1)}_{fi}\right|^2 e^{2\eta t}}{\hbar^2 \omega_{fi}^2} \,. \tag{14.53}$$

On the other hand, from Eq. (14.50) we obtain

$$a_f(t) = -\frac{H^{(1)}_{fi} e^{i(\omega_{fi} - i\eta)t}}{\hbar (\omega_{fi} - i\eta)} \,. \tag{14.54}$$

---

[1]According to Max Born and Vladimir Fock [3] *a physical system remains in its instantaneous eigenstate if a given perturbation acting on it is slowly enough and if there is a gap between the eigenvalue and the rest of the Hamiltonian's spectrum.*

Now,

$$P_f(t) = \frac{|H_{fi}^{(1)}|^2 e^{2\eta t}}{\hbar^2(\omega_{fi}^2 + \eta^2)} . \tag{14.55}$$

Since $0 < \eta \ll 1$ we can neglect η^2 in the above equation. Then Eq. (14.55) becomes the Eq. (14.53).

The adiabatic approximation is applicable also to scattering processes in cases for which the number of excited states of the target contributing significantly to the wave function is limited, and for which the time of traversal of the scattered particle through the region of interaction is small compared with the period of the target motion excitable in the collision.

Solved Problem 5:

A harmonic oscillator in the state ϕ_i is subjected to the perturbation $V_p(x,t) = \alpha x e^{\eta t}$, $0 < \eta \ll 1$ from $t = -\infty$. Calculate the transition probability to a state ϕ_f in time t.

We apply the adiabatic perturbation theory since the perturbation $V_p(x,t)$ is a slowly varying function of t. $P_f(t)$ is given by Eq. (14.55) where

$$\omega_{fi} = \frac{E_f - E_i}{\hbar} = \frac{1}{\hbar}\left[\left(f + \frac{1}{2}\right)\hbar\omega - \left(i + \frac{1}{2}\right)\hbar\omega\right] = (f - i)\omega \tag{14.56a}$$

and

$$
\begin{aligned}
H_{fi}^{(1)} &= \langle\phi_f|\alpha x|\phi_i\rangle \\
&= \alpha\left(\frac{\hbar}{2m\omega}\right)^{1/2}\langle\phi_f|a + a^\dagger|\phi_i\rangle \\
&= \alpha\left(\frac{\hbar}{2m\omega}\right)^{1/2}\left[\sqrt{i+1}\,\delta_{f,i+1} + \sqrt{i}\,\delta_{f,i-1}\right] .
\end{aligned}
\tag{14.56b}
$$

For the upward transition i to $i + 1$ $(\omega_{fi} = \omega)$

$$|H_{fi}^{(1)}|^2 = \frac{\alpha^2\hbar(i+1)}{2m\omega} \tag{14.57}$$

and

$$P_f(t) = \frac{\alpha^2(i+1)e^{2\eta t}}{2m\omega\hbar(\omega^2 + \eta^2)} . \tag{14.58}$$

For the downward transition i to $i - 1$ $(\omega_{fi} = -\omega)$

$$|H_{fi}^{(1)}|^2 = \frac{\alpha^2\hbar i}{2m\omega} \tag{14.59}$$

and

$$P_f(t) = \frac{\alpha^2 i\, e^{2\eta t}}{2m\omega\hbar(\omega^2 + \eta^2)} . \tag{14.60}$$

14.6   Sudden Approximation

In the previous sections, we assumed that the perturbation is applied during a time interval T. Suppose a time-independent perturbation is switched-on suddenly at $t = t'$ and is left on forever. This is known as *sudden* or *impulse approximation*. In the sudden approximation, the Hamiltonian of the system is $H = H^{(0)}$ for $t < t'$ and $H = H^{(0)} + \lambda H^{(1)}$ for $t > t'$. Also, for such problems it is preassumed that the eigenstates of both Hamiltonians $H^{(0)}$ and $H = H^{(0)} + \lambda H^{(1)}$ are known. Sudden approximation theory has been used in the study of single and multiple electron losses of high and intermediate energy projectiles in their collisions with neutral targets, pressure broadening of rotational transitions, core hole ionization in CO, multielectron transition in complex atoms, collisions of fast multicharged ions with atoms and in the collisions of charged particles with highly-excited atoms.

What is $\psi(t)$ in the sudden approximation? For $t < t'$ to get $\psi(t)$ we use $H^{(0)}$ and for $t > t'$ we use the Hamiltonian $H^{(0)} + \lambda H^{(1)}$. Usually t' is chosen as $-\infty$ or 0 for simplicity. Let

$$H^{(0)}\phi_n^< = E_n^< \phi_n^<, \quad t < 0 \tag{14.61a}$$

$$H\phi_n^> = E_n^> \phi_n^>, \quad t > 0. \tag{14.61b}$$

For $t < 0$ and $t > 0$ the wave functions of the system are

$$\psi^< = \sum_n a_n^< \phi_n^< \, \mathrm{e}^{-iE_n^< t/\hbar}, \quad \psi^> = \sum_m a_m^> \phi_m^> \, \mathrm{e}^{-iE_m^> t/\hbar}, \tag{14.62}$$

respectively, where $\phi_m^>$ and $E_m^>$ are the eigenfunctions and eigenvalues, respectively, of the perturbed system. *How do we determine $a_m^>$?* We use the continuity of $\psi^<$ and $\psi^>$ at $t = 0$ to calculate $a_m^>$. At $t = 0$, $\psi^< = \psi^>$, that is,

$$\sum_n a_n^< \phi_n^< = \sum_m a_m^> \phi_m^>. \tag{14.63}$$

Multiplying both sides by $\phi_f^{>*}$ and integrating over all space we get

$$\sum_n a_n^< \langle \phi_f^> | \phi_n^< \rangle = \sum_m a_m^> \langle \phi_f^> | \phi_m^> \rangle = \sum_m a_m^> \delta_{fm}. \tag{14.64}$$

That is, $a_f^> = \sum_n a_n^< \langle \phi_f^> | \phi_n^< \rangle$.

If we assume that for $t < 0$ the system is in the state i then $a_i^< = 1$ and $a_n^< = 0$ for $n \neq i$. Then

$$a_f^> = \sum_n \delta_{ni} \langle \phi_f^> | \phi_n^< \rangle = \langle \phi_f^> | \phi_i^< \rangle. \tag{14.65}$$

To calculate $a_f^>$ we require both initial and final eigenstates. $a_f^{>*} a_f^>$ gives the probability for the transition from the ith state to the fth state of the perturbed system. An interesting consequence of sudden approximation is that the system can undergo an adiabatic change due to the sudden approximation. By adiabatic change we mean a change such that the perturbed system remains in an energy eigenstate with the same quantum number. That is, for example, a system may remain in the nth state of the system for $t < 0$ and $t > 0$ also. However, for $t < 0$ it is in the nth state ($\phi_n^<$) of the unperturbed Hamiltonian while for $t > 0$ it is in the nth state ($\phi_n^>$) of the perturbed system. $\phi_n^<$ and $\phi_n^>$ states have the same quantum number but with different energies $E_n^<$ and $E_n^>$, respectively. This is often illustrated with the following example.

Consider the infinite height potential confining a particle to the region $0 < x < L$. Suppose the right-wall is suddenly moved to the position $x = 2L$ at $t = 0$. If the system is in the ground state for $t < 0$, *what is the probability of it being in the ground state of the new well? What is the probability of it being in the nth state of the new well?*

For $t < 0$ the energy eigenfunctions and eigenvalues are given by

$$\phi_n^< = \sqrt{\frac{2}{L}} \sin \frac{n\pi x}{L}, \quad E_n^< = \frac{n^2 \pi^2 \hbar^2}{2mL^2}. \tag{14.66}$$

The expanded well is also exactly solvable. Its eigenstates are obtained from Eq. (14.66) by simply replacing L by $2L$:

$$\phi_n^> = \sqrt{\frac{1}{L}} \sin \frac{n\pi x}{2L}, \quad E_n^> = \frac{n^2 \pi^2 \hbar^2}{8mL^2}. \tag{14.67}$$

The transition amplitude for finding the expanded system in the ground state is

$$a_1 = \langle \phi_1^> | \phi_1^< \rangle = \int_0^{2L} \phi_1^{>*} \phi_1^< \, dx. \tag{14.68}$$

Since $\phi_1^< = 0$ for $x > L$ we have

$$a_1 = \int_0^L \phi_1^{>*} \phi_1^< \, dx = \frac{\sqrt{2}}{L} \int_0^L \sin\left(\frac{\pi x}{2L}\right) \sin\left(\frac{\pi x}{L}\right) \, dx = \frac{4\sqrt{2}}{3\pi}. \tag{14.69}$$

The transition probability is then $P_1 = a_1^* a_1 = 32/(9\pi^2)$. For the second question,

$$a_n = \langle \phi_n^> | \phi_1^< \rangle = \frac{\sqrt{2}}{\pi} \left[\frac{\sin(n-2)y}{n-2} - \frac{\sin(n+2)y}{n+2} \right]_0^{\pi/2}. \tag{14.70}$$

Since $\sin(m\pi/2) = (-1)^{(m+3)/2}$ for odd m and 0 for even m, we obtain $a_n = 0$ for even n and for odd n

$$a_n = \frac{\sqrt{2}}{\pi} \left[\frac{(-1)^{(n+1)/2}}{n-2} - \frac{(-1)^{(n+1)/2}}{n+2} \right] = \frac{4\sqrt{2}\,(-1)^{(n+1)/2}}{\pi(n^2 - 4)}. \tag{14.71}$$

Then $P_n = a_n^* a_n = 32/(\pi^2(n^2 - 4)^2)$ for odd n and $P_n = 0$ for even n.

14.7 The Semiclassical Theory of Radiation

Based on the considerations of matter in equilibrium with radiation Einstein postulated that the ratio of the probabilities for spontaneous emission, A, and stimulated emission, B, is a constant

$$\frac{A}{B} = \frac{\hbar\omega^3}{\pi^2 c^3}, \tag{14.72}$$

where ω is the frequency of the radiation. Equilibrium is possible and the laws of thermodynamics are satisfied only for the relation (14.72). In 1920, the coefficients A and B were calculated.

Consider two energy levels of an atomic system corresponding to the energies E_1 and E_2, $E_2 > E_1$. Let N_1 and N_2 represent the number of atoms per unit volume in levels 1 and 2, respectively. An atom in the lower energy level is excited to E_2 by absorbing radiation. This process can occur only in the presence of radiation. Such a process is known as *stimulated absorption* or simply *absorption*. The rate of absorption depends on the density $\rho(\omega)$, associated with the radiation field corresponding to the frequency $\omega = (E_2 - E_1)/\hbar$. The rate of absorption is proportional to N_1 and also to $\rho(\omega)$. Thus, the number of absorption per unit volume and per unit time is written as $N_{ab} = CN_1\rho(\omega)$, where C is the coefficient of proportionality.

On the other hand, when the atom is in an excited state, it can make a transition to a lower energy state by emitting electromagnetic radiation. There are two different types of emission process: *spontaneous emission* and *stimulated emission*. In the spontaneous emission, an atom in the excited state emits radiation in the absence of incident radiation. It is not stimulated by incident radiation but occurs spontaneously. The rate of spontaneous emission is proportional to the number of atoms in the excited state. Then the number of spontaneous emissions per unit volume per unit time is given by AN_2.

In the stimulated emission, an incident radiation of appropriate frequency induces an atom in an excited state to emit radiation. For example, a photon of appropriate frequency can cause emission of a photon with the same energy. This is responsible for the working of a laser. The rate of transition to the lower energy level is proportional to N_2 and $\rho(\omega)$. Thus, the number of stimulated emissions per unit volume and per unit time is $BN_2\rho(\omega)$. In recognition of Einstein's insight the quantities A, B and C are called *Einstein coefficients*. The coefficients A and B received important attention in the field of lasers.

At thermal equilibrium, the number of upward transitions must be equal to the number of downward transitions and so A, B and C are related as

$$CN_1\rho(\omega) = AN_2 + BN_2\rho(\omega) \tag{14.73}$$

From this equation we write

$$\rho(\omega) = \frac{A}{B\left(\frac{CN_1}{BN_2} - 1\right)}. \tag{14.74}$$

In statistical mechanics for an ensemble of systems in thermal equilibrium at temperature T, the average number of systems occupying a state of energy E is proportional to the Boltzmann factor $e^{-E/(k_B T)}$, where k_B is the Boltzmann constant. We have $N_1 \propto e^{-E_1/(k_B T)}$ and $N_2 \propto e^{-E_2/(k_B T)}$ and

$$\frac{N_1}{N_2} = e^{(E_2 - E_1)/(k_B T)} = e^{\hbar\omega/(k_B T)}. \tag{14.75}$$

Then Eq. (14.74) becomes

$$\rho(\omega) = \frac{A}{B\left[\frac{C}{B}e^{\hbar\omega/(k_B T)} - 1\right]}. \tag{14.76}$$

This distribution must agree with Planck distribution

$$\rho(\omega) = \frac{\hbar\omega^3}{\pi^2 c^3 \left[e^{\hbar\omega/(k_B T)} - 1\right]}. \tag{14.77}$$

Comparison of Eqs. (14.76) and (14.77) gives

$$B = C, \quad \frac{A}{B} = \frac{\hbar\omega^3}{\pi^2 c^3}. \tag{14.78}$$

Knowing A, B and C we are able to calculate spontaneous emission, stimulated emission and spontaneous absorption. But the Eqs. (14.78) are insufficient to calculate these constants because we have only two equations for three unknowns. The relation $B = C$ indicates that stimulated emission is as strong as absorption. That is, in the presence of incident radiation the probabilities of pumping of electrons into upper states and stimulating the already excited electron to undergo the downward transition are same. Thus, population inversion cannot occur in a two level system. The second equation in (14.78) gives the relation between stimulated and spontaneous emissions.

Suppose we consider an atom in a oscillating electric field $\mathcal{E}(t) = \hat{e}\mathcal{E}_0 \cos\omega t$. The field is switched-on at $t = 0$ and switched-off at $t = T$. $\hat{e}$ represents the unit vector along the direction of the applied field. The interaction energy of the electron with the applied field is given by $H^{(1)}(\mathbf{r}, t) = q\mathcal{E} \cdot \mathbf{r} = q\mathcal{E}_0(\hat{e} \cdot \mathbf{r}) \cos\omega t$, where q represents the magnitude of the electronic charge.

Let $H^{(0)}$ represents the Hamiltonian of the atomic system and ϕ_n denote the eigenfunctions of $H^{(0)}$ belonging to the energy $E_n = \hbar\omega_n$. Writing

$$\psi(\mathbf{r}, t) = \sum a_n(t) e^{-i\omega_n t} \phi_n(\mathbf{r}) \tag{14.79}$$

we obtain

$$i\hbar\dot{a}_f = \sum_n a_n(t) H^{(1)}_{fn}(t) e^{i\omega_{fn} t}, \tag{14.80a}$$

where

$$H^{(1)}_{fn}(t) = \int \phi_f^* H^{(1)} \phi_n \, \mathrm{d}r = q\mathcal{E}_0\hat{e} \cdot \frac{1}{2} \left[e^{i\omega t} + e^{-i\omega t} \right] \int \phi_f^* \mathbf{r}\phi_n \, \mathrm{d}r. \tag{14.80b}$$

We define $D_{fn} = q \int \phi_f^* \mathbf{r}\phi_n \, \mathrm{d}r$ and $\overline{D}_{fn} = \hat{e} \cdot D_{fn}$. Then

$$H^{(1)}_{fn}(t) = \frac{1}{2}\mathcal{E}_0 \left(e^{i\omega t} + e^{-i\omega t} \right) \overline{D}_{fn}. \tag{14.81}$$

The quantity $q \int \phi_f^* \mathbf{r}\phi_n \mathrm{d}r = \rho_{fn}$ is called the *dipole matrix element*. Equation (14.80a) is rewritten as

$$i\hbar\dot{a}_f = \frac{1}{2}\mathcal{E}_0 \sum_n \overline{D}_{fn} a_n(t) \left[e^{i(\omega_{fn}+\omega)t} + e^{i(\omega_{fn}-\omega)t} \right]. \tag{14.82}$$

We assume that at $t = 0$ the state of the atom is ϕ_i and hence $a_i(0) = 1$, $a_n(0) = 0$ for $n \neq i$. Then we have

$$i\hbar\dot{a}_f = \frac{1}{2}\mathcal{E}_0 \overline{D}_{fi} \left[e^{i(\omega_{fi}+\omega)t} + e^{i(\omega_{fi}-\omega)t} \right]. \tag{14.83}$$

Solving this equation we get

$$a_f(t) = -\frac{i\mathcal{E}_0 \overline{D}_{fi}}{2\hbar} \left[\frac{e^{i(\omega_{fi}+\omega)t}}{(\omega_{fi}+\omega)} + \frac{e^{i(\omega_{fi}-\omega)t}}{(\omega_{fi}-\omega)} \right]_0^T. \tag{14.84}$$

The transition may be emission or absorption. In the absorption process, $E_i < E_f$ and ω_{fi} is positive. The radiation field may contain any range of frequencies. However, since $\omega_{fi} - \omega < \omega < \omega_{fi} + \omega$, the term $\omega_{fi} - \omega$ would vanish or become very small. $\omega_{fi} - \omega \approx 0$ means $(E_f - E_i)/\hbar \approx \omega$ or $E_f = E_i + \hbar\omega$. Then the second term in (14.84) is very large compared to the first term. That is, the second term is the resonance term. We neglect the first term, the nonresonant term. The neglect of it is called *rotating wave approximation*, often noticed in magnetic resonance and laser theory. In the emission process, $E_i > E_f$ and ω_{fi} is negative. In this case the first term in Eq. (14.84) is dominant.

14.8   Concluding Remarks

In the previous chapter, we were concerned with the theoretical treatment of time-independent weak perturbation. Because the perturbation is time-independent, the quantum state of the system is time-independent. As the perturbation is weak its effect is only to shift the energy eigenvalues by a fraction. Therefore, we developed a theory to compute corrections to energy eigenvalues and eigenfunctions. In the case of time-dependent perturbations, the system undergoes transition from one state to another during the action of the perturbation. Further, once the perturbation is switched-off the system settles to one of the stationary states. Therefore, we are interested in knowing the probability for the transition from an initial state to another state. One can numerically solve the equation of the transition coefficient and compute transition probability when it is difficult to compute it theoretically. This is described in the last chapter. In the case of adiabatic perturbation where the potential changes slowly with time we obtained a formula for the transition probability for a transition between the initial and final states. When a time-independent perturbation is applied to a system suddenly at an instant of time, say, t_0 the Hamiltonian of the perturbed system is time-independent. For $t > t_0$ the perturbed system will be in one of the stationary states of it. E of nth state of the perturbed and unperturbed systems are different. Consequently, we are interested in finding the transition probability for the system to make a transition from an initial state i (state of the unperturbed system for $t < t_0$) to jth state of the perturbed system for $t > t_0$.

14.9   Bibliography

[1]   P.A.M. Dirac, *Proc. Roy. Soc. London A* 114:243, 1927.

[2]   E. Fermi, *Nuclear Physics.* Univ. of Chicago Press, Chicago, 1950.

[3]   M. Born and V.A. Fock, *Zeitschrift für Physik a Hadrons and Nuclei* 51:165, 1928; http://en.wikipedia.org/wiki/Adiabatic_approximation.

14.10   Exercises

14.1   Evaluate the transition amplitude up to the second-order for the constant perturbation $V(t) = \begin{cases} 0, & t < 0 \\ V_0, & t \geq 0. \end{cases}$

14.2   A harmonic oscillator potential is subjected to the perturbation $\lambda b x^2$ in the time between 0 to T. Obtain the selection rules for the transition from the initial state ϕ_i to ϕ_f in time T and the transition probabilities for the possible transitions.

14.3   If the perturbation added to a harmonic oscillator potential is $\lambda b x^3$ find the selection rules and the transition probabilities for the allowed transitions.

14.4   Assume that the Hamiltonian of a harmonic oscillator is perturbed by the anharmonic term $\lambda b x^4$ during the time interval 0 to T only. Obtain the selection rules for allowed transitions and the corresponding probabilities.

14.5 A one-dimensional harmonic oscillator, originally at rest is acted on by a force $F(t)$. Find the motion of the centre of the wave packet by first-order perturbation theory and compare it with the exact formula.

14.6 A one-dimensional linear harmonic oscillator is acted upon by the force $F(t) = \dfrac{F_0\tau/\omega}{\tau^2 + t^2}$, $-\infty < t < \infty$. At $t = -\infty$, the oscillator is in the ground state. Using the time-dependent perturbation theory to first-order, calculate the probability that the oscillator is found to be in the excited state at $t = \infty$.

14.7 At time $t = 0$ the infinite height potential $V(x) = 0$ for $0 < x < L$ and ∞ otherwise is perturbed by the additional term of the form $V_{\mathrm p}(x) = V_0$ for $L/4 < x < 3L/4$ and 0 otherwise. The perturbation is switched-off at $t = T$. The system is initially in the ground state ϕ_1. What is the probability of finding it in the state ϕ_3 after the time $t = T$?

14.8 A particle in a box potential of width L is perturbed by the term $V_0 \sin(\pi x/L)$ during the time 0 to T. Compute the probability for the transition from the ground state ϕ_1 to the excited state ϕ_3 in time T.

14.9 Assume that at $t = 0$ the infinite height potential of width L is perturbed by the term $V_{\mathrm p}(x) = V_0$ for $0 < x < L/2$ and 0 otherwise. The system is initially in the ground state ϕ_1. Calculate the transition probability P_{31} in first-order if the perturbation is switched-off after $t = T$.

14.10 At time $t = 0$ the infinite height potential is perturbed by the term

$$V_{\mathrm p}(x) = \begin{cases} 0, & 0 < x < L/2 \\ V_0(x - L/2), & L/2 < x < L. \end{cases}$$

The perturbation is switched-off at $t = T$. If the system is initially in the state ϕ_1 calculate the probability of finding it in the ϕ_3 state after time T.

14.11 During the time interval $t = 0$ to T the perturbation $V_{\mathrm p}(x) = V_0$ for $0 < x < L/4$ and 0 for $L/4 < x < L$ is added to the infinite height potential of width L. If the system is initially in its ground state ϕ_1, find the probability for finding the system in the ϕ_3 state for $t > T$.

14.12 A plane rotator is subjected to a constant electric field F' during $0 < t < T$ and then the field is switched-off. The normalized eigenfunctions of the unperturbed system are $\Phi_n = (1/\sqrt{2\pi})\mathrm e^{\mathrm i m\phi}$. Find the probability for the transition from the state l to f.

14.13 A plane rotator is subjected to an electric field $-F\mu\cos\phi$ during $t \in [0, T]$. The system is in the lth state at $t = 0$. The normalized eigenfunctions and energy eigenvalues are $\psi = (1/\sqrt{2\pi})\,\mathrm e^{\mathrm i m\phi}$, $E_m = \hbar^2 m^2/(8\pi^2 I)$. Find the probability for the transition from the state l to f. Obtain the selection rule for the transition.

14.14 A one-dimensional harmonic oscillator of charge to mass ratio e/m and spring constant k is in its ground state. An oscillating uniform electric field $E(t) = 2E_0\cos\omega_0 t$, $\omega_0^2 = k/m$ is applied for t seconds parallel to the motion of the oscillator. What is the probability that the oscillator is excited to the nth state given that $(\omega_{n0} - \omega_0)t \ll 1$?

14.15 Assume that an adiabatic perturbation of the form $H^{(1)} = W(x)\mathrm e^{\alpha t}$ is turned on slowly from $t = -\infty$. Obtain the expression for second-order transition amplitude. Also write the time-independent wave function up to second-order correction.

14.16 Assume that a harmonic oscillator in the ground state is subjected to the perturbation $V_p(x,t) = -\alpha x e^{-\mu t}$ from $t = 0$. Calculate the transition probability to a state ϕ_n in the limit $t \to \infty$.

14.17 Suppose that the Hamiltonian of a system changes as shown below:
$$H = \begin{cases} H_0, & \text{for } t < 0 \\ H_1, & \text{for } t > t_0 \\ H_i, & \text{for } 0 < t < t_0. \end{cases}$$
H_0, H_1 and H_i have complete set of energy eigenfunctions. Show that the error in the sudden approximation is proportional to t_0 for small t_0.

14.18 A harmonic oscillator is in its ground state at $t = -\infty$. Its potential is perturbed by the term $-\alpha x^2 e^{-\mu t^2}$ between $t = -\infty$ and ∞. Find the first-order transition probability to a state E_n.

14.19 A one-dimensional harmonic oscillator consisting of a mass m suspended from a spring with spring constant k is initially in its ground state. At $t = 0$ the upper end of the spring is suddenly raised to a distance d. This disturbance exists during the interval $0 < t < T$ and at $t = T$ the spring is suddenly brought to its original position. Obtain the transition amplitude for the system to be found in its first excited state for $t > T$.

14.20 A harmonic oscillator Hamiltonian is perturbed by the term $m\omega^2(b^2 + 2b)x^2/2$ suddenly at $t = 0$. For $t < 0$ the system is in the ground state. Obtain the energy eigenfunctions and eigenvalues of the perturbed system. Compute the probability of finding the system for $t > 0$ in the ground state of the perturbed system.

14.21 A one-dimensional harmonic oscillator has its spring constant k suddenly reduced by a factor of $1/2$. The oscillator is initially in its ground state. Find the probability for the oscillator to remain in the ground state after the perturbation.

14.22 A one-dimensional harmonic oscillator has its equilibrium point shifted suddenly from $x = 0$ to $x = a$. The oscillator is initially in its ground state. Find the probability for the oscillator to remain in the ground state after the perturbation.

14.23 At $t = 0$ a linear harmonic oscillator potential is suddenly subjected to the time-independent electric field so that the change in the Hamiltonian is $H^{(1)} = -eEx$. At $t = 0$ the system is in its ground state. Find the energy eigenvalues and eigenfunctions of the perturbed system by a suitable change of variable. Then calculate the probability of finding the system in the ground state of the new Hamiltonian.

14.24 For $t < 0$ a hydrogen-like atom with atomic number Z is in its ground state. At $t = 0$ its potential is suddenly perturbed by the additional term $V_p(r) = -1/r$. Determine the eigenstates of the perturbed system. Applying the sudden approximation theory compute the probability for finding the electron in the ground state of the perturbed system.

14.25 An electron which was initially in the hydrogen atom ground state (ψ_{100}, $Z = 1$) has undergone transition to the He atom ($Z = 2$) as a result of beta decay of the nucleus ($Z = 1 \to Z = 2$). Assume that the change in the Hamiltonian is fast compared to the characteristic atomic time so that sudden approximation theory is applied. Find the probability for the transition to the ground state of He.

15

Approximation Methods III: WKB and Asymptotic Methods

15.1 Introduction

The WKB method is due to Gregor Wentzel, Hendrik Anthony Kramers and Leon Nicolas Brillouin [1–3]. The method is applicable for slowly varying potentials, that is, the potentials that change not more than a few wavelengths. This corresponds to states of the systems with large kinetic energy and hence high quantum number. In other words $\hbar$ is small. Therefore, the approximation is known as semiclassical approximation. The method is useful for estimating eigenfunctions and tunnelling probabilities for smooth potentials or for potentials with only a few discontinuities. The WKB approximation is a physically intuitive method and makes many connections with intuition about waves and classical mechanics as well as with results from the early days of quantum theory as it leads to the Bohr quantization condition. The results obtained by the WKB technique are also obtained employing two other different methods, namely, the semiclassical path integral and the instanton procedure [4,5]. In the WKB method, solutions on either side of a turning point are matched with a third solution valid near the turning point. This difficulty is avoided in an asymptotic method of Keller [6]. We discuss the WKB and the asymptotic methods in this chapter.

15.2 Principle of WKB Method

The method essentially consists of three steps:

1. The eigenfunction is written as a series expansion in the powers of $\hbar$. The approximation breaks down at turning points $(E - V(x) = 0)$.

2. A solution valid near turning points is constructed.

3. The third step is the application of boundary conditions.

15.2.1 Asymptotic Solution

We assume the solution of the Schrödinger equation

$$\hbar^2 \psi'' + 2m(E - V(x))\psi = 0 \tag{15.1}$$

as

$$\psi(x) = A(x)\, e^{iS(x)/\hbar}, \tag{15.2}$$

where A and S are real functions. Substitution of Eq. (15.2) in Eq. (15.1) gives

$$\hbar^2 A'' + 2i\hbar A' S' + i\hbar A S'' - A S'^2 + 2m(E - V)A = 0. \tag{15.3}$$

DOI: 10.1201/9781003172178-15

Equating the real and imaginary parts of the above equation separately to zero we obtain

$$\hbar^2 A'' - AS'^2 + 2m(E - V)A = 0\,, \tag{15.4a}$$

$$2A'S' + AS'' = (A^2 S')' = 0\,. \tag{15.4b}$$

The solution of (15.4b) is $A = C/\sqrt{S'}$, where C is an arbitrary constant. Since Eq. (15.4a) is difficult solve exactly, assume a solution in series as

$$A = A_0 + \hbar^2 A_1 + \dots\,, \quad S = S_0 + \hbar^2 S_1 + \dots\,. \tag{15.5}$$

Using Eq. (15.5) in Eq. (15.4a) and equating the coefficient of $\hbar^0$ to zero we get

$$S_0' = \pm[2m(E - V)]^{1/2}\,. \tag{15.6}$$

$A_0 = C/\sqrt{S_0'}$. Next, setting the coefficient of $\hbar^2$ to zero in Eq. (15.4a) gives

$$A_0'' - 2S_0' S_1' A_0 = 0\,. \tag{15.7}$$

Defining $\hbar p(x) = [2m(E - V)]^{1/2}$ Eq. (15.6) is written as $S_0' = \pm p(x)$. Further, Eq. (15.1) now becomes $\psi'' + p^2(x)\psi = 0$. The solution of $S_0' = \pm p(x)$ is $S_0 = \pm \int^x p(x')\,\mathrm{d}x'$. Also, as $A_0 = C/\sqrt{S_0'}$ and $S_1' = A_0''/(2S_0' A_0)$ we obtain

$$A_0 = \frac{C}{\sqrt{p(x)}}\,, \quad S_1' = \frac{\sqrt{p}}{2S_0'}\left[\frac{1}{\sqrt{p}}\right]'' = \frac{\left[(S_0')^{-1/2}\right]''}{2\sqrt{S_0'}}\,. \tag{15.8}$$

In the lowest-order approximation, we consider only A_0 and S_0. Then

$$\psi(x) = A_0(x)\,\mathrm{e}^{\mathrm{i}S_0/\hbar} = \frac{C_1}{\sqrt{p}}\,\mathrm{e}^{\mathrm{i}y_1} + \frac{C_2}{\sqrt{p}}\,\mathrm{e}^{-\mathrm{i}y_1}\,, \tag{15.9}$$

where

$$y_1 = \frac{S_0}{\hbar} = \frac{1}{\hbar}\int_0^x p(x')\,\mathrm{d}x'\,. \tag{15.10}$$

Since $p(x) = \sqrt{2m(E - V)}$, y_1 is real if $V < E$ and imaginary if $V > E$. So, the solution of Eq. (15.9) is oscillating for the region $V < E$ while for $V > E$ it is an exponential form.

The above approximation is valid only if the series is convergent. That is, $\hbar^2 S_1 < S_0$ or $\left|\hbar^2 S_1/S_0\right| < 1$. If $\hbar^2 S_1/S_0$ is small then $\hbar^2 S_1'/S_0'$ will also be small. Therefore, the condition for good approximation is $\left|\hbar^2 S_1'/S_0'\right| \ll 1$. As the de Broglie wavelength is $\lambda = h/p$ we write $S_0' = p = h/\lambda$. Then the condition $\left|\hbar^2 S_1'/S_0'\right| \ll 1$ becomes

$$\left|\lambda'^2 - 2\lambda\lambda''\right| \ll 32\pi^2\,. \tag{15.11}$$

The WKB approximation is valid in the region of x where the condition (15.11) is satisfied. For the region nearer to x_0 where $E - V(x_0) = 0$, the $p(x_0)$ is zero. Thus, the wavelength $\lambda = h/p(x_0)$ is infinity and hence λ' and λ'' also become infinity. The point is that the condition (15.11) is not satisfied near the turning point x_0 and the solution (15.9) is not valid near x_0. At one side of the turning points (15.9) become oscillatory and on the other side, it would be exponential for a potential well. The exponential eigenfunction on one side must be connected to the oscillatory eigenfunction on the other side using connecting formulas.

15.2.2 Solution Near a Turning Point

The solutions of the Schrödinger Eq. (15.1) are regular and analytic. To obtain a solution like (15.9), the Eq. (15.1) is modified slightly so that an exact solution can be constructed.

Assume that $x = 0$ is the turning point. Writing the Taylor series expansion of $E - V$ or p^2 about $x_0 = 0$ as

$$p^2(x) = \rho x^n \left(1 + \alpha x + \beta x^2 + \ldots\right), \quad n > 0 \tag{15.12}$$

and considering the first term alone the wave equation becomes

$$\psi'' + \rho x^n \psi = 0. \tag{15.13}$$

Its solution is

$$\psi(x) = A\sqrt{y/p}\, J_m(y) + B\sqrt{y/p}\, J_{-m}(y), \tag{15.14}$$

where $m = 1/(n+2)$, $y = \int_0^x p(x')\,\mathrm{d}x'$ and $J_m(y)$ is the Bessel function of order m.

15.2.3 Connection at the Turning Point

The turning point $x = 0$ separates $V(x)$ into two region. Region-I is with $x < 0$ and $E > V(x)$. Region-II is with $x > 0$ where $E < V(x)$. Suppose a particle of energy E moves in the potential $V(x)$. Let it approach $x = 0$ from $x_1 < 0$. Classically at $x = 0$, E is $V(x = 0)$ and the kinetic energy is zero. Hence, the velocity is zero and the particle begins to turn back at $x = 0$. But in quantum mechanics because of tunnelling there is a possibility of a particle to be found in the region-II. For this reason, the region-I is called *classical region* and the region-II is called *nonclassical region*. The equations relating the WKB eigenfunctions in these two regions are called *connection formulas*.

In the regions-I and II, we write

$$y_1 = \frac{1}{\hbar} \int_x^0 p(x')\,\mathrm{d}x', \quad (\text{real}) \tag{15.15a}$$

$$y_2 = \frac{1}{\hbar} \int_0^x p(x')\,\mathrm{d}x', \quad (\text{imaginary}), \tag{15.15b}$$

respectively, so that y_1 and y_2 increase as x moves away from the turning point. If we assume the potential near $x > 0$ is linear then the solutions are

$$\psi_1(x) = A_1\sqrt{y_1/p}\, J_{1/3}(y_1) + B_1\sqrt{y_1/p}\, J_{-1/3}(y_1), \quad x < 0 \tag{15.16a}$$

$$\psi_2(x) = A_2\sqrt{y_2/p}\, I_{1/3}(y_2) + B_2\sqrt{y_2/p}\, I_{-1/3}(y_2), \quad x > 0 \tag{15.16b}$$

where I is the Bessel function of imaginary argument. To match ψ_1 and ψ_2 at $x = 0$ we write the solution near $x = 0$. For $x \approx 0$, p^2 is ρx. Then $y_1 \approx y_2 \approx (2/3)\rho^{1/2}|x|^{3/2}$. This condition together with $\psi_1 \approx \psi_2$ at $x = 0$ leads to $A_2 = -A_1$, $B_1 = B_2$.

15.2.4   Asymptotic Connection Formulas

The behaviours of ψ_1 and ψ_2 far from the turning points, that is at $\pm\infty$, is obtained by using the asymptotic forms of Bessel function. These are given by

$$\psi_1^{\text{osc}}(x \to -\infty) \quad \sim \quad \frac{\alpha}{\sqrt{p}} \sin(y_1 + \pi/4), \quad \alpha = \sqrt{2/\pi} \tag{15.17a}$$

$$\psi_2^{\text{osc}}(x \to -\infty) \quad \sim \quad \frac{\alpha}{\sqrt{p}} \cos(y_1 + \pi/4), \tag{15.17b}$$

$$\psi_1^{\text{exp}}(x \to \infty) \quad \sim \quad \frac{\alpha}{2\sqrt{|p|}} e^{-y_2}, \tag{15.17c}$$

$$\psi_2^{\text{exp}}(x \to \infty) \quad \sim \quad \frac{\alpha}{\sqrt{|p|}} e^{y_2}. \tag{15.17d}$$

Here ψ_i^{exp} and ψ_i^{osc} are approximations to the same eigenfunction. ψ_i^{exp} is the continuation into the nonclassical region of ψ_i^{osc}. Thus, the connection between the approximate forms of the eigenfunctions in the classical and nonclassical regions are given by

Classical(osc)	Nonclassical(exp)
$\dfrac{\alpha}{\sqrt{p}} \sin(y_1 + \pi/4)$	$\dfrac{\alpha}{2\sqrt{|p|}} e^{-y_2}$
$\dfrac{\alpha}{\sqrt{p}} \cos(y_1 + \pi/4)$	$\dfrac{\alpha}{\sqrt{|p|}} e^{y_2}.$

15.3   Applications of WKB Method

In the following we discuss a few applications of WKB method.

15.3.1   Energy Levels of a One-Dimensional Potential Well

For any energy level E of a one-dimensional potential well shown in Fig. 15.1, there are two turning points x_- and x_+ such that $E - V(x_-) = E - V(x_+) = 0$. Region-II $(x_- < x < x_+)$ is classically allowed where the eigenfunction is oscillatory. Regions-I and III are nonclassical with exponentially decreasing eigenfunctions. The regions-I, II and III where the WKB

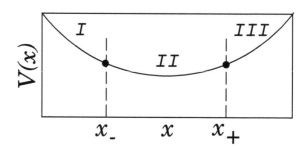

FIGURE 15.1

A one-dimensional potential with two turning points at x_- and x_+. The regions I and III are classically forbidden regions and the region II is the classically allowed region.

approximation is valid are defined by $x \le x_-$, $x_- \le x < x_+$ and $x \ge x_+$, respectively. The eigenfunctions in the regions-I and III are

$$\psi_{\mathrm{I}}(x) \approx \frac{A_1}{2\sqrt{|p|}} \, \mathrm{e}^{-(1/\hbar)\int_x^{x_-}|p(x)|\,\mathrm{d}x} \,, \tag{15.18a}$$

$$\psi_{\mathrm{III}}(x) \approx \frac{A_3}{2\sqrt{|p|}} \, \mathrm{e}^{-(1/\hbar)\int_{x_+}^{x}|p(x)|\,\mathrm{d}x} \,, \tag{15.18b}$$

where $p(x) = \sqrt{2m(E - V(x))}$. Using the connection formula the eigenfunction in the region-II is obtained as

$$\psi_{\mathrm{II}}(x) \approx \frac{A_1}{\sqrt{p}} \sin\left[\frac{1}{\hbar}\int_{x_-}^{x} p(x)\,\mathrm{d}x + \frac{\pi}{4}\right] \,, \tag{15.19a}$$

$$\approx \frac{A_3}{\sqrt{p}} \sin\left[\frac{1}{\hbar}\int_{x}^{x_+} p(x)\,\mathrm{d}x + \frac{\pi}{4}\right] \,. \tag{15.19b}$$

Equation (15.19a) is the solution to the right of x_- and (15.19b) gives the solution to the left of x_+. These two solutions must join smoothly in the interior region-II. So,

$$\frac{A_1}{\sqrt{p}} \sin\left[\frac{1}{\hbar}\int_{x_-}^{x} p(x)\,\mathrm{d}x + \frac{\pi}{4}\right] = \frac{A_3}{\sqrt{p}} \sin\left[\frac{1}{\hbar}\int_{x}^{x_+} p(x)\,\mathrm{d}x + \frac{\pi}{4}\right] \,. \tag{15.20}$$

We rewrite this equation as

$$A_1 \sin\left[\frac{1}{\hbar}\int_{x_-}^{x_+} p(x)\,\mathrm{d}x + \frac{\pi}{2} - \left(\frac{1}{\hbar}\int_{x}^{x_+} p(x)\,\mathrm{d}x + \frac{\pi}{4}\right)\right]$$

$$= A_3 \sin\left[\frac{1}{\hbar}\int_{x}^{x_+} p(x)\,\mathrm{d}x + \frac{\pi}{4}\right] \,. \tag{15.21}$$

Comparison of Eq. (15.21) with $\sin(m\pi - \theta) = (-1)^{m-1}\sin\theta$, $m = 1, 2, \ldots$ gives $A_3/A_1 = (-1)^m$ and

$$\frac{1}{\hbar}\int_{x_-}^{x_+} p(x)\,\mathrm{d}x + \frac{\pi}{2} = m\pi = (n+1)\pi\,, \quad n = 0, 1, \ldots$$

or

$$\phi = \left(n + \frac{1}{2}\right)\pi = \frac{1}{\hbar}\int_{x_-}^{x_+} p(x)\,\mathrm{d}x\,. \tag{15.22}$$

From (15.22) we write

$$2\int_{x_-}^{x_+} p(x)\,\mathrm{d}x = 2\left(n + \frac{1}{2}\right)\hbar\pi\,. \tag{15.23}$$

Now,

$$2\int_{x_-}^{x_+} p(x)\,\mathrm{d}x = \int_{x_-}^{x_+} p(x)\,\mathrm{d}x - \int_{x_+}^{x_-} p(x)\,\mathrm{d}x = \oint p(x)\,\mathrm{d}x\,, \tag{15.24}$$

where $\oint$ represents integration over a complete period of the particle. Substitution of Eq. (15.24) in Eq. (15.23) gives

$$\oint p(x)\,\mathrm{d}x = \left(n + \frac{1}{2}\right)h\,, \tag{15.25}$$

where $p(x) = \sqrt{2m(E - V(x))}$. Integration of the above equation gives E in terms of n and the Eq. (15.25) corresponds to Bohr–Sommerfeld quantum rule.

What is the form of the quantum condition Eq. (15.25) in momentum space? The quantum condition in momentum space is

$$\int_{p_1}^{p_2} \lambda(p)\,\mathrm{d}p = \left(n + \frac{1}{2}\right)\pi\hbar, \quad \lambda(p) = \sqrt{2m(E - V)}, \tag{15.26}$$

where p_1 and p_2 are the turning points.

Using Eq. (15.22) we obtain the quantized energy levels of certain systems.

Solved Problem 1:

Determine the energy levels of one-dimensional harmonic oscillator.

For the harmonic oscillator potential $V(x) = kx^2/2$ the turning points are obtained by equating $p(x) = 0$ in the equation $p(x) = \sqrt{2m(E - V(x))}$. We get the turning points as $x_- = -\sqrt{2E/k}$ and $x_+ = \sqrt{2E/k}$. Then ϕ is obtained as

$$
\begin{aligned}
\phi &= \frac{1}{\hbar}\int_{x_-}^{x_+} p(x)\,\mathrm{d}x \\[2mm]
&= \frac{1}{\hbar}\int_{x_-}^{x_+} \sqrt{2m\left(E - \frac{1}{2}kx^2\right)}\,\mathrm{d}x \\[2mm]
&= \frac{\sqrt{2mE}}{\hbar}\int_{x_-}^{x_+} \sqrt{1 - \frac{kx^2}{2E}}\,\mathrm{d}x \\[2mm]
&= \frac{2E}{\hbar}\sqrt{\frac{m}{k}}\int_{-1}^{1} \sqrt{1 - y^2}\,\mathrm{d}y \\[2mm]
&= \frac{2E}{\hbar\omega}\int_{-\pi/2}^{\pi/2} \cos^2 z\,\mathrm{d}z \\[2mm]
&= \frac{E\pi}{\hbar\omega}.
\end{aligned}
\tag{15.27}
$$

Equating $\phi = E\pi/(\hbar\omega)$ to $(n + 1/2)\pi$ (refer Eq. (15.22)) we get $E = (n + 1/2)\hbar\omega$.

Solved Problem 2:

Applying the WKB method obtain the energy eigenvalues of hydrogen atom.

For a spherically symmetrical potential $V(r)$ the radial equation for $R(r)$ when

$$\psi(r, \theta, \phi) = \frac{1}{r} R(r)\, Y_{lm}(\theta, \phi) \tag{15.28}$$

is

$$R'' + \frac{2m}{\hbar^2}\left[E - V(r) - \frac{\hbar^2 l(l + 1)}{2mr^2}\right]R = 0 \tag{15.29}$$

which has the general form as Eq. (15.1) but the boundary conditions are not at $r = \pm\infty$ but at $r = 0, \infty$. However, the change of variables $r = \mathrm{e}^s$, $R(r) = \mathrm{e}^{s/2}F(s)$ change the boundary conditions into $r = -\infty$ to ∞. Then Eq. (15.29) becomes

$$F'' + \frac{2m}{\hbar^2}\left[(E - V)\mathrm{e}^{2s} - \frac{\hbar^2}{2m}\left(l + \frac{1}{2}\right)^2\right]F = 0. \tag{15.30}$$

The bound state condition is $F = (n + 1/2)\pi$ where

$$F = \frac{1}{\hbar} \int_{s_-}^{s_+} \left[2m(E - V(e^s))e^{2s} - \hbar^2 \left(l + \frac{1}{2} \right)^2 \right]^{1/2} ds. \tag{15.31}$$

Substituting $s = \ln r$ the above equation is rewritten as

$$F = \frac{1}{\hbar} \int_a^b \left[2m(E - V(r)) - \frac{\hbar^2 \left(l + \frac{1}{2} \right)^2}{r^2} \right]^{1/2} dr, \tag{15.32}$$

where a and b are the turning points. We write

$$
\begin{aligned}
F &= \frac{1}{\hbar} \int_a^b \left[2m \left(E + \frac{e^2}{r} \right) - \frac{\hbar^2 \left(l + \frac{1}{2} \right)^2}{r^2} \right]^{1/2} dr \\
&= \frac{\sqrt{2m|E|}}{\hbar} \int_a^b \frac{1}{r} [-(r-a)(r-b)]^{1/2} \, dr
\end{aligned}
\tag{15.33}
$$

where $a + b = e^2/|E|$ and $ab = \hbar^2(l + 1/2)^2/(2m|E|)$. The value of the integral is $\pi \left(a + b - 2\sqrt{ab} \right)/2$, $0 < a < b$. Then

$$F = \frac{\pi \sqrt{m}\, e^2}{\hbar \sqrt{2|E|}} - \left(l + \frac{1}{2} \right) \pi. \tag{15.34}$$

Equating this expression to $(n + 1/2)\pi$ we obtain

$$E = -\frac{me^4}{2\hbar^2(n + l + 1)^2}. \tag{15.35}$$

This is the exact expression for the hydrogen energy eigenvalues provided the principle quantum number is now $n + l + 1$. We note that here n gives the number of nodes in the WKB (radial) eigenfunction while the principle quantum number counts both radial and spherical harmonic nodes and we start from 1 instead of 0.

15.3.2 Penetration Probability of a Barrier

The Schrödinger equation is solved to find the penetration of a square potential barrier. For a barrier of complicated shape, the Schrödinger equation cannot usually be solved exactly and the WKB approximation is often suitable for the problem. The WKB method is applied to obtain a transmission coefficient for penetration through a barrier. Consider a slowly varying barrier, for example, as shown in Fig. 15.2 and assume that the wave is incident from left and $E < V$. The region-III in Fig. 15.2 has transmitted wave alone. Therefore, the solution of the Schrödinger equation is

$$\psi_{\text{III}} = \frac{A}{\sqrt{p_2}} e^{iy_3} = \frac{A}{\sqrt{p_2}} (\cos y_3 + i \sin y_3), \tag{15.36}$$

where $y_3 = (1/\hbar) \int_b^x p_2(x) \, dx$ and $p_2(x) = \sqrt{2m(E - V)}$. Similarly,

$$
\begin{aligned}
\psi_{\text{II}} &= \frac{A}{\sqrt{p_1}} \left[\frac{1}{2} e^{-(1/\hbar) \int_x^b p_1 \, dx} - i e^{(1/\hbar) \int_x^b p_1 \, dx} \right] \\
&= \frac{A}{\sqrt{p_1}} \left[\frac{1}{2} e^{-(1/\hbar) \int_a^b p_1 \, dx} e^{(1/\hbar) \int_a^x p_1 \, dx} \right. \\
&\qquad \left. - i e^{(1/\hbar) \int_a^b p_1 \, dx} e^{-(1/\hbar) \int_a^x p_1 \, dx} \right], \tag{15.37}
\end{aligned}
$$

where $p_1(x) = \sqrt{2m(V - E)}$.

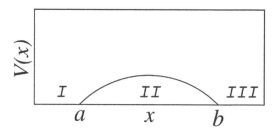

FIGURE 15.2
A slowly varying potential barrier.

In the region-I

$$\psi_{\mathrm{I}} = \frac{A}{\sqrt{p_2}} \left[-\frac{1}{2} e^{-(1/\hbar) \int_a^b p_1 \, dx} \sin \left(\frac{1}{\hbar} \int_x^a p_2 \, dx \right) \right.$$
$$\left. - \frac{2i}{\sqrt{p_2}} e^{(1/\hbar) \int_a^b p_1 \, dx} \cos \left(\frac{1}{\hbar} \int_x^a p_2 \, dx \right) \right]$$
$$= \psi_{\mathrm{I \, inc}} + \psi_{\mathrm{I \, ref}} \,, \tag{15.38}$$

where

$$\psi_{\mathrm{I \, inc}} = -\frac{iA}{\sqrt{p_2}} e^{(i/\hbar) \int_x^a p_2 \, dx} \left[e^{(1/\hbar) \int_a^b p_1 \, dx} - \frac{1}{4} e^{-(1/\hbar) \int_a^b p_1 \, dx} \right], \tag{15.39}$$

$$\psi_{\mathrm{I \, ref}} = -\frac{iA}{\sqrt{p_2}} e^{-(i/\hbar) \int_x^a p_2 \, dx} \left[e^{(1/\hbar) \int_a^b p_1 \, dx} + \frac{1}{4} e^{-(1/\hbar) \int_a^b p_1 \, dx} \right]. \tag{15.40}$$

The transmission coefficient is given by $T = $ transmitted flux/incident flux. The transmitted flux from Eq. (15.36a) is

$$\text{transmitted flux} = \frac{|A|^2}{p_2}. \tag{15.41a}$$

From Eq. (15.39) the incident flux is

$$\text{incident flux} = \frac{|A|^2}{p_2} \left| \exp \left(\frac{1}{\hbar} \int_a^b p_1 \, dx \right) - \frac{1}{4} \exp \left(-\frac{1}{\hbar} \int_a^b p_1 \, dx \right) \right|^2. \tag{15.41b}$$

Then

$$T = \left[\exp \left(\frac{1}{\hbar} \int_a^b p_1 \, dx \right) - \frac{1}{4} \exp \left(-\frac{1}{\hbar} \int_a^b p_1 \, dx \right) \right]^{-2}. \tag{15.42}$$

The WKB approximation is applicable if $(1/\hbar) \int_a^b p_1 \, dx \gg 1$. Then we neglect the second exponential term in Eq. (15.42). The resulting quantity is called *barrier penetration factor* and is

$$\tau = \exp \left(-\frac{2}{\hbar} \int_a^b p_1 \, dx \right) = \exp \left(-\frac{2}{\hbar} \int_a^b \sqrt{2m(E - V(x))} \, dx \right). \tag{15.43}$$

In the next two subsections, we apply the WKB method to α-decay and cold emission of electrons from metals.

15.3.3 Theory of α-Decay

The stability of a nucleus is explained by the short-range attractive force existing between the nucleons inside the nucleus. Inside the nuccleus two neutrons and protons combine and form the α-particle. Since α-particle is a highly stable configuration, it spends an appreciable fraction of time inside the nucleus before being emitted. How they escape from the forces binding the nucleus together was explained by Gamow and others as an example of quantum mechanical penetration of a potential barrier. The existence of such a barrier can be seen if we look at it from either side. An α-particle incident from outside will be repelled by the Coulomb field of the nucleus and on the other hand, it is attracted by a strong nuclear force inside the nucleus.

The potential outside the nucleus arises due to the Coulomb force between the daughter nuclei with atomic number Z plus the centrifugal force. The form of the potential inside the nucleus cannot be given in detail. The energy of the α-particle E is always less than the barrier height. So, the α-particle emission takes place only due to quantum mechanical tunnelling effect. The inner turning point is assumed as the radius of the daughter nucleus and the outer turning point r_0 is obtained from

$$E = \frac{2Ze^2}{r_0} + \frac{l(l+1)\hbar^2}{2\mu r_0^2} . \tag{15.44}$$

The radial part of the Schrödinger equation with $R = u(r)/r$ is

$$\frac{d^2u}{dr^2} = -\left[\frac{2\mu}{\hbar^2}\left(\frac{2Ze^2}{r} + \frac{l(l+1)}{2\mu r^2} - E\right)\right]u . \tag{15.45}$$

From (15.43) the barrier penetration factor is given by

$$T = \exp\left\{-\frac{2}{\hbar}\int_R^{r_0}\left[2\mu\left(\frac{2Ze^2}{r} + \frac{l(l+1)\hbar^2}{2\mu r^2} - E\right)\right]^{1/2}dr\right\} . \tag{15.46}$$

For $l = 0$, from (15.44), we get $r_0 = 2Ze^2/E$. Then the integral in (15.46) is

$$\int_R^{2Ze^2/E}\left[2\mu\left(\frac{2Ze^2}{r} - E\right)\right]^{1/2}dr = \sqrt{2mE}\int_R^{2Ze^2/E}\left(\frac{2Ze^2}{Er} - 1\right)^{1/2}dr . \tag{15.47}$$

Making the substitution $r = 2Ze^2\cos^2\theta/E$ we get

$$\int_R^{2Ze^2/E}\left[2\mu\left(\frac{2Ze^2}{r} - E\right)\right]^{1/2}dr = 2Ze^2\sqrt{\frac{2\mu}{E}}\left[\cos^{-1}s - s\sqrt{1 - s^2}\right] , \tag{15.48}$$

where $s = \sqrt{RE/(2Ze^2)}$. Hence,

$$T = \exp\left\{-\frac{4Ze^2}{\hbar}\sqrt{\frac{2\mu}{E}}\left[\cos^{-1}s - s\sqrt{1 - s^2}\right]\right\} . \tag{15.49}$$

If the speed of the α-particle inside the nucleus is v then the rate of hitting the barrier is v/R. Then the probability of escape per second (which is the inverse of mean-life time τ) is given by

$$\frac{1}{\tau} \approx \frac{v}{R}T . \tag{15.50}$$

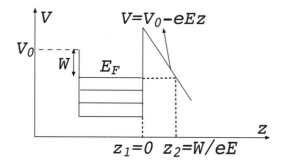

FIGURE 15.3
Potential for electron in a metal with the external electric field.

Since, $v \sim 10^9$ cm/sec and $R \sim 10^{-12}$ cm, we have $v/R \sim 10^{21}$ sec^{-1}. Further, for low energy α-particle, $R \ll 2Ze^2/E$. In the limit $RE/(2Ze^2) \to 0$, we get $\cos^{-1} s = \pi/2$. Therefore, we get from (15.49)

$$T \approx \exp\left[-\frac{2Ze^2}{\hbar}\pi\sqrt{2\mu/E}\right]. \tag{15.51}$$

Substituting (15.51) in (15.50), we write the decay constant as

$$\lambda = \frac{1}{\tau} = \frac{v}{R}\exp\left[-\frac{2Ze^2}{\hbar}\pi\sqrt{2\mu/E}\right]. \tag{15.52}$$

Taking logarithm on both sides, we get the Geiger and Nuttal empirical law

$$\ln\lambda = A - B\frac{Z}{\sqrt{E}}. \tag{15.53}$$

Experimentally, it has been found that a plot of $\ln\lambda$ against $1/\sqrt{E}$ gives fairly a straight-line. The effect of introducing the term $l(l+1)\hbar^2/(2mr^2)$ is not very great because even for $l = 5$ the value of τ changes by a factor not more than 10.

15.3.4   Cold Emission of Electrons From Metals

When a strong electric field is applied metals emit electrons even at room temperature. This phenomenon is called *cold emission*. In a metal, the electrons are filled up to the Fermi level as shown in Fig. 15.3. This is the required energy to remove an electron from the Fermi level. If an external field E is applied then the potential will vary as a function of the distance from the surface of the metal as shown in Fig. 15.3. The two classical turning points are $z_1 = 0$ and $z_2 = W/(eE)$. We have $V(z) = V_0 - eEz$ and

$$E - V(z) = V_0 - eEz - E_F = W - eEz. \tag{15.54}$$

The transmission probability for cold emission is given by

$$\begin{aligned}
T &= \exp\left[-\frac{2}{\hbar}\int_{x_1}^{x_2}\sqrt{2m(V(z) - E)}\,dz\right] \\
&= \exp\left[-\frac{4\sqrt{2mW^3}}{3\hbar eE}\right].
\end{aligned} \tag{15.55}$$

The electron penetrates the barrier due to the applied electric field.

15.4 WKB Quantization with Perturbation

Earlier we have seen the usefulness of a perturbation theory in obtaining correction to energy eigenvalues and eigenfunctions for perturbed systems whose unperturbed part is exactly solvable. In this section, we examine how the quantized energies, E_n, arising from the WKB quantization condition change when the exactly solvable potential is modified slightly. Starting from the WKB phase integral we obtain a formula for first-order correction to energy. We apply the method to some simple potentials and compare the result with perturbation theory.

15.4.1 Formula for First-Order Shift in Energy

Let us obtain a formula for first-order shift in energy [7]. We write the WKB quantization condition as

$$\frac{\sqrt{2m}}{\hbar} \int_{x_-}^{x_+} \sqrt{E_n - V(x)} \, dx = (n + c_l + c_r)\,\pi \,, \tag{15.56}$$

where $c_l = c_r = 1/4$ for linear functions, $c_l = c_r = 1/2$ for infinite walls and other choices for radial and Coulomb-type problems. Let us write $V(x) \to V(x) + \lambda V'(x)$, where $V'(x)$ is the perturbation part of the potential and the change in the quantized energies up to first-order correction as $E_n \to E_n^{(0)} + \lambda E_n^{(1)}$. Now, the integral in Eq. (15.56) takes the form

$$I_n(\lambda) = \int_{x_-(\lambda)}^{x_+(\lambda)} \sqrt{E_n^{(0)} - V(x) + \lambda \left(E_n^{(1)} - V'(x) \right)} \, dx \,. \tag{15.57}$$

Since $E_n^{(1)} - V'(x)$ is assumed to be small we write the Taylor series expansion of the right-side of Eq. (15.57). We obtain

$$I_n(\lambda) = I_n(0) + \lambda \frac{dI_n}{d\lambda} \bigg|_{\lambda=0} + \ldots \,, \tag{15.58a}$$

where

$$I_n(0) = \int_{x_-(0)}^{x_+(0)} \sqrt{E_n^{(0)} - V(x)} \, dx \,, \tag{15.58b}$$

$$\frac{dI_n}{d\lambda} \bigg|_{\lambda=0} = \frac{1}{2} \int_{x_-(0)}^{x_+(0)} \frac{\left(E_n^{(1)} - V'(x) \right)}{\sqrt{E_n^{(0)} - V(x)}} \, dx \,. \tag{15.58c}$$

Even though we have expanded about $\lambda = 0$, for a given value of n, I_n must be independent of λ since Eq. (15.56) is satisfied always. Therefore, any change in $V(x)$ and in the $x_\pm$ must be accompanied by a corresponding change in E_n so as to make $I_n(\lambda) = I_n(0)$. This implies that $dI_n/d\lambda|_{\lambda=0} = 0$. Then the substitution

$$v(x) = \sqrt{2 \left(E_n^{(0)} - V(x) \right) \big/ m} \tag{15.59}$$

in Eq. (15.58c) gives

$$E_n^{(1)} \int_{x_-(0)}^{x_+(0)} \frac{1}{v(x)} \, dx = \int_{x_-(0)}^{x_+(0)} \frac{V'(x)}{v(x)} \, dx \,. \tag{15.60}$$

Since $\int_{x_-(0)}^{x_+(0)} 1/v(x)\, \mathrm{d}x = \tau/2$, where τ is the period of the classical motion we have

$$E_n^{(1)} = \frac{2}{\tau} \int_{x_-(0)}^{x_+(0)} \frac{V'(x)}{v(x)}\, \mathrm{d}x \,. \qquad (15.61)$$

Further, $P_{\mathrm{CM}}^{(0)}(x) = 2/(\tau v(x))$. Therefore,

$$E_n^{(1)} = \int_{x_-(0)}^{x_+(0)} P_{\mathrm{CM}}^{(0)}(x) V'(x)\, \mathrm{d}x = \langle V'(x) \rangle \,. \qquad (15.62)$$

Solved Problem 3:

Find the first-order correction to the energy eigenvalues of the linear harmonic oscillator perturbed by $\lambda b x^2$.

For the unperturbed harmonic oscillator with $V(x) = kx^2/2 = m\omega^2 x^2/2$ the unperturbed energy eigenvalues obtained by solving the Schrödinger equation are $E_n^{(0)} = (n + 1/2)\hbar\omega$. The same result is obtained by WKB method. Now, we consider the effect of the additional quadratic potential $V'(x) = bx^2$. The perturbed potential is $V(x) = \frac{1}{2}m\omega^2 x^2 + \lambda b x^2$. By defining $\bar{\omega} = \omega\sqrt{1 + \dfrac{2b\lambda}{m\omega^2}}$ we write $V(x) = m\bar{\omega}^2 x^2/2$ so

$$E_n = \left(n + \frac{1}{2}\right)\hbar\omega \left(1 + \frac{2b\lambda}{m\omega^2}\right)^{1/2} = E_n^{(0)} \left(1 + \frac{\lambda b}{m\omega^2} + \dots\right) \qquad (15.63)$$

which is an exact result. For the unperturbed harmonic oscillator $P_{\mathrm{CM}}(x)$ is given by (refer Eq. (4.35)) $P_{\mathrm{CM}}(x) = 1/(\pi\sqrt{A_n^2 - x^2})$. Then

$$E_n^{(1)} = \frac{\lambda b}{\pi} \int_{-A_n}^{A_n} \frac{x^2}{\sqrt{A_n^2 - x^2}}\, \mathrm{d}x \,. \qquad (15.64)$$

Introducing the change of variable $x = A_n \sin\theta$ the above integral is easily evaluated. We obtain $E_n^{(1)} = b\lambda A_n^2/2$. The turning points of the unperturbed system are (obtained by equating $p = 0$ and $H = (n + 1/2)\hbar\omega$ in the unperturbed Hamiltonian) are $A_n = \pm[2\hbar(n + 1/2)/(m\omega)]^{1/2}$. Therefore,

$$E_n^{(1)} = \frac{\lambda b \hbar}{m\omega}\left(n + \frac{1}{2}\right) = \frac{\lambda b E_n^{(0)}}{m\omega^2} \qquad (15.65)$$

which is same as the first-order in λ in Eq. (15.63).

For the perturbation of the form $V'(x) = -eEx$ it is easy to check that the WKB first-order correction to E_n is 0. The same result is obtained earlier using time-independent perturbation theory.

15.5   An Asymptotic Method

The WKB method has an intrinsic defect making it inconvenient to use in many situations. In this method the different expressions for the solution must be used on each side of a turning point and further they have to be matched to a third expression, which is valid near the turning point. Consequently, in a problem with two turning points, five different expressions for the solution must be used and matched to one another. This defect can be overcome by using one spatially uniform asymptotic representation of the solution instead of the WKB five-part nonuniform representation. In this section, we present the treatment of the asymptotic method [6] for the harmonic oscillator potential which can be generalized to a general potential.

Let us consider the Schrödinger equation in the form

$$\psi_{xx} + (E' - V'(x))\psi = 0 \,, \tag{15.66}$$

where $E' = 2mE/\hbar^2$ and $V'(x) = 2mV(x)/\hbar^2$ and $V(x)$ is the potential of the system. For simplicity we set $\hbar = 1$, $m = 1$ and $\omega = 1$. For the harmonic oscillator $V'(x) = x^2/4$ and the Schrödinger equation is

$$\psi_{xx} + \left(E' - \frac{x^2}{4} \right)\psi = 0 \,. \tag{15.67}$$

The above equation admits bound state solution only for $E' = (n + 1/2)$. In this case

$$\psi(x) = e^{-x^2/4} H_n(x) \,. \tag{15.68}$$

For a general potential we write the solution of Eq. (15.66) as

$$\psi(x) = \phi_x^{-1/2} e^{-\phi_x^2/4} H_n(\phi_n(x)) \,. \tag{15.69}$$

We substitute Eq. (15.69) in Eq. (15.66) and use the result that Eq. (15.68) obeys Eq. (15.67) with $E' = n + 1/2$. Then Eq. (15.69) is found to satisfy Eq. (15.66) provided

$$\phi_x^2 \left(n + \frac{1}{2} - \frac{\phi^2}{4} \right) = E' - V' + \phi_x^{1/2} \left(\phi_x^{-1/2} \right)_{xx} \,. \tag{15.70}$$

We assume that $V'(x)$ changes very slowly in the distance $2\pi\sqrt{E'}$, which is the wavelength. Then the last term in Eq. (15.70) is negligible compared to the other terms. Now,

$$\phi_x^2 \left(n + \frac{1}{2} - \frac{\phi^2}{4} \right) = E' - V' \,. \tag{15.71}$$

Suppose $E' - V' = 0$ at x_- and x_+ with $x_- < x_+$. Then Eq. (15.71) gives $\phi^2 - 4(n+1/2) = 0$. Its roots are $\phi(x_\pm) = \pm 2\sqrt{n + 1/2}$. One of these conditions can be viewed as an initial condition and the other to determine E'. Integration of Eq. (15.71) from x_- to x gives

$$\int_{x_-}^x \phi_x \left(n + \frac{1}{2} - \frac{\phi^2}{4} \right)^{1/2} dx = \int_{x_-}^x \sqrt{E' - V'} \, dx$$

or

$$\left(n + \frac{1}{2} \right) \frac{\pi}{2} + \frac{\phi}{4} \left(n + \frac{1}{2} - \frac{\phi^2}{4} \right)^{1/2} + \left(n + \frac{1}{2} \right) \sin^{-1} \left(\frac{\phi}{2\sqrt{n + 1/2}} \right)$$

$$= \int_{x_-}^x \sqrt{E' - V'} \, dx \,. \tag{15.72}$$

Setting $x = x_+$ in the above equation we get

$$\left(n + \frac{1}{2} \right) \pi = \int_{x_-}^{x_+} \sqrt{E' - V'} \, dx \,. \tag{15.73}$$

The above result is same as the one given by the WKB method. But the present derivation is simpler. Further, the representation (15.69) for the wave function is valid for all x.

15.6 Concluding Remarks

A novel WKB approach for calculating the life-time of quasi-stationary states in the potential wells of the form $V(x) = P(x) - \mu Q(x)$ has been suggested [8]. An alternative formulation of WKB approximation is proposed [9]. An exact quantization rule of the form

$$\int_{x_a}^{x_b} K(x)\,\mathrm{d}x = (n+1)\pi + \int_{x_a}^{x_b} K'(x)\,[\phi(x)/\phi'(x)]\,\mathrm{d}x,\qquad(15.74)$$

where $K(x) = \sqrt{2m(E_V(x))}/\hbar$, x_a and x_b are two turning points determined by $E = V(x)$ and $\phi(x) = (1/\psi)\mathrm{d}\psi/\mathrm{d}x$ has been proposed [10]. Here $n+1$ is the number of the nodes of $\phi(x)$ in the region $E \geq V(x)$. The second term in (15.74) is found to be independent of the number of nodes of the wave function. Thus, it is sufficient to make use of the ground state in computing it. A proper quantization rule of the form $\int_{x_a}^{x_b} K(x)\,\mathrm{d}x - \int_{x_a}^{x_b} K_0(x)\,\mathrm{d}x = n\pi$, where n is the number of nodes of $\psi(x)$ has been proposed [11]. Using this rule all the bound state eigenvalues can be calculated from the ground state energy only. Evaluation of WKB integral for the special cases (i) n (radial quantum number) $\to \infty$ and l (azimuthal quantum number)-fixed and (ii) $l \to \infty$ and n-fixed have been discussed in refs. [12,13].

15.7 Bibliography

[1] G. Wentzel, *Z. Phys.* 38:518, 1926.

[2] H.A. Kramers, *Z. Phys.* 39:828, 1926.

[3] L. Brillouin, *Comptes Rendus* 183:24, 1926.

[4] C. Itzykson and J. Zuber, *Quantum Field Theory*. McGraw–Hill, New York, 1980.

[5] B.R. Holstein, *Am. J. Phys.* 56:338, 1988.

[6] J.B. Keller, *Am. J. Phys.* 54:546, 1986.

[7] R.W. Robinett, *Am. J. Phys.* 65:320, 1997.

[8] J. Zamastil, J. Cizek and L. Skala, *Phys. Rev. Lett.* 84: 5683, 2000.

[9] M.A. Vandyck, *Eur. J. Phys.* 12:112, 1991.

[10] Z.Q. Ma and B.W. Xu, *Europhys. Lett.* 69:685, 2005.

[11] W.C. Qiang and S.H. Dong, *Europhys. Lett.* 89:10003, 2010.

[12] G. Paiano and M. Pellicoro, *Nuovo Cimento B* 119:239, 2004.

[13] G. Paiano, *Europhys. Lett.* 83:30013, 2008.

15.8 Exercises

15.1 Show that the gradient of the phase S in $\psi(\mathbf{X}) = A(\mathbf{X})e^{iS(\mathbf{X})/\hbar}$ characterizes the probability flux.

15.2 What is the equation we get by applying the semiclassical approximations $\hbar \to 0$ and $\hbar \left| \nabla^2 S \right| \ll (\nabla S)^2$ to the time-dependent Schrödinger equation?

15.3 Using the Hamilton–Jacobi equation

$$\frac{1}{2m} \left| \nabla S(x,t) \right|^2 + V(x) + \frac{\partial S}{\partial t} = 0$$

show that the semiclassical approximation $\left| \nabla^2 S \right| \ll \left| \nabla S \right|^2$ for a stationary state is valid if the de Broglie wavelength is very small compared with the characteristic distance over which the potential varies appreciably (Hint: For stationary states, the Hamilton's principal function S and Hamilton's characteristic function W are related by $S = W(x) - Et$).

15.4 Given $\phi = (1/\hbar) \int_{x_1}^{x_2} p(x)\, \mathrm{d}x$ find $\partial \phi / \partial E$.

15.5 Using the WKB quantization rule find the eigenvalues of the anharmonic oscillator with the Hamiltonian $H = -(\hbar^2/2m)\mathrm{d}^2/\mathrm{d}x^2 + \lambda x^4$.

15.6 Use the WKB approximation to determine the bound state energies of a particle in the potential

$$V(x) = \begin{cases} \dfrac{V_0}{a}|x|, & |x| \le a \\ V_0 = \dfrac{1}{m}\left(\dfrac{h}{a}\right)^2, & |x| \ge a. \end{cases}$$

15.7 Use the WKB method to find the first-order correction to the energy of a harmonic oscillator perturbed by a small potential $V' = \lambda x^4$. Compare it with the result of the perturbation theory.

15.8 The nonrelativistic expansion of the energy of a free particle in one-dimension can be written in the form

$$E = \sqrt{p_x^2 c^2 + m^2 c^4} \approx mc^2 + \frac{1}{2m}p_x^2 - \frac{1}{8m^3 c^2}p_x^4.$$

Treating $-p_x^4/(8m^3 c^2)$ as a perturbation to the harmonic oscillator, estimate the correction to the energy using WKB method.

15.9 A neutron falling in a gravitational field and bouncing-off a horizontal mirror exhibits quantized energy levels. The potential of the problem is $V(z) = mgz$ for $z > 0$ and ∞ for $z < 0$. Applying the WKB method obtain the energy levels of the system.

15.10 For a particle in a box potential, $V(x) = 0$ for $0 < x < L$ and ∞ otherwise, determine the energy eigenvalues by WKB method.

16

Approximation Methods IV: Variational Approach

16.1 Introduction

In chapters 13 and 14 we focused on perturbation theories for time-independent and time-dependent perturbations. For time-independent perturbations we obtained first-order and second-order corrections to energy eigenvalues and eigenfunctions. For time-dependent perturbations we obtained the probability for a system to make a transition from an ith initial state to fth final state. We have also discussed how to study the effect of adiabatic and sudden perturbations. In the present chapter, we consider the variational method. The variational method can be used to construct an approximate ground state eigenfunction. In some cases the method can be used for finding approximately the energy of first few discrete states of a bound system. The perturbation theory uses a set of unperturbed states and the perturbation Hamiltonian $H^{(1)}$ to construct the eigenfunctions and eigenvalues through highly formalized approximations. In contrast, the variational method is based on a *guess* for the state – a trial state is used to determine the corresponding energy. Any well behaved trial state introduced will yield an appropriate energy, although some trial states will be better than other, but whatever the guess, we can always make systematic improvements. We can construct a reasonably accurate eigenfunction often by a physical insight and ingenuity.

16.2 Calculation of Ground State Energy

Let us briefly describe the basic idea of the variational method of calculating ground state energy.

16.2.1 Basic Principle

Let ϕ_n are the eigenfunctions of the eigenvalue equation

$$H\phi_n = E_n\phi_n, \quad n = 0, 1, \ldots. \tag{16.1}$$

DOI: 10.1201/9781003172178-16

An arbitrary wave function is written as $\psi = \sum_n a_n \phi_n$. Now, consider the integral $\int_{-\infty}^{\infty} \psi^* H \psi \, d\tau$. We obtain

$$\int_{-\infty}^{\infty} \psi^* H \psi \, d\tau = \int_{-\infty}^{\infty} \sum_m a_m^* \phi_m^* H \sum_n a_n \phi_n \, d\tau$$

$$= \sum_n |a_n|^2 E_n$$

$$\geq \sum_n |a_n|^2 E_0, \tag{16.2}$$

where E_0 is the ground state energy. For normalized ψ the requirement is $\sum_n |a_n|^2 = 1$. Then

$$E_0 \leq \int_{-\infty}^{\infty} \psi^* H \psi \, d\tau. \tag{16.3}$$

Equation (16.3) is known as *variational theorem*. For unnormalized ψ

$$E_0 \leq \int_{-\infty}^{\infty} \psi^* H \psi \, d\tau \Big/ \int_{-\infty}^{\infty} \psi^* \psi \, d\tau. \tag{16.4}$$

Equation (16.3) states that whatever be the form of wave function ψ the upper bound for the ground state energy E_0 is $\int_{-\infty}^{\infty} \psi^* H \psi \, d\tau$. Thus, we can choose a suitable trial function $\psi = \phi(c_i)$ which depends on the parameters c_i and determine the best upper bound for ground state energy by minimizing the expression $\int_{-\infty}^{\infty} \phi^* H \phi \, d\tau$ with respect to the parameters c_i.

The conditions for $E(c_i) = \int_{-\infty}^{\infty} \phi^*(c_i) H \phi(c_i) d\tau$ to be a mininum are

$$\frac{\partial E}{\partial c_i} = 0 \quad \text{and} \quad \frac{\partial^2 E}{\partial c_i^2} > 0. \tag{16.5}$$

The above conditions are used to determine the minimum value of E.

Solved Problem 1:

Determine the error in the ground state energy given by the variational method.

Suppose ϕ is a trial ground state eigenfunction while ϕ_0 is the true ground state eigenfunction. The difference between them is $\phi - \phi_0 = \lambda \Delta \phi$, where λ is a small constant and $\Delta \phi$ is orthogonal to ϕ_0. Now, E_ϕ is given by

$$E_\phi = \frac{\int_{-\infty}^{\infty} (\phi_0 + \lambda \Delta \phi)^* H(\phi_0 + \lambda \Delta \phi) \, d\tau}{\int_{-\infty}^{\infty} (\phi_0 + \lambda \Delta \phi)^* (\phi_0 + \lambda \Delta \phi) \, d\tau}. \tag{16.6}$$

We rewrite it as

$$E_\phi = \frac{\int_{-\infty}^{\infty} \phi_0^* H(\phi_0 + \lambda \Delta \phi) + \lambda (\Delta \phi)^* H(\phi_0 + \lambda \Delta \phi) \, d\tau}{\int_{-\infty}^{\infty} (\phi_0^* \phi_0 + \lambda \phi_0^* \Delta \phi + \lambda (\Delta \phi)^* \phi_0 + \lambda^2 (\Delta \phi)^* \Delta \phi) \, d\tau}. \tag{16.7}$$

That is,

$$
\begin{aligned}
E_\phi &= \frac{E_0 + \lambda^2 \int_{-\infty}^{\infty} (\Delta\phi)^* H \Delta\phi \, \mathrm{d}\tau}{1 + \lambda^2 \int_{-\infty}^{\infty} (\Delta\phi)^* \Delta\phi \, \mathrm{d}\tau} \\
&= \left[E_0 + \lambda^2 \int_{-\infty}^{\infty} (\Delta\phi)^* H \Delta\phi \, \mathrm{d}\tau \right] \left[1 - \lambda^2 E_0 \int_{-\infty}^{\infty} (\Delta\phi)^* \Delta\phi \, \mathrm{d}\tau \right] \\
&= E_0 + O(\lambda^2),
\end{aligned}
\tag{16.8}
$$

where we have neglected terms containing λ^4 and higher powers of λ. Thus, the error of the variational energy is of the order of λ^2.

16.2.2   Applications

In the following we apply the variational method to some interesting quantum systems.

16.2.3   Harmonic Oscillator

We calculate the ground state energy of the harmonic oscillator with

$$
H = -\frac{\hbar^2}{2m} \frac{\mathrm{d}^2}{\mathrm{d}x^2} + \frac{1}{2} m\omega^2 x^2.
\tag{16.9}
$$

We choose the trial eigenfunction as $\phi = A e^{-\alpha x^2}$. First, we normalize the ϕ:

$$
1 = \int_{-\infty}^{\infty} \phi^* \phi \, \mathrm{d}x = A^2 \int_{-\infty}^{\infty} e^{-2\alpha x^2} \, \mathrm{d}x = 2A^2 \int_{0}^{\infty} e^{-2\alpha x^2} \, \mathrm{d}x.
\tag{16.10}
$$

Using the integral relation $\int_0^{\infty} e^{-\beta x^2} \, \mathrm{d}x = \sqrt{\pi/(4\beta)}$ we obtain $A = (2\alpha/\pi)^{1/4}$.
    Next,

$$
\begin{aligned}
E &= \int_{-\infty}^{\infty} \phi^* H \phi \, \mathrm{d}x \\
&= A^2 \int_{-\infty}^{\infty} e^{-\alpha x^2} \left(-\frac{\hbar^2}{2m} \frac{\mathrm{d}^2}{\mathrm{d}x^2} + \frac{1}{2} m\omega^2 x^2 \right) e^{-\alpha x^2} \, \mathrm{d}x \\
&= \frac{2\hbar^2 A^2 \alpha}{m} \int_0^{\infty} (1 - 2\alpha x^2) e^{-2\alpha x^2} \, \mathrm{d}x + m\omega^2 A^2 \int_0^{\infty} x^2 e^{-2\alpha x^2} \, \mathrm{d}x.
\end{aligned}
\tag{16.11}
$$

Using the result

$$
\int_0^{\infty} x^{2n} e^{-\beta x^2} \, \mathrm{d}x = \frac{(2n-1)!! \, \sqrt{\pi/\beta}}{2^{n+1} \beta^n}.
\tag{16.12}
$$

Using the identity $(2n-1)!! = (2n)!/(2^n n!)$ we get

$$
E = \frac{\hbar^2 \alpha}{2m} + \frac{m\omega^2}{8\alpha}.
\tag{16.13}
$$

The minimization condition

$$
0 = \frac{\partial E}{\partial \alpha} = \frac{\hbar^2}{2m} - \frac{m\omega^2}{8\alpha^2}
\tag{16.14}
$$

gives $\alpha = m\omega/(2\hbar)$. Further, we note that $\partial^2 E/\partial \alpha^2 = m\omega^2/(4\alpha^3) > 0$. Substitution of $\alpha = m\omega/(2\hbar)$ in Eq. (16.13) gives $E = \hbar\omega/2$, which is the well-known expression for zero-point energy of harmonic oscillator. The obtained value of ground state energy is exact because the trial eigenfunction chosen is the correct functional form with the adjustable parameter α.

16.2.4 Hydrogen Atom

The Hamiltonian of the hydrogen atom is $H = -(\hbar^2/2m)\nabla^2 - e^2/r$. We choose the trial eigenfunction as $\phi = Ae^{-ar}$. The normalization of ϕ gives

$$
\begin{aligned}
1 &= \int_0^\infty \int_0^\pi \int_0^{2\pi} \phi^* \phi\, r^2 \sin\theta\, dr\, d\theta\, d\phi \\
&= 4\pi A^2 \int_0^\infty r^2 e^{-2ar}\, dr .
\end{aligned}
\tag{16.15}
$$

Using the relation $\int_0^\infty x^n e^{-\beta x}\, dx = n!/\beta^{n+1}$ we get $A = \sqrt{a^3/\pi}$.

Next, we calculate E, $\int \phi^* H\phi\, d\tau$. Since ϕ is a function of r only, it is sufficient to consider the radial component of H which is given by

$$
H_{\rm rad} = -\frac{\hbar^2}{2m}\frac{1}{r^2}\frac{\partial}{\partial r}\left(r^2\frac{\partial}{\partial r}\right) - \frac{e^2}{r} = H_1 + H_2 .
\tag{16.16}
$$

We write $E = \int \phi^*(H_1 + H_2)\phi\, d\tau = E_1 + E_2$. We obtain

$$
\begin{aligned}
E_1 &= A^2 \int_0^\infty \int_0^\pi \int_0^{2\pi} e^{-ar}\left(H_1 e^{-ar}\right) r^2 \sin\theta\, dr\, d\theta\, d\phi \\
&= \frac{2\pi\hbar^2 A^2 a^2}{m}\int_0^\infty e^{-ar}\left(2re^{-ar} - ar^2 e^{-ar}\right) dr \\
&= \frac{\pi\hbar^2 A^2}{2ma}
\end{aligned}
\tag{16.17a}
$$

and

$$
E_2 = \int \phi^* H_2 \phi\, d\tau = -4\pi A^2 e^2 \int_0^\infty r e^{-2ar}\, dr = -\frac{\pi A^2 e^2}{a^2} .
\tag{16.17b}
$$

Then $E = E_1 + E_2 = \hbar^2 a^2/(2m) - ae^2$. The minimization condition $\partial E/\partial a = 0$ yields $a = e^2 m/\hbar^2$. Substituting this value for a in the above expression for E we get $E = -me^4/(2\hbar^2)$ which is the exact ground state energy.

16.2.5 Helium Atom

The helium atom consists of a nucleus of charge $2e$ and two electrons each of charge $-e$ at distances r_1 and r_2 from the nucleus. The Hamiltonian is

$$
H = -\frac{\hbar^2}{2m}\left(\nabla_1^2 + \nabla_2^2\right) - \frac{2e^2}{r_1} - \frac{2e^2}{r_2} + \frac{e^2}{r_{12}} .
\tag{16.18}
$$

If e^2/r_{12} is neglected then Eq. (16.18) simply represents Hamiltonian of two hydrogen atoms whose ground state energy is $E_1 = -Z^2\beta$, where $Z = 1$, $\beta = me^4/2\hbar^2$. Therefore, we choose the trial eigenfunction as

$$
\phi = Ae^{-Z'(r_1+r_2)/a_0} .
\tag{16.19}
$$

The normalization condition $1 = \int \phi^*\phi d\tau$ becomes

$$
\begin{aligned}
1 &= A^2 \int_{r_1,r_2}\int_{\theta_1,\theta_2}\int_{\phi_1,\phi_2} e^{-2Z'(r_1+r_2)/a_0} r_1^2 \sin\theta_1\, dr_1\, d\theta_1\, d\phi_1 \\
&\qquad \times r_2^2 \sin\theta_2\, dr_2\, d\theta_2\, d\phi_2 .
\end{aligned}
\tag{16.20}
$$

The result is $A = Z'^3/(\pi a_0^3)$. We proceed to calculate E.

The eigenfunction ϕ given by Eq. (16.19) is the ground state energy of hydrogen-like atom with charge $Z'e$. The corresponding ground state energy is $E_1 = -Z'^2\beta$. Therefore, we have

$$\left(-\frac{\hbar^2}{2m}\nabla_1^2 - \frac{Z'^2 e^2}{r_1}\right)\phi = E_1\phi = -Z'^2\beta\phi. \tag{16.21}$$

That is,

$$-\frac{\hbar^2}{2m}\nabla_1^2 = -Z'^2\beta + \frac{Z'e^2}{r_1}. \tag{16.22a}$$

Similarly,

$$-\frac{\hbar^2}{2m}\nabla_2^2 = -Z'^2\beta + \frac{Z'e^2}{r_2}. \tag{16.22b}$$

Then

$$-\frac{\hbar^2}{2m}\left(\nabla_1^2 + \nabla_2^2\right) = -2Z'^2\beta + Z'e^2\left(\frac{1}{r_1} + \frac{1}{r_2}\right). \tag{16.23}$$

The Hamiltonian, Eq. (16.18), is rewritten as $H = H_1 + H_2 + H_3$, where

$$H_1 = -2Z'^2\beta, \quad H_2 = e^2(Z'-2)\left(\frac{1}{r_1} + \frac{1}{r_2}\right), \quad H_3 = \frac{e^2}{r_{12}}. \tag{16.24}$$

The energy E is given by

$$E = \int \phi^*(H_1 + H_2 + H_3)\phi\, d\tau. \tag{16.25}$$

We obtain

$$\int \phi^* H_1 \phi\, d\tau = -2Z'^2\beta, \quad \int \phi^* H_2 \phi\, d\tau = 4Z'(Z'-2)\beta, \tag{16.26a}$$

$$\int \phi^* H_3 \phi\, d\tau = \frac{5}{4}Z'\beta. \tag{16.26b}$$

Then

$$E = 2\beta\left(Z'^2 - \frac{27}{8}Z'\right). \tag{16.27}$$

The minimization condition $\partial E/\partial Z' = 0$ gives $Z' = 27/16$. With this value of Z' the ground state energy is obtained as

$$E = -2\beta\left(\frac{27}{16}\right)^2 = -\frac{me^4}{\hbar^2}\left(\frac{27}{16}\right)^2, \tag{16.28}$$

whereas the exact value of energy is $E_{\text{exact}} = -2^2 me^4/\hbar^2$.

16.3 Trial Eigenfunctions for Excited States

Now, we consider the excited states.

16.3.1 Trial Eigenfunctions Orthogonal to All Lower Levels

The variational method can also be applied to compute the excited state energy values by choosing the trial eigenfunctions to be orthogonal to all the lower state eigenfunctions. For example, if a trial eigenfunction

$$\psi_t = \psi - \phi_0 \int \phi_0^* \psi \, d\tau \tag{16.29}$$

is chosen then

$$\int \phi_0^* \psi_t \, d\tau = \int \phi_0^* \psi \, d\tau - \int \phi_0^* \phi_0 \, d\tau \int \phi_0^* \psi \, d\tau. \tag{16.30}$$

Since $\int \phi_0^* \phi_0 \, d\tau = 1$ we get $\int \phi_0^* \psi_t \, d\tau = 0$. ψ_t is orthogonal to the ground state eigenfunctions. Hence, the expansion $\psi_t = \sum a_n \phi_n$ will not contain a_0. We obtain

$$\langle E_t \rangle = \int \psi_t^* H \psi_t \, d\tau = \sum_{n=1}^{\infty} |a_n|^2 E_n. \tag{16.31}$$

Therefore,

$$
\begin{aligned}
\langle E_t \rangle &\geq \sum_{n=1}^{\infty} |a_n|^2 E_1 \\
&\geq E_1.
\end{aligned}
\tag{16.32}
$$

Therefore, the trial wave function ψ_t gives an upper limit for the first excited state. Extending the above, we can prove that if ψ_t is orthogonal to ϕ_i, $i = 0, 1, \ldots, n$ then ψ_t will give an upper limit of the $(n+1)$th energy eigenvalue.

16.3.2 Finite Basis Set Expansion

A simple procedure for obtaining approximate excited state energies is to expand the trial wave function ψ using states taken from a known complete set $\{\phi_n\}$. The trial wave function is chosen as linear combination of known functions as

$$\psi = \sum_{i=1}^{r} a_i \phi_i, \tag{16.33}$$

where a_i's are variational parameters. The ϕ_i may not be normalized or mutually orthogonal. The energy is

$$E = \int_{-\infty}^{\infty} \psi^* H \psi \, d\tau \Big/ \int_{-\infty}^{\infty} \psi^* \psi \, d\tau. \tag{16.34}$$

Substitution of (16.33) in (16.34) yields

$$E = \frac{\sum_{i=1}^{r} \sum_{j=1}^{r} a_i^* a_j H_{ij}}{\sum_{i=1}^{r} \sum_{j=1}^{r} a_i^* a_j S_{ij}}, \tag{16.35a}$$

where

$$H_{ij} = \int_{-\infty}^{\infty} \phi_i^* H \phi_j \, d\tau, \quad S_{ij} = \int_{-\infty}^{\infty} \phi_i^* \phi_j \, d\tau. \tag{16.35b}$$

From the above equation we write

$$\sum_{i=1}^{r}\sum_{j=1}^{r} Ea_i^* a_j S_{ij} - \sum_{i=1}^{r}\sum_{j=1}^{r} a_i^* a_j H_{ij} = 0. \qquad (16.36)$$

Differentiating the above equation with respect to a_i^* and substituting $\partial E/\partial a_i^* = 0$ we obtain

$$\sum_{j=1}^{r} (H_{ij} - ES_{ij})\, a_j = 0, \quad i = 1, 2, \ldots, r. \qquad (16.37)$$

Differentiation of Eq. (16.36) with respect to a_i and using $\partial E/\partial a_i = 0$ gives the complex conjugate of Eq. (16.37). Equations (16.37) form a set of r homogeneous linear equations for the r unknowns a_i. For nontrivial solution, the determinant of the coefficients matrix of Eq. (16.37) must vanish. That is,

$$\begin{vmatrix} H_{11} - ES_{11} & H_{12} - ES_{12} & \cdots & H_{1r} - ES_{1r} \\ H_{21} - ES_{21} & H_{22} - ES_{22} & \cdots & H_{2r} - ES_{2r} \\ \vdots & & & \vdots \\ H_{r1} - ES_{r1} & H_{r2} - ES_{r2} & \cdots & H_{rr} - ES_{rr} \end{vmatrix} = 0. \qquad (16.38)$$

This will give an algebraic equation of degree r in E. If the r roots of E are arranged in increasing order $E_0, E_1, \ldots, E_{r-1}$ then E_0 will be the ground state energy, E_1 will be the first excited state energy and so on. Substituting these roots in Eq. (16.37) and solving it we are able determine a_j corresponding to each roots. These a_j when substituted in Eq. (16.33) will give the corresponding energy eigenfunction. If the basis chosen as an orthonormal basis then $S_{ij} = \delta_{ij}$. Then Eq. (16.38) will become

$$\begin{vmatrix} H_{11} - E & H_{12} & \cdots & H_{1r} \\ H_{21} & H_{22} - E & \cdots & H_{2r} \\ \vdots & & & \vdots \\ H_{r1} & H_{r2} & \cdots & H_{rr} - E \end{vmatrix} = 0. \qquad (16.39)$$

The accuracy and convergence of the energy values are determined by including one more function ϕ_{r+1} to the basis and solving the resulting $(r+1)$ equations. If $|E_n(r) - E_n(r+1)|$ is very small then the convergence is very good. The basis set is expanded till good convergence is realized.

Although any well-behaved set of basis states may be used to perform the diagonalization of H, one set is noteworthy. If H is partitioned as $H_0 + H_1$, where H_1 need not be weak, then the eigenfunctions of H_0 can be chosen as the basis if they are known.

Solved Problem 2:

If the first $n-1$ eigenfunctions of a particular Hamiltonian are known, write a formal expression for a variational method trial function that could be used to get the upper limit on the nth energy level.

Let $u_1, u_2, \ldots, u_{n-1}$ are known eigenfunctions. If ϕ is any arbitrary trial function then using the Schmidt orthogonalization process, a trial eigenfunction that could be used to get the upper limit on the nth energy level can be expressed as

$$\psi = \phi - \sum_{i=1}^{n-1} u_i \int_{-\infty}^{\infty} u_i^* \phi\, d\tau. \qquad (16.40)$$

Now,

$$
\begin{aligned}
\int u_j^* \psi \, d\tau &= \int u_j^* \phi \, d\tau - \sum_{i=1}^{n-1} \int u_j^* u_i \, d\tau \int u_i^* \phi \, d\tau , \quad j \le n-1 \\
&= \int u_j^* \phi \, d\tau - \sum_{i=1}^{n-1} \delta_{ji} \int u_i^* \phi \, d\tau \\
&= \int u_j^* \phi \, d\tau - \int u_j^* \phi \, d\tau \\
&= 0 .
\end{aligned}
\tag{16.41}
$$

Since ψ is orthogonal to the first $n-1$ eigenfunctions, it can be used as a trial function for the n energy level.

16.4   Application to Hydrogen Molecule

As an example for the application of the linear combination method we shall estimate the ground state energy of the hydrogen molecule. Consider the coordinate system of the hydrogen molecule as shown in Fig. 16.1.

We assume that the nuclei of the two hydrogen atoms A and B are fixed in space at a distance $\mathbf{R}$ apart. The position of the first electron is given by the position vector $\mathbf{r}_{1A}$ with respect to A and that of the second electron is given by $\mathbf{r}_{2B}$ with respect to the second nuclei B. The Hamiltonian for the hydrogen molecule is given by

$$
H = H^{(0)} + H^{(1)} ,
\tag{16.42a}
$$

where

$$
H^{(0)} = \left(-\frac{\hbar^2}{2\mu} \nabla_1^2 - \frac{e^2}{r_{1A}} \right) + \left(-\frac{\hbar^2}{2\mu} \nabla_2^2 - \frac{e^2}{r_{2B}} \right) ,
\tag{16.42b}
$$

$$
H^{(1)} = \frac{e^2}{R} + \frac{e^2}{r_{12}} - \frac{e^2}{r_{1B}} - \frac{e^2}{r_{2A}} .
\tag{16.42c}
$$

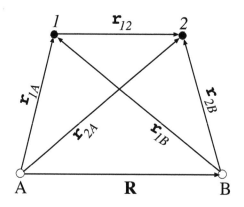

FIGURE 16.1
Coordinate system of hydrogen molecule.

If the interaction between the two atoms ($H^{(1)}$) is ignored then the ground state eigenfunction can be either $\phi_1 = u_A(1)u_B(2)$ or $\phi_2 = u_A(2)u_B(1)$ when the second electron is considered as belonging to the nuclei A and the first electron is belonging to the nuclei B where

$$u_A(1) = \left(\frac{1}{\pi a_0^3}\right)^{1/2} e^{-r_{1A}/a_0} \tag{16.43}$$

and so on. We have

$$H^{(0)}\phi_1 = 2E_0\phi_1, \quad H^{(0)}\phi_2 = 2E_0\phi_2, \tag{16.44}$$

where $E_0 = -e^2/(2a_0)$. Construct a trial eigenfunction as a linear combination of ϕ_1 and ϕ_2 as

$$\psi = a_1\phi_1 + a_2\phi_2. \tag{16.45}$$

From Eq. (16.45) the determinantal equation is

$$\begin{vmatrix} H_{11} - ES_{11} & H_{12} - ES_{12} \\ H_{21} - ES_{21} & H_{22} - ES_{22} \end{vmatrix} = 0, \tag{16.46}$$

where

$$H_{ij} = \int\int \phi_i^* H \phi_j \, d\tau_1 \, d\tau_2, \quad i,j = 1,2 \tag{16.47a}$$

$$S_{ij} = \int\int \phi_i^* \phi_j \, d\tau_1 \, d\tau_2, \quad i,j = 1,2. \tag{16.47b}$$

It can be proved that

$$H_{11} = H_{22}, \quad H_{12} = H_{21}, \quad S_{11} = S_{22} = 1, \quad S_{12} = S_{21}. \tag{16.48}$$

Using (16.48) in (16.46) we get

$$\begin{vmatrix} H_{11} - E & H_{12} - ES_{12} \\ H_{12} - ES_{12} & H_{11} - E \end{vmatrix} = 0. \tag{16.49}$$

The roots of E are

$$E_S = \frac{H_{11} + H_{12}}{1 + S_{12}}, \quad E_A = \frac{H_{11} - H_{12}}{1 - S_{12}}. \tag{16.50}$$

From the homogeneous equation

$$\begin{pmatrix} H_{11} - E & H_{12} - ES_{12} \\ H_{12} - ES_{12} & H_{22} - E \end{pmatrix} \begin{pmatrix} a_1 \\ a_2 \end{pmatrix} = 0 \tag{16.51}$$

we get the normalized eigenvectors as

$$\frac{1}{\sqrt{2}}\begin{pmatrix} 1 \\ 1 \end{pmatrix} \text{ and } \frac{1}{\sqrt{2}}\begin{pmatrix} 1 \\ -1 \end{pmatrix} \tag{16.52}$$

for E_S and E_A, respectively. Then

$$\psi_S = \frac{1}{\sqrt{2}}(\phi_1 + \phi_2), \quad \psi_A = \frac{1}{\sqrt{2}}(\phi_1 - \phi_2). \tag{16.53}$$

The exchange of the electrons changes $\phi_1 \to \phi_2$ and $\phi_2 \to \phi_1$ and hence $\psi_S \to \psi_S$ and $\psi_A \to -\psi_A$. ψ_S is a symmetric eigenfunction and ψ_A is antisymmetric. We have not considered the spin part of the eigenfunction as H does not have spin operators. Since the

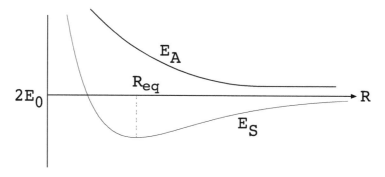

FIGURE 16.2
Energy of the hydrogen molecule as a function of the interatomic distance.

total eigenfunction of fermions must be antisymmetric, ψ_S must be multiplied by the anti-symmetric spin function (singlet state) and ψ_A must be multiplied by the symmetric spin functions (triplet-state).

We obtain

$$S_{12} = \int u_A^*(1)u_B(1)\,d\tau_1 \int u_B^*(2)u_A(2)\,d\tau_2 = S^2\,, \qquad (16.54)$$

where $S = \int u_A^*(r)u_B(r)\,d\tau$. Next,

$$H_{11} = \int\int \phi_1^*\left(H^{(0)} + H^{(1)}\right)\phi_1\,d\tau_1\,d\tau_2 = 2E_0 + k\,, \qquad (16.55a)$$

where

$$k = \int\int e^2\,|u_A(1)|^2\,|u_B(2)|^2\left[-\frac{1}{r_{2A}} - \frac{1}{r_{1B}} + \frac{1}{R} + \frac{1}{r_{12}}\right]d\tau_1\,d\tau_2, \qquad (16.55b)$$

and

$$H_{12} = 2E_0 S^2 + J\,, \qquad (16.56a)$$

where

$$\begin{aligned}
J &= \int\int e^2\,(u_A(1)u_B(1))\,(u_A(2)u_B(2)) \\
&\quad \times \left[-\frac{1}{r_{2A}} - \frac{1}{r_{1B}} + \frac{1}{R} + \frac{1}{r_{12}}\right]d\tau_1\,d\tau_2\,.
\end{aligned} \qquad (16.56b)$$

Using (16.54), (16.55a) and (16.56a) we get

$$E_S = 2E_0 + \frac{K+J}{1+S^2}\,, \quad E_A = 2E_0 + \frac{K-J}{1-S^2}\,. \qquad (16.57)$$

The integrals K, J and S can be evaluated. The variational of E_S and E_A with respect to the interaction distance R is shown in Fig. 16.2. From this figure we note that the ground state is a symmetric and bound state. The antisymmetric state has energy greater than that of two separated hydrogen atoms ($2E_0$) and hence it is not bound. The theoretical equilibrium bond length R_{eq} is found to be 0.77 Å. The difference between E_A and E_S arises since J and S are not zero. K is the energy of the classical Coulomb electrostatic interaction between the electronic charge distribution. For this reason K is called the *Coulombic integral*. J and S are due to the fact that ψ_A and ψ_S have the terms ϕ_1 and ϕ_2 wherein the position of the electrons are interchanged. For this reason J is called the *exchange integral* and S is called the *overlap integral*. Though the exchange interaction arises because of the basic Coulomb interaction, it is purely a quantum mechanical phenomenon which cannot be explained classically. This exchange interaction is mainly responsible for the binding of atoms into molecules. It is also responsible for the ordering of spins in ferromagnetic materials.

16.5 Hydrogen Molecule Ion

Consider the coordinate system of a H_2^+ ion shown in Fig. 16.3. The Hamiltonian of H_2^+ ion neglecting the nucleon motion and spin-orbit interaction is

$$H = -\frac{\hbar^2}{2m}\nabla^2 - \frac{e^2}{r_A} - \frac{e^2}{r_B} + \frac{e^2}{R}, \tag{16.58a}$$

where

$$r_A = |\mathbf{r} - \mathbf{R}_A|, \quad r_B = |\mathbf{r} - \mathbf{R}_B|, \quad R = |\mathbf{R}_A - \mathbf{R}_B|, \tag{16.58b}$$

with $\mathbf{r}$ being the position of the electron, $\mathbf{R}_A$ and $\mathbf{R}_B$ being the positions of the two protons. If R is sufficiently large then depending upon whether the electron is attached to the nucleon A or B, it will be described by u_A or u_B, where u_A and u_B are the hydrogen ground state eigenfunctions centred at A and B, respectively. This suggests that we can use a linear combination of atomic orbitals (LCAO) to construct a molecular orbitals (MO) as

$$\psi = C_A u_A + C_B u_B, \tag{16.59a}$$

where

$$u_A = \left(\frac{1}{\pi a_0^3}\right)^{1/2} e^{-r_A/a_0}, \quad u_B = \left(\frac{1}{\pi a_0^3}\right)^{1/2} e^{-r_B/a_0}. \tag{16.59b}$$

To find the energy levels, we solve the secular determinant

$$\begin{vmatrix} H_{AA} - ES_{AA} & H_{AB} - ES_{AB} \\ H_{BA} - ES_{BA} & H_{BB} - ES_{BB} \end{vmatrix} = 0, \tag{16.60}$$

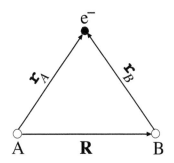

FIGURE 16.3
The coordinate system of a hydrogen ion.

where

$$S_{AA} = S_{BB} = \frac{1}{\pi a_0^3} \int_0^\infty \int_0^\pi \int_0^{2\pi} e^{-2r_A/a_0} r_A^2 \sin\theta_A \, dr_A \, d\theta_A \, d\phi_A$$

$$= 1\,, \tag{16.61a}$$

$$S_{AB} = S_{BA} = \frac{1}{\pi a_0^3} \int \left(e^{-r_A/a_0} - e^{-r_B/a_0} \right) d\tau\,, \tag{16.61b}$$

$$H_{AA} = H_{BB} = \frac{1}{\pi a_0^3} \int e^{-r_A/a_0} H e^{-r_A/a_0} \, d\tau\,, \tag{16.61c}$$

$$H_{AB} = H_{BA} = \frac{1}{\pi a_0^3} \int e^{-r_A/a_0} H e^{-r_B/a_0} \, d\tau\,. \tag{16.61d}$$

The evaluation of the integrals (16.61b-d) is not simple as they are two-centre integrals, since they involve the distance between the point of integration (x, y, z) and two different fixed points A and B. They are evaluated by a change of coordinate system from the usual Cartesian or polar coordinates to a new system in which r_A and r_B appear explicitly. We choose the new coordinates

$$\frac{r_A + r_B}{R} = \xi\,, \quad \frac{r_A - r_B}{R} = \eta \quad \text{and} \quad \phi\,. \tag{16.62}$$

We find the ranges of ξ, η and ϕ as

$$1 \le \xi < \infty\,, \quad -1 \le \eta \le 1\,, \quad 0 \le \phi \le 2\pi\,. \tag{16.63a}$$

The volume element $d\tau$ is

$$d\tau = \frac{1}{8} \left(\xi^2 - \eta^2 \right) R^3 \, d\xi \, d\eta \, d\phi\,. \tag{16.63b}$$

Then

$$r_A = R(\xi + \eta)\,, \quad r_B = R(\xi - \eta)\,. \tag{16.64}$$

Substituting (16.63) and (16.64) in Eqs. (16.61), after straight-forward lengthy calculations we arrive the result

$$S_{AB} = S_{BA} = \left(1 + \frac{R}{a_0} + \frac{1}{3}R^2\right) e^{-R/a_0}\,, \tag{16.65a}$$

$$H_{AA} = H_{BB} = \left[-\frac{1}{2} + \left(1 + \frac{a_0}{R}\right) e^{-2R/a_0} \right] \frac{e^2}{a_0}\,, \tag{16.65b}$$

$$H_{AB} = H_{BA} = \left[\left(\frac{a_0}{R} - \frac{1}{2}\right) S_{AB} - \left(1 + \frac{R}{a_0}\right) e^{-R/a_0} \right] \frac{e^2}{a_0}\,. \tag{16.65c}$$

Now, Eq. (16.60) becomes

$$\begin{vmatrix} H_{AA} - E & H_{AB} - E S_{AB} \\ H_{AB} - E S_{AB} & H_{AA} - E \end{vmatrix} = 0 \tag{16.66}$$

which gives

$$E_\pm = \frac{H_{AA} \pm H_{AB}}{1 \pm S_{AB}}\,. \tag{16.67}$$

The corresponding eigenfunctions are found as

$$\psi_\pm = \frac{1}{\sqrt{2(1 \pm S_{AB})}} (u_A \pm u_B)\,. \tag{16.68}$$

ψ_+ is symmetric with respect to the interaction of the nuclei and ψ_- is antisymmetric. Substituting (16.65) in (16.67) we obtain the $E_\pm$ as a function of R. ψ_+ will give more charge density for the electron at the middle of the two protons compared to ψ_- which will give zero charge density at the centre of the bond.

16.6 Concluding Remarks

The variational method is sensitive to the trial eigenfunction, because the method does not tell us how to choose the trial function. In fact, for any arbitrary function we can solve the equation $\partial E / \partial c_i = 0$. In practice we choose the trial function by inspecting the nature of the potential. In Eq. (16.8) because the lowest-order contribution to $E_\phi - E_0$ is of second-order in the difference $\phi_\alpha - \phi_0$, use of any unsophisticated trial eigenfunction which encompasses the symmetries, boundary conditions, angular momentum properties, etc appropriate to the system can often give a value of E_0 much close to E_ϕ. It is very difficult and some times impossible to access the accuracy of the results as E_0 is not known. That is, it is impossible to judge whether the relative error is small enough or not. Therefore, we cannot be sure whether the chosen trial function is sufficiently accurate.

16.7 Exercises

16.1 For a one-dimensional box of dimension L with eigenfunction ϕ show that $\langle E \rangle = (\hbar^2 / 2m) \int_0^L |\mathrm{d}\phi / \mathrm{d}x|^2 \, \mathrm{d}x$. Using this relation estimate the ground state energy for a particle in the one-dimensional box with trial eigenfunction

$$\phi(x) = \begin{cases} x/(\beta L), & 0 \le x \le \beta L \\ (L-x)/((1-\beta)L), & \beta L \le x \le L. \end{cases}$$

Taking β as the variational parameter compare it with the exact result.

16.2 Estimate the ground state of the infinite-well (one-dimensional box) problem defined by $V = 0$ for $|x| < L$ and ∞ for $|x| > L$ using the trial eigenfunction $\phi = |L|^\alpha - |x|^\alpha$ with α the trial parameter and compare it with the exact energy value.

16.3 Using the calculation of variation, show that the time-independent Schrödinger equation is the Euler–Lagrange equation that minimizes the functional $I(c) = \int_{-\infty}^\infty \phi_\alpha^*(x, c) H(x) \phi_\alpha(x, c) \, \mathrm{d}x$ with respect to variation of $\phi_\alpha(x)$ and $\phi_\alpha^*(x)$ with $\int_{-\infty}^\infty \phi_\alpha^*(x) \phi_\alpha(x) \, \mathrm{d}x = 1$ and the Lagrange multiplier being the energy eigenvalue.

16.4 For a quartic oscillator $H = -(\hbar^2 / 2m) \mathrm{d}^2 / \mathrm{d}x^2 + \lambda x^4$. Choosing the normalized trial eigenfunction as $\phi_0 = (\alpha^2 / \pi)^{1/4} \mathrm{e}^{-\alpha^2 x^2 / 2}$ with α as the variational parameter estimate the ground state energy. Assume that $\int_0^\infty \mathrm{e}^{-(ax)^m} x^n \, \mathrm{d}x = \Gamma\left[(n+1)/m\right]/m a^{n+1}$.

16.5 Consider the triangular potential $V(x) = Fx$, if $x > 0$ and ∞, if $x < 0$. This is used as a model for an electron trapped on the surface of liquid helium by an electric field due to two capacitor plates bracketing the helium and vacuum above it or for the MOSFET. Applying the variational method calculate ground state energy. Use the trial eigenfunction as $\phi = x \mathrm{e}^{-ax}$.

16.6 Estimate the ground state energy of the three-dimensional symmetric harmonic oscillator with $H = -(\hbar^2 / 2m) \nabla^2 + m\omega^2 r^2 / 2$ by taking the 1s hydrogenic normalized orbital $\psi(r, \beta) = (2/\beta^{3/2}) \mathrm{e}^{-r/\beta} Y_0^0$ as the trial eigenfunction with β as the variation parameter. Compare the result with the exact ground state energy and find the error.

16.7   A hydrogen atom is subjected to an electric field E. Estimate the ground state energy of the atom by taking $\psi = N(1 + qEz)u_{100}$ as the trial eigenfunction with q as the variational parameter. Find its polarizability by assuming E is small.

16.8   A particle of mass m is acted on by the three-dimensional potential $V(r) = -V_0 e^{-r/a}$, where $\hbar^2/(V_0 a^2 m) = 3/4$. Use the trial function $e^{-r/\beta}$ to obtain a bound on the energy.

16.9   Let the total Hamiltonian of a system is of the form $H^0 + H'$ with H' small compared to H^0. By choosing the eigenfunction of H^0 as the trial eigenfunction, show that the first-order perturbation is always in excess of the exact value.

16.10   The nonrelativistic hydrogen has the Hamiltonian $H = -(\hbar^2/2m)\nabla^2 - e^2/r$. Estimate its ground state energy by taking the normalized trial eigenfunction

$$\psi(r, \beta) = \left(\frac{2}{\pi}\right)^{1/4} \left(\frac{2}{\beta}\right)^{3/2} e^{-r^2/\beta^2} Y_0^0$$

with β as the variational parameter. Compare it with the exact ground state energy and find the relative error.

17

Scattering Theory

17.1 Introduction

Consider a pond in which water is very still. Suppose a pole is stuck in the pond and is projecting out of the surface of water. When a small stone is dropped into the pond we notice ripples. When the ripples reach the pole, secondary waves are generated. We say that the pole scatters the incident waves. Experimentally, scattering is a process where a particle (or a system) with a certain energy is sent toward another system (target) and the final motion of the particle is studied after it interact with the system. When the kinetic energy is conserved then the scattering is said to be *elastic*, otherwise *inelastic*.

Why are we interested in scattering processes? One of the best ways to explore the structure of particles and the forces between them is to allow them to scatter. This is practically true at the microscopic level where the systems cannot be seen in the literal sense and must be probed by indirect ways. A general scattering event is of the form

$$A_1(\alpha_1) + A_2(\alpha_2) + \ldots \rightarrow B_1(\beta_1) + B_2(\beta_2) \ldots ,$$

where $\{A_1, A_2, \ldots, B_1, B_2, \ldots\}$ are names of the particle and $\{\alpha_1, \beta_1, \ldots\}$ are the variables specifying their states such as spin, momentum, etc. Many of the most important break-throughs in physics have been made by means of scattering.

There are two different ways of developing the theory of scattering processes in quantum mechanics. They give the same results but they proceed from different formulations. The first way is to consider the scattering of a particle as a transition out of an initial state of given momentum into a final state of different momentum but with the same energy. Let the time-dependent free particle wave function representing the initial state be $\psi_i^{(0)}$ and that of one of the final states be $\psi_f^{(0)}$. Then using the time-dependent perturbation theory the probability of finding an incident particle scattered out of the state $\psi_i^{(0)}$ into $\psi_f^{(0)}$ is obtained from the Fermi formula (see sec. 14.3)

$$\Gamma = \frac{2\pi}{T} |\langle f|H'|i\rangle|^2 \rho_f(E) , \tag{17.1}$$

where $\rho_f(E)$ is the density of states in the neighbourhood of the energy E. The second approach is to regard the whole process as a scattering of waves by a potential and set up a differential equation. Its solution will give the scattered wave function. The first approach is more general while the second is more convenient and we will follow the second approach in this book.

If the incident particle and the scattered particle have same energy but different direction of the wave vector $\mathbf{k}$, then it is called *elastic scattering*. On the other hand, if the incident particle excites the target nuclei of atoms then the energy of the scattered particle will be different from that of the incident particle. Such a process is called *inelastic scattering*. In inelastic scattering, one has to find the reaction cross-section in addition to scattering cross-section to obtain the total cross-section.

DOI: 10.1201/9781003172178-17

The general idea of scattering is to shoot projectiles at the system (target) under study and see how these projectiles bounce-off from the target. Historically, scattering has been used as an important tool for exploring the structure of matter. The search for fundamental particles is done entirely using scattering experiments. We note that Rutherford discovered the nucleus of the atom via scattering experiments. Some importance of scattering experiments are as follows:

1. Scattering of nucleons of various energies from a nucleus provides an idea of the strength and range of nuclear forces and structure of nucleus.

2. Scattering of high energy electrons reveals details about charge distribution in nucleus.

3. Scattering of neutrons from a nucleus gives magnetic properties of the nucleus.

What is the aim of quantum scattering theory? Our interest in scattering process is to calculate the probability of certain final states, given the initial state and the Hamiltonian of the system. The aim of the scattering theory in quantum mechanics is to deduce details about the force or the interaction responsible for the scattering. The theory achieves this by establishing a relationship between the cross-section and the wave function of the system. The wave function is obtained by solving the Schrödinger equation. Thus, in a bound state problem the emphasis is on the energy eigenvalues whereas in the scattering problem the focus is on the wave function. The quantum mechanical analysis of scattering of particles deals with calculations of incident, scattered and total wave functions, amplitude of the scattered wave, phase-shift and differential and total scattering cross-sections.

In this chapter, first we review the classical definition of scattering cross-section and derive the relation between the scattering cross-sections of centre of mass and laboratory frame of reference. We establish the relationship between scattering cross-section and the wave function and introduce scattering amplitude. Then we get a Green's function for scattering amplitude and discuss Born's approximation for scattering amplitude. We present partial wave analysis of scattering for spherically symmetric systems. Also, we consider the collisions between identical particles and the case of inelastic scattering.

17.2   Classical Scattering Cross-Section

Consider a scattering process. Suppose a parallel beam of particles of certain energy or momentum is sent towards a target (often a thin metal foil). The target deflects or scatters the particles in various directions. The scattered particles diverge. We choose the position of the target as the origin of the coordinate system (laboratory system). Only a fraction of the incident particles are scattered and others are transmitted and reflected. The direction of the scattered particles is described by the polar angles θ and ϕ as shown in Fig. 17.1.

Suppose call the number of incident particles crossing unit area of cross-section per unit time as the incident flux N_i. Let ΔN_s is the number of particles scattered into a small solid angle $d\Omega$. Evidently, $\Delta N_s \propto N_i \, d\Omega$. We write

$$\Delta N_s = \frac{d\sigma}{d\Omega} N_i \, d\Omega, \quad \sigma \implies \sigma(\theta, \phi) \tag{17.2}$$

where $d\sigma/d\Omega$ is the proportionality constant and is called *differential scattering cross-section*. $d\sigma/d\Omega$ has units of area/steradian (SI units of solid angle): $\Delta\sigma = (d\sigma/d\Omega)/d\Omega$

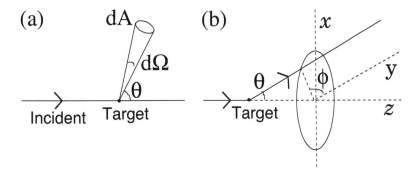

FIGURE 17.1
Representation of a scattering event.

is the area an incident particle must strike per target nucleus, in order to scatter into the
solid angle $d\Omega(= 2\pi \sin\theta \, d\theta)$.

From (17.2) we write

$$\frac{d\sigma}{d\Omega}d\Omega = \frac{\Delta N_s}{N_i} = \frac{\text{Number of particles scattered through } d\Omega \text{ per unit time}}{\text{Number of particles incident per unit time per unit area of cross-section}}. \qquad (17.3)$$

It gives nothing but the probability of a particle to be scattered into the angle $d\Omega$. The total
number of particles scattered per unit time is then calculated by integrating Eq. (17.2) over
the range of Ω:

$$\int \Delta N_s \, d\Omega = \int N_i \frac{d\sigma}{d\Omega} \, d\Omega \, d\Omega = N_i \int d\sigma \, d\Omega. \qquad (17.4)$$

That is, $N_s = N_i\sigma$, where σ is called *total scattering cross-section*. We write $\sigma = N_s/N_i$.
The cross-sectional area of the nucleus is $\pi R^2 = 7.07 \times 10^{-26} A^{2/3}$, where $R = 1.5 \times 10^{-15}$ m
is radius of the nucleus and A is the mass number. For Al^{27} the cross-sectional area is
0.588×10^{-28} m^2 and for Au^{197} it is 2.4×10^{-28} m^2. The nuclear cross-sections are typically
around 10^{-28} m^2. Cross-sections are often measured in barns (b), where 1 barn $= 10^{-28}$ m^2.
Cross-sections run from 0 to several 10^5 barns, but a few barns are typical.

17.3 Centre of Mass and Laboratory Coordinates Systems

Often the scattering cross-section is experimentally measured and compared with the cross-
section derived using theory. There is generally a correction which must be made before the
results of the theory and experiment is compared. This correction arises from the necessity
of taking into account the recoil of target particle.

Let us recall the scattering process. In a laboratory experiment the target is usually
a thin metal foil and a narrow beam of particles is directed to fall on it. The particles
are scattered by nuclei in the target and their number is counted by suitable detectors
kept at various angles. This is shown in Fig. 17.2a. The target is assumed to be stationary
after interaction with the incident particle. However, in reality the target particles will not

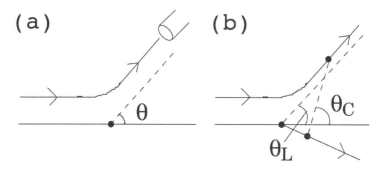

FIGURE 17.2
Scattering angle in the laboratory system and centre of mass system.

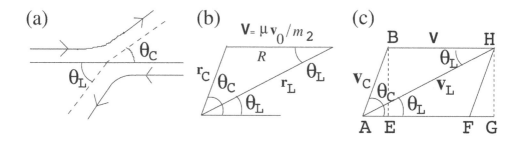

FIGURE 17.3
Scattering process to an observer in the centre of mass system. Here $\mu = m_1 m_2 / (m_1 + m_2)$.

remain stationary, unless they are very massive than the incident particles. It is convenient to work out the theory of scattering in a frame of reference in which the centre of mass of the two particles is regarded as the origin with respect to which the coordinates of the incident particles are defined. Note that the experimental cross-sections are measured with respect to the laboratory coordinates. Therefore, we transform the theoretically calculated quantities to the laboratory frame of reference for comparing the results of the theory with the experimental observation.

When the target particle is in motion after interaction then the scattering angle is defined as the angle between the initial and the final relative vectors of the positions of the two particles as depicted in Fig. 17.2b. θ_L measured in the laboratory frame is different from the actual scattering angle θ_C. We must therefore find out the relation between θ_L and θ_C and only then we can transform the expression for the cross-section to the laboratory frame.

The relationship between θ_L and θ_C are determined by examining the way in which scattering takes place in a coordinate system moving with the centre of mass of both particles. An advantage of centre of mass coordinate system is that here the total linear momentum is zero and two particles always move with equal and opposite momenta.

Figure 17.3a depicts scattering process to an observer in the centre of the mass system. Here the particles execute opposite courses and the angle between the initial and final directions of the relative vector is the scattering angle. It is the scattering angle for both the particles. We define the following:

- $\mathbf{r}_L$ and $\mathbf{v}_L$ are the position and velocity after scattering of the incident particle in the laboratory system.

- $\mathbf{r}_C$ and $\mathbf{v}_C$ are the position and velocity after scattering of the incident particle in the centre of the mass system.

- $\mathbf{R}$ and $\mathbf{V}$ are the position and velocity of the centre of mass in the laboratory system.

- $\mathbf{v}_0$ is the velocity of the incident particle.

At any time $\mathbf{r}_L = \mathbf{R} + \mathbf{r}_C$ and $\mathbf{v}_L = \mathbf{V} + \mathbf{v}_C$ as shown in Fig. 17.3b. Since the target is at rest in the laboratory system the relative velocity of the particle in the laboratory system is $\mathbf{v}_0$ (before scattering). According to conservation of linear momentum we have $m_1\mathbf{v}_0 = m_1\mathbf{V} + m_2\mathbf{V}$ which gives

$$\mathbf{V} = \frac{m_1\mathbf{v}_0}{m_1 + m_2} . \tag{17.5}$$

From Fig. 17.3c we have

$$\tan\theta_L = \frac{\sin\theta_L}{\cos\theta_L} = \frac{HG/AH}{AG/AH} = \frac{BE}{AE + BH} . \tag{17.6}$$

In ABE triangle $BE = \mathbf{v}_C \sin\theta_C$ and $AE = \mathbf{v}_C \cos\theta_C$. Therefore,

$$\tan\theta_L = \frac{\mathbf{v}_C \sin\theta_C}{\mathbf{v}_C \cos\theta_C + \mathbf{V}} = \frac{\sin\theta_C}{\cos\theta_C + \rho} , \tag{17.7}$$

where $\rho = \mathbf{V}/\mathbf{v}_C$. Also, $\mathbf{v}_L \cos\theta_C = \mathbf{v}_C \cos\theta_L + \mathbf{V}$. In the centre of mass system

$$\mathbf{r}_C = \frac{m_2}{m_1 + m_2}\mathbf{r} , \quad \mathbf{v}_C = \frac{m_2}{m_1 + m_2}\mathbf{v}_0 = \frac{m_2}{m_1}\mathbf{V} . \tag{17.8}$$

That is, $\mathbf{V}/\mathbf{v}_C = m_1/m_2$. comparison of this with $\rho = \mathbf{V}/\mathbf{v}_C$ gives $\rho = m_1/m_2$. Further,

$$\mathbf{v}_L^2 = \mathbf{v}_C^2 + \mathbf{V}^2 + 2\mathbf{v}_L \cdot \mathbf{V} \cos\theta_L . \tag{17.9}$$

Now,

$$\cos\theta_C = \frac{\cos\theta_L + \rho}{(1 + 2\rho\cos\theta_L + \rho^2)^{1/2}} . \tag{17.10}$$

Then we obtain

$$\frac{d(\cos\theta_L)}{d(\cos\theta_C)} = \frac{\left(1 + 2\rho\cos\theta_L + \rho^2\right)^{3/2}}{1 + \rho\cos\theta_L} . \tag{17.11}$$

We notice that in general the values of $\cos\theta_L$ and $\cos\theta_C$ are different.

The number of particles scattered per unit time into a given solid angle measured in terms of (i) $\cos\theta_L$ and (ii) $\cos\theta_C$ must be the same. Then

$$2\pi \left(\frac{d\sigma_\theta}{d\Omega}\right)_L \sin\theta_L \, d\theta_L = 2\pi \left(\frac{d\sigma_\theta}{d\Omega}\right)_C \sin\theta_C \, d\theta_C \tag{17.12}$$

from which we get

$$\left(\frac{d\sigma_\theta}{d\Omega}\right)_C = \left(\frac{d\sigma_\theta}{d\Omega}\right)_L \frac{d(\cos\theta_L)}{d(\cos\theta_C)} . \tag{17.13}$$

Using (17.11) in (17.13) we obtain

$$\left(\frac{\mathrm{d}\sigma_\theta}{\mathrm{d}\Omega}\right)_\mathrm{C} = \frac{\left(\frac{\mathrm{d}\sigma_\theta}{\mathrm{d}\Omega}\right)_\mathrm{L}\left(1 + 2\rho\cos\theta_\mathrm{L} + \rho^2\right)^{3/2}}{1 + \rho\cos\theta_\mathrm{L}}. \tag{17.14}$$

When $\rho = 1$ ($m_1 = m_2$) we find $\theta_\mathrm{C} = \theta_\mathrm{L}/2$. Therefore,

$$\left(\frac{\mathrm{d}\sigma_\theta}{\mathrm{d}\Omega}\right)_\mathrm{C} = 4\cos\theta_\mathrm{L}\left(\frac{\mathrm{d}\sigma_\theta}{\mathrm{d}\Omega}\right)_\mathrm{L}. \tag{17.15}$$

These formulas are nonrelativistic and therefore of no use in particle physics. In relativity, the centre of mass is not a useful concept as one would have to know the simultaneous position of two particles with different motions.

We can also express the differential cross-section relations between the two frames in terms of $\cos\theta_\mathrm{C}$. From (17.13) we obtain

$$\left(\frac{\mathrm{d}\sigma_\theta}{\mathrm{d}\Omega}\right)_\mathrm{L} = \left(\frac{\mathrm{d}\sigma_\theta}{\mathrm{d}\Omega}\right)_\mathrm{C}\frac{\left(1 + 2\rho\cos\theta_\mathrm{C} + \rho^2\right)^{3/2}}{1 + \rho\cos\theta_\mathrm{C}}. \tag{17.16}$$

Though the differential cross-sections are different in both laboratory and centre of mass systems the total cross-section is same in both systems. *Why?*

Solved Problem 1:

Show that for proton-proton scattering $\theta_\mathrm{L} = \theta_\mathrm{C}/2$.

We have

$$\tan\theta_\mathrm{L} = \frac{\sin\theta_\mathrm{C}}{\dfrac{m_1}{m_2} + \cos\theta_\mathrm{C}}. \tag{17.17}$$

For proton-proton scattering $m_1 = m_2 = m_\mathrm{p}$ then

$$\begin{aligned}
\tan\theta_\mathrm{L} &= \frac{\sin\theta_\mathrm{C}}{1 + \cos\theta_\mathrm{C}} \\
&= \frac{2\sin(\theta_\mathrm{C}/2)\,\cos(\theta_\mathrm{C}/2)}{2\cos^2(\theta_\mathrm{C}/2)} \\
&= \tan(\theta_\mathrm{C}/2).
\end{aligned} \tag{17.18}$$

That is, $\theta_\mathrm{L} = \theta_\mathrm{C}/2$.

17.4   Scattering Amplitude

The total scattering cross-section $\sigma = N_\mathrm{s}/N_\mathrm{i}$ is a classical result. Now we ask: *What is the corresponding situation in quantum mechanics?* In other words, *what is the relation between σ and the wave function ψ?*

Suppose the particles involved in a scattering event are quantum mechanical particles. They must be described by ψ. The Schrödinger equation of the scattering problem is

$$-\frac{\hbar^2}{2m}\nabla^2\psi + V\psi = E\psi. \tag{17.19}$$

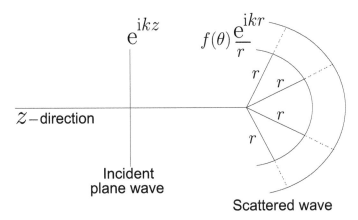

FIGURE 17.4
Spherical wavefront formed by the scattered particles.

We seek solution to (17.19) subjected to the following conditions:

1. Far before scattering ψ represents a homogeneous parallel beam of particles moving in the positive z-direction. Therefore, $\psi_{\text{inc}} \sim \mathrm{e}^{ikz}$. Generally, ψ is a function of r, θ, ϕ. Since particles move along z-direction, ψ is independent of ϕ.

2. Far after scattering ψ represents particles moving in various directions from the target. We consider elastic scattering where the kinetic energy $(mv^2/2)$ is conserved. This means that all particles move with the same constant v. Therefore, at any time the scattered particles form a spherical wave front. This is depicted in Fig. 17.4. The area of the surface of the sphere increases with r. Consequently, the probability of finding a particle in a fixed area $\mathrm{d}A$ decreases with increase in r. Thus, we write

$$\psi_{\text{sca}} \sim f(\theta, \phi) \frac{\mathrm{e}^{i\mathbf{k}\cdot\mathbf{r}}}{r}, \quad r \to \infty \qquad (17.20)$$

where $f(\theta, \phi)$ is called *scattering amplitude*. It is the amplitude of the outgoing spherical wave. The general solution of (17.19) is written as

$$\psi_{r \to \infty} \sim \mathrm{e}^{ikz} + f \frac{\mathrm{e}^{i\mathbf{k}\cdot\mathbf{r}}}{r}. \qquad (17.21)$$

To determine scattering cross-section $\sigma = N_{\text{s}}/N_{\text{i}}$ we need to find number of particles scattered through $\mathrm{d}\Omega$ and number of particles incident per unit time per unit area of cross-section. These are determined as follows.

For the uniform monochromatic beam of particles of given energy, the incident density $\mathbf{J}_{\text{inc}}$ in the z-direction is evaluated from $\psi_{\text{inc}} \sim \mathrm{e}^{ikz}$ as

$$|\mathbf{J}_{\text{inc}}| = \frac{\hbar k}{m}. \qquad (17.22)$$

Let the scattered current density be $\mathbf{J}_{\text{sca}}$ along the $\mathbf{r}$ direction. Consider a solid angle $\mathrm{d}\Omega$ in the direction of $\mathbf{r}$, subtended by an area $\mathrm{d}\mathbf{A}$. Then the number of particles per unit time scattered through $\mathrm{d}\mathbf{A}$ is $\mathbf{J}_{\text{sca}} \cdot \mathrm{d}\mathbf{A}$. The number of particles passing through $\mathrm{d}\mathbf{A}$ per second is $\Delta N_{\text{s}} = \mathbf{J}_{\text{sca}} \cdot \mathrm{d}\mathbf{A} = r^2 \mathbf{J}_{\text{sca}} \cdot \mathrm{d}\Omega$. Since the number of such scattered particles will be proportional to the incident current density $\mathbf{J}_{\text{inc}}$, we write $\Delta N_{\text{s}} = |\mathbf{J}_{\text{inc}}|\mathrm{d}\sigma$. From the above two expressions for ΔN_{s} we get

$$r^2 \mathbf{J}_{\text{sca}} \cdot \mathrm{d}\Omega = |\mathbf{J}_{\text{inc}}|\mathrm{d}\sigma \text{ and } \mathbf{J}_{\text{sca}} \cdot \mathrm{d}\Omega = J_{\text{sca},r}\mathrm{d}\sigma, \tag{17.23}$$

where $J_{\text{sca},r}$ is the r-component of $\mathbf{J}_{\text{sca}}$. From (17.23) we get

$$\frac{\mathrm{d}\sigma}{\mathrm{d}\Omega} = r^2 \frac{J_{\text{sca},r}}{|\mathbf{J}_{\text{inc}}|}. \tag{17.24}$$

The r-component of the current density $\mathbf{J}_{\text{sca}}$ is

$$J_{\text{sca},r} = \frac{\hbar}{2mi}\left(\psi^*_{\text{sca}}\frac{\partial}{\partial r}\psi_{\text{sca}} - \psi_{\text{sca}}\frac{\partial}{\partial r}\psi^*_{\text{sca}}\right). \tag{17.25}$$

Substituting for ψ_{sca} from (17.20) we get

$$J_{\text{sca},r} = \frac{\hbar k}{mr^2}|f(\theta,\phi)|^2. \tag{17.26}$$

Substituting (17.22) and (17.26) in (17.24) we obtain

$$\frac{\mathrm{d}\sigma}{\mathrm{d}\Omega} = |f(\theta,\phi)|^2. \tag{17.27}$$

The total scattering cross-section is then given by $\sigma = \int |f|^2 \sin\theta\,\mathrm{d}\theta\,\mathrm{d}\phi$. If f is independent of ϕ then

$$\sigma = 2\pi \int_0^\pi |f|^2 \sin\theta\,\mathrm{d}\theta. \tag{17.28}$$

In classical mechanics, collisions of two particles are entirely determined by their velocities and impact parameter. In quantum mechanics, we cannot consider a path for a particle. We are thus concerned with the probability of the particle scattered through any given angle after collision.

Solved Problem 2:

The wave function of a scattering process is given by $\psi = e^{ikz} + \dfrac{\cos\theta}{5}\dfrac{e^{i\mathbf{k}\cdot\mathbf{r}}}{r}$. Calculate total scattering cross-section.

The scattering amplitude is $f = (\cos\theta)/5$. Then

$$\sigma = 2\pi \int_0^\pi |f|^2 \sin\theta\,\mathrm{d}\theta = \frac{2\pi}{25}\int_0^\pi \cos^2\theta\,\sin\theta\,\mathrm{d}\theta.$$

Carrying out the integration we get

$$\sigma = \frac{2\pi}{25}\left[-\frac{\cos^3\theta}{3}\right]_0^\pi = \frac{2\pi}{75}(1+1) = \frac{4\pi}{75}.$$

17.5   Green's Function Approach

It is possible to find a solution for the scattering amplitude by transforming the Schrödinger equation into an integral equation employing Green's function method.

Suppose L be a linear differential operator. If $Ly = f(x)$ then its solution can be obtained by Green's function approach. Green's function is the solution of $Ly = \delta(x - x')$. Once the Green's function is known then the solution of $Ly = f(x)$ in the interval $x \in [a, b]$ is given by

$$y(x) = \int_a^b G(x, x') f(x') \, dx' \,. \tag{17.29}$$

Further, if $y_0(x)$ is a solution of $Ly = 0$ then

$$y(x) = y_0(x) + \int_a^b G(x, x') f(x') \, dx' \,. \tag{17.30}$$

This can be easily verified by substituting (17.30) in $Ly = f(x)$.

In a quantum mechanical scattering we wish to solve the Schrödinger equation

$$\left(-\frac{\hbar^2}{2m} \nabla^2 + V \right) \psi = E\psi = \frac{\hbar^2 k^2}{2m} \psi \tag{17.31}$$

which is rewritten as

$$\left(\nabla^2 + k^2 \right) \psi = \frac{2mV}{\hbar^2} \psi \,. \tag{17.32}$$

$\nabla^2 + k^2$ is the Helmholtz operator We are justified in writing $E = \hbar^2 k^2/(2m)$ in (17.31) as the interaction V produces no energy shift in elastic scattering processes. Defining $U(\mathbf{r}) = 2mV(\mathbf{r})/\hbar^2$ we rewrite the above equation as

$$\left(\nabla^2 + k^2 \right) \psi = U(\mathbf{r})\psi(\mathbf{r}) = F(\mathbf{r}) \tag{17.33}$$

which is of the form $Ly = f(\mathbf{X})$ with $L = \nabla^2 + k^2$. The Green's function of Eq. (17.33) satisfies the equation

$$\left(\nabla^2 + k^2 \right) G(\mathbf{r}, \mathbf{r}') = \delta(\mathbf{r} - \mathbf{r}') \,. \tag{17.34}$$

Let us assume that

$$G \propto \frac{e^{\pm ik|\mathbf{r}-\mathbf{r}'|}}{|\mathbf{r} - \mathbf{r}'|} \,. \tag{17.35}$$

Then the left-side of Eq. (17.33) is

$$\begin{aligned} \left(\nabla^2 + k^2 \right) G &= \left(\nabla^2 + k^2 \right) \frac{e^{\pm ik|\mathbf{r}-\mathbf{r}'|}}{|\mathbf{r} - \mathbf{r}'|} \\ &= -k^2 \frac{e^{\pm ik|\mathbf{r}-\mathbf{r}'|}}{|\mathbf{r} - \mathbf{r}'|} + e^{\pm ik|\mathbf{r}-\mathbf{r}'|} \nabla^2 \frac{1}{|\mathbf{r} - \mathbf{r}'|} + k^2 \frac{e^{\pm ik|\mathbf{r}-\mathbf{r}'|}}{|\mathbf{r} - \mathbf{r}'|} \\ &= -4\pi\delta(\mathbf{r} - \mathbf{r}')e^{\pm ik|\mathbf{r}-\mathbf{r}'|} \,, \end{aligned} \tag{17.36}$$

where we have used $\nabla^2(1/|\mathbf{r} - \mathbf{r}'|) = -4\pi\delta(\mathbf{r} - \mathbf{r}')$. From Eq. (17.36) we write

$$\left(\nabla^2 + k^2 \right) \frac{e^{\pm ik|\mathbf{r}-\mathbf{r}'|}}{-4\pi|\mathbf{r} - \mathbf{r}'|} = \delta(\mathbf{r} - \mathbf{r}')e^{\pm ik|\mathbf{r}-\mathbf{r}'|} \,. \tag{17.37}$$

Comparison of Eqs. (17.34) and (17.37) gives

$$G^\pm(\mathbf{r}, \mathbf{r}') = \frac{e^{\pm ik|\mathbf{r}-\mathbf{r}'|}}{-4\pi|\mathbf{r} - \mathbf{r}'|} \tag{17.38}$$

and

$$\delta(\mathbf{r} - \mathbf{r}') = \delta(\mathbf{r} - \mathbf{r}')e^{\pm ik|\mathbf{r}-\mathbf{r}'|} \,. \tag{17.39}$$

Equation (17.39) is true for $\mathbf{r} = \mathbf{r}'$ and $\mathbf{r} \neq \mathbf{r}'$. Thus, G given by Eq. (17.38) is a solution of Eq. (17.34). Moreover, the solution of $(\nabla^2 + k^2)\psi = 0$ is $\psi_0(\mathbf{r}) = e^{ikz}$. Hence,

$$\psi(\mathbf{r}) = e^{ikz} - \frac{1}{4\pi} \int \frac{e^{ik|\mathbf{r}-\mathbf{r}'|}}{|\mathbf{r}-\mathbf{r}'|} U(\mathbf{r}')\psi(\mathbf{r}')\,d\tau', \qquad (17.40)$$

where we have used only G^+ as $r \to \infty$, it must give Eq. (17.21) as the solution. The above equation indicates that the scattered wave is made of spherical waves arising at each point $\mathbf{r}'$ of space. The amplitude of each contribution is proportional to $U(\mathbf{r}')\psi(\mathbf{r}')$, proportional jointly to the strength of the interaction and the amplitude of ψ at $\mathbf{r}'$. All these waves are combined at $\mathbf{r}$ and form the total scattered wave. This is then combined to the incident wave to produce the total wave function at $\mathbf{r}$. In this sense the Green's function G^+ is thought of as a *propagator* carrying the waves from their last point of scattering to the next.

In the limit $r \to \infty$, we write

$$\psi(\mathbf{r}) = e^{ikz} - \frac{1}{4\pi} \frac{e^{ik\cdot r}}{r} \int e^{-i\mathbf{k}\cdot\mathbf{r}'} U(\mathbf{r}')\psi(\mathbf{r}')\,d\tau'. \qquad (17.41)$$

Comparing this with the equation

$$\psi(\mathbf{r}) = e^{ikz} + f(\theta, \phi) \frac{e^{i\mathbf{k}\cdot\mathbf{r}}}{r} \qquad (17.42)$$

we get

$$f(\theta, \phi) = -\frac{1}{4\pi} \int e^{-i\mathbf{k}\cdot\mathbf{r}} U(\mathbf{r})\psi(\mathbf{r})\,d\tau. \qquad (17.43)$$

The above equation is not a straight-forward formula to calculate f. Because $\psi(\mathbf{r})$ which appears in the integral is still not known. However, it provides a very useful approximation called the Born approximation [1] to be discussed in the next section.

We can use the process of interaction to evaluate $\psi(\mathbf{r})$. If $\mathbf{r}$ is replaced by $\mathbf{r}'$ in Eq. (17.40), we get

$$\psi(\mathbf{r}') = e^{ikz'} - \frac{1}{4\pi} \int \frac{e^{ik|\mathbf{r}'-\mathbf{r}''|}}{|\mathbf{r}'-\mathbf{r}''|} U(\mathbf{r}'')\psi(\mathbf{r}'')\,d\tau''. \qquad (17.44)$$

Substituting the above $\psi(\mathbf{r}')$ in (17.40) we obtain

$$\begin{aligned}\psi(\mathbf{r}) = {}& e^{ikz} - \frac{1}{4\pi} \int \frac{e^{ik|\mathbf{r}-\mathbf{r}'|}}{|\mathbf{r}-\mathbf{r}'|} U(\mathbf{r}') e^{\mathbf{k}\cdot\mathbf{r}'}\,d\tau' \\ & - \left(\frac{-1}{4\pi}\right)^2 \int\int \frac{e^{ik|\mathbf{r}-\mathbf{r}'|}}{|\mathbf{r}-\mathbf{r}'|} U(\mathbf{r}') \frac{e^{ik|\mathbf{r}'-\mathbf{r}''|}}{|\mathbf{r}'-\mathbf{r}''|} U(\mathbf{r}'')\psi(\mathbf{r}'')\,d\tau'\,d\tau''. \end{aligned} \qquad (17.45)$$

The above equation gives the first iterated solution. This process is repeated indefinitely, resulting an infinite series called *Neumann series*. If the series converges, then it represents the solution to the Schrödinger equation. The first term in Eq. (17.45) is the incident beam. The second term represents single scattering by the interacted potential $V(r')$ in the volume element $d\tau'$. The third term represents scattering of the first scattered wave at $d\tau'$ by interacted potential $V(r'')$ at volume $d\tau''$. The total effect of such double scattering is obtained by integrating with respect to τ' and τ''. The Neumann series will converge rapidly if the interaction is weak.

17.6   Born Approximation

We have seen that ψ can be given in an infinite series. The equation obtained by cutting-off the series at the nth term is called the nth Born approximation.

17.6.1   First Born Approximation for Scattering Amplitude

Suppose ψ is approximated by a plane wave function e^{ikz}. In doing this we assume that the deviation of ψ from e^{ikz} is infinitely small:

$$g(\mathbf{r}) = \left| \psi - e^{ikz} \right| \ll \left| e^{ikz} \right| = 1 . \tag{17.46}$$

Then the first Born approximation for the scattering amplitude is obtained by setting $\psi = e^{ik_0 z}$, where k_0 is the magnitude of the vector $\mathbf{k}_0$ in the incident direction. We obtain

$$f(\theta, \phi) = -\frac{1}{4\pi} \int e^{-i\mathbf{k} \cdot \mathbf{r}} U(r) e^{i\mathbf{k}_0 \cdot \mathbf{r}} \, d\tau = -\frac{1}{4\pi} \int e^{-i\mathbf{s} \cdot \mathbf{r}} U \, d\tau , \tag{17.47}$$

where $\mathbf{s} = \mathbf{k} - \mathbf{k}_0$ and $U = U(r)$. Thus, the scattering amplitude in the Born approximation is the Fourier transform of the potential apart from the constant factor. This is the first Born approximation.

From Fig. 17.5 we find $s = 2k \sin(\theta/2)$. In spherical polar coordinates $\mathbf{s} \cdot \mathbf{r} = sr \cos \theta$. Then

$$
\begin{aligned}
f &= -\frac{1}{4\pi} \int_0^\infty \int_0^\pi \int_0^{2\pi} e^{-isr' \cos \theta'} U(r') r'^2 \sin \theta' \, dr' \, d\theta' \, d\phi' \\
&= -\frac{1}{s} \int_0^\infty r \sin sr \, U(r) \, dr .
\end{aligned} \tag{17.48}
$$

We note the following important features of Eq. (17.48):

1. f is a function of θ only (since s is a function of θ).
2. f is always real.
3. $\hbar \mathbf{s} = \hbar(\mathbf{k} - \mathbf{k}_0)$ is the momentum transfer. f depends on momentum transfer and not on incident momentum.

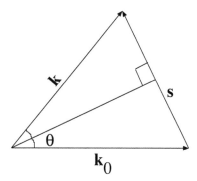

FIGURE 17.5

Vectors $\mathbf{s}$, $\mathbf{k}_0$ and $\mathbf{k}$ and the scattering angle θ.

4. For small k, s is small, then $f = -\int_0^\infty r^2 U(r)\,\mathrm{d}r$. f is independent of θ.

5. f is small for large s.

Solved Problem 3:

Obtain the validity condition for the Born approximation.

The Born approximation is valid only if (17.46) is satisfied. For the ψ given by Eq. (17.40) the validity condition (17.46) takes the form

$$g = -\frac{1}{4\pi}\int \frac{e^{ik|\mathbf{r}-\mathbf{r}'|}}{|\mathbf{r}-\mathbf{r}'|} U(r')\psi(r')\,\mathrm{d}\tau'. \tag{17.49}$$

g takes the largest value near the origin. We wish to have, at the origin

$$\frac{1}{4\pi}\left| \int \frac{e^{ik|-\mathbf{r}'|}}{|-\mathbf{r}'|} U(r')\psi(r')\,\mathrm{d}\tau' \right| \ll 1. \tag{17.50}$$

That is,

$$\frac{1}{4\pi}\left| \int \frac{e^{ikr}}{r}\frac{2mV}{\hbar^2}\psi\,\mathrm{d}\tau \right| \ll 1. \tag{17.51}$$

Replacing ψ by $e^{ikr\cos\theta}$, $\mathrm{d}\tau$ by $r^2\sin\theta\,\mathrm{d}r\,\mathrm{d}\theta\,\mathrm{d}\phi$ and integrating with respect to θ and ϕ we get

$$\frac{m}{\hbar^2 k}\left| \int_0^\infty V\left(e^{2ikr}-1\right)\mathrm{d}r \right| \ll 1 \tag{17.52}$$

if V is spherically symmetric. For a given potential the validity condition can be obtained by evaluating the above integral.

Born approximation gives good result only when the second integral in the infinite series is negligibly small. If the energy is sufficiently high, the wave function ψ is a rapidly oscillating function within the region of interaction, and it can be expected that the integral will be small because of cancellation of contributions from different parts of the region of interaction. Hence, although the interaction may be quite strong, the scattered wave is small and the Born approximation is valid provided the energy is sufficiently high.

17.6.2   Application

Now, we apply the Born approximation to square-well potential and Coulomb potential systems.

17.6.2.1   Scattering from a Square-Well Potential

An attractive square-well potential is represented by

$$V(\mathbf{r}) = \begin{cases} -V_0, & \text{for } r \le a \\ 0, & \text{for } r > a\,. \end{cases} \tag{17.53}$$

For this potential $f(\theta)$ given by Eq. (17.48) is

$$f(\theta) = \frac{2mV_0}{\hbar^2 s}\int_0^a r\sin sr\,\mathrm{d}r = \frac{2mV_0}{\hbar^2 s^3}(\sin sa - as\cos sa). \tag{17.54}$$

Now, the differential scattering cross-section is

$$\frac{d\sigma}{d\Omega} = |f^2| = \frac{4m^2 V_0^2 a^6 g(sa)}{\hbar^4}, \quad g(sa) = \frac{1}{(sa)^6} (\sin sa - sa \cos sa)^2. \tag{17.55}$$

When $sa \ll 1$ (low energy case)

$$\lim_{sa \to 0} g(sa) = \lim_{sa \to 0} \frac{1}{(sa)^6} \left[sa - \frac{(sa)^3}{3!} + \dots - sa \left(1 - \frac{(sa)^2}{2!} + \dots \right) \right]^2$$

$$= \frac{1}{9}. \tag{17.56}$$

Then

$$\frac{d\sigma}{d\Omega} = \frac{4m^2 V_0^2 a^6}{9\hbar^4}. \tag{17.57}$$

Next, the total scattering cross-section is

$$\sigma = \int_0^\pi \int_0^{2\pi} \frac{d\sigma}{d\Omega} \sin\theta \, d\theta \, d\phi = \frac{16\pi m^2 V_0^2 a^6}{9\hbar^4}. \tag{17.58}$$

For the square-well potential the validity condition (17.52) becomes

$$\frac{mV_0}{\hbar^2 k} \left| \int_0^a \left(e^{2ikr} - 1 \right) dr \right| \ll 1. \tag{17.59}$$

Evaluation of the integral in the above equation gives

$$\frac{mV_0}{2\hbar^2 k^2} \left| e^{2ika} - 1 - 2ika \right| \ll 1. \tag{17.60}$$

Defining $2ka = A$ and after simple mathematics we obtain

$$\frac{mV_0}{2\hbar^2 k^2} (2 + A^2 - 2\cos A - 2A\sin A)^{1/2} \ll 1. \tag{17.61}$$

When $A \ll 1$, that is, $2ka \ll 1$ (low-energy limit) we get

$$\left| mV_0 a/(\hbar^2 k) \right| \ll 1. \tag{17.62}$$

Therefore, a and V_0 should be small. In other words, the potential must be weak. Alternatively, if the potential is not weak but the denominator is very large then the above condition is satisfied and the Born approximation is still applicable. Because $\hbar k$ is the momentum, we require the particle momentum should be large. Since momentum is proportional to velocity this is equivalent to saying that the velocity of the incident particle should be large. Therefore, the Born approximation is valid if either the potential is weak or the velocity or the momentum of the incident particle is sufficiently large. Substituting $k = \sqrt{2mE/\hbar^2}$ the condition (17.62) becomes $E \gg 2\hbar^2/(ma^2 V_0^2)$. Since the particle energy must be high, Born approximation is a high energy approximation.

17.6.2.2 Scattering From a Screened Coulomb Potential

If the nucleus is considered as a point charge and orbital electrons are absent then the potential energy of an electron at a distance r from the nucleus is $V(r) = -Ze^2/r$. The effect of an orbital electron is to screen the nucleus from projectiles (electrons). In this case $V(r) = -Ze^2 e^{-r/a}/r$, where a is the atomic radius.

The scattering amplitude is

$$f = -\frac{imZe^2}{\hbar^2 s} \int_0^\infty e^{-r/a} \left(e^{isr} - e^{-isr} \right) \, dr = \frac{2mZe^2 a^2}{\hbar^2 (a^2 s^2 + 1)} \, . \tag{17.63}$$

Then

$$\frac{d\sigma}{d\Omega} = f^2 = \frac{4m^2 Z^2 e^4 a^4}{\hbar^4 (a^2 s^2 + 1)^2} \, . \tag{17.64}$$

The total scattering cross-section is

$$
\begin{aligned}
\sigma &= \int_0^\pi \int_0^{2\pi} \frac{d\sigma}{d\Omega} \sin\theta \, d\theta \, d\phi \\
&= \frac{8\pi m^2 Z^2 e^4 a^4}{\hbar^4} \int_0^\pi \frac{2\sin(\theta/2)\cos(\theta/2)}{[1 + 4k^2 a^2 \sin^2(\theta/2)]^2} \, d\theta \\
&= \frac{16\pi m^2 Z^2 e^4 a^4}{\hbar^4 (1 + 4k^2 a^2)} \, .
\end{aligned}
\tag{17.65}
$$

For the validity condition of Born approximation see the exercise 17.7 at the end of the chapter. If we set $a \to \infty$, then the screened Coulomb potential $V = -Ze^2 e^{-r/a}/r$ becomes the Coulomb potential. The differential scattering cross-section for Coulombic potential can be obtained from (17.64) in the limit $a \to \infty$. We obtain

$$\left(\frac{d\sigma}{d\Omega} \right)_{\text{Coulomb}} = \frac{4m^2 Z^2 e^4}{\hbar^4 s^4} = \frac{m^2 Z^2 e^4}{4\hbar^4 k^4 \sin^4(\theta/2)} \tag{17.66}$$

which is the famous *Rutherford scattering formula* for Coulomb scattering.

17.7   Partial Wave Analysis

When the Born approximation is not valid and the potential is spherically symmetric the partial wave analysis is much useful. In this approximation, a series solution for ψ is written. Each term in the solution is called *partial wave* characterized by a particular angular momentum.

17.7.1   Scattered Wave and Phase-Shift

For a spherically symmetric potential ∇^2 in Eq. (17.33) becomes

$$\nabla^2 = \frac{1}{r^2} \frac{\partial}{\partial r} \left(r^2 \frac{\partial}{\partial r} \right) + \frac{1}{r^2 \sin\theta} \frac{\partial}{\partial \theta} \left(\sin\theta \frac{\partial}{\partial \theta} \right) + \frac{1}{r^2 \sin^2\theta} \frac{\partial^2}{\partial \phi^2} \, . \tag{17.67}$$

The square of the angular momentum $\mathbf{L}^2$ is given by

$$\mathbf{L}^2 = -\hbar^2 \left[\frac{1}{\sin\theta} \frac{\partial}{\partial \theta} \left(\sin\theta \frac{\partial}{\partial \theta} \right) + \frac{1}{\sin^2\theta} \frac{\partial^2}{\partial \phi^2} \right] \, . \tag{17.68}$$

Then

$$\nabla^2 = \frac{1}{r^2} \frac{\partial}{\partial r} \left(r^2 \frac{\partial}{\partial r} \right) - \frac{\mathbf{L}^2}{\hbar^2 r^2} \, . \tag{17.69}$$

Using Eq. (17.69) in (17.33) we get

$$\left[\frac{1}{r^2}\frac{\partial}{\partial r}\left(r^2\frac{\partial}{\partial r}\right) - \frac{\mathbf{L}^2}{\hbar^2 r^2} + k^2 - U\right]\psi = 0\,. \tag{17.70}$$

We assume the solution of the form

$$\psi = R_l(r)Y_l(\theta,\phi)\,. \tag{17.71}$$

If the polar axis (z-axis) is chosen parallel to the direction of incident beam then ψ is independent of ϕ. Since Eq. (17.70) is a linear equation we write the general solution as

$$\psi = \sum_{l=0}^{\infty} R_l(r)Y_l(\theta)\,. \tag{17.72}$$

The solution corresponding to a particular l is called *l-th partial wave*. R_l and Y_l are determined by substituting (17.71) in (17.70) and then solving the corresponding differential equations for them. We get

$$\frac{1}{R_l}\frac{\partial}{\partial r}\left(r^2\frac{\partial}{\partial r}\right)R_l + \left(k^2 - U\right)r^2 - \frac{1}{Y_l\hbar^2}\mathbf{L}^2 Y_l = 0\,. \tag{17.73}$$

The first and second terms are functions of r only while the last term is function of θ only. The above equation is thus valid only if the first two terms and the last term are equal to a same constant. Therefore, we write

$$\frac{1}{\hbar^2 Y_l}\mathbf{L}^2 Y_l = l(l+1)\,. \tag{17.74}$$

This is a Legendre differential equation with argument $\cos\theta$. Its solution is $P_l(\cos\theta)$.
  The radial equation is

$$\frac{\mathrm{d}^2 R_l}{\mathrm{d}r^2} + \frac{2}{r}\frac{\mathrm{d}R_l}{\mathrm{d}r} + \left[k^2 - U - \frac{l(l+1)}{r^2}\right]R_l = 0\,. \tag{17.75}$$

Substituting $R_l = X_l/r$ we get for scattering

$$\frac{\mathrm{d}^2 X_{l,\mathrm{s}}}{\mathrm{d}r^2} + \left[k^2 - U - \frac{l(l+1)}{r^2}\right]X_{l,\mathrm{s}} = 0\,. \tag{17.76}$$

In the limit $r \to \infty$, the above equation becomes

$$\frac{\mathrm{d}^2 X_{l,\mathrm{s}}}{\mathrm{d}r^2} + k^2 X_{l,\mathrm{s}} = 0\,. \tag{17.77}$$

Its solution is

$$X_{l,\mathrm{s}}(r) = C_l'\sin(kr + \Delta_l)\,, \tag{17.78}$$

where C_l' and Δ_l are constants.
  For the incident wave the equation for $X_{l,\mathrm{in}}$ is

$$\frac{\mathrm{d}^2 X_{l,\mathrm{in}}}{\mathrm{d}r^2} + \left(k^2 - \frac{l(l+1)}{r^2}\right)X_{l,\mathrm{in}} = 0\,. \tag{17.79}$$

It is a spherical Bessel equation with the solution

$$X_{l,\mathrm{in}}(r) = \sin\left(kr - \frac{l\pi}{2}\right)\,. \tag{17.80}$$

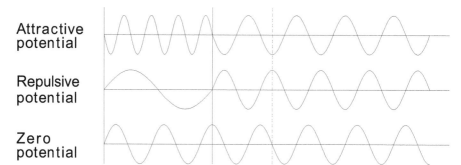

FIGURE 17.6
An example of phase-shifts induced by scattering potentials.

The difference between (17.78) and (17.80) is in the phase of the waves. The effect of the potential is thus to change the phase of the wave from $-l\pi/2$ to Δ_l. The difference in the phase is called *phase-shift* δ_l:

$$\delta_l = kr + \Delta_l - \left(kr - \frac{l\pi}{2}\right) = \Delta_l + \frac{l\pi}{2}. \tag{17.81}$$

δ_l is called the *phase-shift in the lth partial wave* and it depends on l (or the momentum of the motion $\sqrt{l(l+1)}\hbar$) and k (or the momentum of the relative motion $\hbar k$). It measures the amount by which $X_{l,s}(r)$ is displaced in r relative to the $X_{l,in}(r)$. If $V = 0$ then (17.78) and (17.80) are identical. The entire effect of scattering force is therefore described in terms of the set of phase-shifts. Figure 17.6 shows an example of shift in the phase of the scattered wave by repulsive and attractive potentials.

In elastic scattering the number of particles per second entering the scattering region must be equal to the number of particles per second leaving that region. Therefore, the moduli of the amplitudes of the incoming and outgoing waves must be the same. As a result, the effect of the potential is only to produce a phase difference between the two waves. Scattering of quantum particles and the underlying phase-shifts are important in many areas of atomic physics, including Bose–Einstein condensation, degenerate Fermi gases, frequency shifts in atomic clocks and magnetically tuned Feshbach resonances [2].

17.7.2   Scattering Amplitude in Terms of Phase-Shift

From (17.72) and (17.78), we write

$$\psi = \sum \frac{C_l'}{r} P_l \sin(kr + \Delta_l). \tag{17.82}$$

Replacing C_l' by C_l/k (so that everywhere the dependence is kr) we get

$$\psi = \sum \frac{C_l}{kr} P_l \sin\left(kr - \frac{l\pi}{2} + \delta_l\right). \tag{17.83}$$

We need to determine C_l. We compare the above wave function with the wave function given by Eq. (17.21). For this purpose we express e^{ikz} in series form as

$$e^{ikz} = e^{ikr\cos\theta} = \sum_{l=0}^{\infty} i^l (2l+1) P_l(\cos\theta) j_l(kr), \tag{17.84}$$

where j_l is spherical Bessel function. Since we are interested in the wave function far away from the scattering centre, we consider only the asymptotic value of r. Because $\lim_{r\to\infty} j_l(kr) \approx \sin(kr - l\pi/2)/(kr)$ we get

$$e^{ikz} = \sum_{l=0}^{\infty} i^l (2l+1) P_l(\cos\theta) \frac{\sin\left(kr - \dfrac{l\pi}{2}\right)}{kr}. \tag{17.85}$$

Now, comparing (17.83) and (17.21) we obtain

$$\sum \frac{C_l}{kr} P_l \sin\left(kr - \frac{l\pi}{2} + \delta_l\right) = \sum i^l (2l+1)\frac{P_l}{kr} \sin\left(kr - \frac{l\pi}{2}\right)$$
$$+ f\frac{e^{ikr}}{r}. \tag{17.86}$$

We note that the plane wave $\sin(kr - l\pi/2)/(kr)$ is a superposition of the outgoing spherical waves $e^{i(kr-l\pi/2)}/(kr)$ and the incoming spherical waves $e^{-i(kr-l\pi/2)}/(kr)$ converging on origin. Using the relation $\sin x = (e^{ix} - e^{-ix})/(2i)$ Eq. (17.86) is rewritten as

$$e^{ikr}\left[\sum C_l P_l e^{-il\pi/2} e^{i\delta_l} - \sum i^l (2l+1) P_l e^{-il\pi/2} - i2kf\right]$$
$$+ e^{-ikr}\left[-\sum C_l P_l e^{il\pi/2} e^{-i\delta_l} + \sum i^l (2l+1) P_l e^{il\pi/2}\right]$$
$$= 0. \tag{17.87}$$

Since e^{ikr} and e^{-ikr} are two linearly independent solutions the coefficients of $e^{\pm ikr}$ must be identically equal to zero. Equating the coefficients of e^{-ikr} to zero in Eq. (17.87) gives

$$C_l = i^l (2l+1) e^{i\delta_l}. \tag{17.88}$$

Then from the coefficients of e^{ikr} we get

$$f = \frac{1}{2ik}\left[\sum C_l P_l i^{-l} e^{i\delta_l} - \sum (2l+1) P_l\right]$$
$$= \frac{1}{k}\sum (2l+1) P_l e^{i\delta_l} \sin\delta_l. \tag{17.89}$$

We call the f_l's given by

$$f_l = \frac{1}{k}(2l+1) P_l e^{i\delta_l} \sin\delta_l \tag{17.90}$$

as partial scattering amplitudes. $l = 0$ gives f of the wave with $\mathbf{L}^2 = 0$.

The significance of the phase-shifts $\delta_l(k)$ can be seen in a different way by rewriting Eq. (17.83) using (17.88) as

$$\psi = -\sum_{l=0}^{\infty} i^l \frac{(2l+1)}{2ik} P_l \left[\frac{e^{-i\left(kr-\frac{l\pi}{2}\right)}}{r} - \frac{S_l(k) e^{i\left(kr-\frac{l\pi}{2}\right)}}{r}\right], \tag{17.91}$$

where $S_l(k) = e^{2i\delta_l(k)}$ and is called *scattering coefficient* of lth partial wave. In the square-bracket of (17.91), the first and the second terms represent incoming and outgoing spherical waves, respectively. The effect of the potential appears in the outgoing spherical wave as scattering coefficient $S_l(k)$ in each partial wave. The scattering potential can only alter the outgoing wave. We express f_l in terms of $S_l(k)$ as $f_l = \dfrac{1}{2ik}(2l+1)(S_l(k) - 1)$. We see from $S_l(k) = e^{2i\delta_l(k)}$ that the quantity $S_l(k)$ can be generalized to describe inelastic scattering processes (see sec. 17.10). Conservation of probability is then expressed by requiring $|S_l|^2 = 1$.

17.7.3   Total Scattering Cross-Section

We obtain

$$
\begin{aligned}
\sigma &= 2\pi \int_0^\pi |f|^2 \sin\theta \, \mathrm{d}\theta \\
&= \frac{2\pi}{k^2} \sum_l \sum_{l'} (2l+1)(2l'+1) \sin\delta_l \, \sin\delta_{l'} \, \mathrm{e}^{\mathrm{i}\delta_l} \mathrm{e}^{-\mathrm{i}\delta_{l'}} \int_0^\pi P_l P_{l'} \sin\theta \, \mathrm{d}\theta \\
&= \frac{2\pi}{k^2} \sum_l \sum_{l'} (2l+1)(2l'+1) \sin\delta_l \, \sin\delta_{l'} \, \mathrm{e}^{\mathrm{i}\delta_l} \mathrm{e}^{-\mathrm{i}\delta_{l'}} \frac{2\delta_{ll'}}{(2l'+1)} \\
&= \frac{4\pi}{k^2} \sum_l (2l+1) \sin^2\delta_l \,.
\end{aligned}
\tag{17.92}
$$

We write

$$
\sigma_l = \frac{4\pi}{k^2}(2l+1)\sin^2\delta_l
\tag{17.93}
$$

so that $\sigma = \sum_l \sigma_l$. σ_l is zero for $\delta_l = 0$ or π and is maximum when $\delta_l = \pm\pi/2, \pm3\pi/2$.

It is possible to express σ in terms of f also. From Eq. (17.89) the imaginary part of f when $\theta = 0$ is

$$
f_{\mathrm{im}}(0) = \frac{1}{k} \sum_l (2l+1) \sin^2\delta_l \,.
\tag{17.94}
$$

Then Eq. (17.92) gives

$$
\sigma = \frac{4\pi}{k} f_{\mathrm{im}}(0) \,.
\tag{17.95}
$$

This equation is known as *optical theorem* and is the statement of conservation of probability. It relates the imaginary part of scattering amplitude at $\theta = 0$ to total scattering cross-section. *Does the optical theorem applicable to Born approximation? Why?*

The optical theorem given by Eq. (17.95) is applicable for plane wave. Generalization of the optical theorem for certain other types of waves has been reported [3–8].

17.7.4   Relation Between δ_l, $V(r)$ and X_l

To obtain an equation for δ_l consider the Eqs. (17.76) and (17.79). Multiplying Eq. (17.76) by $X_{l,\mathrm{in}}$ and Eq. (17.79) by $X_{l,\mathrm{s}}$ and subtracting one from another and then replacing X_l by rR_l result in

$$
rR_{l,\mathrm{in}}\frac{\mathrm{d}^2}{\mathrm{d}r^2}(rR_{l,\mathrm{s}}) - rR_{l,\mathrm{s}}\frac{\mathrm{d}^2}{\mathrm{d}r^2}(rR_{l,\mathrm{in}}) = r^2 U R_{l,\mathrm{in}} R_{l,\mathrm{s}} \,.
\tag{17.96}
$$

Integrating the above equation with respect to r once gives

$$
\left[rR_{l,\mathrm{in}}\frac{\mathrm{d}}{\mathrm{d}r}(rR_{l,\mathrm{s}}) - rR_{l,\mathrm{s}}\frac{\mathrm{d}}{\mathrm{d}r}(rR_{l,\mathrm{in}})\right]_0^\infty = \int r^2 U R_{l,\mathrm{in}} R_{l,\mathrm{s}} \, \mathrm{d}r \,.
\tag{17.97}
$$

Substituting the solutions $R_{l,\mathrm{in}}$ and $R_{l,\mathrm{s}}$ we obtain

$$
\sin\delta_l = -\frac{1}{k} \int r^2 U(r) R_{l,\mathrm{in}} R_{l,\mathrm{s}} \, \mathrm{d}r \,.
\tag{17.98}
$$

An interesting feature is that the phase-shifts are smooth function of energy. Therefore, the dependence of δ_l on energy and $V(r)$ can be easily analyzed.

We can use Born approximation to Eq. (17.98) for weak potentials $V(r)$. Such a weak potential causes small or less scattering and therefore approximate $R_{l,s}$ by $R_{l,\text{in}}$ in Eq. (17.98) so that

$$\sin \delta_l = -\frac{1}{k} \int r^2 U(r) \left|R_{l,\text{in}}\right|^2 \, dr \,. \tag{17.99}$$

From Eq. (17.99), we conclude that if the potential is predominantly attractive, then $\sin \delta_l$ is positive and if $V(r)$ is repulsive then $\sin \delta_l$ is negative. The nucleon-nucleon scattering system exemplifies this quite nicely. From the analysis of nucleon-nucleon scattering experiments, we are able to determine the phase-shifts. For laboratory energies of less than 300 MeV, a particular n-p phase shift is positive implying that the potential experienced by the n-p system is attractive, but around 300 MeV this phase-shift goes to zero and then for larger energies it becomes negative. Interpretation of this observation leads to the conclusion that for energies greater than 300 MeV, n-p scattering probe the short range behaviour of the potential and the negative value of phase-shift means that there is a repulsive core to the n-p interaction. Hence, for the relevant total angular momentum, the short range nucleon potential is attractive until a very small core radius r_c is reached. For $r < r_c$, the potential changes from attractive to repulsive. The form of the Eq. (17.92) for total cross-section suggests that, for a given k, there should be a maximum value of l, say $l_{\max}(k)$ such that

$$\delta_l(k) = 0, \quad l > l_{\max}(k) \,. \tag{17.100}$$

If Eq. (17.100) is not true and if all the δ_l were nonzero, then Eq. (17.92) would diverge and become infinity. Therefore,

$$\sigma \approx \frac{4\pi}{k^2} \sum_{l=0}^{l_{\max}} (2l+1) \sin^2 \delta_l \tag{17.101}$$

and

$$f(\theta) \approx \frac{1}{k} \sum_{l=0}^{l_{\max}} (2l+1) P_l(\cos\theta) e^{i\delta_l} \sin \delta_l \,. \tag{17.102}$$

Even if $l_{\max}$ is as small as 3 the angular distribution obtained from Eq. (17.102) will be complicated. We can get a rough estimate of $l_{\max}$ using the classical argument. If s is the impact parameter then the angular momentum is $\hbar ks$. Hence, we write $\sqrt{l_{\max}(l_{\max}+1)}\,\hbar \approx \hbar ks$. That is, $\sqrt{l_{\max}(l_{\max}+1)} \approx ks$.

For a given range of potential, $l_{\max}$ increases with k. In general, for low energy (small k), s-wave ($l=0$) scattering dominates (isotropic scattering). In n-p scattering with energies below 10 MeV, the scattering is essentially due to neutrons with $l=0$ and the angular distribution of the scattered neutrons is isotropic in the centre of mass system, and this has been verified experimentally at low energies.

17.7.5 Scattering Length

We have noticed that for low energy scattering only s-state is important. In this case, from Eqs. (17.89) and (17.92) we get

$$f(\theta) = \frac{1}{k} e^{i\delta_0} \sin \delta_0 \,, \quad \sigma = \frac{4\pi}{k^2} \sin^2 \delta_0 \,. \tag{17.103}$$

The scattering is the result of the modification of the incident wave produced by the potential energy inside the range of the force. For s-wave, the behaviour of the wave function in this region is governed by $E - V(r)$ in the Schrödinger equation. Consequently, if E is small compared to $V(r)$, that is, to the depth of the potential well, the scattering is practically independent of E. The limiting value of f as E approaches zero is called the *scattering length* $l_{\rm s}$:

$$\lim_{E \to 0} f(\theta) = -l_{\rm s} \,. \tag{17.104}$$

Then $\sigma = 4\pi l_{\rm s}^2$. If the interaction is weak and the phase-shift is small, then it follows from the Eqs. (17.103) and (17.104) that $\lim_{k \to 0} \delta_0/k = -l_{\rm s}$. Hence, for small energy $\delta_0 = -k l_{\rm s}$.

17.8   Scattering from a Square-Well System

As an example of partial wave analysis, we analyze the case of the attractive square-well potential

$$V(r) = \begin{cases} -V_0, & r < a \\ 0, & r > a. \end{cases} \tag{17.105}$$

This potential can be used as a model to study the scattering of an electron by an atom. In this case the nuclear attraction is essentially takes place within the atom only.

The radial Eq. (17.76) with $R_l = X_l/r$ and $l = 0$ is given by

$$\ddot{X} + (k^2 - U)X = 0\,, \quad U = \frac{2mV}{\hbar^2}\,. \tag{17.106}$$

We write the Eq. (17.106) separately for $r < a$ and $r > a$:

$$\ddot{X}_> + k^2 X_> \;=\; 0\,, \quad r > a \tag{17.107a}$$
$$\ddot{X}_< + k_1^2 X_< \;=\; 0\,, \quad r < a\,, \tag{17.107b}$$

where $k_1^2 = k^2 + k_0^2$ and $k_0^2 = 2mV_0/\hbar^2$. The solutions of (17.107) are

$$X_< = \sin k_1 r\,, \quad X_> = C\sin(kr + \delta_0)\,. \tag{17.108}$$

The constants C and δ_0 must be such that $X_<$, $X_>$, $\dot{X}_<$ and $\dot{X}_>$ are continuous at $r = a$. Then $X_< = X_>$ and $\dot{X}_< = \dot{X}_>$ at $r = a$ give

$$\sin k_1 a = C\sin(ka + \delta_0)\,, \quad k_1\cos k_1 a = Ck\,\cos(ka + \delta_0)\,. \tag{17.109}$$

From the Eqs. (17.109) we obtain

$$\delta_0 = \tan^{-1}\left(\frac{k}{k_1}\tan k_1 a\right) - ka\,. \tag{17.110}$$

For $\delta_0 \ll k_0 a$ we get

$$\sin \delta_0 = ka\left(\frac{\tan k_0 a}{k_0 a} - 1\right)\,. \tag{17.111}$$

For $l = 0$ the total scattering cross-section is

$$\sigma = \frac{4\pi}{k^2}\sin^2\delta_0 = 4\pi a^2\left(\frac{\tan k_0 a}{k_0 a} - 1\right)^2\,. \tag{17.112}$$

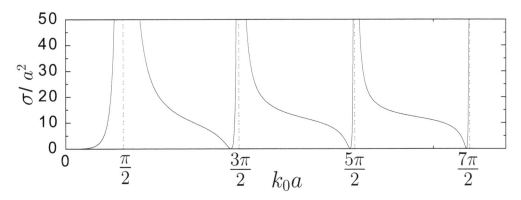

FIGURE 17.7
Variation of σ/a^2 with $k_0 a$ for the square-well potential given by Eq. (17.105). The dashed lines are $k_0 a = \pi/2$, $3\pi/2$, $5\pi/2$ and $7\pi/2$.

Figure 17.7 shows the plot σ/a^2 versus $k_0 a$. We notice that for low bombarding energies whenever $k_0 a \approx (2n+1)\pi/2$, n-integer or

$$V_0 a^2 = \frac{\pi^2 \hbar^2}{8m}, \ \frac{9\pi^2 \hbar^2}{8m}, \ \frac{25\pi^2 \hbar^2}{8m}, \ \cdots \qquad (17.113)$$

the scattering cross-section is very large.

The values of $V_0 a^2$ given by Eq. (17.113) are the values for which new energy levels with $l = 0$ appear. We say that an incident particle that has nearly the correct energy to be bound by the potential tends to concentrate there and give a large distortion in the wave function resulting in a large scattering cross-section. Further, σ becomes ∞ for $k_0 a = \pi/2$, $3\pi/2, \ldots$. Since it was derived from the assumption that $\delta_0 \ll k_0 a$, the infinite cross-section is inadmissible. A detailed analysis gives $\sigma_{\text{max}} = 4\pi/k^2$. Such maxima are called *resonances*.

When $\tan k_0 a \approx k_0 a$ (that is $k_0 a \ll 1$) we have $\sigma \approx 0$. If the values of V_0 and a are so chosen that σ is close to zero or minimum for a nonzero potential then no scattering occurs. This was first observed by Carl Ramsauer in 1921 and is called *Ramsauer effect*. (Similar effect was observed a year later by John Sealy and Victor Albert Bailey. This effect is also known as Ramsauer–Townsend effect). The Ramsauer effect is a very strong evidence for the wave nature of the electron. It was found in the scattering of electrons from atoms of noble (rare) gases, such as neon and argon. Rare gas atoms have a closed shell electronic configuration so that the orbital electrons are fully packed in a well defined spherical region. The combined electrostatic forces due to the nucleus and the orbital electrons are almost zero beyond the radius and hence an incident electron realizes a strong attractive potential-well of finite range. *Can Ramsauer effect occur with a repulsive potential? Why?*

If the sign of V in (17.105) is changed, we get the low-energy scattering cross-section. Solution for the corresponding scattering problem is obtained by simply replacing k_0 in Eq. (17.106) to $i\kappa$ in all the relations. Then we obtain

$$\sigma = 4\pi a^2 \left(\frac{\tanh \kappa a}{\kappa a} - 1 \right)^2. \qquad (17.114)$$

Solved Problem 4:

Find the expression for the scattering length for the square-well potential in terms of the potential and its range.

For a square-well potential we have

$$\sin \delta_0 = ka \left(\frac{\tan k_0 a}{k_0 a} - 1 \right), \quad k_0^2 = U. \tag{17.115}$$

For low energy $\delta_0 \to 0$ and $\sin \delta_0 \to \delta_0$ we get

$$\frac{\delta_0}{k} = a \left(\frac{\tan k_0 a}{k_0 a} - 1 \right). \tag{17.116}$$

Then the scattering length l_s is given by

$$l_s = -\frac{\delta_0}{k} = a \left(1 - \frac{\tan k_0 a}{k_0 a} \right) = a \left(1 - \frac{\tan a\sqrt{U}}{a\sqrt{U}} \right). \tag{17.117}$$

17.9   Phase-Shift of One-Dimensional Case

The partial wave analysis developed for spherically symmetric system has been formulated for one-dimensional systems also [9,10]. In the following we outline this.

Far from scattering centre we write

$$\psi_{\text{scat}\pm}(x) = e^{ikx} + f(\varepsilon)e^{i\varepsilon kx}, \tag{17.118}$$

where $\varepsilon = x/|x| = \pm 1$. The conventional form of the $\psi_\pm$ with the particle incident on the potential from left is

$$\psi = \begin{cases} \psi_-(x) = e^{ikx} + Be^{-ikx}, & x \to -\infty \\ \psi_+(x) = Ce^{ikx}, & x \to \infty. \end{cases} \tag{17.119}$$

We write

$$\psi_+(x) = e^{ikx} + (C - 1)e^{ikx} \tag{17.120}$$

and identify $f(+) = C - 1$ and $f(-) = B$. As an analytic function can be uniquely expressed as a sum of even and odd parts, the general solution of the one-dimensional Schrödinger equation is expanded as

$$\psi(x) = \sum_{l=0,1} \varepsilon^l B_l \cos \left(\varepsilon kx + \frac{1}{2}l\pi + \delta_l \right), \tag{17.121}$$

where B_l and δ_l are determined by solving the Schrödinger equation and the phases $l\pi/2$ are a matter of convention. Comparison of the coefficients of incoming and outgoing waves in (17.120) and (17.121) we get

$$f(\varepsilon) = \sum_{l=0,1} \varepsilon^l f_l, \tag{17.122}$$

where $f_l = (e^{i\delta_l} - 1)/2$ and $B_l = (-1)^l e^{i\delta_l}$. Then the total scattering cross-section is given by

$$\sigma_{\text{tot}} = |f(+)|^2 + |f(-)|^2 = 2 \sum_{l=0,1} \sin^2 \delta_l. \tag{17.123}$$

17.10   Inelastic Scattering

In this section, we obtain total scattering cross-section for inelastic scattering. A plane wave can be treated as a partial wave expansion of incoming and outgoing spherical waves $e^{\pm ikr}/kr$. Due to the scattering object the outgoing spherical waves e^{ikr}/r becomes $S_l e^{ikr}/r$, where $S_l = e^{2i\delta_l}$ and $|S_l| = 1$. For inelastic scattering we write $S_l = \eta_l(k)e^{2i\delta_l}$, where $0 < \eta_l < 1$. There is a loss of flux between the incoming spherical waves and the outgoing spherical waves due to particle decay, particle capture and the excitation of internal states of the target.

The integrated radial flux of the incoming part in Eq. (17.91) can be calculated. The incoming part is written as $f e^{ikr}/r$ where

$$f = \frac{1}{2ik} \sum_{l=0}^{\infty} (2l+1) P_l(\cos\theta) \,. \tag{17.124}$$

For a spherical wave of the form $f e^{ikr}/r$ the radial flux is given by $[\hbar k |f|^2/(mr^2)]\hat{r}$. Then the total integrated flux over all direction is

$$J_{\text{inc}} = \frac{\hbar k}{m} \int |f|^2 \, d\Omega = \frac{\hbar k}{m} \frac{\pi}{k^2} \sum_{l=0}^{\infty} (2l+1) \,. \tag{17.125}$$

For inelastic scattering $|S_l|^2 \neq 1$ and

$$J_{\text{out}} = \frac{\hbar k}{m} \frac{\pi}{k^2} \sum_{l=0}^{\infty} (2l+1)|S_l(k)|^2 \,. \tag{17.126}$$

Then the inelastic cross-section is given by

$$\sigma_{\text{inel}} = \frac{J_{\text{inc}} - J_{\text{out}}}{\hbar k/m} = \frac{\pi}{k^2} \sum_{l=0}^{\infty} \left(1 - |S_l(k)|^2\right) \,. \tag{17.127}$$

The total cross-section is

$$\sigma_{\text{total}} = \sigma_{\text{el}} + \sigma_{\text{inel}} = \frac{2\pi}{k^2} \sum_{l=0}^{\infty} (2l+1)(1 - \text{R.P.}S_l(k)) \,. \tag{17.128}$$

17.11   Concluding Remarks

Several advancements have been made both in the theory and experiments of quantum scattering. For example, using supercomputers a complete solution of the ionization of the hydrogen atom by collision with an electron was obtained. This interaction is responsible for the glow of fluorescent lights and for the ion beams that engrave silicon chips [11]. A quantum scattering theory based on the impact approximation is applied to the level shift and the spectral line broadening in plasmas [12]. A self-contained discussion of scattering is presented for the case of central potentials in one space dimension, which facilitates the understanding of the more complex scattering theory in two and three dimensions [13]. Detection of quantum scattering phase-shifts of individual atoms in a quantum scattering experiment using a novel atom interferometer has been reported. By performing an atomic clock measurement using only the scattered part of wave function of each atom the difference of the s-wave phase-shifts for the two clock states has been measured [2]. An expression for the cross-section in the case of non-plane-wave scattering was obtained [14]. For more details on quantum scattering one may refer to refs.[15–18].

17.12   Bibliography

[1]   M. Born, *Z. Phys.* 38:803, 1926.

[2]   R.A. Hart, X. Xu, R. Legere and K. Gibble, *Nature* 446:892, 2007 and references therein.

[3]   J.A. Lock, J.T. Hodges and G. Gouesbet, *J. Opt. Soc. Am. A* 12:2708, 1995.

[4]   G. Gouesbet, C. Letellier, G. Grehan and J.T. Hodgen, *Opt. Commun.* 125:137, 1996.

[5]   P.S. Camey, E. Wolf and G.S. Agarwal, *J. Opt. Soc. Am. A*14:3366, 1997.

[6]   P.S. Camey and E. Wolf, *Opt. Commun* 155:1, 1998.

[7]   P.S. Comey, *J. Mod. Opt.* 46:891, 1999.

[8]   G. Gouesbet, *J. Math. Phys.* 50:112302, 2009.

[9]   J.H. Eberly, *Am. J. Phys.* 33:771, 1965.

[10]   B. Stec and C. Jedrzejek, *Eur. J. Phys.* 11:75, 1990.

[11]   T.N. Rescigno, M. Baertschy, W.A. Isaacs and C.W. McCurdy, *Science* 286:2474, 1999.

[12]   K. Yamamoto and H. Narumi, *J. Phys. Soc. Jpn.* 47:1383, 1979.

[13]   V.E. Barlette, M.M. Leite and S.K. Adhikari, *Eur. J. Phys.* 21:435, 2000.

[14]   G. Gouesbet, *J. Math. Phys.* 50:112302, 2009.

[15]   L.D. Faddeev and S.P. Merkuriev, *Quantum Scattering Theory For Several Particle Systems*. Kluwer, Dorolrect, 1993.

[16]   D.G. Truhlar and B. Simon, *Multi-Particle Quantum Scattering with Applications to Nuclear, Atomic and Molecular Physics*. Springer, Berlin, 1997.

[17]   D. Belkic, *Principles of Quantum Scattering Theory*. IOP, Bristol, 2004.

[18]   J. Derezinski and C. Gerard, *Scattering Theory of Classical and Quantum N-Particle Systems*. Springer, Berlin, 1997.

17.13   Exercises

17.1   Consider a particle of mass m_1 approaching a target particle of mass m_2 with a velocity v_0. Using the energy conservation theorem show that the kinetic energies in the centre of mass (E_C) and in the laboratory coordinate system (E_L) are related as $E_C = [m_2/(m_1 + m_2)]E_L$, where $E_L = m_1 v^2/2$, $E_C = \mu v^2/2$ and $\mu = m_1 m_2/(m_1 + m_2)$.

17.2 Find the largest laboratory scattering angle θ_L for (a) $m_2 < m_1$ and (b) $m_2 \geq m_1$.

17.3 Show that $\dfrac{d(\cos\theta_L)}{d(\cos\theta_C)} = \dfrac{1 + \rho\cos\theta_C}{(1 + 2\rho\cos\theta_C + \rho^2)^{3/2}}$.

17.4 Assume that the differential cross-section $d\sigma/d\Omega$ for a given interaction potential is isotropic in the centre of mass frame. For mass ratio $\rho \ll 1$, what is the ratio of the differential cross-section in the forward direction to that in the $\theta = \pi/2$ direction in the laboratory frame?

17.5 Using $\psi_{\text{sca}} \sim f(\theta,\phi)e^{i\mathbf{k}\cdot\mathbf{r}}/r$ show that $J_{\text{sca},r} = \hbar k|f|^2/(mr^2)$.

17.6 Find the value of $\nabla_r^2 \dfrac{1}{|\mathbf{r} - \mathbf{r}'|}$.

17.7 For the screened Coulomb potential $V(r) = -Ae^{-\alpha r}/r$ obtain the validity condition for Born approximation and discuss it for both lower and upper velocity limits.

17.8 Calculate the differential cross-section for a central Gaussian potential $V(r) = (V_0/\sqrt{4\pi})e^{-r^2/4a^2}$ under Born approximation.

17.9 For the Yukawa potential $V = V_0\,e^{-r/a}$ obtain the total scattering cross-section by Born approximation.

17.10 For a delta potential $V_0\delta(r - a)$ obtain scattering cross-section for $sa \ll 1$ by Born approximation.

17.11 The scattering amplitude for the Coulombic interaction of two identical bosons of charge Ze is given by

$$f(\theta) = \frac{\gamma}{2k\sin^2(\theta/2)}\,e^{-i2\gamma\ln[\sin(\theta/2)]+i\pi+2i\eta_0},$$

where

$$\gamma = \frac{Z^2e^2}{4\pi\epsilon_0\hbar v}, \quad \eta_0 = \arg\Gamma(1 + i\gamma).$$

Obtain the differential cross-section for the Coulomb scattering (Mott's formula) of two identical spinless bosons.

17.12 A perfectly rigid sphere is represented by a potential $V(r) = \infty$ for $r < a$ and 0 for $r > a$. Particles cannot penetrate into the region $r < a$ because $V(r) = \infty$ in this region. Therefore, the wave function $\psi(r)$ is zero for $r < a$. Because of infinite potential within the radius a, Born approximation is not appropriate. Applying the partial wave analysis calculate the scattering cross-section.

17.13 For the hard sphere considered in the previous exercise find the total scattering cross-section for $l_{\max} = k\alpha$.

17.14 Show that in an elastic scattering between two spinless particles, the centre of mass differential cross-section may be represented by an expression of the type $d\sigma/d\Omega = A + BP_1(\cos\theta) + \dots$.

17.15 Consider the $d\sigma/d\Omega = A + BP_1(\cos\theta) + \dots$ given in the previous exercise. Express the coefficients A and B in terms of the phase shifts δ_l. Use the orthogonality relation

$$\int_0^\pi P_l(\cos\theta)P_{l'}(\cos\theta)\sin\theta\,d\theta = \frac{2}{2l+1}\,\delta_{ll'}$$

and the recurrence relation

$$(l+1)P_{l+1}(\cos\theta) = (2l+1)\cos\theta P_l(\cos\theta) - lP_{l-1}(\cos\theta).$$

18

Identical Particles

18.1 Introduction

Two particles are said to be *identical* when it is not possible to distinguish one particle from the other particle with their inherent physical properties such as mass, charge, spin, etc. Study of identical particles is important for two reasons:

1. Many-body systems contain usually identical particles.

2. They give rise quantum mechanical phenomena that have no classical analogues.

In classical mechanics, we can able to know the trajectory of any particle given its initial position, momentum and the forces acting on it. When we consider a system with two or more identical particles, in principle, we can distinguish them at any time as each one of the particles follows a definite predictable path. This is because classical mechanics is a deterministic theory. For example, when two identical particles collide, we can determine the paths of these particles after collision as their paths are known before the collision. So, in classical mechanics, it is possible to distinguish two identical particles. Also, any particle can be observed without affecting its behaviour.

In quantum mechanics the situation is different because the uncertainty principle does not allow us to know the initial positions and momenta of the particles with desired accuracy. The individual particle is described by a wave packet. Since the wave packet is not localized and spreading with time, it is impossible to distinguish two identical particles as their wave functions always overlap at any point in space. Unless the two particles are well separated with no overlap of their wave functions, quantum identical particles cannot be distinguishable. In quantum mechanics the finite extent of the wave functions associated with each identical particle may result in overlapping of these wave functions. This makes impossible to state the wave function associated with each particle. Thus, a quantum mechanical description of a system of identical particles must be formulated in such a way that the indistinguishability of identical particles is explicitly taken into account [1–3].

18.2 Permutation Symmetry

Quantum theory describes any object at energies below its excitation threshold with four degrees of freedom. For example, these degrees of freedom could be the coordinate vector $\mathbf{r}$ and its spin σ or the momentum vector $\mathbf{p}$ and the helicity. Let us treat the variables corresponding to an N-particles system as $(1, 2, \ldots, N)$. Here 1 corresponds to $\mathbf{r}_1$ and σ_1 of particle 1 and correspondingly for the other particles. So, the Hamiltonian for a system of N identical particles is $H = H(1, 2, \ldots, N)$ and the wave function of the system is

DOI: 10.1201/9781003172178-18

$\psi = \psi(1, 2, \ldots, N)$. The Hamiltonian $H(1, 2, \ldots, N)$ of the N indistinguishable particles is invariant with respect to all their interchanges or permutations.

Let P_{ij} be the operator that permutes i and j and its eigenvalue be ω. P_{ij} is called the *permutation operator*. The effect of the P_{ij}, which interchanges i and j, on an N-particle wave function is

$$P_{ij}\psi(1, 2, \ldots, i, \ldots, j, \ldots, N) = \psi(1, 2, \ldots, j, \ldots, i, \ldots, N). \tag{18.1}$$

We have

$$\begin{aligned} P_{ij}\psi(1, 2, \ldots, i, \ldots, j, \ldots, N) &= \psi(1, 2, \ldots, j, \ldots, i, \ldots, N) \\ &= \omega\psi(1, 2, \ldots, i, \ldots, j, \ldots, N) \end{aligned} \tag{18.2}$$

and

$$\begin{aligned} P_{ij}^2\psi(1, 2, \ldots, i, \ldots, j, \ldots, N) &= \omega P_{ij}\psi(1, 2, \ldots, j, \ldots, i, \ldots, N) \\ &= \omega^2\psi(1, 2, \ldots, i, \ldots, j, \ldots, N), \end{aligned} \tag{18.3}$$

where $\omega^2 = 1$, that is, $\omega = \pm 1$. So, the eigenvalues of the permutation operator are ± 1. Since the eigenvalues are real, P_{ij} is a Hermitian operator. Because $P_{ij}^2 = I$ we have $P_{ij} = P_{ij}^{-1}$. Further, P_{ij} is unitary (see the exercise 18.1 at the end of this chapter).

An operator O changes to O' under the unitary transformation as $UOU^{-1} = O'$. Since the Hamiltonian H is symmetrical in its arguments $(1, 2, \ldots, N)$, because of the indistinguishability of identical particles, we write $PHP^{-1} = H$. Therefore, for every permutation operator $PH = HP$. In fact, for any symmetric operator $O(1, 2, \ldots, N)$ we have $[P, O] = 0$. It is to be noted that all observables O must be permutation invariant. *Why?*

The permutation P has the following properties:

1. If $\psi(1, 2, \ldots, N)$ is an eigenfunction of H with eigenvalue E then $P\psi$ is also an eigenstate with the same eigenvalue. This type of degeneracy is called *exchange degeneracy*.

2. For every permutation $\langle\phi|\psi\rangle = \langle P\phi|P\psi\rangle$.

3. The matrix elements of all observables are same in ψ and $P\psi$.

The principle of indistinguishability with respect to permutation requires at least one simultaneous eigenfunction of all the $N!$ exchange operators and H. Let ψ denotes such a function. Then

$$H\psi = E\psi, \quad P_{ij}\psi = \omega_{ij}\psi, \tag{18.4}$$

where $\omega_{ij} = \pm 1$. It can be proved that

$$P_{ij}P_{ik} = P_{jk}P_{ij} = P_{ik}P_{jk} \tag{18.5}$$

for any three different values of i, j and k. Then from Eq. (18.5) we get

$$\omega_{ij}\omega_{ik} = \omega_{jk}\omega_{ij} = \omega_{ik}\omega_{jk} \tag{18.6}$$

which gives $\omega_{ij} = \omega_{ik}$ and $\omega_{jk} = \omega_{ik}$ or $\omega_{ij} = \omega_{ik} = \omega_{jk}$. That is, the common eigenfunction ψ belongs to the same eigenvalue of all the exchange operators. If $\omega_{ij} = 1$ then we say that ψ is symmetric under the interchange of any pair of particles and denote it by $\psi_{\text{s}}(1, 2, \ldots, N)$. If $\omega_{ij} = -1$, ψ is antisymmetric and denote it by $\psi_{\text{a}}(1, 2, \ldots, N)$. We get

$$\begin{aligned} P_{ij}\psi_{\text{s}}(1, 2, \ldots, i, \ldots, j, \ldots, N) &= \psi_{\text{s}}(1, 2, \ldots, j, \ldots, i, \ldots, N) \\ &= \psi_{\text{s}}(1, 2, \ldots, i, \ldots, j, \ldots, N) \end{aligned} \tag{18.7}$$

and

$$P_{ij}\psi_a(1,2,\ldots,i,\ldots,j,\ldots,N) = \psi_a(1,2,\ldots,j,\ldots,i,\ldots,N)$$
$$= -\psi_a(1,2,\ldots,i,\ldots,j,\ldots,N). \qquad (18.8)$$

For any arbitrary permutation operator P, ψ_s and ψ_a satisfy

$$P\psi_s = \psi_s, \qquad (18.9a)$$

$$P\psi_a = (-1)^P \psi_a, \quad (-1)^P = \begin{cases} 1, & \text{for even permutation} \\ -1, & \text{for odd permutation.} \end{cases} \qquad (18.9b)$$

For ψ_s, every P of the $N!$ elements are assigned the number 1 and for ψ_a, every even (odd) element is assigned the number $1\,(-1)$.

The two-particle case is trivial. There is only one permutation operator, P_{12} (apart from the identity operator I). For three or more particles, the P_{ij} do not commute with one another and in addition to ψ_s and ψ_a, there are states for which not all P_{ij} are diagonal. These states are bases functions of higher-dimensional representation of the permutation group. It is an experimental fact that these states are not realized in nature. Only symmetric and antisymmetric states are realized in nature.

In consequence, the Hilbert space of N identical particles breaks up into subspaces H_s and H_a spanned by totally symmetric and antisymmetric states and a remainder unphysical $(N! - 2)$ states H_r which can be further broken in terms of multi-dimensional irreducible representations. Since P commutes with H, the symmetric character of a wave function remains the same in time and cannot be changed by any perturbation because of the principle of indistinguishability. As a consequence, H_s and H_a are two disconnected subspaces, because there would be no observables whatever with the matrix elements between these subspaces. If a nonstationary state is initially in one subspace, it will stay forever in that subspace. It is an experimental fact that both subspaces exist in nature. The wave functions of all the states of a given system of identical particles must have the same symmetry; otherwise, the wave function of a state formally a superposition of states of different symmetry would be neither symmetrical nor antisymmetrical.

The symmetrical or antisymmetrical property of wave functions depends on the nature of the particle. This dependence arises from the spin of the particle is given by the *spin-statistics theorem*:

The wave functions for any system of N identical particles having half-odd-integer spin must be antisymmetric under interchange of the labels of any pair of the N particles. No others are allowed. Correspondingly, the wave functions of any system of N identical, zero-spin or integer-spin particles must be symmetric under the interchange of the labels of any pair of the N particles.

This theorem gives the connection between spin and statistics. Due to the principle of indistinguishability, the quantum statistics is different from the classical Maxwell–Boltzmann statistics. According to the above theorem, there are two types of particles, with symmetric and antisymmetric wave functions depending upon their spin. Therefore, we have two quantum statistics. The statistics of particles with integer spin is known as *Bose–Einstein statistics* and that of particles with half-integer spin is known as *Fermi–Dirac statistics*. Particles obeying Bose–Einstein statistics are called *bosons* while those obeying Fermi–Dirac statistics are referred to as *fermions*. This correlation between spin and statistics applies to elementary particles and also to composite particles (such as atoms and nuclei). This is quite clear since a composite particle composed of fermions will have integral or half-integral spin as the number of fermions is even or odd. For example, He^3 nucleus is fermion as it has two proton and one neutron. α-particle is a boson because it contains two protons and two neutrons. Examples of fermions are electrons, protons, neutrons, quarks, neutrinos etc. and examples of bosons are photons, π mesons, ^{12}C nuclei etc.

Solved Problem 1:

Prove that $P_{ij}P_{ik} = P_{jk}P_{ij}$.

We obtain

$$P_{ij}P_{ik}\psi(1,2,\ldots,i,\ldots,j,\ldots,k,\ldots,N)$$
$$= \psi(1,2,\ldots,k,\ldots,i,\ldots,j,\ldots,N). \tag{18.10}$$

Also,

$$P_{ij}\psi(1,2,\ldots,i,\ldots,j,\ldots,k,\ldots,N)$$
$$= \psi(1,2,\ldots,j,\ldots,i,\ldots,k,\ldots,N) \tag{18.11a}$$
$$P_{jk}P_{ij}\psi(1,2,\ldots,i,\ldots,j,\ldots,k,\ldots,N)$$
$$= \psi(1,2,\ldots,k,\ldots,i,\ldots,j,\ldots,N). \tag{18.11b}$$

From Eqs. (18.10) and (18.11) we find $P_{ij}P_{ik} = P_{jk}P_{ij}$. Similarly, it can be proved that $P_{ik}P_{ij} = P_{ik}P_{jk}$.

18.3 Symmetric and Antisymmetric Wave Functions

The many-particle Schrödinger equation

$$i\hbar\frac{\partial}{\partial t}\psi(1,2,\ldots,N,t) = H(1,2,\ldots,N)\psi(1,2,\ldots,N,t) \tag{18.12}$$

gives the wave function $\psi(1,2,\ldots,N,t)$ which is a function of time, the spin and the spatial coordinates of the N particles. The solution of Eq. (18.12) may or may not have a well-defined symmetry. Out of all possible solutions, we must choose antisymmetric functions for fermion systems and symmetric functions for boson systems. We have to construct symmetric or antisymmetric functions from such unsymmetrized functions.

We know that there are $N!$ solutions to Eq. (18.12) corresponding to same energy eigen-value as $P\psi$ is also a solution to Eq. (18.12) if ψ is a solution. The sum of all these $N!$ functions gives the unnormalized symmetric wave function ψ_s, since the interchange of any pair of particles changes any one of the component functions into another of them as $N!$ permutation operators form a group. An antisymmetric unnormalized wave function ψ_a is set up by adding all the permuted functions arising from the original solution by means of an even number of interchanges of pairs of particles and subtracting the sum of all the permuted functions arising by means of an odd number of interchanges of pairs of particles in the original solution.

Let us consider, for example, a system of two identical particles. If $\psi(1,2)$ is a solution of the Schrödinger Eq. (18.12) then $P_{12}\psi(1,2) = \psi(2,1)$ is also a solution. Then apart from the normalizing factors, the symmetric and antisymmetric wave functions are of the form

$$\psi_s = A(\psi(1,2) + \psi(2,1)), \quad \psi_a = B(\psi(1,2) - \psi(2,1)). \tag{18.13}$$

If $\psi(1,2)$ and $\psi(2,1)$ are normalized then $A = B = 1/\sqrt{2!}$. For a N-particle system, the symmetrizing and antisymmetrizing operators S_s and S_a are

$$\psi_\text{s} = S_\text{s}\psi(1,\ldots,N,t) = \frac{1}{\sqrt{N!}}\sum_P (1)^P P\psi(1,\ldots,N,t), \tag{18.14a}$$

$$\psi_\text{a} = S_\text{a}\psi(1,\ldots,N,t) = \frac{1}{\sqrt{N!}}\sum_P (-1)^P P\psi(1,\ldots,N,t). \tag{18.14b}$$

Solved Problem 2:

For a three particle system write the symmetric and antisymmetric wave functions.

For a three-particle system

$$\begin{aligned}
\psi_\text{s} &= \frac{1}{\sqrt{6}}\,[\psi(1,2,3) + \psi(1,3,2) + \psi(2,1,3) \\
&\quad + \psi(2,3,1) + \psi(3,2,1) + \psi(3,1,2)], \tag{18.15a} \\
\psi_\text{a} &= \frac{1}{\sqrt{6}}\,[\psi(1,2,3) + \psi(2,3,1) + \psi(3,1,2) \\
&\quad - \psi(2,1,3) - \psi(1,3,2) - \psi(3,2,1)]\,. \tag{18.15b}
\end{aligned}$$

The exchange of two particles in ψ_s does not change the function, whereas the exchange of two particles in ψ_a changes the sign of ψ_a. Hence, for both the symmetric and antisymmetric total eigenfunctions, $\psi_\text{s}^*\psi_\text{s}$ and $\psi_\text{a}^*\psi_\text{a}$ are unchanged under an exchange of the particle. Therefore, any measurable quantity obtainable from the total eigenfunctions of a system of identical particles is unaffected by an exchange of the particle labels. The labels 1 and 2 appear in the expressions for ψ_s and ψ_a, but this labelling does not violate the requirements of indistinguishability because the value of any observable obtained from the eigenfunctions is independent of the assignment of the labels.

For the wave functions given by Eq. (18.13) we have

$$|\psi_\text{s,a}|^2 = \frac{1}{2}\left[|\psi(1,2)|^2 + |\psi(2,1)|^2\right] \pm \text{R.P.}\,[\psi(1,2)\psi^*(2,1)]\,. \tag{18.16}$$

The first term is the joint probability distribution for two noninteracting distinguishable particles. The second term results from the interference between ψ_s and ψ_a and is due to the indistinguishability of identical particles. It should be noted that this interference term vanishes in the regimes where the two one-particle wave functions do not overlap. That is, the wave function $\psi(1,2)$ is different from zero only when the coordinate 1 is in some region R_1, the coordinate 2 is in a region R_2, and R_1 and R_2 have no common domain. Under such a situation, $\psi(1,2)$ vanishes whenever 1 is not in R_1 and 2 is not in R_2. Now, R_1 and R_2 do not overlap and the interference term is zero everywhere. So, the two identical particles can be distinguished as the particles are well separated. Hence, for a system of identical particles, at high energies ($E \gg k_\text{B}T$) and low concentration, probable number of particles per quantum state will be very small. Wave functions of the particles are well separated. The particles can now be distinguished and we can apply the classical statistics instead of quantum statistics.

18.4   The Exclusion Principle

From the analysis of data concerning the energy levels of atom, Pauli in 1925, discovered his famous *exclusion principle*:

> *In a multielectron atom there can never be more than one electron in the same quantum state.*

This principle is applicable not for electron system alone. All fermions follow this principle.

An exact solution of a many-body problem in quantum mechanics cannot be found directly. It is usually found by using the method of successive approximations. In this method in the zeroth-order approximation the particles are noninteractive while in the higher-order approximations the interaction is included through perturbation theory. Hence, the Hamiltonian in zeroth-order approximation is given as sum of the Hamiltonians of the separate particles:

$$H_0(1, 2, \ldots, N) = H_0(1) + H_0(2) + \ldots + H_0(N). \tag{18.17a}$$

The zeroth-order eigenfunction is the product of one-particle eigenfunctions:

$$
\begin{aligned}
\psi(1, 2, \ldots, N) &= \phi_{\alpha_1}(1)\phi_{\alpha_2}(2)\ldots\phi_{\alpha_N}(N), & (18.17b)\\
E_0 &= E_{\alpha_1} + E_{\alpha_2} + \ldots + E_{\alpha_N}, & (18.17c)\\
H_0(j)\phi_{\alpha_j}(j) &= E_{\alpha_j}\phi_{\alpha_j}(j). & (18.17d)
\end{aligned}
$$

$\phi_{\alpha_j}(j)$ are the energy eigenfunctions of the jth particle and α_j are the quantum numbers specifying its state. The eigenfunctions corresponding to E_0 will then be linear combination of the functions $\psi(1, 2, \ldots, N)$.

We write ψ_a in the form of a determinant (Slater determinant)

$$
\psi_a = \frac{1}{\sqrt{N!}}
\begin{vmatrix}
\phi_{\alpha_1}(1) & \phi_{\alpha_1}(2) & \ldots & \phi_{\alpha_1}(N) \\
\phi_{\alpha_2}(1) & \phi_{\alpha_2}(2) & \ldots & \phi_{\alpha_2}(N) \\
\vdots & \vdots & \ldots & \vdots \\
\phi_{\alpha_N}(1) & \phi_{\alpha_N}(2) & \ldots & \phi_{\alpha_N}(N)
\end{vmatrix}.
\tag{18.18}
$$

The change in the sign of the function (18.18) under a permutation of any two particles follows immediately from the change of sign of a determinant when two of its columns are interchanged.

From (18.18), the Pauli exclusion principle follows. According to the Pauli exclusion principle, a system of identical fermions cannot be in a state described by a wave function (18.18) with two single-particle states which are the same. If any of the two single-particle states among $\alpha_1, \alpha_2, \ldots, \alpha_N$ are the same, the determinant vanishes. Thus, on a system of identical fermions, two- or more-particles cannot be in the same state. For a boson system (described by symmetric wave function) any number of particles can occupy a single state. So, the occupation number of bosons in a state n can be $0, 1, 2, \ldots$ whereas it is either 0 or 1 for fermions. Hence, the statistics for bosons and fermions are different and we have two different quantum statistics: Bose–Einstein statistics for bosons and Fermi–Dirac statistics for fermions.

18.5 Spin Eigenfunctions of Two Electrons

The Hamiltonian of a system of identical particles

$$H = \frac{1}{2m} \sum_{i=1}^{N} p_i^2 + V(\mathbf{r}_1, \mathbf{r}_2, \ldots, \mathbf{r}_N) \tag{18.19}$$

does not contain the spin operators of the particles in the nonrelativistic limit if no external applied magnetic field and the spin-orbit interaction is neglected. In this case, we solve the Schrödinger equation and obtain $\Phi(\mathbf{r}_1, \mathbf{r}_2, \ldots, \mathbf{r}_N)$. This is the spatial part of the total wave function. In order to take into account the spin degrees of freedom, the total wave function is written as a product of Φ and the spin eigenfunction $\chi(\mathbf{r}_1, \mathbf{r}_2, \ldots, \mathbf{r}_N)$ as

$$\psi(1, 2, \ldots, N) = \Phi(\mathbf{r}_1, \mathbf{r}_2, \ldots, \mathbf{r}_N)\chi(\mathbf{r}_1, \mathbf{r}_2, \ldots, \mathbf{r}_N) \tag{18.20}$$

or as a linear combination of such products. The above wave function is the first approximation when dealing with a system with a Hamiltonian containing spin-orbit interaction.

The symmetry requirements of the wave function with permutations of particles is applicable to the total wave function $\psi(1, 2, \ldots, N)$ since a permutation of particles corresponds to a permutation of both the spatial and the spin variables. The quantum numbers α_j, corresponding to the wave function of jth particle, contains three space quantum numbers and one spin quantum number for a three-dimensional case. We can construct symmetric and antisymmetric wave functions for both the spatial and spin parts of the total wave function. If both the spatial and spin part of the wave functions are symmetric or antisymmetric then the total wave function will be symmetric. If they differ in their symmetries then the total wave function will be antisymmetric. But the total wave function for a system of bosons must be symmetric and that for a system of fermions must be antisymmetric.

For two systems with angular momentum $\mathbf{J}_1$ with eigenstates $|j_1 m_1\rangle$ and $\mathbf{J}_2$ with eigenstates $|j_2 m_2\rangle$, the combined system with total angular momentum $\mathbf{J} = \mathbf{J}_1 + \mathbf{J}_2$ is found from the $(2j_1 + 1)(2j_2 + 1)$ eigenstates $|j_1 m_1\rangle |j_2 m_2\rangle$ using the Clebsch–Gordan (CG) transformation matrix. This also holds for the addition of spin angular momenta of two electrons. The spin quantum number of the electron $s = 1/2$ gives two eigenfunction $\left|\frac{1}{2}\frac{1}{2}\right\rangle = \alpha$ and $\left|\frac{1}{2} - \frac{1}{2}\right\rangle = \beta$. For the case of two electrons, we get four eigenfunctions $\alpha(1)\alpha(2)$, $\beta(1)\beta(2)$, $\alpha(1)\beta(2)$ and $\beta(1)\alpha(2)$.

In the coupled representation, we can construct eigenstates of the four commuting operators $\mathbf{S}^2 = (\mathbf{S}_1 + \mathbf{S}_2)^2 = |\mathbf{S}_1|^2 + |\mathbf{S}_2|^2 + 2\mathbf{S}_1 \cdot \mathbf{S}_2$, $S_z = S_{1z} + S_{2z}$, S_1^2 and S_2^2. The simultaneous eigenstates of these operators are written as $|s m_s\rangle$, where s takes the values 1 and 0. m_s takes the values $1, 0, -1$ for $s = 1$ and 0 for $s = 0$. If $s = 0$ then the spins are antiparallel and for $s = 1$ the two spins are parallel. Thus, there are four eigenkets $|s m_s\rangle$ in the coupled representation. We have found in sec. 11.6, the CG transformation matrix between these two representations for the case $j_1 = 1/2$ and $j_2 = 1/2$. If we consider the two spin degrees of freedom alone then $s_1 = j_1 = 1/2$ and $s_2 = j_2 = 1/2$. So, using the CG matrix we obtain the spin wave function for two electrons in the coupled representation as

$$\chi_s^{(1)} = |11\rangle = \alpha(1)\alpha(2), \tag{18.21a}$$

$$\chi_s^{(0)} = |10\rangle = \frac{1}{\sqrt{2}}[\alpha(1)\beta(2) + \beta(1)\alpha(2)], \tag{18.21b}$$

$$\chi_s^{(-1)} = |1-1\rangle = \beta(1)\beta(2), \tag{18.21c}$$

$$\chi_a = |00\rangle = \frac{1}{\sqrt{2}}[\alpha(1)\beta(2) - \beta(1)\alpha(2)]. \tag{18.21d}$$

Since $S^2|sm_s\rangle = s(s+1)\hbar^2|sm_s\rangle$ and $S_z|sm_s\rangle = m_s\hbar|sm_s\rangle$ we can easily prove that Eqs. (18.21b) and (18.21d) are eigenstates of S^2 with eigenvalues $2\hbar^2$ and 0 corresponding to s values of 1 and 0, respectively. The eigenvalues of S_z for the three states given by Eqs. (18.21a-c) are found to be $\hbar$, 0 and $-\hbar$, respectively. The eigenvalues of S_z for the state given by Eq. (18.21d) is 0. The triplet states corresponding to $s = 1$ are all symmetric functions and the singlet state corresponding to $s = 0$ is antisymmetric.

Solved Problem 3:

Show that the spin matrices for $s = 1/2$ can be defined by

$$S_x = \frac{\hbar}{2}[(|\alpha\rangle\langle\beta|) + (|\beta\rangle\langle\alpha|)], \quad S_y = \frac{i\hbar}{2}[(|\beta\rangle\langle\alpha|) - (|\alpha\rangle\langle\beta|)],$$
$$S_z = \frac{\hbar}{2}[(|\alpha\rangle\langle\alpha|) - (|\beta\rangle\langle\beta|)].$$

We have $|\alpha\rangle = \begin{pmatrix} 1 \\ 0 \end{pmatrix}$, $|\beta\rangle = \begin{pmatrix} 0 \\ 1 \end{pmatrix}$. Then

$$S_x = \frac{\hbar}{2}\left[\begin{pmatrix} 1 \\ 0 \end{pmatrix}\begin{pmatrix} 0 & 1 \end{pmatrix} + \begin{pmatrix} 0 \\ 1 \end{pmatrix}\begin{pmatrix} 1 & 0 \end{pmatrix}\right] = \frac{\hbar}{2}\begin{pmatrix} 0 & 1 \\ 1 & 0 \end{pmatrix}. \tag{18.22}$$

Next,

$$\begin{aligned} S_y &= \frac{i\hbar}{2}\left[\begin{pmatrix} 0 \\ 1 \end{pmatrix}\begin{pmatrix} 1 & 0 \end{pmatrix} - \begin{pmatrix} 1 \\ 0 \end{pmatrix}\begin{pmatrix} 0 & 1 \end{pmatrix}\right] \\ &= \frac{\hbar}{2}\begin{pmatrix} 0 & -i \\ i & 0 \end{pmatrix}, \end{aligned} \tag{18.23}$$

$$\begin{aligned} S_z &= \frac{\hbar}{2}\left[\begin{pmatrix} 1 \\ 0 \end{pmatrix}\begin{pmatrix} 1 & 0 \end{pmatrix} - \begin{pmatrix} 0 \\ 1 \end{pmatrix}\begin{pmatrix} 0 & 1 \end{pmatrix}\right] \\ &= \frac{\hbar}{2}\begin{pmatrix} 1 & 0 \\ 0 & -1 \end{pmatrix}. \end{aligned} \tag{18.24}$$

18.6   Exchange Interaction

The nonrelativistic Schrödinger equation does not take into account the spin of the particle as in the nonrelativistic approximation the electrical interaction of the particle does not depend on their spins. However, there is a peculiar dependence of the energy of the system on its total spin, due to the principle of indistinguishability of identical particles.

Let us consider a system with two identical particles and with no external interaction. The inter-particle interaction is spin-independent. The relevant Hamiltonian for such a system will be

$$H(\mathbf{r_1}, \mathbf{r_2}) = K_1(\mathbf{r_1}) + K_2(\mathbf{r_2}) + V(|\mathbf{r_1} - \mathbf{r_2}|). \tag{18.25}$$

In the centre of mass coordinates $\mathbf{R}$ and $\mathbf{r} = \mathbf{r_1} - \mathbf{r_2}$, Eq. (18.25) becomes

$$H(\mathbf{r}, \mathbf{R}) = H_{rel}(\mathbf{r}) + K_{cm}(\mathbf{R}), \tag{18.26}$$

where

$$H_{rel} = \frac{1}{2\mu}P_{rel}^2(\mathbf{r}) + V(r), \quad K_{cm}(\mathbf{R}) = \frac{1}{2M}P_{cm}^2(\mathbf{R}). \tag{18.27}$$

Since P_{12} commutes with $H(\mathbf{r}_1, \mathbf{r}_2)$ we have $[P_{12}, K_{\mathrm{cm}}] = 0 = [P_{12}, H_{\mathrm{rel}}]$. Hence, the eigenstates of K_{cm} and H_{rel} can be either symmetric or antisymmetric under the exchange of the two particles. The eigenstates of $K_{\mathrm{cm}} = e^{i\mathbf{K}_{\mathrm{cm}} \cdot \mathbf{R}} = e^{i\mathbf{K}_{\mathrm{cm}} \cdot (\mathbf{r}_1 + \mathbf{r}_2)}$ are already symmetric under the interchange of the labels 1 and 2 and thus the symmetry character of the overall eigenstate of $H(\mathbf{r}_1, \mathbf{r}_2)$ is determined by the eigenstates of H_{rel}. The eigenstates of $H_{\mathrm{rel}}(\mathbf{r})$ are the products of the spatial and spin states as given by Eq. (18.20).

Suppose that the particles have zero spin. Then, we have only the spatial wave function $\Phi(\mathbf{r}_1, \mathbf{r}_2)$ as the spin wave function $\chi(s_1, s_2)$ is absent for such particles. Since spin zero particle is a boson, the total wave function $\Phi(\mathbf{r}_1, \mathbf{r}_2)$ must be symmetric. The spatial eigenstates of H_{rel} are the usual $\psi_{nlm}(r, \theta, \phi)$. The interchange of two similar particles is equivalent to the inversion operation with the origin as the centre of the line joining the two particles. We know that the effect of inversion is to multiply the wave function $\psi_{nlm}(r, \theta, \phi)$ by $(-1)^l$. Therefore, for even values of l, the function $\psi_{nlm}(r, \theta, \phi)$ is symmetric with respect to interchange of the two particles and antisymmetric for odd values of l. The conclusion is that a system of two identical particles with zero spin can have only an even orbital angular momentum.

Next, consider a system consisting of two particles of spin-$\frac{1}{2}$ (say, two electrons), Then the complete wave function must be antisymmetric. As given by Eqs. (18.21b) and (18.21d), the spin wave functions $\chi(s_1, s_2)$ can be either symmetric (the triplet state with $s = 1$) or antisymmetric (the singlet state with $s = 0$), respectively. Since the total wave function must be antisymmetric, the triplet state $\psi_{nlm}(r, \theta, \phi)$ must be antisymmetric and symmetric for the singlet state. Hence, for the triplet state l must be odd and even for the singlet state. So, the antisymmetry requirement leads to $l + s = $ even. That is, $l = $ even, $s = 0$ (the singlet state) and $l = $ odd, $s = 1$ (the triplet state). This is a very important conclusion which will be applied for the case of He-atom in the next section.

If ϕ_{α_1} and ϕ_{α_2} are the spatial eigenfunctions in the zeroth-order approximation then the symmetric spacial eigenfunction is given by

$$\Phi_{\mathrm{s}}(1,2) = \frac{1}{\sqrt{2}} \left[\phi_{\alpha_1}(1)\phi_{\alpha_2}(2) + \phi_{\alpha_2}(1)\phi_{\alpha_1}(2) \right] \tag{18.28}$$

and the antisymmetric spatial function is

$$\Phi_{\mathrm{a}}(1,2) = \frac{1}{\sqrt{2}} \left[\phi_{\alpha_1}(1)\phi_{\alpha_2}(2) - \phi_{\alpha_2}(1)\phi_{\alpha_1}(2) \right] . \tag{18.29}$$

Thus, for the singlet state of two electrons, the spatial wave function is given by Eq. (18.28). Equation (18.29) gives the spatial wave function for the triplet state.

Let us consider a case in which the space variable of the two electrons have almost the same value. Then $\phi_{\alpha_1}(1) \approx \phi_{\alpha_1}(2)$ and $\phi_{\alpha_2}(1) \approx \phi_{\alpha_2}(2)$. Then we find from Eq. (18.29), $\Phi_{\mathrm{a}}(1,2) \approx 0$. This shows that when the triplet state electrons are close together then the probability density will be very small . Since there is almost zero probability of the two electrons coming closer, the triplet electrons act as if they repel each other. This type of repulsion does not arise due to their charges. It is due to the spin-orientation of the two electrons and the requirement of antisymmetric wave functions for fermions system.

Symmetric space eigenfunction given by Eq. (18.28) has inverse property as $\Phi_{\mathrm{s}} \approx \sqrt{2}\phi_{\alpha_1}(1)\phi_{\alpha_2}(2)$ when $\phi_{\alpha_1}(1) \approx \phi_{\alpha_1}(2)$ and $\phi_{\alpha_2}(1) \approx \phi_{\alpha_2}(2)$. This is twice the average value of over all space of the probability density of the symmetric wave function $\Phi_{\mathrm{s}}(1,2)$. Hence, there is a particularly large chance of finding the two noninteracting electrons close together if their space eigenfunction is symmetric. Thus, if the spins of the two electrons are antiparallel and the spin eigenfunction is the antisymmetric singlet ($s = 0$) then the two electrons act as if they attract each other since there is a large chance of finding them close together. The requirement that the fermions systems must have antisymmetric wave functions gives rise a coupling between their spin and space variables. They act as if they move under the influence of a force whose spin depends on their relative orientation. In other words, the possible values of the energy of a system of electrons depend on the total spin. This peculiar interaction arising due to the symmetry requirement is known as *exchange interaction*. It is a purely quantum effect which vanishes in the limit of classical mechanics. Exchange force arises because of the quantum mechanical indistinguishability of identical particles.

18.7   Excited States of the Helium Atom

In the zeroth-order approximation, the two electrons are in the ground state with the electronic configuration $1s^2$. Both electrons are in the hydrogenic state ψ_{100}. We neglect the interaction between the two electrons and write its spatial wave function as

$$\psi_0(1,2) = \psi_{100}(1)\psi_{100}(2) = \frac{1}{\pi}\left(\frac{Z}{a_0}\right)^3 e^{-Z(r_1+r_2)/a_0} \qquad (18.30)$$

with $Z = 2$ for helium atom. The above wave function is symmetric under permutation of the spatial coordinates of the two particles. To obtain an antisymmetric wave function, we multiply (18.30) by the spin antisymmetric wave function χ_a given by Eq. (18.21d). So, the ground states of helium-like atoms correspond to states with total spin zero. We have treated the electron-electron interaction as a perturbation in sec. 16.2 and found the ground state energy of the helium-like atoms. In this section, we shall consider the excited states of helium-like atoms.

The space part of the wave function for the case of one of the electrons is in the ground state (1s) and the other is in an excited state (characterized by nlm) is given by

$$\psi(1,2) = \frac{1}{\sqrt{2}}\left[\psi_{100}(1)\psi_{nlm}(2) \pm \psi_{nlm}(1)\psi_{100}(2)\right], \qquad (18.31)$$

where + sign gives symmetric spatial wave function and − sign gives the antisymmetric spatial wave function. Since the total wave function must be antisymmetric, the symmetric wave function is associated with the singlet state with the antisymmetric spin wave function χ_a. The antisymmetric wave function corresponds to the triplet state with the spin wave functions χ_s with electrons having parallel spins. States with antiparallel spins (singlet states) are called *para-states* and those with parallel spins (triplet states) are termed as *ortho-states*.

The first excited state of helium atom corresponds to the electron configuration $1s^1 2s^1$. The wave functions for the para- and ortho-states are given, respectively, as

$$\Phi_s(1,2) = \frac{1}{\sqrt{2}}\left[\psi_{100}(1)\psi_{200}(2) + \psi_{200}(1)\psi_{100}(2)\right], \qquad (18.32)$$

$$\Phi_a(1,2) = \frac{1}{\sqrt{2}}\left[\psi_{100}(1)\psi_{200}(2) - \psi_{200}(1)\psi_{100}(2)\right]. \qquad (18.33)$$

The Hamiltonian for the He-atom is given by

$$\begin{aligned} H(1,2) &= -\frac{\hbar^2}{2\mu}\left(\nabla_1^2 + \nabla_2^2\right) - Ze^2\left(\frac{1}{r_1} + \frac{1}{r_2}\right) + \frac{e^2}{|\mathbf{r}_1 - \mathbf{r}_2|} \\ &= H_0 + H'. \end{aligned} \qquad (18.34)$$

The electron-electron interaction $H' = e^2/|\mathbf{r}_1 - \mathbf{r}_2|$ can be considered as a perturbation to the unperturbed part of the Hamiltonian H_0 which has exact solutions. To find the energies of the ortho- and para-states in the first-order perturbation theory, it is sufficient to determine $\langle H(1,2) \rangle$ in these states. Since ψ_{100} and ψ_{200} are hydrogen-like functions corresponding to energies E_{1s} and E_{2s} of the hydrogen-like atom with $n=1$ and $n=2$ we get the energy of the para-state as

$$E_s = \int \Phi_s^* H \Phi_s d\tau = E_{1s} + E_{2s} + I + J \tag{18.35}$$

and the energy of the ortho-state as

$$E_s = \int \Phi_a^* H \Phi_a d\tau = E_{1s} + E_{2s} + I - J, \tag{18.36}$$

where

$$I = \int [\psi_{100}(1)]^2 [\psi_{200}(2)]^2 \frac{e^2}{|\mathbf{r}_1 - \mathbf{r}_2|} d^3\mathbf{r}_1 d^3\mathbf{r}_2 \tag{18.37}$$

$$J = \int [\psi_{100}(1)]^* [\psi_{200}(2)]^* \frac{e^2}{|\mathbf{r}_1 - \mathbf{r}_2|} \psi_{100}(2)\psi_{200}(1) d^3\mathbf{r}_1 d^3\mathbf{r}_2. \tag{18.38}$$

The integral I is usually called the *Coulomb integral* and the integral J is called as *exchange integral*. It can be shown that both I and J are positive. So, we find that helium in spin-singlet state (para-state) lies higher than the helium in spin-triplet state (ortho-state). As already discussed, in the singlet case the space function is symmetric and the electrons have a tendency to come close to each other. The effect of the electrostatic repulsion is more significant, hence, a higher energy results. In the triplet case, the space function is antisymmetric and the electrons tend to avoid each other. Each configuration splits into the para-state and the ortho-state with the para-state lying higher. For the ground state, only para-helium is possible. Similar to $1s^1 1s^2$ configuration, other configurations like $1s^1 2p^1$ can be considered for the excited state. Then also there will be para- and ortho-states, giving two different energy levels as shown in Fig. 18.1.

Though the spin orientations of the two electrons lead to two different energy levels for the excited states of helium, it must be remembered that this difference of energy between the two states occurs only due to Coulombic interactions as the Hamiltonian (18.34) contains only the Coulombic terms and no spin dependent term. This type of exchange interaction is called *Heisenberg exchange interaction*, which is considered to be responsible

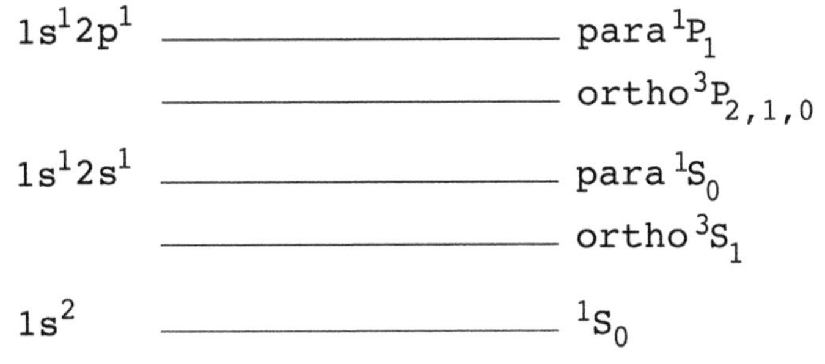

FIGURE 18.1
Energy levels for low-lying configuration of the helium atom.

for spin-ordering in ferromagnets. Para-helium atoms have no magnetic moment and form a diamagnetic gas. Ortho-helium atoms have a magnetic moment and form a paramagnetic gas.

18.8   Collisions Between Identical Particles

Collisions between identical particles are particularly interesting as a direct illustration of the fundamental differences between classical and quantum mechanics.

Consider the collision of two identical particles in the centre of mass frame as shown in Fig. 18.2. When the two particles are identical, it is impossible to decide from the observation of a particle reaching a detector whether it is the incident particle 1 or the recoiling target particle 2. A detector counts the particles scattered into the direction characterized by the polar angles (θ, ϕ). Since the particles 1 and 2 are identical there is no way of deciding whether a particle recorded by the detector results from a collision event in which particle 1 is scattered in the direction (θ, ϕ) (Fig. 18.2a) or from a collision process in which particle 2 is scattered in that direction, so that particle 1 is scattered in the opposite direction $(\pi - \theta, \phi + \pi)$ (Fig. 18.2b).

In classical mechanics, the differential cross-section for scattering in the direction (θ, ϕ) is the sum of the differential cross-sections for observation of particle 1 and particle 2 as the detection does not distinguish the incident and target particles since they are identical. Therefore, the classical differential scattering cross-section for two identical particles is given by

$$\frac{d\sigma_{\mathrm{CM}}}{d\Omega} = |f(\theta, \phi)|^2 + |f(\pi - \theta, \phi + \pi)|^2\,, \tag{18.39}$$

where the scattering amplitude is defined by the scattered wave function

$$\psi_{r \to \infty} \to e^{i\mathbf{k}\cdot\mathbf{r}} + f(\theta, \phi)\frac{e^{i\mathbf{k}\cdot\mathbf{r}}}{r}\,. \tag{18.40}$$

We note that the total wave function (product of the spatial and spin part of the wave function) must be symmetric for bosons and antisymmetric for fermions. As the wave function (18.40) is asymmetric it must be properly symmetrized with respect to permutations

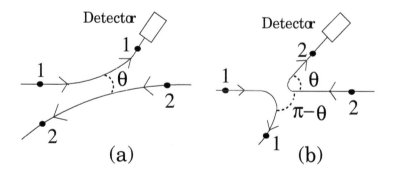

FIGURE 18.2
The scattering of identical particles in the centre of mass frame.

of identical particles. For example, in the case of two identical spinless bosons, the wave function must be symmetric under the interchange of the spatial coordinates of two particles. Now, the interchange of $\mathbf{r}_1 \to \mathbf{r}_2$ corresponds to replacing the relative position vector $\mathbf{r} = \mathbf{r}_1 - \mathbf{r}_2$ by $-\mathbf{r}$, which in polar coordinates corresponds to (r, θ, ϕ) being replaced by $(r, \pi - \theta, \phi + \pi)$. So, in place of Eq. (18.40) we use

$$\psi_S(r \to \infty) = e^{i\mathbf{k}\cdot\mathbf{r}} + e^{-i\mathbf{k}\cdot\mathbf{r}'} + [f(\theta, \phi) + f(\pi - \theta, \phi + \pi)]e^{i\mathbf{k}\cdot\mathbf{r}} . \tag{18.41}$$

This results in a differential cross-section for spinless boson particles

$$\begin{aligned} \frac{d\sigma_{QM}}{d\Omega} &= |f(\theta, \phi) + f(\pi - \theta, \phi + \pi)|^2 \\ &= |f(\theta, \phi)|^2 + |f(\pi - \theta, \phi + \pi)|^2 \\ &\quad + 2\text{R.P.}\left[f(\theta, \phi)f^*(\pi - \theta, \phi + \pi)\right] . \end{aligned} \tag{18.42}$$

It is important to note that this expression differs from the classical expression (18.39) by the presence of the third term on the right-side. The third term represents a truly quantum mechanical effect, which arises due to the interference between the amplitude for scattering of the projectile and target in the direction (θ, ϕ).

If the interaction potential is central then $f(\theta, \phi)$ is independent of ϕ. Then, we get from Eq. (18.42),

$$\frac{d\sigma_{QM}}{d\Omega} = |f(\theta)|^2 + |f(\pi - \theta)|^2 + 2\text{R.P.}\left[f(\theta)f^*(\pi - \theta)\right] \tag{18.43}$$

so that the scattering is symmetric about the angle $\theta = \pi/2$ in the centre of mass system, since the partial wave expansion of the symmetrized scattered amplitude $f_s(\theta) = f(\theta) + f(\pi - \theta)$ contains only even values of l. Moreover, at $\theta = \pi/2$,

$$\frac{d\sigma_{QM}}{d\Omega} = 4|f(\theta = \pi/2)|^2 \quad \text{and} \quad \frac{d\sigma_{CM}}{d\Omega} = 2|f(\theta = \pi/2)|^2 . \tag{18.44}$$

Hence, the quantum mechanical cross-section is twice as large as the classical result.

If we consider scattering by two spin-1/2 identical fermions, then the total wave function must be antisymmetric. Since the spin wave function is either symmetric (triplet state) or antisymmetric (singlet state), the spatial part of the singlet state must be symmetric and that for the triplet state must be antisymmetric. If we assume that the interaction between the two fermions is spin independent and central, then for the singlet state the scattering amplitude is $f_s = f(\theta) + f(\pi - \theta)$ and for the triplet state the scattering amplitude is $f_A = f(\theta) - f(\pi - \theta)$. Therefore, the quantum mechanical scattering cross-sections are given by

$$\left(\frac{d\sigma_{QM}}{d\Omega}\right)_{\text{singlet}} = |f(\theta)|^2 + |f(\pi - \theta)|^2 + 2\text{R.P.}\left[f(\theta)f^*(\pi - \theta)\right] \tag{18.45}$$

and

$$\left(\frac{d\sigma_{QM}}{d\Omega}\right)_{\text{triplet}} = |f(\theta)|^2 + |f(\pi - \theta)|^2 - 2\text{R.P.}\left[f(\theta)f^*(\pi - \theta)\right] . \tag{18.46}$$

If the incident and target particles are unpolarized (UP) (that is, their spins are randomly oriented), the probability of obtaining triplet states is three times that of a singlet state, so that the differential cross-section is given by

$$\left(\frac{d\sigma_{QM}}{d\Omega}\right)_{\text{UP}} = \frac{1}{4}\left(\frac{d\sigma_{QM}}{d\Omega}\right)_{\text{singlet}} + \frac{3}{4}\left(\frac{d\sigma_{QM}}{d\Omega}\right)_{\text{triplet}} . \tag{18.47}$$

From (18.45) and (18.46) we get

$$\left(\frac{d\sigma_{\text{QM}}}{d\Omega}\right)_{\text{UP}} = |f(\theta)|^2 + |f(\pi - \theta)|^2 - \text{R.P.}\,[f(\theta)f^*(\pi - \theta)]\,. \tag{18.48}$$

At $\theta = \pi/2$, Eq. (18.48) gives

$$\left.\frac{d\sigma_{\text{QM}}}{d\Omega}\right|_{(\theta=\pi/2)} = |f(\theta = \pi/2)|^2\,. \tag{18.49}$$

Comparing it with Eq. (18.44), we note that at $\theta = \pi/2$, the quantum mechanical differential cross-section is equal to one-half of the classical result.

18.9   Uncertainty Principle for a System of Identical Particles

The uncertainty relation (8.33) is applicable for a system of single particle. Now, we obtain an uncertainty relation for a system of N identical particles [4]. The Hilbert space $\mathcal{H}$ for a system of our interest is given by

$$\mathcal{H} = \mathcal{H}_1 \otimes \mathcal{H}_2 \otimes \ldots \otimes \mathcal{H}_N = \overset{N}{\underset{i=1}{\otimes}} \mathcal{H}_i, \tag{18.50}$$

where $\mathcal{H}_i$ is associated with the ith particle. We introduce the extended position and momentum observables as

$$x_i = \mathcal{I}_1 \otimes \ldots x_i \otimes \ldots \otimes \mathcal{I}_n, \tag{18.51a}$$
$$p_i = \mathcal{I}_1 \otimes \ldots p_i \otimes \ldots \otimes \mathcal{I}_n, \tag{18.51b}$$

where $\mathcal{I}_i$ is the identity operator and write

$$x = \sum_{i=1}^{N} x_i, \quad p = \sum_{i=1}^{N} p_i. \tag{18.52}$$

For the above (x, p)

$$\begin{aligned}(\Delta x)^2(\Delta p)^2 &\geq \frac{1}{4}|\langle[x,p]\rangle|^2 \\ &\geq \frac{1}{4}|\langle[x_1,p_1]+[x_2,p_2]+\ldots+[x_N,p_N]\rangle|^2 \\ &\geq \frac{1}{4}|\langle iN\hbar\rangle|^2 \\ &\geq \frac{1}{4}N^2\hbar^2. \end{aligned} \tag{18.53}$$

Equation (18.53) can be written as

$$\sum_{i,j=1}^{N} C_x(i,j) \sum_{i,j=1}^{N} C_p(i,j) \geq \frac{1}{4}N^2\hbar^2, \tag{18.54a}$$

where

$$C_\alpha(i,j) = \langle \alpha_i \alpha_j \rangle - \langle \alpha_i \rangle \langle \alpha_j \rangle, \quad \alpha = x, p \tag{18.54b}$$

and are called *quantum covariance functions*. As $C_x(i,i) = (\Delta x_i)^2$ and $C_p(i,i) = (\Delta p_i)^2$ Eq. (18.54a) becomes

$$\left(\sum_{i=1}^{N}(\Delta x_i)^2 + \sum_{i \neq j=1}^{N} C_x(i,j)\right)\left(\sum_{i=1}^{N}(\Delta p_i)^2 + \sum_{i \neq j=1}^{N} C_p(i,j)\right) \geq \frac{N^2 \hbar^2}{4}. \qquad (18.55)$$

In Eq. (18.55) for each observable there are two terms. The first term is the sum of the squares of the dispersions in position (momentum) while the second term describes quantum correlations. When all the quantum covariance functions become zero we recover the usual uncertainty product.

18.10   Concluding Remarks

The symmetry of the Hamiltonian with respect to permutations of its arguments implies the existence of a group of transformations on the state function. The permutation group S_N which consists of all permutations of N identical objects has $N!$ elements. Since the states ψ and $P\psi$ are experimentally indistinguishable, the question arises whether all these $N!$ states are realized in nature. It can be proved that the exchange operators do not commute among themselves. We can prove that, in general $[P_{ij}, P_{kl}] \neq 0$ and they will commute if the labels (i,j) both differ from the labels (k,l) and both are identical. Hence, only a limited set of permutation operators can be diagonalized simultaneously. Due to noncommutivity of P_{ij} operators, a complete set of common eigenfunctions of all P_{ij} cannot exist.

With reference to Eqs. (18.17) it must be noted that one can speak about the states of separate particles like $\phi_{\alpha_1}, \phi_{\alpha_2}, \dots$ only if the particles do not interact among themselves or weakly interact among themselves. So, Pauli's exclusion principle can be formulated only for weakly interacting system of identical fermions. Further, the function (18.18) characterizes a state of the system in which different particles are in single-particle states, $\alpha_1, \alpha_2, \dots, \alpha_N$, it is impossible to say just which particle is in each of these states due to the indistinguishability of identical particles.

18.11   Bibliography

[1]   A.S. Davydov, *Quantum Mechanics*. Pergamon Press, Oxford, 1965.

[2]   L.I. Schiff, *Quantum Mechanics*. McGraw Hill, New York, 1968.

[3]   J.J. Sakurai, *Modern Quantum Mechanics*. Addison–Wesley, New York, 1994.

[4]   G. Rigolin, *Commun. Theor. Phys.* 66:201, 2016.

18.12   Exercises

18.1   Prove the following properties of the permutation operator P.

(a) If $\psi(1, 2, \ldots, N)$ is an eigenfunction of H with an eigenvalue E then $P\psi$ is also an eigenstate of H with the same eigenvalue.

(b) $\langle\phi|\psi\rangle = \langle P\phi|P\psi\rangle$. (c) $\langle P\psi_i|O|P\psi_j\rangle = \langle\psi_i|O|\psi_j\rangle$.

18.2 Does the operator P_{ij} defined through
$$P_{ij}\psi(1, 2, \ldots, i, \ldots, j, \ldots, N) = \psi(1, 2, \ldots, j, \ldots, i, \ldots, N)$$
a unitary operator?

18.3 Show that the symmetric characters of the wave function does not change with time.

18.4 Find whether $[P_{ij}, P_{kl}]$ is zero or not if any of the labels (i, j) do not differ from the labels (k, l).

18.5 The S_3 permutation group contains six permutation operators. Find the two-dimensional representation of S_3 in the basis

$$|\psi_1\rangle = \frac{1}{\sqrt{12}}[2\phi_{\alpha_1}(1)\phi_{\alpha_2}(2)\phi_{\alpha_3}(3) + 2\phi_{\alpha_2}(1)\phi_{\alpha_1}(2)\phi_{\alpha_3}(3)$$
$$-\phi_{\alpha_1}(1)\phi_{\alpha_3}(2)\phi_{\alpha_2}(3) - \phi_{\alpha_3}(1)\phi_{\alpha_2}(2)\phi_{\alpha_1}(3)$$
$$-\phi_{\alpha_3}(1)\phi_{\alpha_1}(2)\phi_{\alpha_2}(3) - \phi_{\alpha_2}(1)\phi_{\alpha_3}(2)\phi_{\alpha_1}(3)]$$

$$|\psi_2\rangle = \frac{1}{\sqrt{2}}[\phi_{\alpha_3}(1)\phi_{\alpha_2}(2)\phi_{\alpha_1}(3) + \phi_{\alpha_3}(1)\phi_{\alpha_1}(2)\phi_{\alpha_2}(3)$$
$$-\phi_{\alpha_1}(1)\phi_{\alpha_3}(2)\phi_{\alpha_2}(3) - \phi_{\alpha_2}(1)\phi_{\alpha_3}(2)\phi_{\alpha_1}(3)] .$$

18.6 Determine the nature of the wave function obtained by applying (a) the symmetrising and (b) the antisymmetrising operators to an arbitrary linear combination of $\psi(1, 2)$ and $\psi(2, 1)$.

18.7 Show that the exchange operator for the spin can be given as
$$P_{12} = \frac{1}{2}\left(I + \frac{4}{\hbar^2}\mathbf{S}_1 \cdot \mathbf{S}_2\right).$$

18.8 The spatial eigenfunctions and eigenvalues of a particle confined in a one-dimensional box located between $x = -a$ and $x = a$ are given by
$$\psi_n^{(e)} = \frac{1}{\sqrt{a}}\cos\frac{(2n+1)\pi x}{2a},$$
$$E_n^{(e)} = \left(n + \frac{1}{2}\right)^2\frac{\hbar^2\pi^2}{2ma^2}, \quad n = 0, 1, 2, \ldots$$
$$\psi_n^{(o)} = \frac{1}{\sqrt{a}}\sin\frac{n\pi x}{a}, \quad E_n^{(o)} = \frac{n^2\hbar^2\pi^2}{2ma^2}, \quad n = 1, 2, \ldots$$

Determine the lowest three energy levels and the corresponding eigenfunctions if there are two identical spin-0 bosons each in spin-up state α in the box.

18.9 Consider the previous exercise. Determine the first few energy spectra if there are two identical spin-1/2 fermions each in spin-up state α in the box.

18.10 Show that for a system of two identical particles the uncertainty relation (18.55) becomes
$$\left((\Delta x_1)^2 + (\Delta x_2)^2\right)\left((\Delta p_1)^2 + (\Delta p_2)^2\right) \geq \hbar^2/4.$$

19

Relativistic Quantum Theory

19.1 Introduction

Most of the applications of quantum mechanics are in the realm of the three phases of matters: gases, liquids and solids. In these cases a nonrelativistic quantum theory is sufficient. However, in principle, a strictly correct treatment should include relativity. For relativistic particles, that is, for particles travelling at speeds close to the speed of light, the nonrelativistic Schrödinger equation is not applicable. Oskar Klein and Walter Gordon[1] proposed a relativistic wave equation for a free particle. Their equation fell into dispute because the probability density function is not positive definite, thereby preventing the probability interpretation.

In 1928 Paul Adrian Maurice Dirac (shared Nobel Prize with Erwin Schrödinger in 1933) introduced a wave equation for a relativistic electron. His equation is considered as one of the highest achievements of 20th century physics. The Dirac equation led to a positive definite probability density function, thereby allowing a probability interpretation as in the nonrelativistic Schrödinger theory. It accounted the spin of the electron. That is, a relativistic electron has the spin property introduced by Goudsmit and Uhlenbeck [1] in order to explain anomalous Zeeman effect. Further, the Dirac equation accounted gyromagnetic ratio, magnetic moment and provided satisfactory explanation for the fine structure of the hydrogen atom. However, the equation predicted the existence of negative energy states which implied that matter was unstable. There are three unusual features of the negative energy states:

1. The hydrogen atom spectrum obtained from the Dirac equation has no ground state.

2. *Zitterbewegung* (trembling motion): The effect is that the expectation value of the position of a free particle displays rapid oscillatory motion, that is, the velocity is not a constant.

3. *Klein paradox*: For a certain range of incident energy the amplitude of the reflected wave from a step potential is larger than the incident wave.

In this chapter, we introduce the relativistic Klein–Gordon and Dirac equations and present their various features in detail.

---

[1] Walter Gordon was born on 13 August 1893 Apolda, Germany and died on 24 December 1939 Stockholm, Sweden.

DOI: 10.1201/9781003172178-19

19.2   Klein–Gordon Equation

For simplicity, we start with a relativistic free particle.

19.2.1   Construction of Klein–Gordon Equation

The nonrelativistic Hamiltonian $H = \mathbf{p}^2/(2m)$ is not valid for relativistic case. The above H is generalized by considering $E^2 = \mathbf{p}^2c^2 + m^2c^4$, where E is the total energy, m is the rest mass and $\mathbf{p}$ is the momentum. Then

$$H = E = \sqrt{\mathbf{p}^2c^2 + m^2c^4}\,. \tag{19.1}$$

This is a classical result. The relativistic wave equation is obtained by replacing H and $\mathbf{p}$ by their corresponding operators in $H\psi = E\psi$. This gives

$$i\hbar\frac{\partial \psi}{\partial t} = \left(-\hbar^2 c^2 \nabla^2 + m^2 c^4\right)^{1/2}\psi\,. \tag{19.2}$$

Right-side of Eq. (19.2) contains a square-root of an operator. We can avoid the square-root by expanding it in a series which will then contain number of terms of even powers of ∇ making the equation difficult to solve. So, consider the squared Hamiltonian and the equation $H^2\psi = E^2\psi$. That is,

$$-\hbar^2\frac{\partial^2 \psi}{\partial t^2} = -\hbar^2 c^2 \nabla^2 \psi + m^2 c^4 \psi\,. \tag{19.3}$$

The above equation is rewritten as

$$\left(\Box^2 - \frac{m^2 c^2}{\hbar^2}\right)\psi = 0\,, \quad \Box^2 = \nabla^2 - \frac{1}{c^2}\frac{\partial^2}{\partial t^2}\,, \tag{19.4}$$

where $\Box^2$ is known as D'Alembertian operator. Equation (19.4) was proposed by Oskar Klein and Walter Gordon [2] and is known as *Klein–Gordon* (KG) equation. For a particle in a potential $V(x)$ the time-independent KG equation is

$$\frac{d^2\psi}{dx^2} + \frac{1}{c^2\hbar^2}\left[(E - V(x))^2 - m^2 c^4\right]\psi = 0\,. \tag{19.5a}$$

In the Schrödinger equation form the above equation is written as

$$\frac{d^2\psi}{dx^2} + (E_{\mathrm{eff}} - V_{\mathrm{eff}})\psi = 0\,, \tag{19.5b}$$

where $E_{\mathrm{eff}} = (E^2 - m^2 c^4)/(c^2\hbar^2)$, $V_{\mathrm{eff}} = (2EV(x) - V^2(x))/(c^2\hbar^2)$. Since $\partial^\mu\partial_\mu = \Box^2$, the KG Eq. (19.4) in covariant form takes the form

$$\left(\partial^\mu\partial_\mu - \frac{m^2 c^2}{\hbar^2}\right)\psi = 0\,. \tag{19.6}$$

Only four-component potentials can be added to the KG equation as it will keep the equation in covariant form.

   What is the probability density P in relativistic case? We calculate it in the next subsection. The KG equation without mass corresponds to elastic stretching of an isolated vortex filament in the ideal fluid. The wave function has the meaning of the curve's position vector. The mass part of the KG equation describes the rotation of the helical curve about the screw axis due to the hydrodynamic self-inductor of the bent vortex filament [3].

19.2.2 Continuity Equation – Probability and Current Densities

Multiplication of Eq. (19.4) by ψ^* from left gives

$$\psi^* \left(\Box^2 - \frac{m^2 c^2}{\hbar^2} \right) \psi = 0 . \tag{19.7}$$

Multiplication of complex conjugate of Eq. (19.4) by ψ from right gives

$$\left(\Box^2 \psi^* \right) \psi - \frac{m^2 c^2}{\hbar^2} \psi^* \psi = 0 . \tag{19.8}$$

Subtracting Eq. (19.8) from Eq. (19.7) we obtain

$$\psi^* \nabla^2 \psi - \left(\nabla^2 \psi^* \right) \psi + \frac{1}{c^2} \left(\frac{\partial^2 \psi^*}{\partial t^2} \psi - \psi^* \frac{\partial^2 \psi}{\partial t^2} \right) = 0 . \tag{19.9}$$

Using $\nabla \cdot (\psi^* \nabla \psi) - \nabla \cdot (\psi \nabla \psi^*) = \psi^* \nabla^2 \psi - \psi \nabla^2 \psi^*$ Eq. (19.9) is written as

$$\nabla \cdot \left[\frac{\hbar}{2mi} (\psi^* \nabla \psi - \psi \nabla \psi^*) \right] + \frac{\partial}{\partial t} \left[\frac{\hbar}{2mc^2 i} \left(\frac{\partial \psi^*}{\partial t} \psi - \psi^* \frac{\partial \psi}{\partial t} \right) \right] = 0 , \tag{19.10}$$

where the term $\hbar/(2mi)$ is introduced for convenience. Defining

$$\rho = \frac{\hbar}{2mc^2 i} \left(\psi \frac{\partial \psi^*}{\partial t} - \psi^* \frac{\partial \psi}{\partial t} \right) = \frac{1}{mc^2} E \psi^* \psi \tag{19.11a}$$

and

$$\mathbf{J} = \frac{\hbar}{2mi} \left[\psi^* \nabla \psi - \psi \nabla \psi^* \right] \tag{19.11b}$$

Eq. (19.10) becomes the continuity equation

$$\frac{\partial \rho}{\partial t} + \nabla \cdot \mathbf{J} = 0 . \tag{19.12}$$

Relativistic $\mathbf{J}$ coincides with the nonrelativistic $\mathbf{J}$ given by Eq. (2.39). However, ρ is different. ρ of nonrelativistic particle is positive definite at all points for all times. But ρ given by the Eq. (19.11a) is negative if $E < 0$.

Since the values of ψ and $\partial \psi / \partial t$ at a given time can be arbitrary, ρ defined by Eq. (19.11a) can be positive, negative or zero. It is thus impossible to interpret ρ as the probability density of finding the particles at a point. In the relativistic quantum mechanics, a particle cannot be localized in space within a volume of dimension less than $\hbar/(4mc)$, since otherwise the energy of the particle $\mathbf{p}^2/(2m)$ will by virtue of the Heisenberg uncertainty principle be larger than $2mc^2$, which would suffice for the production of a pair of particles. So, in relativistic quantum mechanics, we cannot speak of a single particle theory. The idea of a single particle cannot be retained in relativistic quantum mechanics if there are interactions which lead to creation and annihilation of pair of particles. Therefore, the number of particles is not conserved in the relativistic theory. It is thus impossible to trace at high energies (relativistic region) the motion of a single particle. The total charge is however conserved. Therefore, ρ given in Eq. (19.11a) is multiplied by the electric charge e of the particle and is interpreted as volume charge current density. Multiplying (19.11b) with e will give the electric current density. Then Eq. (19.12) will represent the continuity equation for charge density. In this case Eq. (19.12) gives the conservation of total charge as $\int \rho \, \mathrm{d}^3 r = $ constant.

Solved Problem 1:

Discuss the nonrelativistic limit of KG equation.

Consider Eq. (19.4) with $\Psi(\mathbf{X}, t)$ in place of $\psi(\mathbf{X}, t)$. We go over from the relativistic KG Eq. (19.4) to the nonrelativistic Schrödinger equation through the unitary transformation

$$\Psi(\mathbf{X}, t) = \psi(\mathbf{X}, t) e^{-imc^2 t/\hbar} . \tag{19.13}$$

In the nonrelativistic case, the total energy is nearly equal to mc^2. Hence, $E = E' + mc^2$, where $E' \ll mc^2$ is the nonrelativistic energy of the particle. Therefore, $E'\psi = |i\hbar\partial\psi/\partial t| \ll mc^2\psi$. We can thus write

$$\frac{\partial \Psi}{\partial t} = \left(\frac{\partial \psi}{\partial t} - \frac{imc^2}{\hbar}\psi \right) e^{-imc^2 t/\hbar} \approx -\frac{imc^2}{\hbar}\psi e^{-imc^2 t/\hbar} \tag{19.14}$$

and

$$\frac{\partial^2 \Psi}{\partial t^2} \approx -\left[\frac{2imc^2}{\hbar}\frac{\partial \psi}{\partial t} + \frac{m^2 c^4}{\hbar^2}\psi \right] e^{-imc^2 t/\hbar} . \tag{19.15}$$

Using (19.13) and (19.15) in Eq. (19.4) we get the nonrelativistic Schrödinger equation for a free particle. Substituting Eq. (19.14) in (19.11a) we get $\rho = \psi^*\psi$ which is the probability density corresponding to the nonrelativistic Schrödinger equation. Substituting (19.14) in (19.11b) gives the nonrelativistic current density.

19.2.3 Spin of Klein–Gordon Equation

To know the spin of the particles described by KG equation we must consider the transformation laws for the wave function under the original coordinate transformation $x'_\mu = \sum_\nu \Lambda_{\mu\nu} x_\nu$ and $\sum_\mu \Lambda_{\mu\nu}\Lambda_{\mu\nu'} = \delta_{\nu\nu'}$ with $\mu, \nu = 1, 2, 3, 4$. This transformation does not change the length of a four-vector and corresponds to a rotation in three-dimensional space, a proper Lorentz transformation, or an inversion. According to the special theory of relativity, the relativistic wave equation must retain its form under the above transformation. Hence, from the covariant form of the KG equation it follows that the wave function can be multiplied only by a factor with modulus unity. The wave function of Eq. (19.6) thus transform as $\psi(x) \rightarrow \psi'(x') = \lambda\psi(x)$ with $|\lambda| = 1$. $\lambda = 1$ if the transformation corresponds to proper rotations (continuous transformations).

Application of $\mathbf{X} \rightarrow \mathbf{X}' = -\mathbf{X}$, $t' = t$ twice will give identical transformation. Hence, $\lambda^2 = 1$, $\lambda = \pm 1$. For $\lambda = 1$, $\psi(\mathbf{X}', t') = \psi'(-\mathbf{X}, t) = \psi(\mathbf{X}, t)$ and ψ transforms like a scalar. For $\lambda = -1$, $\psi(\mathbf{X}', t') = \psi'(-\mathbf{X}, t) = -\psi(\mathbf{X}, t)$ and ψ transforms like a pseudo-scalar. So, the KG wave functions transform like a scalar or a pseudo-scalar under orthogonal transformation or proper Lorentz transformation. The spin of a particle is determined by the transformation characteristics of wave functions under coordinate transformations. If a wave function transforms like a scalar then it describes a particle of spin zero. Thus, the KG equation describes particles of zero spin like pions and kaons.

19.2.4 Plane Wave Solution

Assume the solution of Eq. (19.4) as $\psi(\mathbf{X}, t) = f(t)e^{i\mathbf{k}\cdot\mathbf{X}}$. Substituting it in Eq. (19.4) leads to the following differential equation for f:

$$f'' + \omega^2 f = 0, \quad \omega^2 = k^2 c^2 + (m^2 c^4/\hbar^2) . \tag{19.16}$$

Using $E = \hbar\omega$, $\hbar k = p$, in the second equation of (19.16) we obtain

$$E = \pm \left(\mathbf{p}^2 c^2 + m^2 c^4\right)^{1/2} . \tag{19.17}$$

Note that, E can be negative. Solution of Eq. (19.16) is $f(t) = e^{\pm i\omega t}$. Hence, $\psi(\mathbf{X}, t) = e^{i(\mathbf{k}\cdot\mathbf{X}\pm\omega t)}$. From this solution we write the wave functions of two kinds for well-defined values of momentum as

$$\psi_{\pm} = e^{i(\mathbf{k}\cdot\mathbf{X}\pm\omega t)} = e^{i(\mathbf{p}\cdot\mathbf{X}\pm Et)/\hbar} . \tag{19.18}$$

Substituting (19.18) in the charge density expression we get

$$\rho_{\pm} = \frac{ie\hbar}{2mc^2} \left(\psi_{\pm}^* \frac{\partial\psi_{\pm}}{\partial t} - \psi_{\pm} \frac{\partial\psi_{\pm}^*}{\partial t}\right) = \pm\frac{eE}{mc^2}\psi_{\pm}^*\psi_{\pm}^* . \tag{19.19}$$

Thus, the solutions of the type ψ_+ correspond to free motion of a particle with momentum $\mathbf{p}$ and charge e while solutions of the type ψ_- correspond to the free motion of a particle with the opposite sign of the charge.

19.2.5 Inadequacies of KG Equation

When the KG equation was reported many physicists thought that the problem of relativistic quantum theory of electron was solved. But Dirac did not think so and very much worried with the negative probability. Some of the inadequacies of the KG equation are:

1. There appears negative energy states. For a free particle $E = \text{K.E}+\text{P.E} = \text{K.E} = \frac{1}{2}mv^2 > 0$; E cannot be negative. If $E < 0$ then $P = E\psi^*\psi/(mc^2) < 0$ but we require $0 \le P \le 1$.

2. Equation (19.4) is second-order in time. This equation can be solved provided if ψ and $\partial\psi/\partial t$ are known at $t = 0$. In nonrelativistic case ψ at $t = 0$ is sufficient to solve the time-dependent Schrödinger equation.

19.3 Dirac Equation for a Free Particle

Dirac set out in seeking for an equation of first-order in t, x, y, z.

19.3.1 Dirac Hamiltonian and Dirac Matrices

Dirac in 1928 proposed the following Hamiltonian for a free particle [4]:

$$E = H = c\boldsymbol{\alpha} \cdot \mathbf{p} + \beta mc^2 , \tag{19.20}$$

where $\boldsymbol{\alpha}$ and β are to be determined. The corresponding wave equation is

$$i\hbar\frac{\partial\psi}{\partial t} = -i\hbar c \left(\boldsymbol{\alpha} \cdot \nabla\right)\psi + \beta mc^2\psi \tag{19.21}$$

which is first-order in space and time. In connection with the Hamiltonian given by Eq. (19.20), Dirac made the following two assumptions:

1. For a free particle H should be independent of space and time coordinates; otherwise such terms indicate the presence of forces. Therefore, $\boldsymbol{\alpha}$ and β should not depend on time.

2. H is linear in $\mathbf{p}$. This implies that $\boldsymbol{\alpha}$ and β must be independent of $\mathbf{p}$. They should also be independent of space and time coordinates in order to describe a free particle.

The E given by Eq. (19.20) must satisfy the relativistic energy-momentum relationship $E^2 = \mathbf{p}^2c^2 + m^2c^4$. Denoting α_1, α_2 and α_3 for α_x, α_y and α_z, respectively, Eq. (19.20) takes the form

$$E = c\sum_{i=1}^{3} \alpha_i p_i + \beta mc^2 \,. \tag{19.22}$$

Squaring Eq. (19.22) we get

$$
\begin{aligned}
E^2 &= c^2\sum_{i} \alpha_i^2 p_i^2 + c^2 \sum_{i,i\neq j}\sum_{j} (\alpha_i\alpha_j + \alpha_j\alpha_i)\, p_i p_j \\
&\quad + mc^3 \sum_{i} (\alpha_i\beta + \beta\alpha_i)\, p_i + m^2c^4\beta^2 \,.
\end{aligned}
\tag{19.23}
$$

Comparison of $E^2 = \mathbf{p}^2c^2 + m^2c^4$ and (19.23) gives

$$\alpha_i\alpha_j + \alpha_j\alpha_i = 0, \quad i \neq j, \quad \alpha_i\beta + \beta\alpha_i = 0, \quad \alpha_i^2 = \beta^2 = 1 \,. \tag{19.24}$$

If α_i and α_j are nonzero numbers then $\alpha_i\alpha_j + \alpha_j\alpha_i = 0$, $i \neq j$ is not true as numbers do not anticommute. Similarly, from $\alpha_i\beta + \beta\alpha_i = 0$ it is clear that β is not a number. However, they can be matrices.

Since $\alpha_i^2 = \beta^2 = I$, the eigenvalues of these matrices are ± 1. We write $\alpha_i = \alpha_i I = \alpha_i\beta^2 = \alpha_i\beta\beta$. Using the relation $\alpha_i\beta + \beta\alpha_i = 0$ we have $\alpha_i = -\beta\alpha_i\beta$. Taking trace on both sides of $\alpha_i = -\beta\alpha_i\beta$ we get tr$\alpha_i = -$tr$\beta\alpha_i\beta$. Since the trace of ABC is same as that of BCA, we get tr$\alpha_i = tr\alpha_i\beta^2 = -tr\alpha_i$ which implies tr$\alpha_i = 0$. Then because the trace is the sum of all the eigenvalues, the numbers $+1$ and -1 must occur same number of times. Hence, the dimension of the matrices α_i and β must be even. If $n = 2$ then the number of possible anticommuting matrices are found to be only 3. They are the *Pauli matrices*:

$$\sigma_1 = \begin{pmatrix} 0 & 1 \\ 1 & 0 \end{pmatrix}, \quad \sigma_2 = \begin{pmatrix} 0 & -i \\ i & 0 \end{pmatrix}, \quad \sigma_3 = \begin{pmatrix} 1 & 0 \\ 0 & -1 \end{pmatrix}. \tag{19.25}$$

But we require four matrices α_1, α_2, α_3 and β. When $n = 4$ there are four matrices and are Hermitian satisfying the relations (19.24). They are

$$
\begin{aligned}
\alpha_1 &= \begin{pmatrix} 0 & \sigma_1 \\ \sigma_1 & 0 \end{pmatrix}, \quad \alpha_2 = \begin{pmatrix} 0 & \sigma_2 \\ \sigma_2 & 0 \end{pmatrix}, \\
\alpha_3 &= \begin{pmatrix} 0 & \sigma_3 \\ \sigma_3 & 0 \end{pmatrix}, \quad \beta = \begin{pmatrix} I & 0 \\ 0 & -I \end{pmatrix}.
\end{aligned}
\tag{19.26}
$$

19.3.2   Dirac Equation in Covariant Form

Rewrite the Eq. (19.21) as

$$i\hbar\frac{\partial\psi}{\partial t} + i\hbar c \sum_{k=1}^{3} \alpha_k \frac{\partial\psi}{\partial x_k} - \beta mc^2\psi = 0 \tag{19.27}$$

or

$$\beta \frac{\partial \psi}{\partial (ict)} - \sum i\beta\alpha_k \frac{\partial \psi}{\partial x_k} + \frac{mc}{\hbar}\psi = 0 \,. \tag{19.28}$$

Define $x_4 = ict$, $\gamma_4 = \beta$, $\gamma_k = -i\beta\alpha_k$, $k = 1, 2, 3$. Then Eq. (19.28) becomes

$$\gamma_4 \frac{\partial \psi}{\partial x_4} + \sum_{k=1}^{3} \gamma_k \frac{\partial \psi}{\partial x_k} + \frac{mc}{\hbar}\psi = 0 \,.$$

That is,

$$\sum_{k=1}^{4} \gamma_k \frac{\partial \psi}{\partial x_k} + \frac{mc}{\hbar}\psi = 0 \,. \tag{19.29}$$

The above equation is called *covariant form of Dirac equation*. In Einstein summation conversion form the above equation is written as

$$\left(\gamma_k \frac{\partial}{\partial x_k} + \frac{mc}{\hbar} \right)\psi = 0 \,. \tag{19.30}$$

We can easily verify that γ matrices are Hermitian, $\gamma^2 = I$, $\gamma_i\gamma_j + \gamma_j\gamma_i = 2\delta_{ij}$ and $\mathrm{Tr}\gamma_i = 0$.

Solved Problem 2:

Let U be the transformation matrix that transforms the wave function ψ in the S frame to the wave function ψ' in the S' frame. Then show that the condition for the relativistic invariance of the Dirac equation is $U^{-1}\gamma_k U L_{k\mu} = \gamma_\mu$, where $L_{k\mu}$ is the Lorentz transformation matrix.

Consider the Dirac Eq. (19.30). We write $\psi' = U\psi$. Since the γ matrices are c-numbers, they remain unchanged in this transformation. We can write

$$\frac{\partial}{\partial x'_k} = \frac{\partial}{\partial x_\mu} \frac{\partial x_\mu}{\partial x'_k} = L_{k\mu} \frac{\partial}{\partial x_\mu} \,. \tag{19.31}$$

We have the Dirac equation in S' frame as

$$\left(\gamma_k \frac{\partial}{\partial x'_k} + \frac{mc}{\hbar} \right)\psi' = 0 \,. \tag{19.32}$$

Substituting $\psi' = U\psi$ and (19.31) in (19.32) we get

$$\left(\gamma_k L_{k\mu} \frac{\partial}{\partial x_\mu} + \frac{mc}{\hbar} \right)U\psi = 0 \,. \tag{19.33}$$

Multiplying this equation from the left with U^{-1}, we get

$$\left(U^{-1}\gamma_k U L_{k\mu} \frac{\partial}{\partial x_\mu} + \frac{mc}{\hbar} \right)\psi = 0 \,. \tag{19.34}$$

The Lorentz invariance requires that Eq. (19.34) must be same as Eq. (19.30). This is true if $U^{-1}\gamma_k U L_{k\mu} = \gamma_\mu$.

19.3.3   Probability and Current Densities

Consider the Dirac equation

$$i\hbar\frac{\partial\psi}{\partial t} = -i\hbar c\,\boldsymbol{\alpha}\cdot\nabla\psi + mc^2\beta\psi\,. \tag{19.35}$$

Multiplication of Eq. (19.35) by $\psi^\dagger$ (ψ is a column vector) from left gives

$$i\hbar\psi^\dagger\frac{\partial\psi}{\partial t} = -i\hbar c\,\psi^\dagger\boldsymbol{\alpha}\cdot\nabla\psi + mc^2\psi^\dagger\beta\psi\,. \tag{19.36}$$

Taking the Hermitian adjoint of Eq. (19.35) and then multiplying it by ψ from right gives

$$-i\hbar\frac{\partial\psi^\dagger}{\partial t}\psi = i\hbar c\,\nabla\psi^\dagger\cdot\boldsymbol{\alpha}\psi + mc^2\psi^\dagger\beta\psi\,. \tag{19.37}$$

Subtraction of Eq. (19.37) from Eq. (19.36) yields

$$\frac{\partial}{\partial t}\left(\psi^\dagger\psi\right) + c\,\nabla\cdot\left(\psi^\dagger\boldsymbol{\alpha}\psi\right) = 0\,. \tag{19.38}$$

Defining $\rho = \psi^\dagger\psi$ and $\mathbf{J} = c\,\psi^\dagger\boldsymbol{\alpha}\psi$, Eq. (19.38) takes the form

$$\frac{\partial\rho}{\partial t} + \nabla\cdot\mathbf{J} = 0 \tag{19.39}$$

which is the continuity equation. Since ρ is nonnegative the Dirac equation has a positive definite probability density. We note that $\int\psi^\dagger\psi\,\mathrm{d}^3\tau$ is conserved. As the Dirac equation is covariant and the derivatives in the Eq. (19.39) form a four-vector, the probability and current densities form a four-vector: $J^k = (\mathbf{J}, ic\rho)$. Then the covariant form of Eq. (19.39) is $\partial_k J^k = 0$.

19.3.4   Plane Wave Solution

Consider the Dirac wave equation for a free particle in the form

$$\left(c\,\boldsymbol{\alpha}\cdot\mathbf{p} + \beta mc^2\right)\psi = E\psi\,. \tag{19.40}$$

Since $\boldsymbol{\alpha}$ and β are 4×4 matrices, the Eq. (19.40) has meaning only when ψ is a four-component function or represented by a 4×1 matrix. Let us seek a plane wave solution of the form

$$\psi(\mathbf{X}, t) = Nu(\mathbf{p})e^{i(\mathbf{p}\cdot\mathbf{X}-Et)/\hbar}\,, \tag{19.41}$$

where N is the normalization constant and u is linear in $\mathbf{p}$. Substitution of Eq. (19.41) in Eq. (19.40) gives

$$\left(c\,\boldsymbol{\alpha}\cdot\mathbf{p} + \beta mc^2\right)u(\mathbf{p}) = Eu(\mathbf{p})\,. \tag{19.42}$$

Since $\boldsymbol{\alpha}$ and β are 4×4 matrices u will have four components and is called *spinor* Therefore, we write

$$u = \begin{pmatrix} u_1 \\ u_2 \\ u_3 \\ u_4 \end{pmatrix} \tag{19.43}$$

and partition it into

$$u = \begin{pmatrix} v \\ w \end{pmatrix}, \quad v = \begin{pmatrix} u_1 \\ u_2 \end{pmatrix}, \quad w = \begin{pmatrix} u_3 \\ u_4 \end{pmatrix}. \tag{19.44}$$

Using Eqs. (19.26) the Eq. (19.42) becomes

$$E \begin{pmatrix} v \\ w \end{pmatrix} - mc^2 \begin{pmatrix} I & 0 \\ 0 & -I \end{pmatrix} \begin{pmatrix} v \\ w \end{pmatrix} = c \begin{pmatrix} 0 & \boldsymbol{\sigma} \cdot \mathbf{p} \\ \boldsymbol{\sigma} \cdot \mathbf{p} & 0 \end{pmatrix} \begin{pmatrix} v \\ w \end{pmatrix}. \tag{19.45}$$

From the above equation we obtain

$$\left(E - mc^2\right) v = c\left(\boldsymbol{\sigma} \cdot \mathbf{p}\right) w, \tag{19.46a}$$

$$\left(E + mc^2\right) w = c\left(\boldsymbol{\sigma} \cdot \mathbf{p}\right) v. \tag{19.46b}$$

When $\mathbf{p}$ is replaced by $-i\hbar\nabla$ Eqs. (19.46) become a set of coupled first-order differential equations for the unknowns v and w. In the following we obtain the energy spectrum and eigenfunctions $u(\mathbf{p})$ from Eqs. (19.46).

19.3.5   Energy Spectrum

Multiplication of Eq. (19.46a) by $(E + mc^2)$ and replacing $(E + mc^2)w$ by the right-side of Eq. (19.46b) we have

$$\left(E^2 - m^2 c^4\right) v = c^2 \left(\boldsymbol{\sigma} \cdot \mathbf{p}\right)^2 v. \tag{19.47}$$

Using the identity $(\boldsymbol{\sigma} \cdot \mathbf{A})(\boldsymbol{\sigma} \cdot \mathbf{B}) = \mathbf{A} \cdot \mathbf{B} + i\boldsymbol{\sigma} \cdot (\mathbf{A} \times \mathbf{B})$, we get $(\boldsymbol{\sigma} \cdot \mathbf{p})^2 = \mathbf{p}^2$ the above equation becomes $\left(E^2 - m^2 c^4 - c^2 \mathbf{p}^2\right) v = 0$, which gives

$$E_\pm = \pm\sqrt{\mathbf{p}^2 c^2 + m^2 c^4}, \quad E_- = -E_+. \tag{19.48}$$

When $|\mathbf{p}| = 0$ we have $E_- = -mc^2$ and $E_+ = mc^2$. As $|\mathbf{p}|$ increases from zero, E_+ starts from mc^2 and extends to ∞. On the other hand, E_- begins from $-mc^2$ and goes to $-\infty$ as $|\mathbf{p}| \to \infty$. Thus, there are positive and negative energy regions separated by an energy gap $2mc^2$. E_- to E_+ is a forbidden energy gap of width $2mc^2$. Since the energy spectra extends to $-\infty$ there is no ground state.

19.3.6   Determination of Eigenfunctions $u(\mathbf{p})$

Equations (19.46) are two-coupled linear equations and their solutions are v and w. To solve them, first, decide whether E_+ or E_- is to be used in Eqs. (19.46). When E is $E_+ = mc^2$ then left-side of Eq. (19.46a) becomes zero. Therefore, choose $E = E_-$ for Eq. (19.46a). Similarly, the choice $E = E_+$ for Eq. (19.46b) is useful. From Eqs. (19.46) for negative energy states

$$v = -\frac{c\boldsymbol{\sigma} \cdot \mathbf{p}}{E_+ + mc^2} w, \tag{19.49a}$$

and for positive energy states

$$w = \frac{c\boldsymbol{\sigma} \cdot \mathbf{p}}{E_+ + mc^2} v. \tag{19.49b}$$

   Earlier, we assumed that u is linear in $\mathbf{p}$. In Eq. (19.49b) if v is a function of $\mathbf{p}$ then w will contain nonlinear terms (that is, power or degree of $\mathbf{p}$ is greater than 1) and consequently u will also have nonlinear terms in $\mathbf{p}$. Therefore, v's are treated independent of $\mathbf{p}$ and can be a constant. Similarly, in Eq. (19.49a) w's are constant. We choose the constants as 0, 1 taking note of the presence of the normalization constant N. There are four possible v's and are

$$\begin{pmatrix} 0 \\ 0 \end{pmatrix}, \begin{pmatrix} 0 \\ 1 \end{pmatrix}, \begin{pmatrix} 1 \\ 0 \end{pmatrix}, \begin{pmatrix} 1 \\ 1 \end{pmatrix}.$$

Since $\begin{pmatrix} 0 \\ 0 \end{pmatrix}$ leads to trivial solution and $\begin{pmatrix} 1 \\ 1 \end{pmatrix}$ is a linear combination of $\begin{pmatrix} 0 \\ 1 \end{pmatrix}$ and $\begin{pmatrix} 1 \\ 0 \end{pmatrix}$ we consider the remaining two choices. $\boldsymbol{\sigma} \cdot \mathbf{p}$ is computed as

$$\boldsymbol{\sigma} \cdot \mathbf{p} = \sigma_x p_x + \sigma_y p_y + \sigma_z p_z = \begin{pmatrix} p_z & p_- \\ p_+ & -p_z \end{pmatrix}, \tag{19.50}$$

where $p_\pm = p_x \pm i p_y$. Now, Eq. (19.49b) becomes

$$w = \frac{c}{E_+ + mc^2} \begin{pmatrix} p_z & p_- \\ p_+ & -p_z \end{pmatrix} v. \tag{19.51}$$

Using $v = \begin{pmatrix} 0 \\ 1 \end{pmatrix}$ and $\begin{pmatrix} 1 \\ 0 \end{pmatrix}$ the following two solutions are obtained:

$$v = \begin{pmatrix} 0 \\ 1 \end{pmatrix}, \quad w = \frac{c}{E_+ + mc^2} \begin{pmatrix} p_- \\ -p_z \end{pmatrix}, \quad u = \begin{pmatrix} v \\ w \end{pmatrix}. \tag{19.52a}$$

$$v = \begin{pmatrix} 1 \\ 0 \end{pmatrix}, \quad w = \frac{c}{E_+ + mc^2} \begin{pmatrix} p_z \\ p_+ \end{pmatrix}, \quad u = \begin{pmatrix} v \\ w \end{pmatrix}. \tag{19.52b}$$

Similarly, for the choices $w = \begin{pmatrix} 0 \\ 1 \end{pmatrix}$ and $\begin{pmatrix} 1 \\ 0 \end{pmatrix}$ we get two more solutions

$$w = \begin{pmatrix} 0 \\ 1 \end{pmatrix}, \quad v = -\frac{c}{E_+ + mc^2} \begin{pmatrix} p_- \\ -p_z \end{pmatrix}, \quad u = \begin{pmatrix} v \\ w \end{pmatrix}. \tag{19.52c}$$

$$w = \begin{pmatrix} 1 \\ 0 \end{pmatrix}, \quad v = -\frac{c}{E_+ + mc^2} \begin{pmatrix} p_z \\ p_+ \end{pmatrix}, \quad u = \begin{pmatrix} v \\ w \end{pmatrix}. \tag{19.52d}$$

The solutions (19.52a) and (19.52b) are positive energy solutions while (19.52c) and (19.52d) are negative energy solutions. Totally, there are four linearly independent solutions.

We rewrite the Eq. (19.52a) as

$$u_1 = 0, \quad u_2 = 1, \quad u_3 = \frac{c}{E_+ + mc^2} p_-, \quad u_4 = -\frac{c}{E_+ + mc^2} p_z. \tag{19.53}$$

Similarly, we express the Eqs. (19.52b-d) in terms of u_1, u_2, u_3 and u_4. The fours solutions given by Eqs. (19.52) are normalized using the condition $\psi^\dagger \psi = 1$ (*why is this form?*). For the solution given by Eq. (19.53) we obtain

$$\begin{aligned} 1 &= |N|^2 \left[u_1^* u_1 + u_2^* u_2 + u_3^* u_3 + u_4^* u_4 \right] \\ &= |N|^2 \left[0 + 1 + \frac{c^2}{(E_+ + mc^2)^2} \left(p_-^2 + p_z^2 \right) \right] \\ &= |N|^2 \left[1 + \frac{c^2 \mathbf{p}^2}{(E_+ + mc^2)^2} \right]. \end{aligned}$$

Therefore, the normalization constant is

$$N = \left[1 + \frac{c^2 \mathbf{p}^2}{(E_+ + mc^2)^2} \right]^{-1/2}. \tag{19.54}$$

Remember that the four solutions are solutions of Dirac equation for a free particle. We may ask: *How are the negative energy solutions are to be understood and what is the reason for the existence of two positive (or negative) energy solutions?* The answer to the second question has been provided by showing that the Dirac equation has spin-1/2 (such as the electron) so that there are two independent spin orientations for given momentum and energy. The wave function ψ of Dirac equation has four components while that of the KG equation has only one component. That is, for Dirac equation more degrees of freedom are available than the required degrees of freedom to describe the translatory motion. Since the additional components are absent for the KG equation, we say that it describes particles of spin zero like π-mesons.

For a free particle we found the energy states which are nondegenerate and there is a forbidden energy range including the value $E = 0$. Now, for a particle in a potential V we ask: *What is the situation for a particle in a potential V? Does degeneracy possible in nonzero potential? Does $E = 0$ allowed for a scalar (that is time-independent) potential? Can we observe energy level crossing?* We answer these important questions in latter sections.

Solved Problem 3:

Construct the plane wave solution of Dirac's free electron in one-dimension.

In one-dimension the Dirac equation $E\psi = (c\alpha_x p_x + \beta mc^2)\psi$ is rewritten as

$$i\hbar \frac{\partial \psi}{\partial t} + i\hbar c \begin{pmatrix} 0 & \sigma_x \\ \sigma_x & 0 \end{pmatrix} \frac{\partial \psi}{\partial x} - mc^2 \begin{pmatrix} I & 0 \\ 0 & -I \end{pmatrix} \psi = 0. \tag{19.55}$$

Writing ψ in four-components form we have

$$i\hbar \frac{\partial \psi_1}{\partial t} + i\hbar c \frac{\partial \psi_4}{\partial x} - mc^2 \psi_1 = 0, \tag{19.56a}$$

$$i\hbar \frac{\partial \psi_2}{\partial t} + i\hbar c \frac{\partial \psi_3}{\partial x} - mc^2 \psi_2 = 0, \tag{19.56b}$$

$$i\hbar \frac{\partial \psi_3}{\partial t} + i\hbar c \frac{\partial \psi_2}{\partial x} + mc^2 \psi_3 = 0, \tag{19.56c}$$

$$i\hbar \frac{\partial \psi_4}{\partial t} + i\hbar c \frac{\partial \psi_1}{\partial x} + mc^2 \psi_4 = 0. \tag{19.56d}$$

An interesting observation is that the equations for the four components of ψ are coupled in pairs only; ψ_1 with ψ_4; ψ_2 with ψ_3. This suggests that we can introduce the two component spinor $\begin{pmatrix} \psi_u \\ \psi_l \end{pmatrix}$, where $\psi_u = \psi_1$ or ψ_2 and $\psi_l = \psi_4$ or ψ_3 with u and l stand for 'upper' and 'lower', respectively.

For stationary states Eqs. (19.56) become

$$i\hbar c\, \psi_1' + \left(E + mc^2\right) \psi_4 = 0, \quad i\hbar c\, \psi_2' + \left(E + mc^2\right) \psi_3 = 0, \tag{19.57a}$$

$$i\hbar c\, \psi_3' + \left(E - mc^2\right) \psi_2 = 0, \quad i\hbar c\, \psi_4' + \left(E - mc^2\right) \psi_1 = 0, \tag{19.57b}$$

where we have written the last equation of (19.56) as first equation and so on. Equations (19.57a) are rewritten in single equation as

$$i\hbar c\, \psi_u' + \left(E + mc^2\right) \psi_l = 0. \tag{19.58a}$$

Similarly, Eqs. (19.57b) are written as

$$i\hbar c\, \psi_l' + \left(E - mc^2\right) \psi_u = 0. \tag{19.58b}$$

For simplicity we write ψ_u as v and ψ_l as w so that Eqs. (19.58) read as

$$i\hbar c\, v' + \left(E + mc^2\right) w = 0, \tag{19.59a}$$
$$i\hbar c\, w' + \left(E - mc^2\right) v = 0. \tag{19.59b}$$

Differentiating (19.59a) with respect to x and using Eq. (19.59b) we get

$$v'' + \frac{p^2}{\hbar^2} v = 0, \quad p^2 = \frac{1}{c^2}\left(E^2 - m^2 c^4\right). \tag{19.60}$$

Then w is given by $w = -(i\hbar c)v'/\left(E + mc^2\right)$. Now, we need to determine v only. The general solution of Eq. (19.60) is $v(x) = Ae^{ipx/\hbar} + Be^{-ipx/\hbar}$.

19.3.7   Negative Energy States

We have seen that the Dirac's theory admits both positive and negative energy solutions. The energy eigenvalues of the plane wave solutions are

$$E_+ = +\left(c^2\mathbf{p}^2 + m^2 c^4\right)^{1/2} \geq mc^2, \tag{19.61a}$$
$$E_- = -\left(c^2\mathbf{p}^2 + m^2 c^4\right)^{1/2} \leq -mc^2. \tag{19.61b}$$

Thus, the energy of a free particle in the Dirac's theory is continuous and ranges from $-\infty$ to ∞ except for a forbidden gap of width $2mc^2$.

An energy gap of width $2mc^2$ exists in the Dirac's theory of electron. We cannot ignore the negative energy solutions without giving a physical interpretation. As electron can undergo a transition from higher energy states to lower energy states, we must consider the possibility of transitions from E_+ to E_- energy states. Hence, the negative energy states cannot be regarded as independent of positive energy states.

An excited state makes a transition to a lower level by emitting a photon. When it is in its ground state, it cannot make any transition to lower states as there are no further lower energy states. But in Dirac's theory there is no lower bound as the allowed negative energies range from $-mc^2$ to $-\infty$. Therefore, an electron with a positive energy E_+ can make a transition onto the negative energy state E_- emitting a γ-ray. Furthermore, once it reaches a negative energy state, it will continuously lower its energy indefinitely by emitting photons since there is no lower bound to the negative energy spectrum. Ultimately there will be no electron in positive energy states at all.

To resolve this difficulty Dirac proposed that all the negative energy states are filled under normal conditions. The catastrophic transitions mentioned above are then disallowed because of the Pauli's exclusion principle. So, the normal state of the vacuum contains of an infinite density of negative energy electrons. When a negative energy electron from this *Fermi sea* is excited into the positive energy states, a hole is left in the sea. This hole allows another negative-energy electron to make a transition to it, leaving a hole elsewhere in the sea. In semiconductor physics this hole acts as a positive charged particle.

Dirac considered first the particle as proton. But Julius Robert Oppenheimer and Hermann Klaus Hugo Weyl theoretically proved that the particle must have the same mass as electrons. In 1932 Carl David Anderson discovered in a Wilson cloud chamber of cosmic rays a track occurring due to a particle of charge $+e$ and mass equal to that of electron. This was identified as the hole in the Dirac's theory and is named *positron*.

The absorption of a photon in the Fermi sea of energy in excess of $2mc^2$ leads to the creation of an electron-positron pair. To ensure conservation of energy and momentum, this event must take place in the presence of a third body carrying away the momentum.

A positive energy electron can fall to this hole and fill the Fermi sea. The decay energy of this electron-positron annihilation is carried away in two photons. Note that physically observable positron has positive energy values. For the conservation of momentum and spin, the positive energy positron will have opposite momentum and spin of the negative energy electron states. But the helicity $\sigma' \cdot \mathbf{p}$ will not change, since the velocity $\mathbf{v} = c^2 \mathbf{p}/E$ as $-E \to +E$, $-\mathbf{p} \to \mathbf{p}$ will be same for the positive energy positron and negative energy electron.

With an infinite sea of electrons, Dirac's theory is no longer the one-particle theory. A many particle theory of Dirac equation can be obtained using the formulation of quantized field. Even though the energy spectrum of free KG particle is identical to the free Dirac particle, it is not possible to construct a sensible hole theory out of KG particles which obey Bose–Einstein statistics.

19.3.8 Jittery Motion of a Free Particle

The velocity of a free particle is

$$\frac{d\mathbf{X}}{dt} = \frac{1}{i\hbar} [\mathbf{X}, H] = -\frac{c}{i\hbar} \boldsymbol{\alpha} \cdot \mathbf{p}\mathbf{X} = c\boldsymbol{\alpha} . \tag{19.62}$$

The time evolution of $\boldsymbol{\alpha}$ is found to be

$$\frac{d\boldsymbol{\alpha}}{dt} = \frac{1}{i\hbar} [\boldsymbol{\alpha}, H] = \frac{1}{i\hbar} \left(c\mathbf{p} + \boldsymbol{\alpha}\beta mc^2 - H\boldsymbol{\alpha}\right) . \tag{19.63}$$

Adding and subtracting $H\boldsymbol{\alpha}$ to the above equation we get

$$\begin{aligned}
\frac{d\boldsymbol{\alpha}}{dt} &= \frac{1}{i\hbar} \left(c\mathbf{p} + \boldsymbol{\alpha}\beta mc^2 - H\boldsymbol{\alpha} + H\boldsymbol{\alpha} - H\boldsymbol{\alpha}\right) \\
&= \frac{1}{i\hbar} \left(c\mathbf{p} + \boldsymbol{\alpha}\beta mc^2 + c\mathbf{p} + \beta\boldsymbol{\alpha} mc^2 - 2H\boldsymbol{\alpha}\right) \\
&= \frac{2}{i\hbar} \left(c\mathbf{p} - H\boldsymbol{\alpha}\right) .
\end{aligned} \tag{19.64}$$

Since H and $\mathbf{p}$ are time-independent the Eq. (19.64) is written as

$$\frac{d\boldsymbol{\alpha}}{dt} = \frac{2iH}{\hbar} \left(-\frac{c\mathbf{p}}{H} + \boldsymbol{\alpha}\right) . \tag{19.65}$$

Defining $\boldsymbol{\alpha}' = \boldsymbol{\alpha} - c\mathbf{p}/H$ the above equation becomes $d\boldsymbol{\alpha}'/dt = (2iH/\hbar)\boldsymbol{\alpha}'$ whose solution is $\boldsymbol{\alpha}'(t) = \boldsymbol{\alpha}'(0)e^{2iHt/\hbar}$. Then $\boldsymbol{\alpha}(t)$ is obtained as

$$\boldsymbol{\alpha}(t) = \frac{c\mathbf{p}}{H} + e^{2iHt/\hbar} \left(\boldsymbol{\alpha}(0) - \frac{c\mathbf{p}}{H}\right) . \tag{19.66}$$

Now, Eq. (19.62) gives

$$\frac{d\mathbf{X}}{dt} = \frac{c^2\mathbf{p}}{H} + c\,e^{2iHt/\hbar} \left(\boldsymbol{\alpha}(0) - \frac{c\mathbf{p}}{H}\right) . \tag{19.67}$$

Although the momentum $\mathbf{p}$ is a constant of motion, that is independent of time, the velocity is not. The solution of Eq. (19.67) is

$$\mathbf{X}(t) = \mathbf{X}(0) + \frac{c^2\mathbf{p}\,t}{H} + \frac{\hbar}{2iH} \left(e^{2iHt/\hbar} - 1\right) \left(c\boldsymbol{\alpha}(0) - \frac{c^2\mathbf{p}}{H}\right) . \tag{19.68}$$

By analogy with classical free particle we expect the term linear in t. In Eq. (19.68) the first two terms of the right-side describe the uniform motion and it corresponds to the classical dynamics of a free relativistic particle. The last term is a feature of relativistic quantum mechanics. This term oscillates extremely rapidly with high frequency, $\omega \sim mc^2/\hbar$, (or period $\sim \hbar/(mc^2)$) and with amplitude $\hbar/(mc)$ *(how?)*. This high frequency oscillatory or jittery or trembling motion is called *zitterbewegung*. The frequency ω is so high that the deviation from the classical mechanics term $c^2\mathbf{p}t/H \approx \mathbf{v}t$ is undetectable. This paradox arises from the coexistence of particles and antiparticles in Dirac's theory and that only a quantum field theory description predicting pair creation and annihilation could cure this problem. The zitterbewegung has fascinated many physicists [5–7] ever since, because the spin of the electron can be associated with the orbital angular momentum of the internal helical motion.

The zitterbewegung amplitude X is equal to the electron Compton wavelength [8] $X = c/\omega_0 = \hbar/(mc) = 3.8 \times 10^{-13}\,\mathrm{m}$ and $s = XE_0/c = \hbar/2$ and $E_0 = s\omega_0 = mc^2/2$. The last property implies $V_0 = mc^2/2$. Further, since the electron is charged, the zitterbewegung motion generates a magnetic moment of magnitude $\mu = ec\pi X^2/(2\pi Xc) = eX/2$. For more discussion on zitterbewegung one may refer to ref. [8].

Now, we may ask: *Does the zitterbewegung occur only for spin-1/2 particles? Can spin-0 particles exhibit such a phenomenon? What is the cause of zitterbewegung? When does it vanish?* In order to get the answers for these questions zitterbewegung has been examined for spin-0 particles using the Hamiltonian form of the KG equation [9]. For a spin-0 particle with the KG Hamiltonian

$$H = (\sigma_3 + i\sigma_2)\frac{\mathbf{p}^2}{2m} + m\sigma_3 \tag{19.69}$$

$\mathbf{X}$ is obtained as

$$\mathbf{X}(t) = \mathbf{X}(0) + \frac{\mathbf{p}t}{H} + \frac{(e^{2iHt}-1)}{2iH}\left(\mathbf{v}(0) - \frac{\mathbf{p}}{H}\right). \tag{19.70}$$

The last term describes the zitterbewegung.

Next, we evaluate $\langle \mathbf{X}(t)\rangle$ with a wave packet, which is a superposition of eigenstates of $\mathbf{p}$ and H, given by

$$\psi_{\mathbf{k}}^{\sigma}(\mathbf{X}) = \frac{\omega^{\sigma}}{(2\pi)^{3/2}}\,e^{i\mathbf{k}\cdot\mathbf{X}} \tag{19.71a}$$

with

$$\sigma = \pm 1, \quad \omega = \sqrt{k^2 + m^2}, \quad \omega^{\sigma} = \frac{1}{2\sqrt{m\omega}}\begin{pmatrix} m + \sigma\omega \\ m - \sigma\omega \end{pmatrix}. \tag{19.71b}$$

The states with $\sigma = 1$ and -1 have energies ω and $-\omega$, respectively. Writing the wave packet in the form $|\psi\rangle = \sum_{\sigma}\int |\psi_{\mathbf{k}}^{\sigma}\rangle a_{\mathbf{k}}^{\sigma}\,d^3k$ we obtain

$$\begin{aligned}
\langle \mathbf{X}(t)\rangle &= \langle \mathbf{X}(0)\rangle + \sum_{\sigma}\int \frac{\mathbf{k}}{\omega}|a_{\mathbf{k}}^{\sigma}|^2 t\,d^3k \\
&\quad + 2\int \frac{\mathbf{k}}{\omega^2}\sin\omega t\,\mathrm{R.P.}\left(a_{\mathbf{k}}^{(-)*}a_{\mathbf{k}}^{(+)*}e^{-i\omega t}\right)d^3k.
\end{aligned} \tag{19.72}$$

The last term describes oscillatory motion. If the wave packet contains only positive or only negative energy waves then the last term vanishes and hence the absence of zitterbewegung. Thus, the zitterbewegung is caused by the interference between the positive and negative energy states in the wave packet.

The zitterbewegung effect has not been observed experimentally for electrons because the corresponding frequency of the free electron is $\approx 10^{20}$Hz. It was recognized that the zitterbewegung effect can appear in solids where the electron at the boundary between two interacting bands obeys the Dirac equation [10–12]. This effect has been studied through simulations in ultra cold atoms [13], superconductivity [14], a single trapped ion [15,16] and graphene [17,18]. Zitterbewegung was studied in non-Hermitian photonic waveguide systems [19] and in Bogoliubov's system [20]. An optical analog of the effect was found in a two-dimensional photonic crystal composed by a triangular lattice of the dielectric cylinders where two bands touch each other forming a double-cone structure with a linear dispersion [21,22]. The optical zitterbewegung effect was demonstrated by solving Maxwell's equations for the electromagnetic pulses propagating through a negative-zero-positive index metamaterial [23].

Solved Problem 4:

Show that the zitterbewegung effect can be removed by projecting the standard position operator to the particle and antiparticle subspaces of Hilbert space, respectively.

We introduce the projection operators

$$P_0^\pm = \frac{1}{2}\left(I_H + |H|^{-1}H\right), \quad |H| = (H^*H)^{1/2} \tag{19.73}$$

with $P_0^+ P_0^- = 0$ and $P_0^+ + P_0^- = I_H$. The time evolution of the projected position operators $\hat{x}_0^\pm = P_0^\pm \hat{x} P_0^\pm$ are given by

$$\hat{x}_0^\pm(t) = \hat{x}_0^\pm(0) + c^2 \mathbf{p} H^{-1} t P_0^\pm . \tag{19.74}$$

It exactly corresponds to the classical expression. This is valid for particles and as well as for antiparticles. The interpretation is that particles and antiparticles show different time evolutions and the zitterbewegung represents the interference term

$$P_0^+ \hat{x} P_0^- + P_0^- \hat{x} P_0^+ = \frac{i\hbar}{2} H^{-1} \hat{F} \tag{19.75}$$

which is absent in a classical case.

19.4 Minimum Uncertainty Wave Packet

In the nonrelativistic case as shown in sec. 8.4, the condition for a wave function $\phi(x)$ that is normalizable to possess a minimum uncertainty product is given by Eq. (8.53). Suppose we assume that $\langle x \rangle = \langle p_x \rangle = 0$. We can write

$$-i\hbar \frac{d\phi(x)}{dx} \propto x\phi(x), \tag{19.76}$$

where the proportionality factor is pure imaginary. The solution of this equation is

$$\phi(x) = \left(\frac{1}{2\pi\sigma^2}\right)^{1/4} e^{-x^2/(4\sigma^2)} . \tag{19.77}$$

Further,

$$(\Delta x)^2 = \langle x^2 \rangle = \int_{-\infty}^{\infty} x^2 \phi^2(x) \mathrm{d}x = \sigma^2, \tag{19.78a}$$

$$(\Delta p_x)^2 = \langle p_x^2 \rangle = -\hbar^2 \int_{-\infty}^{\infty} \left(\frac{\mathrm{d}\phi(x)}{\mathrm{d}x} \right)^2 \mathrm{d}x = \frac{\hbar^2}{4\sigma^2}. \tag{19.78b}$$

For the Dirac wave function $\psi(x) = \begin{pmatrix} v(x) \\ w(x) \end{pmatrix}$ with $\int_{-\infty}^{\infty} \left(|v|^2 + |w|^2 \right) \mathrm{d}x = 1$ the quantity $\langle x^2 \rangle$ is given by

$$\langle x^2 \rangle = \langle x^2 \rangle_v + \langle x^2 \rangle_w. \tag{19.79}$$

In this equation $\langle x^2 \rangle_v$ and $\langle x^2 \rangle_w$ are with respect to $v(x)$ and $w(x)$, respectively. Similarly,

$$\langle p_x^2 \rangle = \langle p_x^2 \rangle_v + \langle p_x^2 \rangle_w. \tag{19.80}$$

The minimum uncertainty relation is satisfied provided $v(x)$ and $w(x)$ separately satisfy Eq. (19.76). Suppose the relativistic $\psi(x)$ is

$$\psi(x) = \begin{pmatrix} v(x) \\ w(x) \end{pmatrix} = \begin{pmatrix} C_v \\ C_w \end{pmatrix} \phi(x), \tag{19.81}$$

where $\phi(x)$ is given by Eq. (19.77). Then for the coefficients C_v and C_w the condition is $C_v^2 + C_w^2 = 1$ and in Eqs. (19.78) ϕ^2 and $(\mathrm{d}\phi/\mathrm{d}x)^2$ are to be replaced by $|\psi|^2$ and $|\mathrm{d}\psi/\mathrm{d}x|^2$. Now, for this $\psi(x)$ we find $\Delta x \Delta p_x = \hbar/2$.

In the nonrelativistic case, the minimum uncertainty wave function given by Eq. (19.77) is the ground state eigenfunction of the linear harmonic oscillator. Now, one may ask: *What is the potential of the system admitting the solution of the form (19.81)?* To obtain such a potential consider the Dirac Hamiltonian in one-dimension [24]

$$H = c\alpha p + \beta S(x). \tag{19.82}$$

For the choice $\alpha = \sigma_y$, $\alpha p = -\mathrm{i}\hbar\sigma_y \frac{\mathrm{d}}{\mathrm{d}x}$ is real and hence one may chose ψ as real. For simplicity assume $V(x) = 0$. Then $H\psi = E\psi$ becomes

$$(\mathrm{i}cp + S)v = Ew, \quad (-\mathrm{i}cp + S)w = Ev. \tag{19.83}$$

When $E = 0$ elimination of S in the above equations after substituting $p = -\mathrm{i}\hbar\frac{\mathrm{d}}{\mathrm{d}x}$ gives $\frac{\mathrm{d}}{\mathrm{d}x}vw = 0$. That is, $vw = $ constant. ψ is normalizable if $vw = 0$. For $v = 0$ the solution of the second equation $(-\mathrm{i}cp + S)w = 0$ is

$$w(x) = N\mathrm{e}^{1/(\hbar c)} \int_0^x S(x)\mathrm{d}x, \tag{19.84}$$

where N is a normalization constant. For $\psi(x)$ to be normalizable S must be such that $S(\infty) < 0$ and $S(-\infty) > 0$ [25]. Since an one-dimensional state cannot be degenerate the above state is nondegenerate. $S(x)$ can be aribtrary, however, it should satisfy the conditions $S(\infty) < 0$ and $S(-\infty) > 0$. A choice which represents a minimum uncertainty wave packet is $S(x) = -(\hbar c/(2\sigma^2))x$ for which

$$v(x) = 0, \quad w(x) = N\mathrm{e}^{-x^2/(4\sigma^2)} = \phi(x). \tag{19.85}$$

For this wave packet $(\Delta x)(\Delta p) = \hbar/2$.

19.5 Spin of a Dirac Particle

In this section, we show that for a Dirac particle orbital angular momentum is not conserved but the sum of orbital angular momentum and spin angular momentum is conserved.

In the Heisenberg picture, the time evolution of an operator is described by Eq. (7.30). For a Dirac particle with the Hamiltonian (19.20) the time evolution of orbital angular momentum $\mathbf{L}$ is given by

$$-i\hbar \frac{d}{dt}(\mathbf{i}L_1 + \mathbf{j}L_2 + \mathbf{k}L_3) = [H, \mathbf{i}L_1 + \mathbf{j}L_2 + \mathbf{k}L_3], \tag{19.86}$$

where L_1, L_2 and L_3 represent L_x, L_y and L_z, respectively. Consider the component L_3. We write

$$-i\hbar \frac{d}{dt}L_3 = [H, L_3] = \left[\sum_{k=1}^{3} c\,\alpha_k p_k + \beta mc^2, \, x_1 p_2 - x_2 p_1\right]. \tag{19.87}$$

Since β is a matrix and $x_1 p_2$ and $x_2 p_1$ are scalars the quantity $[\beta mc^2, x_1 p_2 - x_2 p_1] = 0$. Therefore,

$$
\begin{aligned}
-i\hbar \frac{d}{dt}L_3 &= \left[\sum c\,\alpha_k p_k, x_1 p_2\right] - \left[\sum c\,\alpha_k p_k, x_2 p_1\right] \\
&= -i\hbar c \sum \alpha_k \delta_{k1} p_2 + i\hbar c \sum \alpha_k \delta_{k2} p_1 \\
&= -i\hbar c\,(\alpha_1 p_2 - \alpha_2 p_1) \\
&= -i\hbar c\,(\boldsymbol{\alpha} \times \mathbf{p})_3 .
\end{aligned}
$$

That is, $dL_3/dt = c(\boldsymbol{\alpha} \times \mathbf{p})_3$. Similarly, $dL_2/dt = c(\boldsymbol{\alpha} \times \mathbf{p})_2$ and $dL_1/dt = c(\boldsymbol{\alpha} \times \mathbf{p})_1$. Therefore,

$$\frac{d}{dt}\mathbf{L} = c(\boldsymbol{\alpha} \times \mathbf{p}) \tag{19.88}$$

which is not zero. Hence, $\mathbf{L}$ is not a constant of motion or a conserved quantity. However, it implies that $\mathbf{L} + \mathbf{A}$, where $\mathbf{A}$ is such that $d\mathbf{A}/dt = -c(\boldsymbol{\alpha} \times \mathbf{p})$ is a conserved quantity. In the following we show that $\mathbf{A} = \hbar\boldsymbol{\sigma}'/2$, where $\boldsymbol{\sigma}' = \begin{pmatrix} \boldsymbol{\sigma} & 0 \\ 0 & \boldsymbol{\sigma} \end{pmatrix}$.

The equation of motion of σ_3' is simplified into

$$
\begin{aligned}
-i\hbar \frac{d\sigma_3'}{dt} &= [H, \sigma_3'] \\
&= \left[\sum c\,\alpha_k p_k + \beta mc^2, \sigma_3'\right] \\
&= \sum c\,p_k\,[\alpha_k, \sigma_3'] \\
&= c\,p_1\,[\alpha_1, \sigma_3'] + c\,p_2\,[\alpha_2, \sigma_3'] \\
&= 2ic\,(\alpha_1 p_2 - p_1 \alpha_2) , \tag{19.89}
\end{aligned}
$$

where we have used the result $[\beta mc^2, \sigma_3'] = 0$, $[\alpha_3, \sigma_3'] = 0$, $\sigma_1'\sigma_3' = -\sigma_3'\sigma_1'$ and $\sigma_1'\sigma_3' = -i\sigma_2'$. That is, $(\hbar/2)d\sigma_3'/dt = -c(\boldsymbol{\alpha} \times \mathbf{p})_3$. Therefore, we write

$$\frac{d}{dt}\frac{\hbar}{2}\boldsymbol{\sigma}' = -c(\boldsymbol{\alpha} \times \mathbf{p}) . \tag{19.90}$$

Combining Eqs. (19.88) and (19.90) we obtain

$$\frac{d}{dt}\left(\mathbf{L} + \frac{\hbar}{2}\boldsymbol{\sigma}'\right) = 0. \tag{19.91}$$

$\hbar\boldsymbol{\sigma}'/2$ is called *spin angular momentum* and $\mathbf{J} = \mathbf{L} + (\hbar/2)\boldsymbol{\sigma}'$ is the *total angular momentum*. Thus, $\mathbf{J}$ is a conserved quantity. Hence, we conclude that the Dirac particle is endowed with spin $\hbar\boldsymbol{\sigma}'/2$.

Solved Problem 5:

Show that $\hbar k = \beta(\boldsymbol{\sigma}' \cdot \mathbf{L} + \hbar) = \beta(\boldsymbol{\sigma}' \cdot \mathbf{J} - \hbar/2)$, where $\boldsymbol{\sigma}' = \begin{pmatrix} \boldsymbol{\sigma} & 0 \\ 0 & \boldsymbol{\sigma} \end{pmatrix}$.

We have $\hbar k = \beta(\boldsymbol{\sigma}' \cdot \mathbf{L} + \hbar)$, $\mathbf{L} = \mathbf{J} - \hbar\boldsymbol{\sigma}'/2$. Hence,

$$\begin{aligned}
\hbar k &= \beta\left[\boldsymbol{\sigma}' \cdot \left(\mathbf{J} - \frac{\hbar}{2}\boldsymbol{\sigma}'\right) + \hbar\right] = \beta\left[\boldsymbol{\sigma}' \cdot \mathbf{J} - \frac{\hbar}{2}|\boldsymbol{\sigma}'|^2 + \hbar\right] \\
&= \beta\left[\boldsymbol{\sigma}' \cdot \mathbf{J} - \frac{\hbar}{2}\right]
\end{aligned} \tag{19.92}$$

since $|\boldsymbol{\sigma}'|^2 = {\sigma'_x}^2 + {\sigma'_y}^2 + {\sigma'_z}^2 = 3$.

19.6   Particle in a Potential

In the following we construct the solution of Dirac equation for a particle in a potential V [26]. We consider the Dirac Hamiltonian

$$\begin{aligned}
H &= c\boldsymbol{\alpha} \cdot \mathbf{p} + \beta mc^2 + V \\
&= -i\hbar c\frac{d}{dz}\begin{pmatrix} 0 & -i \\ i & 0 \end{pmatrix} + \begin{pmatrix} 1 & 0 \\ 0 & -1 \end{pmatrix} mc^2 + V,
\end{aligned} \tag{19.93}$$

where $\boldsymbol{\alpha}$ is chosen as α_3. Now, the equation $H\psi = E\psi$ is written as

$$\left[-i\hbar c\frac{d}{dz}\begin{pmatrix} 0 & -i \\ i & 0 \end{pmatrix} + mc^2\begin{pmatrix} 1 & 0 \\ 0 & -1 \end{pmatrix} + V\right]\begin{pmatrix} v \\ w \end{pmatrix} = E\begin{pmatrix} v \\ w \end{pmatrix}. \tag{19.94}$$

From the above equation we obtain

$$-\hbar c\,w' + v\left(mc^2 + V\right) = Ev, \quad \hbar c\,v' - w\left(mc^2 + V\right) = Ew, \tag{19.95a}$$

where prime denotes differentiation with respect to z. Suppose there are two bound state solutions

$$\psi_1 = \begin{pmatrix} v_1 \\ w_1 \end{pmatrix}, \quad \psi_2 = \begin{pmatrix} v_2 \\ w_2 \end{pmatrix} \tag{19.96}$$

with eigenvalues E_1 and E_2, respectively. Now, the Eqs. (19.95) are

$$\begin{aligned}
-\hbar c\,w'_1 + \left(mc^2 + V\right)v_1 &= E_1 w_1, & (19.97a) \\
\hbar c\,v'_1 - \left(mc^2 + V\right)w_1 &= E_1 v_1, & (19.97b) \\
-\hbar c\,w'_2 + \left(mc^2 + V\right)v_2 &= E_2 v_2, & (19.97c) \\
\hbar c\,v'_2 - \left(mc^2 + V\right)w_2 &= E_2 w_2. & (19.97d)
\end{aligned}$$

Multiplying Eq. (19.97a) by v_2 and Eq. (19.97c) by v_1 and subtracting one from another we get

$$\hbar c \left(v_1 w_2' - w_1' v_2\right) = \left(E_1 - E_2\right) v_1 v_2. \tag{19.98}$$

Similarly, from Eqs. (19.97b) and (19.97d) we have

$$\hbar c \left(v_1' w_2 - w_1 v_2'\right) = \left(E_1 - E_2\right) w_1 w_2. \tag{19.99}$$

Combining of the above two equations gives

$$\hbar c \frac{d}{dz} \left(v_1 w_2 - w_1 v_2\right) = \left(E_1 - E_2\right) \psi_1^* \psi_2. \tag{19.100}$$

If $E_1 = E_2$, that is the eigenstates are degenerate, then

$$\frac{d}{dz} \left(v_1 w_2 - w_1 v_2\right) = 0 \text{ or } v_1 w_2 - w_1 v_2 = \text{constant}. \tag{19.101}$$

Because v and $w \to 0$ as $|z| \to \infty$, this constant must be zero. Hence, $v_1/w_1 = v_2/w_2$ from which we obtain

$$\frac{v'}{v} = \frac{\left(E + mc^2 + V\right)}{\hbar c} \frac{w}{v}. \tag{19.102}$$

That is,

$$\frac{v_1'}{v_1} = \frac{\left(E + mc^2 + V\right)}{\hbar c} \frac{w_1}{v_1} = \frac{\left(E + mc^2 + V\right)}{\hbar c} \frac{w_2}{v_2} = \frac{v_2'}{v_2}. \tag{19.103}$$

Equations $v_1/w_1 = v_2/w_2$ and (19.102) imply that $\psi_1 \propto \psi_2$. Thus, they represent the same state. Therefore, the energy states are nondegenerate irrespective of the potential. Further, if $E_1 \neq E_2$ [26],

$$\langle \psi_1 | \psi_2 \rangle = \int_{-\infty}^{\infty} \psi_1^* \psi_2 \, dz = \frac{1}{\left(E_1 - E_2\right)} \left(v_1 w_2 - w_1 v_2\right) |_{-\infty}^{\infty} = 0. \tag{19.104}$$

That is, ψ_1 and ψ_2 are orthogonal. This condition with the nondegeneracy of energy implies that *level crossing* cannot occur. When the potential is varied smoothly, for example, by varying a parameter in it then the energies and wave functions will also vary smoothly. Here by level crossing we mean that two energies E_1 and E_2 cross each other. This requires a singular behaviour of $\langle \psi_1 | \psi_2 \rangle$ and is zero for $E_1 \neq E_2$ and suddenly jumps to a finite value at $E_1 = E_2$. This is not possible when ψ varies smoothly.

19.7 Klein Paradox

The step potential in the relativistic Dirac equation gives rise to the peculiar situation called *Klein paradox* [27–29].

19.7.1 Scattering by Step Potential

Let us consider the step potential, Eq. (5.45) and Fig. 5.8. For $x < 0$ the Dirac equation is the free particle equation given by Eq. (19.60). For $x > 0$ the corresponding equation is obtained by replacing E in Eq. (19.60) (p is a function of E) by $E - V_0$. From the solution for $x < 0$ the solution for $x > 0$ is obtained by replacing E by $E - V_0$ since $V(x)$ in $x > 0$ region is a constant. Let us obtain the solution [28].

The solution of Eq. (19.60) is $v_<(x) = A\left(e^{ipx/\hbar} + Re^{-ipx/\hbar}\right)$. Then

$$w_<(x) = A\left(ae^{ipx/\hbar} - aRe^{-ipx/\hbar}\right),\tag{19.105}$$

where $a = cp/(E + mc^2)$. Now,

$$\begin{aligned}\psi_< = \begin{pmatrix} v_< \\ w_< \end{pmatrix} &= A\left[\begin{pmatrix} 1 \\ a \end{pmatrix}e^{ipx/\hbar} + R\begin{pmatrix} 1 \\ -a \end{pmatrix}e^{-ipx/\hbar}\right]\\ &= A\left[u_+e^{ipx/\hbar} + Ru_-e^{-ipx/\hbar}\right],\end{aligned}\tag{19.106}$$

where $u_\pm = \begin{pmatrix} 1 \\ \pm a \end{pmatrix}$. For $x > 0$ there is no reflected wave. Hence,

$$\psi_> = \begin{pmatrix} v_> \\ w_> \end{pmatrix} = D\,\bar{u}\,e^{i\bar{p}x/\hbar},\tag{19.107a}$$

where

$$\begin{aligned}\bar{u} &= \begin{pmatrix} 1 \\ b \end{pmatrix},\quad \bar{p} = \frac{1}{c}\left[(E - V_0)^2 - m^2c^4\right]^{1/2},\\ b &= \frac{c\bar{p}}{(E - V_0 + mc^2)}.\end{aligned}\tag{19.107b}$$

To determine R and D we use the boundary conditions: $\psi_<(0) = \psi_>(0)$ and $\psi'_<(0) = \psi'_>(0)$. These conditions lead to $A(u_+ + Ru_-) = D\bar{u}$. That is,

$$A\left[\begin{pmatrix} 1 \\ a \end{pmatrix} + R\begin{pmatrix} 1 \\ -a \end{pmatrix}\right] = D\begin{pmatrix} 1 \\ b \end{pmatrix}.\tag{19.108}$$

We get $A(1 + R) = D$, $Aa(1 - R) = bD$. Solving them results in

$$R = \frac{a - b}{a + b},\quad T = \frac{2a}{a + b}.\tag{19.109}$$

The behaviour of the wave functions given by Eqs. (19.106) and (19.107) depends on V_0. We consider the following cases:

1. $E > V_0 + mc^2$.
2. $V_0 - mc^2 < E < V_0 + mc^2$.
3. $E < V_0 - mc^2$.

Case 1: $E > V_0 + mc^2$

In this case $E^2 > m^2c^4 + V_0^2 + 2mc^2V_0$. Then $E^2 - m^2c^4 > 0$. That is, $p^2c^2 = E^2 - m^2c^4$ is positive and hence p is real. Similarly, $E - V_0 > mc^2$ or $(E - V_0)^2 > m^2c^4$ or $\bar{p} = [(E - V_0)^2 - m^2c^4]^{1/2} > 0$. So, $\bar{p}$ is real. Then in Eq. (19.106) $e^{ipx/\hbar}$ and $e^{-ipx/\hbar}$ represent incident wave moving towards positive x-direction and reflected wave moving towards negative x-direction, respectively. $e^{i\bar{p}x/\hbar}$ in Eq. (19.107) represents transmitted wave, moving in the positive x-direction in region-*II* ($x > 0$). There is no reflected wave in region-*II*. This behaviour is similar to the nonrelativistic Schrödinger equation for $E > V_0$.

Case 2: $V_0 - mc^2 < E < V_0 + mc^2$

In this case $(E - V_0)^2 < m^2c^4$. Hence, $\bar{p}$ in Eq. (19.107b) is imaginary. Then

$$\psi_>(x) = AT\bar{u} + \mathrm{e}^{-\bar{p}x/\hbar}, \quad \bar{p} = \left[\left|(E - V_0)^2 - m^2c^4\right|\right]^{1/2}. \tag{19.110}$$

The wave in the region-II is an exponentially decaying wave. In the region-I, we have the right-moving incoming and the left-moving reflected wave. This situation is also similar to the nonrelativistic case for $0 < E < V_0$.

Case 3: $E < V_0 - mc^2$

We have $E - V_0 < -mc^2$ and $E - V_0$ negative. Further, $(E - V_0)^2 > m^2c^4$ and hence $\bar{p}$ is real leading to an oscillatory solution in region-II (the solution in the region-I is oscillatory). In the nonrelativistic case, for $E < V_0$, no such solution is obtained. Furthermore, $E - V_0 + mc^2$ is negative and thus b is negative. We write

$$|R| = \left|\frac{a - b}{a + b}\right| = \left|\frac{a + |b|}{a - |b|}\right| > 1. \tag{19.111}$$

The paradoxical result is that $|R| > 1$. *The amplitude of the reflected wave is larger than the incident wave!* So, more particles are reflected by the step than the incident on it. This result was first reported by Oskar Klein. The above situation is called the *Klein paradox*. In the nonrelativistic case, for $E < V_0$, there is no oscillatory solution in the potential region $(x > 0)$ and R is unity. In contrast, in the relativistic case, even when E is much smaller than V_0 we have oscillatory waves in region-II. The occurrence of Klein paradox has been verified in a numerical simulation [30].

19.7.2   Interpretation of Klein Paradox in Terms of the Hole Theory

The resolution of the problem is associated with pair production [31]. The potential is strong enough to generate particle-antiparticle pairs from vacuum. That is, multi-particle phenomenon is interpreted using a simple single particle wave function. As shown in Fig. 19.1 if $E + mc^2 < V_0$ then the positive energy continuum in region-I contacts with the negative energy continuum in region-II at the boundary. At the boundary, incident electrons stimulate the vacuum in the region-II, that is, filled by the negative energy electrons. As a result, negative energy electrons are ejected into region-I (being observed as positive energy electrons). Holes remaining in region-II are interpreted as positrons. The incident electrons induce pair creation. Here the reflected current becomes larger than the incident current because it consists of reflected electrons and created ones. The current in the region-II is regarded as the flow of positrons.

The Klein paradox is treated as the simplest model of the decay of the charge vacuum in a field creating electron-positron pairs at the potential barrier [32]. Discussion on Klein paradox using Feynman's picture of antiparticles as negative energy solutions travelling backward in time is presented in ref. [28]. Klein paradox was not found for bosons [33].

19.7.3   Further Works on Klein Paradox

The transmission coefficient for the potential

$$V(x) = \begin{cases} \nu x, & 0 < x < L \\ 0, & x < 0 \\ \nu L, & x > L \end{cases} \tag{19.112}$$

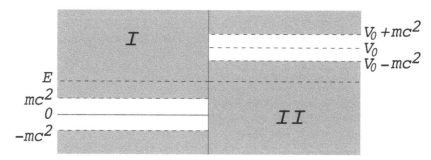

FIGURE 19.1
Positive and negative energy in a step potential.

has been calculated [34]. To obtain negative energy states (holes) to propagate through the barrier, we require $\nu L > 2m$. The potential represents a constant electric field $E = -\nu$ in a finite region of space. After a lengthy calculation involving hypergeometric functions, the obtained R and T reduced to the Klein values of R and T for $|\nu| \sim m^2 c^4$. For weaker fields he obtained $R \approx 1$, $T \approx e^{-\pi mc^2/\nu}$ which shows the exponentially suppressed tunnelling typical of quantum phenomena.

For the square-barrier potential $V(x) = V_0$ for $|x| < a$ and 0 for $|x| > a$ the reflection and transmission coefficients are obtained as [35]

$$R = \frac{(1-k)^2 \sin^2 2pa}{4k^2 + (1-k^2)^2 \sin^2 2pa}, \tag{19.113a}$$

$$T = \frac{4k^2}{4k^2 + (1-k^2)^2 \sin^2 2pa}, \tag{19.113b}$$

where

$$k^2 = \frac{\left(V - E + mc^2\right)\left(E + mc^2\right)}{\left(V - E - mc^2\right)\left(E - mc^2\right)}. \tag{19.113c}$$

$R = 0$ and $T = 1$ when $2pa = n\pi$, where n is an integer. The electron passes through the barrier without reflection and is called a *transmission resonance*.

As a becomes very large for fixed mc^2, E and V the quantity pa becomes very large and $\sin pa$ oscillates very rapidly. In this case we average over the phase angle pa using $\sin^2 pa = \cos^2 pa = 1/2$ and obtain

$$R_\infty = \frac{(1-k^2)^2}{8k^2 + (1-k^2)}, \quad T_\infty = \frac{8k^2}{8k^2 + (1-k^2)^2}. \tag{19.114}$$

For a study on relativistic quantum scattering by locally periodic potentials see ref.[36]. Scattering of a low momentum Dirac particle by an arbitrary potential of finite range is also studied [37]. It has been shown that the amplitude for reflection is 1 and the transmission amplitude is 0 in general. If the potential supports a half-bound state at momentum $k = 0$, this result does not hold. In the case of an asymmetric potential, the transmission coefficient is found to be nonzero and is 1 for a symmetric potential. For discussions on relativistic linear harmonic oscillator one may refer to refs. [38–41]. It has been shown that the Klein paradoxial behaviour can be tested experimentally using electrostatic barriers in single- and bi-layer graphene [42]. For a quantum simulation of Klein paradox one may refer to the ref. [43].

19.8   Relativistic Particle in a Box

In sec. 4.6 the problem of a nonrelativistic particle in a box potential is solved exactly. For the box of width L described by the equation

$$V(x) = \begin{cases} 0, & 0 < x < L \\ V_0, & \text{elsewhere} \end{cases} \tag{19.115}$$

the wave function and energy levels are given by

$$\psi(x) = \begin{cases} A \sin kx, & 0 < x < L \\ 0, & \text{elsewhere} \end{cases} \quad \text{and} \quad E_n = \frac{n^2 \hbar^2 \pi^2}{2mL^2}. \tag{19.116}$$

Now, we consider a relativistic particle in the box potential [44].

19.8.1   Wave Function of a Relativistic Particle

Assume that the mass of the particle is a function of x. We write $m(x) = m$ for $0 < x < L$ and M elsewhere, where M is a constant which will be set to ∞ later. Consider the solutions of the Dirac equation separately in regions I $(x < 0)$, II $(0 < x < L)$ and III $(x > L)$. In each of these regions the solution is of the form of solution of Dirac's free particle, since $m(x)$ is constant in each of them. Inside the well the solution consists of a plane incident wave and a reflected wave on the walls of the well. Outside the well the solution has one wave travelling outwards. For a bound solution $\psi(x) = 0$ in the limit $x \to \pm\infty$. The wave functions in the three regions are given by

$$\psi_{\mathrm{I}}(x) = Ae^{-ik'x} \begin{pmatrix} v \\ -P\sigma_z vk'/k \end{pmatrix}, \tag{19.117a}$$

$$\psi_{\mathrm{II}}(x) = Be^{ikx} \begin{pmatrix} v \\ P\sigma_x v \end{pmatrix} + Ce^{-ikx} \begin{pmatrix} v \\ -P\sigma_z v \end{pmatrix}, \tag{19.117b}$$

$$\psi_{\mathrm{III}}(x) = De^{ik'x} \begin{pmatrix} v \\ P\sigma_z vk'/k \end{pmatrix}, \tag{19.117c}$$

where

$$P = \frac{\hbar kc}{E + mc^2}, \quad k = \frac{\sqrt{E^2 - m^2 c^4}}{\hbar c}, \quad k' = \frac{\sqrt{E^2 - M^2 c^4}}{\hbar c} \tag{19.117d}$$

and A, B, C and D are constants. In the limit $M \to \infty$, E is less than mc^2 and $k' = i\sqrt{M^2 c^4 - E^2}/\hbar c = ik''$. Since k' is an imaginary number, the exponentials in ψ_I and ψ_{II} are $e^{k''x}$ and $e^{-k''x}$, respectively. In the region I in the limit $M \to \infty$, $e^{k''x} \to 0$. In the region-III, $e^{-k''x} \to 0$ as $M \to \infty$. That is, ψ_{I} and $\psi_{\mathrm{III}} \to 0$ as $M \to \infty$.

19.8.2   Relations Between the Constants B and C

To find the relation between the constants B and C and to get the energy eigenvalues use the boundary conditions at $x = 0$ and $x = L$. The boundary condition is that the outward flux of probability at the walls of the well be zero. It is given by $\pm(-i)\beta\alpha_x\psi = \psi$, where the minus sign is for $x = 0$ and plus sign for $x = L$. Let us define $\bar{\psi} = \psi^*\beta$. Multiplication of $\pm(-i)\beta\alpha_x\psi = \psi$ by $\bar{\psi}$ from left and replacing $\bar{\psi}$ by $\psi^*\beta$ in the left-side of the equation and then using the relation $\beta^2 = 1$ we get $\pm(-i)\psi^*\alpha_x\psi = \bar{\psi}\psi$. The above equation states that the probability current density at $x = 0$ and L is $\bar{\psi}\psi$.

The boundary condition at $x = 0$ gives $i\beta\alpha_x\psi_{II}(0) = \psi_{II}(0)$. That is

$$i\begin{pmatrix} 0 & \sigma_x \\ \sigma_x & 0 \end{pmatrix}\left\{ B\begin{pmatrix} v \\ P\sigma_x v \end{pmatrix} + C\begin{pmatrix} v \\ -P\sigma_x v \end{pmatrix}\right\}$$

$$= B\begin{pmatrix} v \\ P\sigma_x v \end{pmatrix} + C\begin{pmatrix} v \\ -P\sigma_x v \end{pmatrix}. \tag{19.118}$$

The first row gives $(iBP - iCP - B - C)v = 0$ from which we obtain

$$C = B\frac{iP - 1}{iP + 1}. \tag{19.119}$$

Next, consider the condition $\psi_{II}^*\beta\psi_{II} = 0$. This gives

$$|B + C|^2 = 0, \quad -P^2|B - C|^2 = 0. \tag{19.120}$$

Combining the two equations of Eq. (19.120) we get $|B + C|^2 - P^2|B - C|^2 = 0$. The same result is obtained from the boundary condition at $x = L$.

19.8.3   Relativistic Probability Density in the Region II

Since the modulus of $(iP - 1)/(iP + 1)$ is 1 let us define

$$\frac{iP - 1}{iP + 1} = e^{i\delta}, \quad \delta = \tan^{-1}\left(\frac{2P}{P^2 - 1}\right). \tag{19.121}$$

Then Eq. (19.119) becomes $C = Be^{i\delta}$. Now, ψ_{II} is written as

$$\psi_{II}(x) = Be^{i\delta/2}\begin{pmatrix} 2\cos(kx - \delta/2)v \\ 2iP\sin(kx - \delta/2)\sigma_x v \end{pmatrix}. \tag{19.122}$$

To compute the energy eigenvalue we consider the boundary condition at $x = L$. The boundary condition together with Eq. (19.117d) gives $\tan kL = -\hbar k/(mc)$. For nonrelativistic momentum $\hbar k \ll mc$ and $P \approx 0$. Then $\delta \sim \tan^{-1}(0/(-1)) = \pi$. Substituting $\delta = \pi$ and $P = 0$ in Eq. (19.122) we get

$$\psi_{II}(x) = Be^{i\pi/2}\begin{pmatrix} 2\sin(kx)v \\ 0 \end{pmatrix} \tag{19.123}$$

which is exactly the nonrelativistic result, Eq. (19.116), apart from a normalization factor and v. The relativistic probability density is obtained as

$$\psi_{II}^*\psi_{II} = 4|B|^2\left[1 + \left(P^2 - 1\right)\sin^2(kx - \delta/2)\right]. \tag{19.124}$$

For $k = 0$ the wave function is zero inside the well. For $k \neq 0$, because $0 < P < 1$, the quantity $\psi_{II}^*\psi_{II}$ never become zero.

19.9   Relativistic Hydrogen Atom

The relativistic hydrogen-like atom is one of the most important applications of the Dirac equation. It is also one of more difficult of the solvable problems. In this section, we take up the Dirac equation of hydrogen atom.

19.9.1   Dirac Equation for Hydrogen Atom

The Dirac equation for the description of electron in hydrogen atom is

$$
E\psi = [c\boldsymbol{\alpha}\cdot\mathbf{p} + \beta mc^2 + V]\,\psi = mc^2\left[\beta + \frac{c\boldsymbol{\alpha}\cdot\mathbf{p}}{mc^2} - \frac{Ze^2}{mc^2 r}\right].
\tag{19.125}
$$

Defining $E = mc^2\mathcal{E}$ and $e^2/(mc^2) = s^2$ Eq. (19.125) is rewritten as

$$
\mathcal{E}mc^2\psi = mc^2\left[\beta + \frac{\boldsymbol{\alpha}\cdot\mathbf{p}}{mc^2} - \frac{Zs^2}{r}\right].
\tag{19.126}
$$

The radial component of the above equation is

$$
mc^2\mathcal{E}R = mc^2\left[\beta - s\alpha_r\sum\left(\frac{\partial}{\partial r} - \frac{\Lambda}{r}\right) - \frac{Zs^2}{r}\right]R,
\tag{19.127a}
$$

where

$$
\beta = \begin{pmatrix} I & 0 \\ 0 & -I \end{pmatrix},
\tag{19.127b}
$$

$$
\alpha_r = \begin{pmatrix} 0 & -i\sigma_r \\ i\sigma_r & 0 \end{pmatrix} = \begin{pmatrix} 0 & -iI \\ iI & 0 \end{pmatrix}\sum_r,
\tag{19.127c}
$$

$$
\sum_r = \frac{\sigma\cdot\mathbf{r}}{r}, \quad \Lambda = \sigma\cdot\mathbf{1}.
\tag{19.127d}
$$

The solutions of Eq. (19.127) are given by [45]

$$
R = \begin{pmatrix} f(r) & R_l^{(a)} \\ ig(r) & R_{l+1}^{(b)} \end{pmatrix} \text{ and } R = \begin{pmatrix} f(r) & R_{l+1}^{(b)} \\ g(r) & R_l^{(a)} \end{pmatrix},
\tag{19.128}
$$

where the behaviour of the angular functions under the angular operators is

$$
\Lambda R_l^{(a)} = l R_l^{(a)}, \quad \Lambda R_l^{(b)} = -(l + D - 2)R_l^{(b)},
$$
$$
\sum R_l^{(a)} = -R_{l+1}^{(b)}, \quad \sum R_{l+1}^{(b)} = -R_l^{(a)},
\tag{19.129}
$$

where D is the dimension. Substitution of Eq. (19.128) into Eq. (19.127) gives two pairs of first-order differential equations for the functions f and g. After much manipulation, these functions are expressed in terms of the solutions of the second-order differential equation [44]

$$
y'' + \left\{\frac{1 + \frac{2\mathcal{E}Zs}{\lambda}}{2\rho} + \left[4Z^2s^2 - \frac{(2l + D)(2l + D - 2)}{4\rho^2}\right] - \frac{1}{4}\right\}y = 0,
\tag{19.130}
$$

where $\lambda = \sqrt{1 - \mathcal{E}^2}$. The good news is that the solution of the above equation is obtained by comparing it with the Eq. (12.78). We obtain $p = \pm\left[(2l + D - 1)^2 - 4Z^2s^2\right]^{1/2}$. Further,

$$
E = mc^2\left[1 + 4Z^2s^2(2n - 2l - 2 + p)^{-2}\right]^{-1/2}
\tag{19.131}
$$

which for $D = 3$ is approximated to

$$
E = mc^2\left[1 - \frac{\gamma^2}{2n^2} - \frac{\gamma^4}{2n^4}\left(\frac{n}{|k|} - \frac{3}{4}\right)\right],
\tag{19.132a}
$$

$$
\gamma = \frac{Ze^2}{\hbar c}, \quad n = n' + |k|, \quad n' = 0, 1, 2, \ldots
\tag{19.132b}
$$

and $|k|$ can take on positive integer values. $E - mc^2$ is the rest energy. The term $-\gamma^2 mc^2/(2n^2)$ gives the nonrelativistic formula.

For $D = 1$, since $l = 0$ the value of p is $\pm(-4Zs^2)^{1/2}$, that is, p is complex. The bad news is that imaginary values of p are not permitted and hence the Dirac equation for the electron in one-dimensional hydrogen atom has no bound state solutions! Although the potential is attractive, the Dirac equation allows no hydrogen in one-dimension. The system has solutions in one-dimension but does not provide bound states. This is the case for $D = 2$, $l = 0$. In contrast to the above, bound states exist for nonrelativistic hydrogen atom in arbitrary dimension. Further, the Schwinger model [46] in quantum field theory produces bound states! *What has caused the failure of relativistic theory?* We shall answer this [47] in the following.

19.9.2   Paradox Resolved

The source of the problem is the Klein paradox [47]. Consider an electron in the central potential. It is attracted to the centre of the potential and the allowed regions are now separated by a V-shaped forbidden region. Let the energy of the electron be $\bar{E} = mc^2 \mathcal{E}$ where $\mathcal{E}$ represents sum of kinetic and potential energies. The Dirac equation has two kinds of solutions—one in the inner allowed region and the another in the sea restricted to the outer region.

In the hole theory, a particle from the sea with an energy $\bar{E}$ is able to penetrate through the forbidden zone into the central region, leaving a hole in the sea. The hole is pushed away from the centre of the potential and behaves like an antiparticle. This creation of pairs discredits the Dirac equation in the case of one-dimensional hydrogen atom. Relativistic equations of motion are single particle equations and they cover only the situations where pulling of particle–antiparticle pairs out of a vacuum is suppressed. In three-dimensional hydrogen, a creation of virtual pairs is a higher-order process and therefore the Dirac equation works well. In contrast, in one-dimensional analysis the creation of pairs is not a small quantum fluctuation. Single particle equations cannot cope with situations in which particle creation is a prominent process. Thus, when the Dirac equation says there are no one-dimensional hydrogen atom, it does not mean that such relativistic bound states are absent but only means that this equation is not applicable to the problem.

19.10   The Electron in a Field

The vector potential $\mathbf{A}$ and the scalar potential ϕ, form a four-vector $A^\mu = (\mathbf{A}, i\phi/c)$. Now, the Dirac equation for an electron in an electromagnetic field is obtained by replacing $\mathbf{p}$ by $\mathbf{p} - e\mathbf{A}/c$ and E by $E - e\phi$, in the equations

$$\left(E - mc^2\right) v = c\left(\boldsymbol{\sigma} \cdot \mathbf{p}\right) w\,, \quad \left(E + mc^2\right) w = c\left(\boldsymbol{\sigma} \cdot \mathbf{p}\right) v\,. \tag{19.133}$$

We obtain

$$\left(E - e\phi - mc^2\right) v = c\boldsymbol{\sigma} \cdot \left(\mathbf{p} - \frac{e\mathbf{A}}{c}\right) w\,, \tag{19.134a}$$

$$\left(E - e\phi + mc^2\right) w = c\boldsymbol{\sigma} \cdot \left(\mathbf{p} - \frac{e\mathbf{A}}{c}\right) v\,. \tag{19.134b}$$

We are interested in positive energy solutions $E = mc^2 + \mathcal{E}$. Substitution of this in Eq. (19.134b) gives

$$\left(2mc^2 + \mathcal{E} - e\phi\right) w = c\boldsymbol{\sigma} \cdot \left(\mathbf{p} - \frac{e\mathbf{A}}{c}\right) v . \tag{19.135}$$

In the case of nonrelativistic motion in a weak field ϕ and $\mathcal{E}$ are so small (compared to mc^2), we neglect $\mathcal{E}$ and $-e\phi$ in Eq. (19.135) so that

$$w \approx \frac{1}{2mc}\boldsymbol{\sigma} \cdot \left(\mathbf{p} - \frac{e\mathbf{A}}{c}\right) v . \tag{19.136}$$

Use of Eq. (19.136) in Eq. (19.134a) gives

$$\frac{1}{2m}\left[\boldsymbol{\sigma} \cdot \left(\mathbf{p} - \frac{e\mathbf{A}}{c}\right)\right]^2 v = (\mathcal{E} - e\phi)v . \tag{19.137}$$

Using $(\boldsymbol{\sigma} \cdot \mathbf{B})(\boldsymbol{\sigma} \cdot \mathbf{C}) = \mathbf{B} \cdot \mathbf{C} + i\boldsymbol{\sigma} \cdot (\mathbf{B} \times \mathbf{C})$ with $B = C = (\mathbf{p} - e\mathbf{A}/c)$ we have

$$\begin{aligned}
\left[\boldsymbol{\sigma} \cdot \left(\mathbf{p} - \frac{e\mathbf{A}}{c}\right)\right]^2 &= \left(\mathbf{p} - \frac{e\mathbf{A}}{c}\right)^2 + i\boldsymbol{\sigma} \cdot \left\{\left(\mathbf{p} - \frac{e\mathbf{A}}{c}\right) \times \left(\mathbf{p} - \frac{e\mathbf{A}}{c}\right)\right\} \\
&= \left(\mathbf{p} - \frac{e\mathbf{A}}{c}\right)^2 - \frac{e\hbar}{c}\boldsymbol{\sigma} \cdot \mathbf{B} ,
\end{aligned} \tag{19.138}$$

where $\mathbf{B} = \nabla \times \mathbf{A}$ is the magnetic field. Thus, Eq. (19.137) becomes

$$\left\{\frac{1}{2m}\left(\mathbf{p} - \frac{e\mathbf{A}}{c}\right)^2 - \frac{e\hbar}{2mc}\boldsymbol{\sigma} \cdot \mathbf{B} + e\phi\right\} v = \mathcal{E}v . \tag{19.139}$$

This equation is known as *Pauli equation* and is similar to the nonrelativistic Schrödinger equation with the additional term $-e\hbar\boldsymbol{\sigma} \cdot \mathbf{B}/(2mc)$. This term suggests that the electron in a magnetic field have an extra energy $-\boldsymbol{\mu} \cdot \mathbf{B} = -e\hbar\boldsymbol{\sigma} \cdot \mathbf{B}/(2mc)$ and behaves as if it has a *magnetic moment* $\boldsymbol{\mu}$ associated with its spin, namely,

$$\boldsymbol{\mu} = \frac{e\hbar\boldsymbol{\sigma}}{2mc} = \mu_B\boldsymbol{\sigma} , \tag{19.140}$$

where $\mu_B = e\hbar/(2mc)$ is called the *Bohr magnetron*.

The Hamiltonian of the Dirac equation contains, in the nonrelativistic approximation, a term which takes the intrinsic spin and magnetic properties of the electron into account. This has been found to be in agreement with experiments. From (19.140) we find $\mu_z = \mu_B\sigma_z = (2/\hbar)\mu_B S_z$. Hence, the ratio of the spin magnetic moment to the spin angular momentum is equal to $e/(mc)$, which is twice the value obtained for orbital motion.

19.11 Spin-Orbit Energy

We show that for central potential $V(r)$ the Dirac equation gives rise spin-orbit interaction. Defining $\mathbf{p}' = \mathbf{p} - (e\mathbf{A}/c)$ the Dirac equation is written as

$$\begin{aligned}
E \begin{pmatrix} v \\ w \end{pmatrix} &= c \begin{pmatrix} 0 & \boldsymbol{\sigma} \cdot \mathbf{p}' \\ \boldsymbol{\sigma} \cdot \mathbf{p}' & 0 \end{pmatrix} \begin{pmatrix} v \\ w \end{pmatrix} + mc^2 \begin{pmatrix} I & 0 \\ 0 & -I \end{pmatrix} \begin{pmatrix} v \\ w \end{pmatrix} \\
&\quad + V \begin{pmatrix} v \\ w \end{pmatrix} .
\end{aligned} \tag{19.141}$$

Write the Eq. (19.141) in the form

$$\left(E - mc^2 - V\right) v = c\boldsymbol{\sigma} \cdot \mathbf{p}' w, \tag{19.142a}$$

$$\left(E + mc^2 - V\right) w = c\boldsymbol{\sigma} \cdot \mathbf{p}' v. \tag{19.142b}$$

Using Eq. (19.142b) for w in Eq. (19.142a) we have

$$\left(E - mc^2 - V\right) v = c^2 (\boldsymbol{\sigma} \cdot \mathbf{p}')^2 f^{-1} v, \quad f = \left(E + mc^2 - V\right). \tag{19.143}$$

Now, consider the result $[A, I] = [A, BB^{-1}] = 0$. From this we obtain $ABB^{-1} = B^{-1}BA$ or $ABB^{-1}B^{-1} = B^{-1}BAB^{-1}$. That is,

$$\begin{aligned} AB^{-1} &= B^{-1}BAB^{-1} - B^{-1}ABB^{-1} + B^{-1}ABB^{-1} \\ &= -B^{-1}[A, B]B^{-1} + B^{-1}A. \end{aligned} \tag{19.144}$$

Using the above relation with $A = \boldsymbol{\sigma} \cdot \mathbf{p}'$ and $B = f^{-1}$ we get

$$\begin{aligned} (\boldsymbol{\sigma} \cdot \mathbf{p}')f^{-1} &= f^{-1}(\boldsymbol{\sigma} \cdot \mathbf{p}') - f^{-1}\left[\boldsymbol{\sigma} \cdot \mathbf{p}', f\right] f^{-1} \\ &= f^{-1}(\boldsymbol{\sigma} \cdot \mathbf{p}') - f^{-2}(-i\hbar)\boldsymbol{\sigma} \cdot \nabla V. \end{aligned} \tag{19.145}$$

Then Eq. (19.143) is

$$fv = c^2 \left\{ f^{-1}(\boldsymbol{\sigma} \cdot \mathbf{p}')^2 + i\hbar f^{-2}(\boldsymbol{\sigma} \cdot \nabla V)(\boldsymbol{\sigma} \cdot \mathbf{p}') \right\}. \tag{19.146}$$

Next, using the relation $(\boldsymbol{\sigma} \cdot \mathbf{B})(\boldsymbol{\sigma} \cdot \mathbf{C}) = \mathbf{B} \cdot \mathbf{C} + i\boldsymbol{\sigma} \cdot (\mathbf{B} \times \mathbf{C})$ we have

$$(\boldsymbol{\sigma} \cdot \mathbf{p}')^2 = \mathbf{p}'^2 - \frac{e\hbar}{c}\boldsymbol{\sigma} \cdot \mathbf{H}_B, \tag{19.147a}$$

$$(\boldsymbol{\sigma} \cdot \nabla V)(\boldsymbol{\sigma} \cdot \mathbf{p}') = \nabla V \cdot \mathbf{p}' + i\boldsymbol{\sigma} \cdot \nabla V \times \mathbf{p}'. \tag{19.147b}$$

For a spherically symmetric potential we use

$$\nabla V = \frac{1}{r}\frac{dV}{dr}\mathbf{r}, \tag{19.148a}$$

$$\nabla V \cdot \mathbf{p} = \frac{1}{r}\frac{dV}{dr}\mathbf{r} \cdot (-i\hbar\nabla) = -i\hbar\frac{dV}{dr}\frac{\partial}{\partial r}, \tag{19.148b}$$

$$\nabla V \times \mathbf{p} = \frac{1}{r}\frac{dV}{dr}(\mathbf{r} \times \mathbf{p}) = \frac{1}{r}\frac{dV}{dr}(\mathbf{r} \times \mathbf{p}) = \frac{1}{r}\frac{dV}{dr}\mathbf{L}. \tag{19.148c}$$

Then Eq. (19.146) becomes

$$\begin{aligned} \left(E - mc^2\right) v = \Bigg\{ V + c^2 f^{-1} &\left(\mathbf{p}'^2 - \frac{e\hbar}{c}\boldsymbol{\sigma} \cdot \mathbf{H}_B\right) \\ &- i\hbar c^2 f^{-2}\left[\nabla V \cdot \mathbf{p} - \nabla V \cdot \frac{e\mathbf{A}}{c}\right. \\ &\left. + i\boldsymbol{\sigma} \cdot \nabla V \times \mathbf{p} - i\boldsymbol{\sigma} \cdot \left(\nabla V \times \frac{e\mathbf{A}}{c}\right)\right]\Bigg\} v. \end{aligned} \tag{19.149}$$

Writing $E' = E - mc^2$, $|E'| \ll mc^2$, $|V| \ll mc^2$ we have

$$f^{-1} = \left(2mc^2 + E' - V\right)^{-1} \approx \frac{1}{2mc^2}\left(1 - \frac{E' - V}{2mc^2}\right) \tag{19.150}$$

and $f^{-2} \approx 1/(4m^2c^4)$. When $\mathbf{A} = 0$, $\mathbf{H}_B = 0$ we have

$$E'\left(1 + \frac{\mathbf{p}^2}{4m^2c^2}\right)v = \left[V + \frac{1}{2m}\left(1 + \frac{V}{2mc^2}\right)\mathbf{p}^2 - \frac{\hbar^2}{4m^2c^2}\frac{dV}{dr}\frac{\partial}{\partial r}\right.$$

$$\left. + \frac{1}{2m^2c^2}\frac{1}{r}\frac{dV}{dr}\mathbf{S}\cdot\mathbf{L}\right]v. \tag{19.151}$$

Defining the terms within the square-bracket as F, it is rewritten as

$$E'v = \left(1 + \frac{\mathbf{p}^2}{4m^2c^2}\right)^{-1}Fv \approx \left(1 - \frac{\mathbf{p}^2}{4m^2c^2}\right)Fv. \tag{19.152}$$

Keeping only terms up to the order $(mc^2)^{-2}$ in the above equation we get

$$E'v \approx \left[V + \frac{\mathbf{p}^2}{2m} - \frac{\mathbf{p}^4}{8m^3c^2} + \frac{\hbar^2}{4m^2c^2}\left\{\nabla^2 V + \frac{dV}{dr}\frac{\partial}{\partial r}\right\}\right.$$

$$\left. + \frac{1}{2m^2c^2}\frac{1}{r}\frac{dV}{dr}\mathbf{S}\cdot\mathbf{L}\right]v. \tag{19.153}$$

In Eq. (19.153) the first two terms in the right-side are the nonrelativistic Hamiltonian. The third terms is the relativistic correction to the kinetic energy $\mathbf{p}^2/2m$. The relativistic kinetic energy is

$$E - mc^2 = \left(c^2\mathbf{p}^2 + m^2c^4\right)^{1/2} - mc^2 = \frac{\mathbf{p}^2}{2m} - \frac{\mathbf{p}^4}{8m^3c^2}. \tag{19.154}$$

The last term in Eq. (19.153) is the spin-orbit energy, that is,

$$H_{\text{SOI}} = \frac{1}{2m^2c^2}\frac{1}{r}\frac{dV}{dr}\mathbf{S}\cdot\mathbf{L}. \tag{19.155}$$

The remaining terms in Eq. (19.153) do not find a physical meaning in terms of classical effects.

19.12 Relativistic Quaternionic Quantum Mechanics

In sec. 6.12 we briefly introduced the topic of quaternionic quantum mechanics. The present section is concerned with relativistic quaternionic quantum mechanics.

Consider a quaternionic wave function $\psi = ((i/c)\psi_0, \boldsymbol{\psi})$ associated with a particle. The evolution equations for $\boldsymbol{\psi}$ and ψ_0 are [48]

$$\nabla \cdot \boldsymbol{\psi} - \frac{1}{c^2}\frac{\partial \psi_0}{\partial t} - \frac{m}{\hbar}\psi_0 = 0, \tag{19.156a}$$

$$\nabla \psi_0 - \frac{\partial \boldsymbol{\psi}}{\partial t} - \frac{mc^2}{\hbar}\boldsymbol{\psi} = 0, \tag{19.156b}$$

$$\nabla \times \boldsymbol{\psi} = 0. \tag{19.156c}$$

These equations are found to be invariant under the transformation

$$t \to it, \quad \mathbf{X} \to i\mathbf{X}, \quad m \to -im. \tag{19.157}$$

From Eqs. (19.156) using appropriate commutator brackets the following wave equations are obtained:

$$\frac{1}{c^2}\frac{\partial^2\psi}{\partial t^2} - \nabla^2\psi + \frac{2m}{\hbar}\frac{\partial\psi}{\partial t} + \frac{m^2c^2}{\hbar^2}\psi = 0, \tag{19.158}$$

$$\frac{1}{c^2}\frac{\partial^2\psi_0}{\partial t^2} - \nabla^2\psi_0 + \frac{2m}{\hbar}\frac{\partial\psi_0}{\partial t} + \frac{m^2c^2}{\hbar^2}\psi_0 = 0. \tag{19.159}$$

We define

$$\boldsymbol{\psi} = \boldsymbol{\alpha}\psi, \quad \psi_0 = -c\psi \tag{19.160}$$

and let

$$m \to im\beta. \tag{19.161}$$

Substitution of Eqs. (19.160) and (19.161) in Eq. (19.156a) gives the Dirac equation [49]. Further, use of Eq. (19.161) in Eq. (19.159) leads to

$$\frac{1}{c^2}\frac{\partial^2\psi}{\partial t^2} - \nabla^2\psi + \frac{2mi\beta}{\hbar}\frac{\partial\psi}{\partial t} - \frac{m^2c^2}{\hbar^2}\psi = 0. \tag{19.162}$$

Making use of the Dirac equation in Eq. (19.162) results in

$$\frac{1}{c^2}\frac{\partial^2\psi}{\partial t^2} - \nabla^2\psi + \frac{2mi}{\hbar}\beta c\boldsymbol{\alpha}\cdot\nabla\psi + \frac{m^2c^2}{\hbar^2}\psi = 0 \tag{19.163}$$

and is a new form of the Dirac equation. When the dissipation term in Eq. (19.158) is neglected then it reduces to the KG equation.

19.13   Concluding Remarks

In nonrelativistic quantum mechanics the central field Hamiltonian commutes with the orbital angular momentum operator $\mathbf{L}$. But in the relativistic theory we have $[H, \mathbf{L}] = -i\hbar c\,\boldsymbol{\alpha}\times\mathbf{p}$. Using the commutation relations of the Pauli matrices, we can show that for the spin angular momentum $\mathbf{S}$ we have $[H, \mathbf{S}] = i\hbar c\,\boldsymbol{\alpha}\times\mathbf{p}$. If we take $\mathbf{J} = \mathbf{L} + \mathbf{S}$ then $[H, \mathbf{J}] = 0$, that is, the total angular momentum $\mathbf{J}$ commutes with the Dirac Hamiltonian H. Since H is linear in $\mathbf{p}$ it does not commute with the ordinary nonrelativistic parity operator P. From the relations $[\boldsymbol{\alpha}, \beta] = 0$, the desired commutation relation is recovered for a parity operator of the form βP and we obtain $[H, \beta P] = 0$. Further, the parity operator commutes with $\mathbf{J}$.

Quantum simulation [50], semiclassical approach [51] and supersymmetry [52] of Dirac equation have been reported. Bound states of the Dirac equation for a pseudoscalar linear plus Coulomb-like potential [53] and solvability of Dirac equation with Lorentz scalar potential [54] and Mie type potentials [55] were investigated. The Dirac equation has been used to describe the electron transport in graphene [56].

A quasi-relativistic wave equation the solutions of which match with those of nonrelativistic quantum mechanics for slow moving particles and valid for particles with relativistic speeds is introduced and studied [57–61]. A relativistic non-Hermitian quantum mechanics is developed [62].

19.14 Bibliography

[1]   G.E. Uhlenbeck and S. Goudsmit, *Nature* 117:264, 1926.

[2]   O. Klein, *Z. Phys.* 37:895, 1926; W. Gordon, *Z. Physik* 40:117, 1926.

[3]   V.P. Dmitriyev, *Aperion* 8:1, 2001.

[4]   P.A.M. Dirac, *Proc. Roy. Soc. Lond. A* 117:610, 1928.

[5]   A.O. Barut and M.G. Cruz, *Eur. J. Phys.* 15:119, 1994.

[6]   B. Thaller, *The Dirac Equation.* Springer, Berlin, 1992.

[7]   J. Bolte and R. Glaser, *J. Phys. A* 37:6359, 2004.

[8]   D. Hestenes, *Found. Phys.* 15:63, 1983.

[9]   M.G. Fuda and E. Furlani, *Am. J. Phys.* 50:545, 1982.

[10] F. Cannata, L. Ferrari and G. Russo, *Solid State Commun.* 74:309, 1990.

[11] F. Cannata and L. Ferrari, *Phys. Rev. B* 44:8599, 1991.

[12] W. Zawadzki, *Phys. Rev. B* 72:085217, 2005.

[13] J.Y. Vaishnav and W.C. Clark, *Phys. Rev. Lett.* 100:153002, 2008.

[14] J.B. Ketterson and S.N. Song, *Superconductivity.* Cambridge University Press, Cambridge, 1999.

[15] L. Lamata, J. Leon, T. Schatz and E. Solano, *Phys. Rev. Lett.* 98:253005, 2007.

[16] Z.Y. Wang and C.D. Xiong, *Phys. Rev. A* 77:045402, 2008.

[17] M.I. Katsnelson, *Eur. Phys. J. B* 51:157, 2006.

[18] B. Trauzettel, Y.A.M. Blanter and A.F. Morpurgo, *Phys. Rev. B* 75:035305, 2007.

[19] G. Wang, H. Xu, L. Huang and Y.Ch. Lai, *New J. Phys.* 19:013017, 2017.

[20] L. Yan, S. Hong, Z. Fu-Lin, C. Jing-Ling, W. Chun-Feng and L.C. Kwek, *Commun. Theor. Phys.* 63:145, 2015.

[21] X.D. Zhang, *Phys. Rev. Lett.* 100:113903, 2008.

[22] F.D.M. Haldane and S. Raghu, *Phys. Rev. Lett.* 100:013904, 2008.

[23] L.G. Wang, Z.G. Wang and S.Y. Zhu, *Eur. Phys. Lett.* 86:47008, 2009.

[24] Y. Nogami and F.M. Toyama, *Am. J. Phys.* 78:175, 2010.

[25] F.A.B. Coutinho, Y. Nogami and F.M. Toyama, *Am. J. Phys.* 56:904, 1988.

[26] F.A.B. Countinho, Y. Nogami and F.M. Toyama, *Am. J. Phys.* 56:904, 1988.

[27] O. Klein, *Z. Phys.* 53:157, 1929.

[28] B.R. Holstein, *Am. J. Phys.* 66:507, 1998.

[29] A. Calogerocos and N. Dombey, *Contemporary Phys.* 40:313, 1999.

[30] H. Nitta, T. Kudo and H. Minowa, *Am. J. Phys.* 67:966, 1999.

[31] A.D. Alhaidari, *Phys. Scr.* 83:025001, 2011.

[32] W. Greiner, B. Muller and J. Rafelski, *Quantum Electrodynamics of Strong Fields.* Springer, Berlin, 1985.

[33] P. Ghose, M.K. Samal and A. Datta, *Phys. Lett. A* 315:23, 2003.

[34] F. Sauter, *Z. Phys.* 69:742, 1931.

[35] H.G. Dosch, J.H.D. Jensen and V.F. Muller, *Phys. Norv.* 5:151, 1971.

[36] D.J. Griffiths and C.A. Steinke, *Am. J. Phys.* 69:137, 2001.

[37] P. Kennedy and N. Dombey, *J. Phys. A* 35:6645, 2002.

[38] J. Guerrero, V. Aldaya, *Mod.Phys.Lett. A* 14:1689,, 1999.

[39] V. Aldaya, J. Bisquert and J.N. Salas, *Phys. Lett. A* 156:381, 1991.

[40] A. Zarzo and A. Martinez, *J. Math. Phys.* 34:2926, 1993.

[41] Y.S. Kim and M.E. Noz, *Am. J. Phys.* 46:484, 1978.

[42] M.I. Katsnelson, K.S. Novoselov and A.K. Geim, *Nature Phys.* 2:620, 2006.

[43] R. Gerritsma, B.L. Lanyon, G. Kirchmair, F. Zahringer, C. Hempel, J. Casanova, J.J.G. Ripoll, E. Solano, R. Blatt and C.F. Roos, *Phys. Rev. Lett.* 106:060502, 2011.

[44] P. Alberto, C. Fiolhais and V.M.S. Gil, *Eur. J. Phys.* 17:19, 1996.

[45] R.E. Moss, *Mol. Phys.* 53:269, 1984.

[46] S. Coleman, *Ann. Phys.* 101:239, 1979.

[47] H. Galic, *Am. J. Phys.* 56:312, 1988.

[48] A.I. Arbab, *Applied Phys. Res.* 3:160, 2011.

[49] A.I. Arbab, *Eur. Phys. Lett.* 92:40001, 2010.

[50] R. Geritsma, G. Kirchmair, F. zahringer, E. Solano, R. Blatt and C.F. Roos, *Nature* 463:68, 2010.

[51] J. Bolte and S. Keppeler, *Annals Phys.* 274:125, 1999.

[52] F. Cooper, A. Khare, R. Musto and A. Wipf, *Annals Phys.* 187:1, 1988.

[53] A.S. de Castro, *Annals Phys.* 311:170, 2004.

[54] C.L. Ho, *Annals Phys.* 321:2170, 2006.

[55] O. Aydogdu and R. Sever, *Annals Phys.* 325:373, 2010.

[56] A.K. Geim, *Science* 324:1530, 2009.

[57] L.G. de Peralta, *Eur. J. Phys.* 41:065404, 2020.

[58] L.G. de Peralta, *Results Phys.* 18:103318, 2020.

[59] L.G. de Peralta, *Sci. Rep.* 10:14925, 2020.

[60] L.G. de Peralta, *J. Mod. Phys.*11:788, 2020.

[61] L.G. de Peralta, L.A. Poveda and B. Poirier, *Eur. J Phys.* 42: 055404, 2021.

[62] K. Jones-Smith and H. Mathus, *Phys. Rev. D* 89:125014, 2014.

19.15   Exercises

19.1 Construct the KG equation for a Coulombic potential and then show that it cannot be applied for hydrogen atom.

19.2 Obtain the dispersion relation for the relativistic free particle.

19.3 What are the similarities and differences between Schrödinger and Dirac equations?

19.4 Find the velocity operator for the Dirac's free particles.

19.5 Find the four linearly independent solutions of the Dirac equation for a free particle moving in the z-direction.

19.6 Write the continuity equation $\partial\rho/\partial t + \nabla \cdot \mathbf{J} = 0$ in covariant form.

19.7 Show that $w = c(\boldsymbol{\sigma}\cdot\mathbf{p})v/(E_+ + mc^2)$ of the free particle Dirac equation gives the same current density as that of the Schrödinger equation under the nonrelativistic limit if the spin contribution is neglected.

19.8 Find $(\boldsymbol{\alpha} - c\mathbf{p}H^{-1})H + H(\boldsymbol{\alpha} - c\mathbf{p}H^{-1})$ for a Dirac's free particle.

19.9 Workout the quantum numbers n', k, l and j for $n = 2$ and $n = 3$ levels for the hydrogen atom in Dirac theory using the selection rules $\Delta l = \pm 1$ and $\Delta j = 0$, ± 1 and find the allowed transitions.

19.10 For $H = c\boldsymbol{\alpha}\cdot(\mathbf{p} + e\mathbf{A}/c) - e\phi + \beta mc^2$ find $d\mathbf{X}/dt$ and $d\mathbf{p}/dt$.

19.11 Show that the matrices α_1, α_2, α_3 and β are linearly independent.

19.12 Find the dimension of the wave function of spin-3/2 particles.

19.13 Taking $\psi(\mathbf{X},t) = \psi(\mathbf{X})\,e^{-iEt/\hbar}$ find the relation between the average values of velocity and momentum for Dirac's free particle. Then show that the average velocity and average momentum are in opposite directions for negative energy states.

19.14 Find $\beta H + H\beta$ for a relativistic free particle.

19.15 Show that $\langle\beta\rangle$ approaches ± 1 in the nonrelativistic limit and vanishes as the speed approaches c.

19.16 Determine $\beta(t)$ and then $\langle\beta(t)\rangle$.

19.17 Show that the helicity operator $\boldsymbol{\sigma}'\cdot\mathbf{n}$ and Dirac's Hamiltonian for a free particle have simultaneous eigenfunction.

19.18 If $H = c\boldsymbol{\alpha}\cdot\mathbf{p} + \beta mc^2 + V(r)$ then find $[H, \beta\boldsymbol{\sigma}'\cdot\mathbf{J}]$ and $[H, \hbar k]$.

19.19 Prove that the Dirac equation can be given as $\left(\gamma'_\mu\dfrac{\partial}{\partial x_\mu} + \dfrac{mc}{\hbar}\right)\psi' = 0$, where $\gamma'_\mu = S^{-1}\gamma_\mu S$ and $\psi = S\psi'$.

19.20 Using the covariant form of Dirac equation find the effect of the gauge transformation of the electromagnetic potential $A_k = A'_k + \partial X/\partial x_k$, where X is an arbitrary function, and the unitary transformation of the wave function $\psi = \psi'e^{ieX/(\hbar c)}$.

20

Mysteries in Quantum Mechanics

20.1 Introduction

Quantum theory is considered as the most successful scientific theory of all time. However, the understanding[1] and interpretation of quantum theory has been subjected to a continuous debate since its formulation. A leading role was played by Einstein in the development of the *old quantum theory* but he was not impressed by the statistical interpretation of the quantum theory. He believed in a world of well defined law and order in which objectivity exists. He never thought that the new quantum theory as a complete theory. In 1926 in a letter to Max Born, Einstein wrote: *Quantum mechanics is very impressive. ... The theory produces good deal but hardly brings us closer to the secret of the old one. I am at all events convinced that God does not play dice.*

In the beginning, *gedankan*[2] (thought) experiments were used to analyze fundamental issues in quantum mechanics. At the end of 20th century, the situation had changed and real experiments were performed. Bohr and Einstein sparred to explain certain gedankan experiments each using very different concepts like causality. The realization of a controversy between quantum mechanics and reasonable assumptions was pointed out by Einstein, Boris Podolsky and Nathan Rosen (EPR) in 1935. Even though their argument is opposite to quantum mechanical results, it contains a truth. David Bohm called the EPR argument a *paradox*. The essence of their criticism is that Schrödinger wave function must be considered as describing an ensemble of systems and cannot attached to be the state of any individual system.

It was John Stewart Bell (1964), an Irish physicist, established what the quantum theory can tell us about the nature of the world. Bell formulated certain inequalities based on the arguments of EPR. They should be valid for any theory satisfying the notions of reality and locality. The striking news is that the Bell's inequality is violated by quantum mechanics but the experimental results agree with the predictions of quantum theory.

In the present chapter, we present the Schrödinger cat paradox and Bohm's version of EPR paradox. We give a brief sketch of Bell's inequality and the quantum mechanical examples violating it.

---

[1]In connection with the understanding of quantum mechanics Feynman once said: *There was a time when the newspapers said that only twelve men understood the theory of relativity. I do not believe there ever was such a time. ... I can safely say that nobody understands quantum mechanics* [R.P. Feynman (1965), *The Character of Physical Law* 129. BBC/Penguin].

[2]A gedankan experiment is a thought experiment which may be imagined, technically not feasible but consistent with the laws of physics.

DOI: 10.1201/9781003172178-20

20.2   The Collapse of the Wave Function

According to *Copenhagen interpretation* of quantum mechanics, the state of a microscopic system is defined as $\psi = a_1\psi_1 + a_2\psi_2 + \ldots$. When a measurement is performed, ψ collapses into any one of the states ψ_1, ψ_2, $\ldots$, with respective probabilities $|a_1|^2$, $|a_2|^2$, $\ldots$. This collapse of the wave function occurs in an unpredictable way.

Consider a simple radioactive nucleus emitting an electron in the decay process. The possible outcomes of the experiment will be either the undecayed nucleus or decayed nucleus. Before the measurement, ψ of the state of the nuclei is $a_{\mathrm{ud}}\psi_{\mathrm{ud}} + a_{\mathrm{d}}\psi_{\mathrm{d}}$, where ψ_{ud} is the undecayed state and ψ_{d} is the decayed state wave functions. At any particular time $|a_{\mathrm{ud}}|^2$ and $|a_{\mathrm{d}}|^2$ are the probabilities of it has not decayed and decayed, respectively. In the actual measurement, we will find either ψ_{d} or ψ_{ud}. No measurement will yield a partially decayed state. The measurement thus caused a discontinuous change of wave function from $a_{\mathrm{ud}}\psi_{\mathrm{ud}} + a_{\mathrm{d}}\psi_{\mathrm{d}}$ to ψ_{ud} or ψ_{d}. This quantum jump phenomenon is called the *collapse of the wave function*.

As the decay of the nuclei is a random process, the emitted electron will travel in any direction with equal probability so it can be represented by a spherical de Broglie wave spreading out from the source. Consider a spherical surface of area $S = S' + S''$, where S' is the area of an electron detector and S'' is the remaining area of the spherical surface. Since the total probability of detecting the electron in the total spherical surface is one, we write

$$\int_S \psi\psi^* \, dV = \int_{S'} \psi\psi^* \, dV + \int_{S''} \psi\psi^* \, dV = 1 \,. \qquad (20.1)$$

Hence, before an experiment is performed to detect the electron, the probability of detection is $\int_{S'} \psi\psi^* \, dV = S'/S$. The probability for nondetection is $\int_{S''} \psi\psi^* \, dV = S''/S$. At a later moment if the detector has detected an electron then the conditions will be $\int_{S'} \psi\psi^* \, dV = 1$ and $\int_{S''} \psi\psi^* \, dV = 0$. If the detector has not recorded the electron then $\int_{S'} \psi\psi^* \, dV = 0$ and $\int_{S''} \psi\psi^* \, dV = 1$.

The discontinuous collapse of the wave function is similar to what happens when the atom decays. Consider the case when the electron is detected. Before the detection of the electron, ψ is spread over a sphere centred on the nucleus. When the wave function collapses, it has to happen everywhere on the sphere. So, the probability of detecting the electron changes suddenly everywhere except at the detector from a finite value to 0. At the detector the probability changes from a small value to a large value at the same instant. Detecting the electron at one place has had an instantaneous effect on the wave function at all other places. This strange *action-at-a-distance* is called *nonlocality*. These examples show how quantum theory entangles the measuring apparatus of the observer in the experiment. The process of detection by the observer has made the probability of detection elsewhere 0. Hence, it is difficult to separate an objective physical reality from the experimental set up.

Let us point out the essentials of Einstein's dissatisfaction with quantum theory [1]. In connection with the above mentioned decay process, we ask: *what is the time of decay?* The quantum theory does not provide the actual time of decay, but only gives the various possible decay times and their corresponding probabilities. Einstein asked the following: *Can this theoretical description be taken as the complete description of the disintegration of a single atom?* The answer is *no*. An atom decays at a definite time, but such a definite time-value is not specified by the wave function ψ. Thus, the description of the individual atom by the wave function is incomplete. The wave function is to be taken as a description, but of an ensemble of systems.

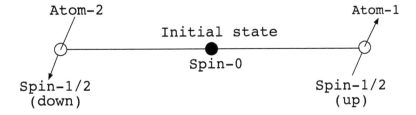

FIGURE 20.1
A spin-0 molecule splits into two atoms.

20.3   Einstein–Podolsky–Rosen (EPR) Paradox

To argue that quantum theory is incomplete in describing the physical reality, EPR formulated a thought experiment. In the following we briefly present the US physicist David Bohm's version of EPR experiment, EPR's arguments and the replies of John von Neumann and Bohr [2,3].

Consider a molecule with two identical atoms. The total spin of the molecule is zero and the spin of each atom is $\hbar/2$. Suppose the molecule is split into two atoms conserving the total angular momentum. Each atom travels in opposite directions. Because the total angular momentum is unchanged if we measure the spin of the atom-1 in some direction then the spin of the atom-2 must be in the opposite direction as shown in Fig. 20.1.

We recall that spinning property of an atom cannot be described in classical terms. That is, the axis of rotation cannot always be defined with certainty. The act of measurement gives the atom a particular axis of rotation. But before the measurement, it cannot be said to spin about a well defined axis. Suppose each atom has the z-component of its spin measured by a Stern–Gerlach apparatus as in Fig. 20.2. The source S sends the atoms 1 and 2 to the vertically oriented magnets P_1 and P_2, respectively. Here P stands for polarization.

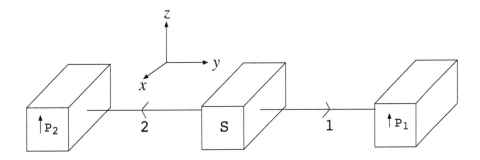

FIGURE 20.2
Measuring the z-component of spin of two atoms. The polarization of the two polarizers is the same. (Reproduced with permission from D. Harrison, *Am. J. Phys.* 50:811, 1982. Copyright 1982, American Association of Physics Teachers.)

20.3.1   Total Wave Function: An Entangled State

The atom-1 has a wave function ϕ^1. It is an eigenvector of the operator P_1. Similarly, the wave function of atom-2 is, say, ϕ^2. For spin-up and spin-down use the subscripts $+$ and $-$, respectively. The corresponding eigenvalues are chosen as $+1$ and -1. Then we have

$$\text{atom-1} - \text{spin-up}\quad : P_1\phi^1_+ = \phi^1_+\,, \tag{20.2a}$$

$$\text{atom-1} - \text{spin-down}\quad : P_1\phi^1_- = -\phi^1_-\,, \tag{20.2b}$$

$$\text{atom-2} - \text{spin-up}\quad : P_2\phi^2_+ = \phi^2_+\,, \tag{20.2c}$$

$$\text{atom-2} - \text{spin-down}\quad : P_2\phi^2_- = -\phi^2_-\,. \tag{20.2d}$$

We know that either atom-1 has spin-up and atom-2 has its spin-down or versa, without knowing which of these two possible states the system has. In such a case the experimental properties are compatible with the idea of the particles possess individual properties but are unknown to us. Suppose we perform superpositions of product states as

$$\psi_{1,2} = \frac{1}{\sqrt{2}}\left(\phi^1_+\phi^2_- - \phi^1_-\phi^2_+\right)\,. \tag{20.3}$$

The minus sign is because the total spin is zero. Here ψ is nonlocal, describes both particles together, even if they are far apart. In $\psi_{1,2}$ specific state is not assigned to any of the particles individually. The point is that a measurement on one particle would collapse the wave function to a definite state of the particles. Schrödinger called such states as *entangled states*[3] [4,5]. The entanglement refers to the quantum description of an ensemble of systems where every system consists of at least-two more subsystems and disallow us to assign a definite quantum state to each of the individual subsystems. *What is unique about the entangled state?* Assigning a state to each particle in individually is not possible. In other words, the individual particles are regarded as not possessing properties in their own.

Another example of entangled states is the following. Let us consider a process of converting a single ultraviolet photon into two red photons using a nonlinear crystal [6] with energy and momentum conservations. In a coordinate system assume that the photons have equal and opposite momenta. We place slits so that photon-1 can be in one of the two states, a or b, and photon-2 is either in a' or b'. Since we do not know the state of the created photon, we have an entangled state

$$\psi \propto \psi_1(a)\psi_2(a') + \psi_1(b)\psi_2(b')\,. \tag{20.4}$$

We note that entanglement is one of the most striking aspects of quantum mechanics. In classical physics, two interacting systems retain their individuality during the interaction process and independent of each other if their coupling is removed. In contrast, a different behaviour is realized in quantum theory. When two quantum systems interact then their identities become entangled. Further, as a result a state vector description of each system is precluded. The state vector of the composite system is inseparable into a product of the states of the subsystems. The entanglement is preserved even after removing the interaction. Consequently, a measurement on one system will affect the other. Entanglement has been confirmed experimentally [7–10]. Manifestation of entanglement in macroscopic properties of a system is noticed in an experiment by Ghosh et al. [11]. They considered an insulating magnetic salt $\text{LiHO}_x\text{Y}_{1-x}\text{F}_4$, which is formed by random substitution of magnetic HO^{3+} ions in place of nonmagnetic Y^{3+} with probability x. The magnetic specific heat and magnetic susceptibility of a single-crystal specimen of the salt as a function of temperature is

---

[3]Schrödinger coined the word *entanglement* in 1935 in a three-part paper on the 'Present Situation in Quantum Mechanics' (E. Schrödinger, *Die Naturwissenschaften* 48:807, 1935; 49:823, 844, 1935).

experimentally determined. The numerical simulations agreed with the experimental result only when the entanglement between the magnetic spins is taken into account.

An application of entanglement is the quantum computers which could allow an exponential increase of computational speed for certain problems such as for example, the factorization of large numbers into primes [12,13]. Quantum entanglement allows to transfer an unknown quantum state of a two-level system from one particle to another distinct particle without actually sending the particle itself. Many other applications of entanglement are now being developed and investigated, for example, in frequency standards [14], distributed quantum computation [15,16], multiparticle swapping [17] and multi-particle entanglement purification [18].

Solved Problem 1:

Obtain the proof of ψ given by Eq. (20.3).

Suppose measure the x-component of the spin. Consider σ_x, σ_z, ϕ_+ and ϕ_- given by

$$\sigma_x = \begin{pmatrix} 0 & 1 \\ 1 & 0 \end{pmatrix}, \quad \sigma_z = \begin{pmatrix} 1 & 0 \\ 0 & -1 \end{pmatrix}, \quad \phi_+ = \begin{pmatrix} 1 \\ 0 \end{pmatrix}, \quad \phi_- = \begin{pmatrix} 0 \\ 1 \end{pmatrix}. \tag{20.5}$$

The eigenvectors of σ_x are $V_+ = \dfrac{1}{\sqrt{2}} \begin{pmatrix} 1 \\ 1 \end{pmatrix}$ and $V_- = \dfrac{1}{\sqrt{2}} \begin{pmatrix} 1 \\ -1 \end{pmatrix}$. In terms of the eigenvalues of σ_z we have

$$V_\pm = \frac{1}{\sqrt{2}} (\phi_+ \pm \phi_-) . \tag{20.6}$$

Substitution of Eq. (20.6) in Eq. (20.3) for the atoms 1 and 2 gives

$$\psi_{1,2} = \frac{1}{\sqrt{2}} \left(V_+^1 V_-^2 - V_-^1 V_+^2 \right) . \tag{20.7}$$

Thus, the measurement of the x-component of the spin for the system described by ψ gives one atom spin-up and other atom spin-down. *What do we get if*

$$\psi' = \frac{1}{\sqrt{2}} \left(\phi_+^1 \phi_-^2 + \phi_-^1 \phi_+^2 \right) \tag{20.8}$$

is used instead of Eq. (20.3)?

20.3.2 Wave Function Before and After the Measurement

Before the measurement of z-component of spin of atom-1, the state vector is

$$\psi_{1,2}^{\text{before}} = \frac{1}{\sqrt{2}} \left(\phi_+^1 \phi_-^2 - \phi_-^1 \phi_+^2 \right) . \tag{20.9}$$

After a measurement, atom-1 is in spin-up state so that

$$\psi_2^{\text{before}} = \psi_2^{\text{after}} = \phi_-^2 . \tag{20.10}$$

Next, suppose we wish to measure the x-component of atom-1. Then

$$\overline{\psi}_{1,2}^{\text{before}} = \frac{1}{\sqrt{2}} \left(\overline{\phi}_+^1 \overline{\phi}_-^2 - \overline{\phi}_-^1 \overline{\phi}_+^2 \right) . \tag{20.11}$$

Finding the atom to be in spin-up, we have

$$\overline{\psi}_2^{\text{before}} = \overline{\psi}_2^{\text{after}} = \overline{\phi}_-^2 . \tag{20.12}$$

The problem now is that different state vectors ϕ_-^2 and $\overline{\phi}_-^2$ are attached to the same reality.

20.3.3   EPR Argument

By measuring the spin of atom-1 we can determine the spin of atom-2 without carrying out a measurement on atom-2. If atom-1 is spin-up about some axis then atom-2 must be spin-down about the same axis. Since the two atoms are noninteracting the spin of atom-2 before and after the measurement of atom-1 must be identical. But, quantum theory denies this information and is unable to give the spin of atom-2 before the measurement of it. According to quantum theory a property of atom-2 is known only if the measurement is performed on it.

The significance of the EPR argument is that it was against quantum mechanics with the principles of locality and reality. The principle of locality implies that the actual situation of atom-2 is independent of what is done with the spatially separated atom-1. EPR pointed out that in a complete theory every element of physical reality must have a corresponding counterpart. Quantum theory does not have this one to one correspondence between reality and theory. This is because it not determines individual events but determine only probabilities of the possible events. EPR proposed that if without disturbing a system, we are able to determine certainly the value of an observable then there exists an element of physical reality to this observable. According to locality, a change recorded by P_1 is due to the change in the orientation of P_1 and independent of the orientation of spatially separated P_2.

As per the above, the spin $S_1(S_2)$ of atom-1(2) can be found definitely without disturbing atom-2(1). Therefore, S_1 and S_2 are elements of reality. But here if S_1 is known then S_2 is also known (without a measurement on atom-2). In this sense the principles of locality and reality are violated and hence the quantum mechanical description is incomplete. In quantum mechanics individual events are thought of simply chance happening and for each individual events there are no corresponding theoretical elements. Then according to EPR, quantum theory is not complete. This is the crux of the EPR argument. EPR did not doubt the correctness of predictions of quantum theory. They were concerned in calling quantum theory as complete theory.

We will consider another example of EPR experiment. As per the conservation of linear momentum, if two particles A and B are emitted in opposite directions, then the magnitude of momenta of A and B must be equal. Thus, a measurement on A or B determines linear momentum of both even though they are separated and no interaction between them. If the linear momentum of A is measured accurately then the linear momentum of B is also known accurately. Now, the position of B can be measured with desired precision and hence that of A can also be known without measurement on A. Since the position and momentum of both A and B have been obtained accurately the uncertainty principle is violated. If this is possible then the position and momentum of A and B are elements of physical reality. But the wave function will not give simultaneously the position and momentum. Hence, according to EPR argument, the wave function does not provide a complete description of physical reality. EPR thought that there were *hidden variables* and their ignorance disallowed us in making exact predictions. These hidden variables are nonlocal. But Bohr strongly disagreed with EPR's argument.

20.3.4   John von Neumann's and Bohr's Reply to EPR

According to von Neumann a measurement of S_z on atom-1 makes the state vector to collapse to $\phi_+^1 \phi_-^2$ or $\phi_-^1 \phi_+^2$. Therefore, if the S_z of atom-1 is measured as $\hbar/2$ then that of atom-2 will be $-\hbar/2$ and vice-versa. Both spins now acquired definite values of S_z. The measurement led an instantaneous effect on the spins of atoms 1 and 2 even though they are spatially separated.

Bohr pointed out that measurements on quantum systems are not like classical measurements. Their results depend on the experimental arrangement which is not local in this sense. The spin of atom-2 is determined with a measurement on it but by the experiment on atom-1. This possibility is because the spin of atom-1 and the spin of atom-2 will have exactly the opposite sign. Further, he argued that, more importantly, the two atoms system is an indivisible even the particles are separated. The system cannot be studied in terms of independent parts. They have instantaneous and nonlocal connections. They are beyond our usual notions of information transfer.

20.4   Hidden Variables

EPR argued that the quantum theory is incomplete. But any method for completing it was not known. EPR arguments raised an intriguing question of whether there is a more refined description of nature than that of quantum mechanics in explaining the action-at-distance. It was proposed that classical-like determinism can be recovered in quantum mechanics by hidden variables. They are assigned to observables before measurement. An average over them gives the statistical values determined by quantum mechanics. Effectively, the hidden variables make the probabilistic nature of the individual system as a system of ensembles of individuals. If we specify the hidden parameters then we can arrive a classical physics type of theory where there would be no uncertainties. As these variables are not realized in actual experiments, they have been named *hidden variables*.

Even before the EPR paper, von Neumann in 1932 shown that hidden variable could not yield the same results as quantum theory. His proof pointed out that hidden variables theories are not useful to complete the quantum theory. The correctness of the proof was questioned. In 1952, David Bohm developed a hidden variables theory and its predictions were in agreement with quantum results. He shown that a hidden variables theory would be able to produce the results of conventional theory and retain the reality of quantum objects. But his model had built-in nonlocality and needed influences to travel faster than the speed of light, thus violating the theory of relativity. In 1964, John Stewart Bell identified the breakdown of proof of von Neumann. He established that the predictions of quantum theory are in agreement with observation while those of hidden variables theories are not.

20.5   The Paradox of Schrödinger's Cat

Let us ask the following question: *Why do we not realize quantum superposition of classical objects?* To answer this let us brief the famous thought experiment of Schrödinger (1935): The paradox of Schrödinger cat [19,20]. This paradox was introduced by Schrödinger when he was not satisfied by the superposition of unobserved states and the discontinuous collapse of the wave function.

Suppose a cat is placed inside a sealed container. Inside it there is a device which can be triggered by a quantum event. If the event happens then the device releases, say poison gas and the cat is killed. The experimental set up is shown in Fig. 20.3. The quantum event is, for example, triggering of a photo-cell by a photon. The photon is emitted by a light source and then reflected-off a half-silvered mirror (it is a mirror which reflects exactly half of the incident light and the remaining half is transmitted directly through the mirror). At

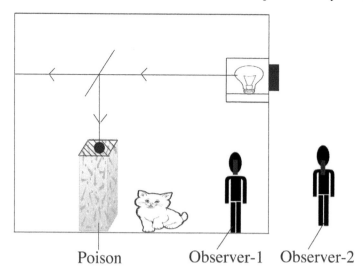

Poison   Observer-1   Observer-2

FIGURE 20.3
Set up of Schrödinger cat experiment.

the mirror the photon is splitted into two parts. One is reflected while the second is focused on a photo cell. When a photon is registered, the gas is released and the cat is killed. If the photon is not registered then it is assumed to be transmitted and the cat is alive.

In the experiment, either the photon is registered and the cat is killed or the photon is not registered (that is, transmitted) and the cat is saved. Any one of the above two events must take place. The probability of each event is $1/2$ because the mirror is half-silvered. Before the measurement, the wave function for the event contains two parts with the amplitude of each part as $1/\sqrt{2}$. The wave function must contain both possibilities-*dead* and *alive* with equal probabilities. That is, the cat is in a linear superposition of being dead and alive:

$$|\psi\rangle = \frac{1}{\sqrt{2}}\left(|\text{dead}\rangle + |\text{alive}\rangle\right). \tag{20.13}$$

When the observation is made the wave function of the cat collapses into dead or alive. *What does the observer inside the container predict?* For him the wave function of the cat collapsed much earlier than the outside observer and the above linear combination has no relevance. But for an outside observer the cat is partially alive and partially dead until he performs the measurement. That is, two different states are assigned to the same reality.

Quantum mechanics says nothing about any individual events but gives all the possibilities. It does not say that before measurement the cat is half-dead and half-alive. When the observer looks to find the state of the cat, there is no implication in quantum mechanics that the death of the cat is caused by *seeing*. Essentially, the wave function is not a physical attribute of the cat.

To Einstein the collapse postulate was precious retreat from realism. It implied that physical quantities have no values until they are observed. This suggests that in the absence of an observer there might be no real world. With this view in mind Einstein asked: *Is the moon there when nobody looks?* [21]. Einstein believed that an objective reality exists in nature.

20.6 Bell's Theorem

Three decades later the EPR experiment, in 1964 John Bell derived a theorem [22] which proved that the existence of local hidden variables is inconsistent with the predictions of quantum theory. Experiments supported Bell's prediction and agreed with quantum mechanical predictions. Now, it appears that local hidden variables theories are incapable and the principle of locality does not hold. In the following we present d'Espagnat's [23] generalization of the proof of Bell's theorem. Its simplified version is given by Harrison [3]. The derivation only involves logic. We follow the description in [3].

Consider a set of macroscopic objects characterized by three two-valued parameters a, b and c. Table 20.1 gives examples of such objects. Let us call these parameters as hidden variables on each object. We assume that the hidden variables are assigned to each object remain unchanged. Further, we do not assume anything about the probabilities of the values of the hidden variables. Assign the $+$ symbol for the properties in the first row of table 20.1 and the $-$ symbol for the properties in the second row. We generate the table 20.2. Now, we denote the following:

1. The number of objects with parameter a but not parameter b as $N(a, -b)$ (c can be anything) and is simply $N_3 + N_4$.

2. The number of objects with parameter b but not a parameter c as $N(b, -c)$, which is $N_2 + N_6$.

3. The number of objects with the parameter a but not parameter c as $N(a, -c)$ and is $N_2 + N_4$.

For any collection of objects with the three different parameters a, b and c the result is the following:

TABLE 20.1
Examples of three double-valued macroscopic objects. (Reproduced with permission from D. Harrison, *Am. J. Phys.* 50:811, 1982. Copyright 1982, American Association of Physics Teachers.)

Object	Colour	Shape
a = Wood	b = Black	c = Sphere
$-a$ = Metal	$-b$ = White	$-c$ = Cube

TABLE 20.2
Various combinations of the objects and their count.

a	b	c	Numbers	a	b	c	Numbers
+	+	+	N_1	−	+	+	N_5
+	+	−	N_2	−	+	−	N_6
+	−	+	N_3	−	−	+	N_7
+	−	−	N_4	−	−	−	N_8

The number of objects with a and not b plus the number of objects with b and not c is greater than or equal to the number of objects with a and not c.

This is written as

$$N(a, -b) + N(b, -c) \geq N(a, -c).$$ (20.14)

In terms of probability the above equation is written as

$$P(a, -b) + P(b, -c) \geq P(a, -c).$$ (20.15)

The above relation is called *Bell's inequality* or *Bell's theorem*[4] which is true for any hidden variables theory that satisfies the notions of reality and locality. This is one version of Bell's inequalities. Let us prove the relation (20.15).

Consider $N(a, -b, c) + N(-a, b, -c)$. This must be ≥ 0 because either some members of the group have these combination or no members do. So,

$$N(a, -b, c) + N(-a, b, -c) \geq 0.$$ (20.16)

Suppose we add $N(a, -b, -c)$ and $N(a, b, -c)$ to (20.16):

$$\begin{aligned} N(a, -b, c) + N(-a, b, -c) + N(a, -b, -c) + N(a, b, -c) \\ > 0 + N(a, -b, -c) + N(a, b, -c). \end{aligned}$$ (20.17)

We note that

$$\begin{aligned} N(a, -b, -c) + N(a, b, -c) &= N(a, -c), & \text{(20.18a)} \\ N(a, -b, c) + N(a, -b, -c) &= N(a, -b), & \text{(20.18b)} \\ N(-a, b, -c) + N(a, b, -c) &= N(b, -c). & \text{(20.18c)} \end{aligned}$$

Using (20.18) in (20.17) we observe

$$N(a, -b) + N(b, -c) \geq N(a, -c).$$ (20.19)

This inequality is obtained without using quantum mechanics but with the principles of locality and reality. For presentations of Bell's theorem without inequalities see refs. [24,25]. It has been shown that the prediction of quantum mechanics violates this inequality. We present another version of Bell's inequality which we use in the next section for a quantum prediction.

Suppose we consider an experiment in which object 1 is picked randomly and a second object 2 is picked with each property opposite of object 1. $P(a, -b)$ for object 1 is equal to the probability that object 1 has property a and the object 2 has the property b (that is, opposite to $-b$). We designate it as $P(1a, 2b)$. Then Eq. (20.19) becomes

$$P(1a, 2b) + P(1b, 2c) \geq P(1a, 2c).$$ (20.20)

20.7 Violation of Bell's Theorem

Let us examine the Bell's theorem, the inequality given by Eq. (20.20), in quantum mechanics [3].

---

[4]Bell's theorem was called as the most profound discovery of science by Stapp [H.P. Stapp, *Nuovo Cimento* 29B:271, 1975] [3].

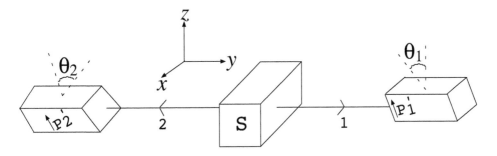

FIGURE 20.4
Measuring the z-component of spin of two atoms. The polarization of the two polarizers is the same. Each polarizer P_i' has been rotated about the y-axis by an angle θ_i. (Reproduced with permission from D. Harrison, *Am. J. Phys.* 50:811, 1982. Copyright 1982, American Association of Physics Teachers.)

20.7.1 Spin Measurement Experiment

We consider Fig. 20.4. The two Stern–Gerlach analyzers P_1' and P_2' are rotated about the y-axis through the angles θ_1 and θ_2, respectively. Now, we have

$$P'\phi'_\pm = \pm\phi'_\pm. \tag{20.21}$$

For Fig. 20.4 we have

$$\phi^1_+ = \cos(\theta_1/2)\phi'^1_+ - \sin(\theta_1/2)\phi'^1_-, \tag{20.22a}$$

$$\phi^1_- = \cos(\theta_1/2)\phi'^1_- + \sin(\theta_1/2)\phi'^1_+ \tag{20.22b}$$

and similarly for atom-2. The $1/2$ in $\theta_1/2$ is because we considered spin-1/2 atoms. Substituting the obtained ϕ's in Eq. (20.3) we have

$$\psi = \frac{1}{\sqrt{2}}\left[\cos\left((\theta_1 - \theta_2)/2\right)\left(\phi'^1_+\phi'^2_- - \phi'^1_-\phi'^2_+\right) \right.$$
$$\left. - \sin\left((\theta_1 - \theta_2)/2\right)\left(\phi'^1_+\phi'^2_+ + \phi'^1_-\phi'^2_-\right)\right]. \tag{20.23}$$

For $\theta_1 = \theta_2$ Eq. (20.23) is identical to Eq. (20.3).

Solved Problem 2:
What are the probabilities for (i) P_1' measuring atom-1 in spin-up and simultaneously P_2' measuring atom-2 also in spin-up and (ii) P_1' measuring atom-1 in spin-up and simultaneously P_2' measuring atom-2 in spin-down?

(i) Let us denote P_{++} be the probability for P_1' measuring atom-1 in spin-up and simultaneously P_2' measuring atom-2 also in spin-up. P_{++} is given by the square of the coefficient of $\phi'^1_+\phi'^2_+$ in Eq. (20.23):

$$P_{++} = -\frac{1}{\sqrt{2}}\sin((\theta_1 - \theta_2)/2) \times -\frac{1}{\sqrt{2}}\sin((\theta_1 - \theta_2)/2) = \frac{1}{2}\sin^2((\theta_1 - \theta_2)/2). \tag{20.24}$$

For $\theta_1 = \theta_2$ we have $P_{++} = 0$ as expected.

TABLE 20.3
Examples of three double-valued atomic spins. (Reproduced with permission from D. Harrison, *Am. J. Phys.* 50:811, 1982. Copyright 1982, American Association of Physics Teachers.)

a = spin-up	b = spin-up	c = spin-up
$\theta = 0°$	$\theta = 45°$	$\theta = 90°$
$-a$ = spin-down	$-b$ = spin-down	$-c$ = spin-down
$\theta = 0°$	$\theta = 45°$	$\theta = 90°$

(ii) In this case the probability P_{+-} is simply the square of the coefficient of $\phi_+'^1\phi_-'^2$ in Eq. (20.23). We obtain

$$P_{+-} = \frac{1}{2}\cos^2((\theta_1 - \theta_2)/2).\qquad(20.25)$$

It becomes maximum ($P_{+-} = 1/2$) when $\theta_1 = \theta_2$.

We use the result in the above solved problem to verify the Bell's theorem. To test the Bell's inequality, first write its quantum version. Assume that pair of the atom emitted by a source have a total spin 0 and they travel in opposite directions. If the right-side atom is spin-up then its companion atom is spin-down provided the Stern–Gerlach filters have the same orientation. We set a, b and c as in table 20.3 and assume the following:

1. The polarizers measure the spins at locations separated by a space interval. Then the polarizers cannot influence each other. That is, we assume that no influence propagates faster than the speed of light. Otherwise, the measurement by one polarizer introduce uncontrollable phase factors into the wave function. This is a kind of uncertainty principle.

2. The objects have properties (such as spin) even if they are not measured.

Suppose the Stern–Gerlach filters allow only spin-up atoms to pass through. We consider the application of Eq. (20.20) to spin measurements. Remember that Eq. (20.20) is for the case of each with properties opposite to the other. In a spin measurement also we consider the same. This is realized if the two polarizers have the same orientation implying that the atoms have opposite spins. We determine whether the right-side atom is spin-up, $0°$, not spin-up $45°$ by measuring its spin at $0°$ and its companion's spin at $45°$. The Bell's inequality, Eq. (20.20), becomes

$$P(1\text{up }0°, 2\text{up }45°) + P(1\text{up }45°, 2\text{up }90°) \geq P(1\text{up }0°, 2\text{up }90°).\qquad(20.26)$$

Evaluate each term in Eq. (20.26) using Eq. (20.24). We have

$$P(1\text{up }0°, 2\text{up }45°) = \frac{1}{2}\sin^2(-45°/2),\qquad(20.27a)$$

$$P(1\text{up }45°, 2\text{up }90°) = \frac{1}{2}\sin^2(-45°/2),\qquad(20.27b)$$

$$P(1\text{up }0°, 2\text{up }90°) = \frac{1}{2}\sin^2(90°/2).\qquad(20.27c)$$

Thus, Eq. (20.26) becomes

$$\frac{1}{2}\sin^2(45°/2) + \frac{1}{2}\sin^2(45°/2) \geq \frac{1}{2}\sin^2(90°/2).\qquad(20.28)$$

That is, $0.0732 + 0.0732 = 0.1464 \geq 0.25$ which is false. The local hidden variables model with the particles having a definite spin orientation before they are detected is wrong. Thus, the prediction of quantum mechanics violates the Bell's inequality. The existence of local hidden variables is inconsistent with the predictions of quantum theory. Thus, Bell discovered that assumption of local realism is not in accordance with quantum physics.

Solved Problem 3:

Verify that in table 20.3 instead of the values $\theta = 0°$, $45°$ and $90°$ if the angles are $0°$, $90°$ and $180°$, respectively, then there is no conflict between the predictions of quantum mechanics and the Bell's inequality.

The Bell's inequality $P(1a, 2b) + P(2b, 2c) \geq P(1a, 2c)$ becomes

$$P(\text{1up } 0°, \text{2up } 90°) + P(\text{1up } 90°, \text{2up } 180°) \geq P(\text{1up } 0°, \text{2up } 180°). \tag{20.29}$$

Since

$$P(\text{1up } 0°, \text{2up } 90°) = \frac{1}{2}\sin^2(-90°/2) = \frac{1}{4}, \tag{20.30a}$$

$$P(\text{1up } 90°, \text{2up } 180°) = \frac{1}{2}\sin^2(-90°/2) = \frac{1}{4}, \tag{20.30b}$$

$$P(\text{1up } 0°, \text{2up } 180°) = \frac{1}{2}\sin^2(-180°/2) = \frac{1}{2} \tag{20.30c}$$

we get the Bell's inequality as $1/4 + 1/4 = 1/2$. So, there is no conflict between the predictions of quantum mechanics and the Bell's inequality.

20.7.2 Experimental Tests of Bell's Theorem

What is the outcome of real experiments? The result is that the inequality is violated. The predictions of quantum mechanics and Bell's theorem were tested experimentally for photons and protons. These experiments helped us to understand the basics of quantum mechanics. Moreover, gave birth to fascinating new topics such as quantum information, quantum teleportation, quantum computation and quantum cryptography. The first experiment was carried out by Clauser and Freedman at Berkely [25]. The most famous experiment was performed by Aspect, Dalibard and Roger in Paris [26]. These experiments are quite difficult. The most striking point is that all the experiments agreed with the quantum mechanical results and Bell's inequality was violated.

Most of the experiments were based on correlations between the polarizations of pairs of photons. Correlated photons can be produced in many ways. An example is annihilation of positronium into two photons. Here initial and final angular momenta are zero. Alternatively, an atom in an excited state with angular momentum zero decays by $J = 0 \to J = 1 \to J = 0$ cascade emitting two photons.

In a typical experiment a source emits polarization-conserved pairs of photons. Each photon is passed through a polarizer. Behind each polarizer, the transmitted photons are registered. Conservation of polarization implies that if the photon-1 is measured and found in horizontal (H) polarization then photon-2 will be vertical (V) and vice-versa. Before measurement the total wave function is

$$|\psi\rangle = \frac{1}{\sqrt{2}}\left(|H\rangle_1|V\rangle_2 - |V\rangle_1|H\rangle_2\right). \tag{20.31}$$

A version of Bell's inequality is

$$|E(a,b) - E(-a,b)| + |E(a,-b) + E(-a,-b)| \leq 2, \qquad (20.32)$$

where

$$E(a,b) = \frac{1}{N} \left[C_{++}(a,b) + C_{--}(a,b) - C_{+-}(a,b) - C_{-+}(a,b) \right] . \qquad (20.33)$$

Here $+$ and $-$ are the outputs of a two-channel polarizer. Then $C_{++}(a,b)$ is the number of coincidences between the $+$ output port of the polarizer measuring photon-1 along a and $+$ output port of the polarizer measuring photon-2 along b. For $a = 0$, $b = 22.5°$ Eq. (20.32) is found to be $2\sqrt{2} < 2$ a clear violation of Bell's inequality.

One may wonder about the contradiction between Bell's theorem and quantum mechanics. A possible reason is the difference between ideal experiment and measurement. But in real experiments we can consider reasonable assumptions about the working of the apparatus, noise, etc. Moreover, in the Bell's theorem the polarizers are assumed to act independently. If this is incorrect then signal (passing faster than light) connecting the two polarizers is possible. Next, the experiments have always involved fixing the angles of polarizers and then taking data. Thus, one may argue that the polarizers have much time to know orientation of each other.

Greenberger et al. [27] have demonstrated Bell's theorem in a new way by analyzing a system consisting of 3 or more correlated spin-1/2 particles. Here again Bell's theorem is violated by quantum mechanical predictions. In 1993, Hardy proposed a situation where nonlocality can be inferred without using inequalities [28].

20.8 Resolving EPR Paradox

The EPR's nonlocality puzzle can be resolved [29] if the correct physical input is used for the calculation of the quantum correlations.

Suppose the result of spin measurement is two-valued. Let A and B denote the outcomes ± 1. a and b denote the settings of the apparatus for the first particle and the second particle, respectively. The statement of locality is $A(a) = \pm 1$, $B(b) = \pm 1$. The existence of reality is that some hidden variables decide the outcomes even before the measurement. This is stated as $A(a, h_1) = \pm 1$, $B(b, h_2) = \pm 1$, where h_1 and h_2 are hidden variables associated with the outcomes. The experimenter can compute a correlation function of the outcomes defined by $P(a,b) = \frac{1}{N} \sum (A_i B_i)$. This is the classical function obtained by averaging the products of the form $(++ = +)$, $(-- = +)$, $(+- = -)$ and $(-+ = -)$. The joint events $(++)$ and $(--)$ are coincidences and the events $(+-)$ and $(-+)$ are anticoincidences. $P(a,b)$ denotes the average of the quantity (number of anticoincidences). The Bell correlation is then $\int dh P(h) A(a, h_1) B(b, h_2)$, where $\int dh P(h) = 1$.

Assume that a measurement on one particle does not alter either the magnitude or phase of the complex amplitude of the other particle. In particular, measurement on one particle does not make the other particle to possess a definite state. At each location a measurement yield two-valued denoted by $+$ and $-$ for each particle. The local amplitudes for the events $+$ and $-$ for the two particles are denoted by the complex functions C_{1+}, C_{1-}, C_{2+} and C_{2-}. Amplitudes C_{1+} and C_{2-} are mutually orthogonal. Then the statement of locality is $C_{1\pm} = C_{1\pm}(a, \phi_1)$ and $C_{2\pm} = C_{2\pm}(b, \phi_2)$, where ϕ_1 and ϕ_2 are the hidden variables.

The correlation function is now of the form $U(a,b) = \text{R.P.}(N C_i C_j^*)$, where N is a normalization factor. Let as $C_{1+} = (1/\sqrt{2})e^{is(\theta_1 - \phi_1)}$ for the first particle at the first polarizer and $C_{2+} = (1/\sqrt{2})e^{is(\theta_2-\phi_2)}$ for the second particle at the second polarizer. θ_1 and θ_2 are the directions of the two polarizers and s is the spin of the particle. We have $C_1 C_1^* = C_2 C_2^* = 1/2$. The correlation function for an outcome of either $(++)$ or $(--)$ of two particles is

$$U(\theta_1, \theta_2, \phi_0) = 2\text{R.P.}(C_1 C_2^*) = \cos\left[s(\theta_1 - \theta_2) + s\phi_0\right]. \tag{20.34}$$

U^2 is the probability for a coincidence detection $(++$ or $--)$ and $1-U^2$ is the probability for an anticoincidence (events of the type $+-$ and $-+$). The average of the quantity (number of coincidences$-$number of anticoincidences) is $U^2 - (1 - U^2) = 2U^2 - 1$. Then

$$P(a, b) = 2U^2 - 1 = 2\cos^2\left[s(\theta_1 - \theta_2) + s\phi_0\right] - 1. \tag{20.35}$$

For the singlet state breaking up into two spin-1/2 particles propagating in opposite directions with total spin 0 the quantities U and P are obtained as

$$U = \cos\left[\frac{1}{2}(\theta_1 - \theta_2) + \frac{\pi}{2}\right] = -\sin((\theta_1 - \theta_2)/2) \tag{20.36}$$

and $P(a, b) = 2\sin^2((\theta_1 - \theta_2)/2) - 1 = -\cos(\theta_1 - \theta_2)$. This is identical to the quantum mechanical prediction obtained from the singlet entangled state and the Pauli spin operator.

20.9 Concluding Remarks

Quantum theory does not give exact outcome of a measurement but it gives various possible outcomes. Even if the event has occurred the outcome is known (according to quantum theory) only a measurement or observation is made. With out observation, the value of a dynamical variable cannot be given any meaning in quantum mechanics. All quantum systems are supposed to exist in some sort of state of probabilistic uncertainty until an observation or measurement is made and the wave function collapses. A measurement process not only makes the wave function collapse into one of the possibilities but in some experiments it is responsible for the wave or particle character of the system. For example, the wave behaviour of light in double-slit experiment and particle-like character in photoelectric experiment are not a property of light but are the property of our interaction with light through measuring process. Schrödinger equation does not tell us which event is going to happen at various times and at various places and how an event will occur at a place at a given time. Its goal is not to predict the actual events going to take place at various times and places but to predict the possible events.

An important point is that given the value of ψ at $t = t_0$ and assuming fixed boundary conditions the Schrödinger equation determines the value of ψ at $t > t_0$. In this sense and because there is no stochastic term in the equation the quantum description of state of a system is deterministic. If the events occur purely by chance then how we are able to find the probabilities of the events. This is because probabilities are given by probability function, and probability function is obtained from wave function. Wave function is the solution of the Schrödinger equation $-$ which is completely deterministic.

20.10 Bibliography

[1] R. Deltete and R. Guy, *Am. J. Phys.* 58:673, 1990.

[2] A. Einstein, B. Podolsky and N. Rosen, *Phys. Rev.* 47:777, 1935.

[3]   D. Harrison, *Am. J. Phys.* 50:811, 1982.

[4]   B.M. Terhal, M.M. Wolf and A.C. Doherty, *Physics Today*, April, pp. 46, 2003.

[5]   T. Leggett, *Physics World*, December pp.73, 1999.

[6]   D.M. Greenberger, M.A. Horne and A. Zeilinger, *Physics Today*, August, pp.22, 1993.

[7]   C. Monroe, D.M. Meekhof, B.E. King and D.J. Wineland, *Science* 272:1131, 1996.

[8]   S. Wallentowitz, R.L. de Matos Filho and W. Vogel, *Phys. Rev. A* 56:1205, 1997.

[9]   D. Bouwmeester, J.W. Pan, M. Daniells, H. Weinfurter and A. Zeilinger, *Phys. Rev. Lett.* 82:1345, 1999; *Nature* 403:515, 2000.

[10] C.A. Sackett, D. Kielpinski, B.E. King, C. Langer, V. Meyer, C.J. Myatt, M. Rowe, Q.A. Turchette, W.M. Itano, D.J. Wineland and C. Monroe, *Nature* 404:256, 2000.

[11] S. Ghosh, T.F. Rosenbaum, G. Aeppli and S.N. Coppersmith, *Nature* 425:48, 2003.

[12] C.P. Williams and S.H. Clearwater, *Explorations in Quantum Computing.* Springer, New York, 1998.

[13] H.K. Lo, S. Popescu and T. Spiller (Eds.), *Introduction to Quantum Computation and Information.* World Scientific, Singapore, 1998.

[14] S.F. Huelga, C. Macchiavello, T. Pellizzari, A.K. Ekert, M.B. Plenio and J.I. Cirac, *Phys. Rev. Lett.* 79:3865, 1997.

[15] L.K. Grover, *Phys. Rev. Lett.* 78:325, 1997.

[16] J.I. Cirac, A.K. Ekert, S.F. Huelga and C. Macchiavello, *Phys. Rev. A* 59:4249, 1999.

[17] S. Bose, V. Vedral and P.L. Knight, *Phys. Rev. A* 57:822, 1998.

[18] M. Murao, M.B. Plenio, S. Popescu, V. Vedral and P.L. Knight, *Phys. Rev. A* 57:R4075, 1998.

[19] R. Penrose, *The Emperor's New Mind.* Vintage, New York, 1990.

[20] A.J. Leggett, *Contemporary Physics* 25:583, 1984.

[21] N.D. Mermin, *Physics Today* April, pp.38, 1985.

[22] J.S. Bell, *Physics* 1:195, 1964.

[23] d'Espagnat, *Am. J. Phys.* 50:811, 1982.

[24] N.D. Mermin, *Am. J. Phys.* 58:731, 1990.

[25] J.J. Freedman and J.S. Clauser, *Phys. Rev. Lett.* 28:938, 1972.

[26] A. Aspect, J. Dalibard and G. Roger, *Phys. Rev. Lett.* 49:1804, 1982.

[27] D.M. Greenberger, M.A. Horne, A. Shimony and A. Zeilinger, *Am. J. Phys.* 58:1131, 1990.

[28] L. Hardy, *Phys. Rev. Lett.* 71:1665, 1993.

[29] C.S. Unnikrishnan, *Current Science* 79:195, 2000.

20.11   Exercises

20.1 List the assumptions used to derive the Bell's inequality.

20.2 When do you get equality in Eq. (20.15)?

20.3 If you make measurements of a physically observable A on a system in rapid succession, do you expect the outcomes to be the same every time?

20.4 For a pair of spin-1/2 particles in the singlet state determine the expectation value of the measurements of the correlated state in the different directions of the unit vectors **a** and **b**, coplanar with σ.

20.5 Assume that at time $t = 0$ a fair coin was tossed by one of the two observers A and B. At $t = t_1$ the coin landed on the ground and shown an outcome. At $t = t_2$, A looked at the coin and observed the outcome as head. B saw the coin at $t = t_3 > t_2$. Assume that the above event is quantum one. Write the wave function of the state of the coin for the two observers at t_1, t_2 and t_3.

20.6 Consider the previous exercise. What would be the outcome of the coin tossing for the observer A during $t_1 < t < t_2$ knowing the outcome at $t = t_2$ (as head) if the event is (i) classical and (ii) quantum?

20.7 What are the eigenvalues of P_1' and P_2' in Eq. (20.21)?

20.8 Evaluate the quantum correlation function $C_{\mathrm{QM}} = P_{++} + P_{--} - P_{+-} - P_{-+}$ corresponding to the wave function (20.23) associated with the measuring the z-component of spin of two atoms in the set up shown in Fig. 20.4.

20.9 Compute the expectation value of $P_1'P_2'$ for the set up shown in Fig. 20.4 using the wave function given by Eq. (20.23).

20.10 Consider the table 20.3. Suppose the angles $45°$ and $90°$ are interchanged. Test the Bell's inequality.

21

Delayed-Choice Experiments

21.1 Introduction

In the 17th century, there existed two theories of light, namely, the wave theory of Huygens and the corpuscular theory of Newton. At the beginning of 19th century Thomas Young demonstrated the wave character of light using diffraction experiments. James Clerk Maxwell developed his electromagnetic wave equation and showed that light is an electromagnetic wave. The wave nature of light could not explain photoelectric effect. Einstein explained the photoelectric effect using the hypothesis that light is a quantum of energy, which is produced or absorbed as a whole like a particle.

In 1924 de Broglie postulated that all particles with mass behave as a wave also with the associated wavelength $\lambda = h/p$. The wave characteristic of electron was confirmed through diffraction experiment with a nickel crystal by Clinton Davisson and Lester Germer in 1927. Claus Jönsson in 1961 performed the first double slit diffraction experiment with electron.

Quantum mechanics differs from classical mechanics as it incorporates with itself strange and weird concepts like wave-particle duality, the uncertainty principle, nonexistence of trajectories and collapse of the wave function due to measurement of observables. The double slit experiments performed with particle and particle detectors demonstrate nonclassical behaviour of microscopic particles. The interference observed in double slit experiments with microscopic particles demonstrate their wave character and detectors of the particles detect them as particles.

Essentially, the arrangement of the double slit experiment appears to induce particle or wave charcter of light. To make things very clear we can consider a Mach–Zehnder interferometer experiment an equivalent experiment of the double slit experiment. To test whether light priorily get the information about the experimental set up in the double slit experiment John Archibald Wheeler proposed a gedanken delayed-choice experiment with the Mach–Zehnder interferometer. This thought experiment was then realized experimentally using various systems. An experiment was proposed to observe simultaneously both particle and wave character of the quantum through quantum entanglement. We discuss these developments in the present chapter.

21.2 Single Slit and Double Slit Experiments

We consider the slit experiments with a beam of particles and with one particle at a time [1].

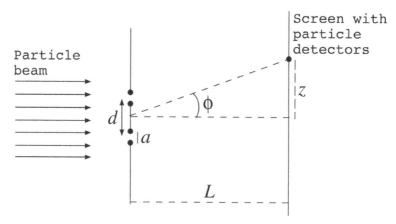

FIGURE 21.1
Schematic of double slit apparatus.

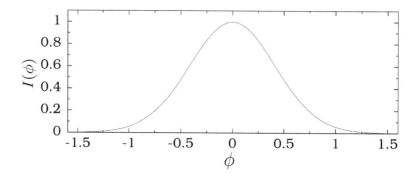

FIGURE 21.2
Intensity of single slit diffraction pattern (Eq. (21.1)) for $I_{\max} = 1$, $L = 1\,\mathrm{m}$, $a = 600\,\mathrm{nm}$ and $\lambda = 632.8\,\mathrm{nm}$.

21.2.1 Experiments With Particle

The experimental set up for the double slit experiment with particle is shown schematically in Fig. 21.1. Let a, d and L are width of a single slit, the distance separation of the two slits and the distance separation of the slits and the screen, respectively. A parallel beam of microscopic particles with constant velocity v is made to fall on a double slit.

Suppose we first close the lower slit allowing the particles to pass only through the upper slit. Then we will get the single slit diffraction pattern at the screen. Classical diffraction theory gives the intensity of the single slit diffraction pattern as [1,2]

$$I(\phi) = I_{\max}\left(\frac{\sin\alpha}{\alpha}\right)^2, \quad \phi = \tan^{-1}\left(\frac{z}{L}\right), \quad \alpha = \frac{\pi a \sin\phi}{\lambda}, \tag{21.1}$$

where ϕ is as defined in Fig. 21.1. The intensity of the single slit diffraction pattern is shown in Fig. 21.2. If the upper slit is closed allowing the particles to pass only through the lower slit we will again get the same diffraction pattern displaced downwards by the slit separation d. Classical theory explains successfully the diffraction pattern of a single slit.

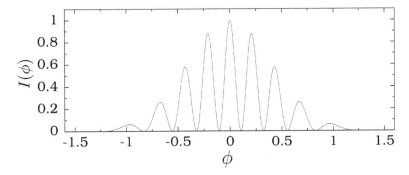

FIGURE 21.3
Intensity of double slit diffraction pattern (Eq. (21.2)) for $I_{\max} = 1$, $L = 1$ m, $a = 600$ nm, $d = 3\,\mu$m and $\lambda = 632.8$ nm.

Suppose we now keep both the slits open then particles can pass through both slits. In this case the intensity of the double slit diffraction pattern is given by [1,3]

$$I(\phi) = I_{\max} \cos^2 \beta \left(\frac{\sin \alpha}{\alpha} \right)^2, \quad \beta = \frac{\pi d \sin \phi}{\lambda}. \tag{21.2}$$

Then we will observe a fringe pattern on the screen as shown in Fig. 21.3. We get the same diffraction pattern for the particles as that of a coherent monochromatic beam of light. The double slit experiment establishes the wave nature of microscopic particles. The maxima and minima of the diffraction, respectively, arise only due to the constructive and destructive interactions of the waves of the particles passing through the upper and lower slits.

21.2.2 Experiment With One Particle At a Time

When a beam of particles is incident on the slits we may think that the interference pattern is observed due to the interferences of waves of some particles going through the upper slit with some other particles passing through the lower slit. In order to test this assumption experiments were done by passing one particle at a time through the double slit apparatus. If the assumption that the interference pattern arises due to the interference arises due to the interference of waves of the particles passing through both the slits then for this case no interference should be observed as a single particle can pass through only in one of the two slits. But it has been experimentally observed by many workers that even if one particle is passed at a time through the double slit apparatus the double slit interference pattern is observed. Though a single particle cannot pass through both the slits at the same time the wave associated with the particle passes through both the slits at the same time thereby producing interference pattern.

We can think of a modification of this experiment in which we try to find the slit through which the particle will pass by keeping detectors at each slit. Then it will be found that each particle passes through one slit or other and never on both slits. Both detectors will not detect the particle simultaneously. If the detectors are of the type which allow the particle to pass through the slits while recording its passage, *will we get the same double slit interference pattern?* We will not get the interference pattern. We will get just the superposition of two single slit diffraction patterns only. That is, the detection of the path of the particle through the slits has destroyed the interference pattern.

The modified experiments have been done with photons and atoms. These experiments were done using *entangled particles* (refer sec. 20.3.1 for some details about entanglement). The which-way information of one particle passing through the slits can be obtained from the measurements done on the other entangled particle. Through this experiment we can conclude that measurement of the path of the particle destroys the interference pattern and the particle shows only its particle characteristics. If no detector detects the path of the particle at the slits an interference pattern is observed and the particle shows only its wave characteristics. The point is that the type of experiments we perform determines either the particle nature or wave nature. No experiment can determine both wave and particle characteristics of a particle simultaneously. As Bohr postulated *the wave and particle nature are complementary to each other.*

21.3 Quantum Mechanical Explanation

We have found from the previous experiment that turning on the slit detectors forced the particle to move in one of the slits whereas turning off the detectors forced the particle to interact with both the slits and produced the interference pattern. As discussed earlier the type of experimental set up decides whether the particle behaves as a particle or a wave. *How does quantum mechanics explain this experiment?* This wave-particle duality experiment is explained by the superposition principle of quantum states.

First, we consider the case when the slit detectors are absent and both slits are kept open. $\psi_1(z)$ and $\psi_2(z)$ are the states localized at the location of slit-1 (upper slit) and slit-2 (lower slit), respectively. Then the particle will emerge from the double slit in a superposition state

$$|\psi\rangle = \frac{1}{\sqrt{2}} \left(|\psi_1\rangle + |\psi_2\rangle \right). \tag{21.3}$$

$|\psi_1\rangle$ and $|\psi_2\rangle$ are orthogonal because of their spatial separation. The probability of finding the particle at position z in the screen is given by

$$|\langle z|\psi\rangle|^2 = \frac{1}{2} \left[|\psi_1(z,t)|^2 + |\psi_2(z,t)|^2 + \psi_1^*(z,t)\psi_2(z,t) + \psi_2^*(z,t)\psi_1(z,t) \right]. \tag{21.4}$$

The last two terms give the interference.

Next, we consider the case where the slit-detectors are present to detect the path of the particles passing through the slits. Let us assume that the which-way detector is a 1-bit detector, $|\uparrow\rangle$ state for particle passing through slit-1 and $|\downarrow\rangle$ state for the particle passing through slit-2. *What will be the state of the particle coming out of the slits?* The which-way detector states get entangled with the states of the two paths and the combined state of the particle and which-way detector is given by

$$|\psi\rangle = \frac{1}{\sqrt{2}} \left[|\psi_1\rangle|\uparrow\rangle + |\psi_2\rangle|\downarrow\rangle \right]. \tag{21.5}$$

$|\uparrow\rangle$ and $|\downarrow\rangle$ are orthogonal states as both particle detectors at the two slits cannot detect the particle at the same time. Now, we get the probability density of the particle falling on the screen at position z as

$$|\langle z|\psi\rangle|^2 = \frac{1}{2} \left[|\psi_1(z,t)|^2 + |\psi_2(z,t)|^2 \right]. \tag{21.6}$$

Equation (21.6) does not have the interference term as in Eq. (21.4). Equation (21.6) corresponds to the classical superpostion of two single slit diffraction pattern. The which-way detection experiment detects the particle nature of the quantum particle.

Solved Problem 1:

Show that the interference terms vanishes for the wave function given by Eq. (21.5).

For the given $|\psi\rangle$ the $\langle\psi|$ is

$$\langle\psi| = \frac{1}{\sqrt{2}}\left(\langle\uparrow|\langle\psi_1| + \langle\downarrow|\langle\psi_2|\right). \tag{21.7}$$

Then

$$
\begin{aligned}
|\psi|^2 &= \langle\psi|\psi\rangle \\
&= \frac{1}{2}\left[\langle\psi_1|\psi_1\rangle\langle\uparrow|\uparrow\rangle + \langle\psi_1|\psi_2\rangle\langle\uparrow|\downarrow\rangle + \langle\psi_2|\psi_1\rangle\langle\downarrow|\uparrow\rangle + \langle\psi_2|\psi_2\rangle\langle\downarrow|\downarrow\rangle\right]. \tag{21.8}
\end{aligned}
$$

As $|\uparrow\rangle$ and $|\downarrow\rangle$ are orthonormal we have $\langle\uparrow|\downarrow\rangle = \langle\downarrow|\uparrow\rangle = 0$ and $\langle\uparrow|\uparrow\rangle = \langle\downarrow|\downarrow\rangle = 1$. Then

$$|\psi|^2 = \frac{1}{2}\left[\langle\psi_1|\psi_1\rangle + \langle\psi_2|\psi_2\rangle\right] = \frac{1}{2}\left(|\psi_1|^2 + |\psi_2|^2\right). \tag{21.9}$$

The interference terms vanish.

21.4 Experiment With Mach–Zehnder Interferometer

We can consider an equivalent double slit experiment with Mach–Zehnder interferometer. The schematic of a Mach–Zehnder interferometer is shown in Fig. 21.4. A quantum particle like photon, electron, neutron enters from left into the interferometer through a semitransparent (50 : 50) beam splitter. We denote the state of the transmitter part in arm b as $|0\rangle$ and the reflected part as $|1\rangle$. Let a phase ϕ be introduced in the reflected arm a. Then the quantum state of the system is a superposition of the two paths

$$|\psi\rangle = \frac{1}{\sqrt{2}}\left(|0\rangle + e^{i\phi}|1\rangle\right). \tag{21.10}$$

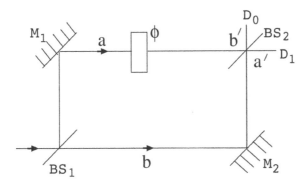

FIGURE 21.4
Schematic of a Mach–Zehnder interferometer.

If the beam splitter BS$_2$ is not present then detector D$_0$ will register the quantum particle travelled in arm b and detector D$_1$ will register the quantum particle travelled in arm b and detector D$_1$ will register the quantum particle travelled in arm a. When BS$_2$ is not present we get the which-way information of the particle. From Eq. (21.10) the probability of detection of the outcome $|0\rangle$ and $|1\rangle$ are $P_a = P_b = 1/2$. Both the detectors will not respond at the same time if a single particle is allowed inside the interferometer. Without BS$_2$ we find that the quantum particle behaves as a particle.

Now, suppose BS$_2$ is present. Then the beam will be splitted with one transmitted and another reflected for both a and b quantum particle beam. The transmitted and reflected states of BS$_2$ can be represented by the states $\frac{1}{\sqrt{2}}(|0\rangle \pm |1\rangle)$. Then the detection probability of a$'$ at D$_1$ detector will be $P_{a'} = \cos^2(\phi/2)$ and the detection probability of b$'$ at D$_0$ will be $P_{b'} = \sin^2(\phi/2)$. We note that the detection probabilities in detectors D$_0$ and D$_1$ are dependent on the phase difference ϕ between the quantum particle travelled in a and b arms. That is, the quantum particle behaves as a wave producing interference pattern on the detectors depending on the phase difference ϕ. In short, we find that the quantum particle behaves as a particle if BS$_2$ is not present and behaves as waves if BS$_2$ is present. Presence and absence of BS$_2$ are two different experiments. Hence, the wave nature and particle nature are revealed by these two different experiments.

21.5   Delayed-Choice Experiment

The two experiments considered in the previous section are mutually exclusive as BS$_2$ cannot be simultaneously inserted or removed. If we have set up any one of the experiment well in advance before the photon enters the interferometer then one could reconcile Bohr's complementary principle with Einstein's local conception of reality. We may then think that the photon could have received some *hidden information* about the chosen experimental set up and it would have travelled either as a particle or a wave according to the experimental set up. Then a hidden variable theory can explain the results of the two experiments.

The next obvious question to be explored is how does a single photon show its wave or particle character when the introduction of BS$_2$ or its removal is delayed until the photon emerges out of BS$_1$. In an interferometer experiment the choice to observe the particle or wave character of a quantum particle can be delayed with respect to the quantum particle entering the interferometer. It is also possible to observe a continuous transformation between these two experimental arrangements before the measurements are done. The history of delayed-choice gedanken experiment can be traced back to 1927 when Heisenberg put forwarded his uncertainty princple using a γ-ray microscope. For a brief history of delayed-choice experiments one may refer to the ref. [4]. The interest on delayed-choice experiment was revived by the thought experiment proposed in the refs. [5,6].

John Archibald Wheeler proposed a delayed-choice experiment with the Mach-Zehnder interferometer in which the choice of which property of the quantum particle will be observed is made after the photon has passed the first beam splitter BS$_1$. The removal or insertion of BS$_2$ is done after the photon has passed through BS$_1$. In this delayed-choice experiment one can decide whether the photon has come by one route (as a particle) or both routes (as a wave) after it has already done its travel.

The complete sheme proposed by Wheeler with single photon was experimentally realized in 2007 [7]. This experiment unambiguously proved Wheeler's supposition that a delayed-choice of experiment affect the past history of the photon. Wheeler's delayed-choice

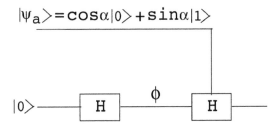

FIGURE 21.5
Schematic of the quantum delayed-choice experiment. For details see the text.

experiment with a single atom was then realized in 2015 [8]. The delayed-choice experiment rules out the assumption that the photon could have known in advance what type of experimental apparatus that it will be confronted with and then decides to behave either as a wave or a particle. Therefore, a hidden variable theory which assumes that the photon knows in advance the type of experimental set up it will face, cannot explain the results of delayed-choice experiment.

A quantum version of Wheeler's delayed-choice experiment based on a quantum-controlled beam splitter has been proposed [9]. This quantum-controlled beam splitter can also be in a superposition state of being present and absent. The schematic of the quantum delayed-choice experiment is shown in Fig. 21.5.

In the quantum delayed experiment, the first beam splitter BS_1 is represented by the Hadamard quantum gate which has the transformation $H = \dfrac{1}{\sqrt{2}} \begin{pmatrix} 1 & 1 \\ 1 & -1 \end{pmatrix}$. If the two photon states of the a and b arms in the standard Mach-Zehnder interferometer are represented by $|1\rangle$ and $|0\rangle$, respectively, then the first H transforms the photon state $|0\rangle$ into the superposition state $\dfrac{1}{\sqrt{2}}(|0\rangle + |1\rangle)$. A phase shifter then introduces a relative phase ϕ between the two states giving the superposition state $\dfrac{1}{\sqrt{2}}(|0\rangle + e^{i\phi}|1\rangle)$. This state is the input state for the second Hadamard gate which is controlled by an ancilla photon.

If the ancilla photon is prepared in state $|0\rangle$ then no beam splitter will be present. With $|0\rangle$ control bit, the second Hadamard operation will be just an identity operation. Hence, the output state will be the input state

$$|\psi_P\rangle = \frac{1}{\sqrt{2}}\left(|0\rangle + e^{i\phi}|1\rangle\right). \tag{21.11}$$

The measurement indicates that the probability detections in the $|0\rangle$ (b-arm) and $|1\rangle$ (a arm) basis give $P_a = P_b = 1/2$. The final measurement indicates the path the photon took. Hence, the ancillary photon state $|0\rangle$ is equivalent to the absence of BS_2 in the conventional Mach–Zehnder interferometer. We detect the *particle nature of the photon* for this experimental condition. If the ancilla photon is prepared in the state $|1\rangle$ then it is equivalent to the presence of BS_2. Applying the Hadamard operation to $|\psi\rangle$ will give the output state

$$|\psi_w\rangle = \cos(\phi/2)|0\rangle - i\sin(\phi/2)|1\rangle \tag{21.12}$$

leaving out a common phase factor. Equation (21.12) gives the probability of detection $P_{a'} = \sin^2(\phi/2)$ and $P_{b'} = \cos^2(\phi/2)$. The final measurement gives information about the phase ϕ that was introduced between the two states. We get the interference pattern corresponding to the *wave nature of the photon* for the experiment done using the ancilla photon prepared in state $|1\rangle$.

Conventional beam splitter BS_2 can only be present or absent. But a quantum controlled beam splitter can be put in a superposition of being present and absent. We get $|\psi_p\rangle$ as output for the ancilla state $|0\rangle_a$ and $|\psi_w\rangle$ for the ancilla state $|1\rangle_a$. Thus, for the ancilla state $\cos\alpha|0\rangle_a + \sin\alpha|1\rangle_a$ we get the output

$$|\psi_f(\alpha, \phi)\rangle = \cos\alpha|\psi_P\rangle_a + \sin\alpha|\psi_w\rangle_a. \tag{21.13}$$

We find that the quantum state and the ancilla photons have now become entangled for $0 < \alpha < \pi/2$. The photon detector D_0 now measures the intensity proability

$$I_b = I_0(\phi, \alpha) = I_P(\phi)\cos^2\alpha + I_w(\phi)\sin^2\alpha \tag{21.14}$$

and the detector D_1 measures the intensity probability

$$I_a = I_1 = 1 - I_0. \tag{21.15}$$

The contrast of an interference pattern is quantified by the term *interferometric visibility* also called *fringe visibility* or simply *visibility* and is given by

$$V = \frac{I_{0max} - I_{0min}}{I_{0max} + I_{0min}}, \tag{21.16}$$

where I_0 is given by Eq. (21.14). I_{0max} and I_{0min} are to be computed by varying ϕ.

From Eqs. (21.11) and (21.12) we find $I_p(\phi) = 1/2$ and $I_w(\phi) = \cos^2(\phi/2)$. So,

$$I_0(\phi, \alpha) = \frac{1}{2}\cos^2\alpha + \cos^2(\phi/2)\sin^2\alpha. \tag{21.17}$$

For this I_0 the visibility is $V = \sin^2\alpha$. We observe that by changing α we can alter continuously the interference pattern from a particle at $\alpha = 0$ and to a wave at $\alpha = \pi/2$.

A quantum beam splitter makes it possible to prepare a photon in a superposition of particles and wave. Such a superposition does not have a classical analog. The presently available hidden variable theories in which particle and wave are realistic properties cannot explain the delayed-choice experiments. Based on quantum-controlled beam splitter experimentally the particle and wave behaviours of photon to be present simultaneously was investigated [10]. Further, this quantum nature of the photon was tested using a Bell inequality. It was found that no model in which the photon knew in advance whether it would behave as particle or wave could account for the observed experimental data.

For more discussion on delayed-choice experiments one may refer to the refs. [11–19].

21.6 Delayed-Choice Quantum Eraser

In a delayed-choice quantum eraser interference fringes are obtained by erasing which-way information after the interfering quantum particle has already been detected [20]. In a quantum eraser the which-way information of the signal photon is encoded in an entangled ancillary photon. One can then perform certain measurement on the ancilla which can make the which-way information unobtainable. The measurement done on the ancillary entangled photon erases the which-way information of the signal photon. This measurement can also be done after the measurement done on the signal photon in the interferometer. When the result of signal measurement done earlier is correlated with the results of the eraser measurement done later we can reconstruct the interference pattern. The temporal order of the measurements on different systems is irrelevant even if the systems are entangled. The measurement on ancilla is performed long after the signal photon has passed through the interferometer. This protocal is known as *delayed-choice quantum eraser*. We give the theory of delayed-choice quantum eraser described in [21].

Consider a photon described by the superposition state $|\psi\rangle = \frac{1}{\sqrt{2}}(|H\rangle + |V\rangle)$, where $|H\rangle$ denotes horizontal polarization state and $|V\rangle$ the vertical polarization state. We let the photon pass through a polarization interferometer. On passing through this interferometer let there be a relative phase ϕ be introduced between the $|H\rangle$ and $|V\rangle$ states. The photon state after the phase change will be

$$|\psi'\rangle = \frac{1}{\sqrt{2}}\left(|H\rangle + e^{i\phi}|V\rangle\right). \tag{21.18}$$

A measurement on the $H - V$ basis will give the probability of $1/2$ for both horizontal and vertical polarizations. The measurement probability $1/2$ for both $|H\rangle$ and $|V\rangle$ states shows that the photon does not show any interference as the paths of the vertical polarized photon and horizontally polarized photon is known. The measurement in $H - V$ basis will detect the photon as a particle.

Now, if we measure in the $\pm45°$ (diagonal) basis with the corresponding eigenstates $|\pm45°\rangle = \frac{1}{\sqrt{2}}(|H\rangle \pm |V\rangle)$ the probability will be

$$
\begin{aligned}
P(+45°) &= |\langle+45°|\psi'\rangle|^2 = \cos^2\phi, \tag{21.19} \\
pP(-45°) &= |\langle-45°|\psi'\rangle|^2 = \sin^2\phi. \tag{21.20}
\end{aligned}
$$

The oscillating probabilities of the detector signals give the interference pattern. Hence, the measurement on the $\pm45°$ basis give the wave nature of the photon.

Next, let us consider two photons in the entangled state

$$|\psi\rangle = \frac{1}{\sqrt{2}}\left(|H\rangle|H\rangle + |V\rangle|V\rangle\right). \tag{21.21}$$

Let one of the photons (called *signal*) pass through the interferometer leading to the composite state

$$|\psi'\rangle = \frac{1}{\sqrt{2}}\left(|H\rangle|H\rangle + e^{i\phi}|V\rangle|V\rangle\right). \tag{21.22}$$

Suppose we measure the signal in the $\pm45°$ basis and the other photon (called *idler*) in the $H - V$ basis then we get the joint probability for the $\pm45°$ polarization measurement of signal and H and V polarization measurements for the idler photon as

$$P(45°, H) = \frac{1}{4}, \quad P(45°, V) = \frac{1}{4}. \tag{21.23}$$

The joint probability of these two independent measurements give results which are independent of ϕ. These two independent measurements give the particle nature of the signal. This result is obtained as the idler measurement has given the which-way information of the signal.

Suppose we measure both the signal and idler polarization in $\pm45°$ basis then the joint probability of finding $+45°$ polarization of the signal and idler photon is given by

$$P(45°, 45°) = \frac{1}{2}\cos^2(\phi/2). \tag{21.24}$$

Finding $+45°$ polarization for signal and $-45°$ polarization for idler photon giving the joint probability. Similarly,

$$P(45°, -45°) = \frac{1}{2}\sin^2(\phi/2). \tag{21.25}$$

We find that the signal photon behaves like a wave after correlating these two independent measurements. Also, the sum of these two probabilities is always $1/2$ and independent of ϕ. Hence, it is found that unless the eraser measurement is actually carried out on the idler photon and the result is correlated with the signal measurement result no interference pattern can be observed. The probabilities given by Eqs. (21.24) and (21.25) do not depend on whether we first measure the idler or the signal. The relative time order of these two independent measurements is immaterial. But the choice between interference and path information is only open until both measurements have been carried out. Quantum eraser experiments have been realized by various groups [4,22–33].

21.7　Concluding Remarks

Quantum mechanics incorporates strange and weird concepts like wave-particle duality, nonexistence of trajectories, uncertainty principle and collapse of the wave function. The quantum particles (usually microscopic particles) behave strangely. For example, they behave as particles in certain experiments and as waves in some other experiments. Davisson and Germer demonstrated the wave nature of electron using crystal diffraction in 1927. In 1963 Richard Feynman described an interference experiment of particles using a double slit. The experiment as described by Feynman was performed in 1973 [34]. Later the double slit interference pattern was also observed even if one particle at a time passed through the double slit apparatus. Experiment with double slit has has also proved that any experiment in which the path of the particle is measured will not show any interference pattern. A slit detector can find whether a particle passed through it or not.

Wheeler raised a question whether we can observe the interference pattern by detecting the path of the particle through the slit after the particle has passed through the slit. The question is whether the particle decides to behave as a particle or wave at the slits itself irrespective of its path determined later. Wheeler's thought experiment has been tested using Mach–Zehnder interferometer using a second beam splitter. The absence of the second beam splitter will reveal the particle nature and the presence of it will show interference pattern corresponding to wave nature of the particle. The removal and inclusion of the second beam splitter can be delayed even after the particle has passed through the first beam splitter. This experiment established that particle behaves as wave or particle depending on the final detection arrangements only.

The above mentioned experiments were done using a quantum beam splitter using Hadamard quantum gate with the control of an ancilla photon. This ancilla photon can be in a superposition state of present and absent and it is possible to prepare a photon state in a superposition of particle and wave. It is also possible to create a pair of entangled photons, a signal photon and an ancillary photon. The ancillary photon carries which-way information of the signal photon. It is possible to erase this information on the ancillary photon after the signal photon has been detected. Such an erasure of the which-way information correlated with the signal information will give interference pattern. The time order of the measurement of signal photon and ancillary photon is of no consequence.

Various delayed-choice experiments serve as striking illustrations of the inherently non-classical feature of quantum mechanics. They play an important role in the foundations of quantum mechanics. These experiments clearly demonstrate that a quantum particle does not decide to behave either as a wave or as a particle by adapting *a priori* on the experimental situation. This is confirmed by experiments with space-like separation of the changing the beam splitter BS_2 from the entry of the photon at BS_1 as no signal can travel with velocity greater than the velocity of light.

We find from the delayed-choice experiments that quantum effects mimic an influence of future actions on past events. The delayed-choice experiments reveal a type of retrocasuality. Actually, it is one of the weird nature of quantum mechanics which presents a kind of causal connection, which is generically different from anything that could be characterized classically since the causal connection cannot be clearly analyzed into a *cause* and *effect* [35] .

In quantum mechanics the relative temporal order of measurement of events is not relevant. No physical interactions or signals are necessary to explain the experimental results. In explaining an individual observation of one system one has to include the whole experimental configuration and also the complete quantum state, potentially describing joint properties with other systems as seen in quantum eraser experiments. The results of the delayed-choice experiments cannot be explained by any hidden variable theory as these experiments highlight certain nonclassical features.

21.8 Bibliography

[1]   R. Rolleigh, *The Double Slit Experiment and Quantum Mechanics*. 2010. https://www.hendrix.edu/uploadedFiles/Departments_and_ Programs/Physics-/Faculty/DS5Quantum.pdf.

[2]   J. Sanny and S. Ling, *University Physics*. Volume 3. OpenStax, Minneapolis, 2016.

[3]   F.A. Jenkins and H.E. White, *Fundamentals of Optics*. McGraw Hill, New York, 1967.

[4]   X.S. Ma, J. Kofler and A. Zeilinger, *Rev. Mod. Phys.* 88:015005, 2016.

[5]   J.A. Wheeler, *The past and the delayed-choice experiment*. In Mathematical Foundations of Quantuam Theory. Editor R. Marlow. Academic Press, New York, 1978.

[6]   J.A. Wheeler and W.H. Zurek (Eds.), *Quantum Theory and Measurement*. Princeton University Press, Princeton, 1984.

[7]   V. Jacques, E. Wu, F. Grosshans, F. Treussart, P. Grangier, A. Aspect and J.F. Roch, *Science* 315:966, 2007.

[8]   A.G. Manning, R.I. Khakimov, R.G. Dall and A.G. Truscott, *Nature Physics* 11:539, 2015.

[9]   R. Ionicioiu and D.R. Terno, *Phys. Rev. Lett.* 107:230406, 2011.

[10] A. Peruzzo, P. Shadbolt, N. Brunner, S. Popescu and J.L. O'Brien, *Science* 338:634, 2012.

[11] F. Kaiser, T. Coudreau, P. Milman, D.B. Ostrowsky and S. Tanzilli, *Science* 338:637, 2012.

[12] D. Ellerman, *Quantum Stud.: Math. Found.* 2:183, 2015.

[13] K. Liu, Y. Xu, W. Wang, S.B. Zheng, T. Roy, S. Kundu, M. Chand, A. Ranadive, R. Vijay, Y. Song, L. Duan and L. Sun, *Sci. Adv.* 3:e1603159, 2017.

[14] F. Vedovato, C. Agnesi, M. Schiavon, D. Dequal, L. Calderaro, M. Tomasin, D.G. Marangon, A. Stanco, V. Luceri, G. Bianco, G. Vallone and P. Villoresi, *Sci. Adv.* 3:e1701180, 2017.

[15] R. Chaves, G.B. Lemos and J. Pienaar, *Phys. Rev. Lett.* 120:190401, 2018.

[16] H.L. Huang, Y.H. Luo, B. Bai, Y.H. Deng, H. Wang, Q. Zhao, H.S. Zhong, Y.Q. Nie, W.H. Jiang, X.L. Wang, J. Zhang, L. Li, N.L. Liu, T. Byrnes, J.P. Dowling, C.Y. Lu and J.W. Pan, *Phys. Rev. A* 100:012114, 2019.

[17] K. Wang, Q. Xu, S. Zhu and X.S. Ma, *Nature Photonics* 13:872, 2019.

[18] E. Polino, I. Agresti, D. Poderini, G. Carvacho, G. Milani, G.B. Lemos, R. Chaves and F. Sciarrini, *Phys. Rev. A* 100:022111, 2019.

[19] M.X. Dong, D.S. Ding, Y.C. Yu, Y.H. Ye, W.H. Zhang, E.Z. Li, L. Zeng, K. Zhang, D.C. Li, G.C. Guo and B.S. Shi, *Quan. Inf.* 6:72, 2020.

[20] M.O. Scully and K. Druhl, *Phys. Rev. A* 25:2208, 1982.

[21] J.M. Ashby, P.D. Schwarz and M. Schlosshaur, *Am. J. Phys.* 84:95, 2016.

[22] S. Durr, T. Nonn and G. Rempe and *Phys. Rev. Lett.* 81:5705, 1998.

[23] A.G. Zajonc, L. Wang, X.Y. Zou and L. Mandel, *Nature* 353:507, 1991.

[24] P.G. Kwiat, A. Steinberg and R. Chiao, *Phys. Rev. A* 45:7729, 1992.

[25] T.J. Herzog, P.G. Kwiat, H. weinfurter and A. Zeilinger, *Phys. Rev. Lett.* 75:3034, 2000.

[26] Y.-H. Kim, R. Yu, S.P. Kulik, Y. Shih and M.O. Scully, *Phys. Rev. Lett.* 84:1, 2000.

[27] S.P. Walborn, M.O. Terra Cunha, S. Padua and C.H. Monken, *Phys. Rev. A* 65:033818, 2002.

[28] H. Kim, J. Ko and T. Kim, *Phys. Rev. A* 67:054102, 2003.

[29] U.L. Anderson, O. Glockl, S. Lorenz, G. Leuchs and R. Fllip, *Phys. Rev. Lett.* 93:100403, 2004.

[30] H. Scarcelli, Y. Zhou and Y. Shih, *Eur. Phys. J. D* 44:167, 2007.

[31] L. Neves, G. Lima, J. Aguirre, F.A. Torres-Ruiz, C. Saavedra and A. Delgado, *New J. Phys.* 11:073035, 2009.

[32] T. Quershi, *Eur. J. Phys.* 41:055403, 2020.

[33] M.B. Schneidera and I.A. LaPuma, *Am. J. Phys.* 70:266, 2020.

[34] O. Donati, D.P. Missiroli and G. Pozzi, *Am. J. Phys.* 41:639, 1973.

[35] A. Shimony, *Conceptual Foundation of Quantum Mechanics.* In The New Physics. Editor P. Davies. Cambridge University Press, Cambridge, 1989. pp.373-395.

21.9 Exercises

21.1 In the quantum version of Wheeler's delayed-choice experiment based on a quantum-controlled beam splitter show that $P_\mathrm{a} = P_\mathrm{b} = 1/2$ when BS_2 is absent and $P_{\mathrm{a}'} = \cos^2(\phi/2)$ and $P_{\mathrm{b}'} = \sin^2(\phi/2)$ when BS_2 is present.

21.2 Find the output states of the Hadamard gate for the input states $|0\rangle$ and $|1\rangle$.

21.3 Show that applying Hadamard operation to $|\psi\rangle = \frac{1}{\sqrt{2}}\left(|0\rangle + e^{i\phi}|1\rangle\right)$ we get a normalized $|\psi_\mathrm{w}\rangle = \cos(\phi/2)|0\rangle - i\sin(\phi/2)|1\rangle$ by omitting a common phase factor.

21.4 Determine the values of visibility when the probability of finding the photon in state $|0\rangle$ is $I_\mathrm{P}(\phi) = 1/2$ for the particle state and when $I_\mathrm{w}(\phi) = \cos^2(\phi/2)$ for the wave state.

21.5 Show that the visibility of the interference pattern with $I_0(\phi, \alpha) = \frac{1}{2}\cos^2\alpha + \cos^2(\phi/2)\sin^2\alpha$ is $V = \sin^2\alpha$.

22

Fractional Quantum Mechanics

22.1 Introduction

A great deal of interest has been paid over the decades to extend the standard quantum mechanics. One extension is the analysis of systems with non-Hermitian Hamiltonians. Another development is the fractional quantum mechanics (FQM) introduced by Nick Laskin in 2000 [1–3]. It is well known that the appearance of the Schrödinger equation of a free particle is like the ordinary diffusion equation

$$\frac{\partial p}{\partial t} = \frac{\sigma}{2}\frac{\partial^2 p}{\partial x^2},\tag{22.1}$$

where p represents the concentration of the diffusion material and σ is the diffusion coefficient. Equation (22.1) is based on the Gaussian distribution for random walk problems. This implies that we can derive the standard Schrödinger equation by considering the Gaussian probability distribution in the space for all possible paths. Richard Phillips Feynman and Albert Roach Hibbs derived the standard Schrödinger equation making use of path integrals over Brownian paths [4]. That is, the consideration of Brownian motion of paths led to the Schrödinger equation.

Use of non-Gaussian distributions gives rise fractional diffusion equation and the corresponding diffusion is known as *anomalous diffusion*. Consider the diffusion equation of the form

$$\frac{\partial^\beta p}{\partial t^\beta} = c\frac{\partial^\alpha p}{\partial x^\alpha}.\tag{22.2}$$

We have the following three cases [5]:

1. Space fractional diffusion equation: $\beta = 1$, $0 < \alpha < 2$.

2. Time fractional diffusion equation: $0 < \beta < 1$, $\alpha = 2$.

3. Space-time fractional diffusion equation: $\beta \neq 1$, $\alpha \neq 2$.

In this direction we can consider different types of Schrödinger equation.

Laskin in 2000 applying the path integral approach of Feynman by considering Lévy distribution instead of Gaussian distribution for the possible paths for quantum mechanical systems shown that the associated equation of motion is

$$i\hbar\frac{\partial\psi}{\partial t} = D_\alpha\left(-\hbar^2\nabla^2\right)^{\alpha/2}\psi + V\psi,\tag{22.3}$$

where $\psi = \psi(x,t)$ (we assume one-dimensional case), $V = V(x,t)$, α is the Lévy index ($1 < \alpha \leq 2$), D_α is a constant and the Laplacian operator is fractional. Equation (22.3) is called *space fractional Schrödinger equation* (space FSE). The probability density $p_{\mathrm{L}}(x,t;,x_0,t_0)$ of the Lévy α-stable distribution is given by

$$p_{\mathrm{L}}(x,t;x_0,t_0) = \frac{1}{2\pi}\int_{-\infty}^{\infty} dk\, e^{ik(x-x_0)}e^{-\sigma_\alpha|k|^\alpha(t-t_0)},\tag{22.4}$$

DOI: 10.1201/9781003172178-22

493

where α is the Lévy index and σ_α is the generalized diffusion coefficient. α defines the index of the distribution and controls the scale properties of the stochastic process $\{x\}$ and σ_α selects the scale unit of the process [2,5,6]. α is also called *stability index* and it takes values between 0 and 2. When $\alpha = 2$ the Lévy distribution becomes the familiar Gaussian distribution. Thus, Eq. (22.3) with $\alpha = 2$ corresponds to the standard Schrödinger equation.

Instead of first-order time derivative term in the Schrödinger equation Mark Naber considered Caputo fractional derivative thereby proposed time FSE [5]. In the time FSE, the time derivative is non-integer order while the space derivative is second-order. The associated Hamiltonian is non-Hermitian and more over nonlocal in time. Space-time FSE was introduced by Shaowei Wang and Mingyu Xu [7].

The behaviour of particles with fractional spin is described by FSE and thus deals with fractional effect in quantum mechanics. Applications of FQM in a condensed matter physics based on Lévy crystals [8], transverse light dynamics in aspherical optical resonators [9], splitting and diffraction of Gaussian beam in the case $1 < \alpha < 2$ for potential free case [10], beam propagation management [11,12], gravitational optics [13] and conical diffraction of input beams [14] have been explored.

Various features of FQM, particularly, properties, determination of solutions, representation in momentum space, physical realization or applications and extension to relativity have been investigated. We present these developments in the present chapter.

22.2   Integer and Fractional Diffusion Equations

The solution of the diffusion Eq. (22.1) is

$$p = \frac{1}{\sqrt{2\pi\sigma(t-t_0)}} e^{-(x-x_0)^2/(2\sigma(t-t_0))} \tag{22.5}$$

and

$$(x - x_0)^2 \propto \sigma(t - t_0). \tag{22.6}$$

The diffusion path is described by the Gaussian function in Eq. (22.5). With $\delta x = x - x_0$, $\delta t = t - t_0$ and $T = N\delta t$ the length of the diffusion path is

$$L = N\delta x = \frac{T\delta x}{\delta t} = \sigma T(\delta x)^{-1}. \tag{22.7}$$

The dimension d_f of a path length is defined through

$$L \propto (\delta x)^{1-d_f}. \tag{22.8}$$

Comparing the Eqs. (22.7) and (22.8) we find that for the diffusion path considered the dimension $d_f = 2$, an integer. The dimensions of a point and a line segment are 0 and 1, respectively. They have integer dimension. There are mathematical and real objects with fractional dimensions also called *fractal dimension*. Examples of objects with fractal dimension include Cantor set, Mandelbrot set, chaotic attractors, lightning and electricity, snowflakes and clouds.

We can realize fractional diffusion equation and FQM by considering Lévy random process with the probability density given by Eq. (22.4). *What is the dimension associated with the Lévy path?* From Eq. (22.4) we have

$$(x - x_0) \propto (\sigma_\alpha(t - t_0))^{1/\alpha}. \tag{22.9}$$

Then

$$L = N\delta x = \frac{T}{\delta t}\delta x = \frac{T}{\delta t}\frac{(\delta x)^\alpha}{(\delta x)^\alpha}\delta x = T\sigma_\alpha(\delta x)^{1-\alpha}. \qquad (22.10)$$

That is, $L \propto (\delta x)^{1-\alpha}$. Comparison of it with $L \propto (\delta x)^{1-d_{\mathrm{f}}}$ gives $d_{\mathrm{f}}^{\mathrm{Lévy}} = \alpha$, $1 < \alpha \le 2$.

The fractional diffusion equation obeying the Lévy distribution is given by

$$\frac{\partial}{\partial t}p_{\mathrm{L}}(x,t;x_0,t_0) = \sigma_\alpha \nabla^\alpha p_{\mathrm{L}}(x,t;x_0,t_0), \qquad (22.11)$$

where $p_{\mathrm{L}}(x,t;x_0,t_0) = \delta(x - x_0)$ and $\nabla^\alpha = \partial^\alpha/\partial x^\alpha$ is the *fractional Riesz derivative* and is defined through the equation

$$\nabla^\alpha p(x,t) = -\frac{1}{2\pi}\int_{-\infty}^{\infty} dk e^{ikx}|k|^\alpha \bar{p}(k,t), \qquad (22.12)$$

where $\bar{p}$ is the Fourier transform of p and

$$\bar{p}(k,t) = \int_{-\infty}^{\infty} dx\, e^{-ikx}p(x,t), \qquad (22.13a)$$

$$p(x,t) = \frac{1}{2\pi}\int_{-\infty}^{\infty} dx\, e^{ikx}\bar{p}(k,t). \qquad (22.13b)$$

22.3   Wave Function and Kernel

Consider the Schrödinger equation $i\hbar\partial\psi/\partial t = H\psi$ in one-dimension. We can define a time evolution operator $U(t)$ that transforms $\psi(0)$ to $\psi(t)$ as $\psi(t) = U(t)\psi(0)$. For the case of H independent of time we write $U(t) = e^{-iHt/\hbar}$. Let the path of a particle starts from x_i at an initial time t_i and reaches x_f at t_f. The probability amplitude A for a particle to be at x_f at t_f is given by

$$A = K(x_f, t_f; x_i, t_i) = \langle x_f| \exp(-iH\delta t/\hbar)|x_i\rangle. \qquad (22.14)$$

K is called the *propagator* from x_i to x_f and is also called *Feynman's kernel* or simply *kernel*. Further,

$$\psi(x,t) = \int dx_i K(x_f, t_f; x_i, t_i)\psi(x_i, t_i). \qquad (22.15)$$

K contains the details about the evolution of ψ.

We can obtain an expression for the kernel K in the form of a summation (or integral) over all paths in the phase space from time t_i to the time t_f, that is, from x_i to x_f. This is known as *path integral representation of kernel* [4,15–18]. We divide the time interval $t_f - t_i$ into N (large) steps with the step size $\delta t = t/N$. Then

$$K(x_f, t_f; x_i, t_i) = \langle x_f| \left(\exp(-iH\delta t/\hbar)\right)^N |x_i\rangle. \qquad (22.16)$$

Using $\int dx_j|x_j\rangle\langle x_j| = 1$ we obtain

$$
\begin{aligned}
K &= \int dx_1 \ldots dx_{N-1}\langle x_f(= x_N)| \exp(-iH\delta t/\hbar)|x_{N-1}\rangle \\
&\quad \times \langle x_{N-1}| \exp(-iH\delta t/\hbar)|x_{N-2}\rangle \ldots \langle x_1| \exp(-iH\delta t/\hbar)|x_0\rangle \\
&= \left(\prod_{j=1}^{N-1}\int dx_j\right)\langle x_f| \exp(-iH\delta t/\hbar)|x_{N-1}\rangle \\
&\quad \times \langle x_{N-1}| \exp(-iH\delta t/\hbar)|x_{N-2}\rangle \ldots \langle x_1| \exp(-iH\delta t/\hbar)|x_0\rangle. \qquad (22.17)
\end{aligned}
$$

Equation (22.17) can be rewritten as

$$K = \left(\prod_{j=1}^{N-1}\int dx_j\right) K_{x_N,x_{N-1}}\ldots K_{x_1,x_0}. \tag{22.18}$$

There exists a connection between K and the classical action S. Let us find it.

Consider one-dimensional free particle with $H = p^2/(2m)$. Its momentum eigenfunction is

$$\langle x|p\rangle = \frac{1}{\sqrt{2\pi\hbar}}\,e^{ipx/\hbar}, \quad \int dp|p\rangle\langle p| = 1. \tag{22.19}$$

Then

$$\begin{aligned}
K_{x_{j+1},x_j} &= \langle x_{j+1}|\exp(-i\delta t(p^2/2m)/\hbar)|x_j\rangle \\
&= \int dp\langle x_{j+1}|\exp(-i\delta t(p^2/2m)/\hbar)|p\rangle\langle p|x_j\rangle \\
&= \frac{1}{2\pi}\int dp\,e^{-i\delta t(p^2/2m)/\hbar}\langle x_{j+1}|p\rangle\langle p|x_j\rangle.
\end{aligned} \tag{22.20}$$

The integral over p is simply a Gaussian integral. Integrating over p gives

$$K_{x_{j+1},x_j} = \left(\frac{m}{2\pi i\hbar\delta t}\right)^{1/2} e^{im/(2\hbar)(x_{j+1}-x_j)^2/\delta t}. \tag{22.21}$$

Using of Eq. (22.21) in Eq. (22.18) gives

$$K = \left(\frac{m}{2\pi i\hbar\delta t}\right)^{N/2}\prod_{j=0}^{N-1}\int dx_j\exp\left\{\frac{im\delta t}{2\hbar}\sum_{j=0}^{N-1}\left[\frac{x_{j+1}-x_j}{\delta t}\right]^2\right\}. \tag{22.22}$$

In the limit $\delta t\to 0$, we write $\dot x_j = (x_{j+1}-x_j)/\delta t$ and replace $\delta t\sum_{j=0}^{N-1}$ by $\int_{t_i}^{t_j}$. Then

$$K = \left(\frac{m}{2\pi i\hbar\delta t}\right)^{N/2}\prod_{j=0}^{N-1}\int dx_j\exp\left\{\frac{i}{\hbar}\int_{t_i}^{t_f}\frac{1}{2}m\dot x^2 dt\right\}. \tag{22.23}$$

We define

$$\int \mathcal{D}x(t) = \lim_{N\to\infty}\left(\frac{m}{2\pi i\hbar\delta t}\right)^{N/2}\prod_{j=0}^{N-1}\int dx_j, \tag{22.24}$$

where $\mathcal{D}$ is called the *functional measure* over the space of $x(t)$ [2]. Now,

$$K = \int \mathcal{D}x(t)\exp\left\{\frac{i}{\hbar}\int_{t_i}^{t_f}\frac{1}{2}p\dot x dt\right\}. \tag{22.25}$$

Consider a particle in three-dimension with $H = \mathbf{p}^2/2m + V(\mathbf{X})$ and the paths in the phase space $(\mathbf{X},\mathbf{p})$. We have

$$\begin{aligned}
K(\mathbf{X}_f,t_f;\mathbf{X}_i,t_i) &= \int_{\mathbf{X}_i}^{\mathbf{X}_f}\mathcal{D}\mathbf{X}(t)\int \mathcal{D}\mathbf{p}(t)\exp\left\{\frac{i}{\hbar}\int_{t_i}^{t_f}dt\left[\frac{1}{2}\mathbf{p}\dot{\mathbf{X}}-V(\mathbf{X})\right]\right\} \\
&= \int \mathcal{D}\mathbf{X}(t)\int \mathcal{D}\mathbf{p}(t)\exp\left\{\frac{i}{\hbar}\int_{t_i}^{t_f}dt\left(\mathbf{p}\dot{\mathbf{X}}-H\right)\right\}.
\end{aligned} \tag{22.26}$$

22.4 Space Fractional Schrödinger Equation

By taking Lévy paths we can realize fractional quantum mechanics [1–3]. In this case the kernel K_L is given by Eq. (22.26) with H replaced by H_α, fractional Hamiltonian and is given by

$$
\begin{aligned}
H_\alpha(\mathbf{p}, \mathbf{X}) &= T_\alpha + V \\
&= D_\alpha(-\hbar^2 \nabla^2)^{\alpha/2} + V(\mathbf{X}) \\
&= D_\alpha |\mathbf{p}|^\alpha + V(\mathbf{X}), \quad 1 < \alpha \le 2,
\end{aligned}
\tag{22.27}
$$

where T_α is the fractional-order kinetic energy operator and D_α is the *quantum diffusion constant* with physical dimension

$$
[D_\alpha] = \mathrm{erg}^{1-\alpha} \times \mathrm{cm}^\alpha \times \sec^{-\alpha}
\tag{22.28}
$$

(*how?*). $D_\alpha = 1/(2m)$ for $\alpha = 2$. $(-\hbar^2 \nabla^2)^{\alpha/2}$ is the *quantum Riez fractional operator* [19] defined through the equation

$$
\begin{aligned}
(-\hbar^2 \nabla^2)^{\alpha/2} \psi(\mathbf{X}, t) &= \frac{1}{(2\pi\hbar)^3} \int_{-\infty}^{\infty} d\mathbf{p} \, e^{i\mathbf{p}\cdot\mathbf{X}/\hbar} |\mathbf{p}|^\alpha \int_{-\infty}^{\infty} e^{-i\mathbf{p}\cdot\mathbf{X}/\hbar} \psi(\mathbf{X}, t) d\mathbf{X} \\
&= \frac{1}{(2\pi\hbar)^3} \int_{-\infty}^{\infty} \phi(\mathbf{p}, t) |\mathbf{p}|^\alpha e^{i\mathbf{p}\cdot\mathbf{X}/\hbar} d\mathbf{p}.
\end{aligned}
\tag{22.29}
$$

Note that

$$
\psi(\mathbf{X}, t) = \frac{1}{(2\pi\hbar)^3} \int_{-\infty}^{\infty} e^{i\mathbf{p}\cdot\mathbf{X}/\hbar} \phi(\mathbf{p}, t) d\mathbf{p},
\tag{22.30a}
$$

$$
\phi(\mathbf{p}, t) = \int_{-\infty}^{\infty} e^{-i\mathbf{p}\cdot\mathbf{X}/\hbar} \psi(\mathbf{X}, t) d\mathbf{X},
\tag{22.30b}
$$

where $\phi(\mathbf{p}, t)$ is the Fourier transform of $\psi(\mathbf{X}, t)$.

With $H = H_\alpha$ and $\epsilon = (t_f - t_i)/N$ the right-side of Eq. (22.26) is defined by [3]

$$
\begin{aligned}
\int \mathcal{D}\mathbf{X}(t) \int \mathcal{D}\mathbf{p}(t) \ldots &= \lim_{N \to \infty} \int_{-\infty}^{\infty} d\mathbf{X}_1 \ldots d\mathbf{X}_N \\
&\times \frac{1}{(2\pi\hbar)^{3N}} \int_{-\infty}^{\infty} d\mathbf{p}_1 \ldots d\mathbf{p}_N \\
&\times \exp\left\{ \frac{i}{\hbar} \mathbf{p}_1(\mathbf{X}_1 - \mathbf{X}_i) - \frac{i}{\hbar} D_\alpha \epsilon |\mathbf{p}_1|^\alpha \right\} \\
&\times \ldots \\
&\times \exp\left\{ \frac{i}{\hbar} \mathbf{p}_N(\mathbf{X}_f - \mathbf{X}_{N-1}) - \frac{i}{\hbar} D_\alpha \epsilon |\mathbf{p}_N|^\alpha \right\} \\
&\times \ldots.
\end{aligned}
\tag{22.31}
$$

One can easily verify that the Hamiltonian H_α is Hermitian, that is, $(\phi, H_\alpha \psi) = (H_\alpha \phi, \psi)$.

22.4.1 Evolution Equation For the Wave Function

We write

$$
\psi_f(\mathbf{X}_f, t_f) = \int d\mathbf{X}_i K_L(\mathbf{X}_f, t_f; \mathbf{X}_i, t_i) \psi_i(\mathbf{X}_i, t_i).
\tag{22.32}
$$

We wish to find the evolution equation for the wave function. For an infinitesimal increment in time, say, from t to $t + \epsilon$, Eq. (22.32) becomes

$$\psi(\mathbf{X}, t + \epsilon) = \int d\mathbf{X}' K_L (\mathbf{X}, t + \epsilon; \mathbf{X}', t) \, \psi(\mathbf{X}', t). \tag{22.33}$$

We use Eq. (22.26) for $K_L(\mathbf{X}, t + \epsilon; \mathbf{X}', t)$. Further, we approximate $\int_t^{t+\epsilon} d\tau V(\mathbf{X}, t)$ by $\epsilon V((\mathbf{X} + \mathbf{X}')/2, t)$. With all these we get

$$
\begin{aligned}
\psi(\mathbf{X}, t + \epsilon) &= \int d\mathbf{X}' \frac{1}{(2\pi\hbar)^3} \int_{-\infty}^{\infty} d\mathbf{p} \exp \left\{ \frac{i}{\hbar} \mathbf{p}(\mathbf{X} - \mathbf{X}') - \frac{i}{\hbar} D_\alpha \epsilon |\mathbf{p}|^\alpha \right. \\
&\quad \left. - \frac{i}{\hbar} \epsilon V \left(\frac{\mathbf{X} + \mathbf{X}'}{2}, t \right) \right\} \psi(\mathbf{X}', t).
\end{aligned}
\tag{22.34}
$$

Expanding both sides of Eq. (22.34) in power series, keeping upto terms containing ϵ only in each term and writing $e^{\epsilon z} \approx 1 + \epsilon z$ Eq. (22.34) becomes

$$
\begin{aligned}
\psi(\mathbf{X}, t) + \epsilon \frac{\partial}{\partial t} \psi(\mathbf{X}, t) &= \int d\mathbf{X}' \frac{1}{(2\pi\hbar)^3} \int_{-\infty}^{\infty} d\mathbf{p} \exp \left\{ \frac{i}{\hbar} \mathbf{p} (\mathbf{X} - \mathbf{X}') \right\} \left(1 - \frac{i}{\hbar} D_\alpha \epsilon |\mathbf{p}|^\alpha \right) \\
&\quad \times \left[1 - \frac{i}{\hbar} \epsilon V \left(\frac{\mathbf{X} + \mathbf{X}'}{2}, t \right) \right] \psi(\mathbf{X}', t) \\
&= \int d\mathbf{X}' \frac{1}{(2\pi\hbar)^3} \int_{-\infty}^{\infty} d\mathbf{p} \exp \left\{ \frac{i}{\hbar} \mathbf{p} (\mathbf{X} - \mathbf{X}') \right\} \\
&\quad \times \left[1 - \frac{i}{\hbar} D_\alpha \epsilon |\mathbf{p}|^\alpha - \frac{i}{\hbar} \epsilon V \left(\frac{\mathbf{X} + \mathbf{X}'}{2}, t \right) \right] \psi(\mathbf{X}', t). \tag{22.35}
\end{aligned}
$$

Using Eq. (22.29) we obtain from Eq. (22.35)

$$\psi + \epsilon \frac{\partial \psi}{\partial t} = \psi - \frac{i}{\hbar} D_\alpha \epsilon \left(-\hbar^2 \nabla^2 \right)^{\alpha/2} \psi - \frac{i}{\hbar} \epsilon V \psi. \tag{22.36}$$

That is,

$$i\hbar \frac{\partial \psi}{\partial t} = D_\alpha \left(-\hbar^2 \nabla^2 \right)^{\alpha/2} \psi + V \psi = H_\alpha \psi. \tag{22.37}$$

This is the *space FSE* as the space deriative in this equation is of fractional (noninteger) order α. In the operator form, Eq. (22.37) can be written as

$$i\hbar \frac{\partial \psi}{\partial t} = H_\alpha(\mathbf{p}, \mathbf{X}) \psi, \quad H_\alpha = D_\alpha |\mathbf{p}|^\alpha + V. \tag{22.38}$$

22.4.2 Time-Independent Equation

Assuming the solution of Eq. (22.37) as $\psi(\mathbf{X}, t) = f(t) \psi(\mathbf{X})$ we obtain

$$f(t) = e^{-iEt/\hbar} \tag{22.39}$$

and

$$D_\alpha \left(-\hbar^2 \nabla^2 \right)^{\alpha/2} \psi(\mathbf{X}) + V(\mathbf{X}) \psi(\mathbf{X}) = E \psi(\mathbf{X}). \tag{22.40}$$

Equation (22.40) is the *time-independent FSE*. For a free particle with $H_\alpha = T_\alpha = (-\hbar^2 \nabla^2)^{\alpha/2}$ and $\psi = e^{i\mathbf{k}.\mathbf{X}}$ from the equation $E\psi = T_\alpha \psi$ we can easily find $E = D_\alpha \hbar^\alpha |\mathbf{k}|^\alpha$, $-\infty < k < \infty$.

In momentum representation we write Eq. (22.40) as

$$D_\alpha |\mathbf{p}|^\alpha \phi(\mathbf{p}) + \frac{1}{(2\pi\hbar)^3}(V * \phi)(\mathbf{p}) = E\phi(\mathbf{p}), \tag{22.41a}$$

where

$$(V * \phi)(\mathbf{p}) = \int_{-\infty}^{\infty} V(\mathbf{p} - \mathbf{X})\phi(\mathbf{X})\mathrm{d}\mathbf{X} \tag{22.41b}$$

and

$$V(\mathbf{p}) = \int_{-\infty}^{\infty} e^{-i\mathbf{p}\cdot\mathbf{X}/\hbar} V(\mathbf{X})\mathrm{d}\mathbf{X}. \tag{22.41c}$$

Here $V(\mathbf{p})$ is the Fourier transform of the potential $V(\mathbf{X})$.

22.4.3 Continuity Equation For Probability Density

Let us obtain the probability continuity equation in the FQM [20]. With $\rho = \psi^*\psi$ we obtain

$$\frac{\partial \rho}{\partial t} = \frac{\partial}{\partial t} \int_V \rho \, \mathrm{d}\tau = \int_V \left(\psi^* \frac{\partial \psi}{\partial t} + \psi \frac{\partial \psi^*}{\partial t}\right) \mathrm{d}\tau. \tag{22.42}$$

We have

$$i\hbar \frac{\partial \psi}{\partial t} = T_\alpha \psi + V\psi, \tag{22.43}$$

$$-i\hbar \frac{\partial \psi^*}{\partial t} = T_\alpha \psi^* + V\psi^*. \tag{22.44}$$

$\psi^* \times$ (22.43) $- \psi \times$ (22.44) gives

$$\begin{aligned}
\psi^* \frac{\partial \psi}{\partial t} + \psi \frac{\partial \psi^*}{\partial t} &= \frac{1}{i\hbar}(\psi^* T_\alpha \psi - \psi T_\alpha \psi^*) \\
&= -iD_\alpha \hbar^{\alpha-1} \left[\psi^* \left(-\nabla^2\right)^{\alpha/2} \psi - \psi \left(-\nabla^2\right)^{\alpha/2} \psi^*\right] \\
&= -\nabla \cdot \mathbf{J} + I_\alpha,
\end{aligned} \tag{22.45a}$$

where

$$\mathbf{J} = -iD_\alpha \hbar^{\alpha-1} \left[\psi^* \left(-\nabla^2\right)^{\frac{\alpha}{2}-1} \nabla\psi - \psi \left(-\nabla^2\right)^{\frac{\alpha}{2}-1} \nabla\psi^*\right], \tag{22.45b}$$

$$I_\alpha = -iD_\alpha \hbar^{\alpha-1} \left[\nabla\psi^* \left(-\nabla^2\right)^{\frac{\alpha}{2}-1} \nabla\psi - \nabla\psi \left(-\nabla^2\right)^{\frac{\alpha}{2}-1} \nabla\psi^*\right]. \tag{22.45c}$$

Substituting (22.45a) in (22.42) we obtain

$$\frac{\partial \rho}{\partial t} + \nabla \cdot \mathbf{J} = I_\alpha. \tag{22.46}$$

Comparing Eq. (22.46) with Eq. (2.43) we note that the probability continuity equation in FQM has an extra source term. $I_\alpha > 0$ implies that there is a source at $\mathbf{X}$ at time t generating the probability. $I_\alpha < 0$ corresponds to the presence of a sink at $\mathbf{X}$ at time t destroying the probability. When $\alpha = 2$ with $D_\alpha = 1/(2m)$

$$\mathbf{J} = \frac{\hbar}{2mi} (\psi^* \nabla\psi - \psi \nabla\psi^*) \tag{22.47}$$

which is same as that of nonfractional quantum mechanics. Further, $I_\alpha = 0$ and hence $\partial\rho/\partial t + \nabla \cdot \mathbf{J} = 0$.

One can easily verify that for $\psi = \psi_1 + \psi_2$, where $\psi_j(x,t) = e^{i(k_j x - E_j t/\hbar)}$, $j = 1,2$ with $k_1 > k_2 > 0$, $E_j = D_\alpha(\hbar k_j)^\alpha$, $j = 1,2$ and the extra source term is

$$I_\alpha = 2D_\alpha\hbar^{\alpha-1}\left(k_1 k_2^{\alpha-1} - k_1^{\alpha-1}k_2\right)\sin\left[(k_2 - k_1)x - (E_2 - E_1)t/\hbar\right] \tag{22.48}$$

and is not zero except for $\alpha = 2$.

22.5 Solutions of Certain Space Fractional Schrödinger Equations

Let us determine the eigenfunctions and eigenvalues of certain fractional one-dimensional quantum mechanical systems [21,22].

22.5.1 Free Particle

The FSE for a free particle ($V(x) = 0$) is $H_\alpha\psi = E\psi$. We can easily verify that $\psi(x) = e^{ikx}$, that is, $\psi(x) = \sin kx$ or $\psi(x) = \cos kx$, $-\infty < k < \infty$ and $E = D_\alpha(\hbar|k|)^\alpha$.

We have

$$\begin{aligned} H_\alpha\psi(x) &= T_\alpha\psi(x) \\ &= D_\alpha\left(-\hbar^2\nabla^2\right)^{\alpha/2} e^{ikx} \\ &= D_\alpha(\hbar|k|)^\alpha\psi(x) \\ &= E\psi(x). \end{aligned} \tag{22.49}$$

We note that for $k > 0$

$$\begin{aligned} H_\alpha\sin kx &= H_\alpha\frac{1}{2i}\left(e^{ikx} - e^{-ikx}\right) \\ &= D_\alpha\left(-\hbar^2\nabla^2\right)^{\alpha/2}\frac{1}{2i}\left(e^{ikx} - e^{-ikx}\right) \\ &= D_\alpha(\hbar|k|)^\alpha\sin kx \\ &= E\sin kx. \end{aligned} \tag{22.50}$$

Similarly, $H_\alpha\cos kx = E\cos kx$. *What are $\psi(x)$ and E for $V(x) = V_0$?*

Solved Problem 1:

The wave function of a free particle with fractional kinetic energy $E = D_\alpha(\hbar k)^\alpha$ is $\psi = e^{i(kx - Et/\hbar)}$. Determine the extra source term I_α in the continuity Eq. (22.46) for $1 < \alpha < 2$.

As $E = D_\alpha(\hbar k)^\alpha$ we write $T_\alpha\psi = D_\alpha(\hbar k)^\alpha\psi$. Since $T_\alpha = D_\alpha(-\hbar^2\nabla^2)^{\alpha/2}$ we write (with $\nabla^2 = \partial^2/\partial x^2$)

$$D_\alpha(-\hbar^2\nabla^2)^{\alpha/2}\psi = D_\alpha(\hbar k)^\alpha\psi \tag{22.51}$$

or $(-\nabla^2)^{\alpha/2}\psi = k^\alpha\psi$. Replacing α by $\alpha - 2$ we get

$$(-\nabla^2)^{\alpha/2-1}\psi = k^{\alpha-2}\psi. \tag{22.52}$$

Also,

$$(-\nabla^2)^{\alpha/2-1}\psi^* = k^{\alpha-2}\psi^*. \tag{22.53}$$

Then Eq. (22.45c) gives

$$
\begin{aligned}
I_\alpha &= -iD_\alpha\hbar^{\alpha-1}\left[\nabla\psi^*\left(-\nabla^2\right)^{\frac{\alpha}{2}-1}\nabla\psi - \nabla\psi\left(-\nabla^2\right)^{\frac{\alpha}{2}-1}\nabla\psi^*\right]\\
&= -iD_\alpha h^{\alpha-1}\left\{\nabla\psi^*\nabla\left[(-\nabla^2)^{\frac{\alpha}{2}-1}\psi\right] - \nabla\psi\nabla\left[(-\nabla^2)^{\frac{\alpha}{2}-1}\psi^*\right]\right\}\\
&= -iD_\alpha h^{\alpha-1}\left\{\nabla\psi^*\nabla\left(k^{\alpha-2}\psi\right) - \nabla\psi\nabla\left(k^{\alpha-2}\psi^*\right)\right\}\\
&= -iD_\alpha h^{\alpha-1}k^{\alpha-2}\left\{\nabla\psi^*\nabla\psi - \nabla\psi\nabla\psi^*\right\}\\
&= 0. \tag{22.54}
\end{aligned}
$$

22.5.2   Dirac-Delta Function Potential

Consider the Dirac-delta function potential $V(x) = V_0\delta(x)$, V_0 is a constant. The FSE in momentum representation is

$$D_\alpha|p|^\alpha\phi(p) + \frac{1}{2\pi\hbar}(V * \phi)(p) = E\phi(p), \tag{22.55}$$

where

$$(V * \phi)(p) = \int_{-\infty}^\infty V(p-x)\phi(x)\mathrm{d}x \tag{22.56}$$

and $V(p)$ is the Fourier transform of $V(x)$. As the Fourier transform of $V(x) = V_0\delta(x)$ is V_0 we write

$$(V * \phi)(p) = V_0\int_{-\infty}^\infty \phi(x)\mathrm{d}x = V_0K, \quad K = \int_{-\infty}^\infty \phi(x)\mathrm{d}x. \tag{22.57}$$

Then Eq. (22.55) is written as

$$\left(|p|^\alpha - \frac{E}{D_\alpha}\right)\phi(p) = -\gamma K, \quad \gamma = \frac{V_0}{2\pi\hbar D_\alpha}. \tag{22.58}$$

Energy Eigenvalue

For $E < 0$ we write $E/D_\alpha = -\lambda^\alpha$, $\lambda > 0$. From Eq. (22.58) we find

$$\phi(p) = -\frac{\gamma K}{|p|^\alpha + \lambda^\alpha}. \tag{22.59}$$

Use of this expression for $\phi(p)$ in $K = \int_{-\infty}^\infty \phi(p)\mathrm{d}p$ gives

$$
\begin{aligned}
1 &= -\gamma\int_{-\infty}^\infty \frac{\mathrm{d}p}{|p|^\alpha + \lambda^\alpha}\\
&= -2\gamma\int_0^\infty \frac{\mathrm{d}p}{p^\alpha + \lambda^\alpha}\\
&= -2\gamma\lambda^{-\alpha}\int_0^\infty \frac{\mathrm{d}p}{1 + (p/\lambda)^\alpha}. \tag{22.60}
\end{aligned}
$$

Defining $p/\lambda = q$ the above equation is rewritten as

$$\begin{aligned}
1 &= -2\gamma\lambda^{1-\alpha}\int_0^\infty \frac{dq}{1+q^\alpha} \\
&= -2\gamma\lambda^{1-\alpha}\frac{\pi}{\alpha}\mathrm{cosec}(\pi/\alpha).
\end{aligned} \tag{22.61}$$

This gives

$$\lambda = \left(-\frac{2\gamma\pi\mathrm{cosec}(\pi/\alpha)}{\alpha}\right)^{1/(\alpha-1)}. \tag{22.62}$$

Using (22.62) in $E/D_\alpha = -\lambda^\alpha$ we find E as

$$E = -\lambda^\alpha D_\alpha = -\left(-\frac{2\gamma\pi\mathrm{cosec}(\pi/\alpha)}{\alpha}\right)^{\alpha/(\alpha-1)}D_\alpha. \tag{22.63}$$

Substitutions of $\gamma = V_0/(2\pi\hbar D_\alpha)$ and $V_0 = -g$, $g > 0$ in the above equation give

$$E = -\left(\frac{g\mathrm{cosec}(\pi/\alpha)}{\alpha\hbar D_\alpha^{1/\alpha}}\right)^{\alpha/(\alpha-1)}. \tag{22.64}$$

For $\alpha = 2$, $D_2 = 1/2m$ the result is $E = -mg^2/(2\hbar^2)$ which is correct. We note that there is only one bound state energy.

Eigenfunction

The inverse Fourier transofrm of $\phi(p)$ gives $\psi(x)$. That is,

$$\psi(x) = \frac{1}{2\pi\hbar}\int_{-\infty}^\infty \phi(p)dp = -\frac{\gamma K}{2\pi\hbar}\int_{-\infty}^\infty \frac{e^{ipx/\hbar}}{|p|^\alpha + \lambda^\alpha}dp. \tag{22.65}$$

The value of the integral is given by the Fox's H-function as

$$\int_{-\infty}^\infty \frac{e^{ipx/\hbar}}{|p|^\alpha + \lambda^\alpha}dp = \frac{2\pi\hbar}{\lambda^\alpha|x|}H_{2,3}^{2,1}\left[\left(\frac{\lambda}{\hbar}\right)^\alpha|x|^\alpha\Big|_{(1,\alpha),(1,1),(1,\alpha/2)}^{(1,1),(1,\alpha/2)}\right]. \tag{22.66}$$

We need to normalize the wave function. We obtain

$$\begin{aligned}
1 &= \int_{-\infty}^\infty \psi^*\psi dx \\
&= \frac{1}{2\pi\hbar}\int_{-\infty}^\infty \phi^*(p)\phi(p)dp \\
&= \frac{(\gamma K)^2(\alpha-1)\lambda^{1-2\alpha}\mathrm{cosec}(\pi/\alpha)}{\hbar\alpha^2}.
\end{aligned} \tag{22.67}$$

This gives

$$(\gamma K)^2 = \frac{\hbar\alpha^2\lambda^{2\alpha-1}\sin(\pi/\alpha)}{\alpha-1}. \tag{22.68}$$

Then

$$\psi(x) = \frac{\alpha}{|x|}\sqrt{\frac{\hbar\sin(\pi/\alpha)}{\lambda(\alpha-1)}}H_{2,3}^{2,1}\left[\left(\frac{\lambda}{\hbar}\right)^\alpha|x|^\alpha\Big|_{(1,\alpha),(1,1),(1,\alpha/2)}^{(1,1),(1,\alpha/2)}\right]. \tag{22.69}$$

The determination of $\psi(x)$ for $E \geq 0$ involves much complex calculation. Interested readers may refer to the ref. [22].

22.5.3 Linear Harmonic Oscillator

For the linear harmonic oscillator potential $V(x) = kx^2/2$ the FSE is

$$D_\alpha \left(-\hbar^2 \frac{d^2}{dx^2} \right)^{\alpha/2} \psi(x) + \frac{1}{2}kx^2\psi(x) = E\psi(x). \tag{22.70}$$

Taking Fourier transform of Eq. (22.70) gives

$$-\frac{1}{2}k\hbar^2 \frac{d^2}{dp^2}\phi(p) + D_\alpha|p|^\alpha \phi(p) = E\phi(p). \tag{22.71}$$

The ordinary Schrödinger equation for the linear harmonic oscillator is

$$-\frac{\hbar^2}{2m}\frac{d^2}{dx^2}\psi(x) + \frac{1}{2}kx^2\psi(x) = E\psi(x). \tag{22.72}$$

In Eq. (22.71) the term $-\frac{1}{2}k\hbar^2\frac{d^2}{dp^2}$ is the potential energy while in Eq. (22.72) $-\frac{\hbar^2}{2m}\frac{d^2}{dx^2}$ is the kinetic energy term. Similarly, in Eq. (22.71) the term $D_\alpha|p|^\alpha$ (kinetic energy term) is analogous to the term $\frac{1}{2}kx^2$ (potential energy) in Eq. (22.72). The roles of kinetic and potential energies got reversed in Eq. (22.72) with $k = 1/m$. Considering the above we can make use of WKB approximation in the momentum space to obtain E.

Referring to Eq. (22.71) in momentum space the quantization condition is

$$\int_{p_-}^{p_+} \sqrt{(2/k)(E - D_\alpha|p|^\alpha)}\, dp = \left(n + \frac{1}{2} \right) \hbar\pi, \tag{22.73}$$

where the classical turning points are $p_\pm = \pm(E/D_\alpha)^{1/\alpha}$. Evaluating the integral we finally obtain

$$E_n = \left[\frac{\left(n + \frac{1}{2}\right)\hbar\pi\sqrt{k}(D_\alpha)^{1/\alpha}\Gamma\left(\frac{3}{2} + \frac{1}{\alpha}\right)}{2\sqrt{2}\,\Gamma(3/2)\Gamma(1 + \frac{1}{\alpha})} \right]^{2\alpha/(2+\alpha)}. \tag{22.74}$$

22.6 Fractional Schrödinger Equation in Relativistic Quantum Mechanics

In this section, we present the realization of FSE in relativistic quantum mechanics [21]. From the relativistic energy-momentum relation we write the kinetic energy T_R as

$$T_R = \sqrt{\mathbf{p}^2 c^2 + m^2 c^4}. \tag{22.75}$$

For $v \ll c$ and $\alpha = 2$

$$T_R \approx mc^2 + \frac{1}{2m}\mathbf{p}^2 = mc^2 + T_2. \tag{22.76}$$

In the case of extremely high speed ($\alpha = 1$), we can neglect the rest energy and write

$$T_R \approx mc^2 = |\mathbf{p}|c = T_1. \tag{22.77}$$

The point is that the variation of the speed of a particle from low value to high value approximately equivalent to the variation of α from 2 to 1. That is, we can regard the relativistic kinetic energy as an approximate realization of the fractional kinetic energy.

In nonrelativistic FQM the operator T_α is defined through

$$T_\alpha \psi(\mathbf{X}) = \frac{1}{(2\pi\hbar)^3} D_\alpha \int_{-\infty}^{\infty} \phi(\mathbf{p})|\mathbf{p}|^\alpha e^{i\mathbf{p}\cdot\mathbf{X}/\hbar} d\mathbf{p}. \tag{22.78}$$

Referring to this equation in relativistic FQM we define the operator T_R as

$$T_R \psi(\mathbf{X}, t) = \frac{1}{(2\pi\hbar)^3} D_\alpha \int_{-\infty}^{\infty} \phi(\mathbf{p}, t)\sqrt{\mathbf{p}^2 c^2 + m^2 c^4}\, e^{i\mathbf{p}\cdot\mathbf{X}/\hbar} d\mathbf{p}. \tag{22.79}$$

The Schrödinger equation is $H_R \psi = E\psi$, $H_R = T_R + V$.

Next, we obtain the solutions of free particle and a particle in a delta function potential [21].

1. Free Particle

For the free particle $V(x) = 0$, $H_R \psi = E\psi$ with $\psi = e^{ikx}$ gives

$$\sqrt{p^2 c^2 + m^2 c^4}\, e^{ikx} = \sqrt{\hbar^2 k^2 c^2 + m^2 c^4}\, e^{ikx}. \tag{22.80}$$

For $\psi = \sin kx$ or $\cos kx$ we have $E = \sqrt{\hbar^2 k^2 c^2 + m^2 c^4}$.

2. The Delta Potential Well

For the potential $V = -V_0 \delta(x)$, $V_0 > 0$ the relativistic Schrödinger equation $H_R \psi = E\psi$ is

$$\left(p^2 c^2 + m^2 c^4\right)^{1/2} \psi(x) - V_0 \delta(x)\psi(x) = E\psi(x). \tag{22.81}$$

The Fourier transform of Eq. (22.81) gives

$$\left(p^2 c^2 + m^2 c^4\right)^{1/2} \phi(p) - \frac{V_0}{2\pi\hbar} \int_{-\infty}^{\infty} \phi(p)dp = E\phi(p). \tag{22.82}$$

As $E < 0$ we rewrite Eq. (22.82) as

$$\left[\left(p^2 c^2 + m^2 c^4\right)^{1/2} + |E|\right] \phi(p) = \frac{V_0}{2\pi\hbar} \int_{-\infty}^{\infty} \phi(p)dp. \tag{22.83}$$

We take the limits of the integral in the above equation as $-p_0$ to $p_0 > mc$ and finally apply the limit $p_0 \to \infty$.

From Eq. (22.83) we write

$$\phi(p) = \frac{V_0}{2\pi\hbar \left[\left(p^2 c^2 + m^2 c^4\right)^{1/2} + |E|\right]} \int_{-p_0}^{p_0} \phi(p)dp. \tag{22.84}$$

Integrating both sides with respect to p over the interval $-p_0$ to p_0 and assuming that $\int_{-p_0}^{p_0} \phi(p)dp \neq 0$ we can cancel the term $\int_{-p_0}^{p_0} \phi(p)dp$ on both sides. The result is

$$\begin{aligned}
1 &= \frac{V_0}{2\pi\hbar} \int_{-p_0}^{p_0} \frac{1}{\left[\left(p^2 c^2 + m^2 c^4\right)^{1/2} + |E|\right]} dp \\
&= \frac{V_0}{\pi\hbar} \int_{0}^{p_0} \frac{1}{\left[\left(p^2 c^2 + m^2 c^4\right)^{1/2} + |E|\right]} dp.
\end{aligned} \tag{22.85}$$

If we consider the range of the integration as $[mc, p_0]$ instead of $[0, p_0]$ then

$$1 > \frac{V_0}{\pi\hbar} \int_{mc}^{p_0} \frac{1}{\left[(p^2c^2 + m^2c^4)^{1/2} + |E|\right]} \, dp. \tag{22.86}$$

Evaluation of the integral for $p > mc$ gives

$$1 > \frac{V_0}{\sqrt{2}\pi\hbar c} \ln \left[\frac{p_0 + \left(|E|/\left(\sqrt{2}\, c\right)\right)}{mc + |E|/\left(\sqrt{2}\, c\right)} \right]. \tag{22.87}$$

That is,

$$|E| > \frac{\sqrt{2}\, p_0 c - \sqrt{2}\, mc^2 e^{\sqrt{2}\,\pi\hbar c/V_0}}{e^{\sqrt{2}\,\pi\hbar c/V_0} - 1}. \tag{22.88}$$

As $p_0 \to \infty$ we have $|E| \to \infty$, that is $E = -\infty$. *What are the expression for $\phi(p)$ and its value as $p_0 \to \infty$? What is the form of $\psi(x)$ for $p_0 \to \infty$?*

22.7 Time Fractional Schrödinger Equation

Similar to space FSE a time FSE can be obtained by considering non-Gaussian evolution. Properties of time FSE were analyzed by Naber [5].

22.7.1 Construction of Time Fractional Schrödinger Equation

The units of the wave function need to be preserved when considering the time FSE. For this purpose consider Planck units for length, time, mass and energy given by

$$L_P = \sqrt{G\hbar/c^3} = 1.61625 \times 10^{-35}\,\text{m}, \tag{22.89a}$$

$$T_P = \sqrt{G\hbar/c^5} = 5.39125 \times 10^{-44}\,\text{s}, \tag{22.89b}$$

$$M_P = \sqrt{\hbar c/G} = 2.17643 \times 10^{-8}\,\text{kg}, \tag{22.89c}$$

$$E_P = M_P c^2 = 1.95950 \times 10^{9}\,\text{J}, \tag{22.89d}$$

where $G = 6.67408 \times 10^{-11}\,\text{m}^3\text{kg}^{-1}\text{s}^{-2}$ is the gravitational constant. The time-dependent Schrödinger equation can be written in Planck units as

$$iT_P \frac{\partial\psi}{\partial t} = -\frac{1}{2m} L_P^2 M_P \frac{\partial^2\psi}{\partial x^2} + \frac{V}{E_P}\psi. \tag{22.90}$$

This is the *time-dependent Schrödinger equation* in dimensionless form.

Introducing $N_m = m/M_P$ and $N_V = V/E_P$ Eq. (22.90) takes the form

$$iT_P \frac{\partial\psi}{\partial t} = -\frac{L_P^2}{2N_m} \frac{\partial^2\psi}{\partial x^2} + N_V\psi. \tag{22.91}$$

The time FSE can be considered as

$$(iT_P)^\beta \frac{\partial^\beta\psi}{\partial t^\beta} = -\frac{L_P^2}{2N_m} \frac{\partial^2\psi}{\partial x^2} + N_V\psi. \tag{22.92}$$

Here $\partial^\beta/\partial t^\beta$ is the *Caputo fractional derivative of order* β defined by [23,24]

$$\frac{\partial^\beta}{\partial t^\beta} f(t) = \frac{1}{\Gamma(1-\beta)} \int_0^t \frac{df(\tau)/d\tau}{(t-\tau)^\beta} d\tau, \ \ 0 < \beta < 1. \tag{22.93}$$

For simplicity define

$$a = \frac{N_{\mathrm{V}}}{T_{\mathrm{P}}^\beta}, \quad b = \frac{L_{\mathrm{P}}^2}{2N_{\mathrm{m}}T_{\mathrm{P}}^\beta}. \tag{22.94}$$

Then Eq. (22.92) becomes

$$\frac{\partial^\beta \psi}{\partial t^\beta} = -\frac{b}{\mathrm{i}^\beta} \frac{\partial^2 \psi}{\partial x^2} + \frac{a}{\mathrm{i}^\beta} \psi. \tag{22.95}$$

22.7.2   Probability Current Equation

We now obtain the proability current equation [25]. With the probability density $\rho = \psi^*\psi$ we have

$$\frac{\partial \rho}{\partial t} = \psi^* \frac{\partial \psi}{\partial t} + \psi \frac{\partial \psi^*}{\partial t}. \tag{22.96}$$

We need to find $\partial\psi/\partial t$ and $\partial\psi^*/\partial t$. A property of the Caputo fractional derivative for $y(t)$ is [23,24]

$$\frac{\partial^{-\beta}}{\partial t^{-\beta}} \left(\frac{\partial^\beta y}{\partial t^\beta} \right) = y(t) - y(0). \tag{22.97}$$

Differentiating Eq. (22.97) with respect to t once we get

$$\frac{\partial^{1-\beta}}{\partial t^{1-\beta}} \left(\frac{\partial^\beta y}{\partial t^\beta} \right) = \frac{\partial y}{\partial t}. \tag{22.98}$$

Making use of it we can find $\partial\psi/\partial t$ and $\partial\psi^*/\partial t$. From Eq. (22.95) we write

$$\frac{\partial^{1-\beta}}{\partial t^{1-\beta}} \left(\frac{\partial^\beta \psi}{\partial t^\beta} \right) = -\frac{b}{\mathrm{i}^\beta} \frac{\partial^2}{\partial x^2} \left(\frac{\partial^{1-\beta}\psi}{\partial t^{1-\beta}} \right) + \frac{a}{\mathrm{i}^\beta} \frac{\partial^{1-\beta}\psi}{\partial t^{1-\beta}}. \tag{22.99}$$

Then use of Eq. (22.98) in Eq. (22.99) gives

$$\frac{\partial\psi}{\partial t} = -\frac{b}{\mathrm{i}^\beta} \frac{\partial^2}{\partial x^2} \left(\frac{\partial^{1-\beta}\psi}{\partial t^{1-\beta}} \right) + \frac{a}{\mathrm{i}^\beta} \frac{\partial^{1-\beta}\psi}{\partial t^{1-\beta}}. \tag{22.100}$$

Defining $\overline{\psi} = \partial^{1-\beta}\psi/\partial t^{1-\beta}$ the above equation is rewritten as

$$\frac{\partial\psi}{\partial t} = -\frac{b}{\mathrm{i}^\beta} \frac{\partial^2\overline{\psi}}{\partial x^2} + \frac{a}{\mathrm{i}^\beta} \overline{\psi}. \tag{22.101}$$

Using Eq. (22.101) and its complex conjugate in Eq. (22.96) we obtain the fractional probability current equation as

$$\begin{aligned}
\frac{\partial\rho}{\partial t} + \frac{\partial J}{\partial x} &= b\left[\frac{1}{\mathrm{i}^\beta} \frac{\partial\psi^*}{\partial x} \frac{\partial\overline{\psi}}{\partial x} + \frac{1}{(-\mathrm{i})^\beta} \frac{\partial\psi}{\partial x} \frac{\partial\overline{\psi}^*}{\partial x} \right] + a\left[\frac{1}{\mathrm{i}^\beta} \psi^*\overline{\psi} + \frac{1}{(-\mathrm{i})^\beta} \psi\overline{\psi}^* \right] \\
&= 2\,\mathrm{R.P.}\left[\frac{1}{\mathrm{i}^\beta}(bA + aB) \right] \\
&= S, \tag{22.102}
\end{aligned}$$

where the fractional probability current density J is given by

$$J = b \left[\frac{1}{i^\beta} \psi^* \frac{\partial \overline{\psi}}{\partial x} + \frac{1}{(-i)^\beta} \psi \frac{\partial \overline{\psi}^*}{\partial x} \right],$$ (22.103a)

$$A = \frac{\partial \psi^*}{\partial x} \frac{\partial \overline{\psi}}{\partial x}, \quad B = \psi^* \overline{\psi}$$ (22.103b)

and R.P. stands for the real part. As $S \neq 0$ the probability is not conserved for the solutions of time FSE.

For a free particle with $a = 0$ and $\beta \to 1$ we have $\overline{\psi} = \partial^{1-\beta} \psi / \partial t^{1-\beta} = \psi$, $\overline{\psi}^* = \psi^*$. Then Eq. (22.102) becomes

$$\frac{\partial \rho}{\partial t} + \frac{\partial J}{\partial x} = b \left[\frac{1}{i} \frac{\partial \psi^*}{\partial x} \frac{\partial \psi}{\partial x} - \frac{1}{i} \frac{\partial \psi}{\partial x} \frac{\partial \psi^*}{\partial x} \right] = 0$$ (22.104)

and

$$\begin{aligned} J &= \frac{b}{i} \left[\psi^* \frac{\partial \psi}{\partial x} - \psi \frac{\partial \psi^*}{\partial x} \right] \\ &= \frac{L_P^2}{2 N_m T_P i} \left[\psi^* \frac{\partial \psi}{\partial x} - \psi \frac{\partial \psi^*}{\partial x} \right] \\ &= \frac{\hbar}{2mi} \left[\psi^* \frac{\partial \psi}{\partial x} - \psi \frac{\partial \psi^*}{\partial x} \right]. \end{aligned}$$ (22.105)

That is, Eq. (22.102) reduces to the continuity equation and Eq. (22.103a) reduces to the probability density of standard quantum mechanics.

22.7.3 Solution of Time Fractional Schrödinger Equation

Assume that the potential V is independent of time. We seek a solution of Eq. (22.92) of the form $\psi(x, t) = f(t)\psi(x)$. Substitution of this solution in Eq. (22.92) yields

$$\frac{(iT_P)^\beta}{f} \frac{\partial^\beta f}{\partial t^\beta} = -\frac{L_P^2}{2N_m \psi} \psi_{xx} + N_V.$$ (22.106)

As the left-side is a function of t only and the right-side is a function of x only we equate both sides to a same constant, say, λ. The equation for ψ is

$$\psi_{xx} + \left(\frac{2N_m \lambda}{L_P^2} - \frac{2N_m N_V}{L_P^2} \right) \psi = 0.$$ (22.107)

The solution of Eq. (22.107) can be determined once the potential term $N_V = V/E_P$, that is, $V(x)$ is known.

For $f(t)$ the equation is

$$(iT_P)^\beta \frac{\partial^\beta f}{\partial t^\beta} = \lambda f \text{ or } \frac{\partial^\beta f}{\partial t^\beta} = (-i)^\beta \omega f, \quad \omega = \frac{\lambda}{T_P^\beta},$$ (22.108)

where $\partial^\beta / \partial t^\beta$ is fractional derivative in t. The solution of Eq. (22.108) can be obtained by employing Laplace transform. First, consider the equation

$$\frac{\partial^\beta f}{\partial t^\beta} = \lambda f$$ (22.109)

with $f(t = 0) = f_0$. Taking Laplace transform on Eq. (22.109) with $\mathrm{L}[f(t)] = \bar{f}(s)$ we get

$$s^\beta \bar{f} - s^{\beta-1} f_0 = \lambda \bar{f}. \tag{22.110}$$

That is,

$$\bar{f} = \frac{s^{\beta-1} f_0}{s^\beta - \lambda} = \frac{f_0}{s} \frac{1}{1 - \frac{\lambda}{s^\beta}} = f_0 \sum_{n=0}^{\infty} \frac{\lambda^n}{s^{1+\beta n}}. \tag{22.111}$$

Taking inverse Laplace transform gives

$$f(t) = f_0 \sum_{n=0}^{\infty} \frac{\lambda^n t^{\beta n}}{\Gamma(1 + \beta n)} = f_0 E_\beta (\lambda t^\beta), \tag{22.112}$$

where E_β is called the *Mittag–Leffler function* [26,27]. Then the solution of Eq. (22.108) is

$$f(t) = f_0 E_\beta \left(\omega(-\mathrm{i} t)^\beta \right). \tag{22.113}$$

Solved Problem 2:

Show that for $\beta = 1$ the function $f(t)$ given by Eq. (22.113) coincides with that of the nonfractional Schrödinger equation.

When $\beta = 1$ we obtain

$$
\begin{aligned}
f(t) &= f_0 E_1(-\mathrm{i}\omega t) \\
&= f_0 \sum_{n=0}^{\infty} \frac{(-\mathrm{i}\omega t)^n}{\Gamma(1 + n)} \\
&= f_0 e^{-\mathrm{i}\omega t} \\
&= f_0 e^{-\mathrm{i}\lambda t / T_\mathrm{P}}.
\end{aligned} \tag{22.114}
$$

As λ is the energy of the time-independent quantum system we replace λ by E/E_P. Then

$$f(t) = f_0 e^{-\mathrm{i}Et/\hbar} \tag{22.115}$$

which is same as the Eq. (2.18a) obtained for nonfractional Schrödinger equation with E being the energy of time-independent quantum system.

Instead of writing series expansion for $\bar{f}$ we can directly take inverse Laplace transform [6]. For the case of Eq. (22.108)

$$\bar{f} = \frac{s^{\beta-1}}{s^\beta - \omega(-\mathrm{i})^\beta} f_0. \tag{22.116}$$

Then

$$f(t) = \frac{1}{2\pi \mathrm{i}} \int_{\gamma-\mathrm{i}\infty}^{\gamma+\mathrm{i}\infty} \frac{e^{st} s^{\beta-1} f_0}{s^\beta - \omega(-\mathrm{i})^\beta} \, \mathrm{d}s. \tag{22.117}$$

The integral in the above equation can be evaluated using residues [5]. With $\rho = \omega(-\mathrm{i})^\beta$ the result is $f = f_0 E_\beta$ where

$$E_\beta \left(\rho t^\beta \right) = \frac{1}{\beta} e^{\mathrm{i}\omega^{1/\beta} t} - F_\beta \left(\rho, t \right), \tag{22.118a}$$

where

$$F_\beta \left(\rho, t \right) = \frac{1}{\pi} \rho \sin(\beta\pi) \int_0^{\infty} \frac{e^{-rt} r^{\beta-1}}{r^{2\beta} - 2\rho \cos(\beta\pi) r^\beta + \rho^2} \, \mathrm{d}r. \tag{22.118b}$$

22.7.4 Schrödinger Picture

In chapter 7 we discussed three pictures of nonfractional quantum mechanics. In the present, subsection, we consider the Schrödinger picture of time FQM.

In sec. 7.2 the evolution operator U is introduced and then the evolution equation of an operator A representing an observable is obtained. We now consider these for time FSE [28].

Let us write $\psi(x,t) = U_\beta(t, t_0)\psi(x, t_0)$, $i^\beta \partial^\beta \psi/\partial t^\beta = H\psi$ and assume that V is independent of time. Use of the above ψ in the time FSE gives the fractional operator equation for U_β as

$$i^\beta \frac{\partial^\beta}{\partial t^\beta} U_\beta(t, t_0) = H U_\beta(t, t_0). \tag{22.119}$$

Taking Laplace transform on both sides of the above equation with $U_\beta(t, t_0) = 1$ and $t_0 = 0$ gives

$$s^\beta \overline{U}_\beta - s^{\beta-1} = (-i)^\beta H \overline{U}_\beta. \tag{22.120}$$

That is,

$$\overline{U}_\beta = \frac{s^{\beta-1}}{s^\beta - (-i)^\beta H} = \frac{1}{s\left(1 - \frac{(-i)^\beta H}{s^\beta}\right)} = \sum_{n=0}^{\infty} \frac{(-i)^{\beta n} H^n}{s^{1+\beta n}}. \tag{22.121}$$

Taking inverse Laplace transform yields

$$U_\beta(t, t_0) = \sum_{n=0}^{\infty} \frac{(-i)^{\beta n} H^n t^{\beta n}}{\Gamma(1 + \beta n)} = E_\beta\left((-it)^\beta H\right), \tag{22.122}$$

where $E_\beta(z)$ is the Muttag–Leffler function. Note that for $\beta = 1$ we find $E_\beta(z) = e^z$ and $U_1 = e^{-iHt}$ which is the standard evolution operator. One can easily verify that $U_\beta^\dagger U_\beta \neq 1$, that is, U_β is not unitary [28] and $U_1^\dagger U_1 = 1$. For the determination of $\psi(x,t)$ of certain time fractional quantum systems using $U_\beta(t, t_0)$ one may refer to the ref. [28].

Next, we find the equation of motion of an operator A. The expectation value of A is

$$\langle A \rangle_\beta = \int_{-\infty}^{\infty} \psi^* A\psi \, dx = \langle \psi|A|\psi\rangle. \tag{22.123}$$

The time variation of $\langle A \rangle_\beta$ is

$$\begin{aligned} \frac{\partial}{\partial t}\langle A \rangle_\beta &= \frac{\partial}{\partial t}\langle \psi|A|\psi\rangle \\ &= \langle \partial\psi/\partial t|A|\psi\rangle + \langle \psi|A|\partial\psi/\partial t\rangle \\ &= i^\beta \langle \partial^{1-\beta}\psi/\partial t^{1-\beta}|AH|\psi\rangle + (-i)^\beta \langle \psi|AH|\partial^{1-\beta}\psi/\partial t^{1-\beta}\rangle. \end{aligned} \tag{22.124}$$

The right-side of Eq. (22.124) is time-dependent and is nonlocal in time. Therefore, $\frac{\partial}{\partial t}\langle A \rangle_\beta$ is not zero and hence $\langle A \rangle_\beta$ is not a constant of motion. When $\beta = 1$

$$\begin{aligned} \frac{\partial}{\partial t}\langle A \rangle_\beta &= i\langle \psi|HA|\psi\rangle - i\langle \psi|AH|\psi\rangle \\ &= -i\langle \psi|[A, H]|\psi\rangle \\ &= \frac{1}{i}\langle \psi|[A, H]|\psi\rangle \end{aligned} \tag{22.125}$$

and is the case of standard Schrödinger equation of motion.

22.8 Solutions of Certain Time Fractional Schrödinger Equations

We can determine the eigenvalues and eigenfunctions of quantum mechanicals systems in the framework of fractional time derivative by solving the associated Schrödinger equation or by applying the time fractional evolution operator $U_\beta(t, t_0)$ on a given initial state of the system. Here we consider the case of solving time FSE for a free particle and the linear harmonic oscillator [6,29].

22.8.1 Free Particle

For a free particle with $V = 0$ the time FSE is

$$(iT_P)^\beta \frac{\partial^\beta \psi}{\partial t^\beta} = -\frac{L_P^2}{2N_m} \frac{\partial^2 \psi}{\partial x^2}. \tag{22.126}$$

Assume that $\psi(x, t) = f(t)\psi(x)$. Substitution of this solution in Eq. (22.126) gives the equation for $\psi(x)$ as

$$\psi_{xx} + k^2 \psi = 0, \quad k^2 = \frac{2N_m \lambda}{L_P^2}. \tag{22.127}$$

Its solution is (refer sec. 4.3)

$$\psi(x) = A e^{\pm ikx}, \quad A = \frac{1}{\sqrt{2\pi}}. \tag{22.128}$$

The equation for f is

$$\frac{\partial^\beta f}{\partial t^\beta} = (-i)^\beta \omega f, \quad \omega = \frac{\lambda}{T_P^\beta}. \tag{22.129}$$

Referring to sec. 22.7.3 the solution of the above equation is

$$f(t) = E_\beta \left(\omega(-it)^\beta \right), \tag{22.130}$$

where E_β is given by Eq. (22.118).

The time-dependent energy is obtained as

$$\begin{aligned} E(t) &= \int_{-\infty}^{\infty} \psi^* i\hbar \frac{\partial}{\partial t} \psi dx \\ &= \frac{i\hbar}{2\pi} \int_{-\infty}^{\infty} \psi^*(x)\psi(x)dx \times E_\beta \left(\omega(it)^\beta \right) \frac{\partial}{\partial t} E_\beta \left(\omega(-it)^\beta \right) \\ &= i\hbar E_\beta \left(\omega(it)^\beta \right) \frac{\partial}{\partial t} E_\beta \left(\omega(-it)^\beta \right). \end{aligned} \tag{22.131}$$

In the function E_β given by Eq. (22.118) as $t \to \infty$ the integral term vanishes. Therefore, for $t \to \infty$

$$E(\infty) = i\hbar \frac{1}{\beta^2} e^{i\omega^{1/\beta} t} \left(-i\omega^{1/\beta} \right) e^{-i\omega^{1/\beta} t} \Big|_{t \to \infty} = \frac{\hbar \omega^{1/\beta}}{\beta^2}. \tag{22.132}$$

When $\beta = 1$ the result is $E = \hbar\omega = \hbar\lambda/T_P = \hbar^2 k^2/(2m)$ and is the E given by Eq. (4.5).

22.8.2 Linear Harmonic Oscillator

For the linear harmonic oscillator with $V = kx^2/2$ the time FSE is

$$(iT_P)^\beta \frac{\partial^\beta \psi}{\partial t^\beta} = -\frac{L_P^2}{2N_m} \frac{\partial^2 \psi}{\partial x^2} + \frac{kx^2}{2E_P}\psi. \tag{22.133}$$

Substitution of $\psi(x,t) = f(t)\psi(x)$ in this equation gives

$$\frac{(iT_P)^\beta}{f} \frac{\partial^\beta f}{\partial t^\beta} = -\frac{L_P^2}{2N_m\psi} \frac{\partial^2 \psi}{\partial x^2} + \frac{kx^2}{2E_P}. \tag{22.134}$$

Equating both sides of the above equation to a same constant, say, ϵ gives the equation for $\psi(x)$ as

$$\psi_{xx} + \left(\frac{2N_m\epsilon}{L_P^2} - \frac{kN_m}{E_P L_P^2}x^2 \right)\psi = 0. \tag{22.135}$$

Define $\xi = \alpha x$. Then Eq. (22.135) becomes

$$\psi_{\xi\xi} + \left(\frac{2N_m\epsilon}{L_P^2\alpha^2} - \frac{kN_m}{E_P L_P^2\alpha^4}\xi^2 \right)\psi = 0. \tag{22.136}$$

Choose

$$\alpha^4 = \frac{kN_m}{E_P L_P^2}, \quad \alpha^2 = \sqrt{\frac{kN_m}{E_P L_P^2}}, \quad \lambda = \frac{2N_m\epsilon}{L_P^2\alpha^2}. \tag{22.137}$$

Then Eq. (22.136) is written as

$$\psi_{\xi\xi} + (\lambda - \xi^2)\psi = 0. \tag{22.138}$$

This is the Eq. (4.23) in sec. 4.4.2. Referring to sec. 4.4.2 the solution of Eq. (22.138) is

$$\psi_n(x) = N_n H_n(\alpha x)e^{-\alpha^2 x^2/2}, \quad N_n = \left(\frac{\alpha}{\sqrt{\pi}2^n n!} \right)^{1/2}. \tag{22.139}$$

As $\lambda = 2n + 1$ the energy ϵ_n is obtained as

$$\epsilon_n = \left(n + \frac{1}{2} \right)\frac{L_P^2\alpha^2}{N_m} = \left(n + \frac{1}{2} \right)T_P\omega, \quad \omega = \sqrt{\frac{k}{m}}. \tag{22.140}$$

Next, consider the equation for f given by

$$\frac{\partial^\beta f}{\partial t^\beta} = \frac{(-i)^\beta \epsilon_n f}{T_P^\beta}. \tag{22.141}$$

As done for the free particle with $f(t = 0) = 1$ the solution of the above equation is obtained as

$$\begin{aligned} f(t) &= E_\beta\left(\Omega_n(-it)^\beta\right), \quad \Omega_n = \frac{\epsilon_n}{T_P^\beta} \\ &= \frac{1}{\beta}\left[e^{-i\Omega_n^{1/\beta}t} - \beta F_\beta\left((-i\Omega_n)^\beta, t\right) \right], \end{aligned} \tag{22.142}$$

where F_β is given by Eq. (22.118b). Then

$$\psi_n(x,t) = N_n H_n(\alpha x)e^{-\alpha^2 x^2/2} f(t). \tag{22.143}$$

Next, $E_n(t)$ is obtained as

$$
\begin{aligned}
E_n(t) &= \int_{-\infty}^{\infty} \psi^* i\hbar \frac{\partial}{\partial t} \psi \mathrm{d}x \\
&= i\hbar N_n^2 \int_{-\infty}^{\infty} H_n^2(\alpha x)e^{-\alpha^2 x^2} \mathrm{d}x f^* f' \\
&= \frac{i\hbar}{\alpha} N_n^2 (2^n n! \sqrt{\pi}) f^* f' \\
&= i\hbar f^* f'.
\end{aligned}
\tag{22.144}
$$

Note that in the time fractional case the energy eigenvalues are time-dependent. However, in $f(t)$ the term F_β monotonically decays with increase in time. Therefore, in the long time limit the first term in f contributes to $E_n(\infty)$. In the limit of $t \to \infty$

$$E_n(\infty) = \frac{\hbar \Omega_n^{1/\beta}}{\beta^2} = \frac{\hbar \epsilon_n^{1/\beta}}{\beta^2 T_{\mathrm{P}}}. \tag{22.145}$$

When $\beta = 1$

$$E_n = \frac{\hbar \epsilon_n}{T_{\mathrm{P}}} = \left(n + \frac{1}{2} \right) \hbar \omega \tag{22.146}$$

which are the energy eigenvalues of linear harmonic oscillator in the standard quantum mechanics.

22.9   Space-Time Fractional Schrödinger Equation

Consider the dimensionless time-dependent Schrödinger equation given by Eq. (22.90) Fractionalization of both space and time derivatives in this equations gives the space-time FSE as

$$(iT_{\mathrm{P}})^\beta \frac{\partial^\beta}{\partial t^\beta} \psi(x,t) = -\frac{L_{\mathrm{P}}^\alpha}{2N_m} \nabla_x^\alpha \psi(x,t) + N_{\mathrm{V}} \psi(x,t), \tag{22.147}$$

where $\partial^\beta/\partial t^\beta$ is the Caputo fractional derivative given by Eq. (22.93) and ∇_x^α is the Riesz fractional derivative given by

$$\nabla_x^\alpha f(x) = \frac{\mathrm{d}^2}{\mathrm{d}x^2} \frac{1}{\Gamma(2-\alpha)} \int_{-\infty}^{x} \frac{f(u)}{(x-u)^{3-\alpha}} \mathrm{d}u. \tag{22.148}$$

Equation (22.147) was proposed by Shaowei Wang and Mingyu Xu [30]. Jiamping Dong and Mingya Xu [31] introduced a space-time FSE from the space FSE and is written as

$$(iT_{\mathrm{P}})^\beta \frac{\partial^\beta}{\partial t^\beta} \psi(x,t) = \frac{D_\alpha T_{\mathrm{P}}^{2-2\alpha}}{M_{\mathrm{P}}^{1-\alpha} E_{\mathrm{P}}^\alpha L_{\mathrm{P}}^{2-2\alpha}} \left(-\hbar^2 \frac{\partial^2}{\partial x^2} \right)^{\alpha/2} \psi(x,t) + \frac{V}{E_{\mathrm{P}}} \psi(x,t). \tag{22.149}$$

Laskin [29] considered the space-time FSE as

$$i^\beta \hbar_\beta \frac{\partial^\beta}{\partial t^\beta}\psi(x,t) = D_{\alpha,\beta}\left(-\hbar_\beta^2 \frac{\partial^2}{\partial x^2}\right)^{\alpha/2}\psi(x,t) + V\psi(x,t), \qquad (22.150)$$

where $\hbar_\beta$ and $D_{\alpha,\beta}$ are appropriate scale coefficients. In Eq. (22.150) $\hbar_\beta$ and $D_{\alpha,\beta}$ have the physical dimensions as

$$[\hbar_\beta] = \text{erg} \cdot \sec^\beta, \quad [D_{\alpha,\beta}] = \text{erg}^{1-\alpha} \cdot \text{cm}^\alpha \cdot \sec^{\alpha-\beta}. \qquad (22.151)$$

The quantum fractional operator (fractional quantum Riesz derivative) $\left(-\hbar_\beta^2 \partial^2/\partial x^2\right)^{\alpha/2}$ is defined as

$$\left(-\hbar_\beta^2 \frac{\partial^2}{\partial x^2}\right)^{\alpha/2}\psi(x,t) = \frac{1}{2\pi\hbar_\beta}\int_{-\infty}^{\infty} dp e^{ipx/\hbar_\beta}|p|^\alpha \phi(p,t) \qquad (22.152)$$

with

$$\psi(x,t) = \frac{1}{2\pi\hbar_\beta}\int_{-\infty}^{\infty} dp e^{ipx/\hbar_\beta}\phi(p,t) \qquad (22.153)$$

and

$$\phi(p,t) = \int_{-\infty}^{\infty} dx e^{-ipx/\hbar_\beta}\psi(x,t). \qquad (22.154)$$

The operator p is given by $-i\hbar_\beta \partial/\partial x$.

In the operator form, Eq. (22.150) is expressed as

$$i^\beta \hbar_\beta \frac{\partial^\beta}{\partial t^\beta}\psi(x,t) = H_{\alpha,\beta}(p,x)\psi(x,t), \quad H_{\alpha,\beta} = D_{\alpha,\beta}|p|^\alpha + V. \qquad (22.155)$$

In three-dimension

$$i^\beta \hbar_\beta \frac{\partial^\beta}{\partial t^\beta}\psi(\mathbf{X},t) = D_{\alpha,\beta}\left(-\hbar_\beta^2 \nabla^2\right)^{\alpha/2}\psi(\mathbf{X},t) + V(\mathbf{X},t)\psi(\mathbf{X},t) \qquad (22.156)$$

and

$$\left(-\hbar_\beta^2 \nabla^2\right)^{\alpha/2}\psi(\mathbf{X},t) = \frac{1}{(2\pi\hbar_\beta)^3}\int_{-\infty}^{\infty} d\mathbf{p} e^{i\mathbf{p}\cdot\mathbf{X}/\hbar_\beta}|\mathbf{p}|^\alpha \phi(\mathbf{p},t). \qquad (22.157)$$

In the rest of this section, we consider the space-time FSE given by Eq. (22.150).

22.9.1   Probability Current Equation

Rewrite Eq. (22.150) as

$$\frac{\partial^\beta}{\partial t^\beta}\psi(x,t) = \frac{D_{\alpha,\beta}}{i^\beta \hbar_\beta}\left(-\hbar_\beta^2 \frac{\partial^2}{\partial x^2}\right)^{\alpha/2}\psi(x,t) + \frac{1}{i^\beta \hbar_\beta}V\psi(x,t). \qquad (22.158)$$

Replacing y in Eq. (22.97) by ψ and then differentiation of the resulting equation with respect to t give

$$\frac{\partial\psi}{\partial t} = \frac{\partial^{1-\beta}}{\partial t^{1-\beta}}\left(\frac{\partial^\beta \psi}{\partial t^\beta}\right) = \frac{D_{\alpha,\beta}}{i^\beta \hbar_\beta}\left(-\hbar_\beta^2 \frac{\partial^2}{\partial x^2}\right)^{\alpha/2}\frac{\partial^{1-\beta}\psi}{\partial t^{1-\beta}} + \frac{V}{i^\beta \hbar_\beta}\frac{\partial^{1-\beta}\psi}{\partial t^{1-\beta}},$$

$$(22.159)$$

where V is assumed to be time-independent. Defining $\overline{\psi} = \partial^{1-\beta}\psi/\partial t^{1-\beta}$, $\mu = D_{\alpha,\beta}/\hbar_\beta$ and $b = V/\hbar_\beta$ Eq. (22.159) is rewritten as

$$\frac{\partial\psi}{\partial t} = \frac{\mu}{i^\beta}\left(-\hbar_\beta^2\frac{\partial^2}{\partial x^2}\right)^{\alpha/2}\overline{\psi} + \frac{b}{i^\beta}\overline{\psi}. \tag{22.160}$$

The rate of change of $\rho = \psi^*\psi$ is

$$\begin{aligned}\rho_t &= \psi^*\psi_t + \psi\psi_t^* \\ &= \psi^*\frac{\mu}{i^\beta}\left(-\hbar_\beta^2\frac{\partial^2}{\partial x^2}\right)^{\alpha/2}\overline{\psi} + \frac{b}{i^\beta}\psi^*\overline{\psi} + \psi\frac{\mu}{(-i)^\beta}\left(-\hbar_\beta^2\frac{\partial^2}{\partial x^2}\right)^{\alpha/2}\overline{\psi}^* \\ &\quad + \frac{b}{(-i)^\beta}\psi\overline{\psi}^*. \end{aligned} \tag{22.161}$$

We find

$$\frac{b}{i^\beta}\psi^*\overline{\psi} + \frac{b}{(-i)^\beta}\psi\overline{\psi}^* = b\left[\frac{\psi^*\overline{\psi}}{i^\beta} + \left(\frac{\psi^*\overline{\psi}}{i^\beta}\right)^*\right] = 2\,\text{R.P.}\left[\frac{b\psi^*\overline{\psi}}{i^\beta}\right] \tag{22.162}$$

and

$$\begin{aligned}&\psi^*\frac{\mu}{i^\beta}\left(-\hbar_\beta^2\frac{\partial^2}{\partial x^2}\right)^{\alpha/2}\overline{\psi} + \psi\frac{\mu}{(-i)^\beta}\left(-\hbar_\beta^2\frac{\partial^2}{\partial x^2}\right)^{\alpha/2}\overline{\psi}^* \\ &= \mu\left(-\hbar_\beta^2\right)^{\alpha/2}\frac{\partial}{\partial x}\left[\frac{1}{i^\beta}\psi^*\frac{\partial^{\alpha-1}\overline{\psi}}{\partial x^{\alpha-1}} + \frac{1}{(-i)^\beta}\psi\frac{\partial^{\alpha-1}\overline{\psi}^*}{\partial x^{\alpha-1}}\right] \\ &\quad - \mu\left(-\hbar_\beta^2\right)^{\alpha/2}2\,\text{R.P.}\left[\frac{1}{i^\beta}\frac{\partial\psi^*}{\partial x}\frac{\partial^{\alpha-1}\overline{\psi}}{\partial x^{\alpha-1}}\right]. \end{aligned} \tag{22.163}$$

Define

$$a = -\mu\left(-\hbar_\beta^2\right)^{\alpha/2}, \quad A = \frac{\partial\psi^*}{\partial x}\frac{\partial^{\alpha-1}\overline{\psi}}{\partial x^{\alpha-1}}, \quad B = \psi^*\overline{\psi}, \tag{22.164a}$$

$$J = a\left[\frac{1}{i^\beta}\psi^*\frac{\partial^{\alpha-1}\overline{\psi}}{\partial x^{\alpha-1}} + \frac{1}{(-i)^\beta}\psi\frac{\partial^{\alpha-1}\overline{\psi}^*}{\partial x^{\alpha-1}}\right]. \tag{22.164b}$$

Then Eq. (22.161) becomes

$$\rho_t + J_x = 2\,\text{R.P.}\left[\frac{aA + bB}{i^\beta}\right] = S, \tag{22.165}$$

where S is the source term. Compare this equation with that of time FQM.

Solved Problem 3:

Show that the probability current equation of space-time FQM reduces to that of standard quantum mechanics for $\alpha = 2$ and $\beta = 1$.

For $\alpha = 2$ and $\beta = 1$ we find $\overline{\psi} = \partial^{1-\beta}\psi/\partial t^{1-\beta} = \psi$, $\overline{\psi}^* = \psi^*$, $A = \psi_x^*\psi_x$, $B = \psi^*\psi$ and $a = -\mu(-\hbar_\beta^2)^{\alpha/2} = D_{2,1}\hbar_1 = \hbar_1/(2m)$. Then

$$S = 2\,\text{R.P.}\left[-i\left(\frac{\hbar_1}{2m}\psi_x^*\psi_x + \frac{V}{\hbar_1}\psi^*\psi\right)\right]. \tag{22.166}$$

As $\psi^*\psi$ and $\psi_x^*\psi_x$ are real the term inside the square-bracket is pure imaginary. So, the source term S becomes zero.

Next,

$$
\begin{aligned}
J &= a\left[\frac{1}{i^\beta}\psi^*\frac{\partial^{\alpha-1}\psi}{\partial x^{\alpha-1}} + \frac{1}{(-i)^\beta}\psi\frac{\partial^{\alpha-1}\psi^*}{\partial x^{\alpha-1}}\right] \\
&= \frac{\mu\hbar_1^2}{i}\left[\psi^*\psi_x - \psi\psi_x^*\right] \\
&= \frac{D_{2,1}\hbar_1}{i}\left[\psi^*\psi_x - \psi\psi_x^*\right] \\
&= \frac{\hbar_1}{2mi}\left[\psi^*\psi_x - \psi\psi_x^*\right].
\end{aligned}
\tag{22.167}
$$

Then $\rho_t + J_x = 0$. Expression for J and the continuity equation obtained coincide with those of nonfractional quantum mechanics.

22.9.2 Solution of Space-Time Fractional Schrödinger Equation

For the Eq. (22.155) with time-independent Hamiltonian $H_{\alpha,\beta}$ assume the solution as $\psi(x,t) = f(t)\psi(x)$. Substitution of this solution in Eq. (22.155) gives

$$
\frac{i^\beta\hbar_\beta}{f}\frac{\partial^\beta f}{\partial t^\beta} = \frac{1}{\phi}H_{\alpha,\beta}\psi.
\tag{22.168}
$$

Equating both sides of this equation to a constant, say, λ gives

$$
H_{\alpha,\beta}\psi(x) = \lambda\psi(x)
\tag{22.169}
$$

and

$$
i^\beta\hbar_\beta\frac{\partial^\beta f}{\partial t^\beta} = \lambda f(t).
\tag{22.170}
$$

λ is the eigenvalue of $H_{\alpha,\beta}$ and it depends on V. Referring to sec. 22.7.3 the solution of (22.170) is

$$
f(t) = f_0 E_\beta\left(\omega(-it)^\beta\right), \quad \omega = \lambda/\hbar_\beta,
\tag{22.171a}
$$

where

$$
E_\beta\left(\sigma t^\beta\right) = \sum_{n=0}^{\infty}\frac{\sigma^n t^{\beta n}}{\Gamma(1+\beta n)}
\tag{22.171b}
$$

is the Mittag–Leffler function.

22.9.3 A Free Particle Wave Function

Let us find the wave function of a free particle in one-dimension. The space-time FSE for a free particle is

$$
i^\beta\hbar_\beta\frac{\partial^\beta}{\partial t^\beta}\psi(x,t) = D_{\alpha,\beta}\left(-\hbar_\beta^2\frac{\partial^2}{\partial x^2}\right)^{\alpha/2}\psi(x,t).
\tag{22.172}
$$

Define the Fourier transform of $\psi(x,t)$ with respect to x as $\phi(p,t)$ (momentum wave function). Applying the Fourier transform to Eq. (22.172) gives

$$i^\beta \hbar_\beta \frac{\partial^\beta}{\partial t^\beta} \phi(p,t) = D_{\alpha,\beta} |p^2|^{\alpha/2} \phi(p,t) \tag{22.173}$$

with $\phi_0(p) = \phi(p, t=0)$ as

$$\phi_0(p) = \int_{-\infty}^{\infty} e^{-ipx'/\hbar_\beta} \psi_0(x') dx'. \tag{22.174}$$

Referring to the Eqs. (22.170) and (22.171) the solution of Eq. (22.173) is written as

$$\phi(p,t) = E_\beta \left(\sigma t^\beta\right) \phi_0(p), \quad \sigma = \frac{D_{\alpha,\beta} |p^2|^{\alpha/2}}{i^\beta \hbar_\beta}. \tag{22.175}$$

Then from Eq. (22.153) and (22.174) $\psi(x,t)$ is obtained as

$$
\begin{aligned}
\psi(x,t) &= \frac{1}{2\pi\hbar_\beta} \int_{-\infty}^{\infty} dp e^{ipx/\hbar_\beta} \phi(p,t) \\
&= \frac{1}{2\pi\hbar_\beta} \int_{-\infty}^{\infty} dp e^{ipx/\hbar_\beta} E_\beta \left(\sigma t^\beta\right) \phi_0(p) \\
&= \frac{1}{2\pi\hbar_\beta} \int_{-\infty}^{\infty} dx' \int_{-\infty}^{\infty} dp e^{ip(x-x')/\hbar_\beta} E_\beta \left(\sigma t^\beta\right) \psi_0(x'). \tag{22.176}
\end{aligned}
$$

22.10   Concluding Remarks

The topic FQM forms an attractive application of fractional calculus to quantum mechanics. The developments in this field leads to further understanding of the fundamentals of quantum theory. The space-time FSE has two fractality parameters α and β. The choice $1 < \alpha < 2$ and $\beta = 1$ gives space FSE. $\alpha = 1$ and $0 < \beta < 1$ gives time FSE. The standard nonfractional Schrödinger equation is obtained for $\alpha = 2$ and $\beta = 1$. For the time FSE and space-time FSE the time-dependence of the solution of the form $\psi(x,t) = f(t)\psi(x)$ is given by the Mittag–Leffler function.

In FQM the probability is not conserved and the probability current equation has a nonzero source term. The eigenfunctions and the eigenvalues of FSE are known only for a very few physical systems. A prime difficulty of finding these for FSE is nonlocal character of the fractional derivatives.

Generalization of FSE considering ultradistribution [32], laser implementation of space FSE [9] and numerical simulation of FSE [33–35] are reported. Research on physical realization and implementation of FQM is in an initial state. Optics is one promising field in this aspect [9–12]. The extension of FQM to quantum cosmology has been set [36].

22.11   Bibliography

[1]   N. Laskin, *Phys. Lett. A* 268:298, 2000.

[2]   N. Laskin, *Phys. Rev. E* 62:3135, 2000.

[3] N. Laskin, *Phys. Rev. E* 66:056108, 2002.

[4] R.P. Feynman and A.R. Hibbs, *Quantum Mechanics and Path Integrals*. McGraw-Hill, New York, 1965.

[5] M. Naber, *J. Math. Phys.* 45:3339, 2004.

[6] R.N. Mantegna, *Phys. Rev. E* 49:4677, 1994.

[7] S. Wang and M. Xu, *J. Math. Phys.* 48:043502, 2007.

[8] B.A. Sticker, *Phys. Rev. E* 88:012120, 2013.

[9] S. Longhi, *Opt. Letts.* 40:1117, 2015.

[10] Y. Zhang, H. Zhong, M.R. Belic, N. Ahmed, Y. Zhang and M. Xiao, *Sci. Rep.* 6:23645, 2016.

[11] Y. Zhang, X. Liu, M.R. Belic, W. Zhong, Y. Zhang and M. Xiao, *Phys. Rev. Lett.* 115:180403, 2015.

[12] C. Huang and L. Dong, *Sci. Rep.* 7:5442, 2017.

[13] A. Iomin, *Mod. Phys. Lett. A* 36:2140003, 2021.

[14] Y. Zhang, H. Zhong, M.R. Belic, Y. Zhu, W. Zhong, Y. Zhang, D.N. Christodoulides and M. Xiao, *Laser & Photo. Rev.* 10:526, 2016.

[15] R.P. Feynman, *Statistical Mechanics*. Benjamin, Reading, MA, 1972.

[16] R.P. Feynman, *Rev. Mod. Phys.* 20:367, 1948.

[17] D.C. Khandekar, S.V. Lawande and K.V. Bhagwat, *Path Integral Methods and Their Applications*. World Scientific, Singapore, 2002.

[18] A. Das, *Field Theory: A Path Integral Approach*. World Scientific, Singapore, 2006.

[19] A.A. Kilbas, H.M. Srivastava and J.J. Truiillo, *Theory and Applications of Fractional Differential Equations*. Elsevier, Amsterdam, 2006.

[20] Y. Wei, *Phys. Rev. E* 93:066103, 2016.

[21] Y. Wei, *Int. J. Theor. Math. Phys.* 5:87, 2015.

[22] E. Capetas de Oliveira, F. Silva Costa and J. Vaz Jr. *J. Math. Phys.* 51:123517, 2010.

[23] A. Carpinteri and F. Mainardi (Eds.), *Fractals and Fractional Calculus in Continuum Mechanics*. Springer, Vienna, 1997.

[24] B.N. Narahari Achar, B.T. Yale and J.W. Hanneken, *Adv. Math. Phys.* 2013:290216, 2013.

[25] A. Tofighi, *Acta Physica Polonica A* 116:114, 2009.

[26] I. Podlubny, *Fractional Differential Equations*. Academic, New York, 1999.

[27] F. Mainardi and R. Gorenflo, *J. Comput. Appl. Math.* 118:283, 2000.

[28] H. Ertik, D. Demirhan, H. Sirin and F. Buyukkilic, *J. Math. Phys.* 51:082102, 2010.

[29] N. Laskin, *Chaos, Solitons & Fractals* 102:16, 2017.

[30] S. Wang and M. Xu, *J. Math. Phys.* 48:043502, 2007.

[31] J. Dong and M. Xu, *J. Math. Anal. Appl.* 344:1005, 2008.

[32] A.L. De Paoli and M.C. Rocca, *Physica A* 392:111, 2013.

[33] S. Duo and Y. Zhang, *Commun. Comput. Phys.* 18:321, 2015.

[34] M. Al-Raeei and M.S. El-Daher, *Heliyon* 6:e044953, 2020.

[35] M. Al-Raeei and M.S. El-Daher, *AIP Adv.* 10:035305, 2020.

[36] P.V. Moniz and S. Jalalzadeh, *Maths.* 8:313, 2020.

[37] A. Liemert and A. Kienle, *Mathematics* 4:31, 2016.

[38] M. Jeng, S.L.Y. Xu, E. Hawkins and J.M. Schwarz, *J. Math. Phys.* 51:062102, 2010.

22.12   Exercises

22.1 Prove that the fractional Hamiltonian $H_\alpha = D_\alpha \left(-\hbar^2 \dfrac{\partial^2}{\partial x^2} \right)^{\alpha/2} + V(x,t)$ is Hermitian.

22.2 For a free particle with fractional kinetic energy if the wave function is $\psi = \psi_1 + \psi_2$, $\psi_j = e^{i(k_j x - E_j t/\hbar)}$, $j = 1,2$ with $k_1 > k_2 > 0$, $E_j = D_\alpha(\hbar k_j)^\alpha$, $j = 1,2$ then determine the extra source term I_α [21].

22.3 Prove that if the state of a closed fractional quantum mechanical system has a parity (odd or even) then the parity is conserved [3].

22.4 Write the Schrödinger equation $i\hbar\partial\psi/\partial t = (-\hbar^2/2m)\partial^2\psi/\partial x^2 + V\psi$ in Planck units $L_P = \sqrt{G\hbar/c^3}$, $T_P = \sqrt{G\hbar/c^5}$, $M_P = \sqrt{\hbar c/G}$ and $E_P = M_P c^2$. Then obtain the space FSE by defining the fractional derivative $\partial^\alpha/\partial x^\alpha$, where $\partial^\alpha/\partial x^\alpha$ is the Riesz fractional derivative

$$\frac{\partial^\alpha}{\partial x^\alpha} f(x) = \frac{d^2}{dx^2} \frac{1}{\Gamma(2-\alpha)} \int_{-\infty}^{x} \frac{f(u)du}{(x-u)^{3-\alpha}}, \quad 1 < \alpha < 2.$$

22.5 Consider the Schrödinger equation

$$iT_P \psi_t = -\frac{1}{2m} L_P^2 M_P \psi_{xx} + \frac{V}{E_P}\psi.$$

Substituting the expressions of L_P, T_P, M_P and E_P in the above equation verify that the result is the Schrödinger equation

$$i\hbar\psi_t = -\frac{\hbar^2}{2m}\psi_{xx} + V\psi.$$

22.6 If $\psi(x) = N(1+\cos\pi x)$ for the periodic potential $V(x) = D_\alpha(\hbar\pi)^\alpha \frac{1}{1+\cos\pi x}$, where $-1 \le x \le 1$ determine N and the average total energy.

22.7 Show that the solution of the FSE (with $\alpha = 1$)

$$i\hbar\frac{\partial\psi}{\partial t} - \frac{1}{2}\left(-\hbar^2\frac{\partial^2}{\partial x^2}\right)^{1/2}\psi = 0$$

can be written as $\psi_1(x - t/2) + \psi_2(x + t/2)$.

22.8 Determine the momentum wave function of a one-dimensional fractional quantum system with the potential $V(x) = ax$ by solving the space FSE [37].

22.9 For the space FSE with $\alpha = 1$ obtain the eigenfunctions and energy eigenvalues of a particle in the linear harmonic oscillator potential $V(x) = kx^2$ [38].

22.10 Evaluate the integral in

$$\frac{V_0}{\pi\hbar} \int_{mc}^{p_0} \frac{\mathrm{d}p}{\left[(p^2c^2 + m^2c^4)^{1/2} + |E|\right]} < 1$$

with $p > mc$ and determine an expression for E [21].

22.11 Show that (a) $\dfrac{\psi^*\overline{\psi}}{\mathrm{i}^\beta} + \dfrac{\psi\overline{\psi}^*}{(-\mathrm{i})^\beta} = 2\,\text{R.P.}\left[\dfrac{\psi^*\overline{\psi}}{\mathrm{i}^\beta}\right]$ and

(b) $\dfrac{\psi_x^*\overline{\psi}_x}{\mathrm{i}^\beta} + \dfrac{\psi_x\overline{\psi}_x^*}{(-\mathrm{i})^\beta} = 2\,\text{R.P.}\left[\dfrac{\psi_x^*\overline{\psi}_x}{\mathrm{i}^\beta}\right]$.

22.12 Show that the time fractional evolution operator $U_\beta(t, t_0)$ is not unitary.

22.13 Determine the eigenvalues and eigenfunctions of a particle in the potential

$$V(x) = \begin{cases} 0, & -a < x < a \\ \infty, & \text{elsewhere} \end{cases}$$

governed by the time FSE [5].

22.14 Obtain the space-time FSE in momentum space [29].

22.15 For the space-time FSE

$$(\mathrm{i}T_P)^\beta \frac{\partial^\beta}{\partial t^\beta}\psi(x, t) = -\frac{L_P^\alpha}{2N_m}\nabla_x^\alpha\psi(x, t) + N_V\psi(x, t),$$

find the associated probability current equation and the probability current density for the case of time-independent potential.

23

Numerical Methods for Quantum Mechanics

23.1 Introduction

In this book so far we are concerned with the various fundamentals of quantum mechanics. In the latter part of the century the subject has started growing on multi direction. Newer quantum phenomena were identified. Quantum version of certain classical phenomena and concepts have been formulated. New topics in quantum physics have born. Application of quantum physics to other branches of science and technology has started. For example, a great deal of research has been focused on the topics such as supersymmetry, Berry's phase, Aharonov–Bohm effect, Wigner distribution function, coherent and squeezed states, quantum computing, quantum cryptography, quantum imaging and quantum lithography to name a few. Theoretical and experimental progress have been made in these topics. The fundamentals and the progresses made in these advanced topics are covered in the volume-II of this book. Before we move on to these advanced topics we wish to mention that in classical physics several fascinating newer phenomena were first realized in numerical simulation and then necessary theory and mathematical tools were developed. Thus, it is important to develop numerical schemes to study quantum systems.

We have seen that solving the Schrödinger equation is the central problem of nonrelativistic quantum mechanics. Accurate solutions of the one-dimensional Schrödinger equation with an arbitrary potential are important for understanding many phenomena in physics. But exactly solvable Schrödinger equations are very limited. Most of the Schrödinger equations cannot be solved analytically. However, number of simple and efficient numerical techniques are developed to study bound and scattering states of quantum mechanical systems. Particularly, there are techniques which allow us to compute eigenvalues and eigenfunctions of stationary states and transmission and reflection probabilities in a very accurate way [1–24]. Such numerical methods are important for the study of complex systems, design and analysis of quantum effect devices [25,26]. With the advent of modern computer technology, robust numerical study of various quantum phenomena are now common.

The present chapter is concerned with the study of numerical computation of the following:

1. Bound state eigenvalues.
2. Bound state eigenfunctions.
3. Ground state eigenvalue and eigenfunction.
4. Transmission and reflection probabilities.
5. Transition probabilities.
6. Electronic distribution in atoms.

There is a simple method developed by van der Maelen Uria, Garcia-Granda and Menendez-Velazquez [17] to compute stationary states eigenvalues and eigenfunctions for arbitrary

DOI: 10.1201/9781003172178-23

analytical or numerical potentials. In their method, the stationary state equation is transformed into a set of linear finite-difference equations. When these equations are expressed in a matrix form, $(A - \lambda I)\psi = 0$, the problem becomes an algebraic eigenvalue problem. The eigenvalues and eigenfunctions of A can be determined numerically and are related to the energy eigenvalues and eigenfunctions of the original quantum mechanical problem. Chen [3] extended the matrix method to time-dependent Schrödinger equation which is useful to analyze time evolution of an initial wave function. The final finite-difference formula derived by Chen is equivalent to the one obtained by Goldberg et al. [1]. A finite-difference time domain method has been developed by Sudiarta and Geldart [23] for finding stationary state eigenvalues and eigenfunctions. Their method is efficient for the calculation of ground state eigenvalue and it uses random wave function as an initial wave function.

For a scattering potential Pang [16] described a numerical method which uses numerical integration of Schrödinger equation and function minimization technique to obtain the solution in the barrier region satisfying appropriate boundary conditions. Wang and Champagne [24] described a method called coupled channel method for numerical calculation of transition probabilities for systems with time-dependent perturbations switched-on at $t = 0$ and switched-off at $t = \tau$. A simple Monte Carlo simulation for easy visualization of electronic distribution in atoms is demonstrated by de Aquino et al. [18].

In this chapter, the above methods are presented in detail. The methods are applied to a number of physically interesting potentials. Numerical results are compared with exact results wherever possible.

23.2 Matrix Method for Computing Stationary State Solutions

Let us rewrite the time-independent Schrödinger equation into the dimensionless form

$$\psi''(x) + \frac{2m}{\hbar^2}[E - V(x)]\psi(x) = 0 , \quad x \in [x_{\min}, x_{\max}] . \tag{23.1}$$

A Schrödinger equation containing a first derivative term can be transformed into the form (23.1) by a change of variable. In the numerical simulation, we choose the boundary condition $\psi \to 0$ as $x \to \pm\infty$ as $\psi(x_{\min}) = \psi(x_{\max}) = 0$ and we set $\hbar = 1$ and $m = 1$. ψ is zero outside the interval $[x_{\min}, x_{\max}]$. $|x_{\min}|$ and $|x_{\max}|$ are sufficiently large. In the following we illustrate the method of van der Maelen Uria et al. [17] for finding eigenvalues and eigenfunctions of (23.1).

23.2.1 Discrete Equivalent of Schrödinger Equation

We discretize the space variable x in the interval $[a, b]$ by choosing the grid of points $x_0, x_1, \ldots, x_N$ with $x_{i+1} - x_i = h$, where h is a constant. The eigenfunctions will take values only at these grid of points. That is, ψ will be a set of numbers $\psi_0, \psi_1, \ldots, \psi_N$, where $\psi_i = \psi(x_i)$. Usually $\psi_0 = \psi(x_{\min}) = 0$ and $\psi_N = \psi(x_{\max}) = 0$. Now, Eq. (23.1) becomes the following discrete equation

$$\psi_i'' + 2[E - V(x_i)]\psi_i = 0, \quad i = 1, 2, \ldots, N - 1. \tag{23.2}$$

We convert the above equation into a difference equation by replacing ψ_i'' by a central-difference formula. A three-point central-difference formula for ψ_i'' is

$$\psi''|_{x=x_i} = \frac{1}{h^2}[\psi_{i+1} - 2\psi_i + \psi_{i-1}] + O(h^2) . \tag{23.3}$$

Neglecting the error term and substituting Eq. (23.3) in Eq. (23.2) we obtain

$$\psi_{i-1} + 2\left[h^2 E - h^2 V\left(x_i\right) - 1\right]\psi_i + \psi_{i+1} = 0, \quad i = 1, 2, \ldots, N-1. \qquad (23.4)$$

Let us rewrite Eq. (23.4) as

$$-\psi_{i-1} + 2\left[1 + h^2 V\left(x_i\right) - h^2 E\right]\psi_i - \psi_{i+1} = 0, \quad i = 1, 2, \ldots, N-1. \qquad (23.5)$$

Suppose say $N = 4$. Then we have the following set of equations

$$
\begin{aligned}
-\psi_0 + 2\left[1 + h^2 V_1 - h^2 E\right]\psi_1 - \psi_2 &= 0, & (23.6a) \\
-\psi_1 + 2\left[1 + h^2 V_2 - h^2 E\right]\psi_2 - \psi_3 &= 0, & (23.6b) \\
-\psi_2 + 2\left[1 + h^2 V_3 - h^2 E\right]\psi_3 - \psi_4 &= 0. & (23.6c)
\end{aligned}
$$

In the above set of equations $\psi_0 = \psi_4 = 0$. In matrix form (23.6) is written as

$$
\begin{pmatrix}
d_{11} - \lambda & -1 & 0 \\
-1 & d_{22} - \lambda & -1 \\
0 & -1 & d_{33} - \lambda
\end{pmatrix}
\begin{pmatrix}
\psi_1 \\
\psi_2 \\
\psi_3
\end{pmatrix} = 0, \qquad (23.7)
$$

where $d_{ii} = 2\left[1 + h^2 V(x_i)\right]$ and $\lambda = 2h^2 E$. This is of the form $(A - \lambda I)\psi = 0$ with A being symmetric tridiagonal. An $N \times N$ matrix A is said to symmetric tridiagonal if

$$
\begin{aligned}
d_{ij} &= 0, \quad j \neq i - 1, i, i + 1, \quad i = 1, 2, \ldots, N \\
d_{i(i+1)} &= d_{(i+1)i}, \quad i = 1, 2, \ldots, N-1.
\end{aligned} \qquad (23.8)
$$

For arbitrary value of N the $N - 1$ coupled linear finite-difference Eqs. (23.4) are expressed in a compact matrix form

$$(A - \lambda I)\psi = 0, \qquad (23.9a)$$

with the elements of A and λ are given by

$$
\begin{aligned}
\lambda &= 2h^2 E, \quad d_{ii} = 2\left[1 + h^2 V\left(x_i\right)\right], \quad i = 1, 2, \ldots, N-1 \\
d_{i(i+1)} &= d_{(i+1)i} = -1, \quad i = 1, 2, \ldots, N-2 \\
d_{ij} &= 0, \quad j \neq i - 1, i, i + 1, \quad i = 1, 2, \ldots, N-1
\end{aligned} \qquad (23.9b)
$$

and I is the unit matrix. That is, the original differential eigenvalue problem now becomes an algebraic eigenvalue problem. Note that the calculation of E and ψ of the stationary states of a given quantum mechanical system is reduced to the calculation of λ and ψ of the matrix A. For nontrivial solution of $(A - \lambda I)\psi = 0$ we require $|A - \lambda I| = 0$ from which we can compute the eigenvalues of A and then the energy eigenvalues E. Then we can compute ψ from $A\psi = \lambda\psi$.

23.2.2 Computation of Eigenvalues of the Matrix A

How do we compute the eigenvalues of the matrix A? Interestingly, the matrix A is a symmetric tridiagonal. For a symmetric tridiagonal matrix efficient algorithms are available to calculate λ. An iterative method is the Jacobi method, but it is very slow for matrices of order greater than, say, ten. An alternate and more efficient technique is QL algorithm [27, 28]. This algorithm is presented in appendix F.

We note that the truncation error in the finite-difference approximation of the second derivative term, Eq. (23.3), is proportional to h^2. The error can be made proportional to h^4 using Richardson's extrapolation. This technique is the following. Suppose $E_i^{(1)}(h)$ and $E_i^{(2)}(\alpha h)$ are two approximate values of E_i obtained by a method of order p with step size h and αh, respectively. Then we can write

$$E_i^{(1)}(h) = E_i(0) + ch^p + O(h^{2p}) , \tag{23.10a}$$

$$E_i^{(2)}(\alpha h) = E_i(0) + c\alpha^p h^p + O(h^{2p}) . \tag{23.10b}$$

Multiplying of Eq. (23.10a) by α^p and subtracting it from (23.10b) gives

$$E_i(0) = \frac{\alpha^p E_i^{(1)} - E_i^{(2)}}{\alpha^p - 1} + O(h^{2p}) . \tag{23.11}$$

The above formula is known as Richardson's extrapolation. For $h = h_1$, $\alpha h = h_2$, $\alpha = h_2/h_1$ and $p = 2$ Eq. (23.11) becomes

$$E_i(\text{more accurate}) = \frac{1}{h_2^2 - h_1^2} \left(h_2^2 E_i^{(1)} - h_1^2 E_i^{(2)} \right) . \tag{23.12}$$

The eigenvalues of the given quantum system may be infinite in number. Since A is $(N-1) \times (N-1)$ matrix in the numerical computation we get only $N-1$ eigenvalues, a subset of the possible eigenvalues.

23.2.3   Computation of the Eigenfunctions of the Matrix A

An eigenfunction of the matrix A for an eigenvalue is obtained by solving the algebraic eigenvalue problem, Eq. (23.9), where λ ($= 2h^2E$) is a given eigenvalue. We write the Eq. (23.9a) (after redefining $N-1$ as N) and $A_i = d_{ii} - \lambda$ as

$$\begin{pmatrix} A_1 & -1 & 0 & 0 & \cdots & 0 & 0 & 0 \\ -1 & A_2 & -1 & 0 & \cdots & 0 & 0 & 0 \\ 0 & -1 & A_3 & -1 & \cdots & 0 & 0 & 0 \\ & & & \vdots & & & & \\ 0 & 0 & 0 & 0 & \cdots & -1 & A_{N-1} & -1 \\ 0 & 0 & 0 & 0 & \cdots & 0 & -1 & A_N \end{pmatrix} \begin{pmatrix} \psi_1 \\ \psi_2 \\ \psi_3 \\ \vdots \\ \psi_{N-1} \\ \psi_N \end{pmatrix} = 0. \tag{23.13}$$

We have to normalize any admissible eigenfunction. Let $C\psi$ is the normalized eigenfunction, where C is the normalization constant. C can be determined from the normalization condition

$$C^2 \int_a^b \psi^* \psi \, dx = 1 . \tag{23.14}$$

Therefore, keeping this in mind, in the numerical scheme we choose $\psi_1 = 1$ and compute other ψ_i's from (23.13). The first row in Eq. (23.13) gives

$$\psi_2 = A_1 \psi_1 \text{ (from row 1)} . \tag{23.15}$$

That is, use row 1 to compute ψ_2, use row 2 to compute ψ_3 and so on. From rows 2 to $N-2$ we have

$$\psi_3 = -\psi_1 + A_2\psi_2 \text{ (from row 2)} ,$$
$$\psi_4 = -\psi_2 + A_3\psi_3 \text{ (from row 3)} ,$$
$$\vdots$$
$$\psi_{N-1} = -\psi_{N-3} + A_{N-2}\psi_{N-2} \text{ (from row } N-2) . \tag{23.16}$$

Note that $(N-2)$th row gives ψ_{N-1}. Since, $(N-1)$th and Nth rows are not yet used and only ψ_N to be determined, we add these last two rows and obtain a single new equation. From this new equation we compute ψ_N:

$$\psi_N = \frac{1}{1-A_N} \left[-\psi_{N-2} - (1-A_{N-1})\,\psi_{N-1} \right] . \tag{23.17}$$

To summarize, with $\psi_1 = 1$, $\psi_2 = A_1 \psi_1$, the expressions for $\psi_3, \ldots, \psi_{N-1}$ are

$$\psi_i = -\psi_{i-2} + A_{i-1}\psi_{i-1} , \quad i = 3, 4, \ldots, N-1 \tag{23.18}$$

and ψ_N is given by the Eq. (23.17). The computed eigenfunction is unnormalized. To normalize it we calculate the normalization constant C from Eq. (23.14). We can evaluate the integral in Eq. (23.14), for example, using the composite trapezoidal rule:

$$\frac{1}{C^2} = \int_a^b |\psi|^2 \, \mathrm{d}x = h \left[\frac{1}{2}\psi_1^2 + \psi_2^2 + \psi_3^2 + \ldots + \psi_{N-1}^2 + \frac{1}{2}\psi_N^2 \right] . \tag{23.19}$$

23.2.4 Application

We apply the method to three quantum mechanical systems and compare the numerically computed eigenvalues and eigenfunctions with the exact values.

23.2.4.1 Harmonic Oscillator

For the linear harmonic oscillator potential $V(x) = kx^2/2$ with $k=1$ the exact eigenvalues are $E_e = (n+1/2)$, $n = 0, 1, \ldots$. It is easy to develop a program to compute the eigenvalues by the QL method (see the exercise 23.1 at the end of the present chapter). In order to get a more accurate result, the eigenvalues are computed for two values of h, h_1 and h_2. Then using the Richardson's extrapolation more accurate values of E are obtained. Table 23.1 presents first few eigenvalues computed with $h_1 = 0.1$ and $h_2 = 0.05$, by extrapolation and exact values. Here $x_{\min} = -10$ and $x_{\max} = 10$. Accuracy can be improved by reducing the values of h_1 and h_2 and choosing large values of $x_{\min}$ and $x_{\max}$.

The exact ground state eigenfunction is given by $\psi(x) = (1/\pi)^{1/4}\,\mathrm{e}^{-x^2/2}$. Table 23.2 gives the numerically computed ground state eigenfunction values at some selected grid points for three values of h where $x_{\min} = -5$ and $x_{\max} = 5$ (for a complete program to compute a normalized eigenfunction see the exercise 23.3 at the end of the present chapter). $\psi(x)$ approaches the exact result as h decreases. Very good agreement with exact eigenfunction is observed for other values of E also. The numerical solution is symmetric with respect to the origin. This solution is called an *even-parity solution*. An odd-parity solution can be obtained by choosing $\psi(x_{\min}) = -1$. Figure 23.1 shows the numerically computed eigenfunctions for $E = 0.5$, 1.5, 2.5 and 3.5. For $E = 0.5$ and 2.5 the value of $\psi(x_{\min})$ is set as 1 whereas for $E = 1.5$ and 3.5 the value of $\psi(x_{\min})$ is set as -1.

23.2.4.2 Square-Well Potential

As a second example consider the square-well potential (5.23) which admits only a finite number of bound states. Its exact energy eigenvalues and eigenfunctions are determined in sec. 5.3. In the numerical computation, we fix V_0 as 35 and $a = 1$ for which only 6 bound states exist. The numerical (exact) energy eigenvalues are computed as 0.98336 (0.98336), 3.92084 (3.92084), 8.77009 (8.77009), 15.43900 (15.43901), 23.71532 (23.71539) and 32.84380 (32.84389). In the numerical simulation, we get eigenvalues above 35 also which are closely spaced and we can treat them as forming continuous spectrum. Figure 23.2 shows the numerically computed and exact energy eigenfunctions ψ_0, ψ_1 and ψ_5 corresponding to the energy eigenvalues $E_0 = 0.98336$, 3.92084 and 32.8438, respectively.

TABLE 23.1
Numerically computed first few eigenvalues (E_n) of the linear harmonic oscillator potential $V(x) = kx^2/2$. Here $x_{\min} = -10$ and $x_{\max} = 10$ and $k = 1$.

E_n ($h = 0.1$)	E_n ($h = 0.05$)	Extra- polated E_n	E_e
9.44309	9.48584	9.50009	9.5
8.45444	8.48866	8.50007	8.5
7.46452	7.49116	7.50004	7.5
6.47333	6.49335	6.50002	6.5
5.48087	5.49523	5.50002	5.5
4.48715	4.49679	4.50000	4.5
3.49217	3.49805	3.50001	3.5
2.49593	2.49898	2.50000	2.5
1.49844	1.49961	1.50000	1.5
0.49969	0.49992	0.50000	0.5

TABLE 23.2
Numerically computed ground state energy eigenfunction values at some grid points for three values of h. The last column is the exact result.

	Numerical ψ			
x	$h = 0.020$	$h = 0.010$	$h = 0.001$	Exact ψ
± 4.0	0.00024	0.00025	0.00025	0.00025
± 3.0	0.00803	0.00832	0.00834	0.00833
± 2.5	0.03175	0.03292	0.03300	0.03300
± 2.0	0.09779	0.10138	0.10165	0.10165
± 1.5	0.23459	0.24321	0.24385	0.24385
± 1.0	0.43828	0.45438	0.45558	0.45558
± 0.5	0.63770	0.66112	0.66287	0.66287
0.0	0.72261	0.74915	0.75113	0.75113

23.2.4.3 Hydrogen Atom

The radial Schrödinger equation for hydrogen-like atom is

$$\frac{1}{r^2}\frac{d}{dr}\left(r^2\frac{dR}{dr}\right) + \left[\frac{2m}{\hbar^2}(E - V(r)) - \frac{l(l+1)}{r^2}\right]R = 0 . \qquad (23.20)$$

Introducing the change of variable $\rho(r) = rR(r)$ and using the Hartree's atomic units, $\hbar = 1$ and $m = 1$, Eq. (23.20) becomes

$$\frac{d^2\rho}{dr^2} + \left[2(E - V(r)) - \frac{l(l+1)}{r^2}\right]\rho = 0 . \qquad (23.21)$$

The Coulomb potential is given by $V(r) = -Z/r$. Eigenvalues and eigenfunctions of Eq. (23.21) can be computed numerically. First, we consider the case of $Z = 1$ and $l = 0$.

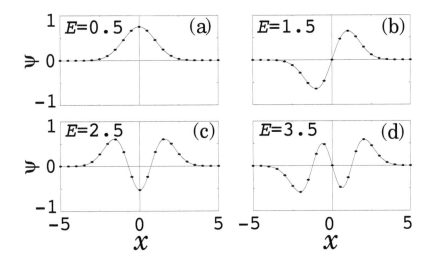

FIGURE 23.1
Numerically computed eigenfunctions of the harmonic oscillator (continuous curve) for first few values of E. The exact values of the eigenfunctions for some values of x are represented by solid circles.

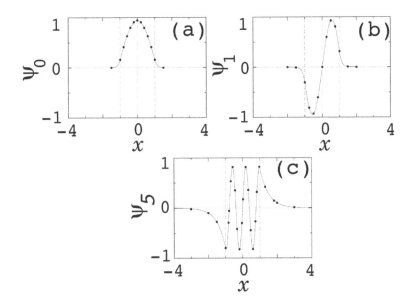

FIGURE 23.2
Numerically computed eigenfunctions of the square-well potential (continuous curve) for a few values of E. The exact values of the eigenfunctions for some values of x are represented by solid circles.

TABLE 23.3
Numerically computed first few energy eigenvalues (E_n) of the hydrogen atom. Here, $r \in [0, 100]$.

E_n ($h = 0.1$)	E_n ($h = 0.05$)	Extra-polated E	E_e
−0.00960	−0.00960	−0.00960	−0.01020
−0.01387	−0.01387	−0.01387	−0.01389
−0.02000	−0.02000	−0.02000	−0.02000
−0.03125	−0.03125	−0.03125	−0.03125
−0.05554	−0.05555	−0.05555	−0.05555
−0.12492	−0.12498	−0.12500	−0.12500
−0.49876	−0.49969	−0.50000	−0.50000

The exact energy eigenvalues are $-1/(2n^2)$. Table 23.3 gives the first few energy values calculated numerically with $h_1 = 0.1$, $h_2 = 0.05$ and by extrapolation. The last column is the exact E. Here, r range is fixed as $r \in (0, 100)$. E_1 to E_5 are identical to the exact values within 5 decimal accuracy. Other values deviate from the exact energy values. As mentioned earlier, accuracy can be improved by choosing large value for maximum value of r and small values for h_1 and h_2. Table 23.4 illustrates the influence of $b = r_{\max}$ on the energy values E_1 to E_5. As expected these values converge to the exact values as the value of b increases. Moreover, the number of negative values of E increases with the increase in the values of b. For $b = 100, 200, 300, 400$ and 500 the number of negative values of E are 8, 12, 15, 17 and 19, respectively. Figure 23.3a shows the numerically computed radial eigenfunction for $l = 0$ and for first three energy eigenvalues.

TABLE 23.4
Numerically computed first few eigenvalues of the hydrogen atom as a function of $b = r_{\max}$. The last row gives exact values. The numerical eigenvalues are obtained from the Richardson's extrapolation of the eigenvalues for $h_1 = 0.1$ and $h_2 = 0.05$.

$r_{\max}$	E_1	E_2	E_3	E_4	E_5
10	−0.50000	−0.11305	0.08974	0.40122	0.81953
20	−0.50000	−0.12499	−0.04999	0.01640	0.11220
30	−0.50000	−0.12500	−0.05542	−0.02469	0.01317
40	−0.50000	−0.12500	−0.05555	−0.03056	−0.01110
50	−0.50000	−0.12500	−0.05555	−0.03120	−0.01788
100	−0.50000	−0.12500	−0.05555	−0.03125	−0.02000
∞	−0.50000	−0.12500	−0.05555	−0.03125	−0.02000

Suppose we choose $l \neq 0$. The possible values of l are $0, 1, \ldots, n - 1$. If we fix $l = 0$ then n must assume the values $1, 2, \ldots$. On the other hand, for $l = 1$ and 2 the quantum number n starts from 2 and 3, respectively. That is, if l is fixed as 1 then the numerically computed eigenvalue spectrum should not contain the energy value E_1. This is also realized

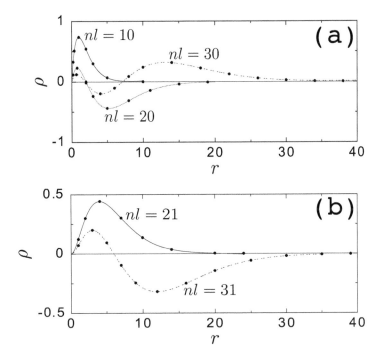

FIGURE 23.3
Numerically computed eigenfunctions (represented by continuous curves) and exact eigenfunctions (marked by solid circles at selected values of r) of hydrogen atom for (a) $nl = 10$, 20 and 30 and (b) $nl = 21$ and 31. For E_1, E_2 and E_3 the range of values of r used are $[0, 10]$, $[0, 20]$ and $[0, 40]$, respectively.

in the numerical computation of E. Figure 23.3b shows the numerically computed radial eigenfunctions ρ_{21} and ρ_{31}.

23.3 Finite-Difference Time-Domain Method

There is another method called finite-difference time-domain (FDTD) method introduced by Sudiarta and Geldart [20] for computation of energy eigenvalues and eigenfunctions of time-independent Schrödinger equation. Their method can be easily extended for higher-dimensional systems. In the FDTD method, the time-dependent Schrödinger equation is discretized using a central-difference formula in both time and space. Using different initial wave function we can obtain various eigenvalues and the corresponding eigenfunctions. The method with random initial wave function is efficient in obtaining ground state eigenvalue and eigenfunction.

23.3.1   Basic Idea

We consider the three-dimensional time-dependent Schrödinger equation

$$i\hbar\frac{\partial}{\partial t}\psi(x,y,z,t) = \left[-\frac{\hbar^2}{2m}\nabla^2 + V(x,y,z)\right]\psi(x,y,z,t)\,. \tag{23.22}$$

The solution of (23.22) in terms of stationary state eigenfunctions is written as

$$\psi(x,y,z,t) = \sum_{n=0}^{\infty} C_n\psi_n(x,y,z)e^{-iE_n t/\hbar}\,, \tag{23.23}$$

where ψ_n's are bound state eigenfunctions of the equation

$$\nabla^2\psi(x,y,z) + \frac{2m}{\hbar^2}[E - V(x,y,z)]\psi(x,y,z) = 0\,. \tag{23.24}$$

The substitution $\tau = \mathrm{i}t$ (called *Wick's rotational transformation*), $\hbar = 1$ and $m = 1$ in Eqs. (23.22) and (23.23) gives

$$\frac{\partial}{\partial\tau}\psi(x,y,z,\tau) = \frac{1}{2}\nabla^2\psi(x,y,z,\tau) - V(x,y,z)\psi(x,y,z,\tau) \tag{23.25}$$

and

$$\psi(x,y,z,\tau) = \sum_{n=0}^{\infty} C_n\psi_n(x,y,z)e^{-E_n\tau}\,. \tag{23.26}$$

In the long time limit $\lim_{\tau\to\infty}\psi \approx C_0\psi_0 e^{-E_0\tau}$. Thus, in an initial wave function $\psi(x,y,z,0)$ if $C_0 \neq 0$ then after a sufficiently long time, the wave function $\psi(x,y,z,\tau) \to \psi_0(x,y,z)$. The obtained ψ_0 is generally unnormalized. The corresponding energy E_0 is obtained from

$$E_0 = \frac{\displaystyle\int_{-\infty}^{\infty}\psi_0^* H\psi_0\,\mathrm{d}^3 X}{\displaystyle\int_{-\infty}^{\infty}\psi_0^*\psi_0\,\mathrm{d}^3 X}\,, \quad \mathrm{d}^3 X = \mathrm{d}x\,\mathrm{d}y\,\mathrm{d}z\,. \tag{23.27}$$

23.3.2   Finite-Difference Algorithm [23]

Because the bound state wave functions are localized, in the numerical simulation we consider finite region of space. Let $x \in [x_{\min}, x_{\max}]$ and similarly for the y and z variables. We divide the x, y and z ranges into discrete lattice sites with spacing between the sites given by $\Delta x = (x_{\max} - x_{\min})/N_x$, where $N_x + 1$ is the number of lattice points along the x-direction. The lattice points x_i are given by $x_i = x_{\min} + i\Delta x$, $i = 0, 1, \ldots, N_x$. Like-wise, define Δy, Δz, y_i and z_i. We denote the temporal increment as $\Delta\tau$. The wave function $\psi(x,y,z,\tau) = \psi(x_i, y_j, z_k, \tau_n)$ or $\psi(i\Delta x, j\Delta y, k\Delta z, n\Delta\tau)$ is denoted as $\psi^n(i,j,k)$.

To discretize the time and space derivative in Eq. (23.25) we use the forward- and central-difference schemes, respectively, given by

$$\frac{\partial\psi}{\partial\tau} \approx \frac{1}{\Delta\tau}\left[\psi^{n+1}(i,j,k) - \psi^n(i,j,k)\right],$$

$$\frac{\partial^2\psi}{\partial x^2} \approx \frac{1}{(\Delta x)^2}\left[\psi^n(i+1,j,k) - 2\psi^n(i,j,k) + \psi^n(i-1,j,k)\right],$$

$$\frac{\partial^2\psi}{\partial y^2} \approx \frac{1}{(\Delta y)^2}\left[\psi^n(i,j+1,k) - 2\psi^n(i,j,k) + \psi^n(i,j-1,k)\right],$$

$$\frac{\partial^2\psi}{\partial z^2} \approx \frac{1}{(\Delta z)^2}\left[\psi^n(i,j,k+1) - 2\psi^n(i,j,k) + \psi^n(i,j,k-1)\right]. \tag{23.28}$$

Further, $\psi(x, y, z) \approx \left[\psi^{n+1}(i, j, k) + \psi^n(i, j, k)\right]/2$. Substitution of (23.28) in (23.25) leads to

$$
\begin{aligned}
\psi^{n+1}(i, j, k) = {} & \alpha\psi^n(i, j, k) \\
& +\beta\gamma_1 \left[\psi^n(i+1, j, k) - 2\psi^n(i, j, k) + \psi^n(i-1, j, k)\right] \\
& +\beta\gamma_2 \left[\psi^n(i, j+1, k) - 2\psi^n(i, j, k) + \psi^n(i, j-1, k)\right] \\
& +\beta\gamma_3 \left[\psi^n(i, j, k+1) - 2\psi^n(i, j, k) + \psi^n(i, j, k-1)\right], \quad (23.29a)
\end{aligned}
$$

where

$$
\alpha = \frac{1 - \dfrac{\Delta\tau}{2}V(i, j, k)}{1 + \dfrac{\Delta\tau}{2}V(i, j, k)}, \quad \beta = \frac{1}{1 + \dfrac{\Delta\tau}{2}V(i, j, k)}, \quad (23.29b)
$$

$$
\gamma_1 = \frac{\Delta\tau}{2(\Delta x)^2}, \quad \gamma_2 = \frac{\Delta\tau}{2(\Delta y)^2}, \quad \gamma_3 = \frac{\Delta\tau}{2(\Delta z)^2}. \quad (23.29c)
$$

At each iteration the energy can be computed using (23.27). The discrete version of (23.27) is

$$
\begin{aligned}
E_0 = {} & \frac{1}{\sum_{i,j,k}\psi(i, j, k)^2}\sum_{i,j,k}\left[V(i, j, k)\psi(i, j, k)^2\right. \\
& -\frac{\psi(i, j, k)}{2(\Delta x)^2}\left[\psi(i+1, j, k) - 2\psi(i, j, k) + \psi(i-1, j, k)\right] \\
& -\frac{\psi(i, j, k)}{2(\Delta y)^2}\left[\psi(i, j+1, k) - 2\psi(i, j, k) + \psi(i, j-1, k)\right] \\
& \left.-\frac{\psi(i, j, k)}{2(\Delta z)^2}\left[\psi(i, j, k+1) - 2\psi(i, j, k) + \psi(i, j, k-1)\right]\right].
\end{aligned}
$$

$$
(23.30)
$$

The boundary condition is $\psi(x, y, z)_{\text{boundary}} = 0$. We note that since the final wave function $\psi_0(x, y, z, \tau)$ is proportional to $e^{-E_0\tau}$, the magnitude of the final wave function may generally be quite small. For this reason we normalize the initial wave function and the wave function after each iteration. If ψ is a solution of the Schrödinger equation then $-\psi$ is also a solution. The normalized $\psi(x, y, z, \tau = \text{large})$ can be either $+\psi_S$ or $-\psi_S$, where ψ_S is a stationary state wave function.

Solved Problem 1:
Obtain the stability condition for the FDTD scheme.

The stability of the method, that is, the condition for the error occurred at a stage of computation to decay, can be analyzed by the Fourier method (also called von Neumann analysis) [29]. To determine the stability condition consider the FDTD algorithm for the one-dimensional Schrödinger equation. In this case (23.29) becomes

$$
\psi_i^{n+1} = \alpha\psi_i^n + \beta\left[\psi_{i+1}^n - 2\psi_i^n + \psi_{i-1}^n\right], \quad (23.31a)
$$

where

$$
\alpha = \frac{1 - (\Delta\tau/2)V_i}{1 + \Delta\tau/2)V_i}, \quad \beta = \frac{\Delta\tau}{2\Delta x^2\left(1 + (\Delta\tau/2)V_i\right)}. \quad (23.31b)
$$

Now, we assume the numerical solution as $\psi = \psi_e + \delta\psi$, where $\delta\psi = e^{iax}e^{b\tau}$ where ψ_e is the exact solution. Substituting $x = i\Delta x$, $\tau = n\Delta\tau$ we get $\psi = \psi_e + e^{iai\Delta x}e^{bn\Delta\tau}$. Define $\xi = e^{b\Delta\tau}$. Then $\psi = \psi_e + e^{iai\Delta x}\xi^n$. Substituting this in the finite-difference formula (23.31a) we get

$$\psi_e^{n+1}(i) + e^{iai\Delta x}\xi^{n+1} = \alpha\psi_e^n(i) + \alpha e^{iai\Delta x}\xi^n + \beta\left[\psi_e^n(i+1) - 2\psi_e^n(i) + \psi_e^n(i-1)\right]$$
$$+\beta\left[e^{ia(i+1)\Delta x} - 2e^{iai\Delta x} + e^{ia(i-1)\Delta x}\right]\xi^n. \tag{23.32}$$

That is,

$$e^{iai\Delta x}\xi^{n+1} = \alpha e^{iai\Delta x}\xi^n + \beta\left[e^{ia(i+1)\Delta x} - 2e^{iai\Delta x} + e^{ia(i-1)\Delta x}\right]\xi^n. \tag{23.33}$$

The above equation gives

$$\xi = \alpha + \beta\left[e^{ia\Delta x} - 2 + e^{-ia\Delta x}\right] = \alpha - 4\beta\sin^2(a\Delta x/2). \tag{23.34}$$

For bound states the necessary and sufficient condition for stability is $|\xi| \leq 1$. Thus, for stability we require $\Delta\tau \leq \left[1/(\Delta x)^2\right]^{-1}$. For the Eq. (23.29) the stability condition becomes

$$\Delta\tau \leq \left[\frac{1}{(\Delta x)^2} + \frac{1}{(\Delta y)^2} + \frac{1}{(\Delta z)^2}\right]^{-1}. \tag{23.35}$$

23.3.3   Application

In the following we apply the FDTD method to certain one- and two-dimensional systems.

23.3.3.1   One-Dimensional Linear Harmonic Oscillator

The ground and first excited states eigenvalues and eigenfunctions of the time-independent Schrödinger equation of the linear harmonic oscillator with the potential $V(x) = x^2/2$ are $E_0 = 1/2$, $E_1 = 3/2$ and

$$\psi_0 = \frac{1}{\pi^{1/4}}e^{-x^2/2}, \quad \psi_1 = \frac{2}{(4\pi)^{1/4}}x\,e^{-x^2/2}. \tag{23.36}$$

We choose $x \in [-5, 5]$, $\Delta x = 0.01$, $\Delta\tau = (\Delta x)^2/5$ and the initial wave function as a random wave function, that is $\psi(x_i, 0) = \xi_i$, where ξ_i's are uniform random numbers in the interval $[-0.5, 0.5]$ generated using Park–Miller method (see appendix G). The boundary condition is $\psi(-5, 0) = \psi(5, 0) = 0$. $\psi(x, \tau)$ is normalized after every iteration. Figure 23.4a shows energy E versus τ. As τ increases E approaches the ground state energy value $E_0 = 0.5$. In Fig. 23.4b an arbitrary wave function evolves to the ground state wave function $(-\psi_0)$. Figures 23.4c and 23.4d show E versus τ and $\psi(x, \tau)$ versus τ, respectively, where the initial wave function is chosen as $\psi(x, \tau = 0) = N\sin 6x$, N being the normalization constant. $E \to E_1 = 1.5$ and $\psi(x, \tau) \to -\psi_1$. Other excited energy eigenvalues and eigenfunctions can be obtained with different initial wave function. The values of E_0 and E_1 obtained are 0.4999969 and 1.499984, respectively.

23.3.3.2   One-Dimensional Particle in a Box

As a second example, consider a particle in a box potential $V(x) = 0$ for $0 \leq x \leq L$ and ∞ elsewhere. Its stationary state eigenvalues and eigenfunctions are $E_n = n^2\pi^2\hbar^2/(2mL^2)$ and $\psi_n(x) = \sqrt{2/L}\sin(n\pi x/L)$. We fix $L = 1$, $\Delta x = 0.01$, $\Delta\tau = (\Delta x)^2/5$, $\psi(x, \tau = 0)$ as a

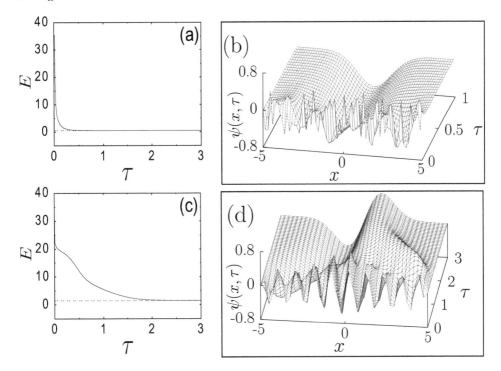

FIGURE 23.4
Variation of (a) ground state energy and (c) first excited state energy of linear harmonic oscillator with τ. The horizontal dashed line indicates the exact energy value. (b) Evolution of initial random wave function into the ground state wave function. The final wave function is $\psi(x, \tau = 1) = \psi_0$. (d) Evolution of $\psi(x, \tau = 0) = N \sin 6x$ into the first excited wave function.

random function and use the boundary condition $\psi(0, \tau) = \psi(L, \tau) = 0$. In Fig. 23.5a the numerically calculated energy is 4.934396 while the exact value of the ground state energy is 4.934802. The numerically calculated ψ at $\tau = 1$ and the exact ground state ψ are plotted in Fig. 23.5b.

23.3.3.3 One-Dimensional Anharmonic Oscillator

Another important quantum system is the anharmonic oscillator with the potential $V(x) = (x^2 - 2)^2$. It is a double-well system and is used to describe structural phase transitions, tunnelling of protons in hydrogen bonded systems and many other systems. Exact eigenvalues and eigenfunctions of the anharmonic oscillator is not known. Its ground state energy obtained by an another method [30] is 1.80081. The ground state energy eigenvalue obtained by the FDTD method is 1.80078, where we used $\Delta x = 0.01$, $\Delta \tau = (\Delta x)^2/5$ and $x \in [-5, 5]$. Figures 23.5c and 23.5d present the results of the FDTD method.

23.3.3.4 Two-Dimensional Systems

Next, we consider the two-dimensional harmonic oscillator with the potential $V(x, y) = (x^2 + y^2)/2$. Its exact ground state energy is 1. In the numerical simulation, we choose $x, y \in [-5, 5]$, $\Delta x = \Delta y = 0.05$, $\Delta \tau = (\Delta x)^2/5$ and $\psi(x, y, \tau = 0)$ as a random function. In Fig. 23.6a for large τ, $E = 0.9998437$. As expected the ground state wave function

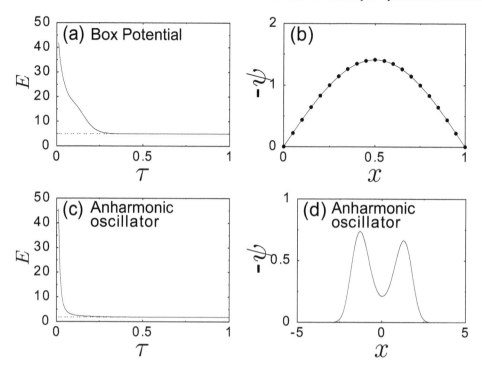

FIGURE 23.5
(a) Variation of ground state energy with τ and (b) the final wave function computed by
the FDTD method (continuous line) and the exact wave function (solid circles) of a particle
in a box potential. The horizontal dashed line indicates the exact energy value. The results
for the anharmonic oscillator are presented in (c) and (d).

obtained (see Fig. 23.6b) is the two-dimensional Gaussian function. Figures 23.6c and 23.6d
present the results for the two-dimensional box potential $V(x,y) = 0$ for $0 \leq x, y \leq 1$ and
∞ otherwise. The exact eigenvalues and eigenfunctions are $E(n_x, n_y) = \left(n_x^2 + n_y^2\right)\pi^2/2$,
$n_x, n_y = 1, 2, \ldots$ and $\psi_{n_x,n_y}(x,y) = 2\sin(n_x\pi x)\sin(n_y\pi y)$ for $0 \leq x, y \leq 1$ otherwise 0.
We fix $\Delta x = \Delta y = 0.02$, $\Delta\tau = (\Delta x)^2/5$. The numerically computed ground state energy
is 9.866358 while the exact value is 9.869604. For $\Delta x = 0.01$ the obtained energy value is
9.868793. That is, accuracy can be improved by decreasing Δx, Δy and $\Delta\tau$.

23.4   Time-Dependent Schrödinger Equation

In sec. 23.2 we described the finite-difference matrix method for computing energy eigenvalues and eigenfunctions of bound states of time-independent Schrödinger equation in one-
dimension. In the present section, we consider the problem of solving the time-dependent
Schrödinger equation. Interestingly, the finite-difference scheme can be extended to the
study of time development of certain initial wave profile [3]. We note that solving the
time-independent Schrödinger equation is a boundary-value problem while solving the time-
dependent Schrödinger equation is essentially an initial-value problem.

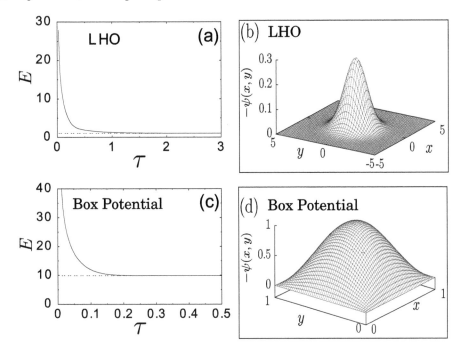

FIGURE 23.6
(a) Variation of ground state energy with τ and (b) the final wave function of two-dimensional linear harmonic oscillator computed by the FDTD method. The results for a particle in a two-dimensional box potential are presented in (c) and (d). The horizontal dashed lines in (a) and (c) indicate the exact energy values.

Setting $\hbar = 1$ and $m = 1$ the time-dependent Schrödinger equation becomes

$$i\frac{\partial \psi}{\partial t} = -\frac{1}{2}\frac{\partial^2 \psi}{\partial x^2} + V(x)\psi . \tag{23.37}$$

We assume the boundary conditions $\psi(x,0) = \phi(x)$ and $\psi(\pm\infty, t) = 0$. As usual we discretize the space variable $x \in [x_{\min}, x_{\max}]$ as $x_0, x_1, \ldots, x_N$ with $x_{i+1} - x_i = \Delta x$ and $t \in [0, T]$ as $0, t_1, t_2, \ldots, t_m$ with $t_{j+1} - t_j = \Delta t$.

The solution $\psi(x,t)$ at the grid point (i,j) is denoted as $\psi_{i,j}$. We approximate ψ, $\partial\psi/\partial t$, $\partial^2\psi/\partial x^2$ as

$$\psi(x,t) = \psi(x_i, t_j) = \frac{1}{2}\left(\psi_{i,j} + \psi_{i,j-1}\right) , \tag{23.38a}$$

$$\frac{\partial}{\partial t}\psi(x,t) = \frac{1}{\Delta t}\left(\psi_{i,j} - \psi_{i,j-1}\right) , \tag{23.38b}$$

$$\frac{\partial^2}{\partial x^2}\psi(x,t) = \frac{1}{2(\Delta x)^2}\left[(\psi_{i+1,j} - 2\psi_{i,j} + \psi_{i-1,j}) + (\psi_{i+1,j-1} - 2\psi_{i,j-1} + \psi_{i-1,j-1})\right]. \tag{23.38c}$$

Using Eqs. (23.38) the time-dependent Schrödinger equation is rewritten as

$$a\psi_{i-1,j} + d\psi_{i,j} + c\psi_{i+1,j} = b_i .\tag{23.39}$$

where

$$a = \frac{1}{2(\Delta x)^2}, \quad d_i = \left(\frac{2i}{\Delta t} - \frac{1}{(\Delta x)^2} - V_i\right),\tag{23.40a}$$

$$b_i = \left(\frac{2i}{\Delta t} + \frac{1}{(\Delta x)^2} + V_i\right)\psi_{i,j-1} - \frac{1}{2(\Delta x)^2}\left(\psi_{i+1,j-1} + \psi_{i-1,j-1}\right).\tag{23.40b}$$

With $\psi_{0,j} = 0$ and $\psi_{N,j} = 0$ Eq. (23.39) is written in matrix form as

$$\begin{pmatrix} d_1 & a_1 & & & & & \\ a_1 & d_2 & a_2 & & & & \\ & a_2 & d_3 & & & & \\ & & & \ddots & & & \\ & & & & d_{N-3} & a_{N-3} & \\ & & & & a_{N-3} & d_{N-2} & a_{N-2} \\ & & & & & a_{N-2} & d_{N-1} \end{pmatrix} \begin{pmatrix} \psi_{1,j} \\ \psi_{2,j} \\ \psi_{3,j} \\ \vdots \\ \psi_{N-3,j} \\ \psi_{N-2,j} \\ \psi_{N-1,j} \end{pmatrix} = \begin{pmatrix} b_1 \\ b_2 \\ b_3 \\ \vdots \\ b_{N-3} \\ b_{N-2} \\ b_{N-1} \end{pmatrix}.\tag{23.41}$$

The above equation is of the form $A\psi = B$. An interesting observation on the matrix A is that it is a symmetric tridiagonal form and can be solved easily. The system given by Eq. (23.41) can be solved by the QL algorithm described in appendix F.

Solved Problem 2:

The initial wave function of a free particle is $\psi(x,0) = (1/\sqrt{2\pi})\,e^{-x^2}$. Numerically compute $\psi(x,t)$ and state its behaviour.

For numerical computation we choose $\Delta x = 0.05$, $\Delta t = 0.01$, $x \in [-5,5]$. Since the Schrödinger equation is a complex equation, $\psi(x,t)$ is complex even if $\psi(x,0)$ is a pure real. Therefore, $\psi(x,t)$ is declared as a complex variable in the numerical algorithm of solving the Eq. (23.41). Real and imaginary parts of complex ψ are obtained as

$$\psi_R = [\psi + \text{conjg}(\psi)]/2, \quad \psi_I = -i[\psi - \text{conjg}(\psi)]/2,\tag{23.42}$$

where $\text{conjg}(\psi)$ denotes complex conjugate of ψ, that is ψ^*. $|\psi|^2 = \psi^*\psi$ is obtained through

$$|\psi|^2 = \psi \times \text{conjg}(\psi) .\tag{23.43}$$

The wave is centred at $x = 0$. As expected the numerically computed $|\psi(x,t)|^2$ shown in Fig. 23.7 spreads over the space variable.

23.4.1 Application

In the following we solve the system (23.41) for certain quantum mechanical systems and capture the time evolution of an initial wave function.

23.4.2 Harmonic Oscillator

For the harmonic oscillator with the potential $V(x) = kx^2/2$ the stationary state energy eigenvalues and eigenfunctions are given by

$$E_n = \left(n + \frac{1}{2}\right)\hbar\omega, \quad \psi_n(x) = N_n H_n(\alpha x)\,e^{-\alpha^2 x^2/2} ,\tag{23.44}$$

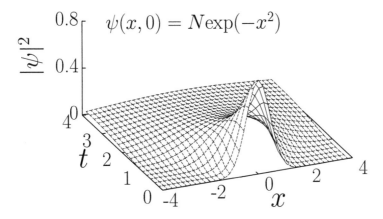

FIGURE 23.7
Numerically computed time evolution of $|\psi|^2$ of free particle.

where $\alpha = \sqrt{m\omega/\hbar}$, H_n are Hermite polynomials and $N_n = (\alpha/(\sqrt{\pi}\,2^n n!))^{1/2}$. The exact time evolution of $\psi_n(x)$ is given by

$$\psi_n(x,t) = \psi_n(x)\,e^{-iE_n t/\hbar}\;. \tag{23.45}$$

We solve the Schrödinger equation with $\psi_0(x,0)$ as the initial wave function. We choose $\Delta x = \Delta t = 0.05$ and $x \in [-5,5]$. Further, we fix $m = 1$, $\omega = 1$ and $\hbar = 1$. Figure 23.8 shows the numerically computed real and imaginary parts of ψ_0 and $|\psi_0|^2$. $\psi_0(x,t) \to 0$ as $x \to \pm\infty$. ψ_R and ψ_I evolve periodically in time while $|\psi(x,t)|^2$ is independent of time.

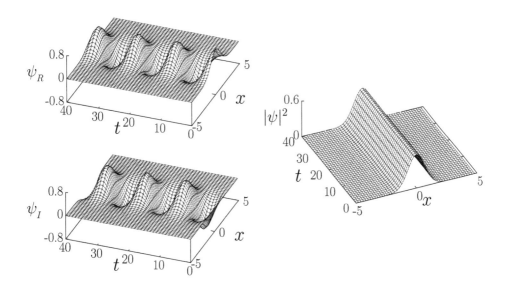

FIGURE 23.8
Numerically computed time evolution of real and imaginary parts of ground state eigenfunction of harmonic oscillator. $|\psi(x,t)|^2$ is also plotted.

The period of the wave function is 4π. Figure 23.9 depicts the numerically computed time evolution of the following two initial states

$$\psi(x,0) \;=\; \sqrt{0.1}\psi_0 + \sqrt{0.9}\psi_1 \,, \tag{23.46a}$$
$$\psi(x,0) \;=\; \sqrt{0.35}\psi_0 + \sqrt{0.15}\psi_1 + \sqrt{0.5}\psi_2 \,. \tag{23.46b}$$

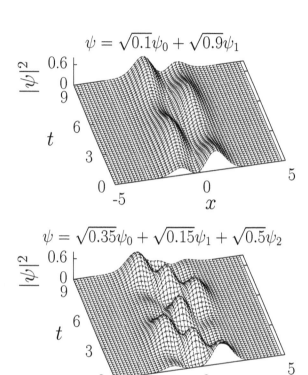

FIGURE 23.9
Numerically computed evolution of $|\psi(x,0)|^2$ given by (23.46).

23.4.3   Particle in the Presence of a Step Potential

Next, we consider a step potential $V(x) = 0$ for $x \le x_0$ and V_0 for $x > x_0$ which has no bound states. The incident wave function undergoes reflection and refraction.

Suppose the initial wave function is $\psi(x,0) = \left(1/\sqrt{2\pi}\right)^{1/2} e^{-x^2} e^{ik_0 x}$. In the numerical simulation, we fix $x_0 = 5$, $k_0 = 5$, $\Delta x = \Delta t = 0.05$ and $x \in [-10, 20]$. Figures 23.10 and 23.11 show, respectively, ψ_R and $|\psi|^2$ for $V_0 = 5$, 10 and 50. In Figs. 23.10a and 23.10b for $V_0 = 5$ and 10 we note that part of the incident wave is reflected while the remaining part is refracted. The amplitude of the reflected wave is relatively larger for $V_0 = 10$. We expect complete reflection for sufficiently large value of V_0. This is the case shown in Figs. 23.10c and Figs. 23.11c for $V_0 = 50$.

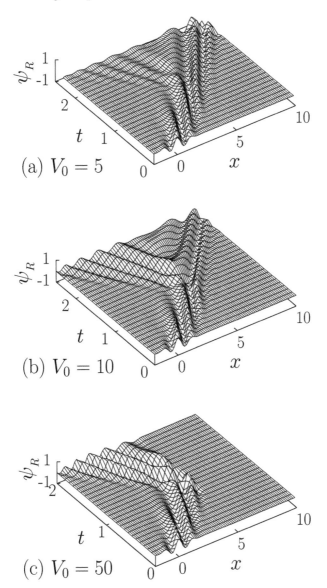

(a) $V_0 = 5$

(b) $V_0 = 10$

(c) $V_0 = 50$

FIGURE 23.10
Time evolution of real part of ψ of a particle in the presence of a step potential.

23.4.4 Tunnelling

Now, we show that tunnelling phenomenon can also be captured by numerically solving the time-dependent Schrödinger equation by the finite-difference method. We consider the rectangular barrier potential with height V_0 and width 0.1. That is, $V(x) = V_0$ for $x \in [5, 5.1]$ and 0 elsewhere. We choose the initial wave function as $e^{-x^2 + i5x}$ and fix $\Delta x = \Delta t = 0.05$ and $x \in [-10, 20]$. The time evolution of ψ_R is shown in Fig. 23.12 for $V_0 = 5$, 20 and 500. For small values of V_0, ψ_R is nonzero in the classically forbidden region $x > 5.1$.

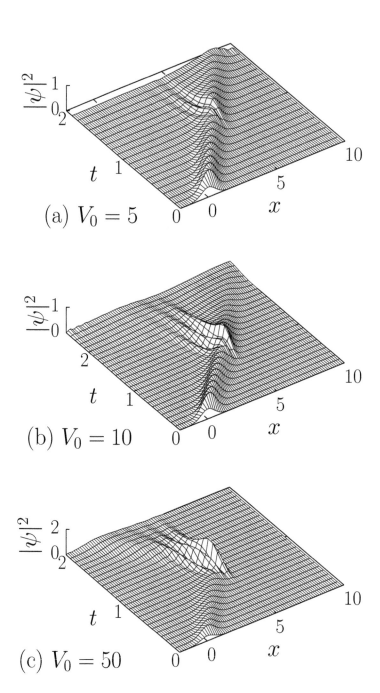

FIGURE 23.11
Time evolution of $|\psi|^2$ of a particle in the presence of a step potential.

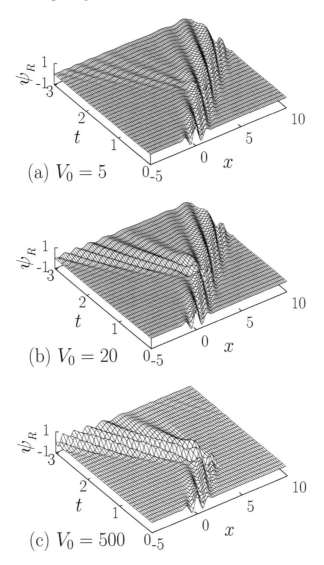

FIGURE 23.12
Numerically computed time evolution of real part of wave function ψ in the presence of a rectangular barrier of height V_0 and width 0.1 at $x = 5$.

The amplitude of the wave in this region decreases with increase in V_0. For $V_0 = 500$ (see Fig. 23.12c) ψ_R is almost negligible in magnitude for $x > 5.1$.

23.4.5 Wave Packet Revivals

In the final example, we consider a particle bouncing on a perfectly reflecting surface under the influence of gravity. The potential of the particle is $V(x) = mgx$ for $x > 0$ and ∞ for $x < 0$. We illustrated the wave packet revival phenomenon in this system in sec. 10.3. Now, we show that this feature can also be reproduced through numerical simulation. Figure 23.13

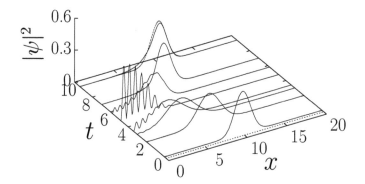

FIGURE 23.13
Numerically computed time evolution of $|\psi|^2$ of the Gaussian wave function given by
Eq. (23.47) in the presence of the potential $V(x) = mgx$ for $x > 0$ and ∞ for $x < 0$.

shows the plot of $|\psi|^2$ versus t of the Gaussian wave function

$$\psi(x,0) = \left(2/(\pi\sigma^2)\right)^{1/4} e^{-(x-x_0)^2/\sigma^2} \tag{23.47}$$

at selected values of t where $x_0 = 10$, $\sigma = 2$, $g = 1$ and $m = 1$. The localized wave packet
solution spreads in time but at a later time T it reforms and appears as its initial state (the
last continuous curve). (The dashed curve is the form of the initial wave.) This is called
wave packet revival.

23.5 Quantum Scattering

Having studied how to compute eigenvalues and eigenfunctions of stationary states of quantum mechanical systems we turn our attention to scattering potentials. A systematic numerical procedure for computing transmission coefficient was described by Pang [16]. We show
how transmission probability of quantum mechanical barriers can be computed numerically.

Quantum mechanical scattering and tunnelling problems in one-dimension are described
by the Schrödinger equation

$$\frac{d^2\psi(x)}{dx^2} + 2[E - V(x)]\psi(x) = 0 , \tag{23.48}$$

where $\hbar^2$ and m are set unity. Let us rewrite the above equation as

$$\frac{dy_1}{dx} = y_2 , \qquad \frac{dy_2}{dx} = 2[V(x) - E]y_1 , \tag{23.49}$$

where $y_1 = \psi$. Assume that $V(x)$ is nonzero only in the region $x \in [0, a]$ and the incident
particle comes from the left with a unit flux. The general solution of Eq. (23.48) is written
as

$$\psi(x) = \begin{cases} e^{ikx} + Be^{-ikx}, & x < 0 \\ Ce^{ik(x-a)}, & x > a \end{cases} \tag{23.50}$$

where the parameters B and C are generally complex and are to be determined. The wave vector k is $\sqrt{2mE/\hbar^2} = \sqrt{2E}$. The solution $\psi(x)$ in the region $x \in [0, a]$ is to be obtained numerically.

The complex solution and its first derivative satisfy the continuity conditions at $x = 0$ and a. These conditions are

$$\psi(0) = 1 + B, \quad \psi(a) = C, \tag{23.51a}$$
$$\psi'(0) = ik(1 - B), \quad \psi'(a) = ikC. \tag{23.51b}$$

Once the constants B and C are known the transmission and reflections probabilities are obtained from the relations $T = |C|^2$, $R = |B|^2$ and $T + R = 1$.

23.5.1 Numerical Solution in the Region $0 \leq x \leq a$

Now, the intricate problem is to determine the solution of Eq. (23.48) in the region $x \in [0, a]$ obeying the continuity conditions, Eq. (23.51). The correct solution is obtained using the numerical procedure outlined below. The procedure consists of four main steps.

1. For a given potential barrier and energy value make an initial guess for the parameter $B = B_r + iB_i$ with $|B_r|$, $|B_i| \leq 1$. We can choose $B_r = \alpha$, $B_i = \sqrt{1 - \alpha^2}$, $|\alpha| \leq 1$.

2. Integrate Eq. (23.49) from $x = 0$ to a using, say, the fourth-order Runge–Kutta method with the initial condition $y_1(0) = 1 + B$ and $y_2(0) = ik(1 - B)$. This is only a trial solution and not the solution satisfying the conditions, Eqs. (23.51).

3. From Eqs. (23.51) we have two expressions for C:

$$C = C_1 = y_1(a), \quad C = C_2 = -\frac{i}{k}y_2(a) \,. \tag{23.52}$$

These two values depend on the values of B_r and B_i. Compute C_1 and C_2 from Eq. (23.52).

4. We require $C_1 = C_2$. This will happen when the trial solution is exact. Introduce the function

$$F = |C_1 - C_2|^2 \,. \tag{23.53}$$

Minimize the function F by varying $y_1(a)$ and $y_2(a)$ so that $F \approx 0$. When this is achieved C_1 and C_2 are identical.

23.5.2 Minimization of F

A simple method of minimization of a multivariable continuous function is the Nelder–Mead simplex (NMS) method [28]. The method is simple, effective and computationally compact. For the basic idea of the method with an example see the exercise 23.12 at the end of this chapter. The method starts with $(N + 1)$ initial guess for an N-dimension function.

For the function F given by Eq. (23.53), $C_1 = C_{1R} + iC_{1I}$, $C_2 = C_{2R} + iC_{2I}$. If real and imaginary components are treated as separate variables then F is a function of 4 variables, that is $N = 4$. Therefore, 5 initial guesses on B are needed. B_r and B_i are chosen as $B_R = 0.2, 0.4, 0.6, 0.8, 1$ and $B_I = \sqrt{1 - B_R^2}$. The variables B, C, y_1 and y_2 are complex functions. Therefore, the Eqs. (23.49), (23.51), (23.52) and (23.53) are rewritten as follows by treating real and imaginary parts separately so that the variables involved are all real:

$$\frac{dy_{1R}}{dx} = y_{2R}, \quad \frac{dy_{1I}}{dx} = y_{2I}, \tag{23.54a}$$

$$\frac{dy_{2R}}{dx} = 2[V(x) - E]y_{1R}, \quad \frac{dy_{2I}}{dx} = 2[V(x) - E]y_{1I}, \tag{23.54b}$$

with the boundary conditions

$$y_{1R}(0) = 1 + B_R, \quad y_{1I}(0) = B_I, \tag{23.55a}$$
$$y_{2R}(0) = kB_I, \quad y_{2I}(0) = k(1 - B_R), \tag{23.55b}$$

and

$$C_{1R} = y_{1R}(a), \quad C_{1I} = y_{1I}(a), \quad C_{2R} = \frac{1}{k}y_{2I}(a), \quad C_{2I} = -\frac{1}{k}y_{2R}(a), \tag{23.56a}$$

$$F = \left(y_{1R} - \frac{y_{2I}}{k}\right)^2 + \left(y_{1I} + \frac{y_{2R}}{k}\right)^2. \tag{23.56b}$$

For the chosen 5 set of values of (B_R, B_I) the corresponding y's at $x = a$ are obtained by solving the Schrödinger equation by the fourth-order Runge–Kutta method. After this step, NMS method is used to find correct $y_1(a)$ and $y_2(a)$ for which $F \approx 0$. Minimization of F is performed actually by varying $y_1(a)$ and $y_2(a)$ and not by directly varying B_R and B_I. The transmission probability T and the reflection probability R are then computed from the expressions $T = |C|^2 = C_{1R}^2 + C_{1I}^2$ and $R = 1 - T$.

Solved Problem 3:

A standard example for a quantum tunnelling is the square-barrier potential $V(x) = V_0$ for $0 \le x \le a$ and 0 otherwise. Applying the NMS method compute numerically the transmission probability T and compare it with the exact T.

The Schrödinger equation for this potential is exactly solvable. The analytic expression for T_{exa} is given by

$$T_{exa} = \begin{cases} 4\mu\left[(1 + \mu)^2 \sin^2 ka + 4\mu \cos^2 ka\right]^{-1}, & E > V_0 \\ 4\mu'\left[(1 - \mu)^2 \sinh^2 k'a + 4\mu' \cosh^2 k'a\right]^{-1}, & E < V_0 \end{cases} \tag{23.57a}$$

where

$$\mu = (E - V_0)/E, \quad \mu' = E/(V_0 - E), \tag{23.57b}$$
$$k = \sqrt{2(E - V_0)}, \quad k' = \sqrt{2(V_0 - E)}. \tag{23.57c}$$

Suppose $a = 5$ and $V_0 = 1$. Step size in the Runge–Kutta method is $h = 0.01$. Tolerance in the NMS method is fixed as 10^{-6}. Developing a program to compute T as a function of energy is given as an exercise (see the exercise 23.13 at the end of this chapter).

Figure 23.14 shows numerically computed T, T_{num}, as a function of E. Table 23.5 presents the values of T_{num}, T_{exa}, $T_{exa} - T_{num}$ and percentage of error, $100 \times |(T_{num} - T_{exa})/T_{exa}|$ for several values of E. Very good agreement between numerical and analytical T is observed. The accuracy can be improved by choosing smaller values of step size h in the Runge–Kutta method and the tolerance value in the NMS method.

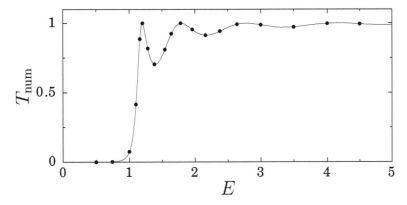

FIGURE 23.14
Numerically computed transmission probability T_{num} (continuous line) as a function of energy for the square-barrier potential with $V_0 = 1$ and $a = 5$. The exact values of T for a few values of E are marked by solid circles.

TABLE 23.5
Comparison of numerically computed transmission probability with the exact transmission probability for the square-barrier potential with $V_0 = 1$ and $a = 5$.

E	Transmission probability		$T_{exa} - T_{num}$	% of error
	T_{num}	T_{exa}		
0.5	0.000180	0.000182	0.1240E-05	0.683
1.0	0.074227	0.074074	−0.1528E-03	0.206
1.5	0.766294	0.765397	−0.8973E-03	0.117
2.0	0.940954	0.940901	−0.5265E-04	0.006
2.5	0.968169	0.969047	0.8779E-03	0.091
3.0	0.988881	0.987819	−0.1063E-02	0.108
3.5	0.973517	0.973134	−0.3829E-03	0.039
4.0	0.997733	0.997956	0.2231E-03	0.022
4.5	0.994056	0.994032	−0.2387E-04	0.002
5.0	0.988368	0.987655	−0.7132E-03	0.072

Solved Problem 4:

T for a locally periodic potential with the unit cell being Dirac delta-function exhibits a band structure character. This property is found to occur for the potential consisting of finite number of identical square-barriers. The potential is given by

$$V(x) = \begin{cases} V_0, & (i-1)2b \le x \le (2i-1)b \\ 0, & \text{otherwise} \end{cases} \tag{23.58}$$

where $i = 1, 2, \ldots, N$. Numerically compute T for a few values of N and point out the features of T.

For the numerical simulation we fix $b = 1$ and $V_0 = 10$. Figure 23.15 shows T_{num} versus E for $N = 1, 2, 4$ and 10. One of the most striking features of this figure is the appearance

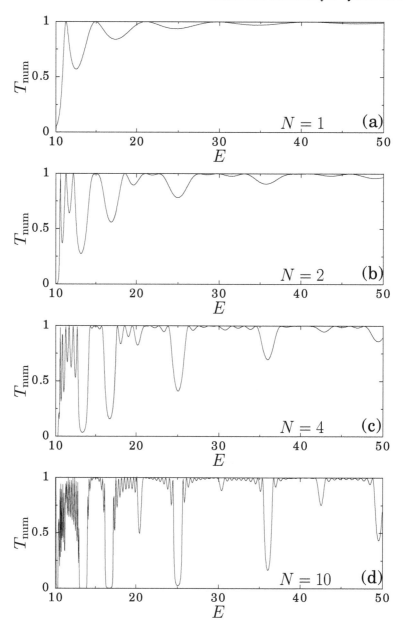

FIGURE 23.15
Numerically computed transmission probability as a function of energy E for a locally periodic potential with square-barrier as a unit cell. (a) For one barrier, (b) for two barriers, (c) for four barriers and (d) for ten barriers. Here $V_0 = 10$ and $b = 1$.

of energy band structure similar to the Dirac comb. The number of maxima with $T_{num} = 1$ increases with increase in N. For $N = 1$, 2 and 4 the T_{num} is always greater than zero. But for large N, for example $N = 10$ (Fig. 23.15d), T_{num} is zero or ≈ 0 for a range of E. Thus, for large N, by suitably choosing E, V_0 and a we can have almost complete transmission or complete reflection.

23.6   Electronic Distribution of Hydrogen Atom

The Monte Carlo technique can be used to represent electronic distribution of hydrogen atom [18], particles distribution in harmonic oscillator, tunnelling through a potential barrier and scattering of particles from a potential. In this section, the electronic distribution of hydrogen atom is considered.

Let us consider the ground state of hydrogen atom. It is given by $\psi_{nlm} = \psi_{100} = (1/\sqrt{\pi})e^{-r}$. The probability density for the electron is $P_{100}(r) = |\psi_{100}|^2 = e^{-2r}/\pi$. If we consider Cartesian coordinate system then $r^2 = x^2 + y^2 + z^2$. For the present discussion the multiplicative factor $1/\pi$ in P_{100} is dropped so that $P_{100} = e^{-2r}$. Suppose we wish to obtain electronic distribution in $x - y - z$ plane. Monte Carlo technique can be employed to this problem. For convenience let us choose the plane $x - z$ with $y = 0$. Essentially, a sequence of N pair of points (x, z) with the distribution $P_{100} = e^{-2r}$ represent the electronic distribution in ground state. Thus, the problem is to generate a set of numbers $\{(x, z)\}$ with the probability distribution P_{100}.

How do we generate $\{(x, z)\}$ *with* $P_{100} = e^{-2r}$? A sequence of numbers with a desired distribution can be obtained using the simple and efficient random sampling technique called *accept-reject* or simply *rejection method*. This method is described in appendix G for a function $f(x)$. This method can be easily extended for $f(x, z)$.

23.6.1   Rejection Technique to Generate Random Numbers with the Distribution $e^{-(x^2+z^2)}$

Let $f(x, z)$ be the desired distribution defined over the interval $x \in [a_x, b_x]$, $z \in [a_z, b_z]$. Select a suitable bounding function $cg(x, z)$ with maximum of $cg(x, z) = cg_m$ is $\geq$ maximum of $f(x, z)$. For the exponential function $e^{-(x^2+z^2)}$ we may choose $x \in [-10, 10]$, $z \in [-10, 10]$ and maximum of $cg(x, z)$ as $cg_m = 1$ since the maximum of $e^{-(x^2+z^2)}$ is 1 (cg_m must be ≥ 1). Now, treat x and z are as uniformly distributed random numbers in the range $[-10, 10]$ and $cg(x, z)$ as uniformly distributed random numbers in the range $[0, 1]$ since $f(x, z) \in [0, 1]$. Then the scheme is the following:

1.   Choose a value for x and a value for z.

2.   Choose a value for $cg(x, z)$.

3.   Compute $f(x, z)$.

4.   If $f(x, z) \leq cg(x, z)$ accept the pair (x, z). Otherwise reject it.

Repeating the above algorithm we can generate a desired number of (x, z). The obtained pairs of numbers obey the chosen distribution $f(x, z)$. Using the above method we can generate any number of pairs of random numbers (or in general n-dimensional random numbers) satisfying a given distribution.

23.6.2   Electronic Distribution in Various States

A program can be developed to generate a sequence of, say, $N = 2000$ pairs of points (x, z) with $f(x, z) = P(x, y = 0, z) = e^{-2r}$ (see the exercise 17 at the end of this chapter). This corresponds to ground state $nlm = 100$. The values of a, b and cg_m are chosen as $-10, 10$ and 1, respectively. The plot of x versus z represents electronic distribution of hydrogen atom in ground state. To obtain the distribution in other states the corresponding expression for $P(x, y, z)$ is to be used. Further, we have to choose appropriate values for a, b and cg_m. These values for some of the states are mentioned in the program. Figure 23.16 displays the

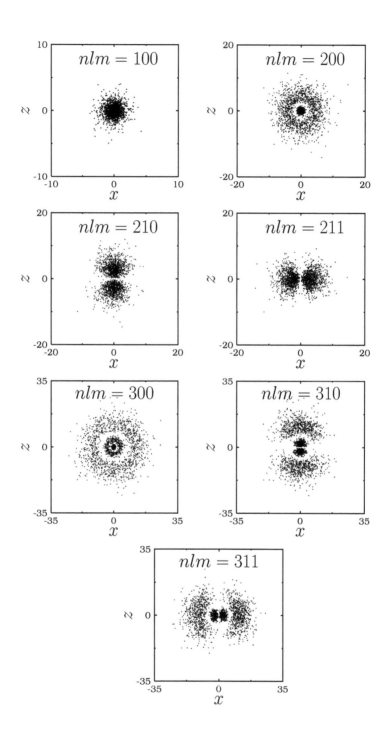

FIGURE 23.16
Electronic distribution of the hydrogen atom in some of its states produced by 2000 points in $x - z$ plane with $y = 0$.

electronic distribution in several states. The distributions corresponding to $nlm = 100, 200$ and 300 exhibit spherical symmetry around their centre. The distributions for $nlm = 210$, 211, 310 and 311 show rotational symmetry around the z-axis. The influence of applied uniform electric field along z-axis on the probability distribution, called *Stark effect* can be visualized by the Monte Carlo method.

23.7   Schrödinger Equation with an External Field

In this section, we present the coupled channel method (CCM) [24] of solving the time-dependent Schrödinger equation with an external field. For

$$i\hbar\frac{\partial\psi}{\partial t} = H\psi \,, \tag{23.59}$$

where $H = H^{(0)} + \lambda H^{(1)}$ we write the solution as

$$\psi(x,t) = \sum_n a_n(t)\phi_n(x)e^{-iE_n t/\hbar} \,. \tag{23.60}$$

In the above solution, $\phi_n(x)$ and E_n are the eigenfunctions and energy eigenvalues, respectively, of the unperturbed Hamiltonian $H^{(0)}$ and $a_n(t)$ are the expansion coefficients. $P_n(t) = |a_n(t)|^2$ is the probability of finding the system in the state n at time t and $\sum_n |a_n(t)|^2 = 1$.

To determine a_n's substitute (23.60) in Eq. (23.59) and obtain

$$i\hbar \sum_n \frac{da_n}{dt}\phi_n e^{-iE_n t/\hbar} = \lambda \sum_n a_n H^{(1)}\phi_n e^{-iE_n t/\hbar} \,. \tag{23.61}$$

Multiplying Eq. (23.61) by ϕ_m^* and integrating over all space we get

$$\frac{da_m}{dt} = \frac{\lambda}{i\hbar}\sum_n a_n H^{(1)}_{mn}(t)e^{i\omega_{mn}t} \,, \tag{23.62}$$

$$H^{(1)}_{mn}(t) = \int_{-\infty}^{\infty}\phi_m^*(x)H^{(1)}(x,t)\phi_n(x)\,dx \tag{23.63}$$

and $\omega_{mn} = (E_m - E_n)/\hbar$. The above equation is exact. This equation is essentially a system of coupled integro-differential equation which is difficult to solve analytically. Therefore, we pose the problem in a slightly different manner. Suppose the perturbation is switched-on at $t = 0$ and switched-off at $t = T$. The system is, say, initially in the nth state and after the time T it settles on to the state m. In numerical simulation we choose a small number of states N, for example, $1 < N < 5$. In this case Eq. (23.62) becomes a closed system of N coupled ordinary differential equations which we can solve by the fourth-order Runge–Kutta method by treating $a_n(t)$ as complex or writing $a_n(t) = R_n(t) + iI_n(t)$. In the latter case we have

$$\frac{dR_m(t)}{dt} = \frac{1}{\hbar}\sum_n [R_n \sin\omega_{mn}t + I_n \cos\omega_{mn}t]\, H^{(1)}_{mn}(t) \,, \tag{23.64a}$$

$$\frac{dI_m(t)}{dt} = \frac{1}{\hbar}\sum_n [I_n \sin\omega_{mn}t - R_n \cos\omega_{mn}t]\, H^{(1)}_{mn}(t) \,. \tag{23.64b}$$

For a given system with an interaction term $H^{(1)}(x,t)$ consider first, say, five states, that is $N = 5$ (if $n = 0$ is not a quantum number, otherwise $N = 4$). $H_{mn}^{(1)}(t)$ given by Eq. (23.63) can be calculated either analytically or numerically. With an initial condition $a_n(0) = R_n(0) + iI_n(0)$, the $2N$ coupled Eqs. (23.64) can be solved for $0 \le t \le \tau$. Then, $|a_n(t)|^2 = [R_n(t)]^2 + [I_n(t)]^2$.

Solved Problem 5:

Consider the linear harmonic oscillator with $H^{(0)} = p_x^2/(2M) + M\omega^2 x^2/2$ interacting with a laser field. M denotes mass of the particle. Suppose the particle is subjected to a laser pulse of electric field $\mathcal{F}(t)$ given by

$$\mathcal{F}(t) = f_0 \sin^2(\pi t/\tau) \cos\omega_L t , \quad 0 \le t < \tau \tag{23.65}$$

where f_0 is the field amplitude, τ is the laser duration and ω_L is the laser centre frequency. Study the transition behaviour by the CCM.

The energy eigenvalues and eigenfunctions of the unperturbed harmonic oscillator are

$$E_n = \left(n + \frac{1}{2}\right)\hbar\omega, \quad n = 0, 1, 2, \ldots \tag{23.66a}$$

$$\phi_n(x) = \left(\frac{\alpha}{2^n n! \sqrt{\pi}}\right)^{1/2} H_n(\alpha x) e^{-\alpha^2 x^2/2} , \tag{23.66b}$$

where $H_n(\alpha x)$ are Hermite polynomials and $\alpha^2 = M\omega/\hbar$.

The laser-oscillator interaction term $H^{(1)}(x,t)$ is $H^{(1)}(x,t) = \mathcal{F}(t)x$. $H_{mn}^{(1)}(t)$ is worked out as $H_{mn}^{(1)}(t) = \mathcal{F}(t)x_{mn}$ where

$$x_{mn} = \sqrt{\frac{\hbar}{M\omega}} \left[\sqrt{\frac{m}{2}}\delta_{m,n+1} + \sqrt{\frac{m+1}{2}}\delta_{m,n-1}\right] . \tag{23.67}$$

In the numerical study fix $\hbar = 1$, $M = 1$, $\omega_L = \omega = 0.2\pi$ and $f_0 = 0.2$. At $t = 0$, assume that the system is in its ground state. In this case, $a_n(0) = \delta_{n0}$. We consider five channels, that is, $N = 4$. $\omega_{mn} = (m - n)\omega$, $m, n = 0, 1, 2, 3, 4$. Equations (23.64) (totally 10 coupled equations) are integrated numerically using the fourth-order Runge–Kutta method with step size $\Delta t = 0.01$. Figure 23.17 shows $\log_{10}|a_n|^2$ versus t/τ for four values of τ. For $\tau = 10$, $|a_0|^2 \gg |a_n|^2$, $n = 1, 2, 3$ and 4 for all values of t. As τ increases $|a_0(t)|^2$ decreases while other $|a_n|^2$ increases. When $\tau = 40$, for $t > \tau$, we notice $|a_0|^2 < |a_n|^2$, $n = 1, 2, 3, 4$. That is, the probabilities for the transitions from ground state to neighbouring excited states are larger than finding the system in the ground state for $t > \tau$. For all values of τ, the excitation probability increases with time rapidly and shows a plateau character near the end of the laser pulse, that is $t \approx \tau$.

In Fig. 23.18 τ is kept as 10 while $a_n(0)$ is varied. For $a_n(0) = \delta_{n0}$, $|a_0|^2 \gg |a_n|^2$, $n = 1, 2, 3, 4$. $|a_m|^2$ with $a_m(0) = \delta_{nm}$, for sufficiently large t, decreases when the number of neighbouring energy levels to mth level increases.

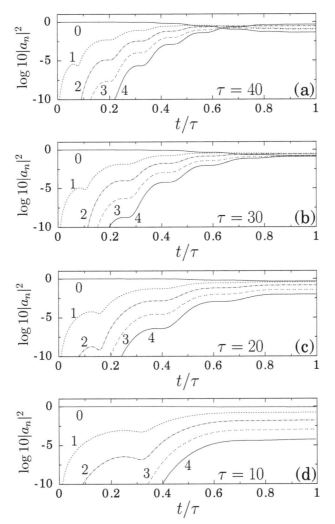

FIGURE 23.17
Transition probability $|a_n|^2$ in $\log_{10}$ scale versus t/τ for the harmonic oscillator subjected to a laser field during a time interval τ. Here $\hbar = 1$, $M = 1$, $\omega_{\mathrm{L}} = \omega = 0.2\pi$, $f_0 = 0.2$ and $a_n(0) = \delta_{n0}$. The numbers $0 - 4$ in each subplot refer to the values of n.

23.8 Concluding Remarks

Accurate numerical methods are essential for studying quantum phenomena for the systems with complicated potentials. It has been found that accurate determination of ionization rates require a very careful numerical study. The numerical methods considered in this chapter are very general and can be applied to more complicated potentials. The methods described in this chapter are useful for the design and analysis of properties and functions of nanoelectronic devices which utilize the nonlinear current-voltage relation of the quantum tunnelling of the electrons.

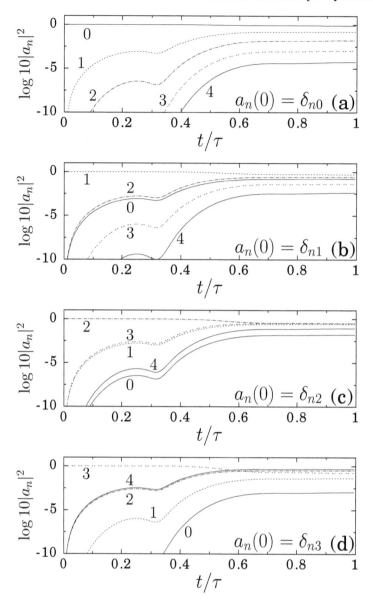

FIGURE 23.18

$|a_n|^2$ in $\log_{10}$ scale versus t/τ for the harmonic oscillator subjected to a laser field during a time interval $\tau = 10$. Here $\omega_L = \omega = 0.2\pi$, $f_0 = 0.2$ and for various $a_n(0)$. The numbers $0-4$ in each subplot refer to the values of n.

A finite-element method [15] and matrix method [31] were developed for solving the one-dimensional quantum tunnelling problem. A (t,t') method was introduced for analyzing propagation of an initial state in explicit time-dependent Hamiltonians [32,33]. A numerical algorithm employing Crank–Nicholson approximation for the time dependence along with a Numerov extension for the discrete transparent boundary conditions was described in [34]. Numerical methods are available for the computation of phase shift and total scattering cross-section. Numerical visualization of quantum matrix diagonalization was reported [35].

Monte Carlo simulation can also be used for ground state energy calculation. For a detailed discussion on such numerical schemes one may refer to [36].

23.9   Bibliography

[1] A. Goldberg, H.M. Schey and J.L. Schwartz, *Am. J. Phys.* 35:77, 1967.

[2] R. Guardiola and J. Ros, *J. Comput. Phys.* 45:374, 1982.

[3] R.L.W. Chen, *Am. J. Phys.* 50:902, 1982; 51:570, 1983.

[4] M.D. Milkhailov and N.L. Vulchanov, *J. Comput. Phys.* 50:323, 1983.

[5] U. Blukis and J.M. Howell, *J. Chem. Edu.* 60:207, 1983.

[6] V. Fack and G.V. Berghe, *Comput. Phys. Commun.* 39:187, 1986.

[7] H. Kobeissi, M. Kobeissi and A. El Hajj, *J. Comput. Chem.* 9:844, 1988.

[8] J.R. Cash, A.D. Raptis and T.E. Simos, *J. Comput. Phys.* 91:413, 1990.

[9] S.E. Koonin and D.C. Meredith, *Computational Physics.* Addison-Wesley, Redwood City, 1990.

[10] L. Greengard and V. Rokhlin, *Commun. Pure Appl. Math.* 44:419, 1991.

[11] A.K. Ghatak and S. Lokanathan, *Quantum Mechanics.* MacMillan, New Delhi, 1992.

[12] M.E. Hosea and L.F. Shampine, *IMA J. Numer. Anal.* 13:397, 1993.

[13] T.E. Simos, *Comput. Phys. Commun.* 74:63, 1993.

[14] D. Stauffer, F.W. Hehl, V. Winkelmann and J.G. Zabolitzky, *Computer Simulation and Computer Algebra.* Springer, Berlin, 1993.

[15] R. Goloskie, J.W. Kramer and L.R. Ram Mohan, *Comput. Phys.* 8:679, 1994.

[16] T. Pang, *Comput. Phys.* 9:602, 1995.

[17] J.F. van der Maelen Uria, S. Garcia-Granda and A. Menendez-Velazquez, *Am. J. Phys.* 64:327, 1996.

[18] V.M. de Aquino, V.C. Aguilera-Navarro, M. Goto and H. Iwamoto, *Am. J. Phys.* 69:788, 2001.

[19] D.M. Sullivan and D.S. Citrin, *J. Appl. Phys.* 91:3219, 2002.

[20] Z. Wang, Y. Ge, Y. Dai and D. Zhao, *Comput. Phys. Commun.* 160:23, 2004.

[21] D.M. Sullivan, *J. Appl. Phys.* 98:084311, 2005.

[22] A.J. Zakrzewski, *Comput. Phys. Commun.* 175:397, 2006.

[23] I.W. Sudiarta and D.J.W. Geldart, *J. Phys. A* 40:1885, 2007.

[24] J. Wang and J.D. Champagne, *Am. J. Phys.* 76:493, 2008.

[25] R.T. Bates, *Sci. Am.* 258:78, 1988.

[26] F. Capasso, *Physics of Quantum Electron Devices.* Springer, Berlin, 1990.

[27] W.H. Press, S.A. Teukolsky, W.L. Vetterling, B.P. Flannery, *Numerical Recipes in Fortran.* Cambridge University Press, Cambridge, 1992.

[28] J.H. Mathews, *Numerical Methods for Mathematics, Science and Engineering.* Printice Hall of India, New Delhi, 1998.

[29] M. Pal, *Numerical Analysis for Scientists and Engineers: Theory and C Programs.* Narosa, New Delhi, 2007.

[30] M. Jafarpour and D. Afshar, *J. Phys. A* 35:87, 2002.

[31] V.A. Fedirko, S.V. Polyakov and D.A. Zenyuk, *Math. Models Comp. Sim.* 2:704, 2010.

[32] U. Peskin, and N. Moiseyev, *J. Chem. Phys.* 99:4590, 1993.

[33] U. Peskin, R. Kosloff and N. Moiseyev, *J. Chem. Phys.* 100:8849, 1994.

[34] C.A. Moyer, *Am. J. Phys.* 72:51, 2004.

[35] K. Randles, D.V. Schroeder and B.R. Thomas, *Am. J. Phys.* 87:857, 2019.

[36] J.M. Thijssen, *Computational Physics.* Cambridge University Press, Cambridge, 1999.

23.10 Exercises

23.1 Develop a Fortran-90 program for the numerical computation of eigenvalues of linear harmonic oscillator by solving its time-independent Schrödinger equation by the QL method.

23.2 Compute the first few energy eigenvalues of the perturbed hydrogen atom whose potential is $V = -1/r + 1/r^2$ by the matrix method. Do the computation for two values of step size h and obtain the result for $h \to 0$ by Richardson extrapolation. Tabulate the results. The exact energy eigenvalues are $E_n = -1/2(n+1)^2$, $n = 1, 2, \ldots$.

23.3 Develop a Fortran-90 program for the numerical computation of the eigenfunction of linear harmonic oscillator for a given energy eigenvalue by solving its time-independent Schrödinger equation by the QL method.

23.4 Consider the box potential $V(x) = 0$ for $0 < x < a$ and ∞ otherwise. Its exact energy eigenvalues are $E_n = n^2\pi^2\hbar^2/(2ma^2)$ and the ground state eigenfunction is $\phi_1 = \sqrt{2}\sin \pi x$. Applying the matrix method determine the energy eigenvalues with values < 100 units and the ground state eigenfunction for $a = 1$ and $m = 1$ units.

23.5 Applying the matrix method compute the ground state eigenfunction ($E = -0.125$) of the perturbed hydrogen atom with the potential $V = -1/r + 1/r^2$ by the matrix method. Tabulate the results. The exact eigenfunction is $\psi_1 \sim r^2 e^{-r/2}$.

23.6 For the shifted harmonic oscillator potential $V(x) = x^2/2 + m$, $m = 0,1,2,\ldots$ the exact energy eigenvalues are $E_n^{(m)} = (m+n+1/2)$. By the matrix method

(a) compute numerically first few energy eigenvalues for $m = 1$.

(b) When $n = 0$, $E_0^{(m)} = (m+1/2)$, $m = 0,1,2,\ldots$. Verify that the eigenfunctions $\phi_m(x)$ all are identical even though the eigenvalues are distinct.

23.7 Consider the harmonic oscillator with $V = x^2/2$ perturbed by the term $V_{\text{per}} = -\lambda eFx$, where e is the electron charge, F is the strength of the applied electric field and λ is the perturbation parameter. Assume that $\lambda = 0.5$, $e = 1$ and $F = 1$. Applying the FDTD method with $\phi_0(x, \tau = 0) = \sin 6x$ determine the ground state energy eigenvalue and the eigenfunction. Compare the computed ϕ_0 with the exact $\phi_0 = (1/\pi^{1/4})e^{-(x-\lambda)^2/2}$.

23.8 For the system with the potential $V(x) = F|x|$ with $F = \sqrt{2}$ find the energy eigenvalues for $\phi(x, \tau = 0) = \cos x$ and $\sin x$. Plot E versus τ and $\phi(x, \tau)$.

23.9 Numerically verify whether tunnelling occurs in the case of the potential $V(x) = 3e^{-0.1x^2}$ for $|x| < 1$ and 0 for $|x| > 1$ with the initial Gaussian wave function $\psi(x, 0) = e^{-x^2}$.

23.10 Consider a system with $V(x) = x$ for $x > 0$ and ∞ for $x < 0$. Study the time evolution of the initial Gaussian wave packet $\psi = (1/2\pi)^{1/4}e^{-(x-10)^2/4}$.

23.11 For a box potential $V(x) = 0$ for $|x| < 1$ and ∞ elsewhere plot the time evolution of $|\psi(x, t)|^2$ for the initial wave functions $\psi(x, 0) = \cos(\pi x/2)$, $\psi(x, 0) = \sin \pi x$ and $\psi(x, 0) = \sqrt{15/16}\,(x^2 - 1)$.

23.12 Prepare a note on the Nelder–Mead simplex (NMS) method of finding a minimum of a two-dimensional function.

23.13 Develop a Fortran-90 program for the numerical computation of transmission probability for a one-dimensional scattering problem.

23.14 The exact transmission probability T of the potential $V = V_0\delta(x - x_0)$ is $T = 1/(1+\beta^2)$, where $\beta^2 = mV_0^2/(2E\hbar)$. Numerically compute T and compare it with the exact T. What is the value of T for large E?

23.15 Compute the transmission probability T as a function of energy E for the truncated Gaussian potential
$$V(x) = \begin{cases} V_0\,e^{-(x-a/2)^2}, & 0 \le x \le a \\ 0, & \text{otherwise.} \end{cases}$$

23.16 Consider the sech potential
$$V(x) = \begin{cases} V_0\,\text{sech}[b(x - a/2)], & 0 \le x \le a \\ 0, & \text{otherwise.} \end{cases}$$
Sketch the potential. Numerically computing T describe its dependence on the energy E of the incident particles.

23.17 Develop a Fortran-90 program for the numerical computation of electronic distribution of hydrogen atom by Monte Carlo method.

23.18 Obtain the image representation of electronic distribution in x–z plane with $y = 0$ of hydrogen atom in the states (a) $(\psi_{200} + \psi_{210})/\sqrt{2}$ and (b) $(\psi_{200} - \psi_{210})/\sqrt{2}$.

23.19 A three-dimensional harmonic oscillator potential is $V(r) = r^2/2 = (x^2 + y^2 + z^2)/2$. In spherical polar coordinates its eigenfunctions are given by $\psi_{nlm} = Ne^{-r^2/2}r^l L_k^{l+\frac{1}{2}}(r^2)Y_l^m(\theta, \phi)$, where N is the normalization constant, $L_k^{l+\frac{1}{2}}(r^2)$ are Laguerre polynomials with $k = (n - l)/2$ and Y_l^m are spherical harmonics. Obtain the distribution of particles in the $x - z$ plane with $y = 0$ for the ψ_{000}, ψ_{110}, ψ_{200} and ψ_{220} states.

23.20 Consider a particle of mass M and charge e in an infinite height potential well centred at $x = L/2$ and $V(x) = 0$ for $0 < x < L$ and ∞ otherwise. The particle is subjected to an applied laser field $\mathcal{F}(t) = f_0 \sin^2(\pi t/\tau) \cos \omega_L t$, $0 \le t < \tau$. Fix $\tau = 40$, $\omega = \omega_L = 0.2\pi$, $e = 1$, $M = 1$ and $L = 2$. Assume that the system is initially in the ground state. Applying the coupled channel method compute $|a_n(t)|^2$ as a function of t/τ for a few values of f_0 and analyze the transitions between the states.

A

Calculation of Numerical Values of h and k_B

Planck, in his novel paper entitled *On the energy distribution in the black body spectrum* [1] presented the calculation of the values of the universal constants h and k_B. Planck calculated the values of them with the help of measurements which have already been performed.

Kurlbaum [2] found that the total energy E_T, which is radiated from $1\,\mathrm{cm}^2$ in $1\,\mathrm{sec}$ by a black body at $T^\circ\,\mathrm{C}$ satisfies the relation

$$E_{100} - E_0 = 0.0731\,\mathrm{watts/cm}^2 = 0.731 \times 10^6\,\mathrm{ergs/(cm^2 sec)}\,. \tag{A.1}$$

Then the spatial density of the total radiant energy at $1\,\mathrm{K}$ is given by

$$u = \frac{4\,(E_{100} - E_0)}{3c\,(373^4 - 273^4)} = 7.061 \times 10^{-15}\,\mathrm{ergs/(cm^2 sec)}\,. \tag{A.2}$$

But u for $T = 1\,\mathrm{K}$ is obtained as

$$
\begin{aligned}
u &= \int_0^\infty U\,\mathrm{d}f \\
&= \frac{8\pi h}{c^3} \int_0^\infty \frac{f^3}{\mathrm{e}^{hf/k_\mathrm{B}} - 1}\,\mathrm{d}f \\
&= \frac{8\pi h}{c^3} \int_0^\infty f^3 \left(\mathrm{e}^{-hf/k_\mathrm{B}} + \mathrm{e}^{-2hf/k_\mathrm{B}} + \cdots \right)\,\mathrm{d}f \\
&= \frac{8\pi k_\mathrm{B}^4}{c^3 h^3} \left(1 + \frac{1}{2^4} + \frac{1}{3^4} + \cdots \right) \\
&= \frac{48\pi k_\mathrm{B}^4}{c^3 h^3} \times 1.0823\,.
\end{aligned} \tag{A.3}
$$

Comparison of Eqs. (A.2) and (A.3) gives

$$\frac{k_\mathrm{B}^4}{h^3} = 1.1682 \times 10^{15}\,. \tag{A.4}$$

Lummer and Pringsheim [3] measured $\lambda_\mathrm{m} T$, where λ_m is the wavelength for the maximum of E at the temperature $T\,\mathrm{K}$, as

$$\lambda_\mathrm{m} T = 0.2940\,\mathrm{cm.deg.K}\,. \tag{A.5}$$

Differentiating

$$E = \frac{8\pi hc}{\lambda^5}\frac{1}{\mathrm{e}^{hc/(k_\mathrm{B}\lambda T)} - 1} \tag{A.6}$$

with respect to λ and equating it to zero we get

$$\left(1 - \frac{hc}{5k_\mathrm{B}\lambda_\mathrm{m} T} \right) \mathrm{e}^{hc/(k_\mathrm{B}\lambda_\mathrm{m} T)} = 1 \tag{A.7}$$

DOI: 10.1201/9781003172178-A

which on simplification gives

$$\lambda_m T = \frac{hc}{4.9651 k_B} \tag{A.8}$$

and

$$\frac{h}{k_B} = \frac{4.9651 \times 0.294}{c} = 4.866 \times 10^{-11}\,\text{sec.deg.K}. \tag{A.9}$$

From Eqs. (A.4) and (A.9) we obtain

$$h = 6.55 \times 10^{-27}\,\text{erg.sec}, \tag{A.10a}$$
$$k_B = 1.346 \times 10^{-16}\,\text{erg/deg.K}. \tag{A.10b}$$

Bibliography

[1]   M. Planck, *Annalen der Physik* 4:553, 1901; also see pages 38–39 in I. Duck and E.C.G. Sudarshan, *100 Years of Planck's Quantum*. World Scientific, Singapore, 2000.

[2]   F. Kurlbaum, *Wied Ann.* 65:759, 1898.

[3]   O. Lummer and E. Pringsheim, *Verhandl. der Deutsch. Physikal. Gesellsch.* 2:176, 1900.

B

A Derivation of the Factor $h\nu/(e^{h\nu/k_{\mathrm{B}}T} - 1)$

Let us consider a box with n identical photons and an atom inside it [1]. The probability that the atom will emit a photon is $(n+1)|a|^2$, where $|a|^2$ is the probability the atom would emit if no photons were present. The probability that the atom would absorb a photon is $n|a|^2$.

Suppose we consider a cavity with N atoms and each atom has, say, two energy levels. We denote N_l and N_h be the average number of atoms in the lower and higher states, respectively. Then at temperature T the ratio N_h/N_l is given by

$$\frac{N_h}{N_l} = \frac{e^{-E_h/k_{\mathrm{B}}T}}{e^{-E_l/k_{\mathrm{B}}T}} = e^{-(E_h-E_l)/k_{\mathrm{B}}T} \,. \tag{B.1}$$

Now, each atom in the lower state can absorb a photon and jump to the higher state E_h. Like-wise each atom in the energy level E_h can go into E_l by emitting a photon. In equilibrium the rate of emission and absorption must be same. If $\bar{n}$ is the average number of photons in a given state then the rate of absorption and emission for photons are $N_l\bar{n}|a|^2$ and $N_h(\bar{n}+1)|a|^2$, respectively. Then $N_l\bar{n} = N_h(\bar{n}+1)$ or

$$\frac{N_h}{N_l} = \frac{\bar{n}}{(\bar{n}+1)} \,. \tag{B.2}$$

Using Eq. (B.1) for N_h/N_l we get

$$\frac{\bar{n}}{(\bar{n}+1)} = e^{-(E_h-E_l)/k_{\mathrm{B}}T} \,. \tag{B.3}$$

From the above equation we find

$$\bar{n} = \frac{1}{e^{(E_h-E_l)/k_{\mathrm{B}}T} - 1} \tag{B.4}$$

which is the average number of photons in any state with frequency ν. Since each photon has an energy $h\nu$, the average energy of the photons of frequency ν in the cavity is $\bar{n}h\nu$. That is,

$$\langle E \rangle = \bar{n}h\nu = \frac{h\nu}{e^{(E_h-E_l)/k_{\mathrm{B}}T} - 1} \,. \tag{B.5}$$

Bibliography

[1] G. Venkataraman, *Bose and His Statistics*. University Press, Hyderabad, 1992.

C

Bose's Derivation of Planck's Law

In this appendix we present Bose's derivation [1] of Planck's law without using the wave aspects of light quanta. Bose made the following three assumptions:

1. Light quanta are indistinguishable.

2. For light quanta the number of particles is conserved.

3. There exists a two-fold degeneracy for each momentum state.

Let a radiation of total energy is enclosed in a volume V and each light quanta is characterized by the number N and energy $h\nu_s$ ($s = 0$ to ∞). We write

$$E = \sum_s N_s h\nu_s = V \int \rho_\nu d\nu \,. \tag{C.1}$$

We need to find N_s. Bose reasoned that if one could state the probability for any distribution characterized by a set of N_s then it could be determined by the requirement that the probability be maximum satisfying (C.1).

The momenta of the quanta satisfy $p_x^2 + p_y^2 + p_z^2 = h^2\nu^2/c^2$. Then

$$\int dx dy dz dp_x dp_y dp_z = 4\pi \left(h^3 \nu^2/c^3 \right) V \, d\nu \,. \tag{C.2}$$

Equation (C.2) indicates that if the phase space is divided into cells of magnitude h^3 then the number of cells in the frequency range $d\nu$ is $4\pi V(\nu^2/c^3)d\nu$. In order to take into account the polarization the above quantity is multiplied by 2 then it is $8\pi V(\nu^2/c^3)d\nu$.

Now, define N^s be the number of quanta belonging to the frequency range $d\nu^s$. We want to find the number of different ways we can distribute these quanta over the cells belonging to the frequency range $d\nu^s$. This is given by $A^s!/(p_0^s!p_1^s!\ldots)$, where p_n^s denotes the number of cells with n quanta only and $A_s = (8\pi\nu^2/c^3)d\nu^s$. Then $N_s = \sum_{r=0}^{\infty} r p_r^s$. The probability of the state defined by all the p_r^s is

$$\prod \frac{A^s!}{p_0^s!p_1^s!\ldots} \,. \tag{C.3}$$

Defining $A^s = \sum_r p_r^s$ then for large p_r^s we have

$$\ln W = \sum_s A^s \ln A^s - \sum_s \sum_r p_r^s \ln p_r^s \,. \tag{C.4}$$

We want to make it maximum with $E = \sum_s N^s h\nu^s$. Performing the variation we have

$$\sum_s \sum_r \delta p_r^s \left(1 + \ln p_r^s \right) = 0 \,, \qquad \sum_s \delta N^s h\nu^s = 0 \,, \tag{C.5a}$$

$$\sum_r \delta p_r^s = 0 \,, \qquad \delta N^s = \sum_r r \delta p_r^s \,. \tag{C.5b}$$

DOI: 10.1201/9781003172178-C

We obtain

$$\sum_s \sum_r \delta p_r^s \left(1 + \ln p_r^s + \lambda^s\right) + \frac{1}{\beta} \sum_s h\nu^s \sum_r r\delta p_r^s = 0 \,. \tag{C.6}$$

p_r^s is found to be $p_r^s = B^s \mathrm{e}^{-rh\nu^s/\beta}$. Then

$$A^s = \sum_r B^s \mathrm{e}^{-rh\nu^s/\beta} = B^s \left[1 - \mathrm{e}^{-h\nu^s/\beta}\right]^{-1}, \quad B^s = A^s \left[1 - \mathrm{e}^{-h\nu^s/\beta}\right]. \tag{C.7}$$

Now,

$$N^s = \sum_r r p_r^s = \frac{A^s \mathrm{e}^{-h\nu^s/\beta}}{1 - \mathrm{e}^{-h\nu^s/\beta}} \,. \tag{C.8}$$

Then

$$E = \sum_s \frac{8\pi h(\nu^s)^3 \mathrm{d}\nu^3}{c^3} \frac{V\mathrm{e}^{-h\nu^s/\beta}}{1 - \mathrm{e}^{-h\nu^s/\beta}} \,, \tag{C.9}$$

$$S = k_{\mathrm{B}} \left[\frac{E}{\beta} - \sum_s A^s \ln\left(1 - \mathrm{e}^{h\nu^s/\beta}\right)\right] \,. \tag{C.10}$$

From the above we get $\beta = k_{\mathrm{B}}T$ because of the condition $\partial S/\partial E = 1/T$. Then

$$E = \sum_s \frac{8\pi h(\nu^s)^3}{c^3} V \left[\mathrm{e}^{h\nu^s/k_{\mathrm{B}}T} - 1\right]^{-1} \mathrm{d}\nu^s \tag{C.11}$$

which is equivalent to Planck's formula.

Bibliography

[1]   S.N. Bose, *Current Science* 66:943, 1994. This paper is an English translation of Bose's German article *Z. Phys.* 26:178, 1924.

D

Distinction Between Self-Adjoint and Hermitian Operators

As mentioned in sec. 3.2 an action and domain constitute an operator. The reason for the specification of domain is that in quantum mechanics the operators representing the observables should be Hermitian. Let us bring out the difference between self-adjoint and Hermitian operators with an example [1,2].

For simplicity we consider the one-dimensional free particle confined in the region $[0, \infty]$. Its Hamiltonian is $H = -\mathrm{d}^2/\mathrm{d}x^2$, where we have set $\hbar^2/(2m) = 1$. The particle is confined to the interval $0 \le x \le \infty$ and hence there must be an infinite barrier at $x = 0$. Hence, $\phi(0) = 0$. Suppose H acts on $\phi(x)$. We are interested in the functions $\phi(x)$ with

$$\int_0^\infty |\phi(x)|^2 \, \mathrm{d}x = \text{finite}, \quad \phi(x) \to 0 \ \text{as} \ x \to \infty \ . \tag{D.1}$$

Such a set of functions specify the domain (domain is the set of specified functions on which an operator acts) of the operator H and we call it as $\mathcal{D}$. $\mathcal{D}$ can be a subset of a domain $\mathcal{L}[0, \infty]$ which includes above square-integrable functions and other types of functions.

Find the adjoint of H. We obtain

$$
\begin{aligned}
(\phi_m, H\phi_n) &= -\int_0^\infty \phi_m^* \frac{\mathrm{d}^2}{\mathrm{d}x^2} \phi_n \, \mathrm{d}x \\
&= -\phi_m^*(0)\phi_n'(0) + \int_0^\infty {\phi_m^*}' \phi_n' \, \mathrm{d}x. \\
&= {\phi_m^*}'(0)\phi_n(0) - \phi_m^*(0)\phi_n'(0) - \int_0^\infty \phi_n \frac{\mathrm{d}^2}{\mathrm{d}x^2} \phi_m^* \, \mathrm{d}x.
\end{aligned}
\tag{D.2}
$$

That is,

$$
\begin{aligned}
(\phi_m, H\phi_n) &= {\phi_m^*}'(0)\phi_n(0) - \phi_m^*(0)\phi_n'(0) \\
&\quad + \int_0^\infty \left(H^\dagger \phi_m\right)^* \phi_n \, \mathrm{d}x \ ,
\end{aligned}
\tag{D.3}
$$

where $H^\dagger = -\mathrm{d}^2/\mathrm{d}x^2$. $H^\dagger$ is the adjoint of H provided the boundary expression in Eq. (D.3) is zero:

$$ {\phi_m^*}'(0)\phi_n(0) - \phi_m^*(0)\phi_n'(0) = 0 \ . \tag{D.4}$$

Now, consider the set of functions ϕ with $\phi(x) \to 0$ as $x \to \infty$ and $\phi(0) = 0$. Then $\phi_n(0) = 0$, $\phi_m^*(0) = 0$ and the Eq. (D.4) is satisfied so that

$$(\phi_m, H\phi_n) = \left(H^\dagger \phi_m, \phi_n\right) \ . \tag{D.5}$$

Hence, $H^\dagger = -\mathrm{d}^2/\mathrm{d}x^2$ is the adjoint of $H = -\mathrm{d}^2/\mathrm{d}x^2$. In Eq. (D.5) the action of H and $H^\dagger$ is defined on the set of functions with $\phi(0) = 0$. Thus, in Eq. (D.5) the action and the domain of the operator H on ϕ_n in left-side is equal to the action and the domain of $H^\dagger$

DOI: 10.1201/9781003172178-D

on ϕ_m in the right-side. Precisely, $H^\dagger = H$ and $\mathcal{D}(H) = \mathcal{D}(H^\dagger)$. An operator A with the property $\mathcal{D}(A) = \mathcal{D}(A^\dagger)$ is called a *self-adjoint operator*. For the set of functions ϕ with $\phi(0) = 0$ (and $\phi(x) \to 0$ as $x \to \infty$) H is a self-adjoint as shown above.

Next, we consider the same operator $H = -\mathrm{d}^2/\mathrm{d}x^2$ but with the conditions $\phi_n(0) = 0$, $\phi_n'(0) = 0$, $\phi_m(x)$, $\phi_n(x) \to 0$ as $x \to \pm\infty$. Is the operator H with the above conditions self-adjoint? The domain of H is the set of functions with $\phi(0) = 0$, $\phi'(0) = 0$. The action of $H^\dagger$ is on ϕ_m and the boundary expression (D.4) is satisfied for arbitrary values of $\phi_m^*(0)$ and $\phi_m^{*}{}'(0)$ still giving the result $H^\dagger = H$. Thus, the domain of $H^\dagger$ is larger than the domain of H and $\mathcal{D}(H) \subset \mathcal{D}(H^\dagger)$. Such an operator is called a *Hermitian operator*.

From the above discussion we define Hermitian and self-adjoint operators as follows [1,2]: Consider an operator with a dense domain $\mathcal{D}(A)$ and denote the domain of its adjoint $A^\dagger$ as $\mathcal{D}(A^\dagger)$ such that for all ϕ and $\psi \in \mathcal{D}(A)$ we have

$$\left(\phi, A^\dagger \psi\right) = \left(A^\dagger \phi, \psi\right) . \tag{D.6}$$

An operator A which has same action of $A^\dagger$ is said to be (i) *Hermitian if* $\mathcal{D}(A) \subset \mathcal{D}(A^\dagger)$ and (ii) *self-adjoint if* $\mathcal{D}(A) = \mathcal{D}(A^\dagger)$.

Bibliography

[1] G. Bonneau, J. Farant and G. Valent, *Am. J. Phys.* 69:322, 2001.

[2] V.S. Araujo, F.A.B. Coutinho and J. Fernando Perez, *Am. J. Phys.* 72:203, 2004.

E

Proof of Schwarz's Inequality

If ψ_1 and ψ_2 are two states then

$$||\psi_1|| \cdot ||\psi_2|| \geq \left| \int_{-\infty}^{\infty} \psi_1^* \psi_2 \, d\tau \right| . \tag{E.1}$$

To prove this we consider

$$||\psi_1||^2 \cdot ||\psi_2||^2 \geq \left| \int_{-\infty}^{\infty} \psi_1^* \psi_2 \, d\tau \right|^2 \tag{E.2}$$

which can also be written as

$$(\psi_1, \psi_1)(\psi_2, \psi_2) \geq \left| (\psi_1, \psi_2) \right|^2 . \tag{E.3}$$

For $\psi_1 = 0$, the inequality is obviously satisfied. For $\psi_1 \neq 0$, we decompose ψ_2 into a part parallel to ψ_1 and a part orthogonal to ψ_1:

$$\psi_2 = z\psi_1 + \chi \ \text{ with } \ (\psi_1, \chi) = 0 . \tag{E.4}$$

Then

$$(\psi_1, \psi_2) = z(\psi_1, \psi_1) \tag{E.5}$$

which gives

$$z = \frac{(\psi_1, \psi_2)}{(\psi_1, \psi_1)} . \tag{E.6}$$

Further,

$$
\begin{aligned}
(\psi_2, \psi_2) &= (z\psi_1 + \chi, z\psi_1 + \chi) \\
&= z^* z (\psi_1, \psi_1) + (\chi, \chi) \\
&\geq z^* z (\psi_1, \psi_1)
\end{aligned} \tag{E.7}
$$

because (χ, χ) is always positive. Substituting Eq. (E.6) in Eq. (E.7) we obtain

$$(\psi_2, \psi_2) \geq \frac{\left| (\psi_1, \psi_2) \right|^2}{(\psi_1, \psi_1)} . \tag{E.8}$$

That is,

$$(\psi_1, \psi_1)(\psi_2, \psi_2) \geq \left| (\psi_1, \psi_2) \right|^2 . \tag{E.9}$$

In other words,

$$||\psi_1||^2 \cdot ||\psi_2||^2 \geq \left| \int_{-\infty}^{\infty} \psi_1^* \psi_2 \, d\tau \right|^2$$

or

$$||\psi_1|| \cdot ||\psi_2|| \geq \left| \int_{-\infty}^{\infty} \psi_1^* \psi_2 \, d\tau \right| . \tag{E.10}$$

DOI: 10.1201/9781003172178-E

F

Eigenvalues of a Symmetric Tridiagonal Matrix – QL Method

Suppose we have a system of linear simultaneous equations of the form $(A - \lambda I)X = 0$ where A be a $N \times N$ matrix, λ is a parameter and X be a vector of dimension N. This linear system has a nontrivial solution if and only if the determinant of $A - \lambda I$ is zero: $\det(A - \lambda I) = 0$. This can be written as

$$
\begin{vmatrix}
a_{11} - \lambda & a_{12} & \cdots & a_{1N} \\
a_{21} & a_{22} - \lambda & \cdots & a_{2N} \\
& & \cdots & \\
a_{N1} & a_{N2} & \cdots & a_{NN} - \lambda
\end{vmatrix} = 0. \tag{F.1}
$$

Expanding the above determinant, one gets a polynomial of degree N:

$$
P(\lambda) = (-1)^N \left(\lambda^N + \alpha_1 \lambda^{N-1} + \ldots + \alpha_{N-1}\lambda + \alpha_N \right) . \tag{F.2}
$$

This equation will have N roots for λ. $(A - \lambda I)X = 0$ will have a nontrivial solution only for the N roots of λ. The λ's are called *eigenvalues* of A.

For a symmetric $N \times N$ matrix all the eigenvalues can be computed numerically, by first reducing it into a symmetric tridiagonal matrix using Householder method and then employing QL method to the reduced matrix [1,2]. Here, we describe the QL method for finding all the eigenvalues of symmetric tridiagonal matrix of the form

$$
A = \begin{pmatrix}
a_1 & d_1 & 0 & \cdots & & & & 0 \\
d_1 & a_2 & d_2 & \cdots & & & & 0 \\
0 & d_2 & a_3 & \cdots & & & & 0 \\
& & & \cdots & & & & \cdot \\
& & & & a_{N-2} & d_{N-2} & 0 & \\
& & & & d_{N-2} & a_{N-1} & d_{N-1} & \\
& & & & 0 & d_{N-1} & a_N &
\end{pmatrix} . \tag{F.3}
$$

In the QL method the matrix $A_1 = A$ is factorized into the form $A_1 = Q_1 L_1$, where $Q_1 = Q$ is an orthogonal matrix and $L_1 = L$ is a lower triangular matrix. Actually, a matrix P_{N-1} is constructed so that the element in the location $(N-1, N)$ in the matrix $P_{N-1}A$ is zero. That is,

$$
P_{N-1}A = \begin{pmatrix}
a_1 & d_1 & 0 & & & & \\
d_1 & a_2 & d_2 & & & & \\
& & & \cdots & & & \\
& & & & q_{N-2} & p_{N-1} & 0 \\
& & & & r_{N-2} & q_{N-1} & p_N
\end{pmatrix} . \tag{F.4}
$$

DOI: 10.1201/9781003172178-F

Similarly, with a suitable P_{N-2} the element in the position $(N-2, N-1)$ of $P_{N-1}A$ can be set into zero. Repeating this procedure $N-1$ times we obtain

$$P_1 P_2 \ldots P_{N-1} A = \begin{pmatrix} p_1 & 0 & 0 & & & & \\ q_1 & p_2 & 0 & & & & \\ & & & \cdots & & & \\ & & & r_{N-3} & q_{N-2} & p_{N-1} & 0 \\ & & & 0 & r_{N-2} & q_{N-1} & p_N \end{pmatrix}. \tag{F.5}$$

Now, we define a matrix Q as $Q_1 = P_{N-1}^T P_{N-2}^T \ldots P_1^T$. Then we form the product $A_2 = L_1 Q_1$. We note that the matrix Q_1 is orthogonal. Therefore, $Q_1^T A_1 = Q_1^T Q_1 L_1 = L_1$. Thus, $A_2 = Q_1^T A_1 Q_1 = Q_1^{-1} A_1 Q_1$. A_2 is similar to A_1 and have the same eigenvalues. In general, $A_k = Q_k L_k$ and $A_{k+1} = Q_k^T A_k Q_k$. A_{k+1} is similar to A_k and hence have the same structure.

Next, consider the case of setting the element A_{pq} and A_{qp} to zero. In this case P_1 has the form

$$P_1 = \begin{pmatrix} 1 & \cdots & & & \\ & c & \cdots & s & \\ & & \cdots & & \\ & -s & \cdots & c & \\ & & & & 1 \end{pmatrix}. \tag{F.6}$$

Note that all the diagonal elements of P_1 are 1 except for the element c. The c appears at two places. All off-diagonal elements of P are 0 except for the elements s which appears at two places.

The tridiagonal form of A_2 indicates that it also has zeros below the lower diagonal. From a detailed calculation one can find that the terms r_j are used only to compute these zero elements. The point is that in writing a computer program the r_j's need not be stored. For each P_j it is sufficient to store the coefficients s_j and c_j. Moreover, we need not compute and store Q explicitly. In fact, we can use s_j's and c_j's to find the product

$$A_2 = LQ = L P_{N-1}^T P_{N-2}^T \ldots P_1^T . \tag{F.7}$$

We can speed up the process employing the *shifting technique*. Here, we can make use of the fact that if λ_j is an eigenvalue of A then $\lambda_j - s_i$ is an eigenvalue of $A - s_i I$. This results in

$$A_i - s_i I = Q_i L_i, \quad A_{i+1} = L_i Q_i, \quad \text{for } i = 1, 2, \ldots, k_j \tag{F.8}$$

with $\lambda_j = s_1 + s_2 + \ldots + s_{k_j}$.

We can determine the correct shift required at each step using the four elements in the upper left corner of the matrix. For example, for the matrix $\begin{pmatrix} a_1 & d_1 \\ d_1 & a_2 \end{pmatrix}$ the eigenvalues x_1 and x_2 are the roots of the equation $x^2 - (a_1 + a_2) x + a_1 a_2 - d_1^2 = 0$. The root that is closer to a_1 is the value of s_i in Eq. (F.8). The QL iteration with shifting is repeated until $d_1 \approx 0$ which produce $\lambda_1 = s_1 + s_2 + \ldots + s_{k_1}$. Repeating the process with the lower $(n-1)$ rows give $d_2 \approx 0$ and the next eigenvalue λ_2. Next iteration gives $d_{n-2} \approx 0$ and the eigenvalue λ_{n-2}. The quadratic formula can be used to find the last two eigenvalues.

Example:

We find the eigenvalues of the matrix $A = \begin{pmatrix} 1 & -1 & 0 \\ -1 & 2 & -1 \\ 0 & -1 & 1 \end{pmatrix}$ employing the QL method. The exact eigenvalues of the above matrix are $\lambda = 0, 1, 3$.

For the given matrix $a_1 = 1$, $a_2 = 2$, $d_1 = -1$ and the quadratic equation is $x^2 - 3x + 1 = 0$. Its roots are $x_1 = 2.61803$ and $x_2 = 0.38197$. The root close to $a_1 = 1$ is x_2. The required shift is $s_1 = 0.38197$. Then the first shifted matrix is obtained as

$$A_1 - s_1I = \begin{pmatrix} 0.61803 & -1 & 0 \\ -1 & 1.61803 & -1 \\ 0 & -1 & 0.61803 \end{pmatrix}. \tag{F.9}$$

The LQ matrix is computed as

$$A_2 = L_1Q_1 = \begin{pmatrix} 0 & 0.52573 & 0 \\ 0.52573 & 0.61803 & -0.85065 \\ 0 & -0.85065 & 2.23607 \end{pmatrix}. \tag{F.10}$$

$|d_1| = 0.52573$ in A is not less than the preassumed small value, say, $\delta = 10^{-5}$. Therefore, we perform second iteration with A_2. Now, $a_1 = 0$, $a_2 = 0.61803$, $d_1 = 0.52573$ and the roots of the quadratic equation are $x_1 = 0.91884$, $x_2 = -0.30081$. The second shift is worked out as $s_2 = -0.30081$. The second shifted matrix is

$$A_2 - s_2I = \begin{pmatrix} 0.30081 & 0.52573 & 0 \\ 0.52573 & 0.91884 & -0.85065 \\ 0 & -0.85065 & 2.53687 \end{pmatrix}. \tag{F.11}$$

Then we obtain the LQ matrix as

$$A_3 = L_2Q_2 = \begin{pmatrix} -0.07668 & -0.06711 & 0 \\ -0.06711 & 0.94706 & -0.25379 \\ 0 & -0.25379 & 2.88615 \end{pmatrix}. \tag{F.12}$$

In the third iteration the shift is $s_3 = -0.08107$. Then we get

$$A_4 = L_3Q_3 = \begin{pmatrix} -0.00009 & 0.00001 & 0 \\ 0.00001 & 1.00358 & -0.08565 \\ 0 & -0.08565 & 2.99623 \end{pmatrix}. \tag{F.13}$$

The next shift s_4 is -0.00009. The LQ matrix is obtained as

$$A_5 = L_4Q_4 = \begin{pmatrix} 0 & 0 & 0 \\ 0 & 1.00041 & -0.02860 \\ 0 & -0.02860 & 2.99959 \end{pmatrix}. \tag{F.14}$$

In the next iteration $s_5 = 0$ and hence an eigenvalue is obtained as $\lambda_1 = s_1 + s_2 + s_3 + s_4 + s_5 = 0$. Next, we replace the first diagonal element in A_5 by λ_1 and write

$$A_5 = \begin{pmatrix} 0 & 0 & 0 \\ 0 & 1.00041 & -0.02860 \\ 0 & -0.02860 & 2.99959 \end{pmatrix}. \tag{F.15}$$

The LQ matrix is

$$A_6 = L_5Q_5 = \begin{pmatrix} 0 & 0 & 0 \\ 0 & 1.00005 & -0.00953 \\ 0 & -0.00953 & 2.99995 \end{pmatrix}. \tag{F.16}$$

The last two eigenvalues are determined using the 2×2 right-side corner matrix. The roots of the quadratic equation are $x_1 = 3$ and $x_2 = 1$. Then $\lambda_2 = \lambda_1 + x_1 = 3$ and $\lambda_3 = \lambda_1 + x_2 = 1$. Therefore, the eigenvalues are 0, 1, 3.

Bibliography

[1]   W.H. Press, S.A. Teukolsky, W.L. Vellerling and B.P. Flannery, *Numerical Recipes in Fortran*. Cambridge University Press, Cambridge, 1998.

[2]   J.H. Mathews, *Numerical Methods for Mathematics, Science and Engineering*. Printice Hall of India, New Delhi, 1998.

G

Random Number Generators for Desired Distributions

A sequence of numbers generated by a random physical process is called *random numbers*. Physical processes such as radioactivity decay, thermal noise in electronic devices, cosmic ray arrival times etc., give rise to a sequence of random numbers. Set of random numbers can be generated employing simple iterative algorithms. Such numbers are called *pseudo random numbers* because they are predictable and reproducible.

A. Uniform Random Number Generator

Consider random numbers in the interval say $[\alpha, \beta]$. Suppose we divide the interval $[\alpha, \beta]$ into N subintervals with $N \to \infty$. If the random numbers are distributed with equal probability in each interval then they are called *uniform random numbers*. In this case the probability for the random numbers taking values between x and $x + dx$ is $dx/(\beta - \alpha)$. The mean and variance of uniform random numbers are $x_{\mathrm{m}} = (\alpha + \beta)/2$ and $\sigma^2 = (\beta - \alpha)^2/12$. For a uniform random numbers in the interval $[0, 1]$ we have $x_{\mathrm{m}} = 0.5$ and $\sigma^2 = 1/12$.

Computers themselves have a library routine usually with the name like *rnd* or *ran* or *rand* to generate uniformly distributed random numbers in the interval $[0, 1]$. Such a routine uses the recurrence relation

$$I_{j+1} = aI_j + c, \quad \mathrm{mod}(m), \quad j \geq 1 \tag{G.1}$$

where m is the modulus with $m > 0$; a is the multiplier with $0 < a < m$; c is the increment with $0 \leq c \leq m$; I_1 is the starting value with $0 \leq I_1 < m$. We start with an integer I_1 and calculate $aI_1 + c$, which is also an integer. We divide it by m, obtain the remainder and denote it as I_2. Repeating the above procedure we can generate a sequence of integers $\{I_1, I_2, \ldots\}$. The sequence obtained with $m = 1000$, $I_1 = 10$, $a = c = 89$ is 979, 220, 669, 630, 159, ... Such a sequence generated from Eq. (G.1) will repeat itself with a period less than m. If we divide the obtained numbers by m we get a sequence in the interval $[0, 1]$. We can convert it into any suitable range $[\alpha, \beta]$ through $x_i \to (\beta - \alpha)x_i + \alpha$.

Park and Miller proposed a generator based on the choices [1]

$$a = 7^5, \ c = 0, \ m = 2^{31} - 1 = 2147483647 \,. \tag{G.2}$$

When the value of aI_j is greater than the maximum value of, say, a 32-bit integer then we cannot implement the Eqs. (G.1)-(G.2) in a computer. To overcome this problem the following suggestion of Schrage is very helpful. From the approximate factorization of m given by

$$m = aq + r \,, \quad q = \mathrm{int.}[m/a], \ r = m \,(\mathrm{mod} \ a) \,, \tag{G.3}$$

where int denotes integer part. For $r < q$ and $0 < I < m - 1$ both $aI(\mathrm{mod} \ q)$ and $r.\mathrm{int.}[I/q]$ lie in the range $[0, m - 1]$. We can write

$$aI \,(\mathrm{mod} \ m) = \begin{cases} aI \,(\mathrm{mod} \ q) - r.\mathrm{int.}[I/q], & \text{if it is } > 0 \\ aI \,(\mathrm{mod} \ q) - r.\mathrm{int.}[I/q] + m, & \text{otherwise.} \end{cases} \tag{G.4}$$

DOI: 10.1201/9781003172178-G

With

$$P_1 = I \,(\text{mod } q), \quad P_2 = \text{int.}[I/q] \tag{G.5}$$

we rewrite the Eq. (G.4) as

$$aI \,(\text{mod } a) = \begin{cases} aP_1 - rP_2, & \text{if it is } > 0 \\ aP_1 - rP_2 + m, & \text{otherwise.} \end{cases} \tag{G.6}$$

Some useful choices of a, q and r are given below: (i) $a = 7^5$, $q = 127773$, $r = 2836$, (ii) $a = 48271$, $q = 44488$, $r = 3399$ and (iii) $a = 69621$, $q = 30845$, $r = 23702$. The period of the numbers generated using the above procedure is $2^{31} - 2 = 2147483646$. The above procedure generates numbers in the interval $[0, m]$. Note that the random numbers are now integers. The numbers generated can be converted into the range $[0, 1]$ by dividing them by m.

The following steps generate N random numbers in the interval $[0, 1]$.

```
M = 2147483647
A = 7**5
R = 2836
Q = 127773
I = 3
DO J = 1, N
      P1 = MOD(I,Q)
      P2 = INT(I/Q)
      I = A*P1 - R*P2
      IF(I.LE.0) I = I + M
      RAN = FLOAT(I)/FLOAT(M)
END DO
```

Figure G.1a depicts first 100 random numbers of the sequence generated. The mean and variance obtained with 10^4 numbers are 0.5017 and 0.0833, respectively, whereas the exact values are 0.5 and 0.0833..., respectively.

B. Generation of Random Numbers with a Desired Distribution

Using uniform random numbers we can generate random numbers of other types of distribution like Gaussian, exponential, binomial, etc. This can be achieved by employing the methods such as direct inversion, rejection method and Metropolis. We describe the rejection technique [2].

Suppose $f(x)$ be the desired distribution defined over the interval $[\alpha, \beta]$. We select a suitable bounding function $g(x)$ such that $cg(x) \geq f(x)$ for all values of $\alpha \leq x \leq \beta$. For the exponential function $f(x) = e^{-x^2}$ for $\alpha \leq x \leq \beta$ and 0 otherwise we can choose the range $[\alpha, \beta]$ as $[-5, 5]$. Since $f(x) \leq 1$ for $x \in [\alpha, \beta]$ we have $cg(x) \geq 1$. We set $c = 1$ so that $g(x) \in [0, 1]$. We choose $g(x)$ and x as uniformly distributed random numbers in the intervals $[0, 1]$ and $[-5, 5]$, respectively. The scheme is the following.

1. Choose a value of x randomly in the interval $[-5, 5]$ and calculate $f(x)$.

2. Choose a value of $cg(x)$, the value of $g(x)$ is between 0 and 1, and call it y.

3. If $y > f(x)$ then go to step (1) otherwise x is a member of the chosen distribution number.

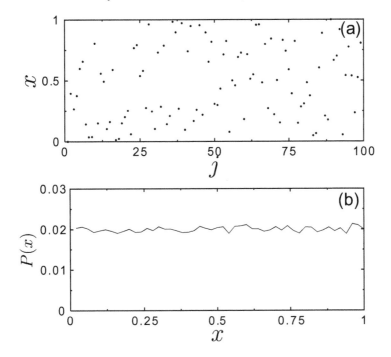

FIGURE G.1
(a) Plot of first 100 uniform random numbers (x) generated. (b) Computed probability distribution of 5×10^4 numbers. The interval $[0, 1]$ is divided into 50 subintervals. $P(x)$ is almost constant.

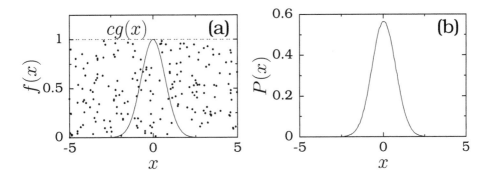

FIGURE G.2
(a) Description of rejection technique. The x-component of a point falling below the curve $f(x)$ is a number with the chosen exponential distribution. (b) Probability distribution (unnormalized) of 5×10^5 random numbers generated.

We can obtain a sequence of random numbers with the distribution $f(x)$ by repeating the above steps. The procedure is simply the following. Consider the Fig. G.2a. The dots are the points with coordinate values (x_i, y_i) with $x_i \in [-5, 5]$ and $y_i \in [0, cg]$ $(= [0, 1])$. Discard the points falling outside the curve $f(x)$. The x values of the points lying within the curve form random numbers with the distribution e^{-x^2}. Here $cg(x)$ not necessarily defined in $[0, 1]$. It can be $[0, 1.5]$ or $[0, 2]$ or $[0, > 1]$. The number of points discarded to generate N numbers with a required distribution depends on the choice of $cg(x)$. The number of discarded points will be minimum if the maximum value of $cg(x)$ is chosen as the maximum value of $f(x)$. Figure G.2b shows the probability distribution function $P(x)$ of generated random numbers.

Bibliography

[1] W.H. Press, S.A. Teukolsky, W.T. Vetterling and B.P. Flannery, *Numerical Receipes in Fortran*. Foundation Books, New Delhi, 1998.

[2] K.P.N. Murthy, *Monte Carlo Basics*. Indian Society for Radiation Physics, Kalpakkam, 2000.

Solutions to Selected Exercises

1.2 $3.106\,\text{Å}$.

1.6 $1674\,\text{nm}$.

1.8 $2.7756 \times 10^{-19}\,\text{J}$.

1.10 $24847.5\,\text{V}$.

1.12 $\lambda = 2481.37\,\text{Å}$.

1.14 130797×10^{18} photons.

1.16 $\lambda' = 0.76224\,\text{Å}$ and $E = 2.60784 \times 10^{-15}\,\text{J}$.

1.18   3.

1.20   From $n\lambda = 2a\sin\theta$ with $n = 1$ $\lambda = 1.64948\,\text{Å}$ and from $\lambda = h/(mv)$
       $\lambda = 1.67093\,\text{Å}$.

2.2 $i\hbar\partial\psi/\partial t = -(\hbar^2/2m)\nabla^2\psi - (q\mathbf{E}(t) \cdot \mathbf{X})\psi$.

2.6   Not equal to $E_1 + E_2$.

2.8 $V(\mathbf{X}, t) = V^*(\mathbf{X}, -t)$.

2.14 $u_x - u^2 - \lambda + g = 0$, $\lambda = 2mE/\hbar^2$ and $g = 2mV/\hbar^2$.

2.18 $1/(\sigma\sqrt{\pi})^{1/2}$.

2.20 $\sqrt{2(1+\alpha^2)/\alpha^3}$.

2.22 $1 - e^{-b}\left(b^2 + 2b + 2\right)/4$.

2.26   0.

2.28 $3\hbar\omega/2$.

2.30 $1/2$.

2.34 $\hbar\pi/(E_2 - E_1)$.

2.38 $5\hbar^2/(mL^2)$.

2.44 $\dfrac{\sqrt{3}}{2}\phi_1 + \dfrac{1}{2}\phi_2$.

2.48 $\sum_n |C_n|^2 E_n$.

2.50 $n^2\hbar^2\pi^2/(2ma^2)$.

3.2 $x^6 + 10x^4 + 25x^2$.

3.6   Not linear.

3.8   To shift t by $t + \tau$.

3.12 $y = A\sin nx$ and $\lambda = n^2$, $n = 1, 2, \ldots$.

3.16 $\psi(x) = Ne^{ikx}$, $E = V_0 + (h^2k^2/(2m))$, $k^2 = 2m\left(E - V_0\right)/\hbar^2$.

3.18   (a) $f(0)$. (b) $\hat{I}^2$.

3.22 $[x^n, p_x] = \mathrm{i}\hbar n x^{n-1}$ and $[x, p_x^n] = \mathrm{i}\hbar n p_x^{n-1}$.

3.26 (a) $\mathrm{i}\hbar$. (b) 0.

3.30 $[H, x] = -\mathrm{i}\hbar p_x/m$ and $[H, p_x] = 0$.

3.34 0.

3.38 No.

3.46 To change A to $A' = \mathcal{T} A \mathcal{T}^{-1}$.

4.2 $\psi(x) = A\mathrm{e}^{\pm \mathrm{i}k'x}$, $k'^2 = (2m/\hbar^2)(E - a)$ and $E = a + (\hbar^2 k'^2/(2m))$.

4.6 $\langle x \rangle = \langle p_x \rangle = 0$.

4.8 $\langle E \rangle = \hbar\omega$.

4.10 0.888.

4.14 $E_n = \left(n + \dfrac{1}{2}\right)\hbar\omega - \dfrac{e^2 E^2}{2m\omega^2}$.

4.18 $a = -1/(2b)$, $b = \sqrt{mk}/\hbar$, $N = (4b^5/\pi)^{1/4}$ and $E = 5\hbar\omega/2$.

4.22 $E_1 = 1.50574 \times 10^{-20}$ Js.

4.26 $\langle x \rangle = \langle p_x \rangle = 0$, $\langle x^2 \rangle = a^2\left(n^2\pi^2 - 6\right)/(3n^2\pi^2)$, $\langle p_x^2 \rangle = \hbar^2 n^2 \pi^2/(4a^2)$.

4.30 $\langle E \rangle = 5\hbar^2/(4ma^2)$.

4.32 $c/(2a)$.

4.34 $P_1(0 < x < 0.5) = P_2(0 < x < 0.5) = 0.5$.

4.38 (a) $P = 64/(9\pi^2)$.

4.42 $\psi(t) = C\sqrt{\pi/2}\left(1 - \cos 2\phi\, \mathrm{e}^{-\mathrm{i}E_2 t/\hbar}\right)$.

4.44 $\triangle l = \pm 1$ and $\triangle m = 0, \pm 1$.

4.46 $P_{\mathrm{CM}} \propto 1/x$ and $P_{\mathrm{QM}}(y) \propto \dfrac{\sin^2(y^2/2)}{y}$, where $x = \sqrt{\hbar/(m\omega)}\, y$.

4.54 $\ddot{\eta} - \eta\dot{\gamma}^2 + d\dot{\eta} + \omega_0^2\eta = 0$, $\eta\ddot{\gamma} + 2\dot{\eta}\dot{\gamma} + d\eta\dot{\gamma} = 0$.

4.60 $C_n = 8\sqrt{15}/(n^3\pi^3)$ for odd n and 0 for even n.

5.2 $T_{\min} = 1 - \dfrac{V_0^2}{(2E - V_0)^2}$.

5.4 $T = \left[1 + (7/3)\sinh^2(3/2)\right]^{-1}$.

5.6 $T = [16/(4 + 3.2)]\,\mathrm{e}^{-160\sqrt{20}/\hbar}$.

5.10 $J_{\mathrm{inc}} = \dfrac{\hbar}{m}\alpha|A|^2$ and $J_{\mathrm{tra}} = \dfrac{\hbar}{m}\alpha|F|^2$.

5.12 $T = \dfrac{1}{1 + 4(E/V_0)((E/V_0) - 1)\sin^2(2\sqrt{2}\,\sqrt{mE}\,a/\hbar)}$.

5.14 $E = 9V_0/8$.

5.20 $E = mV_0^2/(2\hbar)$.

6.2 $\langle \chi|A|\phi \rangle = \sum_i \sum_j D_i^* C_j \langle i|A|j \rangle$.

6.4 Normalized.

6.6 (b) $\dfrac{1}{\sqrt{2}}\begin{bmatrix} 1 & 1 \\ 1 & -1 \end{bmatrix}$. (c) $2|v_1\rangle - 3|v_2\rangle$ and $-\dfrac{1}{\sqrt{2}}|v_1'\rangle + \dfrac{5}{\sqrt{2}}|v_2'\rangle$.

6.8 $\partial U/\partial t = (1/\mathrm{i}\hbar)HU$.

6.10 $[C]$.

6.12 $A = A^\dagger = A^{-1}$.

6.14 $\lambda^* B |\psi_2\rangle \langle \psi_1| A^\dagger$.

6.16 $|c_i|^2$.

6.18 $\begin{bmatrix} e^\theta & 0 \\ 0 & e^{-\theta} \end{bmatrix}$.

6.28 $[L_q, R_p] = 0$.

7.2 $\dot{f} A + f \dot{A}$.

7.4 (a) $\langle x p_x + p_x x \rangle / m$. (b) $\langle p_x^2 \rangle / m + \langle x F \rangle$.

7.6 (a) $P_x + X$. (b) $-P_x - X$.

7.8 $\dot{x} = f + p_x$ and $\dot{p}_x = -g(t)$.

7.10 $e^{-iHt/\hbar} \dfrac{dA_H}{dt} e^{iHt/\hbar}$.

7.12 $\dfrac{d}{dt} x_{\text{int}} = \dfrac{1}{m} p_{x\text{int}}$ and $\dfrac{d}{dt} p_{x\text{int}} = 0$.

7.14 $da/dt = -i\omega a$ and $da^\dagger/dt = i\omega a^\dagger$.

7.16 (a) p_x/m, (b) $q E_x$ and (c) $q\mathbf{E}$.

7.18 $x(0) \cos \omega t + (1/m\omega) p_x(0) \sin \omega t$.

7.20 $\dfrac{3}{4} |0\rangle\langle 0| + \dfrac{1}{4} |1\rangle\langle 1|$.

8.2 5.7748×10^{-12} m.

8.4 ≈ 0.01157536 m.

8.6 $v/(4\pi)$.

8.8 h.

8.10 $\hbar^2/(me^2)$.

8.12 0.5×10^{11} Hz.

8.14 $\Delta E = 6.59 \times 10^{-9}$ eV and $\Delta\nu = 159.1475 \times 10^4$ cycles/sec.

8.16 $\dfrac{\hbar\pi}{2a} \sqrt{\dfrac{1}{3} - \dfrac{2}{\pi^2}}$.

8.20 $\approx -e^2/(8\pi\epsilon_0 a_0)$.

8.24 $\Delta T = \langle T^2 \rangle - \langle T \rangle^2 = 0$.

9.2 It is diagonal with the continuum eigenvalues.

9.4 0.

9.6 $\dfrac{A\sqrt{2}}{\sqrt{\pi}} \dfrac{\sin ka}{k}$.

9.8 $\dfrac{1}{\sqrt{\pi L}} \dfrac{\sin(k - k_0)L}{(k - k_0)}$.

9.10 $\dfrac{(2r_0)^{3/2} r_0}{\pi(r^2 + r_0^2)^2}$, $r_0 = \hbar/\Delta p$.

9.12 $(\sigma/\sqrt{\pi})^{1/2} e^{-\sigma^2 k^2/2}$.

9.14 $\dfrac{i\sqrt{2}\,Ax}{\sqrt{\pi}\,(\alpha^2 + x^2)}.$

10.2 $v_{\mathrm{p}} = 0.5v.$

10.4 $\hbar/2.$

10.6 $|\psi(x,t)|^2 = \dfrac{1}{\sqrt{\pi}\sigma(t)}\,\exp\left\{-\left[x - (x_0 + \dfrac{p_0 t}{m} + \dfrac{Ft^2}{2m})\right]^2 \Big/ \sigma^2(t)\right\}.$

11.6 0 and $i\hbar z.$

11.8 0.

11.10 $2i\hbar\mathbf{r}.$

11.12 0.

11.14 0, 1 and $-1.$

11.16 (a) 0, 1, 2 and 3. (b) 27.

11.18 (a) $(2j_1 + 1)(2j_2 + 1)\ldots(2j_N + 1).$ (b) 105.

11.20 (a) $E_{lm} = \dfrac{l(l+1)\hbar^2}{2I_1} + \left(\dfrac{1}{2I_2} - \dfrac{1}{2I_1}\right)m^2\hbar^2.$ (b) $Y_{lm}.$

11.22 (a) 2/3. (b) $(\sqrt{2}+1)/\sqrt{6})^2.$

11.26 $2\hbar^2[l(l+1) - m^2].$

11.28 0 or $\sqrt{2}\,\hbar$ or $-\sqrt{2}\,\hbar.$

11.34 $\langle J_z\rangle = \hbar/3$, $\langle J_z^2\rangle = 2\hbar^2/3$ and $\Delta J_z = \sqrt{5/9}\,\hbar.$

12.2 0.3233.

12.4 $e^{-\alpha r}$, $\alpha = \sqrt{2\mu E/\hbar^2}.$

12.6 $|R_{nl}|^2 r^2.$

12.8 (a) 1/2, (b) 1/2 and (c) 2/3.

12.10 $3a_0/2.$

12.12 The electron is located near the surface of a sphere of radius $n^2 a_0.$

12.14 $E_5 = -\mu e^4/(50\hbar^2)$ and $r_\pm = [25\hbar^2 \pm 5\sqrt{5}\,\hbar]/(\mu e^2).$

12.16 $\langle\mathbf{r}\rangle = 0.$

12.18 $E = \dfrac{me^4}{\hbar^2\left[1 + 2b + \sqrt{1+4b}\right]}$, $b = \dfrac{2ma}{\hbar^2}.$

12.20 $\alpha > -1/2.$

13.2 $\left|1 - (1/2)\sum_{r\neq n}|H_{rn}^{(1)}|^2 / \left(E_r^{(0)} - E_n^{(0)}\right)^2\right|^2.$

13.4 Both $\mu^2\hbar/(m^3\omega^5)$ and $(30n^2 + 30n + 11)/4(2n+1)$ to be small.

13.6 $-\hbar/2.$

13.8 $aL^2(6\pi^2 - 1)/(18\pi^2).$

13.10 $eEL/2.$

13.12 $3eE_{\mathrm{e}}\hbar^2/(2ma).$

13.14 $8V_0 n^2/(\pi(4n^2 - 1)).$

13.16 $\lambda L^2/4$.

13.18 $-3e^2(a_0 - \rho_0)/(2a_0\rho_0)$.

13.20 $E_1^{(1)} = -\alpha/a_0$ and $E_2^{(1)} = -\alpha/4a_0$.

13.22   (a) $E_{1,2} = (1/2)\left[E_1^{(0)} + E_2^{(0)} \pm \sqrt{\left(E_1^{(0)} - E_2^{(0)}\right)^2 + 4\lambda^2\Delta^2}\right]$.

13.24 $E_{lm}^{(1)} = 0$ and $E_{10}^{(2)} = 0$, $E_{11}^{(2)} = -E_{1-1}^{(2)} = C^2\hbar^2/(2B)$.

14.2   The selection rules are $f = i \pm 2$.

14.4   The selection rules are $f = i \pm 2,\ i \pm 4$.

14.6 $[F_0^2\pi^2/(2m\hbar\omega^3)]\,\mathrm{e}^{-2\omega\tau}$.

14.8 $[256V_0^2/(225\hbar^2\pi^2\omega_{31}^2)]\sin^2(\omega_{31}T/2)$, $\omega_{31} = 4\pi^2\hbar/(mL^2)$.

14.10 $[m^2L^6V_0^2/(16\pi^8\hbar^4)]\sin^2(\omega_{31}T/2)$, $\omega_{31} = 4\pi^2\hbar/(mL^2)$.

14.12   0.

14.14 $[e^2E_0^2/(2m\omega_0\hbar)]\delta_{n1}$.

14.16 $[\alpha^2/(2m\omega\hbar(\mu^2 + \omega^2))]$.

14.18 $\left[\alpha\sqrt{\pi}\mathrm{e}^{-\omega^2/\mu^2}/(\sqrt{2}\,m\omega\mu)\right]^2$.

14.20 $8(1 + b)/(2 + b)^4$.

14.22 $\left|a_f^>\right|^2 = \mathrm{e}^{-a^2/2}$.

14.24 $\left|a_{100,100}^>\right|^2 = \dfrac{64[Z(Z + 1)]^3}{(2Z + 1)^6}$.

15.2 $\dfrac{1}{2m}|\nabla S|^2 + V + \dfrac{\partial S}{\partial t} = 0$.

15.4 $1/(2\hbar\nu)$, ν-classical frequency of the motion of the particle.

15.6 $E_0 = [3/(16\sqrt{2})]^{2/3}V_0$, $E_1 = 3^{2/3}E_0$, $E_2 = 5^{2/3}E_0$ and $E_3 = 7^{2/3}E_0$.

15.8 $-[3/(16m^3c^2)](\hbar m\omega)^2(n + 1/2)^2$.

15.10 $n^2\hbar^2\pi^2/(2mL^2)$.

16.2 $\dfrac{(\alpha + 1)(2\alpha + 1)}{2\alpha - 1}\left(\dfrac{\hbar^2}{4mL^2}\right)$, $\alpha \approx 1.72$.

16.4 $1.082(\hbar^2/(2m))^{2/3}\lambda^{1/3}$.

16.6 $\sqrt{3}\,\hbar\omega$.

16.8 $-V_0/32$.

16.10 $-me^4/(3\pi\hbar^2)$.

17.2   (a) $\theta_\mathrm{L}^{\max} = \sin^{-1}(m_2/m_1)$. (b) θ_L can take any angle.

17.4 $\approx 1 + 2\rho$.

17.6 $-4\pi\delta(|\mathbf{r} - \mathbf{r}'|)$.

17.8 $(4m^2V_0^2a^6/\hbar^4)\,\mathrm{e}^{-2s^2a^2}$.

17.10 $16\pi m^2a^4V_0^2/\hbar^4$.

17.12 $\sigma = 4\pi a^2$.

18.2 P_{ij} is a unitary operator.

18.4 $[P_{ij}, P_{il}] \neq 0$.

18.6   (a) Symmetric wave function. (b) Antisymmetric wave function.

18.8 $\epsilon/2$, $5\epsilon/4$ and $5\epsilon/2$, where $\epsilon = \hbar^2\pi^2/(2ma^2)$.

19.2 $\hbar\omega = (c^2\hbar^2k^2 + m^2c^4)^{1/2}$.

19.4 $c\alpha$.

19.6 $\partial^\mu J_\mu = 0$, where J_μ is the four-vector.

19.8   0.

19.10 $d\mathbf{X}/dt = c\alpha$ and $d\mathbf{p}/dt = e\nabla\phi - e\nabla(\alpha \cdot \mathbf{A})$.

19.12   8 components column vector.

19.14 $2mc^2$.

19.16 $\beta(t) = (mc^2/H) + \beta'(0)e^{2iHt/\hbar}$ and $\langle\beta(t)\rangle = (1 - (c^2\mathbf{p}^2/E^2))^{1/2}$.

19.18 $[H, \hbar k] = 0$.

19.20   The Dirac equation is unchanged.

20.2   If there are no objects in certain category.

20.4 $-\cos(\theta_1 - \theta_2)$.

20.8 $C_{\mathrm{QM}} = -\cos(\theta_1 - \theta_2)$.

21.2 $H|0\rangle = \frac{1}{\sqrt{2}}(|0\rangle + |1\rangle)$ and $H|1\rangle = \frac{1}{\sqrt{2}}(|0\rangle - |1\rangle)$.

21.4   For the particle state $V = 0$ and for the wave state $V = 1$.

22.2 $I_\alpha = 2D_\alpha\hbar^{\alpha-1}\left(k_1k_2^{\alpha-1} - k_1^{\alpha-1}k_2\right)\sin\left[(k_2 - k_1)x - (E_2 - E_1)t/\hbar\right]$.

22.4 $iT_P\dfrac{\partial\psi}{\partial t} = -\dfrac{L_P^2}{2N_m}\dfrac{\partial^2\psi}{\partial x^2} + N_V\psi$.

22.6 $N = 1/\sqrt{3}$ and $\langle H_\alpha\rangle = \langle T_\alpha\rangle + \langle V\rangle = D_\alpha(\hbar\pi)^\alpha$.

22.8 $\psi(\omega,t) = \psi_0(\omega + at)\exp\left[i\dfrac{\omega|\omega|^\alpha - (\omega + at)|\omega + at|^\alpha}{2a(1+\alpha)}\right]$.

22.10   The value of the integral is $\dfrac{V_0}{\sqrt{2}\,\pi\hbar c}\ln\left[\dfrac{\sqrt{2}\,p_0c + |E|}{\sqrt{2}\,mc^2 + |E|}\right]$.

22.14 $i^\beta\hbar_\beta\dfrac{\partial^\beta}{\partial t^\beta}\phi(p,t) = D_{\alpha,\beta}|p|^\alpha\phi(p,t) + \int dp'\phi(p',t)U_{p,p'}$.

23.2 $E = -0.12500, -0.05556, -0.03125, -0.02000$ and -0.01389.

23.4 $E = 4.92822, 19.71288, 44.35400$ and 78.85155.

23.6   (a) $E = 1.5, 2.5, 3.5, 4.50001$ and 5.50000.

23.8 $E = 1.0189$ and 2.33765 for $\phi(x, \tau = 0) = \cos x$ and $\sin x$, respectively.

23.14 $T = 1$ for large E.

Index

γ matrices, 433

adiabatic
    approximation, 346
    perturbation, 346
angular momentum
    spin, 444
    total, 444
anti-
    linear operator, 178
    unitary
        operator, 178
        transformation, 178
anticommuting operators, 59

Baker–Campbell–Hausdorff formula, 256
barrier potential, 129, 544
    tunnelling probability, 132, 134, 544,
        545
    wave function, 130, 133
Bell's inequality, 470
black body
    definition, 2
    Planck's radiation formula, 5
    Rayleigh–Jeans' law, 3
    spectrum, 3
    Stefan's law, 3
    Wien's law, 3
Bohr
    –Sommerfeld quantum rule, 11, 359
    magnetron, 307, 453
    radius, 9, 298
Born approximation, 395
    validity condition, 396
bound state
    definition, 32, 79, 84
    of a linear harmonic oscillator, 84
    of a particle in a box, 96
    of a quantum pendulum, 104
    of a rigid rotator, 120
    of Morse oscillator, 100
    of Pöschl–Teller potentials, 102
box normalization, 33

bra vector, 171
bremsstrahlung, 7

Clebsch–Gordan coefficients, 276
coherent state, 235, 244
commutator, 59
commuting operators, 59
complete set, 65
Compton
    effect, 16
    experiment, 15
    wavelength, 16, 440
constant perturbation, 339
continuity equation, 35, 47, 429, 434, 499,
    506
Copenhagen interpretation, 31, 462
correspondence principle, 13, 14
coupled channel method, 549
current density, 35, 36, 429

Dalgarno and Lewis method, 319
de Broglie
    wavelength, 19
    waves, 18
degeneracy, 112
degenerate states, 46
delayed-choice
    experiment, 484, 485
    quantum eraser, 487
density matrix, 38, 197
differential cross-section, 386, 392
diffraction pattern
    double slit, 481
    single slit, 480
dipole matrix element, 351
Dirac equation
    for a free particle, 431
    for hydrogen atom, 451
    in covariant form, 433
    in four components, 437
Dirac-delta normalization, 34
Doppler effect, 214

Ehrenfest's theorem, 41

eigenfunction(s), 44, 56
    even parity, 69
    odd parity, 69
    of L_z, 217
    of a linear harmonic oscillator, 84
    of a particle in a box, 96, 98, 449
    of a quantum pendulum, 104
    of a rigid rotator, 120
    of fractional quantum systems
        Dirac-delta function potential
            system, 502
        free particle, 500
        linear harmonic oscillator, 511
    of Morse oscillator, 100
    of Pöschl–Teller potentials, 102
    simultaneous, 61
eigenvalue(s), 56
    degenerate, 66, 316
    equation, 56, 259
    nondegenerate, 66, 316
    of L, 259
    of L_z, 259
    of S_x, 268
    of S_y, 268
    of S_z, 268
    of $\mathbf{L}^2$, 259
    of $\mathbf{S}^2$, 268, 283
    of a linear harmonic oscillator, 84
    of a particle in a box, 96, 449
    of a quantum pendulum, 104
    of a rigid rotator, 120
    of a square-well potential, 140
    of fractional quantum systems
        Dirac-delta function potential
            system, 502
        free particle, 510
        linear harmonic oscillator, 503, 512
    of Morse oscillator, 100
    of Pöschl–Teller potentials, 102
    of relativistic hydrogen atom, 451
    spectrum, 44, 56
Einstein coefficients, 349, 350
electric dipole transition, 45
energy
    eigenfunction, 45
    levels, 44
entangled states, 464
EPR
    argument, 466
    paradox, 463
exchange

    degeneracy, 412
    interaction, 420, 421
expectation value
    in matrix form, 166
    of A, 38
    of $f(\mathbf{X}, t)$, 37
    of L_x, 261
    of L_x^2, 262
    of L_y, 261
    of $\mathbf{L}$, 262
    of $\mathbf{S}$, 271
    of an operator, 55

Fermi's golden rule, 340
Feynman's kernel, 495
fine structure, 323
fractional
    Caputo derivative, 506
    diffusion equation, 495
    Riesz derivative, 495, 497, 512
    Schrödinger equation
        in momentum representation, 499
        space, 493, 498
        space-time, 512, 513
        time, 505, 506
        time-independent, 498
    time evolution operator, 509
Franck–Hertz experiment, 10

Gaussian wave
    function, 49
    packet, 240
Green's function, 393

harmonic perturbation, 342
Heisenberg
    equation of motion, 194
    picture, 189, 193
    uncertainty
        principle, 208
        relation, 208
Hellmann–Feynman theorem, 332
Hermitian operator, 62
hidden variables, 467
Hilbert space, 65, 176
hydrogen atom
    Dirac equation, 451
    electronic distribution, 547, 548
    energy eigenfunction, 301
    energy eigenvalues, 299, 528
    momentum eigenfunction, 305

radial eigenfunction, 301
radial equation, 297, 526
relativistic energy, 451
hydrogenic systems, 311
hyperfine structure, 323
hypervirial
    relations, 332
    theorem, 332

interaction picture, 189

ket vector, 171
Klein paradox, 427
Klein–Gordon equation, 428
    inadequacies, 431
    plane wave solution, 431

ladder operators, 91
Lie group, 256
linear harmonic oscillator
    eigenfunctions, 84
    eigenvalues, 84
    momentum eigenfunction, 230
    Schrödinger equation
      in momentum space, 230
    selection rule, 89
    time-dependent wave function, 107
    transition probability, 551, 552
linear vector space, 160

magnetic
    moment, 440, 453
    quantum number, 259, 296
Mathieu equation, 104
matrix
    anti-Hermitian, 163
    Hermitian, 163
    method, 522
momentum
    eigenfunction, 223
      of a free particle, 224
      of a linear harmonic oscillator, 230
      of a particle in a box, 231
    eigenvalue, 224
    representation, 225
    space, 223
    wave function, 225
Monte Carlo method, 547
Morse oscillator, 100

non-Hermitian Hamiltonians, 72
noncommuting operators, 59

nonlocality, 462
normalization condition, 32
    in matrix form, 166
number
    operator, 92, 99
    state, 92

observable, 38, 65
old quantum theory, 23
operator(s)
    adjoint, 62
    anti-Hermitian, 62
    anti-linear, 178
    anti-self-adjoint, 62
    anti-unitary, 178
    anticommuting, 59
    commuting, 59
    creation and annihilation, 91, 92, 99
    D'Alembertion, 428
    displacement, 180
    Hermitian, 62, 564
    identity, 71
    invariant, 191
    ladder, 91
    linear, 56
    noncommuting, 59
    number, 92
    orbital angular momentum, 258
    permutation, 412
    projection, 179
    raising and lowering, 91, 258
    self-adjoint, 62, 564
    spin, 267
    time reversal, 69
    unitary, 71, 178, 180
optical theorem, 402
orbital
    angular momentum operator
      raising and lowering, 258
    quantum number, 297
orthogonal eigenfunctions, 46
orthogonality condition, 46
    in matrix form, 166
orthonormal condition, 46

paradox
    EPR, 463
    Klein, 427, 447
    zitterbewegung, 440
parity
    even, 69

odd, 69
  operator, 68
  time reversal, 69
partial wave, 398, 399
Pauli
  equation, 453
  exclusion principle, 416
  matrices, 270, 432
phase-shift, 400
photoelectric
  effect, 1, 5
  equation, 7
photons, 6
Planck
  constant, 4
  radiation formula, 5
polarization vector, 273
principal quantum number, 299
probability current density, 36, 499, 507, 514
projection operator, 179
propagator, 495
Pöschl–Teller potential, 151
  eigenfunctions, 102
  eigenvalues, 102, 152

QL method, 567
quanta, 5
quantization
  first, 29
  of frequency, 96
  of space, 12, 259
  of spin, 12
  second, 29
quantum
  Hamilton–Jacobi equation, 47
  number, 84, 259
  potential, 47
  state, 84
quaternion, 181

Ramsauer effect, 405
reflection amplitude
  for a barrier potential, 131
  for step potential, 144
reflectionless potentials, 151
rejection method, 572
rigid rotator
  eigenfunctions, 120
  eigenvalues, 120
  selection rule, 121

rotating wave approximation, 351
Rutherford scattering formula, 398

scattering
  amplitude, 391
  cross-section, 387, 392
  elastic, 385
  inelastic, 385
  length, 404
  numerical scheme, 542
  state, 32, 79
Schrödinger
  cat experiment, 467
  equation, 27, 28
    discrete equivalence, 522, 523
    for many particle system, 414
    fractional, 493, 498, 505, 506
    in dimensionless form, 505
    in matrix form, 166
    in momentum representation, 226
    time-independent, 30
  picture, 189, 190
Schwarz's inequality, 207
selection rule, 45
  for a linear harmonic oscillator, 89
  for a forbidden transition, 45
  for a particle in a box, 97
  for a rigid rotator, 121
  for an allowed transition, 45
space quantization, 12
specific heat capacity, 18
spherical harmonics, 120
spin, 267
  -orbit energy, 455
  -statistics theorem, 413
  angular momentum, 444
  of Dirac particles, 444
  operator, 267
spinor, 270, 434
spontaneous emission, 350
square-well potential, 137
  energy eigenfunctions, 140, 141
  energy eigenvalues, 140
  transmission probability, 142, 143
Stark effect, 327
state vectors, 176
stationary state, 9
step potential, 142, 143
  reflection amplitude, 144
  transmission amplitude, 144
  wave function, 144

Stern–Gerlach experiment, 11
stimulated
    absorption, 350
    emission, 350
sudden approximation, 348
superoscillatory function, 244
symmetry, 177
    Wigner, 177

transition probability
    for a linear harmonic oscillator, 551,
        552
    for adiabatic perturbation, 347
    for constant perturbation, 340
    for harmonic perturbation, 342–344
    numerical scheme, 549
transmission
    amplitude
        for a barrier potential, 134
        for a Dirac comb, 150
        for a locally periodic potential, 546
        for a square-well potential, 142, 143
        for a step potential, 144
    probability
        for a barrier potential, 131
tunnelling
    dynamical, 153
    effect, 132
    numerical simulation, 539
    probability for a barrier potential,
        132
turning point, 80, 355

ultraviolet catastrophe, 3
uncertainty
    definition, 205, 206
    in L_x, 262
    in L_y, 262
    minimum product, 208

relation, 208
    for E and time, 209, 341
    for x and k, 209
    for x and p_x, 209
    for classical statistics, 206
    modified, 218, 219
unitary
    operator, 178
    transformation, 178

variational theorem, 372
visibility, 486

wave
    -particle duality, 20
    function, 68
        collapse, 462
        definition, 27
        dimension, 32
        nonnormalizable, 33
        norm, 32
        normalization condition, 32
        square-integrable, 32
    packet
        definition, 235, 236
        fractional revival, 248
        Gaussian minimum uncertainty, 240
        mirror revival, 246
        of a free particle, 242
        revival, 542
Wigner's theorem, 178
Wigner–Eckart theorem, 287
WKB method, 355
work function, 6, 141

Young's double slit experiment, 20

zero-point energy, 84, 96
zitterbewegung, 427, 440